PRECESSION, NUTATION, AND WOBBLE
OF THE EARTH

Earth rotation has long been used as a measure of time, together with using the stars as reference points, to determine travellers' whereabouts on the globe. However, the rotation of the Earth is not uniform and its orientation with respect to the sky goes on changing.

Covering both astronomical and geophysical perspectives, this book describes changes in the Earth's orientation, specifically precession and nutation, and how they are observed and computed in terms of tidal forcing and models of the Earth's interior. Following an introduction to key concepts and elementary geodetic theory, the book describes how precise measurements of the Earth's orientation are made using observations of extragalactic radio sources by very long baseline interferometry (VLBI) techniques. It demonstrates how models are used to accurately pinpoint the location and orientation of the Earth with reference to the stars and how to determine variations in its rotation speed (length-of-day variations). A theoretical framework is also presented that describes the role played by the structure and properties of the Earth's deep interior.

Incorporating suggestions for future developments in nutation theory for the next generation models, this book is ideal for advanced-level students and researchers in the fields of solid Earth geophysics, planetary science, and astronomy.

PROFESSOR V. DEHANT is Senior Research Scientist at the Royal Observatory of Belgium, responsible for the Operational Direction *Reference Systems and Planetology*, and is a part-time professor at the Université Catholique de Louvain. Her research is focused on the interior, rotation, and habitability of the terrestrial objects of our Solar System, mostly in relation to the evolution of their interiors and atmospheres. Professor Dehant is also Principal Investigator and Co-Investigator of several instruments in ESA and NASA missions, and has received several awards including the Descartes Prize of the European Union. She is involved in several advisory committees such as the ESA Space Science Advisory Committee (SSAC) and is a member of the Royal Academy of Belgium (Science Class). She is the author of more than 400 publications.

PROFESSOR P. M. (SONNY) MATHEWS served as Senior Professor of Theoretical Physics at the University of Madras in India, and is now retired. His main research interests are in the Earth's nutation and precession and other concomitant phenomena, as well as the properties of, and processes in, the Earth's deep interior. Professor Mathews is the lead originator of the Mathews–Herring–Buffett (MHB) theory of nutation adopted by the IAU and IUGG. In recognition of his work on the MHB model of nutation, he was elevated to the Fellowship of the American Geophysical Union (2004). He was elected a Fellow of the Indian Academy of Sciences in 1975 and of the Indian National Science Academy in 1983. He has published 120 papers.

PRECESSION, NUTATION, AND WOBBLE OF THE EARTH

V. DEHANT
Royal Observatory of Belgium

P. M. (SONNY) MATHEWS
Formerly of the University of Madras

CAMBRIDGE
UNIVERSITY PRESS

University Printing House, Cambridge CB2 8BS, United Kingdom

One Liberty Plaza, 20th Floor, New York, NY 10006, USA

477 Williamstown Road, Port Melbourne, VIC 3207, Australia

4843/24, 2nd Floor, Ansari Road, Daryaganj, Delhi - 110002, India

79 Anson Road, #06-04/06, Singapore 079906

Cambridge University Press is part of the University of Cambridge.

It furthers the University's mission by disseminating knowledge in the pursuit of education, learning and research at the highest international levels of excellence.

www.cambridge.org
Information on this title: www.cambridge.org/9781107465824

© V. Dehant and P. M. Mathews 2015

First published 2015
First paperback edition 2017

A catalogue record for this publication is available from the British Library

ISBN 978-1-107-09254-9 Hardback
ISBN 978-1-107-46582-4 Paperback

Contents

Preface

For several years (about 20) both authors have been working in the domain of Earth rotation, and in particular on nutation. The Working Group on nutation, established by the International Astronomical Union and the International Union of Geodesy and Geophysics in 1994, was a starting point for bringing together scientists thinking about what was missing in the nutation series adopted by the International Astronomical Union. Collaboration between the scientists of the WG was very successful and, in particular, the authors' collaboration began at that time. Recently, it has appeared to us that there were no existing books yet dedicated to the subject, and the scientific community is looking for a suitable publication. The literature contains a lot of relevant articles, but many of them rely on previous work and do not give full details. This book aims at bringing everything together for the first time. The book is addressed to students or scientists who want to understand nutations. The aim of this book is to give a reasonably comprehensive introduction to the fundamental concepts, mathematical formalism, and methodology of the Earth's nutation. It is only assumed that the student or reader is familiar with the elementary principles of calculus, although we might have used in some parts short-cuts for reasons of simplicity, and with the underlying physical principles in the foreground. Another important aim of this book is to make a comprehensive list of the geophysical and astronomical processes involved in nutation, in order to be able to investigate the "next decimal place."

The authors wish to take this opportunity to acknowledge all those who have aided in the preparation of this book.

Abbreviations

Notation	Definition
AAM	Atmospheric Angular Momentum
AU	Astronomical Units
BCRF	Barycentric Celestial Reference Frame
BCRS	Barycentric Celestial Reference System
BIH	Bureau International de l'Heure
CEP	Celestial Ephemeris Pole
CHAMP	CHAllenging Minisatellite Payload
CIO	Celestial Intermediate Origin
CIP	Celestial Intermediate Pole
CMB	Core Mantle Boundary
cpsd	cycle per sidereal day
CRF	Celestial Reference Frame
CRS	Celestial Reference System
CW	Chandler Wobble
DExxx	Development Ephemeris where xxx is a number
DSN	Deep Space Network
ECRF	Ecliptic Celestial Reference Frame
ECRS	Ecliptic Celestial Reference System
ELP20yy	Ephéméride Lunaire Parisienne where 20yy is a year
EMB	Earth–Moon Barycenter
EOPs	Earth Orientation Parameters
EOT	Empirical Ocean Tide Model
ERA	Earth Rotation Angle
ESA	European Space Agency

Notation	Definition
ESTRACK	ESA Tracking station
FCN	Free Core Nutation
FES	Finite Element Solution
FICN	Free Inner Core Nutation
FOC	Fluid Outer Core
GCM	General Circulation Model
GCRF	Geocentric Celestial Reference Frame
GCRS	Geocentric Celestial Reference System
GMST	Greenwich Mean Sidereal Time
GNSS	Global Navigation Satellite System
GOCE	Gravity field and steady-state Ocean Circulation Explorer
GOT	Goddard Ocean Tide model
GPS	Global Positioning System
GRACE	Gravity Recovery And Climate Experiment
GSH	Generalized Spherical Harmonics functions
GST	Greenwich Sidereal Time
GTRF	Geocentric Terrestrial Reference Frame
GTRS	Geocentric Terrestrial Reference System
IAG	International Association of Geodesy
IAU	International Astronomical Union
IBO	Inverted Barometric Ocean
ICB	Inner Core Boundary
ICRF	International Celestial Reference Frame
ICRS	International Celestial Reference System
ICW	Inner Core Wobble
IERS	International Earth rotation and Reference frame Service
IMCCE	Institut de Mécanique Céleste et de Calcul des Ephémérides
INPOP20xx	Intégrateur Numérique Planétaire de l'Observatoire de Paris where 20xx is a year
InSIGHT	Interior exploration using Seismic Investigations, Geodesy and Heat Transport
IRF	Inertial Reference Frame
IRF	Intermediate Reference Frame
ITRF	International Terrestrial Reference Frame
ITRS	International Terrestrial Reference System
IUGG	International Union for Geodesy and Geophysics

Notation	Definition
J2000	event (epoch) at date 2000 January 1.5
JGM	Joint Gravity Model
JPL	Jet Propulsion Laboratory
LaRa	Lander Radioscience experiment
LExxx	Lunar Ephemeris where xxx is a number
LLR	Lunar Laser Ranging
LOD	Length-Of-Day
LTE	Laplace Tidal Equations
mas	milliarcsecond
μas	microarcsecond
MHB	Nutation model of Mathews, Herring, and Buffett
MHB2000	Nutation model of Mathews, Herring, and Buffett (Mathews *et al.*, 2002)
MOP	Mars Orientation Parameters
NAIF	Navigation and Ancillary Information Facility
NASA	National Aeronautics and Space Administration
NDFW	Nearly Diurnal Free Wobble
NIBO	Non-Inverted Barometric Ocean
NNRC	No-Net-Rotation Condition
NRO	Non-Rotating Origin
OAM	Ocean Angular Momentum
OD	Ocean Dynamics
PFCN	Prograde Free Core Nutation
PM	Polar Motion
PN	Precession Nutation
POD	Precise Orbit Determination
PREM	Preliminary Reference Earth Model
RMS	Root Mean Square
SIC	Solid Inner Core
SLR	Satellite Laser Ranging
SOS	Sasao, Okubo, Saito (Sasao *et al.*, 1980)
SPICE	Spacecraft ephemeris, Planet, satellite, comet, or asteroid ephemerides, Instrument description kernel, Pointing kernel, Events kernel
TGP	Tide Generating Potential
TIO	Terrestrial Intermediate Origin
TOM	Tilt-Over Mode

Notation	Definition
TRF	Terrestrial Reference Frame
TRS	Terrestrial Reference System
UT	Universal Time
VLBI	Very Long Baseline Interferometry
VSOP	Variations Séculaires des Orbites Planétaires

1

Introduction – Fundamental definitions – Motivation

1.1 Rotation and global shape of the Earth

At a very elementary level, the Earth is considered to be an axially symmetric ellipsoid, rotating with uniform angular speed about its symmetry axis, which is the polar axis passing through the Earth's center and its north and south poles; under the steady rotation, the direction of this axis maintains a fixed direction in space, i.e., relative to the directions of the "fixed stars." (The celestial objects that come closest to the ideal of remaining "fixed in space" are the quasars, the most distant extragalactic celestial objects.) The direction of the symmetry axis is at an inclination of about 23.5° to the direction of the normal to the plane of the Earth's orbit around the Sun (or more precisely, about the solar system barycenter, i.e. the center of mass of the solar system).

An axially symmetric ellipsoidal shape, bulging at the equator and flattened at the poles, and an internal structure with the same symmetry, would result from the centrifugal force associated with uniform Earth rotation around the polar axis, counterbalanced by gravity. The ellipsoidal structure computed on the basis of this balance of forces, assuming that the material of the rotating body behaves like a fluid under the incessant action of forces acting over very long timescales (i.e., that the resistance of even solid regions to shear deformation is overcome under such conditions), is called the "hydrostatic equilibrium ellipsoid." The Earth's figure (shape) does conform quite closely, though not perfectly, to that of such an ellipsoid. The equatorial radius of the Earth exceeds the polar radius by about 21 km; this is often described as the *equatorial bulge*. This bulge is to be viewed in relation to the mean radius of about 6371 km.

1.2 Orbit of the Earth

The force of gravitational attraction of the Sun on the total mass of the Earth holds the Earth in orbital motion around the Sun; similarly, the Earth's gravitational force maintains the orbital motion of the Moon around the Earth. Each of these two-body orbits in the Sun–Earth–Moon system would be elliptical, in accordance with Kepler's third law, if the third body were absent. In reality, small deviations from the planar elliptical nature of the orbits result from the gravitational attraction of the two individual bodies by the third body, and from the much smaller gravitational forces exerted by the planets.

1.3 Earth orientation – precession and nutation

Earth rotation can be separated into the rotation speed around its symmetry axis (figure axis) and the orientation of this axis (or another axis of the Earth) in space. In reality, Earth rotation and orientation are variable and even yield information on its interior structure. Most of us know that the rotation of a boiled egg noticeably differs from that of a raw egg. This simple observation shows that information on the inside of an egg can be obtained from its rotation. The same idea applies to the observation of the rotation and orientation of the Earth, relative to a "space-fixed" reference frame. (A space-fixed frame is one in which the directions of the most distant sources emitting in the radio frequency spectrum in the sky remain unchanged in time.)

Consider, in particular, the motion of the figure axis in space. It is a composite of two types of motion. The first is a secular motion, called precession, wherein the pole of the axis (which is on the celestial sphere, i.e., on a sphere of unit radius (in arbitrary units) in the "space-fixed" reference frame) traces a circular path at a constant rate (to the first order around J2000) on the celestial sphere, around the normal to the *ecliptic plane*. (The ecliptic plane is the plane of the orbit of the Earth–Moon barycenter (EMB) around the Sun. It is often loosely referred to as the Earth's orbital plane as mentioned in Appendix C.) The axis maintains a constant angle (to a first approximation) of about 23.5° to the normal to the ecliptic in the course of this (precessional) part of the motion, and hence it describes a cone with a half-angle of 23.5° with its axis along the normal.

The second part of the motion is a composite of numerous periodic motions, each of which manifests itself as an ellipse on the surface of the celestial sphere, with its center at the "mean pole," i.e., the spot on the precessional path where the pole would be if the short periodic motions did not exist. This elliptical motion can be resolved into oscillatory motions in two orthogonal directions on the celestial sphere: one over the circular precessional path on the celestial sphere, and the other in the orthogonal direction (towards/away from the normal to the ecliptic). The composite of all the periodic motions constitutes the nutation; it takes the pole of the axis up to a maximum of about 10 arcseconds away from the mean pole.

The above features are shared by the motions of the rotation and angular momentum axes too, see Chapter 2.

1.4 Primary cause of precession and nutation

As the Earth's rotation axis is tilted with respect to the orbital plane, the equatorial bulge is out of the equatorial plane during the orbital motion. As a result, the Sun and the Moon exert a gravitational torque on the Earth tending to twist the equator towards the orbital plane of the Sun/Moon relative to the Earth. As the Earth is rotating, it reacts to this torque like a spinning top to the gravitational pull on it. The main effect is precession, which is the slow motion of the rotation axis of the Earth around the normal to its orbital plane.

The precessional and nutational motions of the axes are thus primarily due to torques arising from the action of the gravitational potentials of the Moon and the Sun (and of the planets, to a minor extent) on what is loosely called the "equatorial bulge" of the Earth. The gravitational attraction being higher at points nearer to the celestial body than at more distant points, the mass of the equatorial bulge on the side nearer the celestial body gets pulled with a greater force than the bulge on the opposite side. As a result, the Earth gets subjected to a net torque (except at instants when the celestial body is on the equatorial plane), which tends to tilt the axis of the Earth in space; this is illustrated in Fig. 1.1 for the simplified model of a homogeneous and axially symmetric ellipsoidal Earth. It is the gyroscopic response of the rotating Earth to this torque that causes the directions of the various Earth-related axes to keep on varying.

The torque exerted by each of the celestial bodies on the Earth has a time independent part; it is the sum of these constant torques that generates the precessional motion, which is at a constant rate of about 50 arcseconds per year around the circle over the cone mentioned earlier. This motion is more precisely characterized as *luni-solar precession*. The term "luni-solar" recognizes the dominant roles of the Moon and the Sun in the various phenomena considered, but is generally used in the literature as inclusive of the small effects of the planets. We shall encounter later the "planetary precession," which is a misleading term because it does not involve any motion of the Earth's axes in space; it is only a reflection of the extremely slow tilting, caused by the action of the planets, of the reference plane in space (to be defined later) with respect to which the direction of the Earth's axis and hence its precession is defined (see Chapter 5).

The torque due to the solar system bodies has also a huge number of spectral components with frequencies that are related to those of the orbital motions of these bodies relative to the Earth. Each of these components produces a circular motion of the pole of whichever axis might be specified. The superposition of such motions

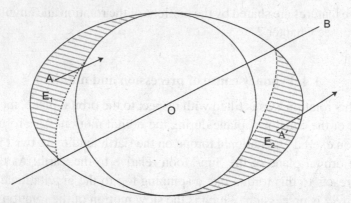

Figure 1.1 Gravitational forcing on the Earth. The figure shows the cross section of a hypothetical homogeneous ellipsoidal Earth with its inscribed sphere, and a celestial body B. Gravitational forces due to B acting on equal mass elements located symmetrically about the line passing through the center (O) and in the direction of B are of equal magnitude and produce equal torques, while in opposite directions, they cancel each other out. But the regions E_1 and E_2 outside the sphere are not symmetrical about the line in the direction of B. So there are net torques on these regions (say at A, A'), which do not cancel out: the torque on E_2 is stronger than that on E_1 because the former region is closer to the gravitating body B.

associated with all the spectral components of nonzero frequency constitutes the full nutation of the specified axis.

The maximum angular displacement due to nutation is about 10 arcseconds or 10 000 milliarcseconds (mas), as mentioned earlier. For this angular displacement, the motion in space of the intersection point of the axis with the mean spherical surface of the Earth is about 310 meters (3.1 cm/mas). The largest of the spectral components, by far, is the so-called Bradley nutation or principal nutation, with a period of about 18.6 years (Bradley, 1748); the major and minor axes of the elliptical motion with this period are about 9200 and 6800 mas, respectively. Other major nutations are already considerably smaller than the principal nutation; they have periods of approximately six months, 9.3 years, two weeks, one year, etc. The combined precession–nutation is illustrated schematically in Fig. 1.2 and as elliptical motions in Fig. 1.3 where nutation in longitude $\Delta\psi$ and obliquity $\Delta\epsilon$ are plotted against each other as functions of time.

1.5 Nutation of a non-rigid Earth

Nutation for a rigid Earth can easily be computed once the time dependence of the torque on the Earth is determined using the ephemerides, which give the relative positions of celestial bodies as functions of time (see Chapter 5). Those theoretical

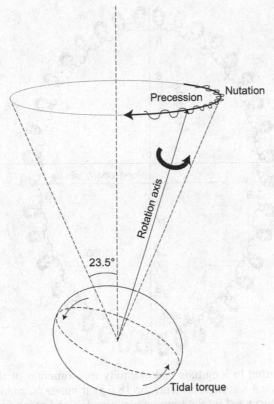

Figure 1.2 Precession and nutation. Precession is a smooth motion of the pole of the Earth's axis along the circle around the normal to the ecliptic plane (shown by the dashed vertical line), and the wiggly excursions from the precessional path represent nutation.

values, however, do not reproduce the observed values since the Earth is deformable and thus non-rigid, and also contains a liquid layer, namely the outer core, as mentioned in the next paragraph (Section 1.6). Thanks to the high precision of the observations by Very Long Baseline Interferometry (see Chapter 4), information on the interior of the Earth can be obtained through appropriate theoretical analyses of the VLBI data, which highly motivates the research in this field. In order to do so, models for the precession and nutation of a non-rigid body have been developed (see Chapter 7). These models are based on knowledge of the Earth's interior gained from seismic studies (see Section 1.6). The response of this Earth to a unit gravitational forcing is then computed. The nutation amplitudes are then obtained using the amplitude of the torque acting on the Earth for any particular forcing period and the response of the deformable Earth to unit forcing at that same forcing period.

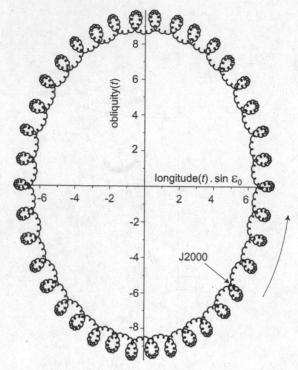

Figure 1.3 Nutation in longitude and obliquity as a function of time over 18.6 years. The elliptical feature represents the 18.6 year nutation, and the wiry loops around it are contributed by the semi-annual nutation and other nutations of still smaller amplitudes. Units: arcseconds.

It is apt to draw attention at this point to the existence of normal modes of the non-rigid Earth, which have a large influence on forced nutations at nearby frequencies (see Section 2.5).

1.6 Models of the Earth's interior

The Earth system consists of what is commonly referred to as the "solid Earth," plus the fluid layers at the surface, namely, the oceans and the atmosphere. The so-called solid Earth is, in reality, far from being wholly solid. It is made up of three major regions or layers: the outermost solid layer called the mantle, extending down from the mean outer radius of 6371 km to a mean radius of about 3480 km; a solid region with a mean radius of about 1220 km around the Earth's center, called the solid inner core (SIC); and a fluid region in between, called the fluid outer core (FOC). In constructing models for the structure and characteristics of the interior of the Earth (density, elastic properties, etc., as functions of position), time dependent deformations of all kinds are of course disregarded.

Information obtained from data on the travel times of seismic waves to various stations spread out over the Earth's surface, and from other data on the frequencies of seismic normal modes, is used to construct spherically symmetric models of the variation of density and elastic parameters of the matter from the center to the surface. This process involves the use of an appropriate procedure to remove the effects of the ellipticity of equidensity surfaces in the Earth's interior from the observational data. Of the many spherically symmetric models constructed in this fashion, the Preliminary Reference Earth Model (PREM; Dziewonski and Anderson, 1981) has been most widely used in recent times, superseding earlier models (e.g., 1066A of Gilbert and Dziewonski, 1975). These models are presented in the form of tables of values of the density, elastic parameters, gravity, etc., at various radial distances from the geocenter. Given such a model, hydrostatic equilibrium theory is employed to obtain the corresponding axially symmetric *ellipsoidal* model, in which the surfaces of equal density and equal elastic parameters (including, in particular, the Earth's exterior surface and internal boundaries such as the Core Mantle Boundary (CMB) and the Inner Core Boundary (ICB)) are axially symmetric and ellipsoidal.

Hydrostatic equilibrium Earth models have been of great importance to nutation theory, though the deviations from the hydrostatic equilibrium structure are not ignorable and have been taken into account in recent theoretical treatments. The deviations are of two kinds. The first kind is revealed through a slightly higher value for the dynamical ellipticity of the Earth, as inferred from precession–nutation studies, than the value that hydrostatic equilibrium theory would lead to. The excess ellipticity may be explained as the effect of the viscous forces resulting from the convective flow within the mantle as a consequence of the temperature gradient between the upper and lower boundaries of the mantle, the latter (the CMB) being at a much higher temperature than the former. The mantle material yields to these forces, persisting over millions of years, as if it were a highly viscous fluid. Secondly, the actual Earth is not strictly ellipsoidal. Deviations from the ellipsoidal structure are reflected in deviations of the Earth's own external gravitational potential from ellipsoidal symmetry, which manifest themselves through their effects on the orbital motions of near-Earth satellites; they are quantified by analyses (1) of Satellite Laser Ranging (SLR) data on these orbits, (2) of microwave data of relative position changes between twin satellites as from the GRACE[1] mission, (3) of data from accelerometers aboard satellites, as in the CHAMP[2] mission, (4) exploiting GPS satellite-to-satellite tracking data as well, and (5) of gradiometer data of acceleration differences over short distances between an ensemble of test masses inside a satellite, as in the GOCE[3] mission.

---

[1] GRACE stands for Gravity Recovery And Climate Experiment.
[2] CHAMP stands for CHAllenging Minisatellite Payload.
[3] GOCE stands for Gravity field and steady-state Ocean Circulation Explorer.

1.7 Precession, nutation, and geodynamics

The three regions/layers of the Earth's interior are not "locked together" in their rotational motions; they have considerable freedom to rotate about differing axes and at different rates. Therefore, it is necessary to clarify here that what is called the rotation vector in Section 1.3 was meant to refer to the angular velocity vector of rotation as observed at the surface of the Earth. It closely reflects the rotation of the mantle region as a whole. There are different angular velocity vectors for the fluid core region and for the SIC. Further elaboration of the concept of the rotation vector becomes necessary in the case of the fluid core region wherein the fluid flow deviates from being purely rotational in nature, primarily because the boundaries of the region are essentially ellipsoidal rather than spherical. This aspect and other refinements will be deferred to Chapter 7. See the paragraphs containing Eqs. (7.71) to (7.74) of that chapter.

1.8 The Earth's normal modes

The free motions are not driven by any external torque and depend on the interior constitution of the Earth. The free normal modes of high interest for nutation are the manifestations of compensatory variations of the rotation vectors of the different layers inside the Earth related to internal physics.

The three-layer Earth has four such modes: the Chandler wobble (CW) with a period of about −435 days in an Earth-fixed reference frame (the period −435 days means that the rotational motion has a period of 435 days in the sense opposite to that of the Earth's rotation); the free core nutation (FCN) with a period of about −430 days in the celestial frame, along with its associated wobble (the nearly diurnal free wobble or NDFW) which has nearly diurnal frequency in the terrestrial frame as its name implies. We use the term wobble in a very broad sense for any periodic or quasi-periodic motion of the Earth's rotation axis with respect to an Earth-fixed reference frame, irrespective of the frequency or the physical origin of the motion (see Section 2.6). For nutation computation it is also necessary to consider the less prominent free inner core nutation (FICN) and inner core wobble (ICW) that are explained in Section 2.20.

There are various mechanisms by which the rotation of one region influences the rotations of the others: the fluid pressure on the ellipsoidally shaped boundaries between regions; the torque due to the action of the gravitational field of the mantle (which is an ellipsoidal shell) on the ellipsoidal inner core when their symmetry axes are out of alignment; electromagnetic forces induced by differential rotation of adjoining regions in the presence of magnetic fields crossing their common boundary; viscous drag exerted by the core fluid on the solid regions; local topography on the solid side of the fluid–solid boundaries; and so on. The coupling of the three regions in this manner is an important aspect of the rotational response of the Earth to external torques as well as of the free rotational motions in the absence of external torques. In particular, while a wholly solid Earth has only a single rotational normal mode, the three-layer Earth has four: the Chandler wobble (which is the counterpart of the single mode of the wholly solid Earth) (CW); the free core nutation (FCN); the free inner core nutation (FICN) also called the prograde free core nutation (PFCN); and the inner core wobble (ICW). These modes are defined in Section 2.20. The resonance associated with the FCN is a prominent feature of the low frequency forced nutations, and the FICN/PFCN resonance too is not insignificant. The FICN and ICW modes would not arise if a SIC were not present.

1.9 Motivation for the book

The motivation for this book stems from the major advances that have taken place, in the very recent past, in the precision of observations of Earth rotation variations, and the advances made, in response to this development, in the theoretical modeling of the variations. The main focus is on developments in the theory of precession–nutation and polar motions. The precision of observations of the time dependent nutation has reached the sub-milliarcsecond level (of the order of 0.2 mas). The matching of these observations by the theoretical model adopted recently by the IAU (International Astronomical Union) and IUGG (International Union for Geodesy and Geophysics) is also at about the same level. (The precision of observations in the case of individual spectral components of the nutation is of the order of 10 microarcseconds or so, except for nutations of very long periods such as the 18.6 year nutation which have larger uncertainties.) Now, the contribution to nutation from the "non-rigidity" of the Earth – a term which is often used conveniently (if rather loosely) to include not only the deformability of the Earth but also the existence of the fluid core – is of the order of 60 mas. This is over a hundred times the current uncertainty in the observational and theoretical results indicated above. The potential for deriving information from nutation studies about aspects of the Earth's interior which influence nutations is then evident. This is a powerful motivation for such studies. Another motivation, not less important, arises from the fact that nutations leave their imprint on a variety of quantities of importance in geophysics, which are now measurable with very high precision through space geodetic and other means, and that the nutation contributions have to be separated out in order to bring out clearly the other geophysical information contained in the data. Examples are movements of observational sites, variations of the geopotential outside the Earth, gravity variations recorded by superconducting gravimeters, etc., caused by phenomena other than Earth rotation variations.

1.10 Organization of the book

The book is organized as follows. In Chapter 2 we introduce the concepts, but we have dedicated the complete Chapter 3 to the reference systems and frames in use in our book. It is indeed of high importance to define these frames precisely, as Earth rotation, precession, nutation, and polar motion relate the body-fixed terrestrial frame to the celestial frame "fixed in space." Recent concepts such as the *non-rotating origin* are shifted to the end of the book (see Chapter 12, more addressed to specialists in the domain).

In order to bring out clearly some of the essential aspects of nutations and wobbles, we provide in Chapter 2 the explicit solutions for a very simple model of the motion of the celestial body (see Section 2.24). This is very helpful for students beginning to study precession and nutation. Chapter 4 is dedicated to the observational methods in the study of precession and nutation as well as polar motion. Once the concepts and observational data are set up, we set the basis for the precession and nutation computation. Chapter 5 summarizes the rigid Earth nutation theories. Parts of this chapter may look quite complicated in addressing the Hamiltonian theory. However, we provide the necessary basis for the understanding of this theory. These parts may be left for specialists, in particular when the reader is a geophysicist. Then, the next chapters (Chapter 6 on the deformation of the Earth, Chapter 7 on the nutation of a non-rigid Earth, and Chapter 8 on anelasticity contribution) provide the theory of the non-rigid Earth as used for the last precession–nutation model adopted by the international community. As the Earth is not only a "solid body" but possesses fluid layers above its surface, it is necessary to correct the nutation for the existence of oceans and of an atmosphere, which is performed in Chapter 9. Chapter 10 provides the reader (specialized in the domain of nutation) with recent refinements of the non-rigid Earth nutation theory, which also provides clues to understanding the future steps in nutation theory (the chapter is addressed to those who want to go further in the modeling of precession and nutation). Chapter 11 treats the comparison of observation and theory.

Precession is a very useful observable for internal geophysics of the Earth because its rate is proportional to its dynamical flattening, reflecting the global mass repartition inside the Earth (different with respect to the equator than with respect to the direction of the rotation axis). Similarly to the Earth, Mars is a terrestrial planet (with an internal structure like the Earth's, with an iron core, a silicate mantle, a lithosphere, and a rocky crust), but smaller (its radius is about half the radius of the Earth), and rotating a little bit slower (24h 37min 23s or 1.026 days). It is thus flattened at the pole and its inclination is 25.19°, which is similar to the 23.439° of the Earth. Mars is further away from the Sun (about 1.5 times the

Earth–Sun distance), but the existence of the gravitational pull of the Sun is still important. As a consequence, Mars undergoes precession and nutation.

Observation of Mars' precession yields an estimate of the dynamical flattening, which, in conjunction with the observation of the gravitational potential of the planet itself, provides a measure of its layer mass distribution. For Mars, precession is one of the main and best-determined constraints on the interior structure. Future observation of the nutations of Mars will be performed with near future space missions such as the InSIGHT mission (Interior exploration using Seismic Investigations, Geodesy and Heat Transport). We have therefore decided to dedicate the complete Chapter 13 to the nutations of Mars.

2

Concepts and elementary theory

2.1 Gravitational potential

Within the Earth itself, the gravitational potential at some point P due to the Sun (or any of the other celestial bodies) varies with the location of P, inversely as its distance from the body. The gravitational attraction on the mass elements of the Earth varies as the inverse square of the same distance. (We are using the word "attraction" here with the specific meaning of "the force exerted per unit mass.") Suppose that the potential, as a function of the location of P, is separated into two parts: one that depends linearly on the position vector from the Earth's center of mass to P (including, in general, a term that is independent of the position) and the remainder of the potential function that is non-linear in the vector. The first part produces a uniform attraction on mass elements throughout the Earth, and is responsible for the relative orbital motion of the celestial body and the Earth. It is the second part that is responsible for the variation of the gravitational attraction over the volume of the Earth. This part of the potential is designated as the *tide generating potential (TGP)* or, for brevity, as the *tidal potential*. The reason for this nomenclature is simple: it generates the ocean tides as well as deformations of the solid Earth, referred to as solid Earth tides. (The term "solid Earth" is generally applied to the whole of the Earth excepting the fluid layers at the surface, namely, the oceans and the atmosphere.) It is the same tidal potential that generates Earth rotation variations too. Thus, the contents of this book relate to the effects of the TGP.

2.2 Axes associated with Earth rotation

Variations in the Earth's rotation are most naturally conceived of as variations in the direction or the magnitude (or both) of the Earth's *angular velocity vector* Ω, also called the *rotation vector*. (Throughout this book we use the bold notation for denoting vectors.) Another vector, namely, the *angular momentum* vector, is what appears most directly in the dynamical effects of the TGP on the Earth: the rate of change of this vector is equal to the torque exerted by the TGP, irrespective of the Earth's internal structure and other properties. The directions of these vectors are referred to as the *rotation axis* and the *angular momentum axis*, respectively. It has been traditional to employ the angular momentum axis as the reference axis in developing the theory of nutation and precession of *rigid* Earth models. For such models, the motions in space of the angular momentum axis determine those of the figure and rotation axes.

Yet another axis, the *figure axis*, is the most relevant one in the context of the observational determination of the temporally varying orientation of the Earth as a whole in space, or more precisely, the orientation of a terrestrial ("Earth-fixed") reference frame relative to the most distant celestial objects that appear to be stationary in space. The figure axis envisioned here and mentioned in the Introduction of this book is the axis of maximum moment of inertia of the "static" figure of the Earth, i.e., of the Earth in its instantaneous orientation, disregarding the deformations caused by the gravitational forces exerted by external bodies and by other forces. The figure axis so defined is a natural choice for the $\hat{z}$-axis of a terrestrial frame. Though it is not realistic to expect an exact coincidence between these two axes in a practical realization of a terrestrial reference frame, the two are extremely close, and we shall treat the two as identical. Thus, the direction of the figure axis gives an accurate reflection of the instantaneous orientation of the equatorial plane of the terrestrial frame in space; this axis is therefore of the greatest interest for Earth orientation studies. (The instantaneous axis of maximum moment of inertia *of the dynamically deforming Earth* has been identified as the figure axis in some of the literature; it moves in relation to the mean figure of Earth itself. Though its motions in space have been studied, they are not of great interest, and will not be considered in this book.)

2.3 Celestial sphere, celestial poles, and equators of axes

We use the term *geocenter* to mean the center of mass of the whole Earth including the oceans and the atmosphere, disregarding all time dependent perturbations of the distribution of mass in the Earth and its fluid layers. It has a fixed position in relation to the solid Earth as defined in Section 2.1 and is denoted by O. Ocean tides and atmospheric mass movements do cause the instantaneous center of mass of the whole Earth to have small motions, of the order of a couple of centimeters, relative to the geocenter as defined above; they are referred to as geocenter motions.

We note, at the outset, that the direction of any axis pointing outwards from the geocenter can be represented by an associated *celestial pole*. The pole is the intersection of the axis with the geocentric *celestial sphere* which is a sphere of unit radius (in arbitrary units); it is taken to be non-rotating in space, i.e., having no rotation relative to the very distant objects in space which appear unmoving in space. Thus the poles which represent the directions from the geocenter to those objects will remain stationary on the celestial sphere. On the other hand, the direction in space of the Earth's angular velocity vector (i.e., the rotation axis) or that of its figure axis keeps varying, reflecting the variations in Earth rotation; correspondingly, the pole of rotation and the pole of the figure axis move over the celestial sphere. Variations in the direction of such axes in space may therefore be pictured equally well as motions of the associated pole on the celestial sphere.

Each Earth-related axis (or its pole) has a corresponding equatorial plane, which is the plane perpendicular to that axis, passing through the geocenter. The great circle along which the equatorial plane intersects the celestial sphere is the equator of the axis. Thus, the pole of the figure axis is also the pole of the geographical equator; the pole of the angular momentum axis is also the pole of the corresponding equatorial plane which is perpendicular to the angular momentum vector and passes through O. Evidently, any motion of an axis, or equivalently that of its pole, is reflected in a corresponding motion of its equator, and vice versa. It is quite common, in the astronomy literature, to view the lunisolar nutation and precession as motions of the equator rather than motions of the axis. The instantaneous equator of the precessing and nutating Earth is called the *true equator* of date; if the precessional motion alone is considered, ignoring nutation, the instantaneous position of the equator is called the *mean equator* of date.

All three of the above alternative pictures – motion of an axis, or motion of its celestial pole or of its equator – will be invoked at various places in this book; which is used in a particular context will depend on its utility for conveying the ideas clearly in that context.

It is necessary to point out here that the changes in directions of geocentric vectors *relative to a frame tied to the Earth* may be represented by the motions of the poles of those vectors on an *Earth-fixed* geocentric sphere of unit radius. The pole, on this sphere, of an axis or of some other radial direction from the geocenter, is the intersection of the axis/direction with this sphere. The Earth's rotation axis, for instance, moves relative to the Earth itself, besides changing direction in space; it remains close to the figure axis (within an arcsecond). Evidently the poles of the two axes too remain at the same small angular distance from each other. Consequently, it is possible, as a close approximation, to take the rotation pole to be the intersection of the rotation axis with the plane that is tangential to the sphere at the pole of the figure axis, and it is a common practice to do so.

The role of the geocentric celestial sphere and of the celestial pole on this sphere in the representation of Earth rotation variations was discussed above. A *barycentric* celestial sphere with its origin O at the solar system barycenter is appropriate for representing the motions of solar system bodies. The circular or elliptical orbits of planets around the Sun (or more accurately, around the solar system barycenter) may be represented by the circles formed by the intersection of the respective orbital planes with the celestial sphere that has the barycenter as its origin O. The positions of planets as well as stars, quasars, etc., in the sky are represented on this sphere by the points of intersection of the sphere with the vectors directed from the solar system barycenter towards the respective objects; the motion (if any) of any one of these objects in the sky is depicted by the motion of the corresponding point over the sphere.

Precession and nutation are two aspects of the motion, relative to a "space-fixed" reference frame, of any of the Earth-related axes considered above.

2.4 Terminology for nutation

As a matter of terminology, we use the plural "nutations" when referring to a class of spectral components of the nutational motion, and reserve the singular (nutation) for the total nutational motion. This terminology is shared by the motions of the rotation and angular momentum axes too.

2.5 Normal modes

The free rotational modes (rotational normal modes) of the Earth constitute another ingredient, independent of any external excitation, of the variations in direction of the axes.

2.6 Wobble and sway

We use the term wobble in a very broad sense for any periodic or quasi-periodic motion of the Earth's *instantaneous rotation axis with respect to the figure of the Earth*, irrespective of the frequency or the physical origin of the motion. The essence of wobble is the separation of the rotation axis (i.e., the direction of the instantaneous angular velocity vector $\mathbf{\Omega}$) from the direction of the mean angular velocity vector $\mathbf{\Omega}_0$ which is also the direction of the figure axis. The components of $\mathbf{\Omega}$ in the terrestrial frame are usually denoted by $\mathbf{\Omega}_0(m_1, m_2, (1 + m_3))$. Wobble is characterized by the equatorial components (m_1, m_2) of $(\mathbf{\Omega}/\mathbf{\Omega}_0)$. (Variations of m_3 represent the fractional variations in the spin-rate, which manifest themselves in length-of-day (LOD) variations.) Wobble is an inescapable accompaniment of nutation; it is, in fact, intimately related to the nutation of the figure axis, and plays an essential role in various approaches to the treatment of nutation.

As in the case of nutation, the plural "wobbles" is used when referring to a class of spectral terms of the time dependent wobble motion (e.g., the low frequency wobbles). Every spectral term of nutation is necessarily accompanied by a corresponding wobble with frequency and amplitude determined by those of the nutation, and vice versa. The most prominent example of a wobble, namely, the Chandler free wobble, is a nearly periodic motion, with variable amplitude, of the rotation axis around the figure axis (see Section 2.20).

"Sway" also is a term used (Ooe, 1973a, 1973b) for the motion of the Earth's rotation axis, but it refers to the motion of this axis in space, *in the absence of any forcing of extraterrestrial origin*, relative to the axis of angular momentum, which remains invariant since external torques are supposed to be absent. Sway is associated with terrestrial phenomena which produce changes in the inertia tensor.

2.7 Length-of-day (LOD) variation

A different aspect of the variations in Earth rotation is the variation in its spin rate, which is represented by the z-component, in the terrestrial reference frame, of the angular velocity of rotation of the Earth in space. The reference value of the spin rate (the so-called mean rate of Earth rotation) is one cycle per sidereal day (cpsd), equivalent to an angular speed

$$\Omega_0 = 2\pi \text{ rad/sidereal day} = 7.292115 \times 10^{-5} \text{ rad/second}. \qquad (2.1)$$

One *sidereal day* is the time that the Earth would take to rotate through 2π radians relative to distant ("unmoving") objects in space if its rotation axis remained strictly aligned to its axis of maximum moment of inertia, and correspondingly its direction in space remained strictly unvarying. (The nutational and wobble motions of this axis give rise to subtle difficulties in conceptualizing the precise meaning of one complete rotation in space; these will be considered later.) The sidereal day is very close to 23 hours 56 minutes. The layman's "day" is the *solar day* which is the length of time needed for the Earth to rotate once in relation to the direction to the "mean" Sun in the sky; this direction varies cyclically with a period of one year because of the orbital motion of the Sun relative to the Earth (or vice versa). The length of 1 solar day is $r = 1.00273781191135448$ times that of one sidereal day (see the value of r in Chapter 5 of IERS Technical Note 21 or 32 or 36). So the above rate is equivalent to $r \times 2\pi$ rad/solar day or $r \times 2\pi/86400$ rad/sec, which reduces to the number given on the right-hand side of the above equation.

Any decrease/increase in the spin rate results in a corresponding increase/decrease in the duration of a day, which is described by the term "length-of-day (LOD) variation."

Deviations of the spin rate from Ω_0 occur for a variety of reasons. Spin rate variations are dominated by the effects of variations in the atmospheric pressure and in wind speeds and wind patterns over the globe, which occur on timescales of up to many years and are not predictable. Smaller variations are produced with a spectrum of periods in the long-period range (from about 5 days to 18 years) by the action of external gravitational potentials on the density perturbations associated with solid Earth tides and ocean tides. The same mechanism gives rise also to a deceleration of Earth rotation (a constant negative rate of change of the spin rate), which corresponds to a secular increase in the LOD. This secular change is often referred to as lunar braking of the Earth's rotation, as the preponderant role in this phenomenon is played by the Moon's gravitation.

The hour angle (arc span of hours) of rotation about the mean rotation axis or figure axis is termed as *Universal Time UT*. It conforms, within a close approximation, to the mean diurnal motion of the Sun. The notation *UT1* designates the hour angle of rotation approximately about the mean diurnal rotation axis. More precisely, it is the hour angle of the Earth's rotation about the celestial intermediate pole (CIP) axis, that will be introduced later (see Section 2.15). The difference between UT and UT1 is that the latter accounts for non-diurnal *polar motion* (i.e., long-term motion of the Earth's pole of rotation relative to the Earth itself). It is conventionally related to the *Greenwich Mean Sidereal Time (GMST)*. GMST is the hour angle between the prime meridian ($0°$ longitude) at Greenwich and the *vernal equinox* (the intersection of the planes of the Earth's equator and the ecliptic) (see Appendix C for precise definitions) measured westward along the celestial equator. The Mean Time indicates that the vernal point used is the intersection of the Earth's mean equator of date (which takes account of precession but not nutation, as explained above) and the ecliptic of date (which changes with precession). UT1 is also related to the conventional *Earth rotation angle (ERA)* that will be defined in Chapter 12, where the conventions involving the non-rotating origin will be introduced. For more details, please see Chapters 5 and 12.

2.8 Solid Earth and ocean tides

The gravitational acceleration of a mass element in the Earth due to the gravitational field of an external body (say, the Moon or the Sun or other solar system body) is not uniform over the volume of the Earth: the farther an Earth point is from the body, the weaker is the acceleration. The differing accelerations of different elements of matter in the Earth cause deformations of the solid Earth ("solid Earth tides"), as well as ocean tides. Both solid Earth and ocean tides play a significant role in all aspects of Earth rotation variations.

2.9 Inertia tensor; principal axes and moments of inertia

The concepts of the inertia tensor and principal axes and moments of inertia are central to the problem of rotational motion of bodies. So we digress a little here to introduce them formally, even though they should be familiar from courses on mechanics.

The moment of inertia I of a body (rigid or deformable) about an axis in the direction of the unit vector $\hat{\mathbf{n}}$ at any instant of time is

$$I = \int \rho(\mathbf{r})[r^2 - (\mathbf{r} \cdot \hat{\mathbf{n}})^2]d^3r = \sum_{i,j} I_{ij}e_i e_j, \qquad (2.2)$$

where $\cdot$ indicates the scalar product between two vectors and where $\rho(\mathbf{r})$ is the density of matter as a function of position $\mathbf{r}$ in the body, and the integration is over the whole volume of the body ($d^3r = dx_1 dx_2 dx_3 = dx\,dy\,dz$); the e_i, ($i = 1, 2, 3$), are the Cartesian components of $\hat{\mathbf{n}}$ in a specified reference frame, and the summations are from 1 to 3. The I_{ij} are the elements of the *inertia tensor* [**I**] in the reference frame employed. This tensor is evidently symmetric, its elements being

$$I_{ij} = I_{ji} = \int \rho(\mathbf{r})(r^2 \delta_{ij} - x_i x_j)d^3r \qquad (2.3)$$

by virtue of (2.2). Written explicitly, they are:

$$I_{11} = \int \rho(\mathbf{r})(y^2 + z^2)d^3r, \; I_{22} = \int \rho(\mathbf{r})(z^2 + x^2)d^3r, \; I_{33} = \int \rho(\mathbf{r})(x^2 + y^2)d^3r, \qquad (2.4)$$

$$I_{12} = -\int \rho(\mathbf{r})xy\,d^3r, \; I_{23} = -\int \rho(\mathbf{r})yz\,d^3r, \; I_{31} = -\int \rho(\mathbf{r})zx\,d^3r. \qquad (2.5)$$

Here, and in the rest of the book, d^3r stands for $dx\,dy\,dz$.

The tensor [**I**], which may be viewed as a symmetric matrix, can be diagonalized by an orthogonal transformation, which is equivalent to a transformation from the original reference frame to a rotated one. The off-diagonal elements of the transformed matrix vanish:

$$I'_{12} = -\int \rho\, x'y'd^3r' = I'_{23} = -\int \rho\, y'z'd^3r' = I'_{31} = -\int \rho\, z'x'd^3r' = 0, \qquad (2.6)$$

where x', y', z' are the coordinates of the position vector in the rotated frame, and the argument $\mathbf{r}$ of ρ has been suppressed. The diagonal elements of the tensor in the rotated frame are

$$I'_{11} = \int \rho\,(y'^2 + z'^2)d^3r', \quad I'_{22} = \int \rho\,(z'^2 + x'^2)d^3r', \quad I'_{33} = \int \rho\,(x'^2 + y'^2)d^3r'.$$
$$(2.7)$$

The ordering of the three axes of this frame is done in such a way that I'_{11} is the smallest and I'_{33} the largest of the three elements:

$$I'_{11} = A \leq I'_{22} = B \leq I'_{33} = C. \qquad (2.8)$$

The non-vanishing elements A, B, C of the diagonalized inertia tensor are the *principal moments of inertia*. The inertia tensor $[\mathbf{I}]$ reduces to:

$$[\mathbf{C}] = \begin{pmatrix} A & 0 & 0 \\ 0 & B & 0 \\ 0 & 0 & C \end{pmatrix}. \qquad (2.9)$$

The axes of the reference frame in which the inertia tensor is diagonal are called the *principal axes* and the frame itself is known as the *principal axis frame*.

2.10 Low and high frequency nutations

Spectral components having frequencies within *the low frequency band* form the dominant and most important class of nutations of the axis of figure (or of rotation or angular momentum). This band, by definition, consists of frequencies of magnitude below 0.5 cpsd. Only a very small part of the total nutational motion is contributed by high frequency spectral components (outside the low frequency band) whose amplitudes are several orders of magnitude smaller than those of the leading low frequency nutations. Therefore, the greatest effort needed in nutation theory is for modeling the low frequency nutations.

It is an important fact that only the axially symmetric part of the Earth's structure plays a role in the generation of this part of the nutation spectrum as well as the precession through the action of the lunisolar gravitational potentials. The most important and most numerous of all the nutations are those low frequency nutations that arise from the action of the lunisolar gravitational potential on the dynamical ellipticity of the Earth. (The Bradley nutation belongs to this class of nutations, which may conveniently be referred to as the *principal class*; the ellipse described by this nutation is larger than that characterizing any of the nutations outside this class by factors of the order of 10^5 or more.) The dynamical ellipticity is defined as

$$H_d = \frac{C - \bar{A}}{C} = \frac{e}{1 + e} \quad \text{or} \quad e = \frac{C - \bar{A}}{\bar{A}} = \frac{H_d}{1 - H_d}, \qquad (2.10)$$

where C is the Earth's maximum moment of inertia (about the polar axis) and $\bar{A} = (A + B)/2$ is the mean of the two principal equatorial moments of inertia. (The astronomy community employs H_d as a measure of the dynamical ellipticity while geophysicists generally use e.) The value of H_d is close to $1/305$. Since the Earth's deviation from axial symmetry, represented by the so-called triaxiality parameter $e' = (B - A/\bar{A})$, is of the order of $e^2 \approx 10^{-5}$, one often employs the approximation $B = A$; in that case H_d and e reduce to $(C - A)/C$ and $(C - A)/A$, respectively. But it is to be borne in mind that the precise values to be used for these parameters are those defined by (2.10).

The magnitude of the torque on an axially symmetric ellipsoidal Earth is proportional to e, and the rate of precession under the action of that torque is proportional to H_d. For accurate computations of the low frequency nutations it is necessary that the value used for e be sufficiently accurate. The values computed for e from the best available Earth models based on seismological data lead to precession rates that differ by more than 1% from observational estimates of the precession rate having uncertainties of less than 0.01%. It has been the practice therefore to use, in nutation theory, the value obtained for e from the observational estimate of H_d from precession.

Much smaller than nutations of the above class, yet far from insignificant, are other low frequency nutations arising from aspects of the matter distribution in the Earth that are not strictly ellipsoidal though still axially symmetric. Finally, there exist high frequency nutations which, by definition, have frequencies of magnitude exceeding 0.5 cpsd; the slight deviations of the Earth's structure from axial symmetry are responsible for these. The magnitudes of such nutations are well below 0.1 mas. In view of the overwhelming dominance of nutations arising from the part of the Earth's structure that is characterized by e or H_d, the treatments of nutation in books deal more or less exclusively with the special case of an axially symmetric ellipsoidal Earth. This book too is concerned, for the most part, with this special case; however calculations of the effects of deviations from this idealized structure are necessary when results of high accuracy are needed, and we shall present those too.

2.11 Interplay of nutation, tides, ocean, and atmosphere

All three layers of the Earth are deformable. This fact is of considerable importance in the context of Earth rotation variations. Tidal deformations of the mantle and the core regions (the so-called body tides) are produced by the spatially non-uniform accelerations of the elements of matter over the volume of the Earth which result from the direct action of the lunisolar gravitational potential. Indirect contributions to solid Earth deformation arise from the loading of the Earth's surface by the ocean tides raised by the same gravitating bodies; further loading is produced by atmospheric pressure changes over the globe caused by solar heating and, to a far smaller extent, by lunisolar gravitation. The deformations, whether of direct or of indirect origin, perturb the inertia tensors of all the regions and produce associated changes in the Earth's gravitational potential. The changes in the inertia tensor affect Earth rotation for obvious reasons. Conversely, rotation variations give rise to perturbations of the centrifugal potential, which in turn produce deformations over and above the deformations already referred to (namely, those due to the direct action of the tidal potential and due to the ocean and atmospheric loading). Thus solid Earth deformations and rotation variations are inextricably linked. At the same time, the ocean response to external gravitational forcing is influenced by movements of the ocean floor due to solid Earth deformation, and also by the associated perturbations of gravity. Thus ocean tides are also linked to solid Earth deformation and hence to Earth rotation variations; a more direct effect on rotation is through the angular momentum associated with tidal ocean currents.

The interplay of the various phenomena outlined above has to be taken into account when seeking a treatment of Earth rotation variations or any of the other phenomena to a high level of accuracy. Other considerations like the effect of mantle anelasticity (which produces a small delayed deformational response to forcing besides the instantaneous elastic response) also enter the picture (see Fig. 2.1).

2.12 Remaining concepts such as fundamental reference planes

The remainder of this chapter is devoted to introducing the concepts and terminology relating to various aspects of Earth rotation variations and to making illustrative calculations on precession, nutations, and wobbles, assuming very simple models for the Earth and for the gravitational forcing of the Earth. Forcing by the Sun, supposed to move with uniform angular speed along a circular orbit around the Earth, is used as the example. We also make simplified presentations of reference frames that are of the most use for these problems, and give an elementary derivation of the kinematical relations between nutations and wobbles; the relations are not exact, but are good enough to have been very widely used, and have played an important role in nutation theory.

The classical description of nutation and precession is in terms of the ecliptic longitude and obliquity variables, which are defined in relation to two fundamental reference planes: (a) the ecliptic plane, introduced earlier in this chapter and in the previous chapter (see also Appendix C), which is the plane defined by the orbital motion of the Earth–Moon barycenter around the Sun; and (b) the equatorial plane pertaining to the Earth's axis (figure, rotation, or angular momentum). Reference frames tied to the ecliptic plane and to the equatorial plane of the figure axis are

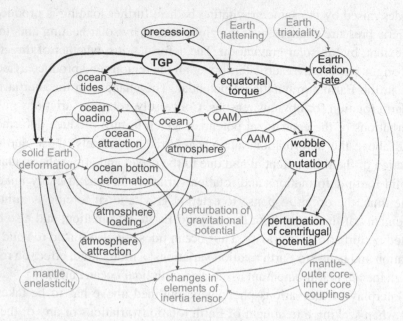

Figure 2.1 Showing the interplay of nutation, solid Earth deformations, and ocean and atmospheric tides (see Dehant and Mathews, 2007). Medium weight line is associated with the external geophysical fluids (where OAM and AAM mean *ocean* and *atmosphere angular momentum*, respectively); light line, with the solid Earth properties; black, with the global rotation/nutation phenomena; bold black with the tide generating potential (TGP).

introduced later in this chapter and dealt with in greater detail in Chapter 3. The intersection of the two fundamental planes is the origin of ecliptic longitude. The transformation between ecliptic and terrestrial reference frames, expressed in terms of the relevant Euler angles, is then presented, and this is followed by a derivation of the kinematical relations in exact form. The difference of these relations from the commonly used approximate form referred to earlier is small but needs to be considered when very accurate results are desired. An alternative description of the transformation between celestial and terrestrial reference frames is through a sequence of transformations which involve the concept of a CIP and its motions relative to the celestial frame as well as relative to the terrestrial frame; the first of these motions of the CIP is identified as precession–nutation, and the second as polar motion. The effect of Earth rotation variations on space geodetic data is, in fact, expressed by means of a representation of this type, which is therefore of obvious practical importance; it is by estimation of the variables appearing in the successive transformations, from the data, that one obtains precise information on the temporal variations of the nutation–precession variables, polar motion variables, and the rate of Earth rotation. We present the essential aspects of this representation of the transformation in the next sections of this chapter and in Chapter 3.

In order to follow the literature on precession and nutation, one has to have some familiarity with the observational aspects, the essentials of which are presented briefly in Chapter 3. The various approaches to theoretical modeling and computation are presented in Chapters 4 to 10. Fitting of the observational data on nutations by theory, and the geophysical inferences that may be made thereby, are treated in Chapter 11. Chapter 12 deals with the recently adopted conventions concerning precession and nutation, and polar motion.

2.13 Sign conventions

A perusal of the literature of Earth rotation variations shows that various authors use different kinds of conventions which make it necessary for the reader to be vigilant in order to avoid misunderstandings. We feel it desirable to specify at the outset the conventions we use in this book, and to clarify some related issues.

1. *Sign convention for potentials*

We deal first with the sign convention related to the gravitational potential $W(\mathbf{r}, t)$ due to one or more bodies, the potential energy $V(\mathbf{r}, t)$ of a mass m at $\mathbf{r}$ in the potential field, and the force $\mathbf{F}(\mathbf{r}, t)$ exerted by the potential on that mass.

The relation between the gravitational *potential energy* V of a mass m and the force $\mathbf{F}$ on the mass is unambiguous: $\mathbf{F}(\mathbf{r}, t) = -\nabla V(\mathbf{r}, t)$. However, there are two conventions in use regarding the sign assigned to the *potential* $W(\mathbf{r}, t)$: the convention followed generally by physicists is in accordance with the relation

$$V(\mathbf{r}, t) = mW(\mathbf{r}, t) \tag{2.11}$$

between the potential energy and the potential, while that found generally in the literature of geophysics corresponds to the relation $V(\mathbf{r}, t) = -mW(\mathbf{r}, t)$. Thus, the gravitational potential W^B due to a body B located at $\mathbf{r}_B(t)$ at time t, and the acceleration $\mathbf{A}(\mathbf{r}, t)$ of a mass at $\mathbf{r}$ that is caused by the potential, are

$$W^B(\mathbf{r}, t) = -\frac{GM_B}{|\mathbf{r} - \mathbf{r}_B|} \quad \text{and} \quad \mathbf{A}(\mathbf{r}, t) = -\nabla W^B(\mathbf{r}, t) \tag{2.12}$$

according to the "physicists' convention," and

$$W^B(\mathbf{r}, t) = +\frac{GM_B}{|\mathbf{r} - \mathbf{r}_B|} \quad \text{and} \quad \mathbf{A}(\mathbf{r}, t) = +\nabla W^B(\mathbf{r}, t)$$

in the "geophysicists' convention," with $\mathbf{F}(\mathbf{r}, t) = m\mathbf{A}(\mathbf{r}, t)$ for both conventions.

We shall follow the physicists' convention throughout this book.

2. *Notation for frequencies*

The term frequency in this book will invariably refer to the angular frequency $(2\pi/T)$ where T is the period.

The symbol ω is reserved for the frequencies, referred to the terrestrial reference frame, of the spectral components of the tidal gravitational potentials acting on the Earth. The Cartwright and Tayler (1971) table of the frequencies and amplitudes of the tidal components has been in very wide use. According to the convention employed in that work and in other works on the spectral representation of the TGP, the tidal frequencies ω are all non-negative.

It is a well-known fact that any elliptical motion of given frequency can be decomposed into two counter-rotating circular motions. In the case of any elliptical motion constituting some spectral component of the nutation of the Earth, its decomposition into two circular parts yields one part in which the motion is in the same sense as the diurnal rotation of the Earth, while motion is in the opposite sense in the other part; the first of these is said to be prograde and the second retrograde. Throughout this book, the prograde component of an elliptical nutation of frequency ω_n will be characterized by the positive frequency $|\omega_n|$ and the retrograde one by $-|\omega_n|$. It is necessary to caution the reader that there are several important papers wherein the opposite convention is used. The origin of the two different conventions is linked to alternative choices (both equally valid) of the complex representation of the spectral components of the tidal potential; stated briefly, a spectral component of the potential, expressed as a real trigonometric function of the time variable, may be written as the real part of either of two complex expressions which are complex conjugates of one another. Consequently, if the time dependent factor is taken as $e^{i\omega_n t}$ in one of the two conventions, it will be replaced by $e^{i(-\omega_n)t}$ in the other convention, thus reversing the sign of the frequency. (The coefficient of $i\,t$ in the exponent is defined as the frequency.) Along with this, the *amplitude* of the spectral component, which multiplies the time dependent factor, will get replaced by its complex conjugate on switching conventions. Further elucidation of these points will have to await the introduction of the complex representation in Chapter 5 and Chapter 8. Suffice it to say, at this stage, that the existence of opposing sign conventions offers scope for much confusion! The reader needs to be alert to the convention followed in each paper that he takes up for study.

As wobble is a motion relative to the terrestrial reference frame, the frequencies of its spectral components are with reference to this frame. A single spectral component of frequency ω of the TGP causes the excitation of a pair of circular wobbles with the frequencies ω and $-\omega$, as will be seen in Chapter 5. We generally use the symbol ω_w for the frequencies of circular wobbles; ω_w is positive (equal to

ω) for prograde wobbles, and negative (equal to $-\omega$) for retrograde wobbles. The signs are reversed when the alternative convention is used, as for instance in Wahr (1981a–d).

There is a significant body of literature in which frequencies are expressed in units of cpsd; we adopt this practice in several chapters of this book. The symbols used for the frequencies in this case are usually ν and σ for circular nutations and wobbles, respectively: $\Omega_0 \nu$ is equal to $|\omega_n|$ for prograde nutations and $-|\omega_n|$ for retrograde ones. Similar relations hold between $\Omega_0 \sigma$ and $\pm|\omega_w|$. The frequencies of the wobbles are the same as those of the spectral components of the torque vectors (referred to the terrestrial frame) that generate them; so the frequencies of the latter may also be denoted by ω_w or $\sigma \Omega_0$.

2.14 Reference frames: basic aspects

Appropriate celestial (space-fixed) and terrestrial reference frames are needed for the description of motion of any given axes in space and relative to the rotating Earth. We start with a brief introduction to reference frames at this point; precise characterizations of such frames are deferred to Chapter 12.

All frames considered here are geocentric, i.e., they have their origin at the geocenter. Each frame is attached to its *principal plane* which passes through the origin. The frame consists of a right-handed triad of orthogonal unit vectors of which the first two lie in the principal plane and the third is perpendicular to the plane; once this plane is specified, the definition of the frame is complete when the first axis too is specified in some way, usually as being along the line of intersection (nodal line) of the principal plane with another plane.

2.14.1 Terrestrial reference frames

A geocentric terrestrial reference system (GTRS) has been defined jointly by IAU and IUGG as "a system of geocentric space-time coordinates within the framework of general relativity, co-rotating with the Earth and related to the geocentric celestial reference system (GCRS) by a spatial rotation which takes into account the Earth orientation parameters." The co-rotation condition is defined as no residual rotation with regard to the Earth's surface, and the geocenter is understood as the center of mass of the whole Earth system, including oceans and atmosphere.

Geocentric terrestrial reference frames (GTRF) are realizations of the GTRS for which the orientation is operationally maintained in continuity with past international agreements and are conceived of as being co-moving with the Earth in space. Their axis orientations are related, ideally, to the Earth's equator and its figure axis, the latter being associated with the $\hat{z}$-axis. In such a system, positions of points attached to the solid surface of the Earth have coordinates which undergo only small variations with time, due to geophysical effects (mainly tectonic/seismic motions and deformations of tidal origin). The time evolution of the orientation is ensured by using a no-net-rotation condition with regard to horizontal tectonic motions over the whole Earth (the no-net-rotation condition (NNRC), which is also called the Tisserand condition; it states that the total angular momentum of all tectonic plates should be zero).

The frame which is the most important for reference purposes is the international terrestrial reference frame (ITRF) which may be thought of as having the Earth's polar axis or axis of maximum moment of inertia as its third axis, and hence the equatorial plane as its principal plane; its first axis is in the direction of the origin of longitudes in the equatorial plane, which lies in the conventional Greenwich or International meridian (a half-plane passing through the north and south poles and through "Greenwich," as the origin of longitudes, or close to the conventional Greenwich for the present-day definition of the prime meridian). We shall have occasion also to use the principal axis frame, already introduced in Section 2.9, having its first axis lying along the axis of least moment of inertia, in the equatorial plane.

A terrestrial reference frame is thus ensured, in practice, using 3-dimensional (Cartesian) coordinates of the positions and velocities of a large number of such stations. In constructing such a frame, one has thus to take account of the fact that the positions of the stations, relative to each other as well as relative to the body of the Earth, keep changing for a number of reasons: the slow relative motions of the distinct "plates" of which the crust is made up over the Earth's surface, tidal and other deformations of the Earth, and so on. In reality, the ITRF is a realization of GTRS by a set of instantaneous coordinates (and velocities) of reference points distributed on the topographic surface of the Earth (mainly space geodetic stations and related markers). Currently the ITRF provides a model for estimating, to high accuracy, the instantaneous positions of these points, which is the sum of conventional corrections provided by the IERS (International Earth Rotation and Reference Frame Service) Convention center (solid Earth tides, pole tides, . . .) and of a regularized position. At present, the latter is modeled by a piecewise linear function, the linear part accounting for such effects as tectonic plate motion, postglacial rebound, and the piecewise aspect representing discontinuities such as seismic displacements. The initial orientation of the ITRF is that of the BIH (Bureau International de l'Heure) Terrestrial System at epoch 1984.0. For continuity with previous terrestrial reference systems, the first alignment was thus close to the mean equator of 1900 and the Greenwich meridian.

The length unit is the SI unit. However, a scale factor is usually evaluated in order to permit each technique to be consistent (at the same scale), with

VLBI as the reference. Such a scale may change from one reference frame realization to the other.

Any direction in the terrestrial frame may be identified by the longitude λ and colatitude θ (or the latitude $(\pi/2 - \theta)$) of the surface point at which the direction vector, originating at the geocenter, intersects the Earth's surface). (As is only too well-known, the latitude goes from $0°$ at the equator to $\pm 90°$ at the north and south poles, respectively; and the colatitude goes from $0°$ at the north pole to $180°$ at the south pole.)

2.14.2 Celestial reference frames

We introduce here the two basic types of celestial reference frames: the ecliptic and the equatorial celestial reference frames. They have the ecliptic plane and the Earth's mean equatorial plane, respectively, as their principal planes. The principal planes are often identified (especially in the astronomy literature) by the circles along which they intersect the celestial sphere: the ecliptic for the ecliptic plane, and the equator for the equatorial plane. The equatorial plane referred to here is that of the Earth's figure axis. As for the ecliptic, it represents the plane of the orbit of the Earth–Moon barycenter around the Sun. (The path of the Earth itself does not stay strictly in this plane because of its motion relative to the Earth–Moon barycenter as each body revolves about the other. But the excursions out of the plane are small, and it is not unusual for the ecliptic to be described loosely as the plane of the Earth's orbit around the Sun.)

2.14.3 Ecliptic celestial reference frame

It is an inconvenient fact that the ecliptic plane is not entirely stationary in space; it undergoes a very slow rotation because of the gravitational pulls of the other planets on the Earth–Moon system. But this rotation does not affect the phenomenon of primary interest to us, namely, the precession–nutation of the Earth's axes in space. So it is sufficient for our purposes to consider a fixed ecliptic and the reference frame attached to it. (Technical points related to the motion of the ecliptic are dealt with at a later stage; see Chapters 3 and 5.) The plane coinciding with the ecliptic of the epoch J2000 is taken as the space-fixed principal plane for the ecliptic celestial reference system (ECRS). J2000 is at 12 hours universal time on 1 January 2000. The direction of the third axis of the system may be loosely characterized as pointing towards the "northern" side of the ecliptic plane (inclined at about $23.5°$ to the direction of the Earth's north pole). The first axis is chosen to be in the direction of the *mean equinox* of J2000 (more specifically, the mean vernal equinox). The mean vernal and autumn equinoxes of any given epoch lie on the line of nodes (intersection) of the ecliptic plane and the mean equatorial plane of that epoch. The vernal (or spring) equinox is on the celestial sphere at that end of the nodal line where the Sun, in its apparent motion around the Earth along the ecliptic, crosses over from south of the Earth's equator to the northern side; the autumn equinox is at the other end.

The direction of a celestial object relative to the ecliptic frame is characterized by its barycentric ecliptic longitude and latitude, defined analogously to the geocentric longitude and latitude in the terrestrial frame. The change in orientation of the direction of the Earth's axis from its mean position is expressed, however, in terms of changes in its longitude and obliquity, as will be explained later. (Obliquity is the angle made by this direction to that of the normal to the ecliptic.)

2.14.4 Equatorial celestial reference frame

A plane close to the mean equatorial plane of J2000 is chosen as the principal plane for the space-fixed equatorial reference frame. (The reason for having to specify an epoch is simply that the mean equatorial plane keeps moving: that motion is the precession of the equator, which has been traditionally called the lunisolar precession.) With the geocenter as the origin, such a frame is called the GCRS. The third axis of the GCRS points towards the north pole, of course. The first axis is chosen, as in the case of the ecliptic frame, to be in the direction of the mean vernal equinox of J2000 (or as close as possible to it).

Directions in relation to the GCRS are specified by the declination δ and the right ascension α. The former is analogous to the latitude in the terrestrial frame and α to the longitude; the origin of α is at the equinox.

The above definitions are simplified presentations of more precise characterizations which are too technical to be of interest at this stage (see Chapter 12).

2.15 Precession and nutation of different axes

Precession and nutation have been already introduced in earlier sections as motions in space of one or the other of various axes related to the Earth's figure or rotation.

The motion of the angular momentum vector in space is governed by the torque equation referred to a space-fixed reference frame, which equates the rate of change of this vector directly to the torque acting on the Earth. The torque is determined solely by the structure of the unperturbed Earth as a whole, meaning that the existence and properties of the core regions, for instance, make no difference to the torque. (We are ignoring here the extremely small contributions involving the Earth's tidal deformations, which have been considered only during the last few years; they will be taken up in Chapter 9.) But, for this very reason, nutations of this axis reveal nothing about the Earth's properties beyond the moments of the density distribution in the Earth, and are therefore only of limited interest in studies of the nutations of the "real" (non-rigid) Earth. In rigid Earth theories, however, except in Hamiltonian theories, it has been the tradition to make computations of the nutations of the angular momentum axis and then to use these results to obtain the nutations of the figure and rotation axes.

The nutation of the figure axis, on the other hand, is very sensitive to the presence and properties of the core regions, and to their interactions with the mantle and between themselves; and the nutation of this axis is a part of the time dependent transformation between the celestial and terrestrial reference frames. Since precise observational estimation of the parameters of this transformation has been done on a regular basis from space geodetic data for over two decades now, correspondingly precise knowledge of the nutation of the figure axis (or more precisely the nutation of the CIP) over this period is now available. Therefore, theoretical studies of the nutation of the figure axis of non-rigid Earth models have become of considerable importance, and have served to shed light on several properties of the interior of the Earth. The nutation of the rotation axis is closely related to that of the figure axis, and may readily be inferred from the latter.

Yet another axis is employed for the definition of nutation in a technical sense in the context of the definition and observational estimation of Earth orientation parameters (EOPs). This axis is in the direction of a conceptual pole, now called the celestial intermediate pole or CIP (see Capitaine, 2000, 2002, 2012, Capitaine *et al.*, 2000, 2002, 2003(a–d), 2004, 2005(a–c), 2006, 2007) which is introduced in Chapter 3, Section 3.2.3. As explained there, the nutational motion of the CIP is the same as that of the pole of the figure axis except for a restriction on the spectral content of the motion to frequencies of magnitude up to 0.5 cpsd. Therefore, a study of the motion of the figure axis suffices to provide the needed information about the motion of the CIP.

2.16 Precession and nutation variables

The classical characterization of the direction of the Earth's axis in space is in relation to the ecliptic plane. The angle between the Earth's axis and the normal to the ecliptic plane (or what is the same thing, the angle between the ecliptic and equatorial planes) is called the *obliquity* and is denoted by ϵ. The mean obliquity (or more precisely, the value of the mean obliquity at the epoch J2000) is $\epsilon_0 \approx 23.5°$. The time dependent deviation

$$\Delta\epsilon = \epsilon - \epsilon_0 \tag{2.13}$$

of the obliquity from the constant value ϵ_0 represents the *nutation in obliquity*, and to a minor extent, also a very slow secular variation in obliquity. (See Fig. 2.2 for

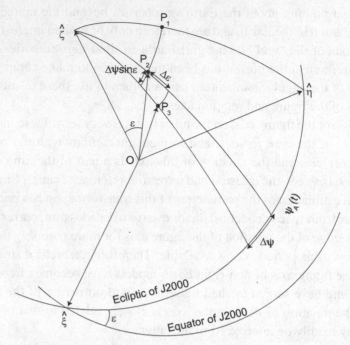

Figure 2.2 Displacement of Earth's pole (P_1, P_2, P_3) on the celestial sphere due to precession–nutation expressed in terms of the precession angle (ψ_A) and the nutation angles ($\Delta\psi$ and $\Delta\epsilon$) in longitude and in obliquity.

a representation of the motion of the pole related to precession and nutation and Fig. 2.3 for a representation of the nutation effects on the coordinates of the pole.) The maximum deviation due to the nutation in obliquity is about ± 10 arcseconds. These motions in obliquity take the pole of the Earth's axis over a very small arc of the great circle on the celestial sphere which passes through this pole and the pole of the ecliptic. (The latter pole lies on the normal to the ecliptic of course.)

Nutation and precession in *longitude* constitute the component of the motion of the Earth's pole that is orthogonal to the motion in obliquity. This motion is over an arc of the circle of constant obliquity ϵ on the celestial sphere. It is measured by the change in ecliptic longitude of the position of the pole, i.e., by the angle through which the point has moved around the pole of the ecliptic. Precession in longitude is the secular part of the motion, which is denoted by $\psi_A(t)$. At the present rate of precession, which is very close to 50 arcseconds per year, the pole of the Earth's axis would take about 25 700 years to complete a full circuit around the pole of the ecliptic. The oscillatory part of the motion on the constant-obliquity circle, denoted by $\Delta\psi(t)$, is the nutation. $\Delta\psi(t)$ could go up to about ± 19 arcseconds. The linear displacement of the Earth's pole along the above circle, corresponding to the nutational increment $\Delta\psi$ to the longitude, is $\Delta\psi \sin\epsilon$ since

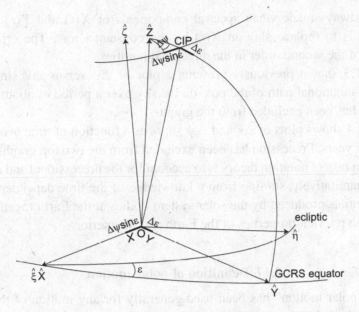

Figure 2.3 Displacement of the Earth's pole (CIP) due to nutation indicated by $\Delta\psi \sin\varepsilon$ and $\Delta\varepsilon$ on the celestial sphere and expressed as coordinates X and Y along $O\hat{X}$ and $O\hat{Y}$. The coordinate systems are indicated with the unit vectors $\hat{X}$, $\hat{Y}$, and $\hat{Z}$ for the frame tied to the GCRS equator, and $(\hat{\xi}, \hat{\eta},$ and $\hat{\zeta})$ for the frame tied to the ecliptic of J2000 (neglecting the ecliptic motion).

the radius of the circle is $\sin\epsilon$; similarly for the precessional motion in longitude (see Fig. 2.3).

The direction in which the pole of the nutating axis has to move in order that the resulting change in $\psi_A(t)$ or $\Delta\psi(t)$ be positive is in the sense of a left-handed rotation around the pole of the ecliptic, *opposite* to that of the motion of the Earth–Moon system around the normal to the ecliptic. Figures 2.2 and 2.3 bring out the fact that, at J2000, the direction of increasing longitude is parallel to the direction of the axis $O\hat{X}$ of the GCRS. It may also be seen that the second axis $O\hat{Y}$ is parallel to the direction of increasing obliquity. Therefore, the Cartesian coordinates of the pole of the Earth's axis in the geocentric celestial reference frame (GCRF) are

$$X(t) = [\psi_A(t) + \Delta\psi(t)] \sin\epsilon, \qquad Y(t) = \Delta\epsilon(t). \qquad (2.14)$$

Note that the first of these relations becomes gradually more and more inaccurate as the secular term $\psi_A(t)$ keeps increasing steadily with the passage of time; this is because ψ_A is measured along a circle, while $X(t)$ is measured along a Cartesian coordinate axis which is a straight line tangential to this circle. Correction terms have to be introduced for sufficiently long times. Another approximation which is almost always made when spectral components of $X(t)$ and $Y(t)$ are to be considered is to replace $\sin \epsilon$ in $X(t)$ by the constant $\sin \epsilon_0$. The error caused thereby is of the second order in the nutation quantities.

Figure 1.3 shown previously presents a plot of $\Delta \epsilon$ versus $\Delta \psi \sin \epsilon_0$ which depicts the nutational path of the pole on the sky over a period of about 19 years. Precession has been excluded from the graph.

Figure 2.4 shows plots of $\Delta \epsilon$ and $\Delta \psi \sin \epsilon_0$ as a function of time over a period of about 35 years. Precession has been excluded from the two top graphs.

The main task of nutation theory is to account for the precessional and nutational motions quantitatively, starting from a knowledge of the time dependent gravitational potentials produced by the solar system bodies at the Earth together with a model of the relevant properties of the Earth and its interior.

2.17 Definition of polar motion

The term "polar motion" has been used generally for any motion of the Earth's pole of rotation relative to the Earth, and in particular for the wobble as well as for secular polar motion. The latter is a steady drift of the mean pole of rotation (with the periodic wobbles averaged out) relative to the Earth; it may be viewed as the present-day trend of the wandering of the Earth's pole of rotation over the surface of the Earth over geological timescales, after taking account of the motion of the tectonic plates.

It is important to recognize, however, that the technical meaning of polar motion or polar motions in the context of the definition and observational estimation of EOPs is quite different from that of wobble or wobbles. Polar motion(s) is defined in that context, as will be elaborated upon in Section 3.2.3, as the motion(s) of the CIP with frequencies outside the retrograde diurnal frequency range (i.e., with frequencies ≤ -1.5 or ≥ -0.5 cpsd) relative to the rotating terrestrial reference frame. The distinction between these polar motions, on the one hand, and wobbles having the same frequencies, on the other, is explained further in the section just mentioned.

In general, polar motion may be conceived of as the motion of a specified pole relative to the terrestrial reference frame. Besides motions of the rotation pole and the CIP, one may, for example, consider those of the pole of the celestial reference frame, for reasons that will become apparent in due course; in fact we shall do so later in this chapter (see Section 2.25).

The position of the time dependent pole which is viewed as undergoing polar motion (whichever pole it might be) is expressed in terms of the *polar motion parameters* denoted by (x_p, y_p). These parameters, which are time dependent, characterize the sequence of two rotations, $R_2(x_p)R_1(y_p)$, which would carry the

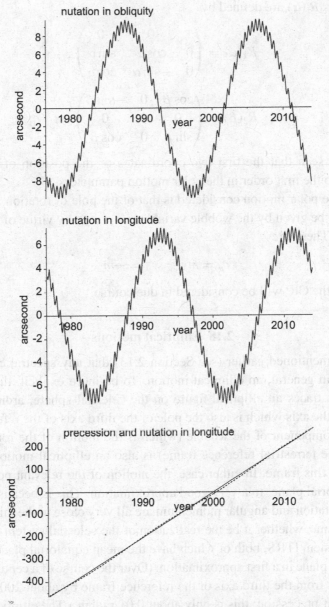

Figure 2.4 Nutation $\Delta\epsilon$ in obliquity (top graph) and $\Delta\psi$ in longitude (middle graph) as a function of time over 35 years. Bottom graph: Precession in longitude (dashed line), with nutation superimposed (solid line).

pole of the terrestrial frame (i.e., the pole of the $\hat{z}$-axis, with coordinates $(0, 0, 1)$) to the instantaneous position of the pole that is undergoing polar motion. Here R_1 and R_2 stand for rotations about the first and second axes, respectively (see Appendix A of the book, Eqs. (A.2) and (A.3)); the angles of rotation are the arguments shown.

The matrices $R_i(\alpha)$ are defined by

$$R_1(\alpha) = \begin{pmatrix} 1 & 0 & 0 \\ 0 & \cos\alpha & \sin\alpha \\ 0 & -\sin\alpha & \cos\alpha \end{pmatrix}, \tag{2.15}$$

$$R_2(\beta) = \begin{pmatrix} \cos\beta & 0 & -\sin\beta \\ 0 & 1 & 0 \\ \sin\beta & 0 & \cos\beta \end{pmatrix}. \tag{2.16}$$

It is readily seen that the first two coordinates of the position of this pole are $(x_p, -y_p)$, to the first order in the polar motion parameters.

In case the polar motion considered is that of the pole of rotation, these coordinates should be given by the wobble variables (m_1, m_2), by virtue of the definition of wobbles. Therefore

$$x_p = m_1, \quad y_p = -m_2. \tag{2.17}$$

The case of the CIP will be considered in due course.

2.18 Elliptical motions

It has been mentioned earlier (see Section 2.13) that any spectral component of nutation is, in general, an elliptical motion. To be more explicit, the pole of the relevant axis traces an elliptical figure on the celestial sphere, around the mean direction of the axis which is also the pole of the third axis of the reference frame. A spectral component of the wobble (which is the motion of the rotation axis as viewed in the terrestrial reference frame) is also an elliptical motion around the third axis of this frame. In either case, the motion of the relevant pole is parallel to the equatorial plane, to a very good approximation. This is because the axes of figure and rotation and angular momentum are all very close to the third axis of the reference frame, whether it be the realization of the celestial system GCRS or the terrestrial system ITRS, both of which have the mean equatorial plane of J2000 as the principal plane in a first approximation. (Over the course of a century, the figure axis deviates from the third axis of the reference frame by about 2000 arcseconds on account of precession; this is only about 0.01 radian.) The elliptical motion of any of the axes that is nutating or wobbling is therefore describable by harmonic variations of the equatorial components of the axis. The reason for the elliptical nature of the motion of the axes lies in the elliptical motion of the torque vector that gives rise to the motion; this fact is illuminated by the simple examples presented in later sections of this chapter. We restrict ourselves at this stage to a general consideration of elliptic motions and the resolution of their spectral components into retrograde and prograde circular motions, primarily in relation to nutations.

The standard representation of the time dependence of the terms $\Delta\psi_n(t)$ and $\Delta\epsilon_n(t)$ of frequency ω_n in the spectral expansion of the nutation $\Delta\psi$ in longitude and $\Delta\epsilon$ in obliquity is through expressions of the following form:

$$\Delta\psi_n(t) = \Delta\psi_n^{\text{ip}} \sin \Xi_n(t) + \Delta\psi_n^{\text{oop}} \cos \Xi_n(t), \qquad (2.18)$$

$$\Delta\epsilon_n(t) = \Delta\epsilon_n^{\text{ip}} \cos \Xi_n(t) + \Delta\epsilon_n^{\text{oop}} \sin \Xi_n(t), \qquad (2.19)$$

where ip and oop stand for "in-phase" and "out-of-phase" as explained below. The coefficients $\Delta\psi_n^{\text{ip}}, \Delta\psi_n^{\text{oop}}, \Delta\epsilon_n^{\text{ip}}, \Delta\epsilon_n^{\text{oop}}$ are called the *coefficients of nutation*, and $\Xi_n(t)$, given by

$$\Xi_n(t) = (\omega_n t + \chi_n), \qquad (2.20)$$

is the *argument of the nutation*. As stated already in Section 2.13, the frequency ω_n is traditionally taken as positive in most cases, but is negative in some cases including that of the largest of the nutations – the 18.6 year nutation.

The argument $\Xi_n(t)$ originates mainly in the lunisolar potential, which drives the nutation. It is a linear combination of a number of fundamental arguments representing the fundamental periodicities present in the orbital motions of the Moon and the Sun relative to the Earth (see Section 5.5.6). Nutations of low frequencies, which constitute the dominant class, are generated by the action of a certain part of this potential (to be precisely identified later) which acts on the equatorial bulge of the Earth or, more precisely, on its dynamical ellipticity e. As viewed from the celestial frame, the generic spectral component of this part of the potential is proportional to $\sin \Xi_n(t)$, according to the conventions that are very widely followed (though the choice of fundamental arguments of the nutations may be such as to necessitate the presence of $\cos \Xi_n(t)$ terms as well, as discussed later). If one employs a rigid Earth model, or if one ignores the dissipative phenomena in the non-rigid Earth (e.g., mantle anelasticity, ocean tidal effects, fluid viscosity), which affect its rotational dynamics and cause the nutational response to be out of phase with the driving potential, $\Delta\psi_n(t)$ will contain only the $\sin \Xi_n$ term, and only the $\cos \Xi_n$ term will be present in $\Delta\epsilon_n(t)$, for this class of nutations (see Chapters 5 and 7). These terms describe the "in-phase" part of the nutation, symbolized by the superscript ip. The remaining terms describe the "out-of-phase" (oop) part of the nutation. Estimates from VLBI data show the ratios $\Delta\psi_n^{\text{oop}}/\Delta\psi_n^{\text{ip}}$ and $\Delta\epsilon_n^{\text{oop}}/\Delta\epsilon_n^{\text{ip}}$ to be very small, of the order 10^{-3} or less for almost all n, with only an exceptional case where one of them is about 10^{-2} (see Table 2.1). Nevertheless the magnitudes of the out-of-phase coefficients are much above the observational uncertainties for many of the nutation components, and their theoretical interpretation through adequate modeling of the various dissipative physical processes poses a challenge to the theorist.

Recall now that the nutational part of the motion of the pole of the axis of interest is described in terms of its equatorial coordinates $(X(t), Y(t))$ in the celestial reference frame, with the secular term proportional to ψ_A removed from $X(t)$: see the relations (2.14) and Fig. 2.4. These relations become linear in $\Delta\psi$ and $\Delta\epsilon$ on replacing $\sin\epsilon = \sin(\epsilon_0 + \Delta\epsilon)$ by $\sin\epsilon_0$; the error introduced thereby is very small, being of order $\Delta\psi\Delta\epsilon$. Subject to this approximation, one may readily write down the spectral components of $X(t)$ and $Y(t)$ as

$$X_n(t) = \Delta\psi_n(t)\sin\epsilon_0 = A_n \sin \Xi_n(t) + A'_n \cos \Xi_n(t), \tag{2.21}$$

$$Y_n(t) = \Delta\epsilon_n(t) = B_n \cos \Xi_n(t) + B'_n \sin \Xi_n(t), \tag{2.22}$$

wherein the unprimed coefficients relate to the in-phase part and the primed to the out-of-phase part of the nutation:

$$A_n = \Delta\psi_n^{\mathrm{ip}}\sin\epsilon_0, \quad A'_n = \Delta\psi_n^{\mathrm{oop}}\sin\epsilon_0, \quad B_n = \Delta\epsilon_n^{\mathrm{ip}}, \quad B'_n = \Delta\epsilon_n^{\mathrm{oop}}. \tag{2.23}$$

Let us now look at the in-phase and out-of-phase parts separately. Since Ξ_n is linear in t, it is evident that the curve traced by the in-phase part of the expressions (2.21) and (2.22) as the time advances is an ellipse with its major and minor axes along the equatorial axes of the reference frame. It is also clear that in the case of a dissipative Earth, the out-of-phase parts of $X_n(t)$ and $Y_n(t)$ may be viewed as constituting a separate elliptical motion of a similar nature.

2.19 Decomposition into circular motions

We have already made note of the well-known fact that any elliptical motion can be decomposed into two counter-rotating circular motions. The decomposition of the two-vector with components $X_n(t)$ and $Y_n(t)$ is done by writing it as

$$\begin{pmatrix} X_n(t) \\ Y_n(t) \end{pmatrix} = A_n^+ \begin{pmatrix} \sin \Xi_n \\ -\cos \Xi_n \end{pmatrix} + A_n^- \begin{pmatrix} -\sin \Xi_n \\ -\cos \Xi_n \end{pmatrix}$$

$$+ A_n'^+ \begin{pmatrix} \cos \Xi_n \\ \sin \Xi_n \end{pmatrix} + A_n'^- \begin{pmatrix} \cos \Xi_n \\ -\sin \Xi_n \end{pmatrix}, \tag{2.24}$$

with

$$A_n^+ = \frac{1}{2}(A_n - B_n), \qquad A_n^- = -\frac{1}{2}(A_n + B_n), \tag{2.25}$$

$$A_n'^+ = \frac{1}{2}(A'_n + B'_n), \qquad A_n'^- = \frac{1}{2}(A'_n - B'_n). \tag{2.26}$$

It is evident that each of the two-component vectors on the right-hand side of Eq. (2.24) executes a circular motion. The two two-component vectors having the coefficients A_n^+, $A_n'^+$ together constitute one elliptical motion, and the other two constitute another elliptical motion.

A little reflection shows that if the frequency ω_n is positive, the first and third of these vectors rotate in the sense from the positive direction of the first axis of the equatorial reference frame towards that of the second axis. This is the same sense in which the Earth rotates, and is therefore said to be *prograde*. The second and fourth vectors rotate in the opposite or *retrograde* sense. If ω_n is negative, it is the second and fourth of the two-vectors that are prograde while the other two are retrograde. Thus the amplitudes of the two prograde parts are A_n^+, $A_n'^+$ and those of the retrograde parts are A_n^-, $A_n'^-$ if ω_n is positive; these connections are reversed if ω_n is negative. One can then write down the following unified expressions, good for both signs of ω_n, for the prograde and retrograde amplitudes:

$$A_n^p = \frac{1}{2}(s_n A_n - B_n), \qquad A_n^r = -\frac{1}{2}(s_n A_n + B_n), \qquad (2.27)$$

$$A_n'^p = \frac{1}{2}(A_n' + s_n B_n'), \qquad A_n'^r = \frac{1}{2}(A_n' - s_n B_n'), \qquad (2.28)$$

wherein the superscripts p and r stand for "prograde" and "retrograde," respectively, and s_n is the sign of ω_n, i.e.,

$$s_n = \omega_n/|\omega_n|. \qquad (2.29)$$

The inverse relations are

$$A_n = s_n(A_n^p - A_n^r), \qquad B_n = -(A_n^p + A_n^r), \qquad (2.30)$$

$$A_n' = (A_n'^p + A_n'^r), \qquad B_n' = s_n(A_n'^p - A_n'^r). \qquad (2.31)$$

Substitution of these relations into Eqs. (2.21) and (2.22) enables one to rewrite $X_n(t)$ and $Y_n(t)$ explicitly in terms of the in-phase prograde and retrograde coefficients A_n^p, A_n^r, and the out-of-phase ones, $A_n'^p$, $A_n'^r$. The resulting expressions are:

$$\begin{pmatrix} X_n(t) \\ Y_n(t) \end{pmatrix} = A_n^p \begin{pmatrix} s_n \sin \Xi_n \\ -\cos \Xi_n \end{pmatrix} + A_n'^p \begin{pmatrix} \cos \Xi_n \\ s_n \sin \Xi_n \end{pmatrix}$$

$$+ A_n^r \begin{pmatrix} -s_n \sin \Xi_n \\ -\cos \Xi_n \end{pmatrix} + A_n'^r \begin{pmatrix} \cos \Xi_n \\ -s_n \sin \Xi_n \end{pmatrix}. \qquad (2.32)$$

The complex combination

$$\tilde{\eta}_n(t) \equiv X_n(t) + i Y_n(t) \qquad (2.33)$$

is extensively used in many theoretical treatments of nutations. It is readily seen from (2.32) that $\tilde{\eta}_n(t)$ reduces to a sum of prograde and retrograde terms with complex amplitudes $\tilde{\eta}_n^{\mathrm{p}}$ and $\tilde{\eta}_n^{\mathrm{r}}$, respectively:

$$\tilde{\eta}_n(t) = -i[\tilde{\eta}_n^{\mathrm{p}} e^{i\,\Xi_n^{\mathrm{p}}(t)} + \tilde{\eta}_n^{\mathrm{r}} e^{i\,\Xi_n^{\mathrm{r}}(t)}], \qquad \Xi_n^{\mathrm{p}} = -\Xi_n^{\mathrm{r}} = |\omega_n|t + s_n\chi_n, \qquad (2.34)$$

with $s_n = \omega_n/|\omega_n|$. The amplitudes of the two parts are

$$\tilde{\eta}_n^{\mathrm{p}} = A_n^{\mathrm{p}} + iA_n^{\prime\mathrm{p}}, \qquad \tilde{\eta}_n^{\mathrm{r}} = A_n^{\mathrm{r}} + iA_n^{\prime\mathrm{r}}. \qquad (2.35)$$

These are related to the coefficients of nutation appearing in (2.18) and (2.19) through Eqs. (2.30) and (2.31) taken together with (2.23). These equations are also a matter of conventions as already mentioned in Section 2.13. The same convention as the one above is used in Defraigne *et al.* (1995b) and in the IERS Conventions. Herring *et al.* (1986a) as well as Wahr and Sasao (1981) define the prograde and retrograde amplitudes differently as they use $\tilde{\eta}_n^{\mathrm{p}} = A_n^{\mathrm{p}} - iA_n^{\prime\mathrm{p}}$, and $\tilde{\eta}_n^{\mathrm{r}} = A_n^{\mathrm{r}} + iA_n^{\prime\mathrm{r}}$. A_n^{p}, $A_n^{\prime\mathrm{p}}$, A_n^{r}, $A_n^{\prime\mathrm{r}}$ are the quantities that are denoted by $A_k^{\mathrm{pro\ ip}}$, $A_k^{\mathrm{pro\ op}}$, $A_k^{\mathrm{retro\ ip}}$, $A_k^{\mathrm{retro\ op}}$, respectively, by Defraigne *et al.* (1995b) and in the IERS Conventions; similarly, A_n, B_n, A_n', B_n' are the quantities that are denoted by $\Delta\psi_n^{\mathrm{p,ip}}\sin\epsilon_0$, $\Delta\psi_n^{r,\mathrm{oop}}\sin\epsilon_0$, $\Delta\epsilon_n^{\mathrm{p,ip}}$, $\Delta\epsilon_n^{\mathrm{p,oop}}$, respectively, by Defraigne *et al.* (1995b) and in the IERS Conventions. The ip, oop designation is appropriate to the dominant class of low frequency nutations driven by the action of the external potential on the Earth's dynamical ellipticity; there are other categories of nutations for which this designation would be incorrect, though those are far less important.

One important feature of the expression (2.34) will be taken for granted throughout the book: the *frequency of a circular nutation*, defined to be the factor multiplying the combination of i and the time variable in the exponent of the time dependent factor, *is positive if the nutation is prograde and negative if the nutation is retrograde.*

A spectral component of the wobble quantity $\tilde{m}(t) = m_1(t) + im_2(t)$ also describes a prograde or retrograde circular motion, with an amplitude that is complex, in general. The amplitudes of the nutations and the associated wobbles are, in fact, intimately related as shown in Section 2.26, and so there is no need to consider the decomposition of the wobbles in detail here.

The complex representation of nutations and wobbles and also of the torque vectors which generate these motions is employed throughout this book, and it would be advantageous for the reader to be well acquainted with it.

Nutations have periods extending from thousands of years down to fractions of a day. The largest of them are the 18.6 year, 9.3 year, annual, semiannual, and 13.66 day nutations. For these five periods, the coefficients $\Delta\psi_n^{\mathrm{ip}}$, $\Delta\epsilon_n^{\mathrm{ip}}$, $\Delta\psi_n^{\mathrm{oop}}$, and $\Delta\epsilon_n^{\mathrm{oop}}$, of the nutations of the figure axis are shown in Table 2.1, and the corresponding prograde and retrograde amplitudes are listed in Table 2.2. The values are from recent computations, and are given both for a non-rigid Earth

Table 2.1 *Coefficients in longitude and obliquity of prominent nutations (in mas).* *"ip" means in-phase, and "oop" out-of-phase.*

	Rigid Earth nutations				Non-rigid Earth nutations			
	In longitude		In obliquity		In longitude		In obliquity	
Period*	ip	oop	ip	oop	ip	oop	ip	oop
−18.6 years	−17281	0.3	9228	0.0	−17208	3.3	9205	1.5
−9.3 years	209	0.0	−90	0.0	207	0.1	−90	0.0
1 year	126	0.0	0	0.0	148	1.2	7	−0.2
0.5 year	−1277	0.0	553	0.0	−1317	−1.4	573	−0.5
13.66 days	−222	0.0	95	0.0	−228	0.3	98	0.1

* The periods shown are the historically assigned values, a few of which are negative.

Table 2.2 *Amplitudes of prograde and retrograde circular nutations (in mas).*

	Rigid Earth nutations				Non-rigid Earth nutations			
	Retrograde		Prograde		Retrograde		Prograde	
Period*	ip	oop	ip	oop	ip	oop	ip	oop
18.6 years	−8051	0.1	−1177	0.1	−8025	1.4	−1180	0.0
9.3 years	87	0.0	4	0.0	86	0.0	4	0.0
1 year	−25	0.0	25	0.0	−33	0.3	26	0.1
0.5 year	−23	0.0	−531	0.0	−25	−0.1	−548	−0.5
13.66 days	−3	0.0	−92	0.0	−4	0.0	−94	0.1

* The periods shown are for the prograde nutations. The retrograde nutations have the corresponding negative values for their periods.

model and for a hypothetical rigid Earth. The entries in the table bring out the fact that the nutations of the figure axis are significantly affected by non-rigidity of the Earth, especially at some frequencies; in particular, the amplitude of the retrograde annual nutation of the non-rigid Earth is over 30% higher than that of the rigid Earth, while that of the prograde annual nutation is hardly affected by non-rigidity. The reason for the differing levels of sensitivity of different spectral components is the existence of certain normal modes and associated resonances (especially the free core nutation (FCN)) which affect the Earth's response to the torque in a strongly frequency dependent fashion, as will be seen from the theory presented in later chapters.

The concepts of elliptical and circular motions and of their interrelations, outlined above, are of importance in the context of the motions, relative to a celestial equatorial reference frame, of the various nutating axes and of the torque vector responsible for the nutations, as well as in the context of the motion of the torque vector relative to the terrestrial reference frame and of the wobble motion which is defined relative to that frame. As mentioned above, the response of the Earth to the retrograde component of the torque is often quite different from its response to the prograde component though both the components are parts of the same elliptical motion of the torque vector.

2.20 Free rotational modes

Before we go on to the study of rotation variations driven by external forcing, we consider the free rotational normal modes of the Earth which exist independently of any external forcing, and the frequencies of these normal modes (the eigenfrequencies). The normal modes play an important role in the forced motions through the resonances that they produce in the response to forcing at frequencies near the eigenfrequencies.

The Earth has an infinite number of modes of *free oscillations* or *normal modes* which are excited by earthquakes and are referred to therefore as the seismic normal modes. A subclass of the free oscillation modes is constituted by the so-called core modes, in which the oscillations are effectively confined to the fluid core region. Three modes of *translational* oscillations of the inner core floating within the core fluid are also worth mentioning; these are called the Slichter modes.

None of the above classes of normal modes is of immediate relevance to variations in Earth rotation. The modes which are of importance in this context are the four *rotational* normal modes characterized by different patterns of differential rotations of the mantle, FOC, and SIC. Their existence is related to the fact that an axially symmetric (but non-spherical) body can rotate "smoothly," with no variations in direction of the rotation axis, only if the rotation axis coincides with the symmetry axis. If the body is set in rotation about any other axis, what results is a "wobbly" motion, as in the familiar example of a symmetrical top. The existence of more than one rotational normal mode for the Earth is due to the presence of the core regions and the consequent possibility of differential rotations of the mantle, the fluid core, and the SIC. It may be noted that only one rotational mode would exist if the Earth did not have a fluid core, and only two if the fluid core extended all the way to the center (i.e., if the inner core were absent).

In each of the rotational modes, the rotation axes of the three regions, as well as the symmetry axis of the SIC, are offset from the symmetry axis of the mantle by different angles. (The magnitude of the angular offset of the rotation axis of each region from the mantle symmetry axis is the *amplitude* of the wobble of that region.) All four axes rotate with a common frequency (the frequency of the normal mode) around the mantle symmetry axis. The rotational normal modes not only contribute directly to Earth rotation variations, but also give rise to resonances in the forced nutations and wobbles when the frequency of forcing is close to one or the other of the normal mode eigenfrequencies. The four rotational modes are the following:

- The Chandler wobble (CW), which is a low frequency mode with a period of about 432 days in a terrestrial reference frame. It is characterized by an offset of the direction of the axis of rotation of the mantle from the figure axis of the ellipsoidal Earth, accompanied by a motion of the rotation axis around the figure axis in the prograde sense. In this mode, the rotation axes of the three regions are nearly coincident. This is the only rotational eigenmode which could exist if the Earth were wholly solid. (The period would of course be different in that case.) The counterpart of the CW for a rigid Earth is the Eulerian wobble with a period close to 300 days, which is determined solely by the Earth's dynamical ellipticity $e \equiv (C - A)/A$, i.e., the fractional difference between the moments of inertia about the polar and equatorial axes. In fact, the period, when expressed in cpsd, is just $(1/e)$.

- The free core nutation (FCN), which is a free mode in which the offset of the rotation axis of the fluid core from the mantle symmetry axis (i.e., the amplitude of the wobble of the FOC) is far larger than the wobble amplitude of the mantle itself. The frequency of this mode is nearly diurnal as seen in a terrestrial frame; for this reason, the manifestation of this mode in the terrestrial frame is referred to as the nearly diurnal free wobble (NDFW). The motion of the rotation axes relative to the body of the Earth is accompanied by a circular motion of the mantle symmetry axis in space; the latter is known as the FCN. It is retrograde, and has a period of about 430 days. Ellipticity of the core mantle boundary (CMB) is essential for the existence of the FCN mode. It causes the equatorial bulge of the fluid mass to impinge on the mantle at the CMB in the course of the rotations of the core and the mantle about non-coincident axes; the resulting pressure of the fluid on the boundary, and the inertial reaction on the fluid core itself, sustain the differential rotation between the core and the mantle in the FCN mode.

- The free inner core nutation (FICN), a mode in which the amplitude of the wobble of the SIC dominates over the wobble amplitudes of the other regions; the figure axis and rotation axis of the SIC are both offset from the mantle symmetry axis by almost the same amount. The FICN has, like the FCN, a nearly diurnal retrograde period in a terrestrial reference frame. The motion of the axes in space is prograde in this mode, unlike the case of the FCN. For this reason, the FICN is also referred to as the prograde free core nutation (PFCN); when it is necessary to emphasize the distinction between this mode and the FCN, the latter is referred to as the retrograde free core nutation (RFCN). The period of the FICN in space, as inferred from the resonance associated with this mode in the forced nutation, is about one thousand days. The ellipsoidal shape of the inner core boundary (ICB) is essential for the generation of this mode. As in the case of the FCN, the fluid pressure acting on the ellipsoidal boundary (the ICB in this case) plays an important role here; but an even larger role is played by the gravitational coupling of the SIC to the mantle due to the offset between the symmetry axes of these two ellipsoidal regions.

- The inner core wobble (ICW), a mode in which the offset of the symmetry axis of the SIC from that of the mantle exceeds the offsets of the three rotation axes by several orders of magnitude. It has a long period (longer than the CW) in a terrestrial frame, and is prograde. The ellipsoidal structure of the SIC and of the rest of the Earth, and the density contrast between the SIC and the fluid at the ICB, are essential elements in the generation of this normal mode.

2.21 Causes of the forced motions

The variations in Earth rotation as seen from a terrestrial reference frame are simply the variations of the angular velocity vector of the mantle relative to this frame, with accompanying variations of the angular velocities of other regions of the Earth (FOC, SIC) in the same frame. Wobbles and spin rate (or LOD) variations are different aspects of the angular velocity variations. Both are governed by the equations of angular momentum balance, i.e., the equations which relate the rate of change of angular momentum of each region to the torque acting on that region. Since angular momentum is the product of the moment of inertia (which is in general a tensor quantity, called the inertia tensor) with the angular velocity vector, variations of this vector can result from the presence of a torque or from variations in the inertia tensor even in the absence of a torque. The free Chandler wobble is the prime example of angular velocity variations associated with variations of the inertia tensor relative to a space-fixed frame that result when the instantaneous axis of rotation does not coincide with the symmetry axis.

For *forced* wobbles and the associated nutations and polar motions, however, the gravitational torques due to solar system bodies are by far the most important cause, though variations in the inertia tensor too play a significant role. (Various mechanisms that produce variations in the inertia tensor have already been mentioned in the introductory section.) The exchange of angular momentum between the geophysical fluid layers at the Earth's surface (the atmosphere and the oceans), on the one hand, and the solid Earth on the other, is another mechanism which has a role in Earth rotation variations.

The dominant role in LOD variations of tidal origin is played by the deformations produced by the tidal potentials, but LOD variations in general are primarily a manifestation of angular momentum exchange between the solid Earth and the atmosphere at day- or year-timescales.

The mutual influences of Earth rotation variations, solid Earth deformations, ocean tides, and atmospheric effects have been alluded to in the introductory section. Recall, in particular, Fig. 2.1 which illustrates the interplay of the various phenomena.

The theoretical treatments of the different phenomena should therefore be developed, at least when high accuracy is aimed at, in such a manner as to take this interplay into account so as to ensure mutual consistency of the treatments of all the phenomena.

2.22 Equations of rotational motion

Now that an overview of the phenomena that we will be studying in this book has been presented, we proceed to the basic equations of motion governing the phenomena.

2.22.1 Equation of rotational motion in a celestial frame

Rotational motion is governed by the equation of angular momentum balance (torque equation) which relates the variations in angular momentum to the applied torque. In an inertial (space-fixed) reference frame, the equation is simply

$$\left(\frac{d\mathbf{H}}{dt}\right)_S = \mathbf{L}, \tag{2.36}$$

where the subscript S indicates that the derivative of the vector in a space-fixed celestial frame is meant. The torque vector $\mathbf{L}$ is the same as $\boldsymbol{\Gamma}$; the changed notation is introduced because the components of $\mathbf{L}$ are to be referred to the celestial frame and are to be distinguished from those of $\boldsymbol{\Gamma}$ which are relative to the terrestrial frame. We now have, trivially,

$$\mathbf{H} = \int \mathbf{L}\, dt \tag{2.37}$$

in a celestial reference frame. This result is independent of whether or not the body which is subjected to the torque is rigid or has any core regions.

The components H_X, H_Y are very small compared to H_Z which is closely approximated by $C\Omega_0$; so $|\mathbf{H}| \approx C\Omega_0$. The X and Y components of the *unit vector* in the direction of $\mathbf{H}$, which characterize the precession and nutation *of the angular momentum axis*, are then

$$X_H = \frac{H_X}{C\Omega_0} = \frac{\int L_X dt}{C\Omega_0}, \qquad Y_H = \frac{H_Y}{C\Omega_0} = \frac{\int L_Y dt}{C\Omega_0}. \qquad (2.38)$$

The variations in longitude and obliquity are related to these quantities through equations analogous to (2.14):

$$X_H = (\psi_{A,H} + \Delta\psi_H)\sin\epsilon, \qquad Y_H = \Delta\epsilon_H, \qquad (2.39)$$

where $\Delta\psi_H$ and $\Delta\epsilon_H$ are the nutations in longitude and obliquity of the angular momentum axis, and $\psi_{A,H}$ has been used to represent the change in longitude due to precession.

Evaluation of the nutation quantities may be done after the torque acting on the Earth and its time dependence are determined. The expression for the torque on an axially symmetric ellipsoidal Earth is obtained in Section 2.23.6 below as a function of the position of the external body, and the treatment of a simplified model may be found in Section 2.24.2. The torque in the case of a more general axially symmetric Earth is derived in Section 5.4, and for a very general Earth structure in Section 5.7. The latter section presents also the spectral representation of the torque, including, in particular, an explicit expression for the axially symmetric ellipsoidal case.

2.22.2 Equation of rotational motion in a terrestrial frame; Euler's equations for a rigid body

Consider now the description of the Earth's rotational motion as seen from an Earth-fixed reference frame. The frame itself is rotating in space, and we denote its angular velocity vector by $\mathbf{\Omega}$. The torque equation governing the variation of the angular momentum vector $\mathbf{H}$ in this reference frame is

$$\frac{d\mathbf{H}}{dt} + \mathbf{\Omega} \wedge \mathbf{H} = \mathbf{\Gamma}, \qquad (2.40)$$

where $\wedge$ indicates the cross product between two vectors, and where the vector $\mathbf{\Gamma}$ is the torque acting on the Earth, referred to the terrestrial frame. On writing the equation out in terms of the Cartesian components, we get

$$\dot{H}_x + \Omega_y H_z - \Omega_z H_y = \Gamma_x,$$
$$\dot{H}_y + \Omega_z H_x - \Omega_x H_z = \Gamma_y, \qquad (2.41)$$
$$\dot{H}_z + \Omega_x H_y - \Omega_y H_x = \Gamma_z,$$

where $\dot{H}_x = dH_x/dt$, and so on.

The above equations, unlike the corresponding ones in an inertial frame, involve not just $\mathbf{H}$ but $\boldsymbol{\Omega}$ too. Therefore it is necessary, in order to solve these equations, to make use of the relationship of the former to the latter. If the Earth were wholly solid, the relation would be $\mathbf{H} = [\mathbf{C}] \cdot \boldsymbol{\Omega}$, where $[\mathbf{C}]$ is the Earth's inertia tensor. This is no longer true if core regions are present. While the mantle rotates with angular velocity $\boldsymbol{\Omega}$, the other regions rotate, in general, with angular velocities differing from $\boldsymbol{\Omega}$; these differences would result in additional contributions to $\mathbf{H}$ from the core regions. Furthermore, the inertia tensors of the different regions will have to include contributions from tidal deformations of the respective regions when the deformability of the Earth is to be taken into account. Thus it is necessary to use as inputs various aspects of the Earth's structure and properties before the torque equation in the terrestrial frame can be formulated; and additional equations which govern the variation of the angular velocities of the core regions will have to be considered simultaneously. We consider here only the simplest of Earth models, namely that of a wholly solid rigid Earth, to illustrate how the wobble motion is determined; more realistic models will be dealt with in later chapters.

For any *rigid* body, Eqs. (2.41) reduce to the Euler equations when coordinate axes parallel to the principal axes of the body are chosen. With such a choice,

$$H_x = A\Omega_x, \qquad H_y = B\Omega_y, \qquad H_z = C\Omega_z, \tag{2.42}$$

where A, B, C stand for the principal moments of inertia in increasing order of magnitude, as usual. Substituting into Eqs. (2.41), we obtain

$$A\dot{\Omega}_x + (C - B)\Omega_y\Omega_z = \Gamma_x,$$
$$B\dot{\Omega}_y + (A - C)\Omega_z\Omega_x = \Gamma_y, \tag{2.43}$$
$$C\dot{\Omega}_z + (B - A)\Omega_x\Omega_y = \Gamma_z.$$

These are the celebrated Euler equations.

2.22.3 Axially symmetric ellipsoidal case: Wobble motion

We specialize now to the Earth considered as an axially symmetric ($B = A$), ellipsoidal, rigid body (with no core), with its instantaneous rotation axis very close to the $\hat{z}$-axis and with an angular rotation rate very close to Ω_0. Then

$$\Omega_x = \Omega_0 m_x, \qquad \Omega_y = \Omega_0 m_y, \qquad \Omega_z = \Omega_0(1 + m_z), \tag{2.44}$$

where m_x, m_y, are the *wobble variables*. Evidently, they characterize the direction of the projection of $\boldsymbol{\Omega}$ on to the x-y plane, which is the equatorial plane since the $\hat{z}$-axis is in the direction of the polar axis (the axis of maximum moment of inertia).

The other component m_z represents the spin rate variation. The magnitudes of m_x, m_y, m_z are very much less than unity (they are typically of order 10^{-8}).

The $(B - A)$ term in the third of the Euler equations (2.43) now drops out because of axial symmetry; and Γ_z vanishes, if it is of gravitational origin, as will be seen later (see the third of Eqs. (2.60)). It follows that H_z remains constant, equal to $C\Omega_0$, i.e., $m_z = 0$ in the above expressions. The first two of the torque equations then become

$$A\Omega_0 \frac{dm_x}{dt} + \Omega_0^2(C - A)m_y = \Gamma_x, \quad A\Omega_0 \frac{dm_y}{dt} + \Omega_0^2(A - C)m_x = \Gamma_y. \quad (2.45)$$

The addition of i times the second of these equations to the first one yields a single equation involving the complex quantities

$$\tilde{m} = m_x + im_y, \qquad \tilde{\Gamma} = \Gamma_x + i\Gamma_y. \qquad (2.46)$$

The combined (complex) equation is

$$\frac{d\tilde{m}}{dt} - ie\Omega_0\tilde{m} = \frac{\tilde{\Gamma}}{A\Omega_0}, \qquad (2.47)$$

where e is the dynamical ellipticity in the sense in which this term is commonly used in the geophysics literature: $e = (C - A)/A$. Solution of the above equation is a trivial matter:

$$\tilde{m}(t) = e^{ie\Omega_0 t} \left[\tilde{m}(0) + \int \frac{\tilde{\Gamma}(t)}{A\Omega_0} e^{-ie\Omega_0 t} dt \right]. \qquad (2.48)$$

Free wobble

The first term, $\tilde{m}(0)e^{ie\Omega_0 t}$, does not involve the torque. It represents a free wobble of frequency $e\Omega_0$ or e cpsd, which is indeed the Eulerian free wobble. In the case of more realistic non-rigid Earth models, it becomes the CW, with a frequency differing from $e\Omega_0$.

Forced wobble

The second term in (2.48) is the forced wobble, which can be evaluated once the time dependence of $\tilde{\Gamma}$ is known. If a single spectral component, of frequency ω_w, of the wobble is considered, the component of the torque which gives rise to it has the same frequency of course. On setting

$$\tilde{\Gamma}(t) = \tilde{\Gamma}(\omega_w)e^{i\omega_w t}, \qquad (2.49)$$

Eq. (2.47) simplifies to

$$i(\omega_w - \Omega_0 e)\tilde{m}(\omega_w) = \tilde{\Gamma}(\omega_w)/(A\Omega_0), \qquad (2.50)$$

so that

$$\tilde{m}(\omega_w) = -i\,\frac{\tilde{\Gamma}(\omega_w)}{A\Omega_0(\omega_w - \Omega_0 e)}. \tag{2.51}$$

The factor $1/(\omega_w - \Omega_0 e)$ represents the *resonance* in the forced wobble that is associated with the Eulerian free wobble which has $e\Omega_0$ as its eigenfrequency. This resonance is well outside the retrograde diurnal band of frequencies and therefore it has little impact on the wobbles in this frequency range or the associated low frequency nutations. Other resonances, which have been referred to already in Section 2.20, appear when the existence of the Earth's fluid core and inner core is taken into account; they do have a profound influence on the Earth's forced wobbles in the retrograde diurnal band.

2.23 Gravitational action of a celestial body on the Earth

The action of the gravitational field of a heavenly body B such as the Moon or the Sun on the matter comprising the Earth produces deformations of the Earth as well as a torque which causes the Earth to nutate and precess. The starting point for calculation of these effects is the expression for the gravitational potential $W^B(\mathbf{r}, t)$ at the position $\mathbf{r}$ of some point within the Earth, due to the presence of a celestial body of mass M_B at the position $\mathbf{d}$ relative to the geocenter (this vector was previously denoted $\mathbf{r}_B$, see Section 2.13, and is now written $\mathbf{d}$ so as to avoid indices). This potential is $-GM_B/|\mathbf{d} - \mathbf{r}|$. The variation of W^B over the volume of the Earth is given by its dependence on $\mathbf{r}$, while its temporal variation is caused by the time dependence of $\mathbf{d}$ which is due to the motion of the celestial body relative to the geocenter.

2.23.1 Temporal variation of the Sun and Moon position vectors

It is useful to take a look at the temporal behavior of $\mathbf{d}$ before going on to consider the spatial variation of the potential. We may refer $\mathbf{d}$ and $\mathbf{r}$ to either of two types of geocentric reference frames: a celestial frame with axes having fixed orientations in space, or a terrestrial frame which is rotating in space. The standard convention is that a rotational motion which is in the prograde/retrograde sense as seen from either of the reference frames is assigned a positive/negative frequency in that frame.

The orbital motions of the Moon and the Sun in space relative to the geocenter are both in the same sense as that of the Earth's rotation, though their orbits lie in different planes: the Sun's orbit in the ecliptic plane, inclined to the equator at $23.5°$, and the Moon's orbit in a plane which is inclined by about $28.5°$ to the equatorial plane and about $5°$ to the ecliptic. Actually, the Moon's orbital plane does not remain fixed in space: it precesses along the ecliptic plane with a period of 18.6 years, i.e., the line of intersection of the two planes rotates in the ecliptic plane with this period, in the sense opposite to that of the apparent motion of the Sun along the ecliptic. The precession of the lunar orbit is caused by the solar torque acting on the Earth–Moon system.

In a celestial frame, the motion of the Sun around the geocenter has a period of one year of T_S days (each of 24 hours duration), where $T_S = 365.242$. This is the so-called "tropical year": the time taken by the Sun to complete the journey along the ecliptic from the equinox back to the equinox. The orbital period of the Moon is $T_M = 27.322$ days. The mean frequency of the Sun's orbital motion is thus $1/T_S$ cycles/day and the mean angular velocity is $2\pi/T_S$ radians/day. The qualification "mean" is necessary because the angular velocity varies somewhat as a result of the eccentricity of the Sun's orbit around the geocenter because the orbit is elliptical (with eccentricity 0.017) rather than circular. (The eccentricity is defined as the ratio between the distance of the foci of the ellipse from its center and the semi-major axis.) When speaking about the Sun's motion around the Earth, we are referring to the apparent motion of the Sun as seen from the geocenter. For similar reasons, the angular velocity of the Moon is variable, with a mean value $2\pi/T_M$ rad/day. The elliptical nature of the orbits causes multiples of the mean frequency to be present in the periodic motions around the respective orbits. Actually, even the picture of elliptical orbits is an oversimplification: it would be valid only if the Sun–Earth system and the Moon–Earth system were not subject to any other forces. In reality, the Sun perturbs the relative motion of the Earth and the Moon, and the Sun–Earth relative motion is affected by the presence of the Moon; over and above these are the gravitational perturbations by the other solar system bodies. These effects add further richness to the spectrum of the frequencies ω_c that are present in the solar and lunar motions in space (the subscript c is to indicate the celestial frame). All these frequencies are much less than 1 cpsd in magnitude. A discussion of the fundamental arguments representing the fundamental frequencies present in the orbital motions of the solar system bodies may be found in Section 5.5.6.

Yet another factor comes into play when the equatorial components of the torque on Earth are considered. As one may easily visualize, the projection of the orbit of the Sun/Moon on to the equatorial plane is elliptical even if the motion in the orbital plane itself were approximated as circular. The consequent elliptical motion in space of the equatorial projection of the position vector **d** decomposes into a pair of prograde and retrograde motions. This fact has the important consequence, which will be illustrated explicitly by a simple example in later sections, that the torque on the Earth as seen in the celestial reference frame (and hence the nutation it causes) has both prograde and retrograde circular components characterized by pairs of frequencies $\pm\omega_c$, despite the motion of either body in its own orbital plane being in one sense only.

Consider now the rotation of the Earth itself. Its mean rate of rotation in space is once per *sidereal day*; in fact one sidereal day is approximately the mean interval between two successive crossings of any given meridian by any one of the very distant "fixed" stars. ("Mean" here is to indicate that variations in the Earth's spin rate are averaged over and that the deviations in direction of the rotation axis from its mean direction are ignored: the direction remains within a fraction of a second of arc from the third axis of the terrestrial frame.) The rate of rotation *relative to the Sun* is once per day, where the day, or more precisely, the *solar day*, is the mean interval between two successive passages of the Sun across any given meridian. This rate is lower than the sidereal rotation rate because of the Sun's orbital motion in the same sense as the Earth's rotation. In fact, 1 cycle per day = 1 cycle per sidereal day (1 cpsd) minus $1/T_S$ cycles per day, whence

$$1 \text{ cpsd} = 1 + 1/T_S = 366.242/365.242 = 1.002\,737\,909 \text{ cycles per day.} \quad (2.52)$$

Thus 1 sidereal day = 365.242/366.242 day, shorter than the solar day by about 4 minutes.

We shift our attention now to frequencies as seen relative to the terrestrial reference frame. Since the frame is rotating in space in the prograde sense with a mean frequency of 1 cpsd, it should be obvious that the frequency of any motion around this axis as seen in the rotating terrestrial frame will be the frequency of the same motion in the celestial frame minus 1 cpsd. Thus the components of the solar and lunar motions parallel to the equatorial plane will have frequencies close to −1 cpsd in the terrestrial frame; and these are retrograde motions. Similarly, the nutational motion of the rotation axis, for example, with frequencies in space that are much less than 1 cpsd in magnitude will appear in the terrestrial frame as a retrograde wobble motion with frequencies close to −1 cpsd. Figure 2.5 represents the prograde and retrograde motions in terrestrial and celestial reference frames. A formal proof of the general relationship between frequencies in the two reference frames comes as part of the proof of the kinematical relations, to be given towards the end of this chapter.

2.23.2 Gravitational potential

We begin with the expression for the gravitational potential $W^B(\mathbf{r})$ at a point P within the Earth due to the external body B which is treated as a point object. If the position vector of P relative to the geocenter is $\mathbf{r}$, and that of B is $\mathbf{d}$, the distance of

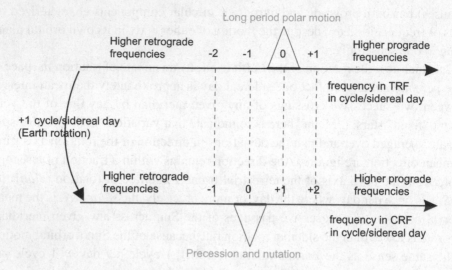

Figure 2.5 Representation of motions in the terrestrial and celestial frames. Note that the frequency scale in the celestial reference system (CRF) is shifted to the left, relative to that in the terrestrial reference frame (TRF), by 1 cpsd.

P from the body B is $|\mathbf{d} - \mathbf{r}|$, and we have

$$W^B(\mathbf{r}, t) = -\frac{GM_B}{|\mathbf{d} - \mathbf{r}|} = -\frac{GM_B}{d}\left(1 - \frac{2\mathbf{d}\cdot\mathbf{r}}{d^2} + \frac{r^2}{d^2}\right)^{-1/2}$$

$$\approx -\frac{GM_B}{d}\left(1 + \frac{(\mathbf{d}\cdot\mathbf{r})}{d^2} - \frac{r^2}{2d^2} + \frac{3(\mathbf{d}\cdot\mathbf{r})^2}{2d^4} + \cdots\right), \qquad (2.53)$$

where $d = |\mathbf{d}|$. Figure 2.6 represents the different position vectors.

Since the Earth's radius is a little under 6400 km, and the distance d to the Moon is about 60 times as large, the terms of successively higher degrees in (r/d) in the potential due to the Moon keep decreasing by a factor $\leq (1/60)$ at points in the Earth. If the potential due to the Sun is considered, the decrease is by a factor $\leq 4.3 \times 10^{-5}$ as the distance d to the Sun is about 150 million km.

2.23.3 Gravitational acceleration and tidal deformation: Tidal potential

The gravitational acceleration of a material element at $\mathbf{r}$ is $-\nabla W^B(\mathbf{r}, t)$, and the force on the element occupying a volume $d^3 r$ is $-\rho(\mathbf{r})\nabla W^B(\mathbf{r}, t)d^3 r$, where $\rho(\mathbf{r})$ is the density of matter at $\mathbf{r}$. The first term in the expansion (2.53) is independent of $\mathbf{r}$ and therefore it does not contribute to the acceleration. As for the second term which is of the first degree in (r/d), its *gradient* is independent of $\mathbf{r}$ and is in the direction of $\mathbf{d}$. This uniform acceleration is clearly the centripetal acceleration of the Earth as

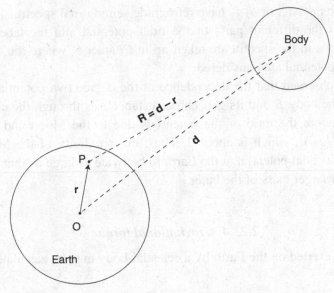

Figure 2.6 Position vectors of a point within the Earth and the center of a celestial body.

a whole towards the Sun (or the Moon), which is responsible for the relative orbital motion of the Earth and the celestial body; it has no other effect on the Earth. The next term, which is of the second degree in (r/d), causes *differential* accelerations of the different parts of the Earth. This is the *tidal* gravitational acceleration, which is responsible for the deformations of the solid Earth (solid Earth tides) and for the ocean tides. (Terms of degrees higher than 2 in the expansion of the potential W^B, which will be considered in later sections, give rise to additional – much smaller – tidal potentials, as was seen in the last paragraph.) The solid Earth and ocean tides are among the important effects of external gravitation on the Earth.

The gravitational potential $W^B(\mathbf{r}, t)$, with the exclusion of the parts which are independent of $\mathbf{r}$ or linear in $\mathbf{r}$, is the tidal potential due to B. (It may be recalled that the term "tidal potential" was defined in the introductory section itself.) In the remainder of this chapter, we consider just the dominant degree two part of the tidal potential:

$$W^B_{n=2}(\mathbf{r}, t) = -\frac{GM_B}{2d^5}[3(\mathbf{d} \cdot \mathbf{r})^2 - d^2 r^2]. \qquad (2.54)$$

Since the equatorial components d_x, d_y of $\mathbf{d}$ in the terrestrial frame have retrograde diurnal spectra and d_z has a low frequency spectrum, the spectrum of the terms involving d_z^2 in the degree two potential are in the low frequency band, and that of terms containing $d_x d_z$ or $d_y d_z$ is in the retrograde diurnal band, while terms involving d_x^2, d_y^2, or $d_x d_y$ have retrograde semidiurnal spectra. The spectral expansions of the different parts of the tidal potential and the labeling of the individual terms in the spectra are taken up in Chapter 5, where the terms of all orders in the potential are considered.

It may be observed that the dependence of the degree two potential (2.54) on the mass of the body B and its geocentric distance d is through the combination M_B/d^3. Therefore the ratio of the potentials due to the Moon and the Sun is $(M_M/M_S)(d_S/d_M)^3$, which is about 2. Thus, the proximity of the Moon to the Earth causes its tidal potential at the Earth to be twice as large as that of the Sun, despite the far larger mass of the latter.

2.23.4 Gravitational torque

The torque $\mathbf{\Gamma}$ exerted on the Earth by a celestial body may be calculated in either of two ways:

1. By expressing the force (and hence the torque) on a mass element in the Earth in terms of the gravitational potential due to the external body at the location of the mass element, and then integrating over the whole Earth;
2. By expressing the force (and hence the torque) exerted by the whole Earth on the external body in terms of the potential due to the Earth at the position of the body. The torque on the Earth due to the body is the negative of the quantity thus computed.

We focus our attention here on the first approach, leaving the presentation of the second one to Chapter 5. The starting point here is the gravitational potential $W^B(\mathbf{r}, t)$ of (2.53). The torque that it produces on an element of matter at $\mathbf{r}$ is $-\rho(\mathbf{r})\mathbf{r} \wedge \nabla W^B(\mathbf{r}, t) d^3 r$. The torque on the whole Earth is therefore

$$\mathbf{\Gamma} = -\int \rho(\mathbf{r})\mathbf{r} \wedge \nabla W^B \, d^3 r \approx \frac{GM_B}{d} \int \rho(\mathbf{r}) \left(\frac{\mathbf{r} \wedge \mathbf{d}}{d^2} + \frac{3(\mathbf{d} \cdot \mathbf{r})(\mathbf{r} \wedge \mathbf{d})}{d^4} \right) d^3 r,$$

(2.55)

where the integration is over the volume of the Earth. Note that the torque does not involve the non-tidal (degrees zero and one) parts of the potential. We ignore at this stage the terms in the torque which arise from higher order terms in the expansion (2.53), as they are of much smaller magnitude for reasons indicated above. The first term in (2.55) involves $\int \rho(\mathbf{r})\mathbf{r} \, d^3 r$, which is simply the Earth's mass times the position vector of the center of mass; this vector is null because the center of mass is the origin (geocenter) itself. The torque is therefore given by the integral of the second term in the above expression.

2.23.5 *Torque components in terrestrial frame*

For the evaluation of the torque integral, we need to express **r** and **d** in terms of its components with respect to a suitable reference frame. It is most convenient to use a geocentric terrestrial reference frame having its axes along the principal axes of the static Earth, which we refer to as the *principal axis frame*. ("Static" means that any variations in the density distribution due to tidal deformations and other causes are not considered.) The moments of inertia about the $\hat{x}$-, $\hat{y}$-, and $\hat{z}$-axes are A, B, and C, in increasing order, the first two axes being in the equatorial plane. The three components of the torque integral are then seen to be

$$\begin{pmatrix} \Gamma_x \\ \Gamma_y \\ \Gamma_z \end{pmatrix} = \frac{3GM_B}{d^5} \int \rho(\mathbf{r})(xd_x + yd_y + zd_z) \begin{pmatrix} (yd_z - zd_y) \\ (zd_x - xd_z) \\ (xd_y - yd_x) \end{pmatrix} d^3r. \qquad (2.56)$$

The components d_x, d_y, d_z of **d** are time dependent because of the motion of the Sun relative to the terrestrial frame. Since we have chosen a principal axis coordinate system,

$$\int \rho(\mathbf{r})xy\, d^3r = \int \rho(\mathbf{r})yz\, d^3r = \int \rho(\mathbf{r})zx\, d^3r = 0, \qquad (2.57)$$

$$\int \rho(\mathbf{r})(y^2 + z^2)d^3r = A,$$

$$\int \rho(\mathbf{r})(z^2 + x^2)d^3r = B, \qquad (2.58)$$

$$\int \rho(\mathbf{r})(x^2 + y^2)d^3r = C,$$

and consequently,

$$2\int \rho(\mathbf{r})x^2 d^3r = B + C - A, \quad 2\int \rho(\mathbf{r})y^2 d^3r = C + A - B,$$

$$2\int \rho(\mathbf{r})z^2 d^3r = A + B - C. \qquad (2.59)$$

When the integral in the expression for Γ_x in (2.56) is simplified using (2.57), we are left with $d_y d_z \int (y^2 - z^2)d^3r$, which may be evaluated using (2.59). The other components of the torque can be evaluated similarly. Thus we obtain

$$\begin{pmatrix} \Gamma_x \\ \Gamma_y \\ \Gamma_z \end{pmatrix} = \frac{3GM_B}{d^5} \begin{pmatrix} (C - B)d_y d_z \\ (A - C)d_z d_x \\ (B - A)d_x d_y \end{pmatrix}. \qquad (2.60)$$

It is known that $(B - A)$ is very small (about 1/300 times) compared to the other two differences, and that its role in Earth rotation is rather minor. Therefore we specialize to the case of an axially symmetric Earth ($B = A$) for the remainder of this section. In this case, $\Gamma_z = 0$, and so the torque vector lies in the equatorial plane; it is seen to be perpendicular to the equatorial projection (d_x, d_y) of **d**. The complex combination $\tilde{\Gamma} = \Gamma_x + i\Gamma_y$ of the equatorial components becomes

$$\tilde{\Gamma} = -i\frac{3GM_B}{d^5}Ae\tilde{d}d_z, \tag{2.61}$$

where $\tilde{d} = d_x + id_y$, and where $e = (C - A)/A$ is the dynamical ellipticity (of the axially symmetric case); e is the Earth parameter which plays the dominant role in gravitationally forced nutation and wobble. For future reference, we write the torque (2.61) in a compact form as

$$\tilde{\Gamma}(t) = -iA\Omega_0^2 e\tilde{\phi}(t), \qquad \tilde{\phi}(t) = \phi_1 + i\phi_2 = \frac{3GM_B}{\Omega_0^2 d^5}(d_x + id_y)d_z. \tag{2.62}$$

We have explicitly indicated here the fact that $\tilde{\Gamma}$ and $\tilde{\phi}$ are functions of time in view of the time dependence of the components d_x, d_y, d_z of the position vector of the body B. Since $\tilde{\Gamma}(t)$ has the same dimensions as $A\Omega_0^2$, it is evident that $\tilde{\phi}(t)$ is dimensionless. It was introduced by Sasao *et al.* (1980) as a convenient dimensionless quantity in terms of which the part of the degree two tidal potential which produces the above torque could be expressed.

The magnitude of $\tilde{\phi}$ is clearly of the same order as that of $3GM_B/(2\Omega_0^2 d^3)$. To estimate its magnitude in the case that B is the Sun, we may take d to be the mean Earth–Sun distance; then one sees from Kepler's Third Law that $GM_S/d^3 = n_S^2$ (with the subscript S denoting the Sun), where n_S is the so-called "mean motion" of the Sun relative to the Earth, i.e., the mean angular velocity of the Earth's orbital motion, corresponding to a period close to 366 sidereal days. Therefore the magnitude of $\tilde{\phi}(t)$ due to the Sun is $\approx (3/2)(1/366)^2 \approx 1.12 \times 10^{-5}$. The corresponding magnitude when $\tilde{\phi}$ is due to the Moon is approximately twice the above value, as already noted in the last paragraph of Section 2.23.3. Thus the dimensionless magnitudes of both the solar and lunar degree two potentials are of the order of 10^{-5}.

Which part of the degree two tidal potential gives rise to the torque $\tilde{\Gamma}$ of Eq. (2.62)? The answer may be found from the observation that (2.62) involves $d_x d_z$ and $d_y d_z$. On examining the potential (2.53), we see that these products come from the part of the potential that involves $(\mathbf{d} \cdot \mathbf{r})^2$, or more specifically, from the part

$$-\frac{3GM_B}{d^5}(d_x d_z\, xz + d_y d_z\, yz) = -\Omega_0^2(\phi_1 xz + \phi_2 yz)$$

$$= \text{Re}\left[-\Omega_0^2\tilde{\phi}(t)(x - iy)z\right]. \tag{2.63}$$

A potential which is a linear combination of xz and yz is called a *tesseral* potential of degree two. Thus the equatorial components of the gravitational torque on an axially symmetric ellipsoidal Earth are produced solely by the degree two tesseral potential, which is responsible for its nutations and wobbles.

2.23.6 Torque components in space-fixed frame

One might ask: What if we employed the components (X, Y, Z) of $\mathbf{r}$ and (d_X, d_Y, d_Z) of $\mathbf{d}$ in a space-fixed equatorial reference frame (say, the GCRF), and wished to compute the torque components (L_X, L_Y, L_Z) in that frame on the same lines as above? The problem that one encounters in attempting such a course is that, though the space-fixed $\hat{X}$-, $\hat{Y}$-, $\hat{Z}$-axes may be chosen so as to coincide with the Earth's principal axes at some instant of time, such coincidence is destroyed by the Earth's axial rotation, nutation, and precession. However, we may ignore nutation and precession for the limited purpose of a lowest order calculation of the torque, and then, if the Earth is assumed to be axially symmetric, the axial rotation does not make any difference since any equatorial axis is a principal axis. With these approximations, one may evaluate integrals of the form (2.56) with (X, Y, Z) replacing (x, y, z) by using the same results as in (2.57) and (2.59). The components of $\mathbf{L}$ in the space-fixed frame then become

$$L_X = \frac{3GM_B}{d^5}(C - A)d_Y d_Z, \quad L_Y = -\frac{3GM_B}{d^5}(C - A)d_Z d_X, \quad L_Z = 0. \quad (2.64)$$

In terms of the polar coordinates (the geocentric distance d, the declination δ, and right ascension α) of the celestial body, we have

$$d_X = d\cos\delta\cos\alpha, \qquad d_Y = d\cos\delta\sin\alpha, \qquad d_Z = d\sin\delta, \quad (2.65)$$

and the torque components become

$$\frac{L_X}{C\Omega_0^2} = E\left(\frac{c}{d}\right)^3 \sin 2\delta \sin\alpha, \quad \frac{L_Y}{C\Omega_0^2} = -E\left(\frac{c}{d}\right)^3 \sin 2\delta \cos\alpha, \quad L_Z = 0, \quad (2.66)$$

where c is the semi-major axis of the elliptical orbit of the body B, and E is the notation used by Melchior and Georis (1968) for the parameter defined by

$$E = \frac{3GM_B}{2\Omega_0^2 c^3}\frac{(C - A)}{C}. \quad (2.67)$$

The factor (c/d) is quite close to unity since the eccentricities of the solar and lunar orbits are small (0.017 and 0.055, respectively).

Values of d, δ, α as functions of time may be taken, for any of the celestial bodies, from the appropriate ephemerides. Time integration of the torque components may then be carried out and thus X_H and Y_H and also $\Delta\psi_H$ and $\Delta\epsilon_H$ may be found

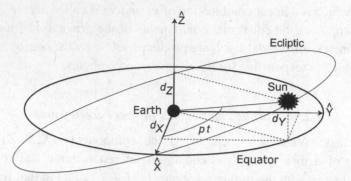

Figure 2.7 The apparent solar orbit around the Earth, in relation to the equatorial plane.

using Eqs. (2.38) and (2.39). Explicit solutions will be worked out in the next section assuming a simple model for the motion of the Sun.

2.24 Solutions for a simple model

In order to bring out clearly some of the essential aspects of nutations and wobbles, we obtain below the explicit solutions for a very simple model of the motion of the celestial body.

2.24.1 Celestial body moving uniformly in a circular orbit

A greatly simplified model of the torquing action of the Sun on the Earth, which is presented in this section, will suffice to illustrate how precession and nutation are generated. For this purpose, we take the Earth to be an axially symmetric ellipsoid, as at the end of the last section. The Sun is taken to be in uniform motion with angular frequency p along a circle of radius d centered at the geocenter and lying in the plane of the ecliptic (see Fig. 2.7). In the literature of astronomy, the mean angular velocity of the apparent orbital motion of the Sun around the Earth goes by the name "mean motion" and is denoted by n_S or n (see Chapter 5); the symbol p is merely a slightly simpler alternative to n_S, which is 2π radians per year. From elementary Newtonian theory one has

$$p^2 = n_S^2 = GM_B/c^3,\tag{2.68}$$

where c is the constant radius of the circular orbit. The position coordinates of the body in the ecliptic plane may then be written as

$$\xi = c\cos pt, \qquad \eta = c\sin pt, \qquad \zeta = 0.\tag{2.69}$$

Let us take both the first axis of the ecliptic frame and the $\hat{X}$-axis of our mean equatorial reference frame fixed in space to be along the intersection of the two principal planes, which is in the direction of the vernal equinox. ($O\hat{X}$ is the direction of increasing ecliptic longitude, i.e., of positive $\Delta\psi$. The motion in obliquity of the figure axis is in the direction of the $\hat{Y}$-axis.) The angle between the principal plane of the equatorial frame and the ecliptic is of course the obliquity ϵ, which we shall treat as a constant, ignoring its small variation due to nutation. Then, with our simplified model of the motion of the Sun in the ecliptic plane, its coordinates in our reference system are:

$$d_X = c \cos pt, \qquad d_Y = c \cos \epsilon \sin pt, \qquad d_Z = c \sin \epsilon \sin pt, \qquad (2.70)$$

the time variable being measured from the equinox (see Fig. 2.7).

Note that (d_X, d_Y) describes the projection of the Sun's circular orbit from the ecliptic plane on to the equatorial plane; the projected orbit is, not surprisingly, an ellipse. The motion around the ellipse, which is described by the evolution of α, the angle made by the projected vector with the $\hat{X}$-axis in the $\hat{X}$-$\hat{Y}$ plane, is not at a uniform rate. This is reflected in the relation $\tan \alpha = (d_Y/d_X) = \cos \epsilon \tan pt$.

2.24.2 Torque in the simple model

The time dependence of the torque (2.64) can now be displayed in explicit form:

$$L_X = F \cos \epsilon \, (1 - \cos 2pt), \qquad L_Y = -F \sin 2pt, \qquad L_Z = 0, \qquad (2.71)$$

where $F = (3p^2/2) \, Ae \sin \epsilon$. Thus, the overall magnitude of the solar torque on the Earth is determined by p and the obliquity ϵ, besides the Earth parameter $(C - A)$; in particular, the torque would vanish if the obliquity ϵ were zero, i.e., if the orbit of the Sun were to lie in the equatorial plane itself.

One may get an idea of the strength of the solar torque from the value of F. For the rotation variations induced by the torque, the non-dimensional quantity $F/(\Omega_0^2 A)$ is more relevant:

$$\frac{F}{\Omega_0^2 A} = \frac{3}{2} \frac{p^2}{\Omega_0^2} e \sin \epsilon. \qquad (2.72)$$

We know that the period of the relative orbital motion of the Sun and the Earth is one year of 365.2422 solar days or 366.2422 sidereal days. So the frequency p of this orbital motion is $\Omega_0/366.2422$ or equivalently, $1/366.2422$ cpsd. We also have, from the IERS Conventions 2010,

$$\epsilon = 23°26'21''.406, \qquad \sin \epsilon = 0.397777, \qquad \cos \epsilon = 0.917482. \qquad (2.73)$$

Therefore $F/(\Omega_0^2 A) = 4.448 \times 10^{-6}\, e$; it is a dimensionless measure of the amplitude of the torque. One finds it to be 1.461×10^{-8} on using the value $e = 0.0032845$ from a recent estimate.

Returning to (2.71) and (2.72), we take note of three important features that are evident: the presence of a constant (time independent) term in the X component of the torque, the semiannual frequency of the periodic part of both the equatorial components of the torque, and the fact that the periodic part of the variation of (L_X, L_Y) is an elliptical motion in the equatorial plane. When the equation of motion is solved, the constant term leads to precession. The semiannual nutation, which is in fact the largest of the nutations of solar origin, is the result of the corresponding periodic term in the torque; it has both prograde and retrograde components since the torque vector has both. A host of nutation terms other than semiannual arise when one takes into account the deviations from the very simple model that we considered: the orbit of the Sun relative to the Earth is not circular but elliptical, so that d is not constant; and the simple harmonic time dependence of d_X and d_Y as portrayed in (2.70) is a simplification – all the more so because the orbit is disturbed by the gravitation of other bodies, especially the Moon's. But the present simplified picture serves to highlight the essential aspects of the nutation–precession phenomenon.

2.24.3 *Prograde and retrograde components of the torque*

The periodic parts of the torque components above have the form

$$L_X = -F \cos \epsilon \cos 2pt, \qquad L_Y = -F \sin 2pt. \tag{2.74}$$

This vector traces an elliptical path in the $\hat{X}$-$\hat{Y}$ plane. It may be resolved, like any periodic elliptical motion, into two counter-rotating circular motions:

$$\begin{pmatrix} L_X \\ L_Y \end{pmatrix} = -\frac{F}{2} \left[(1 + \cos \epsilon) \begin{pmatrix} \cos 2pt \\ \sin 2pt \end{pmatrix} - (1 - \cos \epsilon) \begin{pmatrix} \cos 2pt \\ -\sin 2pt \end{pmatrix} \right]. \tag{2.75}$$

The first of the two columns on the right-hand side of this equation represents a vector in prograde motion while the second is in retrograde motion. The two opposite motions are easily identifiable if one looks at the complex combination:

$$\tilde{L} \equiv L_X + i L_Y = (-F/2)\,[(1 + \cos \epsilon)\,e^{i2pt} - (1 - \cos \epsilon)\,e^{-i2pt}]. \tag{2.76}$$

The prograde part coming from the first term of (2.76) has the time dependence $e^{i\omega t}$ with the positive frequency $\omega = 2p$, and the retrograde part has the dependence $e^{i\omega' t}$ with the negative frequency $\omega' = -2p$.

2.24.4 Nutation and precession of the angular momentum axis

With the components of **L** given in the celestial frame by (2.71), evaluation of $\int L_X dt$ and $\int L_Y dt$ to obtain H_X and H_Y is trivial. The coordinates X_H, Y_H (in the space-fixed reference frame) of the pole of the angular momentum axis on the celestial sphere may then be written down directly using Eqs. (2.38):

$$X_H = (3p^2/2\Omega_0)\, H_d \sin \epsilon \cos \epsilon \, [t - (1/2p) \sin 2pt],$$
$$Y_H = (3p^2/2\Omega_0)\, H_d \sin \epsilon \, [(1/2p) \cos 2pt], \tag{2.77}$$

where $H_d = (C - A)/C = Ae/C = e/(1 + e)$.

Now, X_H is $\sin \epsilon$ times the change in the longitude coordinate ψ of the position of the celestial pole of the angular momentum axis, and Y_H is $\Delta \epsilon$. Therefore the rate of precession in longitude, which comes from the derivative of the secular term in X_H, is

$$\dot{\psi}_{A,H} = \frac{1}{\sin \epsilon} \frac{d X_{H sec}}{dt} = \frac{3p^2}{2\Omega_0} H_d \cos \epsilon, \tag{2.78}$$

while the nutations in longitude and obliquity are obtained from the periodic parts:

$$\Delta \psi_H = -(3p/4\Omega_0)\, H_d \cos \epsilon \, \sin 2pt,$$
$$\Delta \epsilon_H = (3p/4\Omega_0)\, H_d \sin \epsilon \cos 2pt. \tag{2.79}$$

The coefficients of $\sin 2pt$ and $\cos 2pt$ in these expressions are known as the *coefficients of nutation* in longitude and obliquity. With the simplified picture of solar motion that has been employed in this section, the only frequency present is the semiannual one. In a realistic treatment, numerous nutation frequencies appear, each with its own coefficients of nutation.

It may be noted that both the precession rate and the coefficients of nutation of the angular momentum axis are proportional to H_d. We shall see later, when considering the precession and nutation of the figure axis in Section 2.26, that the precession of that axis too is proportional to H_d, but the nutation amplitudes are not.

On using the values given earlier for the various parameters one finds that the precession rate around J2000 due to solar gravitational torque as given by (2.78) is 7.7291×10^{-5} rad/year or equivalently, 15 943 mas/year. Lunar attraction too gives rise to precessional motion, a little more than double that due to the Sun; the greater role of the Moon despite its mass being very much smaller than that of the Sun is a consequence of its being very much closer to the Earth. The total precession rate is about 50 000 mas/year. (This means that if one imagines a geocentric vector of length equal to the Earth's radius in the direction of the angular momentum vector,

the tip of the vector moves in space at the rate of about 615 m/year – not counting
the velocity of the vector attached to the geocenter due to the orbital motion of the
Earth itself.)

As for the semiannual nutation, with angular frequency $2p = 2n_s = 4\pi$ rad/year
(or 1/183.1211 cpsd), one finds the coefficients in longitude and latitude to be
-1269 and 550 mas, respectively (this corresponds to a quasi-circular motion of
the tip of the above-mentioned vector with an amplitude of about 16 m as seen
from space). All these numbers are very close to those obtained from accurate
treatments of the problem. The largest of the nutation terms is of lunar origin. It
has a period of about 18.6 years, which arises from the precession of the lunar orbit
around the ecliptic. The coefficients in longitude and latitude of this nutation are
about $-17\,000$ mas and 9200 mas, respectively.

Returning to the expressions (2.77), we resolve the periodic parts of X and Y
into prograde and retrograde components:

$$\begin{pmatrix} X \\ Y \end{pmatrix} = \frac{3p}{8\Omega_0} H_d \sin \epsilon \left[(1 + \cos \epsilon) \begin{pmatrix} -\sin 2pt \\ \cos 2pt \end{pmatrix} + (1 - \cos \epsilon) \begin{pmatrix} \sin 2pt \\ \cos 2pt \end{pmatrix} \right].$$

(2.80)

Since the frequency $2p$ is positive, we see in the light of the discussion of Section 2.18 that the first term in the above expression represents a prograde motion
while the other term is retrograde, and that their amplitudes are $-(3p^2/4\Omega_0) H_d$
$\sin \epsilon (1 + \cos \epsilon)$ and $-(3p^2/4\Omega_0) H_d \sin \epsilon (1 - \cos \epsilon)$, respectively. These results
are also reflected in the complex combination of X_H and Y_H, which is

$$X_H + iY_H = i(3p/8\Omega_0) H_d \sin \epsilon [(1 + \cos \epsilon) e^{2ipt} + (1 - \cos \epsilon) e^{-2ipt}]. \quad (2.81)$$

Comparison with Eqs. (2.33) and (2.34) shows, on noting that $X_H + iY_H$ has only
a single pair of spectral components here of angular frequencies $\pm 2p$, that the
prograde and retrograde amplitudes in the present case are

$$\tilde{\eta}_H^p = -(3p/8\Omega_0) H_d \sin \epsilon (1 + \cos \epsilon),$$
$$\tilde{\eta}_H^r = -(3p/8\Omega_0) H_d \sin \epsilon (1 - \cos \epsilon).$$

(2.82)

It is worth emphasizing here that the results (2.78), (2.79), (2.81), and (2.82)
are for the angular momentum axis and not for the figure axis. These results are
*independent of whether the Earth is rigid or non-rigid and of whether or not it
has a core*, since there was no need to input any information about the structure
and properties of the Earth (other than the dynamical ellipticity) in order to obtain

the solution for the angular momentum vector. In contrast, the motions of the figure axis and the rotation axis are strongly influenced by the Earth's properties such as the existence of core regions. As a consequence, studies on those motions are the ones that serve to provide glimpses into the properties of the Earth's interior.

2.24.5 Wobble motion in the simple model

Let us consider, for illustration, the spectral terms relating to the simple example considered in Section 2.24.1. We took the source of the potential there to be the Sun, which was assumed to be moving uniformly along a circle of radius c in the ecliptic plane. We can take the expressions (2.70) for the coordinates of the Sun in the space-fixed reference frame, and transform them to terrestrial coordinates (d_x, d_y, d_z) for use here. For a lowest order calculation, which is all that we are interested in at present, we can take the $\hat{z}$- and $\hat{Z}$-axes to be coincident and the Earth rotation to be at a constant rate Ω_0 around this common axis. It is easy to see then that

$$d_x = d_X \cos \Omega_0 t + d_Y \sin \Omega_0 t, \quad d_y = -d_X \sin \Omega_0 t + d_Y \cos \Omega_0 t, \quad d_z = d_Z,$$

(2.83)

if we take the origin of time (rather arbitrarily) to be the instant at which the first axis of the terrestrial frame coincides with that of the celestial frame to within the approximations just mentioned. Then

$$\tilde{d} \equiv d_x + i d_y = (d_X + i d_Y) e^{-i\Omega_0 t}.$$

(2.84)

We may now take the coordinates of the Sun in the celestial frame from the expressions (2.70). Then $d_X + i d_Y = c(\cos pt + i \cos \epsilon \sin pt)$. On introducing this in (2.84) and noting that $d_z = d_Z = c \sin \epsilon \sin pt$, we obtain

$$\tilde{d} d_z = (c^2/2) \sin \epsilon \, [\sin 2pt + i \cos \epsilon (1 - \cos 2pt)] \, e^{-i\Omega_0 t}$$

$$= i(c^2/4) \sin \epsilon \, [2 \cos \epsilon \, e^{-i\Omega_0 t} - (1 + \cos \epsilon) \, e^{i(2p - \Omega_0)t}$$

$$+ (1 - \cos \epsilon) \, e^{i(-2p - \Omega_0)t}].$$

(2.85)

Now, the equatorial torque $\tilde{\Gamma}$ in the terrestrial frame is given by Eq. (2.61); adapted to the present simple case, it is

$$\tilde{\Gamma} = -i \frac{3GM_B}{c^5} A e \tilde{d} d_z.$$

(2.86)

On introducing (2.85) and substituting the resulting expression for $\tilde{\Gamma}$ into the term involving $\tilde{\Gamma}$ in the general solution (2.48) obtained in Section 2.22.3, we find the forced wobble to be

$$\tilde{m}(t) = \frac{3p^2}{4\Omega_0^2} \, ie \sin \epsilon \, [(2 \cos \epsilon)F_1(t) + (1 + \cos \epsilon)F_2(t) + (1 - \cos \epsilon)F_3(t)]$$

(2.87)

with $p^2 = GM_B/c^3$ as before, and with

$$F_1 = \frac{e^{-i\Omega_0 t}}{(1+e)}, \quad F_2 = \frac{\Omega_0 e^{i(2p-\Omega_0)t}}{(2p - \Omega_0 - e\Omega_0)}, \quad F_3 = \frac{\Omega_0 e^{i(-2p-\Omega_0)t}}{(2p + \Omega_0 + e\Omega_0)}. \quad (2.88)$$

Comparison of the last two terms in the expression (2.87) for the wobble $\tilde{m}$ (with F_2 and F_3 given by Eq. (2.88)) with Eqs. (2.81) and (2.82) for the nutation helps to bring out an important fact of quite general validity: with each circular component of the nutation is associated a circular component of the wobble; the frequency of the latter is that of the former minus Ω_0. The first term in (2.87) with frequency $-\Omega_0$ is the wobble corresponding to the precession term in X_H, Eq. (2.77); like the latter, it is excited by the torque term with zero frequency in space.

2.25 Polar motion of the pole of the CRS: relation to nutation

The polar motion considered here is not that of the CIP (introduced in Section 2.17), but that of the pole of the celestial reference system (CRS). Its polar motion is, by definition, the motion of the CRS pole relative to the terrestrial reference system (TRS); it is described by the variation of the vector from the pole of the TRS to the pole of the CRS (referred to hereafter as the *polar motion vector* of the pole of the CRS). This vector is of course the negative of the vector from the pole of the CRS to that of the TRS. The Cartesian coordinates of this pole in the equatorial CRS (the GCRS) are denoted (X, Y). The coordinates of the polar motion vector in this frame are therefore $(-X, -Y)$. Its components in the terrestrial frame are (see Section 2.17) $(x_p, -y_p)$, where x_p, y_p are the polar motion parameters of the pole now under consideration.

Now, the terrestrial frame is in rotation with an angular speed very close to Ω_0 about an axis which deviates from the $\hat{Z}$-axis of the celestial frame only by an angle of the order of the precession–nutation quantities (X, Y) which are very small. When these deviations are ignored, the equatorial planes of the terrestrial and celestial frames are in coincidence, and the equatorial axes of the terrestrial frame rotate relative to those of the celestial frame in this plane. If the small deviations of the rotation rate from Ω_0 are neglected as well, then the angle made by the first axis of the terrestrial frame (defined to be in the Greenwich Meridian) with the first axis of the celestial frame (which is along the mean equinox of J2000) is equal to $\Omega_0(t - t_0)$ where t_0 is an instant at which the two sets of axes were in coincidence. This angle is what is called the Greenwich mean sidereal time (GMST, see Section 2.7),

$$\text{GMST} = \Omega_0(t - t_0). \tag{2.89}$$

The matrix of transformation from the celestial to the terrestrial frame is thus

$$T = \begin{pmatrix} \cos(\text{GMST}) & \sin(\text{GMST}) & 0 \\ -\sin(\text{GMST}) & \cos(\text{GMST}) & 0 \\ 0 & 0 & 1 \end{pmatrix}. \tag{2.90}$$

On applying this transformation matrix to $(-X, -Y)$, we see that

$$x_p = -X \cos(\text{GMST}) - Y \sin(\text{GMST}), \tag{2.91}$$
$$-y_p = X \sin(\text{GMST}) - Y \cos(\text{GMST}).$$

The complex polar motion variable $\tilde{p} = x_p - iy_p$, is then related to $(X + iY)$ through

$$\tilde{p} \equiv x_p - iy_p = -(X + iY)\, e^{-i\,\text{GMST}} = -\tilde{\eta}\, e^{-i\,\text{GMST}}. \tag{2.92}$$

We know that $X + iY$ is made up of the nutation $\tilde{\eta}$ of the pole of the TRS plus a secular term representing precession of that pole. We see from the above equation that the complex polar motion $\tilde{p}$ of the pole of the CRS differs from $(X + iY)$ only by a negative sign and a phase factor. The relations connecting the frequency and amplitude of any spectral component of the nutation with those of the corresponding component of the polar motion of the pole of the CRS follow simply from (2.92). For a circular nutation component of frequency ω_c,

$$\tilde{p}(\omega_p) = -\tilde{\eta}(\omega_c), \qquad \omega_p = \omega_c - \Omega_0. \tag{2.93}$$

Not surprisingly, the frequency ω_p of the polar motion is the same as that of the corresponding wobble. But the amplitude relation in (2.93) is quite different from what is found below for the wobble amplitude (see Eq. (2.105)). The amplitude of the polar motion of the pole of the CRS is of the same magnitude as that of the corresponding nutation of the figure axis for any frequency, while the amplitude of the wobble is in general very different from that of the nutation. The amplitude of the wobble associated with the 18.6 year nutation, for instance, is smaller in magnitude by a factor of 6816 compared to that of the nutation.

We have mentioned earlier that nutation–precession as well as polar motion are supposed to refer, according to resolutions of the IAU, to motions of the CIP. It is shown in the next chapter that though the relations (2.93) were derived for the pole of the CRS, the relations for motions of the CIP are effectively the same, surprising as it may seem. In fact, the relation (2.92) was obtained and applied by Gross (1992), Brzezkiński (1992), and Eubanks (1993) in the context of polar motion of the CEP/CIP (see also Mathews and Bretagnon, 2003).

2.26 Kinematical relations between wobble and nutation

Consider the motion of the Earth's figure axis relative to the equatorial celestial reference frame of J2000 (see Section 2.14). The coordinates of the pole of this axis in this frame at an instant t are $X(t)$, $Y(t)$ with $X(t) = [\psi_A(t) + \Delta\psi(t)] \sin\epsilon$, $Y(t) = \Delta\epsilon(t)$, where $\psi_A(t)$ represents the precession in longitude. The Z coordinate of the pole is of no interest here.

The motion of the pole in the interval $(t, t + dt)$ changes its coordinates by $(dX, dY, 0)$. We transform this infinitesimal vector to the terrestrial reference frame now, in order to establish its relation to the components of the wobble. We seek to obtain the transformed components only to the first order in small quantities. For this limited purpose, we may take the transformation matrix to the zeroth order, which is simply the matrix representing the rotation through the angle $\Omega_0(t - t_0) = \text{GMST}$ around the $\hat{Z}$-axis, given by (2.90). (In doing so, we are ignoring the first-order offsets between the $\hat{Z}$-axis of the celestial frame on the one hand, and the figure and rotation axes of the Earth on the other, and also ignoring the variations in the spin rate, as in Section 2.25.) The components of the infinitesimal displacement vector of the figure axis pole in the terrestrial frame are thus obtained as

$$dx = dX \cos(\text{GMST}) + dY \sin(\text{GMST}),$$
$$dy = -dX \sin(\text{GMST}) + dY \cos(\text{GMST}). \tag{2.94}$$

The errors in these expressions are of the second order in the neglected small quantities.

The displacement (dx, dy) of the figure axis pole in the terrestrial frame is, however, determined directly by the angular velocity vector. The Earth's rotation with angular velocity $\boldsymbol{\Omega}$ has the effect that any mantle fixed vector $\mathbf{V}$ gets displaced by an amount $(\boldsymbol{\Omega} \wedge \mathbf{V}) dt$ in an infinitesimal time interval dt. In particular, the displacement of the $\hat{z}$-axis during the interval $(t, t + dt)$ is $(\boldsymbol{\Omega} \wedge \hat{z}) dt$, with components $(\Omega_y, -\Omega_x, 0) dt = \Omega_0(m_y, -m_x, 0) dt$ relative to the terrestrial frame of the instant t. Since the $\hat{z}$-axis is the figure axis, we have thus

$$dx = \Omega_0 m_y \, dt, \qquad dy = -\Omega_0 m_x \, dt. \tag{2.95}$$

We obtain the relations that we seek on equating the expressions (2.94) and (2.95):

$$
\begin{aligned}
\Omega_0 m_x &= \dot{X} \sin(\text{GMST}) - \dot{Y} \cos(\text{GMST}), \\
\Omega_0 m_y &= \dot{X} \cos(\text{GMST}) + \dot{Y} \sin(\text{GMST}).
\end{aligned}
\tag{2.96}
$$

The familiar complex combination of the pair of equations now yields

$$\Omega_0 \tilde{m} = i\,(\dot{X} + i\dot{Y}) e^{-i\,\text{GMST}} = i\,\frac{d\tilde{\eta}}{dt}\, e^{-i\,\text{GMST}}, \tag{2.97}$$

where

$$\tilde{\eta} = X + iY. \tag{2.98}$$

It is important to note that no dynamical considerations have been invoked while deriving the result (2.97). Therefore it is valid for rigid and non-rigid Earth models alike. Since it is of purely kinematical origin, it is known as the *kinematical relation*; it relates the wobble to the nutation–precession (in the time domain).

The relations connecting (X, Y) to the nutation–precession variables in longitude and obliquity have been given already in Eqs. (2.14). They express X in terms of the precession part ψ_A and the nutation part $\Delta\psi$ of the variation ψ in longitude and Y in terms of $\Delta\epsilon$. We may rewrite Eq. (2.98) using those relations as

$$\tilde{\eta} = (\psi_A + \Delta\psi) \sin\epsilon_0 + i\,\Delta\epsilon. \tag{2.99}$$

We have approximated $\sin\epsilon$ by $\sin\epsilon_0$ in the above, as is commonly done in this form of the kinematical relation, primarily to facilitate the transition from the time domain to the frequency domain. The corrections necessitated by this approximation are discussed in Chapter 10.

For a circular nutation of general frequency ω_c (which may be positive or negative), we may write the positive and negative frequency parts of Eqs. (2.34) and (2.35) in the common form

$$
\begin{gathered}
X_c(t) = \tilde{\eta}(\omega_c) \sin \Xi_c(t), \qquad Y_c(t) = -\tilde{\eta}(\omega_c) \cos \Xi_c(t) \\
\tilde{\eta}_c(t) = X_c(t) + iY_c(t) = -i\tilde{\eta}(\omega_c)\, e^{i\,\Xi_c(t)}.
\end{gathered}
\tag{2.100}
$$

The subscripts c in $X_c(t)$, $Y_c(t)$, and $\tilde{\eta}_c(t)$ serve as a reminder that they belong to a *circular* nutation component of frequency ω_c. The argument Ξ_c is related to ω_c in the usual fashion:

$$\Xi_c(t) = \omega_c t + \chi_c. \tag{2.101}$$

The nutation is prograde if $\omega_c > 0$ and retrograde if $\omega_c < 0$. (The representation of circular nutations here is the same as in Section 2.18, except for minor changes in notation.)

With the use of Eq. (2.100) together with (2.101) in (2.97), one obtains the following expression for the spectral component $\tilde{m}_w(t)$ of the wobble associated with the nutation $\tilde{\eta}_c(t)$:

$$\Omega_0 \tilde{m}_w(t) = i\omega_c \tilde{\eta}(\omega_c) e^{i(\Xi_c(t) - \text{GMST})}. \tag{2.102}$$

The frequency ω_w of this wobble, which is the time derivative of $(\Xi_c(t) - \text{GMST})$, is $\omega_c - \Omega_0$. Thus the wobble frequency is the nutation frequency minus 1 cpsd.

Let us now restrict ourselves to the principal class of nutations and the wobbles associated with them. It may be recalled that the nutations and wobbles of this class result from the action of the tidal potential on the dynamical ellipticity e; the nutations have frequencies ω_c in the low frequency band: $|\omega_c| < 0.5$ cpsd. The frequencies $\omega_w = \omega_c - \Omega_0$ of the associated wobbles are in the retrograde diurnal band centered at -1 cpsd ($-1.5\,\Omega_0 < \omega_w < -0.5\,\Omega_0$). We denote the argument of these wobbles by $\Theta_w(t)$, with $d\Theta_w(t)/dt = \omega_w$.

Anticipating the developments in Chapter 5, we note that the amplitude $\tilde{m}(\omega_w)$ of the spectral component $\tilde{m}_w(t)$ of frequency ω_w in the retrograde diurnal band is defined by

$$\tilde{m}_w(t) = i\tilde{m}(\omega_w) e^{i\Theta_w(t)}, \tag{2.103}$$

and that the relation of the argument $\Theta_w(t)$ of these retrograde wobbles to the argument Ξ_c of the associated nutations of this class is

$$\Xi_c = \text{GMST} + \Theta_w + \pi. \tag{2.104}$$

This is a special case of the relation (5.267) of Chapter 5 for Ξ^- with $m = 1$, as will become clear from the treatment there. (Ξ_c and Θ_w here stand for Ξ^- and $-\Theta_\omega$ of that equation.) It will also be seen there that ω_w is the negative of the frequency ω of the spectral component of the tide generating potential that is responsible for this wobble component.

Substituting for $\tilde{m}_w(t)$ from (2.103) into Eq. (2.102) and using the relation (2.104), we obtain the kinematical relation in the spectral domain:

$$\tilde{\eta}(\omega_n) = -\frac{\Omega_0}{\omega_n}\tilde{m}(\omega_w), \qquad \omega_w = \omega_n - \Omega_0. \tag{2.105}$$

An additional relation, which is not usually written down explicitly, pertains to the spectral component of frequency $\omega_c = 0$ in the celestial frame, for which $\Xi_c = 0$, leading to $\Theta_w = -\text{GMST} - \pi$. We introduce this value of Θ_w and the corresponding frequency $\omega_w = -\Omega_0$ into Eq. (2.103), and substitute the resulting expression for $\tilde{m}_w$ into (2.97), which reduces then to $d\tilde{\eta}_{\omega_c=0}/dt = -\Omega_0 \tilde{m}(-\Omega_0)$. This component of $\tilde{\eta}$ having a constant rate of variation with time represents precession. With reference to Eq. (2.99) wherein $\Delta\psi$ and $\Delta\epsilon$ contain periodic components only, we can conclude that

$$\dot{\psi}_A \sin\epsilon_0 = -\Omega_0 \tilde{m}(-\Omega_0). \tag{2.106}$$

So the rate of precession in longitude is $\dot{\psi}_A = -\Omega_0 \tilde{m}(-\Omega_0)/\sin\epsilon_0$. For this equation to be consistent, $\tilde{m}(-\Omega_0)$ has to be real; it may be seen from the results of Section 5.7 that this is indeed the case. This fact implies, as a corollary, that the lunisolar potential does not produce any secular change in obliquity. (Actually, a minuscule rate of change of the obliquity does arise, in the second order, from the torque exerted by the tidal potential on the incremental density distribution produced by the action of the potential on the deformable Earth involving dissipative processes such as anelasticity and ocean tide dynamics, see Chapter 10.)

We see from the first relation in (2.105), that the amplitude $\tilde{\eta}(\omega_n)$ of the nutation contains a factor $1/\omega_n$, which simulates a normal mode at $\omega_n = 0$. Unlike a proper normal mode, it has no connection to any property of the rotating Earth. It has been referred to in the literature as the "tilt-over mode" (TOM), following Smith (1974). The resonance factor $1/\omega_n$ produces a great enhancement of the nutation amplitude compared to the associated wobble amplitude when the frequency is very small, their ratio $-(\Omega_0/\omega_n)$, apart from a minus sign, being the period of the nutation in sidereal days. Thus, the amplitude of the retrograde 18.6 year nutation (period -6798 solar days $= -6816$ sidereal days) is 6816 times as large as that of the corresponding retrograde diurnal wobble. There is no such resonant kinematic factor in other frequency bands, and this is a major reason for the overriding importance of the low frequency band in nutations.

The second equation in (2.105) relates the frequency of the circular wobble associated with a circular nutation to that of the nutation itself. This important general relation is simply a consequence of the Earth's diurnal rotation in space with angular velocity Ω_0. Examples of this relationship were seen in the preceding section.

Finally, we note that for the *rigid Earth*, the amplitude $\tilde{\eta}(\omega_n)$ of the nutation produced by a torque $\tilde{\Gamma}(\omega_w)$ having frequency ω_w in the terrestrial frame may be obtained by introducing (2.51) into the first of relations (2.105):

$$\tilde{\eta}(\omega_n) = i \frac{\tilde{\Gamma}(\omega_w)}{A(\omega_w - \Omega_0 e)(\Omega_0 + \omega_w)}. \tag{2.107}$$

As for the precession rate, one finds from (2.51) at $\omega_w = -\Omega_0$ together with (2.106) that

$$\dot{\psi}_A \sin \epsilon_0 = -i \frac{\tilde{\Gamma}(-\Omega_0)}{A\Omega_0(1 + e)}. \tag{2.108}$$

We have seen earlier, in Eq. (2.62), that the torque $\tilde{\Gamma}$ on an axially symmetric ellipsoidal Earth is proportional to e. Therefore the rate of precession of the figure axis is proportional to $e/(1 + e) = H_d$ (see Eq. (2.10)). The same property holds true also for the angular momentum axis, as was seen in Eq. (2.78).

2.27 Transfer function

Consider two Earth models, one rigid and the other non-rigid, both of which are forced by the same torque $\tilde{\Gamma}(\omega_w)$. (In the simplest case, when both are axially symmetric and ellipsoidal, this means that both must have the same value for the dynamical ellipticity e.) Let the amplitudes of the circular wobbles of frequency ω_w of the rigid and non-rigid Earth models be $\tilde{m}_{\text{Rig}}(\omega_w)$ and $\tilde{m}(\omega_w)$, respectively. (The subscript Rig stands for "Rigid.") Both $\tilde{m}_{\text{Rig}}$ and $\tilde{m}$ are proportional to $\tilde{\Gamma}(\omega_w)$, and therefore the ratio $\tilde{m}(\omega_w)/\tilde{m}_{\text{Rig}}(\omega_w)$ is independent of the amplitude of this torque. Moreover, as was observed in the last section, the kinematical relation (2.97) and its frequency domain version (2.105) are valid for any type of Earth model, rigid or non-rigid. Therefore it follows from the latter equation that

$$T_n(\omega_n) \equiv \frac{\tilde{\eta}(\omega_n)}{\tilde{\eta}_{\text{Rig}}(\omega_n)} = \frac{\tilde{m}(\omega_w)}{\tilde{m}_{\text{Rig}}(\omega_w)} \equiv T_w(\omega_w), \qquad \omega_n = \omega_w + \Omega_0. \tag{2.109}$$

These ratios, which are independent of the forcing torque as noted above, are the *transfer functions* from the rigid to the non-rigid Earth for nutation and wobble; they are defined only for circular motions, and are determined solely by the structure and properties of the non-rigid Earth. The computation of $T_w(\omega_w)$ for wobbles may be done by solving the torque equations in the terrestrial frame, with unit forcing, for the chosen non-rigid Earth model to obtain $\tilde{m}(\omega_w)$, and then dividing by the rigid Earth amplitude $\tilde{m}_{\text{Rig}}(\omega_w)$ per unit forcing. No separate computation is needed for $T_n(\omega_n)$ since $T_n(\omega_n) = T_w(\omega_w)$.

Transfer functions are of considerable utility for the following reason. Nutations of the rigid Earth have been studied extensively in the literature, starting with the orbital motions of the Moon, Sun, and the planets relative to the Earth (taking into account the mutual gravitational perturbations of these bodies); and highly accurate results are available for the rigid Earth nutation amplitudes. Once the transfer function is computed for the chosen non-rigid Earth model for any frequency ω_n of the excitation, the non-rigid Earth amplitude $\tilde{\eta}(\omega_n)$ is simply obtained as $\tilde{\eta}(\omega_n) = T_n(\omega_n)\,\tilde{\eta}_{\mathrm{Rig}}(\omega_n)$. There is no need any longer for one to worry about the intricacies of celestial mechanics, which are all taken care of in the computation of the rigid Earth amplitudes available in the literature. The sole Earth parameter on which the rigid Earth amplitudes (for an axially symmetric ellipsoidal model) depend is the dynamical ellipticity e. In constructing a rigid Earth nutation theory, the value to be employed for e (or rather, the related quantity $H_d = e/(1 + e)$) is inferred from observational data on the rate of precession of the real Earth.

It may be noted that in later chapters, frequencies are usually expressed in units of cpsd: $\nu = \omega_n / \Omega_0$ cpsd in the celestial frame and $\sigma = \omega_w / \Omega_0$ cpsd in the terrestrial frame (see remarks made in the last paragraph of Section 2.13). Accordingly, the transfer functions of the nutations and wobbles will be denoted by $T_n(\nu)$ and $T_w(\sigma)$, respectively; the two transfer functions are equal for $\nu = \sigma + 1$.

2.28 Oppolzer terms

The kinematical relation established in the last section enables us to obtain an equation which relates the nutation of the rotation axis to that of the figure axis. This equation, like the kinematical relation, holds whether the Earth is rigid or non-rigid.

Let the pole of the $\hat{Z}$-axis of the space-fixed (celestial) reference frame be S and the poles of the Earth's rotation axis and figure axis be R and F, respectively. By definition, the temporal variations of the vectors from S to F and from S to R, as seen from the celestial frame constitute the nutations of the figure axis and rotation axis. We denote the complex combinations of the (equatorial) components of the former vector in the celestial frame by $(X + iY)$, as before, and that of the latter, relating to the rotation axis, by $(X_R + iY_R)$. Now it is evident that the difference $(X_R + iY_R) - (X + iY)$ gives the complex representation of the components (in the celestial frame) of the vector from F to R. It is the variation of this same vector, as seen in the *terrestrial* frame, that constitutes the wobble and is represented by $\tilde{m}$; its components in the celestial frame will then be represented by $\tilde{m}e^{i\Omega_0(t-t_0)}$. Consequently we have

$$(X_R + iY_R) - (X + iY) = \tilde{m}e^{i\Omega_0(t-t_0)}. \tag{2.110}$$

We can now use the kinematical relation (2.97) to eliminate $\tilde{m}$ from the above equation. Thus we obtain the relation we seek:

$$X_R + iY_R = (X + iY) + \frac{i}{\Omega_0}(\dot{X} + i\dot{Y}). \tag{2.111}$$

For a spectral component of the nutation with frequency ω_n, one writes

$$X(t) + iY(t) = -i\tilde{\eta}(\omega_n)e^{i(\omega_n t + \chi_n)} \tag{2.112}$$

and

$$X_R(t) + iY_R(t) = -i\tilde{\eta}_R(\omega_n)e^{i(\omega_n t + \chi_n)}. \tag{2.113}$$

Equation (2.110) then leads to the following relation connecting the spectral components:

$$\tilde{\eta}_R(\omega_n) = \left(1 - \frac{\omega_n}{\Omega_0}\right)\tilde{\eta}(\omega_n). \tag{2.114}$$

Thus the fractional difference between the nutation amplitudes of the rotation axis and the figure axis is seen to be proportional to the frequency of the nutation.

It is of interest to note that the precession of the figure axis in longitude, which is represented by a term Pt in X, leads to the precession part $P(t + i/\Omega_0)$ in $(X_R + iY_R)$. Thus one finds that the rotation pole has, expectedly, the same rate of precession in longitude as the pole of the figure axis; but the rotation pole is offset by a constant amount (P/Ω_0) in obliquity.

One may ask whether the nutation of the angular momentum axis is also related in a simple manner to the nutations of the other two axes. The answer is in the negative. The reason is that the relation of the angular momentum vector to the rotation vector depends very much on the structure of the Earth and so the relation between the nutations of the angular momentum and rotation axes differs from one Earth model to another. The particular case of a rigid Earth is of considerable interest, however, because classical treatments of the rotation variations of the rigid Earth have had, as their primary output, the nutations of the angular momentum axis; the nutations of the figure and rotation axes were then inferred. The relations used for this last step will now be derived, taking the Earth to be a rigid axially symmetric ellipsoid. From Eq. (2.42) and Eq. (2.44) the components of the angular momentum vector in a principal axis frame are then

$$H_x = A\Omega_0 m_x, \qquad H_y = A\Omega_0 m_y, \qquad H_z = A\Omega_0(1 + m_z). \tag{2.115}$$

The equatorial components of the unit vector along $\mathbf{H}$ are then $(A/C)m_x$, $(A/C)m_y$, to the lowest order in m_x, m_y, m_z. These are the components of the vector offset of the pole of the angular momentum axis from that of the figure axis (the $\hat{z}$-axis); they are (A/C) times the corresponding components (m_x, m_y) for the pole of the rotation axis. The scalar factor (A/C) is of course independent of whether the offset vectors of the angular momentum axis and the rotation axis are viewed from the terrestrial or the celestial frame. Therefore we have

$$(X_H + iY_H) - (X + iY) = (A/C)[(X_R + iY_R) - (X + iY)]. \qquad (2.116)$$

This means that when considering the offset of the angular momentum axis, a relation of the form (2.110) holds with $\tilde{m}$ replaced by $(A/C)\tilde{m} = \tilde{m}/(1 + e)$. Consequently we have, instead of (2.111), the relation

$$X_H + iY_H = (X + iY) + \frac{i}{(1 + e)\Omega_0}(\dot{X} + i\dot{Y}). \qquad (2.117)$$

The corresponding relation for the nutation amplitudes for the frequency ω_n is

$$\tilde{\eta}_H(\omega_n) = \left(1 - \frac{\omega_n}{(1 + e)\Omega_0}\right)\tilde{\eta}(\omega_n). \qquad (2.118)$$

Therefore

$$\tilde{\eta}(\omega_n) - \tilde{\eta}_H(\omega_n) = \frac{\omega_n}{(1 + e)\Omega_0 - \omega_n}\tilde{\eta}_H(\omega_n). \qquad (2.119)$$

This is called the *Oppolzer term for the figure axis*. It enables one to compute the amplitude of the nutation of the figure axis, given that of the angular momentum axis. The *Oppolzer term for the rotation axis* can be obtained by combining the above equation (Eq. (2.118)) with (2.114). The result is that

$$\tilde{\eta}_R(\omega_n) - \tilde{\eta}_H(\omega_n) = -\frac{e\omega_n}{(1 + e)\Omega_0 - \omega_n}\tilde{\eta}_H. \qquad (2.120)$$

It must be kept in mind that the expressions for the Oppolzer terms (2.119) and (2.120) are valid only for the rigid Earth model considered, unlike the kinematical relation and the relation (2.114) between $\tilde{\eta}_R$ and $\tilde{\eta}$, which are independent of the Earth model.

Traditionally, nutation tables for the rigid Earth have been presented as tables of the coefficients in the spectral expansions of $\Delta\psi_H(t)$ and $\Delta\epsilon_H(t)$ in terms of real simple harmonic (cosine and sine) functions of time, together with the corresponding coefficients for the Oppolzer terms. In the case of non-rigid Earth models, tables are presented, as a rule, for the nutations of the figure axis.

2.29 Notation used in this chapter

For the notation used in this chapter, see Table 2.3.

Table 2.3 Notation in Chapter 2.

Notation	Definition	Note
Ω_0	mean Earth angular velocity	$= 2\pi$ radian/sidereal day
$\Omega_0 = 1$ cpsd	1 cycle per sidereal day	$= 7.292115 \times 10^{-5}$ radian/second
H_d	dynamical ellipticity of the Earth (astronomy community)	$H_d = \frac{C-\bar{A}}{C}, H_d = \frac{e}{(1+e)}$
e	dynamical ellipticity of the Earth (geodesy community)	$e = \frac{C-\bar{A}}{\bar{A}}, e = \frac{H_d}{(1-H_d)}$
I	mean moment of inertia	
I_{ij}	(i, j) element of the tensor of inertia	
I'_{ij}	(i, j) element of the inertia tensor in principal axis frame	
$[\mathbf{I}]$	tensor of inertia	
A, B	equatorial moments of inertia	$I'_{11} = A, I'_{22} = B, B > A$
$\bar{A}$	mean equatorial moment of inertia	$\bar{A} = (A + B)/2$
C	polar moment of inertia	$I'_{33} = C, C > B > A$
$[\mathbf{C}]$	tensor of inertia for ellipsoidal Earth in a frame related to principal moments of inertia	
$\mathbf{\Omega}$	Earth's instantaneous angular velocity vector	
$\mathbf{\Omega_0}$	mean angular velocity vector	
θ	colatitude	
λ	longitude	
δ	declination	
α	right ascension	
ϵ	obliquity	
ϵ_0	mean obliquity at J2000	
$\psi_A(t)$	precession in longitude of the figure axis or CIP	
$\dot{\psi}_A$	precession rate in longitude of the figure axis or CIP	
$\Delta\psi(t)$ or $\Delta\psi$	nutation in longitude of the figure axis or CIP	
$\Delta\epsilon(t)$ or $\Delta\epsilon$	nutation in obliquity of the figure axis or CIP	
$\Delta\psi_n(t)$ or $\Delta\psi_n$	one elliptical component of the nutation in longitude $\Delta\psi$	
$\Delta\epsilon_n(t)$ or $\Delta\epsilon_n$	one elliptical component of the nutation in obliquity $\Delta\epsilon$	
$\Delta\psi_H$	nutation in longitude of the angular momentum axis	

Symbol	Description	Relation
$\Delta\epsilon_H$	nutation in obliquity of the angular momentum axis	$\tilde\eta_H = \Delta\psi_H \sin\epsilon + i\Delta\epsilon_H$
$\Delta\psi_R$	nutation in longitude of the rotation axis	$\tilde\eta_R = \Delta\psi_R \sin\epsilon + i\Delta\epsilon_R$
$\Delta\epsilon_R$	nutation in obliquity of the rotation axis	$\tilde\eta = \Delta\psi \sin\epsilon + i\Delta\epsilon$; often $= X + iY$
$\tilde\eta_H$	nutation of the angular momentum axis	
$\tilde\eta_R$	nutation of the rotation axis	usually containing both precession and nutation, but often referring to nutation only
$\tilde\eta$	nutation of the figure axis or CIP in space	$X(t)+iY(t) = [\psi_A(t) + \Delta\psi(t)]\sin\epsilon + i\Delta\epsilon(t)$
$X(t), Y(t)$ or X, Y	coordinates of the CIP in the GCRF	$\tilde\eta_n(t) = X_n(t) + iY_n(t) = -i\tilde\eta(\omega_n)\, e^{i\,\Xi_n(t)}$
		$= -i\tilde\eta(\omega_n)\, e^{i(\omega_n t + \chi_n)}$
$\tilde\eta_n(t)$ or $\tilde\eta_n$	n-component of the nutation	
$\tilde\eta(\omega_n)$	amplitude of the component of frequency ω_n of $\tilde\eta$	
$\Xi_n(t)$	argument of the nutation of frequency ω_n	$\Xi_n(t) = \omega_n t + \chi_n$
ω_n	frequency of (elliptical) nutation in the celestial frame	
ν	nutation frequency in cpsd; positive ν corresponds to prograde nutation, and negative ν, to retrograde nutation	$\Omega_0\nu = |\omega_n|$ for prograde nutation
		$\Omega_0\nu = -|\omega_n|$ for retrograde nutation
χ_n	phase of the component of frequency ω_n of $\tilde\eta_n$	
ip	in-phase	
oop	out-of-phase	
$\Delta\psi_n^{\mathrm{ip}}$	coefficients of nutation for the sine (ip) component of $\Delta\psi_n$	$\Delta\psi_n(t) = \Delta\psi_n^{\mathrm{ip}} \sin\Xi_n(t) + \Delta\psi_n^{\mathrm{oop}} \cos\Xi_n(t)$
$\Delta\psi_n^{\mathrm{oop}}$	coefficients of nutation for the cosine (oop) component of $\Delta\psi_n$	
$\Delta\epsilon_n^{\mathrm{ip}}$	coefficients of nutation for the cosine (ip) component of $\Delta\epsilon_n$	$\Delta\epsilon_n(t) = \Delta\epsilon_n^{\mathrm{ip}} \cos\Xi_n(t) + \Delta\epsilon_n^{\mathrm{oop}} \sin\Xi_n(t)$
$\Delta\epsilon_n^{\mathrm{oop}}$	coefficients of nutation for the sine (oop) component of $\Delta\epsilon_n$	
A_n	coefficients of nutation for the sine component of $\Delta\psi_n \sin\epsilon_0$	$\Delta\psi_n(t)\sin\epsilon_0 = A_n \sin\Xi_n(t) + A_n' \cos\Xi_n(t)$
A_n'	coefficients of nutation for the cosine component of $\Delta\psi_n \sin\epsilon_0$	$A_n = \Delta\psi_n^{\mathrm{ip}} \sin\epsilon_0;\ A_n' = \Delta\psi_n^{\mathrm{oop}} \sin\epsilon_0$
B_n	coefficients of nutation for the cosine component of $\Delta\epsilon_n$	$\Delta\epsilon_n(t) = B_n \cos\Xi_n(t) + B_n' \sin\Xi_n(t)$
B_n'	coefficients of nutation for the sine component of $\Delta\epsilon_n$	$B_n = \Delta\epsilon_n^{\mathrm{ip}};\ B_n' = \Delta\epsilon_n^{\mathrm{oop}}$
$X_n(t), Y_n(t)$	components of the CIP position	$X_n(t) = \Delta\psi_n(t)\sin\epsilon_0$
		$= A_n \sin\Xi_n(t) + A_n' \cos\Xi_n(t)$
		$Y_n(t) = \Delta\epsilon_n(t) = B_n \cos\Xi_n(t) + B_n' \sin\Xi_n(t)$
$\Xi_n^{\mathrm{p}}(t)$	argument of the prograde nutation of frequency $|\omega_n|$	$\Xi_n^{\mathrm{p}}(t) = |\omega_n|t + \omega_n/|\omega_n|\,\chi_n$

(cont.)

Table 2.3 (cont.)

Notation	Definition	Note
$\Xi_n^r(t)$	argument of the retrograde nutation of frequency $-|\omega_n|$	$\Xi_n^r(t) = -|\omega_n|t - \omega_n/|\omega_n|\chi_n$
A_n^p	amplitude of the in-phase prograde component of the nutation	$A_n^p = \frac{1}{2}\left(\frac{\omega_n}{|\omega_n|}A_n - B_n\right)$
$A_n^{\prime p}$	amplitude of the out-of-phase prograde component of the nutation	$A_n^{\prime p} = \frac{1}{2}\left(A_n^\prime + \frac{\omega_n}{|\omega_n|}B_n^\prime\right)$
A_n^r	amplitude of the in-phase retrograde component of the nutation	$A_n^r = -\frac{1}{2}\left(\frac{\omega_n}{|\omega_n|}A_n + B_n\right)$
$A_n^{\prime r}$	amplitude of the out-of-phase retrograde component of the nutation	$A_n^{\prime r} = \frac{1}{2}\left(A_n^\prime - \frac{\omega_n}{|\omega_n|}B_n^\prime\right)$
s_n	sign of ω_n	$s_n = \frac{\omega_n}{|\omega_n|}$
$\tilde{\eta}_n^p$	prograde nutation amplitude	$\tilde{\eta}_n^p = A_n^p + iA_n^{\prime p}$
$\tilde{\eta}_n^r$	retrograde nutation amplitude	$\tilde{\eta}_n^r = A_n^r + iA_n^{\prime r}$
$\tilde{\eta}_n(t)$ or $\tilde{\eta}_n$	one nutation component of the CIP motion	$\tilde{\eta}_n = X_n + iY_n = \Delta\psi_n\sin\epsilon_0 + i\Delta\epsilon_n$
		$= -i[\tilde{\eta}_n^p e^{i\Xi_n^p(t)} + \tilde{\eta}_n^r e^{i\Xi_n^r(t)}]$
$\tilde{\eta}_c(t)$	circular nutation	$\tilde{\eta}_c(t) = X_c(t) + iY_c(t) = -i\tilde{\eta}(\omega_c)\,e^{i\Xi_c(t)}$
		$= -i\tilde{\eta}(\omega_c)\,e^{i(\omega_c t + \chi_c)}$
		$\tilde{\eta}_c(t) = -i\tilde{\eta}(\omega_c)\,e^{i\Xi_c(t)}$
$\tilde{\eta}(\omega_c)$	prograde/retrograde nutation amplitude of $\tilde{\eta}_c$ at positive/negative circular frequency ω_c	
ω_c	frequency of circular nutation in the celestial frame	$\omega_c = \omega_n$
χ_c	phase of circular nutation in the celestial frame	
$\Xi_c(t)$	argument of a circular nutation of frequency ω_c	$\Xi_c(t) = \omega_c t + \chi_c$
x_p, y_p	polar motion parameters	$x_p = m_1;\ y_p = -m_2$
$(x_p, -y_p)$	coordinates of the CIP in the TRF	
$\tilde{p}(t)$ or $\tilde{p}$	complex sum of the first two components of polar motion	$\tilde{p} = x_p - iy_p = i\tilde{p}(\omega_p)\,e^{i(\omega_p t + \chi_p)}$
		and $\tilde{p} = -(X + iY)e^{-i\Omega_0(t-t_0)}$
		$= -\tilde{\eta}_c(t)e^{-i\Omega_0(t-t_0)} = i\tilde{\eta}(\omega_c)\,e^{i(\Xi_c - \Omega_0(t-t_0))}$
$\tilde{p}(\omega_p)$	amplitude of the component of frequency ω_p of $\tilde{p}$	$\tilde{p}(\omega_p) = -\tilde{\eta}(\omega_c) = -\tilde{\eta}(\omega_n)$
ω_p	frequency of polar motion in the terrestrial frame	$\omega_p = \omega_n - \Omega_0 = \omega_c - \Omega_0 = \omega_w$
χ_p	phase of the component of frequency ω_p of $\tilde{p}$	
$\Theta_p(t)$	argument of polar motion in the terrestrial frame	$\Theta_c(t) = \omega_c t + \chi_c.$
R_1, R_2, R_3	rotations about the first, second, and third axes	

$$\boldsymbol{\Omega} = \boldsymbol{\Omega}_0 + \Omega_0 \mathbf{m}$$
$$\Omega_1 = \Omega_0 m_1, \ \Omega_2 = \Omega_0 m_2, \ \Omega_3 = \Omega_0(1 + m_3)$$
$$\Omega_x = \Omega_0 m_x, \ \Omega_y = \Omega_0 m_y, \ \Omega_z = \Omega_0(1 + m_z)$$
$$\tilde{m} = m_1 + i m_2 = m_x + i m_y$$
$$\Omega_0 \tilde{m}(t) = i i \bar{\eta} \, e^{-i\Omega_0(t-t_0)}$$
$$\tilde{m}_w(t) = i \tilde{m}(\omega_w) e^{i\Theta_w(t)}$$
$$\text{and } \Omega_0 \tilde{m}_w(t) = -\omega_c \tilde{\eta}(t) \, e^{-i\Omega_0(t-t_0)}$$
$$= i \omega_c \tilde{\eta}(\omega_c) e^{i(\Xi_c(t) - \Omega_0(t-t_0))}$$
$$\omega_w = \omega_n - \Omega_0 = \omega_c - \Omega_0 = \omega_p$$
$$\Omega_0 \sigma = |\omega_w| \text{ for a prograde wobble}$$
$$\Omega_0 \sigma = -|\omega_w| \text{ for a retrograde wobble}$$
$$\Theta_w(t) = \omega_w t + \chi_w; \ \Xi_c = \Omega_0(t-t_0) + \Theta_w + \pi$$
$$\text{and } \Theta_w = \Theta_p; \ \Xi_c = \Omega_0(t-t_0) + \Theta_p + \pi$$
$$\tilde{m}(\omega_w) = -\frac{\omega_n}{\Omega_0} \tilde{\eta}(\omega_n) = \frac{\omega_n}{\Omega_0} \tilde{p}(\omega_p)$$
$$T_n(\omega_n) = \frac{\tilde{\eta}(\omega_n)}{\tilde{\eta}_{Rig}(\omega_n)}$$
$$T_w(\omega_w) = \frac{\tilde{m}(\omega_w)}{\tilde{m}_{Rig}(\omega_w)}; \ T_n(\omega_n) = T_w(\omega_w)$$
$$\left(\frac{d}{dt}\right)_S = \frac{d}{dt} + \boldsymbol{\Omega}\wedge$$
$$\mathbf{H} = (H_x, H_y, H_z) \text{ in the TRF}$$
$$\text{and } \mathbf{H} = (H_X, H_Y, H_Z) \text{ in the CRF}$$
$$X_H = \frac{H_x}{C\Omega_0}, \ Y_H = \frac{H_y}{C\Omega_0}$$
$$\boldsymbol{\Gamma} = (\Gamma_x, \Gamma_y, \Gamma_z) \text{ in the TRF}$$
$$\tilde{\Gamma} = -i \frac{3GM_n}{d^5} A e \tilde{d} d_z$$
$$\text{and } \mathbf{L} = \boldsymbol{\Gamma} = (\Gamma_X, \Gamma_Y, \Gamma_Z) \text{ in the CRF}$$
$$\tilde{\Gamma} = \Gamma_x + i \Gamma_y; \ \tilde{L} = L_x + i L_y$$

$\mathbf{m}$	wobble
m_1, m_2 or m_x, m_y	wobble equatorial components
$\tilde{m}$ or $\tilde{m}(t)$	complex sum of the wobble equatorial components
$\tilde{m}_w(t)$	one component of the complex sum of the wobble equatorial components
ω_w	frequency of the equatorial wobble in TRF
σ	wobble frequency in cpsd; positive σ corresponds to prograde wobble, and negative σ, to retrograde wobble
χ_w	phase of the equatorial wobble in TRF
$\Theta_w(t)$	argument of the equatorial wobble in TRF
$\tilde{m}(\omega_w)$	amplitude of the component of frequency ω_w of $\tilde{m}$
m_3 or m_z	third component of the wobble
$\tilde{\eta}_{Rig}(\omega_n)$	rigid Earth nutation
$T_n(\omega_n)$	transfer function for nutation
$\tilde{m}_{Rig}(\omega_w)$	rigid Earth wobble
$T_w(\omega_w)$	transfer function for wobble
$\left(\frac{d}{dt}\right)_S$	time derivative in the CRF
$\frac{d}{dt} =$	time derivative in the TRF
$\mathbf{H}$	angular momentum vector
X_H, Y_H	normalized components of $\mathbf{H}$ in space
$\Gamma(t)$ or $\boldsymbol{\Gamma}$	torque acting on the Earth in the terrestrial frame
$\mathbf{L}(t)$ or $\mathbf{L}$	torque acting on the Earth in the celestial frame
$\tilde{\Gamma}, \tilde{L}$	complex sum of the equatorial components of the torque

(cont.)

Table 2.3 (cont.)

Notation	Definition	Note
$\tilde{\phi}$	normalized $\tilde{\Gamma}$ in the TRF	$\tilde{\Gamma} = -iA\Omega_0^2 e\tilde{\phi}$ and $\tilde{\phi} = \phi_1 + i\phi_2 = \frac{3GM_B}{\Omega_0^2 d^5}\tilde{d}d_z$
E	normalization factor of $\mathbf{\Gamma}$ in the CRF	$E = \frac{3GM_B}{2\Omega_0^2 c^3}\frac{C-A}{C}$
ω	frequency in the terrestrial frame	
$\mathbf{r}$	position vector of a point within the Earth	relative to the geocenter; $\mathbf{r} = (x, y, z)$
$\mathbf{d}$	position vector of a celestial body B	relative to the geocenter; $\mathbf{d} = (d_x, d_y, d_z)$ in the TRF and $\mathbf{d} = (d_X, d_Y, d_Z)$ in the CRF $\mathbf{R} = \mathbf{d} - \mathbf{r}$
d	distance from the geocenter to the celestial body B	
d_X, d_Y, d_Z	components of $\mathbf{d}$ in the CRF	
d_x, d_y, d_z	components of $\mathbf{d}$ in the TRF	
$\tilde{d}$	complex sum of the first two components of $\mathbf{d}$	$\tilde{d} = d_x + id_y$
$W^B(\mathbf{r}, t)$	gravitational potential of a celestial body B, at $\mathbf{r}$ within the Earth	
$\mathbf{r}_B$	position vector of a celestial body B relative to the geocenter	$\mathbf{r}_B = \mathbf{d}$
G	constant of gravitation	$G = 6.67 \times 10^{-11}$
M_E	mass of the Earth	$M_E = 5.97 \times 10^{24}$ kg
M_B	mass of a celestial body B	
$\rho(\mathbf{r})$	density of matter at $\mathbf{r}$	
c	semi-major axis of the elliptical orbit of the body B	
$p = n_S$	mean angular velocity of the apparent orbital motion of the Sun	= angular velocity of the Sun in uniform motion along a circle of radius d centered at the geocenter = 2π radians per year; $p^2 = n_S^2 = GM_B/c^3$
T_S	orbital period of the apparent motion of the Sun around the Earth	
T_M	orbital period of the Moon around the Earth	
$(\,.\,)$	scalar product	

3

Reference systems and frames

3.1 Celestial and terrestrial reference systems and frames

The term "reference systems" applies to the conceptual basis for the definition of "reference frames." The "frames" are realizations of the "systems" with the aid of the best available observations.

Celestial reference systems have their axes defined in relation to objects in space; terrestrial reference systems have axes that are fixed to the Earth, and hence rotating in space. All geocentric reference systems are taken to have a common origin.

3.1.1 Equator, ecliptic, and equinox, and their motions

The motion of an Earth-related axis in space is accompanied by a corresponding motion of the equator of that axis in space. These motions are manifestations of *lunisolar* precession and nutation. The instantaneous position of the equator resulting from just the secular part of the motion, namely precession, is the *mean* equator of date, while the *true* equator of date is the actual instantaneous position of the equator which includes the effect of nutation too. The true/mean equator of epoch (e.g., J2000) represents the position occupied by the true/mean equator at that fixed epoch.

The ecliptic, the plane of the orbit of the Earth–Moon barycenter around the Sun, was introduced in the previous chapters (Chapters 1 and 2, see also Appendix C) while defining precession (see Section 1.3). The plane of the ecliptic does not remain strictly fixed: it undergoes a slow rotation in space about an axis lying in this plane. This rotation is caused by the gravitational pulls exerted on the Earth–Moon system by the other planets. Therefore, the ecliptic at any given instant, known as the ecliptic of date, differs from the ecliptic of J2000 (or other fixed epoch). It is important to recognize that the secular (precessional) part of the motion of the Earth's axis at any instant is around the normal to (i.e., parallel to)

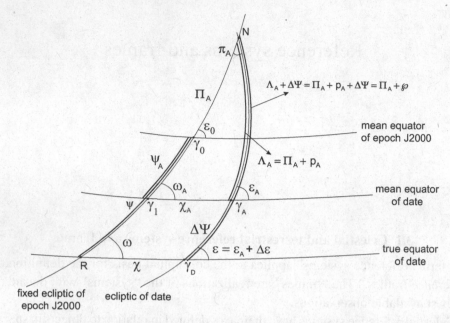

Figure 3.1 Representation of precession motions in longitude and of the nutation in longitude with reference to the ecliptics of epoch and of date, of the mean equators of epoch and of date, and of the true equator of date. This figure is an adaptation of figures in Lieske (1976), Capitaine (1990), and Bretagnon *et al.* (2003).

the ecliptic *of date* (see Fig. 3.1). Therefore the *direction* of the precessional motion in space undergoes a slow variation as the ecliptic plane rotates in space; and the rate of precession gets a small contribution from the motion of the ecliptic.

The intersection of the equatorial and ecliptic planes (the so-called line of nodes of the two planes) determines the equinoxes. The vernal or spring equinox is where the Sun, in its apparent annual motion around the Earth along the ecliptic, crosses the equator, ascending from south of the equator to the north; the Sun crosses the equator again, descending to the south, at the autumn equinox. (These crossing points are referred to in the astronomy literature as the ascending node and the descending node, respectively, of the ecliptic on the equator.) For our purposes, the word "equinox" will always be taken to mean the vernal equinox unless otherwise specified in a given context.

In layman's parlance, the term "equator" refers to the geographical equator associated with the Earth's figure axis. One should keep in mind, however, that also other kinds of axes are employed in describing Earth rotation variations (e.g., the angular momentum axis), and in any given context, the equator would be that which pertains to the chosen axis. The equinox would also be that associated with the specific equator.

One has the true equinox of date γ_D at the intersection of the ecliptic of date and the true equator of date; for the mean equinox of date γ_A, replace "true" in the above by "mean." We use the notation γ_0 for the mean equinox of epoch. The mean equinox moves from its initial position γ_0 to γ_A and γ_D as time evolves.

Motions in longitude and obliquity, with a moving ecliptic

The characterization of precession and nutation in longitude in the literature of astronomy is generally as the motion of the appropriate equinox along the ecliptic of date; and the motions in obliquity are represented by variations of the angle of inclination between the ecliptic of date and the appropriate equator of date. The dominant parts of both these types of motions arise from the motion of the equator (which we have previously represented by motions of the axis perpendicular to the equator) while the motion of the ecliptic contributes a small part to each.

With reference to Fig. 3.1, the general precessional motion in longitude in an infinitesimal interval of time is the motion of the *mean equinox of date*, namely γ_A, along the ecliptic of date in that interval. The lunisolar precession in longitude is represented by the motion, along the fixed ecliptic of J2000, of the intersection point γ_1 of this ecliptic with the mean equator of date. The reason for this name is that nearly all of this motion is caused by the torques exerted by the Moon and the Sun on the ellipsoidal Earth, though a small contribution comes from planetary torques on the Earth and also from geophysical processes and general relativity (see Chapter 5). Lunisolar precession is independent of the motion of the ecliptic and is due solely to the motion of the equator in space; for this reason, it is being increasingly referred to by the name "precession of the equator."

The measure of the accumulated general precession from J2000 till date is $p_A = \Lambda_A - \Pi_A$, where Π_A is the arc from γ_0 to N and Λ_A the arc from N to γ_A, as shown in the figure. The rate of general precession is about $50''.288$ per year. It is made up of $50''.384$ per year due to the precession of the mean equator in space (*lunisolar precession*), represented by the arc from γ_0 to γ_1 along the fixed ecliptic of epoch, and a contribution of $-0''.096$ per year from the motion of the ecliptic in space (*ecliptic motion*). Since the motion of the ecliptic is a consequence of the perturbation of the orbits of the Earth and the Moon by the gravitational pull of the planets, the ecliptic motion contribution is also called "planetary precession."

As for the nutation in longitude, denoted by $\Delta\psi$, it is represented by the arc from γ_A to γ_D.

The motions in obliquity, as pictured in terms of the relevant ecliptics and equators, may also be described with reference to Fig. 3.1. It is convenient to use the term mean obliquity of date for the angle ϵ_A between the mean equator of date and the ecliptic of date at the equinox γ_A, and the term obliquity of date for the angle ϵ between the true equator of date and the ecliptic of date at the equinox γ_D.

(It is not uncommon to use γ for γ_D.) Thus $\Delta\epsilon = (\epsilon - \epsilon_A)$ is the angle between the true equator of date and the mean equator of date, and this is the nutation in obliquity. The difference $(\epsilon_A - \epsilon_0)$ represents the precession in obliquity, ϵ_0 being the angle at γ_0, the mean equinox of J2000, between the mean equator of J2000 and the ecliptic of J2000. Nearly all of it is due to ecliptic motion, as given by $(\epsilon_A - \omega_A)$. There is a tiny contribution from the motion of the equator, which is $(\omega_A - \epsilon_0)$.

Whenever the equinox of J2000 is referred to hereafter, it must be understood to be the *mean equinox* of the epoch J2000, unless otherwise stated ($\gamma_0 = \gamma_{J2000}$).

3.1.2 *Ecliptic celestial reference system (ECRS) of J2000*

An ecliptic reference system that is fixed in space is based on the ecliptic and equinox of J2000.

Its axes $\hat{\xi}$, $\hat{\eta}$, and $\hat{\zeta}$ (also denoted by $O\hat{\xi}$, $O\hat{\eta}$, and $O\hat{\zeta}$) constitute a right-handed orthogonal triad. The first two axes are in the plane of the ecliptic of J2000, with $\hat{\xi}$ pointing in the direction of the mean equinox of J2000. The pole of the ECRS is the point on the celestial sphere in the direction of $O\hat{\zeta}$. The origin O of the reference system will be taken to be at the geocenter for our purposes. The angle between the ecliptic and the mean equator of J2000, which is the mean obliquity ϵ_0 of J2000, is $23°26'21''.406$ according to the IERS Conventions 2010 (while the IAU 2000 or IERS Conventions 2003 value was $23°26'21''.448$ or $84381''.448$).

The position of a celestial object referred to the ECRS may be expressed in Cartesian or spherical coordinates (ecliptic longitude and latitude). The direction of increase of the longitude variable is positive eastwards.

3.1.3 *International celestial reference system (ICRS) and geocentric celestial reference system (GCRS)*

Relativistic descriptions characterizing Earth rotation can take two different approaches: a dynamical description in a suitable relativistic reference system, and a kinematical description related to a set of observed distant celestial objects. Dynamical description of rotational motion in general relativity is complicated but has been done by experts in this field. In a kinematical description the reference is chosen so that the positions of a set of distant celestial objects show no net rotation with respect to the barycentric reference system. Such a reference frame is purely operational. The frame corresponds to a frame defined by these objects in the absence of light deflection, in isolation from the solar system, and when all masses of the solar system would have vanished. However, in reality, the directions so defined do not trace the actual directions of objects, because the light undergoes numerous bendings while crossing the Galaxy and the extragalactic space.

The ICRS is ideally defined as being a space-fixed reference system with its origin at the solar system barycenter, and with its axes $\hat{X}$ and $\hat{Y}$ (also denoted by OX and OY sometimes in the literature) in the plane of the mean equator of J2000, $\hat{X}$ being in the direction of the mean equinox of J2000. The direction of the third axis $\hat{Z}$ at this epoch (also represented by OZ) defines the celestial pole of the ICRS. It is kinematically non-rotating with respect to the ensemble of distant extragalactic objects. It is the idealized barycentric celestial reference system (BCRS).

The GCRS differs from the BCRS in that its origin is at the geocenter. We shall denote the GCRS also by $O\hat{X}\hat{Y}\hat{Z}$, but with O at the geocenter. The axes $O\hat{\xi}$, $O\hat{\eta}$, $O\hat{Z}$, $O\hat{Y}$ of the two reference systems are coplanar, and $O\hat{\xi}$ and $O\hat{X}$ coincide. The angle from $O\hat{\zeta}$ to $O\hat{Z}$ (or from $O\hat{\eta}$ to $O\hat{Y}$) is of course the obliquity ϵ_0 of the equator of the GCRS to the ecliptic at J2000. The GCRS is defined such that the transformation between BCRS (or ICRS) and GCRS spatial coordinates contains no rotation component, so that GCRS is kinematically non-rotating with respect to BCRS (or ICRS).

3.1.4 Geodesic precession and nutation

The equations of motion for computing precession and nutation are written according to the laws of Newtonian physics. In the reference frame introduced above (i.e., the GCRS, BCRS, and ICRS), corrections related to relativity must be introduced. They are called "geodetic" precession and nutation. Geodetic (or geodesic) precession adds to the computed precession in order to be able to be compared to the observed precession and can be treated in the same way as Newtonian precession. Similarly, the geodesic nutation must be added to the computed nutations in order that the result be comparable to the observed ones. These small corrections to precession (about 19 mas/year) and nutations (the most important one being $-0.153\sin(l')$ mas where l' is the mean anomaly of the Sun) are due to the fact that precessional and nutational motions are not in a flat space but rather in curved space, thanks to Einstein's theory of general relativity. The curvature of space is thus changing the way the Earth spin axis moves. In other words, geodetic precession and nutation are due to the fact that an object in rotation (the Earth), orbiting in a gravitational field from another non-rotating body, will undergo forces that modify the direction of its rotation axis (de Sitter, 1916a and b). The geodesic nutation plays no role for the transfer function. The reference frame where precession and nutations are computed is a dynamically non-rotating reference frame, while the geocentric celestial reference frame (GCRF) (as well as the international celestial reference frame (ICRF) and the barycentric celestial reference system (BCRF)) is kinematically non-rotating. They differ from each other by only small rotations referred to above as geodesic precession and nutation.

3.1.5 International terrestrial reference system (ITRS)

The ITRS is "Earth-fixed," as its name implies, and as explained in Section 2.14.1, it has its origin at the geocenter (understood as the center of mass of the whole Earth system, including oceans and atmosphere) and has axes $\hat{x}$ and $\hat{y}$ (also represented by $O\hat{x}$ and $O\hat{y}$) in the plane of the true equator of date, with $\hat{x}$ approximately on the Greenwich meridian, under the condition that there is no residual rotation with regard to the Earth's surface. The point on the celestial sphere in the direction of the Earth's figure axis $\hat{z}$ (also denoted by $O\hat{z}$) is the pole of the ITRS. The angle between $O\hat{z}$ and $O\hat{\zeta}$ is the obliquity ϵ of date, with $(\epsilon - \epsilon_0)$ representing precession and nutation in obliquity.

3.1.6 Reference frames, realization of reference systems

Realizations of the above systems are the ecliptic celestial reference frame (ECRF), the ICRF, the GCRF, and the ITRF. The ICRF, for instance, is constructed with the aid of VLBI determinations of the directions of distant celestial radio sources so as to have its fundamental $(O\hat{X}\hat{Y})$ plane parallel to the mean equator of J2000 as nearly as possible, with $O\hat{X}$ pointing as nearly as possible in the direction of the mean equinox of J2000.

The procedures used for constructing the CRF have been briefly described in Sections 2.14.4 and 3.1.3, and for constructing the TRF in Section 2.14.1.

The axes $(O\hat{X}, O\hat{Y}, O\hat{Z})$ and $(O\hat{x}, O\hat{y}, O\hat{z})$ constitute right-handed coordinate systems, like $(O\hat{\xi}, O\hat{\eta}, O\hat{\zeta})$. The axes of these systems and the corresponding axes of the respective frames may not be in exact coincidence, because of the imperfections of the observations on which the frames were based. The offset of the $O\hat{Z}$-axis of the ICRF from that of the ICRS is determined by the large volume of data of higher precision that has become available in recent years; it corresponds to a fixed set of very small coordinate transformations, and it is called the ICRS pole offset or the frame bias.

3.2 Transformations between reference frames

3.2.1 Transformation from ecliptic frame of fixed epoch to the terrestrial frame

The transformation from the ecliptic frame ECRF of J2000 to the ITRF may be effected through a sequence of three rotations using the notation R_1, R_2, R_3 introduced in Appendix A (see Eq. (A.2)). The first rotation $R_3(\Psi)$ is about the third axis $O\hat{\zeta}$ of the ECRF, i.e., around the pole of the ecliptic of J2000, through the angle Ψ which would take $O\hat{\xi}$ over to $O\hat{x}'$ along the intersection between that ecliptic and the true equator of date which defines the principal plane of the ITRF;

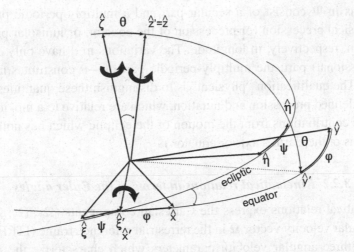

Figure 3.2 Euler angles for going from the ecliptic frame of fixed epoch to a terrestrial reference frame.

$O\hat{x}'$ is the first axis of an intermediate reference frame (IRF$_1$) to which the ECRF is carried over by this rotation. Next comes a rotation $R_1(\theta)$ about $O\hat{x}'$ through a *negative* angle θ so as to bring $O\hat{\zeta}$ into coincidence with the polar axis $O\hat{z}$ of the terrestrial frame. (θ is negative because the rotation which takes $O\hat{\zeta}$ towards $O\hat{z}$ is a "left-handed" rotation about $O\hat{x}'$. It is close to $-23.5°$, and differs only slightly from the negative of the obliquity ϵ.) The second intermediate frame (IRF$_2$) thus arrived at has $O\hat{x}'$ as its first axis and $O\hat{z}$ as its third axis. The final rotation $R_3(\varphi) = R_z(\varphi)$ completes the transformation by taking $O\hat{x}'$ into the position of $O\hat{x}$. (The explicit forms of the matrices R_i are given in Eq. (A.2) of the Appendix.) The full transformation from the ECRF to the TRF, illustrated in Fig. 3.2, is

$$\text{ECRF} \xrightarrow{R_\zeta(\Psi)} \text{IRF}_1 \xrightarrow{R_{x'}(\theta)} \text{IRF}_2 \xrightarrow{R_z(\varphi)} \text{TRF}, \tag{3.1}$$

wherein the subscripts on the rotation operators R indicate the axes about which the successive rotations are made. It is represented by the matrix transformation $R_3(\varphi) R_1(\theta) R_3(\Psi)$.

The angles Ψ, θ, φ are the Euler angles. (The symbols used for these angles by Bretagnon *et al.* (1997) are ψ, ω, φ.) They are time dependent, since the Earth's orientation in space varies with time. The variation of φ is almost entirely due to the diurnal axial rotation of the Earth; it is given by $\Omega_0 t$ plus small contributions from Earth rotation variations, especially the spin rate variations. Both Ψ and $(\theta - \theta_0)$, where θ_0 is the negative of the obliquity at J2000, consist of precessional and nutational motions and are therefore very small.

Variations in Ψ consist of a secular part and a multiply-periodic part, which are the physical precession (or precession of the equator, or lunisolar precession) and nutation, respectively, in longitude. The variations in θ have only a tiny secular (precessional) part; the multiply-periodic part of $-\theta$ constitutes nutation in obliquity. (The qualification "physical" is to distinguish these quantities from the classically defined precession and nutation, which are relative to a moving ecliptic and include contributions from the motion of the ecliptic which has nothing to do with motions of the Earth's axis or equator.)

3.2.2 *Kinematical relations in terms of the Euler angles*

The kinematical relations express the Cartesian components $(\Omega_x, \Omega_y, \Omega_z)$ of the Earth's angular velocity vector $\boldsymbol{\Omega}$ in the terrestrial reference frame (ITRF) in terms of a set of three angular velocity parameters which characterize the motion of the TRF in relation to an ecliptic-based celestial reference frame (ECRF). These parameters are the rates of change of the three Euler angles Ψ, θ, φ introduced in the last section.

The instantaneous angular velocity vector $\boldsymbol{\Omega}$ may be resolved into components $\dot{\Psi}, \dot{\theta}, \dot{\varphi}$ along the successive rotation axes:

$$\boldsymbol{\Omega} = \dot{\Psi}\,\hat{\zeta} + \dot{\theta}\,\hat{x}' + \dot{\varphi}\,\hat{z}, \tag{3.2}$$

where $\hat{\zeta}, \hat{x}', \hat{z}$ are the unit vectors along $O\hat{\zeta}$, $O\hat{x}'$, $O\hat{z}$, respectively. On the other hand one has the usual expression for $\boldsymbol{\Omega}$ in the terrestrial frame:

$$\boldsymbol{\Omega} = \Omega_x\hat{x} + \Omega_y\hat{y} + \Omega_z\hat{z}. \tag{3.3}$$

Note that in terms of the wobble and spin rate variables,

$$\Omega_x = \Omega_0 m_1, \qquad \Omega_y = \Omega_0 m_2, \qquad \Omega_z = \Omega_0(1 + m_3), \tag{3.4}$$

and $\Omega_0\hat{z}$ is the reference (mean) angular velocity of Earth rotation.

Relations connecting the two sets of angular velocity components which appear in Eqs. (3.2) and (3.3) may be obtained by transforming both these expressions for $\boldsymbol{\Omega}$ to a common reference frame. It is most convenient to choose this common frame to be IRF$_2$ having $\hat{x}'$, $\hat{y}'$, and $\hat{z}' = \hat{z}$ as its axes. Since $\hat{\zeta}$, which appears in the first term of (3.2), is the third axis of IRF$_1$ (as well as of the ECRF), it has components $(0,0,1)$ in this frame. Therefore, on transforming to IRF$_2$ through the rotation $R_{x'}(\theta)$, the components in IRF$_2$ of the first term in (3.2), namely $\dot{\Psi}\hat{\zeta} = \dot{\Psi}\hat{\zeta}'$, become

$$R_{\hat{\zeta}'}(\theta)\begin{pmatrix} 0 \\ 0 \\ \dot{\Psi} \end{pmatrix} = \begin{pmatrix} 1 & 0 & 0 \\ 0 & \cos\theta & \sin\theta \\ 0 & -\sin\theta & \cos\theta \end{pmatrix}\begin{pmatrix} 0 \\ 0 \\ \dot{\Psi} \end{pmatrix} = \begin{pmatrix} 0 \\ \sin\theta \\ \cos\theta \end{pmatrix}\dot{\Psi}. \tag{3.5}$$

The axes $\hat{x}'$, $\hat{z}$ involved in the second and third terms of (3.2) are axes of IRF$_2$ itself. Therefore these terms contribute $\dot{\theta}$ and $\dot{\varphi}$, respectively, to the first and third components of the vector (3.2) in IRF$_2$. We obtain the components of $\boldsymbol{\Omega}$ of (3.2) in IRF$_2$ by combining these with the components due to the first term, shown in (3.5):

$$(\dot{\theta}, \ \dot{\Psi} \sin \theta, \ \dot{\varphi} + \dot{\Psi} \cos \theta). \tag{3.6}$$

Consider next the components, in the IRF$_2$, of $\boldsymbol{\Omega}$ of the form (3.3). Since the transformation from the terrestrial reference frame to IRF$_2$ is through the rotation $R_z^{-1}(\varphi)$, the desired components in the latter frame are given by

$$R_z^{-1}(\varphi) \begin{pmatrix} \Omega_x \\ \Omega_y \\ \Omega_z \end{pmatrix} = \begin{pmatrix} \cos \varphi & -\sin \varphi & 0 \\ \sin \varphi & \cos \varphi & 0 \\ 0 & 0 & 1 \end{pmatrix} \begin{pmatrix} \Omega_x \\ \Omega_y \\ \Omega_z \end{pmatrix}$$

$$= \begin{pmatrix} \Omega_x \cos \varphi - \Omega_y \sin \varphi \\ \Omega_x \sin \varphi + \Omega_y \cos \varphi \\ \Omega_z \end{pmatrix}. \tag{3.7}$$

On equating these components to those in Eqs. (3.6), we obtain the relations

$$\Omega_x \cos \varphi - \Omega_y \sin \varphi = \dot{\theta},$$
$$\Omega_x \sin \varphi + \Omega_y \cos \varphi = \dot{\Psi} \sin \theta, \tag{3.8}$$
$$\Omega_z = \dot{\Psi} \cos \theta + \dot{\varphi}.$$

These are the kinematical relations between angular velocity components in space and those in the terrestrial frame.

The first two of these relations can be expressed more compactly by taking their complex combination and denoting $\Omega_x + i\Omega_y$ by $\tilde{\Omega}$:

$$\tilde{\Omega} e^{i\varphi} \equiv \Omega_0 \tilde{m} e^{i\varphi} = i(\dot{\Psi} \sin \theta - i\dot{\theta}), \tag{3.9}$$

where $\tilde{m} = m_1 + im_2$ as usual.

Though we had specified the terrestrial reference frame to be the ITRF at the beginning of this section, relations of the same form hold good with any terrestrial frame which has the true equatorial plane of date as its principal plane. If, for instance, one desired to choose the principal axis frame with the x coordinate axis along the principal axis of the lowest moment of inertia A, the only change needed would be to replace the angle φ (from O$\hat{x}'$ to the Greenwich meridian) by the angle $\tilde{\varphi}$ from O$\hat{x}'$ to the new x-axis. Of course the values of the components of $\boldsymbol{\Omega}$ would be different from those in the ITRF; let the new values be Ω_x', Ω_y', Ω_z'. The kinematical relations will then be obtained from (3.8) by the replacements

$$(\Omega_x, \Omega_y, \Omega_z) \to (\Omega_x', \Omega_y', \Omega_z'), \qquad \varphi \to \tilde{\varphi} = \varphi + \alpha, \tag{3.10}$$

where α is the longitude of the x principal axis in the ITRF.

3.2.3 *The intermediate pole CEP/CIP, and its nutations and polar motions*

Observations as well as the theory of Earth rotation variations show that the spectral content of the motion of the Earth's figure axis in space is almost entirely in the low frequency domain. The spectrum of the motion of the Earth's rotation pole relative to the Earth itself (i.e., of "polar motion" in the classical sense) is also almost exclusively in the low frequency domain. Therefore it made sense to "break up" the transformation between celestial and terrestrial frames into precession–nutation and polar motion parts (apart from an axial rotation) such that each part contains only spectral components of low frequencies (between −0.5 and 0.5 cpsd). This was attempted to be accomplished by the introduction of a conceptual "intermediate pole," named the celestial ephemeris pole (CEP), which was defined (see Seidelmann, 1982) to have only spectral components with frequencies between −0.5 and 0.5 cpsd in its motions relative to the celestial reference frame and also relative to the terrestrial reference frame (see Fig. 3.3). The diurnal Earth rotation was represented as a rotation about the CEP axis, and the angle of rotation was identified with (−GST), where GST, the Greenwich Sidereal Time, is the angle through which the Greenwich Meridian has moved from the *true* equinox.

The above limitation on the frequency spectra in both the celestial and the terrestrial frames amounts, in effect, to ignoring all but these two windows of the frequency spectrum in the relative motion between the celestial and terrestrial reference frames. The rapid improvements in the precision of observations in recent years, as well as advances in theory, made it necessary to take account of the other parts of the frequency spectrum too to achieve a better accounting of the observations. Recognizing this fact, the IAU resolved in 2000 to adopt a new reference pole, named the celestial intermediate pole (CIP), to replace the CEP. The celestial motion of the CIP, like that of the CEP, has spectral components only in the low frequency band (equivalent to frequencies within the retrograde diurnal band in the terrestrial frame); however, unlike the case of the CEP, the spectrum of the motion of the CIP in the terrestrial frame includes all frequencies outside the retrograde diurnal band (see Fig. 3.4).

Another difference is in the angle of rotation around the CIP: it is no longer identified with (−GST) as before, but as the negative of the "Earth rotation angle" ERA (initially named "stellar angle"). It is defined as the arc measured along the equator of the CIP from the "non-rotating origin" (Guinot, 1979, Capitaine, 1986, 1990, Capitaine *et al.*, 2000) related to the ICRF, located on this equator, and called the CIO, the celestial intermediate origin (instead of from the equinox), to the "non-rotating origin" related to the ITRF, located on this equator, and called the TIO, the terrestrial intermediate origin (instead of from the Greenwich meridian). The IAU adopted in 2000 the term celestial ephemeris origin (CEO) for the

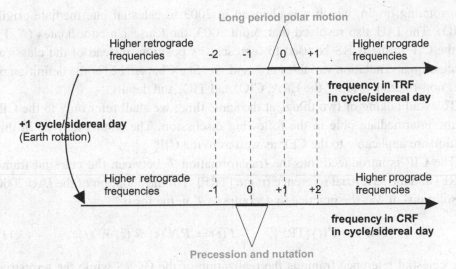

Figure 3.3 Frequency bands involved in the motions of the CEP in the terrestrial reference frame (TRF) and celestial reference frame (CRF) are limited to the sections shown in heavy black. The permitted frequency bands in each reference frame are marked in heavy black.

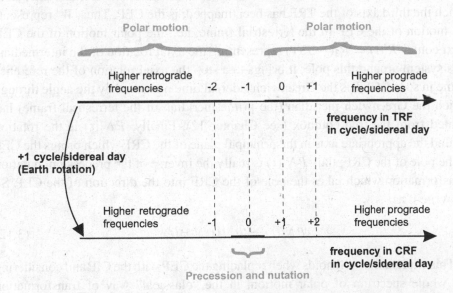

Figure 3.4 Frequency bands involved in the motion of the CIP in the terrestrial reference frame (TRF) and celestial reference frame (CRF) are shown in heavy black. The permitted frequency bands in each reference frame are marked in heavy black. Note that all frequencies in the TRF are allowed, other than a band of width 1 cpsd about −1 cpsd.

non-rotating origin, which was changed in 2003 to celestial intermediate origin (CIO). The IAU also resolved that, from 2003, the Cartesian coordinates (X, Y) of the CIP in the ICRS be used to represent its position, instead of the classical precession and nutation variables $\Delta\psi$ and $\Delta\epsilon$. See Chapter 12 for the definition of the "non-rotating origin," the ERA, CIO, and TIO, and details.

To avoid talking of two things at the same time, we shall refer only to the CIP as the intermediate pole in the following discussion. The results derived in this section are applicable to the CEP as well as to the CIP.

The CIP is introduced into the transformation T between the celestial frame [CRF] and the terrestrial reference frame [TRF]. Till the adoption of the IAU 2000 Resolutions, it was the practice to decompose T in the form

$$[CRF] = T(t)\,[TRF], \qquad T(t) = PN'(t)\,R'(t)\,W'(t). \tag{3.11}$$

The celestial reference frame is the realization of the GCRS while the terrestrial frame TRF is the realization of TRS, which has the true equator of date as its principal plane. The operation $W'(t)$ is a rotation about an axis in the equatorial plane of the TRF, taking the axes of the TRF into an intermediate axis system which rotates in space (like the TRF). The pole of the third axis of this frame, on to which the third axis of the TRF has been mapped, is the CEP. Thus, W' represents the motion of the CEP in the terrestrial frame, i.e., the polar motion of the CEP. Next comes $R'(t) = R_3(-GST)$ representing the axial rotation of the intermediate axis system around this pole; it brings to a stop the axial rotation of the reference frame in space. GST is the Greenwich sidereal time measured by the angle through which the Greenwich meridian (the prime meridian of the terrestrial frame) has rotated from the true equinox (see Chapter 12). Finally, $PN'(t)$ is the rotation around the appropriate axis in the principal plane of the CRF which brings the CEP to the pole of the CRF; thus $PN'(t)$ is really the inverse of the precession–nutation transformation which takes the pole of the CRF into the direction of the CEP. So we write it as

$$PN'(t) = P^{-1}(t)N^{-1}(t). \tag{3.12}$$

This explanation still holds when replacing the CEP with the CIP and considering the whole spectrum of polar motion, in the "classical" way of transformation between the TRF and the CRF.

With the adoption of the IAU 2000 Resolutions, the CIP has replaced the CEP, and GST has been replaced by the Earth rotation angle ERA or $\theta(t)$ (initially referred to as the stellar angle). As has been mentioned earlier, the spectrum of motions of the CIP in the terrestrial reference frame has been defined to include all frequencies other than those in the retrograde diurnal band which are viewed as motions in the celestial frame.

The replacement of GST by the ERA with a relation involving s and s', very small angles determined by the history of the nutation–precession and polar motions, is detailed in Chapter 12. The transformations between the reference frames are also detailed therein.

Referring to Fig. 3.1, the $PN'(t)$ matrix can be written as

$$PN'(t) = R_1(-\epsilon_0)R_3(\psi_A)R_1(\omega_A)R_3(-\chi_A)R_1(-\epsilon_A)R_3(\Delta\psi)R_1(\epsilon_A + \Delta\epsilon).$$

(3.13)

The introduction of polar motion components (x_p, y_p), as given in Chapter 2, enables one to write

$$W'(t) = R_2(x_p)R_1(y_p).$$ (3.14)

With the introduction of the CIP as an intermediate pole, the relative motion between the poles of the CRS and the TRS is viewed as a composite of two parts: a precession–nutation motion of the CIP in space relative to the CRS, with a spectrum confined exclusively to the low frequency band (−0.5 to 0.5 cpsd) as seen in the CRS, and a polar motion of the CIP relative to the pole of the TRS, with a spectrum which has no components in the retrograde diurnal band (-1.5 to -0.5 cpsd) as seen in the TRS. The absence of polar motions with retrograde diurnal frequencies may be expressed alternatively by saying that the amplitudes of polar motions with such frequencies are zero, or equivalently, that the CIP appears coincident with the pole of the TRS when viewed from the terrestrial frame at such frequencies. But we know that this band of frequencies appears from the celestial frame as the low frequency band. Therefore the above-mentioned coincidence of the CIP and the pole of the TRS in this frequency range means, in effect, that the nutational motions of the CIP, which have frequencies restricted to the low frequency band in the celestial frame, are also nutations of the figure axis (the pole of the TRS being, for all practical purposes, on the figure axis). Similar reasoning leads one to the conclusion that the polar motion of the CIP, within the permitted range of frequencies, is effectively the polar motion of the pole of CRS relative to the TRS. But the reverse of this motion (namely, that of the pole of the TRS relative to the CRS) is the nutation of the figure axis in the range of frequencies outside the frequency range in the CRS. (Note that this range is forbidden for the CIP but not for the figure axis.) Therefore the polar motion of the CIP is equivalent to the negative of the nutation of the figure axis, transformed to the terrestrial frame. These observations are of non-trivial significance because they enable us to carry over the results of theories of the nutations of the figure axis (and related results for the polar motions of the pole of the CRS, see Section 2.25) to the motions of the pole of the CIP for which one has no dynamical equations.

The considerations of this section apply equally well to the "old" intermediate pole, though one has to be conscious of the fact that the spectrum of the polar motion of the CEP is restricted to just the low frequency band (see Fig. 3.3 and 3.4).

3.2.4 Representation of precession and nutation referred to moving ecliptic

We introduced in Section 3.1.1 the concepts of precession and nutation in relation to a moving ecliptic. Now we need to see how these concepts enter into the representation of the precession–nutation operator $PN'(t)$ of the last section.

In the transformation from the celestial frame of J2000 to the terrestrial frame of date (which is the inverse of the transformation (3.11) above), the first factors are $N(t)P(t)$. The precession $P(t)$ takes the celestial frame of epoch J2000 which is based on the mean equator and equinox γ_0 of epoch into a frame based on the mean equator and equinox γ_A of date, with a common origin O for both the frames. Nutation $N(t)$ then carries the new frame over into one based on the true equator and equinox γ (or γ_D) of date. The equinox is of course along the intersection of the relevant equator and ecliptic, and gives the direction of the first axis of the reference frame.

The precession $P(t)$ can be written using the previously introduced angles (see Fig. 3.1) as

$$P(t) = R_3(\chi_A)R_1(-\omega_A)R_3(-\psi_A)R_1(\epsilon_0). \tag{3.15}$$

The precession $P(t)$ is however traditionally represented as a product of three rotations (see Fig. 3.5):

$$P(t) = R_3(-z_A)R_2(\theta_A)R_3(-\zeta_A). \tag{3.16}$$

Figure 3.5 shows the mean equator and ecliptic of the epoch J2000 as well as those of date; the poles of the two equators are marked P_0 and P_D, respectively. The great circle containing these two poles is shown; the pole of this circle is Q, the intersection of the two equators. The arcs F_0Q and F_DQ are both $90°$. We note that $R_3(-\zeta_A)$ in (3.16) is a rotation in the mean equatorial plane of J2000, i.e., around the pole P_0, and brings γ_0 to F_0. This rotated intermediate frame has OF_0 as its first axis and OQ as the second. The next rotation $R_2(\theta_A)$ is about OQ, to carry the mean equator of epoch into that of date; it takes F_0 into F_D, making OF_D the first axis of the second intermediate frame, with OQ remaining as its second axis. The last rotation $R_3(-z_A)$ around P_D carries F_D over to γ_A on the first axis of the mean equinox–equator frame of date.

The parameters ζ_A, θ_A, z_A were employed here to characterize precession, which carries the mean equinox of epoch to that of date (γ_0 to γ_A). We have already encountered in Section 3.1.1 an alternative set of parameters, namely, $\chi_A, \psi_A, \omega_A, \Pi_A, \pi_A, \Lambda_A$, which is employed for the same purpose. All these

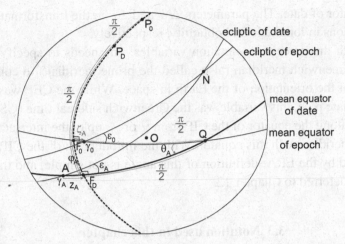

Figure 3.5 Precession transformation based on angles ζ_A, θ_A, z_A. The arcs P_0F_0, P_DF_D, F_0Q, and F_DQ are all $\pi/2$. (Figure adapted from Lieske, 1976.)

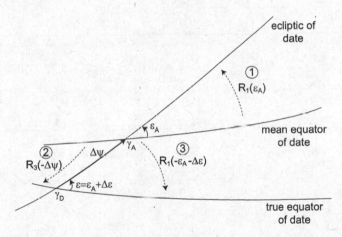

Figure 3.6 Nutation transformation according to Eq. (3.17).

parameters are known as polynomial expressions in the time variable as provided in the IERS Conventions (e.g. IERS Conventions 2010, Petit and Luzum, 2011).

Consider next the nutation $N(t)$ (see Fig. 3.1 and Fig. 3.6). It is represented by the product

$$N(t) = R_1(-\epsilon_A - \Delta\epsilon)R_3(-\Delta\psi)R_1(\epsilon_A). \tag{3.17}$$

The first rotation $R_1(\epsilon_A)$ around γ_A carries the mean equator of date into the ecliptic of date, the next rotation $R_3(-\Delta\psi)$ around the pole of this ecliptic takes γ_A to γ_D, and the final rotation $R_1(-\epsilon_A - \Delta\epsilon)$ around γ_D brings the ecliptic of date over into the true equator of date. The parameters $\Delta\psi$ and $\Delta\epsilon$ in the transformation (3.17) are the nutations in longitude and obliquity, respectively.

Along with the precession–nutation variables, one needs to specify the position of the Greenwich meridian (also called the prime meridian) to complete the description of the orientation of the Earth in space. While the CEP was in use as the intermediate pole, this variable was the Greenwich sidereal time (GST), which is measured along the equator of the CIP from γ or γ_D up to the intersection of the Greenwich meridian with this equator. With the introduction of the CIP, GST has been replaced by the ERA; definition of this angle is not simple, and the detailed definition is deferred to Chapter 12.

3.3 Notation used in this chapter

For the notation used in this chapter, see Table 3.1.

Table 3.1 *Notation in Chapter 3.*

Notation	Definition	Note
$(O\hat{\xi}, O\hat{\eta}, O\hat{\zeta})$	axes of the ecliptic frame	$O\hat{\xi}\hat{\eta}$ is the ecliptic plane with $O\hat{\xi}$ pointing in the direction of the mean equinox of J2000
$(O\hat{X}, O\hat{Y}, O\hat{Z})$	axes of the ICRF	fundamental ($O\hat{X}\hat{Y}$) plane parallel to the mean equator of J2000, with $O\hat{X}$ pointing in the direction of the mean equinox of J2000
$(O\hat{x}, O\hat{y}, O\hat{z})$	axes of the ITRF	$O\hat{x}\hat{y}$ is the plane of the true equator of date, with $O\hat{x}$ on the Greenwich meridian
$(O\hat{x}', O\hat{y}', O\hat{z})$	intermediate non-rotating frame	$O\hat{x}'\hat{y}'$ is the same plane as $O\hat{x}\hat{y}$ with $O\hat{x}'$ pointing in the direction of the intersection between the ecliptic of J2000 and the true equator of date
R_1, R_2, R_3	rotations about the first, second, and third axes	
ψ	longitude Euler angle	angle on fixed ecliptic of epoch J2000 between γ_0 and R
Θ	obliquity Euler angle	angle at R between true equator of date and ecliptic of epoch J2000
φ	rotation Euler angle	angle on the true equator of date between the Greenwich meridian ($O\hat{x}$) and R (or $O\hat{x}'$ or $O\hat{X}'$)
γ_0	mean equinox of epoch	intersection-point between ecliptic of epoch J2000 and mean equator of epoch J2000
γ_A	mean equinox of date	intersection-point between ecliptic of date and mean equator of date
γ_D, γ	true equinox of date	intersection-point between ecliptic of date and true equator of date
γ_1		intersection-point between ecliptic of epoch J2000 and mean equator of date
Q		intersection-point between mean equator of epoch J2000 and mean equator of date
F_D		90° from Q on mean equator of date

(cont.)

Table 3.1 (cont.)

Notation	Definition	Note
F_0		$90°$ from Q on mean equator of epoch J2000
R		intersection-point between ecliptic of epoch J2000 and true equator of date
N		intersection-point between ecliptic of epoch J2000 and ecliptic of date
ϵ_0	mean obliquity of epoch	angle at γ_0 between ecliptic of epoch J2000 and mean equator of epoch J2000
ϵ_A	mean obliquity of date	angle at γ_A between ecliptic of date and mean equator of date
ϵ_D, ϵ	true obliquity of date	angle at γ_D (or γ) between ecliptic of date and true equator of date
$\Delta\epsilon$	nutation in obliquity	$\Delta\epsilon = \epsilon - \epsilon_A = \epsilon_D - \epsilon_A$
ω_A	obliquity at γ_1	angle at γ_1 between ecliptic of epoch J2000 and mean equator of date
π_A		angle at N between ecliptic of epoch J2000 and ecliptic of date
z_A		angle on mean equator of date between γ_A and F_D
ζ_A		angle on mean equator of epoch J2000 between F_0 and γ_0
θ_A		angle at Q between mean equator of epoch J2000 and mean equator of date; also angle on the equator of Q from F_0 to F_D
Π_A		angle on ecliptic of epoch J2000 between N and γ_0
Λ_A		angle on ecliptic of date between N and γ_A
ψ_A	lunisolar precession	angle on ecliptic of epoch J2000 between γ_0 and γ_1
p_A	general precession	angle on ecliptic of date and $=\Lambda_A - |\Pi_A|$
ψ'_1	ecliptic motion	angle on ecliptic of epoch J2000 and $=|p_A| - \psi_A$
χ_A	planetary precession	angle on mean equator of date between γ_1 and γ_A
$\Delta\psi$	nutation in longitude	angle on ecliptic of date between γ_A and γ_D
$\wp$	general precession and nutation in longitude	angle on ecliptic of date and $=\Lambda_A + \Delta\psi - |\Pi_A|$
$\mathbf{\Omega}$	Earth's instantaneous angular velocity vector	

$P(t)$ precession part of the matrix transformation from the CRF to the TRF

$N(t)$ nutation part of the matrix transformation from the CRF to the TRF

ERA Earth rotation angle

$W(t)$ polar motion part of the matrix transformation in non-rotating origin approach from the TRF to the CRF

$W'(t)$ polar motion part of the matrix transformation in classical approach from the TRF to the CRF

$Q(t)$ precession–nutation part of the matrix transformation in non-rotating origin approach from the TRF to the CRF

$PN'(t)$ precession–nutation part of the matrix transformation in classical approach from the TRF to the CRF

$R(t)$ axial rotation part of the matrix transformation in the non-rotating origin approach from the TRF to the CRF

$R'(t)$ axial rotation part of the matrix transformation in the classical approach from the TRF to the CRF

$T(t)$ matrix transformation from the TRF to the CRF

$$P(t) = R_3(-z_A)R_2(\theta_A)R_3(-\zeta_A)$$
$$= R_3(\chi_A)R_1(-\omega_A)R_3(-\psi_A)R_1(\epsilon_0)$$
$$N(t) = R_1(-\epsilon_A - \Delta\epsilon)R_3(-\Delta\psi)R_1(\epsilon_A)$$

$$PN'(t) = P(t)^{-1}N(t)^{-1}$$

$$R(t) = R_3(-ERA)$$

$$R'(t) = R_3(-GST)$$

$$[CRF] = T(t)[TRF], \; T(t) = Q(t)R(t)W(t) = PN'(t)R'(t)W'(t)$$

4

Observational techniques – ephemerides

The angles representing the orientation of the Earth in space are determined, as functions of time, primarily through very long baseline interferometry (VLBI). The temporal variations of these angles are characterized by the terms precession, nutation, and length-of-day (LOD) variations. However global positioning system (GPS) or more generally global navigation satellite system (GNSS) observations have also shown the capability to give useful results for the shorter period nutations, besides LOD variations.

4.1 VLBI

The purpose of this section is not to provide all the detail on the VLBI principle, measurements, and processing and certainly not to provide all the mathematical details. The reader may refer to textbooks including tutorials such as Thompson *et al.* (1986), Felli and Spencer (1989), or Zensus and Napier (1995) for further details.

4.1.1 Principle of observation

This geodetic technique is based on simultaneous recordings of radio emissions received from extragalactic radio sources by a number of large radio telescopes (6 to 16) from which the source is "visible" at the time of observation. Emissions from a radio source arise from the time variable electromagnetic fields over its surface. Electromagnetic waves radiate away from the source in accordance with Maxwell's equations. On arrival at the VLBI stations on the Earth, the radiations within a certain range of frequencies (e.g. in the X-band around 8.4 GHz) are recorded by the VLBI antennas and associated devices with the aid of band-pass filters.

A VLBI network consists of a number of large radio dish antennas, ranging from 3 to 100 meters in diameter, distributed all over the globe, located at distances of hundreds to thousands of kilometers from one another. Electromagnetic waves from a radio source, so far away from the Sun that any motion it has in space is practically undetectable from the Earth, are received and recorded as functions of the time of arrival, at each of a subset of radio antennas in the VLBI network from which the source is simultaneously visible at the time of observation. Up to about 3 minutes are dedicated to each radio source, after which the network of VLBI antennas switches to another radio source. Different subsets of the VLBI network are used for recording signals from different sources, depending on visibility of each source at the time of observation. Such observations continue for about 24 hours, and during this time recordings of radio emissions from a large number of sources (typically about 100) are carried out.

Radio sources are emitting in the form of continuum radiation, the power spectrum of which shows only a slow variation with frequency and may be regarded as constant over the receiver bandwidth. An observer sees random noise if observing one radio source (therefore the radio signal obtained at the antenna is not really a signal; it will however be called signal for convenience). The recorded signal at one radio telescope is essentially identical to the signal recorded by a telescope at a different location, except for the effect of Earth rotation and for the length and orientation of the baseline between the two stations (see Fig. 4.1) (and for local perturbations to a minor extent).

The antenna signal is time-stamped with an extremely precise and stable atomic clock (usually a hydrogen maser). Cross-correlation procedures are used in the processing of VLBI observation data. The signals (signal 1 and signal 2) received at the two antennas (stations) are cross-compared by moving one of the signals (signal 2) along its time-base relative to the other signal (signal 1) as shown in Fig. 4.2 (displacement in time is referred to as "time delay" between each pair of stations). The reason for the delay is that the line-of-sight distance between an individual telescope and a given radio source is different from that between any other telescope and the same radio source.

Each recording electronically captured during the observation session carries alongside the radio signal received at the VLBI station a time signal from the "station clock" that is situated at the station. The station clocks are atomic clocks of very high accuracy and stability, such as a hydrogen maser clock (short-term frequency stability of 10^{-13} at 1 second, stability of 10^{-15} or below at 100 to 10 000 seconds, and an intrinsic frequency drift as low as 10^{-15} per day even for a passive maser, i.e., without implementing the automatic cavity tuning system).

The antennas remain fixed on the ground and the sky appears to rotate above them. Thus the angle marking the line-of-sight in the direction of the radio source changes with time and the antenna tracking the source for several hours necessarily rotates with the Earth. The motion of the antenna associated with this rotation

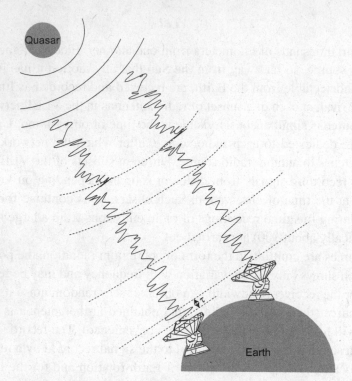

Figure 4.1 Principle of VLBI. The radiation leaving the source (Quasar) at a given time arrives at the two stations at different times, one with a delay of τ with respect to the other.

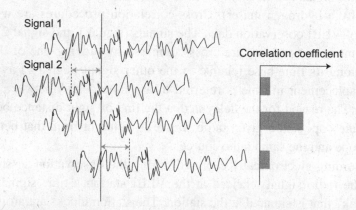

Signal 1

Signal 2

Correlation coefficient

Figure 4.2 Principle of VLBI. The radio signals received at two VLBI antennas are correlated in order to determine the delay. The signals arriving at different radio-telescopes are essentially identical, except for being displaced in time (horizontally in the figure) relative to one another. The correlations between the top signal and the other three signals are shown on the right-hand side of the figure. When the correlation between signal 1 and signal 2 recorded at different stations is determined with a variable displacement in time between the two, the correlation becomes a maximum (as in the second version of signal 2 in the figure) when the displacement is such as to eliminate the delay between the reception of the signals at the two stations.

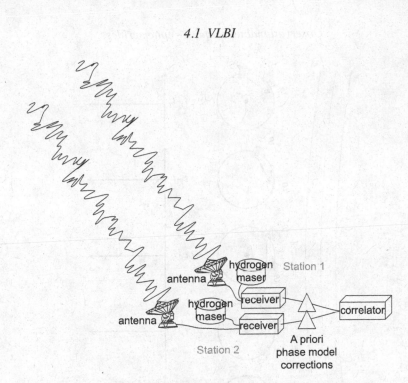

Figure 4.3 Chain of VLBI processing.

causes the frequency of the radiation recorded by the antenna to change on account of the Doppler effect; this change differs from antenna to antenna because of the differences between their rotation-induced velocities on account of their different locations. The difference in frequencies arising from this effect needs to be taken into account while correlating the signal. The cross-correlation function is multiplied by a sinusoid coming from a model phase. It is then integrated over a time interval to produce the so-called stopped fringes.

In processing the VLBI data, one uses a correlator that plays back the recorded data from all the stations simultaneously where the data are aligned and correlated by searching for the maximum of the cross-correlation function. The timing of the playback is adjusted according to the atomic clock information associated with each data stream, and the estimated times of arrival of the radio signal at each of the telescopes. A first determination of the distance between the telescopes and the effect of the Earth rotation provide a first guess of the time delay for correlating the signals. These can easily be taken a priori into account (they are called "phase model corrections") and fine adjustments are made until interference fringes are detected. Constructive interference occurs when the waves are in phase, and destructive interference when they are half a cycle out of phase. Thus, an interference fringe pattern is produced. The outputs of the correlator are the amplitude and phase of the fringes, from which the delays and delay rates of the wavefront can be derived.

The processing of the measurements is sketched in Fig. 4.3.

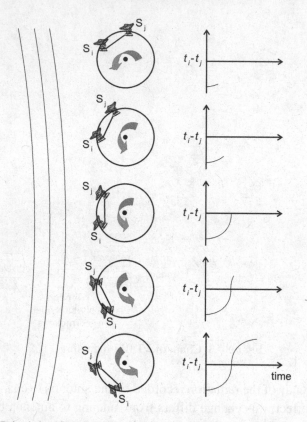

Figure 4.4 Principle of VLBI used for Earth rotation. The radio emission leaving the source at a given time arrives at the two stations i and j at the times t_i and $t_j = t_i + \Delta t_{ij}$. The time delay Δt_{ij} or τ_{ij} between stations i and j is estimated by processing of the records of the emissions received at these stations. Time delays for different configurations of the Earth orientation are shown as a function of time on the right hand side of the figure.

The Earth rotation and orientation are studied using VLBI, as may be easily understood from Fig. 4.4. If two radio telescopes i and j are based at $\mathbf{r}_i$ and $\mathbf{r}_j$ on the Earth's surface, and if the unit vector from the geocenter to the source is $\mathbf{e}$, the time delay is evidently

$$\tau_{ij} = \mathbf{e} \cdot (\mathbf{r}_i - \mathbf{r}_j)/c, \tag{4.1}$$

where c is the velocity of light. The vector $(\mathbf{r}_i - \mathbf{r}_j)$ connecting the sites of the two telescopes is called the baseline vector; it ranges in length from a few tens of meters to nearly the diameter of the Earth for different pairs of telescopes. The time delay τ_{ij} measures thus the projection d_{ij} of the baseline vector on to the direction $\mathbf{e}$: $d_{ij} = \mathbf{e} \cdot (\mathbf{r}_i - \mathbf{r}_j)$. This delay depends mainly on the time of observation because the diurnal rotation of the Earth, and most importantly for our purposes, the variations of the Earth's orientation in space, cause the angle between the baseline vector and the direction **e** to keep changing.

Determination of the delay itself is based on the expected identity of the temporal variations of the emissions recorded at stations i and j when one of these recordings is displaced in time relative to the other by an amount corresponding to τ_{ij}. The determination is done by using a correlator/processor which "displaces" the two recordings relative to one another until the correlation between the records is maximum (the fringes are formed after the observing session). The relative time displacement gives the time delay sought. The electronic recordings of the radio signals from all the VLBI stations in the observing network have to be sent to the correlator station/processing center for this purpose. The delay is measured with a precision at the level of 10 picoseconds.

Given the time delays τ_{ij} and hence the baseline projections d_{ij} for the pairs (ij) of VLBI stations employed in each observation session, it is possible to deduce the baseline vectors themselves. The complete set of time delays for all the pairs of VLBI radio antennas employed in each observing session, together with the observation time to which each time delay pertains, constitute the basic VLBI data set that is used subsequently in the analysis to extract information about various physical processes.

Such data sets are available for the period from the late 1970s onwards, and are being constantly added to with ever-increasing precision. The time delays are presently processed at different Analysis Centers of the International VLBI Service (IVS) (Schuh and Behrend, 2012, Behrend, 2013), hosted by different countries (USA, Europe, Japan, Russia, Ukraine . . .). There is in addition an IVS Coordinating Center which is responsible for coordinating observing programs and scheduling observing sessions. There are also Data Centers that hold the IVS data in storage, and their product files are available from their websites or via the IVS website.

The time delays or equivalently the baseline projections change with time, reflecting the geophysical changes in the distance between the stations due to plate tectonics, tides, atmospheric loading, etc., $\tau_{\text{geophysics}}$, and the changes related to the changes in the Earth rotation and orientation in space, $\tau_{\text{orientation/rotation}}$.

From the accumulation of VLBI data, one finally determines the VLBI station positions and velocities within the terrestrial frame as a function of time, corrected for phenomena such as the tides that can be modeled from suitable theories, as well as the EOPs as a function of time.

Due to delays in the propagation, the relation between the time delays τ_{ij} and the baseline distances d_{ij} is not perfect and τ_{ij} contains a lot of different contributions:

$$\tau_{ij} = \tau_{\text{geom}} + \tau_{\text{clock}} + \tau_{\text{tropo}} + \tau_{\text{iono}} + \tau_{\text{small}}, \qquad (4.2)$$

Table 4.1 *Estimates of errors in the time delay.*

Error source	Order of magnitude on delay (ns)
Antenna phase displacement	0.01
Antenna deformation	0.01
Troposphere	5
Station displacement	10
Gravitation	50
Ionosphere	100
Clocks and cables	1000
Earth rotation (UT1 and polar motion)	10^2
Earth orientation (nutation)	1000

where $\tau_{geom} = \tau_{ij} = (d_{ij}/c)$ and accounts for the geometry of the stations with respect to the radio source and the Earth orientation and it is at the millisecond level; τ_{clock}, which is related to the lack of synchronization of the frequency standards (hydrogen maser) at the different stations, can be determined from clock comparison and is at better than the microsecond (μs) level (clock offsets at better than a tenth of nanosecond, with clock rate estimation at a few nanoseconds per day); τ_{tropo} and τ_{iono} represent the non-negligible contributions of the Earth atmosphere (troposphere and ionosphere, respectively) on the delay and are different from one ground station to another, τ_{small} represents various small effects such as the source structure (which can vary with time), the plasma effect on the radio emission, the local antenna distortions, the antenna phase shift effect, and the cable delays (see IERS Standards 2010). These errors are at different levels, represented in Table 4.1.

These errors are modeled at their best estimation and the remaining errors may be estimated through fitting parameters on the observation.

The classical way to analyze the data is to consider that the celestial objects observed by VLBI are those which materialize the CRF and are fixed in space, and to use this fixed reference frame to determine the variations of the Earth orientation in space from the d_{ij} and their derivatives determined from the observed time delay and the delay rates. As the VLBI stations rotate in space and as the sources are numerous, the projection of the baseline in the direction of the sources d_{ij} changes with time and can in turn be expressed in terms of the position and orientation of the Earth in space, and the positions and velocities of the stations within the Earth. These are in turn functions of parameters such as precession, nutation, plate tectonics, tides, loading effects. (For instance, Earth orientation contributes to the measured delay at about the 100 nanosecond level for UT1 and polar motion and at about the 1 μs level for nutation and precession.) In the computation, one must also consider the effects of the ionosphere and troposphere on the signal, as well as station clock drift, antenna phase center locations, antenna positions and velocities, thermal effects (distortion) on the antenna ... Many of these parameters can be estimated using models. These models are not perfect and some of the parameters may be re-estimated when interpreting the data by the method of least squares. Depending on the time variability, the remaining undetermined parameters may be assigned fixed values or may be estimated on a set of observations or even, when they vary very rapidly, re-estimated for each "arc" of observation (by "arc" of observation we mean the observation during an entire tracking period of two radio sources configured for the processing and during the same session). The parameters are in fact separated into two categories, the "global" parameters, which are evaluated globally on a large data span and which do not depend on the different data arcs, and the "arc" parameters, which are evaluated as a function of time separately from one arc to the other and depend on the individual experiment considered. Haas (2004) explains that arc parameters are parameters of which the value determined by the analysis is only valid during a particular observation session or part of it. An example is the troposphere parameters. Global parameters are valid for a longer time period and not only for one observing session. This is the case for instance for the initial station coordinates and velocities or the radio source coordinates. Time dependence of the global parameters could be envisaged by means of amplitude coefficients, or by weekly solutions. These global parameter determinations use a large number of VLBI session observations. Many different strategies may be applied in order to determine the EOPs. While one strategy is usually described by one author, others are conceivable. The choice of the optimal strategy depends on the emphasis of the different possible outcomes. Strategies may differ with respect to the objectives. For one purpose the models may be sufficient, while for another differences/corrections may be estimated from data. The consideration of their time variability may change from one approach to another. For instance, one can either determine the station positions and velocities globally or on arc-measurements; in the first case, one would have to consider tidal and atmospheric loading effects as determined and estimated for the different arcs; in the second approach one would be able to estimate the tidal motion of the stations. Another example is the choice of the sources that are considered to be "stable" or not, or the choice of estimation or not of the source positions (if the sources are not stable, see Ma and Shaffer, 1991, Eubanks *et al.*, 1994).

The classical perturbation of a radio wave propagation must be considered in VLBI analysis as well. It is particularly important to properly correct for the tropospheric effects on the signal as the varying atmosphere density perturbs the radio signal at a level above the observation precision. These effects are different from station to station. It is assumed that the neutral atmosphere considered here is a mixture of dry air (consisting mainly of CO_2, O_2, and N_2) and water vapor. The first part may be computed from an atmosphere model and the second part from a measurement of the water vapor content of the local atmosphere.

It is easy to determine the effect of the troposphere dry part as there are models for that (as seen in Chapter 9 of the IERS Conventions, these models use only the pressure at the station), but it is not easy to derive the wet part because the water content is not precisely known unless one has a radiometer measuring it. There are models available in the Conventions for the wet part too, but they are not very precise. Usually the tropospheric delay is estimated by means of a simple parameterization, as a function of the elevation of the source in the sky of the observer at the antenna. The hydrostatic part amounts to 90% of the signal and the wet part with the parameterization must be accounted for only when millimeter accuracy is desired. The tropospheric wet zenith delay is usually considered as one of the most limiting factors in VLBI analysis.

There are several software packages used for these computations, which are comparable to one another at the picosecond level in the time delays (see Ma *et al.*, 1998).

There are a few analysis centers in the world sending the results of their analyses to the IVS, which is one of the services of the IERS. These analysis centers provide regular combinations of VLBI data from different baselines and the derived EOPs. The observation campaigns are coordinated by the IVS. The combined results are released as IVS products and are available to the scientific community.

4.2 Construction of the ICRF

Some of the radio sources are known as not being stable and/or not being really point-like (one speaks about compactness). At the beginning of the observational program, a first set of a few tens of radio sources which were bright and compact in S- and X-bands was observed; the number of such sources has now (in 2013) gone up to 3414 objects from which 295 are considered as "defining" sources (the number has increased from 212 to 295 since the IAU or IUGG resolutions in 2003). Their current positions are known to better than a milliarcsecond, the ultimate accuracy being primarily limited by the structural instability of the sources at radio wavelengths (IERS Conventions 2010, Petit and Luzum, 2011). Characteristics of the apparent motions and spatial extensions of the sources have been investigated prior to the definition of the ICRF (see Ma *et al.*, 1998, Feissel *et al.*, 2000, Gontier *et al.*, 2001, and Fey *et al.*, 2004, 2009). Systematic and random behaviours in the time series of individual determinations of coordinates are found. The phenomena which are usually invoked for explaining these behaviours of the radio sources are the changes in structure of the emitting region of the source, and the gravitational lensing by material crossing the line of sight.

A selection process was put in place on the basis of the density of observations and on the compactness and the stability in time of the direction to each source in a source-fixed frame (this, in turn, was determined from the temporal variations of the power spectrum for typical periods ranging from a few months to a few years with a 0.5 year resolution). A ranking of the sources as "defining," "candidate," and "other" radio sources with decreasing confidence in their stability or observation, was used to prepare the final selection. The ICRS is defined on the basis of 295 extragalactic radio sources distributed over the entire sky. These sources obey three criteria: (1) the source must have shown enough positional stability when comparing arc-solution with global-solution; (2) the source must have no important structure in its X-band image; and (3) the position error must be less than three times the formal error (0.15 mas). Only sources which meet all of these criteria are "defining" sources. Sources which meet only one or two of these criteria are "candidate" sources. Sources which do not meet any of the three criteria are "other" sources. In practice, if a "defining" source has a proper motion, it is below 50 μas/year, and the weighted rms of its position variations from one epoch to the next does not exceed 0.5 mas (considered as three times the formal error). A realistic error estimate for the invariant source positions has been made by inflating the formal errors by a factor 1.5 (0.25 mas for the noise floor) (Ma *et al.*, 1998).

The IAU recommended that the origin of the CRS be at the barycenter of the solar system and the directions of the axes be fixed with respect to the quasars. These recommendations further stipulate that the CRS should have its principal plane as close as possible to the mean equator at J2000.0 and that the origin of this principal plane should be as close as possible to the dynamical equinox of J2000.0. This system was prepared by the IERS and was adopted by the IAU General Assembly under the name of the International Celestial Reference System (ICRS).

With the "defining" sources, the final orientation of the axes of what is to become the ICRS has been obtained by a rotation of the radio source positions that are consistent with those of the FK5 J2000 optical system within the limits of the link accuracy; this is done after application of a global three-dimensional rotation to place each pair of catalogs (the catalog with the "defining" sources and the FK5 J2000 catalog) in best coincidence, based on sources commonly observed. There is nevertheless a difference between the mean J2000 pole and the ICRF pole of about 7 mas in longitude on the equator (in $\Delta\psi \sin \epsilon_0$) and about 9 mas in obliquity (in $\Delta\epsilon$). This celestial pole offset can be re-evaluated when new data are obtained.

The maintenance of the ICRF is performed by re-evaluating the sources obeying the criteria for being "defining," "candidate," or "others."

The first realization of the ICRF was constructed in 1995 by using the VLBI positions of 212 "defining" compact extragalactic radio sources (IERS Conventions 1996, Ma *et al.*, 1997, 1998). Following the maintenance process which characterizes the ICRS, two extensions of the frame were constructed: (1) ICRF-Ext.1 by using VLBI data available since 1999 (IERS Conventions, 2000); and (2) ICRF-Ext.2 by using VLBI data available since 2002 (Fey *et al.*, 2004, 2009). The positions and errors of defining sources are unchanged from ICRF1. For candidate and other sources, new positions and errors have been calculated. All of them are listed in the catalogs in order to have a larger, usable, consistent catalog. The total number of objects is 667 in ICRF-Ext.1 and 717 in ICRF-Ext.2.

The generation of a second realization of the ICRF (Fey *et al.*, 2004, 2009) was constructed in 2009 by using positions of 295 new "defining" compact extragalactic radio sources selected on the basis of position stability and the lack of extensive intrinsic source structure. It contains accurate positions of an additional 3119 compact extragalactic radio sources for a total number of 3414 sources, which is more than five times the number of sources of the first realization. The noise floor is now at the level of 40 μas (which is really an improvement with respect to the previously adopted ICRF of which the noise floor was at the level of 0.25 mas) (Gipson, 2006). This stability of the positions of the 295 defining sources, and their more uniform sky distribution, eliminates the two largest weaknesses of the first realization. The present and future VLBI observing program aims at improving these precisions and accuracies. The program is denser than previously, with data analyzed by different centers for redundancy, with more than one VLBI sub-network in parallel, more stations in the southern hemisphere (using mobile stations), and more sources especially in the southern sky.

4.2.1 *Present-day limitations*

The present-day limitations of the observations are mainly related to the fact that the radiation from every quasar is crossing the atmosphere, and consequently is perturbed by the water vapor content in the troposphere (Jacobs *et al.*, 1998, see also Gontier *et al.*, 2001, Souchay *et al.*, 2003b, Charlot *et al.*, 2002, Sovers *et al.*, 1998, 2002, Niell and Tang, 2002, Boehm *et al.*, 2002, Boehm and Schuh, 2004a, 2004b, 2007, Tesmer *et al.*, 2007, Heinkelmann *et al.*, 2007, Steigenberger *et al.*, 2007, Krügel *et al.*, 2007, Kouba, 2008, Boehm *et al.*, 2010, Teke *et al.*, 2011, Heinkelmann *et al.*, 2011, Nilsson *et al.*, 2011, Pany *et al.*, 2011, Nafisi *et al.*, 2012). In order to compute these corrections, one uses a simplified model and a so-called "mapping function" enabling one to compute the tropospheric correction in one particular direction from a sophisticated numerical weather model, knowing the local temperature, pressure, and humidity conditions. The tropospheric delay is in principle modeled like this, but there are uncertainties in the modeling of the propagation delay due to the wet component of the troposphere. Improvements are studied in the scientific community using water vapour radiometer (WVR) measurements or combinations with GPS data (e.g. Haas, 2004, Snajdrova *et al.*, 2006, Krügel *et al.*, 2007, Rioja *et al.*, 2012).

Another concern that has been recently widely investigated in the literature is related to the radio source structure. In the ideal case the radio sources observed for geodetic VLBI would be structureless compact objects and are treated as point sources in the VLBI analysis software. However, Fey and Charlot (1997, 2000), see also MacMillan and Ma (2007), have shown that many sources have considerable structure. Techniques for tackling this problem in VLBI analyses have received much attention in the recent past (see Bourda *et al.*, 2008, 2010, Moor *et al.*, 2011, Fomalont *et al.*, 2011).

A third concern that has also been mentioned in Haas (2004) is the importance of the local deformations due to loading of the atmosphere, the ocean, and the hydrology, in addition to the local deformation of the telescope as a function of the temperature. Some of these corrections are treated routinely and are already incorporated in the VLBI software packages. A whole chapter is devoted to them in the IERS conventions 2010 (see Chapter 7 of IERS conventions 2010, Petit and Luzum, 2011). As mentioned in these conventions, it may be necessary to compute, in addition, the crust-frame translation (geocenter motion) due to the atmospheric tidal mass.

A fourth concern is the thermal deformation of the VLBI antennas caused by temperature variations at VLBI sites. The corresponding displacements of the VLBI reference points typically contain seasonal and daily signatures. The amplitudes of the annual vertical motion of the antenna reference point can reach several millimeters, depending on the design of the antenna structure, on the material, and on environmental effects such as global station position, station height, and climatology effects, as has been shown by Wresnik *et al.* (2007). These authors have applied simple methods to correct for this effect using different parameters such as the difference of the environmental temperature with respect to a defined reference temperature, the antenna dimensions, the elevation of the antenna, the material of the antenna structure, and the antenna temperature, and have shown that it is possible to reduce the effect to several tenths of millimeters. The displacements of the VLBI reference points can indeed be computed from conventional models for VLBI antenna thermal deformation (Nothnagel, 2008, see also IERS Conventions 2010, Petit and Luzum, 2011).

Another source of errors consists in the deficiencies in the network geometry (the majority of stations are in the northern mid-latitudes where the observations are at low elevation, and thus where the tropospheric delay is the strongest). The errors that may be induced in the source positions by such deficiencies have been shown, a few years ago, to amount to as much as 0.5 mas. Not only the poor coverage of the southern hemisphere in the configuration of the observed sources, but also the geometry of the 15 to 20 station networks, organized in sub-networks, have been problems; but these were addressed recently (see e.g. Feissel-Vernier *et al.*, 2005).

One of the most important developments in the determination of precession and nutation using VLBI is the strategy of using a no-net-rotation condition on the set of sources chosen. A complete analysis of these effects has been performed by Lambert and Dehant (2007). These authors have shown that the use of different strategies may affect the estimates obtained for nutation amplitudes by up to several tens of microarcseconds. The differences between the series computed by different authors are indeed at that level (see for instance, Malkin, 2001, Gontier *et al.*, 2001, Gontier and Feissel-Vernier, 2005, Feissel-Vernier *et al.*, 2000, 2005, 2006, Panafidina *et al.*, 2006, Sokolova and Malkin, 2007, 2008, 2009, Titov and Malkin, 2009).

Small inconsistencies in the VLBI software should also be understood. A last potential problem is related to the evaluation of the EOPs using observations over a period of time short compared to long-period nutations and precession. While it was the case for the long-period 18.6 year and 9.3 year nutations previously (see e.g. Williams, 1994, Souchay and Folgueira, 1998), the situation is better now concerning these important nutations for geophysical interpretation.

4.3 GPS observations

GPS observations may also provide information on nutations. Because they are measuring distances from stations to satellites of which the orbits are in principle independent of the Earth orientation (except for second-order effect in the gravity field exerted on the satellite), the precession and nutations appear in the measured distance itself. The distances measured are used in the least square estimation of the orientation parameters. In order to carry out the least square procedure, the partial derivative of the distance with respect to the orientation parameters must be expressed. As the measurements are performed with respect to the Earth, the frame used is the terrestrial frame where the wobbles appear instead of the nutations (for the relation between wobble and nutation, see Chapter 2 and Chapter 3). The uncertainty being the same all along the wobble frequency band, the uncertainties on the nutation components depend on the frequency. The larger the frequency, the higher the nutation precision. For that reason, a nutation series derived from GPS observations is only accurate for periods shorter than 20 days (Rothacher *et al.*, 1999a and b).

A combination at the series level of VLBI and GPS series of precession and nutation has been performed by Vondrák *et al.* (2003). They have shown that the precision of the GPS estimates of the amplitudes of small-period nutations is comparable to that of estimates from VLBI, but for periods above two weeks, the estimates from GPS observations are not of much value. Vondrák *et al.* (2003) concluded also that due to the complicated modeling of satellite orbits in space, the accuracy of the nutation may not match the precision obtained. A combined VLBI/GPS series is expected, however, to perform better in terms of accuracy. This will be especially true in the future as GPS data will be more accurate (modernization of GPS starting in 2004) and as GALILEO European positioning data will be available. Other groups of scientists have also started this work at present on combinations of VLBI and GPS data sets (see Englich *et al.*, 2007, Thaller *et al.*, 2007, Artz *et al.*, 2012, and also a few abstracts of conferences of, e.g., Kudryashova *et al.*, 2008, 2011, Steigenberger *et al.*, 2007, Panafidina *et al.*, 2013).

In parallel to the work on nutations, developments have taken place on methods to determine UT1 rather than LOD from GPS measurements: see Kammeyer (2000), Johnson *et al.* (2001), Ray *et al.* (2005), and Thaller *et al.* (2007). The approaches used in these works can be applied in order to get nutation directly from GPS instead of passing through wobble observations. Neither the true-of-date positions of GPS satellites nor their orbit planes are known from signal arrival times, but rather relative positions with respect to the terrestrial frame where the observing stations are situated. The orbit plane referred to the TRF contains however important information that can be used to determine EOPs, as has been done by Kammeyer (2000) for UT1 rotation angle. GPS measurements provide information on the relative positions of the satellites with respect to the observing stations on Earth. The stations are stable within the ITRF except for the plate tectonic motions and the local geodynamic phenomena (tides, atmospheric loading), which can all be modeled. The spacecraft orbits are stable within the ICRF, in space, except for the orbit changes induced by the gravity field influencing the orbital parameters and possibly small perturbations induced by the non-gravitational forces acting on the spacecraft and perturbing their orbits, which can all be modeled as well. As a consequence, from these data it is possible to obtain the changes in the Earth orientation in space "contaminating" the relative positions of the satellites with respect to the Earth. It is thus possible to get the rotation angle and the nutations from these relative position data. In his study, Kammeyer has shown that it is possible, due to the geometry and periods of the phenomena, to separate the Earth orientation effects from the other effects (mainly from the non-gravitational radiation pressure acting on the satellites). The GPS orbits are given relative to ITRF (they contain thus the Earth orientation effect). In practice, they can be expressed in the ICRF using a-priori EOPs, and the only remaining changes with time are related: (1) to the Earth's gravitational effects on the orbital elements (which are very well known as the Earth's gravity field is very well known); (2) to the errors in the EOPs used; and (3) to the non-gravitational forces acting on the satellites, which can be modeled as well. Accumulation of the data allows determining the rotation angle (UT1-UTC) and nutation residuals. The orbit plane can be determined with a precision of 0.2 mas per day, enough to determine the EOPs. This method has been theoretically proven and used for UT1 determination from GPS by Kammeyer (2000). Kammeyer's UT1 estimates are the best, much more stable than even the IGS combined LOD estimates that have been partially calibrated for biases. Nevertheless, the method has never been routinely used and has never been applied to nutation determination.

On the other hand, due to the duration of the VLBI observation sessions and the number of sessions per week, it is impossible to get subdiurnal nutation, or even nutation of a few days. So there too, the GPS measurements are able to fill the gap. Note that this will provide us with polar motion data rather than nutation as the conventional definition of nutations considers only long periods (greater than 2 days) in space.

4.4 Orbital motions and fundamental arguments

The Keplerian orbit of a body which is gravitationally bound to another (such as the Earth to the Sun or the Moon to the Earth) is characterized, in the absence of any perturbation by other bodies, by five constant elements and a time dependent one relating to the instantaneous position of the body in the orbit as follows:

- the semi-major axis a (i.e., half the length of the largest diameter of an ellipse), typically expressed in AU (astronomical units) in the planetary ephemerides, and in kilometers for the Moon in the lunar ephemerides (1 AU is the mean Earth–Sun distance);
- the eccentricity e (i.e., the ratio between the distance of the foci of the ellipse from its center and the semi-major axis);
- the inclination i (i.e., the angle between the orbital plane of the celestial body and the ecliptic);
- the longitude Ω of the ascending node (i.e., the angle in the ecliptic plane between the reference X-direction – which is typically the vernal equinox – and the point at which the body passes upwards (north) across the ecliptic);
- the longitude of the perihelion ϖ, the angle in the body's orbit plane between the ascending node line and perihelion (or the perigee in the case of the orbit of the Moon or of the apparent solar orbit around the Earth) in the direction of the body's orbit;

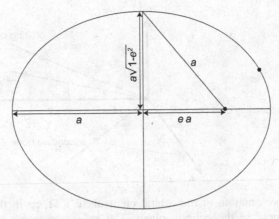

Figure 4.5 Representation of the orbital elements a and e in an osculating plane, i.e. a plane approximating the actual orbit at the specified epoch.

- the mean anomaly M (i.e., the angle from the perihelion to a fictitious body having a constant mean motion equal to the mean of the real motion of the celestial body over one orbit, the angle from the perihelion to the actual position of the body being the true anomaly v). The time rate of change of M is the mean angular velocity of the body; it is commonly referred to as the *mean motion* of the body. In the case of the lunar orbit around the Earth, "perihelion" is to be replaced by "perigee."

The elements a and e characterize the size and the degree of ellipticity of the orbit of either of the bodies around the other; and the inclination i of the orbit relative to the ecliptic, together with the parameters Ω, and ϖ, describe the orientation of the orbit in space relative to the ecliptic. In a Kepler orbit, these five elements being independent of time, the characteristics of the orbit remain unvarying, as does the orientation of the orbit in space. The first two of these orbital elements are displayed in Fig. 4.5 and the others in Fig. 4.6.

The values of these elements and the masses of the bodies determine the mean angular velocity of motion of either body in the orbit through Kepler's third law, and hence the mean anomaly M (angle from the perihelion/perigee to the mean position of either body on the orbit) at an arbitrary time follows from its value at some fixed epoch.

When the two-body system is perturbed by other bodies, the orbit undergoes slow changes and motions in space, which are described by slow variations in the five elements.

For the planets other than the Earth, the motion of each one around the Sun can be characterized by the elements a, e, i, Ω, and ϖ of its own orbit; and M describes the motion of the planet in the orbit. A slow variation of the elements of the orbit

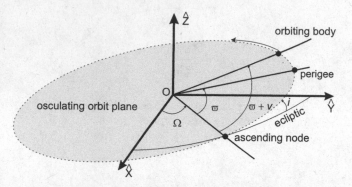

Figure 4.6 Representation of the orbital elements ν, i, Ω, ϖ in the osculating plane with respect to the ecliptic plane. The reference frame visualized by $(O \hat{X}, \hat{Y}, \hat{Z})$ is tied to the ecliptic, with $\hat{Z}$ being perpendicular to the ecliptic. The grey ellipse represents the orbit in the osculating plane.

of each planet takes place over very long periods of time because of interactions with all the other bodies of the solar system.

In the case of the lunar and apparent solar orbits around the Earth, the variation of Ω, which is periodic, is not really slow, since the period is just 18.6 years; as will be seen later, the variation of ϖ too is treated as periodic though with a much longer period; all the other elements vary much more slowly. In addition to the slow variations, the orbit undergoes small short-period deviations from its Keplerian nature (including deviations from planarity and from a strictly elliptical shape). If we ignore the short period variations, what we perceive at any time is a motion along the *osculating* Keplerian ellipse, i.e., the ellipse characterized by the values of the elements at that particular time.

The equations of motion of the bodies of the solar system are the Lagrange equations for the coordinate variables (Cartesian or polar, for instance) of these mutually interacting bodies, or the equivalent Hamiltonian equations. (See Chapter 5 for an elementary introduction to the Lagrangian and Hamiltonian forms of mechanics.) First-order differential equations describing the variation of the six orbital parameters can be constructed for each planet or other celestial body. These equations are called the Lagrange equations after their derivation by Lagrange (1736–1813). They can be found in many celestial mechanics textbooks such as those of Kovalevsky (1967), Beutler (2005), or Capderou (2012). The masses of the bodies are of course essential inputs into these equations. The solution of the set of equations of motion, carried out by analytical methods or by numerical integration, yields the coordinate variables of the various bodies as functions of time. These solutions constitute what is called the *ephemerides*. The ephemerides in analytical form generally take the form of spectral expansions of the coordinate variables of the individual bodies, in which each individual term is a simple harmonic (sine or cosine) function of an *argument* which is essentially linear in time (deviations from linearity are very small). The arguments of the various spectral terms are distinct linear combinations, with small integer coefficients, of a set of *fundamental arguments* representing various fundamental periodicities present in the motions of the bodies. The frequency of any particular term in the spectral expansion is the time derivative of the argument of that term. The ephemerides obtained by numerical integration provide the values of the position coordinates at regular intervals of time over an appropriate interval.

Reverting to the case of a two-body orbit, one notes that the only fundamental argument is the mean anomaly of either of the two bodies on the orbit. For the relative motion of the Earth and the Moon, it is usual to take this argument as the mean anomaly of the Moon in its orbit around the Earth; for the Sun's apparent orbital motion around the Earth, the argument is its mean anomaly in the solar orbit; in each case, the argument is the equivalent of the variable M appearing in the above list of elements of a Keplerian orbit. The time derivative of the argument is the mean angular frequency of the orbital motion.

In the Moon–Earth–Sun system with each body interacting gravitationally with the other two, the relative motions of the bodies may be viewed as consisting of the orbital motion of the Moon and the apparent orbital motion of the Sun around the Earth, with each of these two-body orbits perturbed by the action of the other body. A prominent effect of the Sun's action on the Moon is a precession of the plane of the Moon's orbit in space with an essentially constant angular velocity. This precession has the consequence that the longitude Ω (referred to the equinox) of the node of the lunar orbit on the ecliptic varies essentially at a constant rate. It may be recalled that in the case of the two-body (Earth–Moon) system, this longitude was a constant, represented by the orbit element Ω. In the three-body system the element Ω has gone over into the time dependent *argument* Ω, varying with an angular frequency corresponding to a period of 18.6 years. The motion of the node is in the direction of decreasing ecliptic longitude, and therefore the angular frequency is negative, which means that the period should also be written, strictly speaking, as -18.6 years. It may be noted that in the course of the precession of the lunar orbital plane and of its node, the inclination i of the plane to the ecliptic plane remains essentially constant (see Chapters 1 and 2, as well as Appendix C for more details).

The mean longitude of the Moon along its own orbital plane, referred to the node, is the equivalent of $\varpi + M$ of the two-body case considered above; in the three-body lunar theory, this quantity is represented by the argument F, and the symbol M is replaced by l. Since ϖ (which was one of the constant orbital elements in the two-body case) now corresponds to the combination $(l - F)$ of two fundamental arguments, it is clear that the perigee itself is now viewed as having a periodic motion along the Moon's orbit, its period relative to the node of the lunar orbit being given by the time derivative of $(l - \Omega)$. Thus we have three fundamental arguments l, F, Ω, as well as the elements a_M, e_M, i_M, associated with the lunar orbit.

Besides these, we have other fundamental arguments and elements associated with the apparent solar orbit around the Earth. The first of the arguments is l', the angular distance from the perigee of this orbit to the mean Sun; and a second one is D, called the mean elongation of the Moon from the Sun, which is defined as the difference between F (which was defined above) and its equivalent for the Sun, namely the angular distance, in the plane of the solar orbit (i.e., the ecliptic), from the node of the lunar orbit to the position of the mean Sun. In addition, one has the elements a_S, e_S, and ϖ_S pertaining to this orbit. The element i_S is zero, as the mean solar orbit coincides with the ecliptic.

Thus the orbital motions of the Moon and the Sun relative to the Earth, in the Moon–Earth–Sun system are characterized by six elements, namely, a_M, e_M, i_M, a_S, e_S, ϖ_S, and five fundamental arguments l, l', F, D, Ω. As noted earlier, small deviations with multiple periods take place from the mean orbits as a result of the influence of the third body on each of the two orbits. Nutation frequencies too are determined by the same set of five fundamental arguments; other arguments relating to the planetary motions also need to be included when the very much smaller contributions from those bodies are taken into account.

4.5 Existing ephemerides

The time dependence of the positions of the solar system bodies relative to the Earth needs to be known for a variety of astronomical and geophysical studies. In the latter context, such knowledge is required for determination of the time dependent torques exerted by these bodies on the Earth, from which the precession and nutations of the Earth may be computed. Observations and computations of the orbital motions of the planets and their satellites (including, in particular, the Earth and the Moon) are therefore of wide interest. The results of such studies on orbital motions are presented through the *ephemerides*, in the form of analytical expressions for, or tables of numerical values of, the position coordinates as functions of time. Typically, the expressions or numerical values are given for the Cartesian or spherical polar coordinates of the bodies in a space-fixed heliocentric or geocentric frame.

An ephemeris is a tabulation of computed positions (and/or various derived quantities such as right ascension and declination) of an orbiting celestial body at specific times. The construction of an ephemeris of a celestial body is a computation of its orbital elements referred to a conic (most commonly an ellipse) in inertial space. Ephemerides thus describe celestial object states (equivalent to its Cartesian position and velocity) at a specific epoch. Typically orbital elements are used to express an object's motion on its "osculating orbit," i.e. an orbit tangential to and approximating the actual orbit at the specified epoch. The orbital elements consist of the semi-major axis a, the eccentricity e, the mean anomaly M, the inclination i, the longitude of the ascending node Ω, and the argument of the perihelion ω.

The state vectors or the osculating orbital elements are generated by a numerical integration of the equations of motion of celestial bodies fitted to the best available observations of the positions and velocities of the Sun, the Moon, the planets, their satellites, and some of the largest asteroids. The numerical integration involves stepwise computation (in time) of the position of each of these objects, as determined by the gravitational forces of all of the other objects in the solar system. Adjustments are made to the masses and shapes of the objects to get the best agreement with their observed positions over the last century. These ephemerides have been computed by different groups: (1) DExxx ephemerides constructed at JPL (Jet Propulsion Laboratory, US; see Standish, 1982a and b, 1990a and b, 1998a, b and c, Standish *et al.*, 1995, Standish and Fienga, 2002, Newhall *et al.*, 1983, Kuchynka and Folkner, 2013, Kuchynka *et al.*, 2014, and Folkner *et al.*, 2014), where xxx represents a number chosen by the JPL scientists; the latest is DE431 (Folkner *et al.*, 2014) and the one to which one refers in the IERS Conventions 2010 is DE421 (Folkner *et al.*, 2008); (2) the EPM20yy ephemerides constructed at IAA (Institute of Applied Astronomy of Russian Academy of Science; see Pitjeva, 2001a and b, 2005, 2013, Pitjeva and Standish, 2009); and (3) INPOPyy ephemerides as done recently in IMCCE (Institut de Mécanique Céleste et de Calcul des Ephémérides; see Fienga *et al.*, 2005, 2006, 2008, 2009a and b, 2010, 2011, 2014, Fienga, 2012, Somenzi *et al.*, 2010, Verma *et al.*, 2014) where yy represents the year of the release of the ephemerides.

In parallel to these numerical ephemerides, one finds also analytical computations for ephemerides that are very often used in the domain of precession and nutation computation. This is the case of VSOP ephemerides of Bretagnon (1982), Bretagnon and Francou (1988), Bretagnon *et al.* (1997, 1998), or Moisson and Bretagnon (2001).

The equations used to determine the ephemerides are the basic equations of motion refined to account for fine perturbations such as the Moon's spin–orbit coupling or the relativistic effects, and are integrated numerically using primary dynamical constants such as the GM_i of the planets (relative to the Sun GM_{Sun}) and initial conditions such as present-day positions constrained from the observations.

Several kinds of observation are used to construct these ephemerides; this includes many optical observations, i.e., astrometric observations, as well as transit observations and occultations of planets and their satellites. Radio signal Doppler shifts and ranging observations of landers at the surface of a planet or of spacecraft in orbit around or flying by celestial objects, as well as laser ranging (i.e. enabling measurements of the round trip time (or distance) of ultra-short pulses of light), mainly to the Moon but also to the spacecraft, are used also for the adjustments considered in the process of building the ephemerides. Doppler shift measurements enable measurements of line-of-sight projections of the velocities of the spacecraft; these measurements constrain the dynamics of their orbits and therewith are helping the ranging measurements to be accurately used to build the ephemerides of the celestial bodies. The spacecraft are themselves oriented in space using optical observations of stars. Additional VLBI observations of spacecraft such as the Phobos spacecraft approaching Mars or Magellan spacecraft approaching Venus have been used to determine offsets and rotations between a given planetary ephemeris in an optical frame based on astrometric observation from Earth or from space telescopes and in the ICRF. The ephemerides used presently are referred to the ICRF.

The latest ephemerides (either represented by the highest number or by the year in their denomination) always represent improvements in the sense that the sets of observations are augmented with more recent data and more types of data, and that the data reduction techniques are refined. In the most recent, the asteroid perturbations on the planetary orbits have been modeled.

In practice, the numerical values as a function of time of the solar-system-barycentric positions and velocities of the solar system objects specified in the ephemerides in Cartesian coordinates in the ICRF are given for time steps and time spans of the user's choice. Computer software is available on different websites for reading and using the data files. This is provided for instance for Mercury, Venus, the Earth–Moon barycenter, Mars, Jupiter, Saturn, Uranus, Neptune, Pluto, and the Sun. NASA also provides on-line a toolkit for managing the data files of the high-precision ephemerides for solar-system bodies using JPL's HORIZONS system. This provides an easy access to key solar system data and generates with a high flexibility accurate ephemerides for solar system objects in whatever format is wanted by the user, e.g., in the form of Cartesian vectors (position, velocity, and acceleration) or in the form of osculating orbital elements. Users requiring high-precision ephemerides readable from their own software may also find ephemeris data files in SPK format and a software toolkit (SPICE Kernels) to read those files from JPL's NAIF (navigation and ancillary information facility) web-site.

Even with the accurate recent data, the planetary motions show rather large uncertainties. For instance, the planet Mars is perturbed by the presence of many asteroids whose masses are quite poorly known. Furthermore, it is not possible to solve for the asteroid masses, other than for the largest few, because there are too many of them. This will cause the orbit of Mars to deteriorate at the level of 1–2 km over the span of the observations, and growing at the rate of a few km/decade outside that span. Due to the presence of recent spacecraft around Mars such as Mars Global Surveyor (1999–2006), Mars Odyssey (2001–), Mars Express (2003–), and Mars Reconnaissance Orbiter (2005–), the Martian ephemerides are being constantly improved at present. The future GAIA mission might also help to bring in new precise optical observations (see Fienga, 2012).

For precession and nutation computation, it is important to get ephemerides for the Moon and the Earth (or equivalently the Sun). The Moon is indeed a particular case as, first, the state vectors in the lunar ephemerides are geocentric positions of the Moon and are usually not specified in barycentric coordinates, and second, there are continuous observations of the Earth–Moon distance by lunar laser ranging (LLR).

Analysis of LLR data provides information on the lunar orbit, mass, rotation, solid-body tides, and interior. Lunar ephemerides are a product of the LLR analysis. Extending the data span and improving range accuracy continuously yield improved and new scientific results as well as lunar ephemerides. In particular, the use of LLR for ephemerides enables one to improve the orbital motion of the Moon following the increasing accuracy in the LLR measurements and to better define the dynamical reference frame related to the lunar motion in the ICRS. The computations introduce the inertial mean ecliptic of J2000.0., the position of which can be adjusted with respect to various equatorial systems.

4.6 Ephemerides used for the rigid Earth nutation theory

The ephemerides used for the precession and nutation theory are the so-called VSOPyy ephemerides (Bretagnon, 1982, Bretagnon and Francou, 1988) (VSOP stands for "Variations Séculaires des Orbites Planétaires" and yy for the year of publication) for the planets and the ELP20yy (Chapront-Touzé and Chapront, 1983, 1988, Chapront *et al.*, 1999, 2002, Chapront and Francou, 2003) for the Moon. Numerical integration of orbital motions in the solar system constrained from observations, such as the DE- or LE-kinds of ephemerides solely serve for verification; expansion in the frequency domains is necessary for the computation of precession and nutations. The VSOP- and ELP- ephemerides were built starting in the 1980s by the Bureau des Longitudes (BdL), now IMCCE (replacing the Services des Calculs of the BdL). They are analytical expressions representing the motions of the celestial objects in elliptical, spherical, or rectangular coordinates, reported with respect to the heliocentric ecliptic of J2000 or of date, or the barycentric ecliptic of J2000. They are analytical solutions of the Lagrange equations for the motions generated by the gravitational interactions between the celestial bodies. The Lagrange equations for the problem of N bodies (the Sun and $N-1$ planets) is a system of $3(N-1)$ differential equations of the second order, which can be recast as a system of $6(N-1)$ equations of the first order. If all except two of the bodies (such as the Sun and any single planet, or the Earth and the Moon) are ignored, the solution describes the motion along a Keplerian ellipse.

The variables used in the Lagrange equations are the orbital elements as defined in Section 4.5, such as the semi-major axes of the osculator orbit for instance, and treated as a time dependent variable. The analytical expressions of the solution are then provided in terms of a polynomial of the time T where T is the time expressed in thousands of Julian years ($T = $ (Julian date $-$ 2451 545)/365250), and of series of cosine functions of which the arguments and amplitudes are also polynomials in T. We give here an example for the heliocentric coordinates X, Y, Z of the Earth relative to the dynamical ecliptic and equinox of J2000 (see Chapter 12 for more details on the precise concepts adopted by the international community):

$$X = a_X + b_X T + \sum_{i=1}^{N} \left\{ A_{iX} \cos(a_{iX} + b_{iX}T) + B_{iX}T \cos(a'_{iX} + b'_{iX}T) \right\}, \quad (4.3)$$

$$Y = a_Y + b_Y T + \sum_{i=1}^{N} \left\{ A_{iY} \cos(a_{iY} + b_{iY}T) + B_{iY}T \cos(a'_{iY} + b'_{iY}T) \right\}, \quad (4.4)$$

$$Z = a_Z + b_Z T + \sum_{i=1}^{N} \left\{ A_{iZ} \cos(a_{iZ} + b_{iZ}T) + B_{iZ}T \cos(a'_{iZ} + b'_{iZ}T) \right\}. \quad (4.5)$$

They are given in Poisson series of time where $a_{iX}, a_{iY}, a_{iZ}, b_{iX}, b_{iY}, b_{iZ}, a'_{iX}, a'_{iY}, a'_{iZ}$, and $b'_{iX}, b'_{iY}, b'_{iZ}$ are linear combinations of the Delaunay arguments of the Moon and the Sun ℓ, ℓ', D, F, and Ω, and of the quantities λ_i, with $i = 1$ to 8 representing the mean longitudes of the eight planets. Poisson series are series where both amplitudes and arguments of the terms are low-degree polynomials of time as opposed to the classical Fourier series where the amplitudes and frequencies in the terms are constants. The Delaunay argument Ω is the mean longitude of the ascending node of the Moon; ℓ and ℓ' are the mean anomalies of the Moon and the Sun, respectively, and are often denoted by ℓ_M and ℓ_S; F is $L - \Omega$ where L is the mean longitude of the Moon; and D is the mean elongation of the Moon from the Sun, i.e., the angular distance of the Moon from the Sun in the sky. The Lagrange equations used to obtain these solutions are first-order differential equations for the orbital elements involving partial derivatives of the perturbing function. The orbital elements vary slowly, but the rate may vary quickly, since the perturbing function depends on position in the orbit. They can thus be solved numerically in terms of Poisson series of time in which the long-period arguments associated with the nodes and perihelia longitudes of the planets are developed in polynomials of time (there exists a minimum fixed frequency, i.e. while developing the series, periodic terms with frequencies less than a minimum value chosen for this purpose are replaced by polynomials) (see Simon *et al.*, 1994, 2013).

The precision of the VSOPyy and ELP20yy solutions depends on the precision of the computation of each term in the series, and on the truncation level of the series itself.

It must be noted finally that the solutions are found using Taylor expansions in the small quantities such as the orbit eccentricities and that it is necessary to use an iterative method accounting for the different orders of magnitude of the different parts of the equations in order to reach the final form of the solution (see the example above).

Presently, the last of the ephemerides of the VSOP class is VSOP2013 (Simon *et al.*, 2013), which started from VSOP2000 (Moisson, 2000) including additions to the previous solution from relativistic effects as well as perturbations induced by some asteroids (the more massive ones as well as those which are in resonance with Mars and Jupiter, such as Vesta, Ceres, Pallas, Iris, and Bamberga), the Moon and Pluto. This ephemeride is complemented by the theory of the outer planets (TOP). Simon *et al.* (2013) have increased the precision of the VSOP solutions of Jupiter and Saturn by using TOP solutions and have also improved the construction of TOP by computing the perturbations due to the terrestrial planets from VSOP solutions while constructing VSOP2013 and TOP2013. At present, there are unexplained differences between the DExxx-kind of ephemerides and the VSOP-kind of ephemerides and between the DExxx-ephemerides themselves. Modifications in the integration constants, in the masses assigned, and in the asteroids considered do not explain all the differences.

A last remark concerns the fact that the solutions above for the motion of a celestial body are more complicated than a simple Keplerian orbit. For the Moon and the Sun around the Earth, for instance, there are additional effects such as: (1) the direct planetary effects on their motions; (2) the indirect planetary effects related to the perturbation of the lunar orbit around the Earth by the gravitational force of the Sun, and of the Earth's orbit around the Sun by the Moon's gravitational force (the so-called three-body effects); (3) the J_2-tilt effects due to the perturbation of the Moon's orbit by the non-central part of the Earth's gravitational potential that arises from the Earth's equatorial bulge. These orbital perturbations leave their imprint on the forces and torques exerted by the Moon and the Sun on the Earth. These effects are fully explained in Chapter 5.

4.7 Other kinds of observations needed for nutation studies

Besides the observation of nutation itself, ocean and atmosphere need to be observed
for obtaining the data assimilated in the general circulation models used to compute
the atmosphere and ocean corrections on nutation (see Chapter 9) – mainly pressure
and wind fields for the atmosphere, and altimetry data and tide gauges for the ocean.
Additionally, as explained in Chapter 10, seismic and magnetic field observations
are necessary to constrain the processes studied in the fluid core and the deformable
Earth.

5

Rigid Earth precession and nutation

In the forcing of Earth rotation variations by gravitational action of the solar system bodies, the moments of the matter distribution inside the Earth, which go by the name "geopotential coefficients," play a direct role. The geopotential coefficients of degree two, which are proportional to the elements of the inertia tensor, are by far the most important of these; but coefficients of all higher orders too have their roles. One needs to use the spherical harmonics (or more formally, the surface spherical harmonic functions) for defining the geopotential coefficients. These functions and their properties (to the extent that they are essential for our purposes) are presented in Section 5.3.5, before we go on to construction of the geopotential coefficients.

All the geopotential coefficients remain invariant in time for a rigid Earth which cannot, by definition, deform. The potential energy of the Earth's gravitational interaction with a celestial body is expressible in terms of these coefficients and the time dependent position of that body in relation to an Earth-fixed reference frame. The determination of the rotation variations should then, in principle, be a simple matter, if the values of these coefficients and the time dependence of the positions of the Moon, Sun, and planets relative to the Earth are known. It is generally understood that the values to be used for the geopotential coefficients of the "rigid Earth" are to be the same as for the actual Earth in its undeformed uniformly rotating state, not subject to any perturbations. These values are determined from satellite laser ranging (SLR) or GNSS observations of the orbital motions of dedicated artificial Earth satellites such as LAGEOS, CHAMP, GRACE, GOCE... (see Section 1.6 of Chapter 1), after separating out the contributions from time dependent perturbations.

The orbits of the Moon around the Earth and of the Earth and other planets around the Sun are Keplerian ellipses (see Section 4.4) when these bodies are idealized as point masses and each body is viewed as being subject to the gravitational influence of just one other body (the Sun, in the case of the Earth and other planets, and the Earth in the case of the Moon). The Earth's orbital motion around the Sun is one view of the relative motions of the Earth and Sun. One could describe the relative motion equally well as a motion of the Sun around the Earth. From this point of view, one often uses the term "apparent orbital motion of the Sun around the Earth" as equivalent to the Earth's orbital motion. Perturbation of the lunar orbit by the Sun has the effect of making the Moon's orbital plane precess around the ecliptic with a period of 18.6 years while keeping a constant angle of about 27.5° with the ecliptic plane; it also produces small deviations of the orbit from the Keplerian elliptical shape. The Earth's orbit too ceases to be strictly Keplerian as a result of lunar perturbations.

Determination of the perturbations in Earth rotation produced by the Moon and the Sun, taking into account the influences of all three bodies on the orbital motions of the others, is often referred to as the "main problem." Further perturbations of the lunar and apparent solar orbits around the Earth result from a variety of causes. We mention just two here, leaving others to be introduced later: (1) gravitational perturbation of these orbits by the planets; and (2) the perturbation of the Moon's orbit by the non-central part of the Earth's gravitational potential that arises from the ellipticity of the Earth. These orbital perturbations leave their imprint on the forces and torques exerted by the Moon and the Sun on the Earth. The consequent effects on Earth rotation are referred to as *the indirect planetary effect* and *the J_2-tilt effect*, respectively. (The parameter J_2 was introduced in Section 2.9 and Section 2.10; it is further detailed in Section 5.3.5, where its relation to the dynamical ellipticity is exhibited.) Additional torques on the Earth arise from the direct gravitational action of the planets on the Earth which constitutes *the direct planetary effect*.

5.1 Approaches to the determination of rigid Earth nutations

The rotational dynamics of the rigid Earth may be formulated in a number of different ways which may be classified into two broad groups: formulations based on the Hamiltonian approach and those using the torque approach.

In the Hamiltonian approach, the starting point is the Hamiltonian of the rotating Earth, which is the sum of its rotational kinetic energy and the potential energy of its gravitational interaction with the Moon, Sun, and the planets. The potential energy is determined by the time dependent positions of these bodies relative to an Earth-fixed reference frame, the variables representing the Earth orientation, and the relevant Earth parameters (namely, the geopotential coefficients). The time dependence of the potential energy function on account of its dependence on the position variables of the celestial bodies is usually expressed in terms of its spectral representation. The dependence of both the kinetic and potential energies on Earth orientation is expressed through their functional dependence on suitably chosen coordinates representing Earth orientation and their canonically conjugate momenta. It is a common practice to choose these coordinate variables to be quantities characterizing the orientation in space of the Earth's angular momentum axis. Once the expression for the potential energy is derived, the task is to solve the Hamiltonian equations of motion to obtain the time dependence of the variables representing the orientation of the angular momentum axis and thereafter also that of the variables relating to the orientations of the figure and rotation axes.

The torque approach is based on the solution of the torque equation (equation of angular momentum balance), either in the celestial frame or in the terrestrial frame. The former is an inertial frame and the latter rotates in space; the equations with reference to both these types of frames have been introduced already in Chapter 2. The input needed for determining the externally induced variations in Earth rotation is of course the torque acting on the Earth. The gravitational torque exerted by any of the solar system bodies depends on the time dependent position of the body relative to an Earth-fixed frame, and on the geopotential coefficients characterizing the distribution of matter in the Earth. (As we have seen in Chapter 2, the dynamical ellipticity, to which one of the geopotential coefficients is related, is responsible for all but a small part of the nutational and precessional motions.) The coordinates of the solar system bodies are given as functions of time in the ephemerides, see Chapter 4.

Two ways of computing the general expression for the torque are presented in this chapter, one beginning with the Earth's gravitational potential and its action on the Moon or other celestial body (see Section 5.3.7), and the other starting with the tidal potential of such an external body at points within the Earth (Section 5.5) and the effects of this potential on the Earth. (The tidal potential is that part of the gravitational potential of the body which produces differential accelerations of elements of matter at different locations in the Earth, as noted already in Section 2.23.3 of Chapter 2. The differential accelerations produce torques on the Earth, whether it be rigid or non-rigid, but deformations do not occur unless the Earth is non-rigid.)

An alternative representation of the time dependence is through the use of the spectral expansion of the tidal potential (Section 5.5.4); it enables one to obtain the torque on the Earth in the form of a spectral expansion (Section 5.5.7). Several tables, constructed in different ways, which list the coefficients of the simple harmonic spectral terms in the expansion of the total tidal potential of all the bodies, are available, as mentioned in Section 2.13. They are used for numerical computations.

Expressions for the torque in terms of the geopotential coefficients are in the TRF. Solution of the torque equation in this frame yields the terrestrial components of the rotation vector, i.e., the wobble variables and the spin rate variation. The kinematic relations have to be used then to obtain the nutation variables. One could seek, instead, to write the torque equation directly in relation to a space-fixed reference frame. However, the torque components in such a frame have to be obtained by transformation from those in a terrestrial frame, and this can be done in a reasonably simple manner only in certain special cases and subject to certain approximations (see Section 2.23.6). Once the torque is determined, it is a trivial matter to arrive at the nutation of the angular momentum axis by directly integrating the torque equation in the celestial frame. The transformation from the torque equations in the terrestrial frame to equivalent equations in the celestial frame can be carried out rigorously, taking into account the full range of geopotential coefficients, as was done by Bretagnon *et al.* (1997); the resulting equations are rather complicated, but lead to solutions of high accuracy when solved iteratively, given accurate inputs (values of the geopotential coefficients, ephemerides, etc.). Their method is outlined in Section 5.8.

5.2 Hamiltonian approach

5.2.1 *Lagrangian and Hamiltonian formulations of mechanics*

Mechanics of systems of particles

Elementary treatments of the dynamics of a system of n particles start with Newton's second law of motion which relates the acceleration $\ddot{\mathbf{x}}_a$ of the ath particle of mass m_a to the force $\mathbf{F}_a$ acting on it. It is sufficient for our purposes to consider the forces acting on the particles in the system to be derivable from a potential energy function V which depends on the positions $(\mathbf{x}_1, \mathbf{x}_2, \ldots, \mathbf{x}_n)$ of the particles, and may, in addition, depend explicitly on the time variable t also, but has no dependence on the particle velocities. The Newtonian equations are then

$$m_a \frac{d^2 \mathbf{x}_a}{dt^2} = \mathbf{F}_a = -\nabla_a V, \qquad (a = 1, 2, \ldots, n), \qquad (5.1)$$

where ∇_a is the gradient operator with components $(\partial/\partial x_a, \partial/\partial y_a, \partial/\partial z_a)$.

The equations in the above form are inherently linked to a Cartesian coordinate system. In the case of a particle in a central potential $V(r)$, it would be more natural and efficient to formulate the dynamics in terms of a spherical polar coordinate system. Considering this and other more general situations where it is advantageous to use a set of *generalized coordinates* q_i, $(i = 1, 2, \ldots, N)$ rather than the Cartesian ones, different formulations of the dynamics have been developed. Here N is the number of *degrees of freedom*, which is $3n$ for the n-particle system in three dimensions.

We introduce now the Lagrangian and Hamiltonian formulations.

Lagrangian equations

The first step in going over to the Lagrange formulation is to express the Cartesian coordinates (x_a, y_a, z_a), $(a = 1, 2, \ldots, n)$, as functions of the desired generalized coordinates q_i. (These functions will be taken to have no explicit dependence on time.) If spherical polar coordinates (r, θ, φ) were to be adopted, for example, one would have for the first particle, $q_1 = r_1$, $q_2 = \theta_1$, $q_3 = \varphi_1$, with $x_1 = q_1 \cos q_2 \cos q_3$, $y_1 = q_1 \cos q_2 \sin q_3$, $z_1 = q_1 \cos q_2$; for the second particle, (x_2, y_2, z_2) would be given by similar expressions with (q_1, q_2, q_3) replaced by (q_4, q_5, q_6), and so on. In general, the Cartesian velocity components, expressed in terms of the new variables, are

$$\dot{x}_a = \sum_{i=1}^{N} \frac{\partial x_a}{\partial q_i} \dot{q}_i, \qquad \dot{y}_a = \sum_{i=1}^{N} \frac{\partial y_a}{\partial q_i} \dot{q}_i, \qquad \dot{z}_a = \sum_{i=1}^{N} \frac{\partial z_a}{\partial q_i} \dot{q}_i, \qquad (5.2)$$

where $N = 3n$. The partial derivatives in the above expression are, in general, functions of the q_i. So $\dot{x}_a, \dot{y}_a, \dot{z}_a$ are now replaced by linear combinations of the generalized velocities $\dot{q}_i$ with coefficients that are functions of the coordinates q_i.

The Lagrangian formulation is based on the use of a scalar function L called the Lagrangian which is a function of generalized coordinates q_i and the velocities $\dot{q}_i$ with $i = 1, 2, \ldots N$. The Lagrangian is defined by

$$L = T(q, \dot{q}) - V(q, t), \qquad (5.3)$$

where $V(q, t)$ is the potential energy expressed in terms of the N generalized coordinates, collectively denoted here by q. It is important to note that in view of (5.2), the kinetic energy T of the system, which is $\sum_{a=1}^{n} \frac{1}{2} m_a (\dot{x}_a^2 + \dot{y}_a^2 + \dot{z}_a^2)$, is a homogeneous function of the second degree in the generalized velocity variables $\dot{q}_i$, with coefficients which are functions of the q_i, in general:

$$T = \sum_{a=1}^{n} \frac{1}{2} m_a \sum_{i=1}^{N} \sum_{j=1}^{N} A_{ij}^a(q) \dot{q}_i \dot{q}_j, \quad A_{ij}^a(q) = \left(\frac{\partial x_a}{\partial q_i} \frac{\partial x_a}{\partial q_j} + \frac{\partial y_a}{\partial q_i} \frac{\partial y_a}{\partial q_j} + \frac{\partial z_a}{\partial q_i} \frac{\partial z_a}{\partial q_j} \right).$$
$$(5.4)$$

Note that for each a, the $A_{ij}^a(q)$ constitute a symmetric matrix. The Lagrangian equations of motion are

$$\frac{d}{dt} \left(\frac{\partial L}{\partial \dot{q}_i} \right) - \frac{\partial L}{\partial q_i} = 0. \qquad (5.5)$$

These equations, like the Newtonian ones, consist of $N = 3n$ second-order ordinary differential equations with respect to time. It is easy to verify that they reduce to (5.1) if the familiar expression in terms of Cartesian variables is employed for the kinetic energy.

Hamiltonian formulation

In the Hamiltonian formulation, the N second-order Lagrangian equations are transformed into $2N = 6n$ differential equations of the first order. Hamilton introduced the momentum variables

$$p_i = \frac{\partial L}{\partial \dot{q}_i}, \tag{5.6}$$

as well as the *Hamiltonian function* (or simply, the *Hamiltonian*) defined by

$$H = \sum_{i=1}^{N} p_i \dot{q}_i - L. \tag{5.7}$$

The variables q_i and p_i for any given i are said to form a *canonically conjugate* pair; p_i is the momentum canonically conjugate to q_i.

Now, since V does not involve the generalized velocities $\dot{q}_i$, we have

$$p_i = \frac{\partial L}{\partial \dot{q}_i} = \frac{\partial T}{\partial \dot{q}_i} = \sum_a \sum_j m_a A_{ij}^a(q) \dot{q}_j \tag{5.8}$$

from Eq. (5.4). Consequently $\sum_i p_i \dot{q}_i$ reduces to $2T$, and hence, on using the definition (5.3) of L, we see that the Hamiltonian is the total energy of the system:

$$H(q, p, t) = 2T - L = T(q, p) + V(q, t). \tag{5.9}$$

Here and elsewhere, p (respectively q) stands collectively for the set of variables p_i (respectively q_i), ($i = 1, 2, \ldots, N$).

As is implied by the notation used in the above equation, H is to be expressed purely in terms of q and p, i.e., the $\dot{q}_i$ appearing in the definition (5.7) are to be expressed in terms of q and p. This may be done by treating the Eqs. (5.8) as a set of linear algebraic equations $\sum_j X_{ij}(q) \dot{q}_j = p_i$ for the variables $\dot{q}_j$ (with $X_{ij}(q) = \sum_a m_a A_{ij}^a$), and solving them; the solutions for the $\dot{q}_j$ will evidently be functions of the momentum variables p_i and the coordinate variables q_i.

In the special case of Cartesian coordinates q_i, $T = \sum \frac{1}{2} m_i \dot{q}_i^2$, where the m_i come in sets of three equal numbers: $m_1 = m_2 = m_3$ is the mass of the first particle, and so on. Then $p_i = m_i \dot{q}_i$, and $\sum p_i \dot{q}_i = \sum (p_i^2 / 2m_i)$.

The equations of motion in Hamiltonian form can be obtained in the general case by comparing the differential of H considered as a function of q and p with the differential of the defining relation (5.7). The former is evidently

$$dH = \sum_i \left(\frac{\partial H}{\partial q_i} dq_i + \frac{\partial H}{\partial p_i} dp_i \right), \tag{5.10}$$

while the latter yields

$$dH = \sum_i \left(p_i d\dot{q}_i + \dot{q}_i dp_i - \frac{\partial L}{\partial q_i} dq_i - \frac{\partial L}{\partial \dot{q}_i} d\dot{q}_i \right). \tag{5.11}$$

Now, the Lagrangian equation (5.5) enables us to replace $\partial L/\partial q_i$ by $\dot{p}_i$, while the definition (5.6) tells us that $\partial L/\partial \dot{q}_i = p_i$. Hence the above expression simplifies to $dH = \sum_i (\dot{q}_i dp_i - \dot{p}_i dq_i)$. On equating it to the right-hand side of Eq. (5.10), we are led to the equations

$$\dot{q}_i = \frac{\partial H}{\partial p_i}, \qquad \dot{p}_i = -\frac{\partial H}{\partial q_i}. \tag{5.12}$$

These are the Hamiltonian equations of motion.

Poisson bracket

A very useful role is played in the Hamiltonian formulation of mechanics by the Poisson bracket, which is defined for any pair of functions A and B of p and q by

$$\{A, B\} = \sum_m \left(\frac{\partial A}{\partial q_m} \frac{\partial B}{\partial p_m} - \frac{\partial B}{\partial q_m} \frac{\partial A}{\partial p_m} \right). \tag{5.13}$$

Since all the $6n$ variables are mutually independent at any given time,

$$\frac{\partial q_i}{\partial q_m} = \delta_{im}, \qquad \frac{\partial q_i}{\partial p_m} = 0, \qquad \frac{\partial p_i}{\partial q_m} = 0, \qquad \frac{\partial p_i}{\partial p_m} = \delta_{im}. \tag{5.14}$$

It follows that

$$\{q_i, q_j\} = 0, \qquad \{p_i, p_j\} = 0, \qquad \{q_i, p_j\} = \delta_{ij}, \tag{5.15}$$

and that for any function $F(q, p)$,

$$\{q_i, F\} = \frac{\partial F}{\partial p_i}, \qquad \{p_i, F\} = -\frac{\partial F}{\partial q_i}. \tag{5.16}$$

The equations of motion (5.12) may therefore be written as

$$\dot{q}_i = \{q_i, H\}, \qquad \dot{p}_i = \{p_i, H\}. \tag{5.17}$$

In fact, the time derivative of any function A of the q and p is given by the Poisson bracket $\{A, H\}$:

$$\dot{A} = \sum_i \left(\frac{\partial A}{\partial q_i} \dot{q}_i + \frac{\partial A}{\partial p_i} \dot{p}_i \right) = \{A, H\}, \tag{5.18}$$

where the last step follows on substituting for $\dot{q}_i$ and $\dot{p}_i$ from the Hamiltonian equations of motion (5.12). In arriving at this equation, we have assumed that the dependence of A on t is only through the dependence of the q_i and p_i on t. If A

has an additional explicit dependence on t, i.e., if $A = A(q, p, t)$, then the above equation will get modified to

$$\frac{dA(q, p, t)}{dt} = \{A, H\} + \frac{\partial A}{\partial t}. \qquad (5.19)$$

In the Hamiltonian formulation, the set of coordinate variables q_i and the momentum variables p_i are on an equal footing. All the $6n$ variables are governed by the above set of first-order differential equations. Unlike the Lagrangian formalism which permits only transformations connecting one set of $H = 3n$ coordinate variables q_i to a new set, the Hamiltonian formalism permits *canonical transformations* which connect the whole canonically conjugate set of N pairs of coordinates and momenta (q_i, p_i) to a new set (Q_i, P_i) wherein each member $(Q_i$ or $P_i)$ may be a function of all the $6n$ old variables. The equations of motion in terms of the new variables are of identical form to (5.12), the only difference being that (Q_i, P_i) replace (q_i, p_i) and that the new equations involve a transformed Hamiltonian $K(Q, P, t)$:

$$\dot{Q}_i = \frac{\partial K}{\partial P_i}, \qquad \dot{P}_i = -\frac{\partial K}{\partial Q_i}. \qquad (5.20)$$

It may be possible, by a suitable choice of the canonical transformation, to obtain a form for K such that the new equations of motion (5.20) are much simpler to solve than the old ones; therein lies the reason for the interest in canonical transformations.

In treatments of canonical transformations in textbooks of classical mechanics, it is standard practice to deal with four classes of such transformations. Each one is generated by one type of *generating function*, $F_1(q, Q, t)$, $F_2(q, P, t)$, $F_3(p, Q, t)$, or $F_4(p, P, t)$; each generating function involves either the q or the p from the old canonical variables and either the Q or the P from the new set. The new Hamiltonian K is related to H by

$$K = H + \frac{\partial F}{\partial t}, \qquad (5.21)$$

where F is the generating function employed. We do not pursue these further here. Instead, we will focus our attention on the use of the canonical formalism in the pioneering work of Kinoshita (1977) on the nutations and precession of a rigid Earth.

Lie transformation

The Lie transformation for going over from one set of canonical variables q_i, p_i to another set q_i^*, p_i^* is effected by the operator e^{L_W}, where L_W is defined to be the operator which, acting on an arbitrary function $F(q, p)$, produces its Poisson bracket with the *determining function* $W(q, p)$ of the transformation:

$$L_W F = \{F, W\}. \qquad (5.22)$$

Successive applications of L_W lead to

$$L_W^2 F = L_W\{F, W\} = \{\{F, W\}, W\},$$

$$L_W^r F = L_W L_W^{r-1} F = \{L_W^{r-1} F, W\}. \tag{5.23}$$

The Lie transformation takes functions $F(q, p)$ over into corresponding transformed functions $F^*(q, p)$:

$$F^*(q, p) = e^{L_W} F(q, p) = \sum_{r=0}^{\infty} \frac{1}{r!} L_W^r F(q, p). \tag{5.24}$$

$F^*(q, p)$ is said to be the Lie transform of F with respect to the determining function W. It may be noted that the Lie transform W^* of W is W itself, since $L_W W = \{W, W\} = 0$:

$$W^*(q, p) = e^{L_W} W(q, p) = W(q, p). \tag{5.25}$$

It is often the case, in applications of the Lie theory, that W is of $\mathcal{O}(\epsilon)$ where ϵ is a small parameter. The generic term in the above series is then said to be of order (ϵ^r). We shall use this terminology.

We shall need to use the following properties of the Lie operator:

$$e^{L_W}(F + G) = e^{L_W} F + e^{L_W} G, \tag{5.26}$$

$$e^{L_W}(FG) = (e^{L_W} F)(e^{L_W} G), \tag{5.27}$$

$$e^{L_W}\{F, G\} = \{(e^{L_W} F), (e^{L_W} G)\}. \tag{5.28}$$

The first of these relations is self-evident. The second relation states that the Lie transform of a product of two functions is the product of the Lie transforms of the respective functions. We proceed now to verify this relation, from which the third follows trivially.

Observe first that the application of L_W to a product of two functions $F(q, p)$ and $G(q, p)$ leads, as a direct consequence of the definition of L_W, to the relation

$$L_W(FG) = F(L_W G) + (L_W F)G. \tag{5.29}$$

We see then, by repeated application of L_W and use of the above result at each stage, that

$$L_W^r(FG) = \sum_{s=0}^{r} \binom{r}{s}(L_W^s F)(L_W^{r-s} G), \quad \text{where} \quad \binom{r}{s} = \frac{r!}{s!(r-s)!}. \tag{5.30}$$

The right-hand side is readily identified as $r!$ times the part that is of $\mathcal{O}(\epsilon^r)$ in the product $(e^{L_W} F)(e^{L_W} G)$. Equation (5.27) then follows immediately. The next equation is a corollary.

The Lie transformation associated with a determining function W carries the set of canonical variables (q_i, p_i), $(i = 1, 2, \ldots, n)$, over into the new set (q_i^*, p_i^*):

$$q_i^* = e^{Lw} q_i = q_i + \{q_i, W\} + \frac{1}{2!}\{\{q_i, W\}, W\} + \cdots, \tag{5.31}$$

$$p_i^* = e^{Lw} p_i = p_i + \{p_i, W\} + \frac{1}{2!}\{\{p_i, W\}, W\} + \cdots \tag{5.32}$$

The important fact that the (q_i^*, p_i^*) constitute a canonically conjugate set of variables follows trivially from Eq. (5.28):

$$\{q_i^*, p_j^*\} = \{e^{Lw} q_i, e^{Lw} p_j\} = e^{Lw}\{q_i, p_j\} = e^{Lw}\delta_{ij} = \delta_{ij}, \tag{5.33}$$

and, similarly,

$$\{q_i^*, q_j^*\} = \{p_i^*, p_j^*\} = 0. \tag{5.34}$$

It follows as a consequence of the basic properties (5.26), (5.27), and (5.28) that under a Lie transformation, an arbitrary function $F(q, p)$ goes over into the same function of the Lie-transformed variables, namely $F(q^*, p^*)$:

$$e^{Lw} F(q, p) = F(q^*, p^*). \tag{5.35}$$

It should be remembered that the Lie transform $e^{Lw} F(q, p)$ of $F(q, p)$ is a different function $F^*(q, p)$ of the same variables, as we have noted earlier in Eq. (5.24).

Inverse of the transformation e^{Lw}

Let $F^*(q, p)$ be the Lie transform of $F(q, p)$, with the determining function W. By definition,

$$e^{Lw} F = F^* = F + \{F, W\} + \frac{1}{2!}\{\{F, W\}, W\} + \frac{1}{3!}\{\{\{F, W\}, W\}, W\} + \cdots \tag{5.36}$$

$$e^{L-w} F^* = F^* - \{F^*, W\} + \frac{1}{2!}\{\{F^*, W\}, W\} - \frac{1}{3!}\{\{\{F^*, W\}, W\}, W\} + \cdots \tag{5.37}$$

But the latter expansion is the same as that of $e^{-Lw} F^*$. It follows then that

$$e^{L-w} F^* = e^{-Lw} F^* = e^{-Lw} e^{Lw} F = F. \tag{5.38}$$

Thus e^{L-w} is the inverse of e^{Lw}, i.e., $(-W)$ is the determining function for the inverse of the transformation determined by W.

One could, if necessary, verify the above statement by the "brute force" method of evaluating the terms in (5.37) one by one after substituting for F^* from (5.36) and adding up the resulting expressions.

Transformation of Poisson brackets

The set of relations (5.31) and (5.32) may be inverted, in principle, to obtain the q_i, p_i as functions $q_i(q_i^*, p_i^*)$, $p_i(q_i^*, p_i^*)$ of the new variables. Suppose this has been done. Then any function $F(q, p)$ may be expressed as a function $\tilde{F}$ of the new variables:

$$F(q, p) = F(q(q^*, p^*), p(q^*, p^*)) = \tilde{F}(q^*, p^*). \tag{5.39}$$

Let the function $\tilde{W}(q^*, p^*)$ be also constructed in a similar fashion from $W(q, p)$. Then

$$\{F, W\}_{(q,p)} = \{\tilde{F}, \tilde{W}\}_{(q^*,p^*)}, \tag{5.40}$$

where the subscripts show the variables with respect to which the partial derivatives in the Poisson brackets are taken. To prove the above result, we write $\partial/\partial q_i$ and $\partial/\partial p_i$ as follows:

$$\frac{\partial}{\partial q_i} = \sum_j \left(\frac{\partial q_j^*}{\partial q_i} \frac{\partial}{\partial q_j^*} - \frac{\partial p_j^*}{\partial q_i} \frac{\partial}{\partial p_j^*} \right), \quad \frac{\partial}{\partial p_i} = \sum_k \left(\frac{\partial q_k^*}{\partial p_i} \frac{\partial}{\partial q_k^*} - \frac{\partial p_k^*}{\partial p_i} \frac{\partial}{\partial p_k^*} \right), \tag{5.41}$$

and substitute these expressions for the differential operators in

$$\{F, W\} = \sum_i \left(\frac{\partial F}{\partial q_i} \frac{\partial W}{\partial p_i} - \frac{\partial F}{\partial p_i} \frac{\partial W}{\partial q_i} \right). \tag{5.42}$$

When F and W are replaced by their expressions in terms of (q^*, p^*), namely, $\tilde{F}$ and $\tilde{W}$, respectively, the expression which is obtained for $\{F, W\}$ involves summations over three dummy indices. It can be written, after interchanges of a pair of dummy indices in some of the terms, as

$$\{F, W\}_{(q,p)} = \sum_{ijk} \left(\frac{\partial q_j^*}{\partial q_i} \frac{\partial \tilde{F}}{\partial q_j^*} + \frac{\partial p_j^*}{\partial q_i} \frac{\partial \tilde{F}}{\partial p_j^*} \right) \left(\frac{\partial q_k^*}{\partial p_i} \frac{\partial \tilde{W}}{\partial q_k^*} + \frac{\partial p_k^*}{\partial p_i} \frac{\partial \tilde{W}}{\partial p_k^*} \right)$$

$$- \sum_{ijk} \left(\frac{\partial q_j^*}{\partial p_i} \frac{\partial \tilde{F}}{\partial q_j^*} + \frac{\partial p_j^*}{\partial p_i} \frac{\partial \tilde{F}}{\partial p_j^*} \right) \left(\frac{\partial q_k^*}{\partial q_i} \frac{\partial \tilde{W}}{\partial q_k^*} + \frac{\partial p_k^*}{\partial q_i} \frac{\partial \tilde{W}}{\partial p_k^*} \right)$$

$$= \sum_{jk} \{q_j^*, p_k^*\}_{(q,p)} \left(\frac{\partial \tilde{F}}{\partial q_j^*} \frac{\partial \tilde{W}}{\partial p_k^*} - \frac{\partial \tilde{F}}{\partial p_k^*} \frac{\partial \tilde{W}}{\partial q_j^*} \right)$$

$$+ \sum_{jk} \{q_j^*, q_k^*\}_{(q,p)} \frac{\partial \tilde{F}}{\partial q_j^*} \frac{\partial \tilde{W}}{\partial q_k^*} + \sum_{jk} \{p_j^*, p_k^*\}_{(q,p)} \frac{\partial \tilde{F}}{\partial p_j^*} \frac{\partial \tilde{W}}{\partial p_k^*}$$

$$= \{\tilde{F}, \tilde{W}\}_{(q^*,p^*)}. \tag{5.43}$$

The last step follows on making use of the properties (5.33) and (5.34). The validity of Eq. (5.40) is thus established.

Perturbation expansion of transformed variable

Expand $F^* = e^{L_W} F$ of (5.36) into parts of different orders. Then

$$F_0^* = F_0, \qquad F_1^* = F_1 + \{F_0, W_1\},$$

$$F_2^* = \{F_0, W_2\} + \{F_1, W_1\} + \frac{1}{2!}\{\{F_0, W_1\}, W_1\} + \cdots \qquad (5.44)$$

As we shall see later, this separation of F^* into different orders does not hold up once the equation of motion is invoked to determine W_1 so as to cause F_1^* to be purely secular.

Equations of motion

Let us choose $F(q, p)$ to be the Hamiltonian. Then the equations of motion are

$$\dot{q}_i = \{q_i, F\} = \frac{\partial F}{\partial p_i}, \qquad \dot{p}_i = \{p_i, F\} = -\frac{\partial F}{\partial q_i}. \qquad (5.45)$$

More generally, if $X(q, p)$ is any function of the (q_i, p_i),

$$\dot{X} = \{X, F\}. \qquad (5.46)$$

Let us consider now the rates of change of the new canonical variables:

$$\dot{q}_i^* = \frac{\partial q_i^*}{\partial q_j}\dot{q}_j + \frac{\partial q_i^*}{\partial p_j}\dot{p}_j. \qquad (5.47)$$

Using Eqs. (5.45) we obtain

$$\dot{q}_i^* = \frac{\partial q_i^*}{\partial q_j}\frac{\partial F}{\partial p_j} - \frac{\partial q_i^*}{\partial p_j}\frac{\partial F}{\partial q_j} = \{q_i^*, F\} = \{q_i, F^*\}. \qquad (5.48)$$

The Hamiltonian in terms of the new set of variables is $F^*(q^*, p^*)$. In order to verify this statement, look at the equations of motion.

In applying the Lie transformation to the solution of some perturbation problem where F is the Hamiltonian containing small perturbation parts that are of orders $\epsilon, \epsilon^2, \ldots$, W is taken to be of leading order ϵ plus other parts of higher orders:

$$F = F_0 + F_1 + F_2 + \cdots, \qquad W = W_1 + W_2 + \cdots. \qquad (5.49)$$

Then F^* also has a similar expansion, and one can separate Eq. (5.35) into the different orders. On equating the parts of the same order on the left- and right-hand sides of the equation, we get:

$$F_0^* = F_0, \qquad F_1^* = F_1 + \{F_0, W_1\}, \tag{5.50}$$

$$F_2^* = F_2 + \{F_0, W_2\} + \{F_1, W_1\} + \frac{1}{2!}\{\{F_0, W_1\}, W_1\}, \dots \tag{5.51}$$

The challenge then is to choose the W_i appropriately so as to make it easier to obtain the solution in terms of the transformed variables.

Once the W_i are chosen, the new (Lie transformed) canonical variables may be readily obtained. For example,

$$h^* = h + \{h, W\} + \frac{1}{2}\{\{h, W\}, W\} + \cdots \tag{5.52}$$

If W is expanded into a series of terms of successively higher orders, i.e., if

$$W = W_1 + W_2 + \cdots \tag{5.53}$$

we obtain

$$h^* = h + \{h, W_1\} + \{h, W_2\} + \frac{1}{2}\{\{h, W_1\}, W_1\} + \cdots$$
$$= h + \frac{\partial W_1}{\partial H} + \frac{\partial W_2}{\partial H} + \frac{1}{2}\left\{\frac{\partial W_1}{\partial H}, W_1\right\} + \cdots \tag{5.54}$$

This is one of the Eqs. (4.11) on p. 286 of Kinoshita (1977), with an important difference: the expression for $(h^* - h)$ here is what Kinoshita gives for $(h - h^*)$, and so on.

5.2.2 Rotational motion of rigid bodies

After the general introduction to the Hamiltonian formulation of systems of particles we turn to the mechanics of rigid bodies. By definition, the distance between any two points in a rigid body remains invariant in time. Consequently, the most general motion that a rigid body can undergo consists of a translation of the body as a whole together with a rotational motion of the body about some axis passing through a fixed point of the body. It is convenient to choose this point as the center of mass. The rotational part of the motion is what interests us in the context of the problem of nutation and precession of a rigid model Earth.

When a rigid body is in rotation with an angular velocity ω about a fixed point which is chosen as the origin, the velocity $\mathbf{v}$ of a point of the body at position $\mathbf{r}$ is

$$\mathbf{v} = \omega \wedge \mathbf{r}. \tag{5.55}$$

The kinetic energy of the matter in a volume element $d^3\mathbf{r} = dx\,dy\,dz$ at $\mathbf{r}$ is then $[(1/2)\rho(\mathbf{r})d^3\mathbf{r}](\omega \wedge \mathbf{r}) \cdot (\omega \wedge \mathbf{r})$, and the kinetic energy of the whole body is

$$T = \frac{1}{2} \int \rho(\mathbf{r})[r^2\omega^2 - (\mathbf{r} \cdot \omega)^2]d^3\mathbf{r}$$

$$= \frac{1}{2} \int \rho(\mathbf{r})[r^2\omega_i\omega_j\delta_{ij} - x_i\omega_i x_j\omega_j]d^3\mathbf{r} = \frac{1}{2}I_{ij}\omega_i\omega_j, \qquad (5.56)$$

where the integration is over the volume of the body. We have written $\omega^2 = \omega_j\omega_j\delta_{ij}$, and adopted the summation convention that repeated indices (like i and j here) are to be summed from 1 to 3. The I_{ij} are the elements of the inertia tensor of the body, defined in Section 2.9, and repeated here for convenience:

$$I_{ij} = \int \rho(\mathbf{r})(r^2\delta_{ij} - x_i x_j)d^3\mathbf{r}. \qquad (5.57)$$

The inertia tensor constitutes a real symmetric matrix; so there exists a real orthogonal matrix which transforms it into diagonal form. This orthogonal matrix represents a rotation which brings the coordinate axes into alignment with the principal axes of the body.

When considering the rotation of the model rigid Earth, the origin (the fixed point around which the rotation takes place) is taken to be the geocenter. The principal moments of inertia A, B, C of the Earth ($A \leq B < C$) are the elements of the diagonalized inertia tensor with reference to this origin. In the following, we use the diagonalized form of the inertia tensor.

The angular momentum vector of the body is, by definition,

$$\mathbf{L} = \int \rho(\mathbf{r})[\mathbf{r} \wedge (\omega \wedge \mathbf{r})]d^3\mathbf{r}. \qquad (5.58)$$

The components of $\mathbf{L}$ may be evaluated in a manner similar to that of T. It is readily seen that

$$L_i = I_{ij}\omega_j \qquad \text{and} \qquad T = L_i\omega_i, \qquad (5.59)$$

where the summation convention is again implied. When the inertia tensor is diagonal, we have

$$\mathbf{L} = (A\omega_x, B\omega_y, C\omega_z), \qquad (5.60)$$

and

$$T = \frac{1}{2}(A\omega_x^2 + B\omega_y^2 + C\omega_z^2) = \frac{1}{2}\left(\frac{L_x^2}{A} + \frac{L_y^2}{B} + \frac{L_z^2}{C}\right). \qquad (5.61)$$

The last expression gives the form of the Hamiltonian of a freely rotating rigid body in terms of the Cartesian components of the angular momentum vector. If the body is subject to external forcing, the Hamiltonian will have an additional term, namely the potential energy of interaction with the external force field.

5.2.3 Kinoshita's theory for the nutation of a rigid Earth

In this section and those following presenting the theory of Kinoshita (1977), we try to conform as far as possible to the notation used in that work, to make it easier for the reader who is so inclined to study that work in more detail. In particular, we follow Kinoshita in denoting the universal constant of gravitation (which is generally represented by G) by κ^2; the symbol G is used here, as in Kinoshita's work, for the magnitude of the Earth's rotational angular momentum.

Nutation theory concerns itself primarily with the forced nutations of one or the other of three different axes of the Earth – the angular momentum axis, the figure axis, and the rotation axis. The prime focus of Kinoshita's treatment is on the nutation and precession of the angular momentum axis of the Earth modeled as a rigid body. He obtains the nutations in longitude and obliquity of this axis in analytical form as series of periodic spectral terms arising from the gravitational influences of the Moon and the Sun. (Precession is the secular component of the motion of the axis.) Expressions are then derived for the difference between the nutation series for the angular momentum axis and the corresponding series for each of the two other important axes (namely, the figure axis and the rotation axis). The differences are referred to as the *Oppolzer terms* for the figure axis and for the rotation axis. Such terms have already been considered in Section 2.28.

Reference planes and their relative orientations

Three different reference planes are employed in the development of the theory: the equatorial plane perpendicular to the instantaneous direction of the Earth's figure axis; the plane perpendicular to the angular momentum axis, which we refer to for brevity as the angular momentum plane; and the inertial plane, which is the fixed ecliptic plane of a chosen epoch. When we refer to the ecliptic plane in this section, it must be understood to mean the inertial plane. What we call the angular momentum plane is sometimes referred to as the *Andoyer plane* because of its important role in Andoyer's canonical theory of Earth rotation.

The "absolute" directions of the nutation axes in space are determined in relation to an inertial reference frame. However, the directions of the motions in longitude and obliquity are specified, classically, with reference to a moving plane, the ecliptic of date. Therefore it will be necessary to consider this fourth reference plane too in the process of construction of the nutation series in longitude and obliquity, though we shall restrict our attention to the first three planes for the present. The modifications needed in the treatment of the following sections when the moving

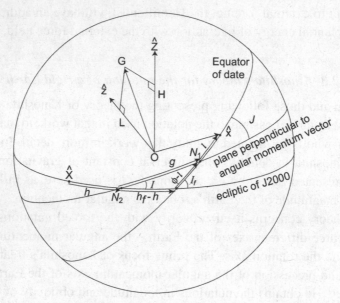

Figure 5.1 The notation for the three planes and circles and for the angles used in Kinoshita's theory (adapted from Kinoshita, 1977).

ecliptic is used as the reference plane instead of the inertial plane will be mentioned briefly at the end.

Longitude and obliquity variables

Figure 5.1 represents the three reference planes by the great circles along which they intersect the geocentric celestial sphere. The directions of the axes of the inertial reference frame relative to the origin O are indicated by points $\hat{X}$, $\hat{Y}$ on the circle pertaining to the inertial plane and by the pole $\hat{Z}$ of the plane (not all of these are shown in the figure). $O\hat{Z}$ is perpendicular to the fixed ecliptic and $O\hat{X}$ is in the direction of the (mean) equinox of epoch. The terrestrial frame employed here is not the usual one having its first axis in the Greenwich meridian, but a frame with axes $O\hat{x}$, $O\hat{y}$, $O\hat{z}$ coinciding with the Earth's principal axes; $O\hat{z}$ is the polar axis perpendicular to the equatorial plane.

The angles I, I_f, J shown in Fig. 5.1 are the inclinations of the three reference planes to one another. It should be obvious that the inclination of one plane to another is also the angle between the axes perpendicular to the two planes. Thus, the angle I between the angular momentum and ecliptic planes is the obliquity of the angular momentum axis, I_f is the obliquity of the figure axis, and J is the angle between the figure axis and the angular momentum axis. Figure 5.1 also shows the arcs h, $h_f - h$, g, l, ϕ which represent the angles between pairs of geocentric vectors in one or the other of these planes. Variations in h are a reflection of the motions of the angular momentum pole in a direction parallel to the ecliptic plane, and therefore represent the motions of the angular momentum axis in longitude. A similar role is played by h_f for the figure axis.

At this point we need to take note of an important fact: the nodes of the angular momentum and equatorial planes on the inertial plane as shown in the figure are the *descending* nodes of the ecliptic on the respective planes, while the reference point for the definition of nutations in longitude and obliquity is the *ascending* node at the diametrically opposite point on the celestial sphere. Consequently, positive increments to the obliquity variable ϵ and to the inclination I require positive rotations about the two opposing axes. Secondly, a positive increment to h is in the sense of a positive rotation around the pole of the ecliptic (see Fig. 5.1) while a positive increment to $\Delta\psi$ is in the sense of a negative rotation about that pole, as explained in Section 2.16. These two observations imply that

$$\Delta\psi = -\Delta h, \quad \text{and} \quad \Delta\epsilon = -\Delta I. \tag{5.62}$$

Variations $-\Delta h$ constitute the precession–nutation in longitude, and $-\Delta I$, that in obliquity, of the angular momentum axis; the corresponding quantities for the figure axis are $-\Delta h_f$ and $-\Delta I_f$. These, along with $-\Delta h_r$ and $-\Delta I_r$ for the rotation axis, are what nutation theory seeks to determine. One word of caution, however: inasmuch as the classical definitions of nutations in longitude and obliquity are in relation to the moving ecliptic and not the fixed ecliptic of epoch, as already mentioned, the theory has to take account of the difference between the two; this too is done in Kinoshita's work.

Andoyer's canonical variables, and Hamiltonian equations

The basic equations from which the variations Δh, ΔI, etc. are to be determined are the dynamical equations for the rotational motion of the rigid Earth. In Hamilton's canonical formulation for the rigid body, the Hamiltonian F is expressed in terms of a chosen set of canonically conjugate pairs of variables which characterize the instantaneous state of the body. One could, for instance, choose L_x, L_y, L_z and the angles of rotation around the x-, y-, and z-axes as the canonically conjugate momentum and coordinate variables, with the Hamiltonian given by T of Eq. (5.61) in the absence of any external perturbations.

However, the canonical variables employed by Kinoshita (1977) are the so-called Andoyer coordinate variables (Andoyer, 1911, 1923), which are angle variables, and their conjugate angular momenta. Andoyer's angle variables are chosen to have a more or less direct relation to the nutation angles in longitude and obliquity. They consist of the angles g, h, and l in the angular momentum, inertial, and equatorial planes, respectively (see Fig. 5.1); their conjugate variables are G, H, and L, which represent the angular momentum components along the axes perpendicular to the

respective planes. Specifically, G is the magnitude of the angular momentum vector
L, H is the Z component of **L**, and L is the z-component of **L**. The above six are
the Andoyer variables.

The Hamiltonian equations of motion for the Andoyer variables are

$$\frac{d}{dt}(g, h, l) = \frac{\partial F}{\partial(G, H, L)}, \qquad \frac{d}{dt}(G, H, L) = -\frac{\partial F}{\partial(g, h, l)}. \qquad (5.63)$$

The task of the theory is to obtain the solutions of these equations, firstly with a
Hamiltonian F_0 which is free of any external influences on the rigid Earth, and then
with the Hamiltonian F which includes the perturbation due to the gravitational
potential of external bodies, primarily the Moon and the Sun. The nutation quantities enumerated above are related to the perturbations produced in the dynamical
variables by $F - F_0$.

Longitude and obliquity: relations to canonical variables

We recall from Section 5.2.3 that variations in the Andoyer variable h constitute
the nutation–precession in longitude. The obliquities I, I_f, J of the different axes
are not among the Andoyer variables; however, I and J are connected to Andoyer's
action variables through geometrical relations. For example, the definitions of H
and L require that

$$\cos I = H/G, \qquad \cos J = L/G. \qquad (5.64)$$

The equatorial components L_x and L_y of **L** may also be expressed in terms of the
Andoyer variables. Noting that the projection of **L** on to the equatorial plane has
to be at $90°$ to the nodal line of the equatorial and angular momentum planes, one
sees that the direction of the projected vector makes an angle $(\pi/2 - l)$ with Ox;
the magnitude of this vector is of course $(L_x^2 + L_y^2)^{1/2} = (G^2 - L^2)^{1/2}$ since L is
L_z. Hence

$$L_x = (G^2 - L^2)^{1/2} \sin l, \qquad L_y = (G^2 - L^2)^{1/2} \cos l, \qquad L_z = L. \quad (5.65)$$

Figure axis

We still need to express the variables h_f and I_f relating to the orientation of the
equatorial plane (and hence that of the figure axis) in terms of the Andoyer variables.
The relations needed for this purpose follow from spherical trigonometric formulae
applied to the spherical triangle formed by the inertial, angular momentum, and
equatorial planes (see Fig. 5.1):

$$\cos I_f = \cos I \cos J - \sin I \sin J \cos g,$$

$$\frac{\sin(h_f - h)}{\sin J} = \frac{\sin(\phi - l)}{\sin I} = \frac{\sin g}{\sin I_f}. \qquad (5.66)$$

The angle J between the angular momentum and equatorial planes is very small, of the order of 10^{-6} rad. Therefore quantities of $\mathcal{O}(J^2)$ may be neglected; in particular, it is sufficient to take $\sin J \approx J$, $\cos J \approx 1$. Furthermore, $I_f - I$ and $h_f - h$ are of $\mathcal{O}(J)$. One may show then from the above relations, with the neglect of quantities of $\mathcal{O}(J^2)$, that

$$h_f = h + \frac{J}{\sin I} \sin g,$$

$$I_f = I + J \cos g, \qquad (5.67)$$

$$\phi = l + g - J \cot I \sin g.$$

The first two of the above equations, taken together with Eq. (5.64), complete the relations of I, I_f, and h_f to the Andoyer variables.

Equations (5.66) and hence (5.67), are simply consequences of the trigonometric formulae which relate the orientations of the three planes, and have nothing to do with the dynamics. So they hold for the Earth subject to external torques as well as in the free case.

Rotation axis

The components of the figure axis in the terrestrial principal axis reference frame are simply $(0, 0, 1)$; those of the rotation axis are $(\omega_x, \omega_y, \omega_z)/\omega$, where $\omega = (\omega_x^2 + \omega_y^2 + \omega_z^2)^{1/2} \approx \omega_z$; the error in this approximation, being of the second order in ω_x and ω_y, is quite negligible. The components of the angular momentum vector of the rigid axially symmetric Earth being $(A\omega_x, A\omega_y, C\omega_z)$, the unit vector characterizing the angular momentum axis has components $(A\omega_x/C\omega_z, A\omega_y/C\omega_z, 1)$ with the neglect of second-order terms as above. The differences of the components of the figure axis and the rotation axis from those of the angular momentum axis are $(A/C)(\omega_x/\omega_z, \omega_y/\omega_z, 0)$ and $(1 - A/C)(\omega_x/\omega_z, \omega_y/\omega_z, 0)$, respectively. (The deviation of the Earth from axial symmetry is so small that its effect on these results is negligible.)

The above observations are independent of whether the Earth's rotation is subject to any perturbation or not. Therefore we can conclude quite generally (for a rigid Earth) that the deviation of the rotation axis from the angular momentum axis is $(-e)$ times that of the figure axis, and hence that

$$h_r - h = -e(h_f - h), \qquad I_r - I = -e(I_f - I), \qquad (5.68)$$

with $(h_f - h)$ and $(I_f - I)$ given by Eqs. (5.67).

Free rotational motion of the axially symmetric rigid Earth

Substantial simplification of the Kinoshita theory becomes possible by ignoring triaxiality, i.e, by assuming axial symmetry ($B = A$), as we shall do in the following. This approximation makes it possible to see the major features of the Hamiltonian approach much more clearly, while producing only entirely negligible errors in the computation of the low frequency nutations. Triaxiality causes excitation of semidiurnal nutations; computation of these by the torque approach is quite straightforward, and is presented elsewhere (see Sections 5.4 and 5.6.1 of this chapter).

The Hamiltonian for free rotation consists of just the kinetic energy. The expression (5.61) for the kinetic energy of the rotating rigid body may now be written in terms of the Andoyer variables by using the relations (5.65) and the fact that L stands for L_z. Setting $B = A$ and taking L_x and L_y from (5.65), we obtain

$$F_0 = \frac{L_x^2 + L_y^2}{2A} + \frac{L_z^2}{2C} = \frac{G^2 - L^2}{2A} + \frac{L^2}{2C}. \tag{5.69}$$

As F_0 does not contain any of the angle variables g, h, l, we see from the second of Eqs. (5.63) that their conjugate momenta G, H, L are all constants and hence that the inclinations I and J are also constants in view of Eqs. (5.64). Constancy of G, H, and L means simply that the magnitude and direction of the angular momentum vector in space remain independent of time, which is expected since no external torque is acting on the body.

Next, we see from the first of the equations of motion (5.63) together with (5.69) that

$$\dot{h} = 0, \qquad \dot{g} = \frac{G}{A}, \qquad \dot{l} = \frac{L}{C} - \frac{L}{A}. \tag{5.70}$$

Thus h is constant; and g and l are linear functions of t since $\dot{g}$ and $\dot{l}$ are both constants. The rates of linear variation may be written in terms of the components of the angular velocity vector by making use of (5.60):

$$n_g = \dot{g} = \left(\frac{C}{A} + \frac{A(\omega_x^2 + \omega_y^2)}{2C\omega_z^2}\right)\omega_z \approx \frac{C}{A}\omega_z, \qquad n_l = \dot{l} = -\left(\frac{C}{A} - 1\right)\omega_z. \tag{5.71}$$

5.2.4 Gravitational perturbation; spectral representation

Hamiltonian including gravitational perturbation

Starting from Eq. (2.54), as will be done later in this chapter (see Section 5.5.2, in particular Eq. (5.200)), the degree two part of the potential energy of the tidal gravitational interaction between the axially symmetric ellipsoidal Earth and a celestial body B is

$$U = \frac{\kappa^2 M}{R^3}(C - A)P_2(\sin \delta) = k'\left(\frac{a}{R}\right)^3 P_2(\sin \delta). \tag{5.72}$$

Here M is the mass of the celestial body, R is its geocentric distance, δ is its latitude in the equatorial reference frame, a is the semi-major axis of its elliptical orbit, and

$$k' = \frac{\kappa^2 M}{a^3}(C - A) = \frac{M}{M + M_E}C\, n^2\, H_d. \tag{5.73}$$

The last form for k', wherein $H_d = (C - A)/C$ and n denotes the mean motion of the body (i.e., mean angular velocity in its Keplerian orbit around the geocenter), is obtained by making use of Kepler's third law:

$$n^2 a^3 = \kappa^2(M + M_E), \tag{5.74}$$

where M_E is the mass of the Earth and M, that of the other body, and where $n_{\mathbb{C}} = 2\pi\, 36525/27.3217 = 8399.7$ rad/cy, $\Omega_0 = 2\pi\, 36525 = 229493$ rad/cy, and $n_{\odot} = 2\pi\, 100 = 628.32$ rad/cy. With $(C - A)/C = 3.2738 \times 10^{-3}$ one finds $k'_{\mathbb{C}} = 7546.72/$cy $= 0.073$ rad/cy and $k'_{\odot} = 3475.41/$cy $= 0.073$ rad/cy.

The total potential energy is the sum of terms of the same form as (5.72), one term for the Moon, one for the Sun, and others for the various planets (which, however, play only very minor roles). We shall continue to show explicitly only one such term, with the express understanding that the contributions from all the other bodies too are to be taken into account. The orbit of each body as seen from the Earth is elliptical in the zeroth order, i.e., when the attraction of third bodies is ignored.

The perturbed Hamiltonian is now

$$F = F_0 + U = \frac{G^2 - L^2}{2A} + \frac{L^2}{2C} + U. \tag{5.75}$$

To make the dependence of the potential U on variables representing the orientation of the Earth in space explicit, it is sufficient to go over from U of (5.72) which involves the latitude δ of the gravitating external body in the TRF to an expression in terms of the coordinates of the body referred to the ecliptic frame. The transformation between these two frames is accomplished by a succession of two rotations (see Fig. 5.1).

The first of these is a negative rotation through the angle J about the node N_1 of the two planes; it carries the equatorial plane into the Andoyer plane. This rotation transforms $P_2(\sin \delta)$ into a linear combination of spherical harmonic functions which have, as their arguments, the latitude of the perturbing body and its longitude reckoned from the node N_1 along the Andoyer plane. The coefficients in the linear combination are functions of J.

The second one is a negative rotation through the angle I between the Andoyer plane and the ecliptic plane of date about the node N_2 of the two planes, to carry the former plane into the latter. It transforms each of the just-mentioned spherical harmonics, with their longitude argument now reckoned from N_2, into a linear combination, with coefficients that are functions of I, of spherical harmonics having the ecliptic latitude β and the ecliptic longitude reckoned from this node as arguments. The first of these longitudes, represented by an arc along the Andoyer plane, now includes the arc from N_2 to N_1 which is the Andoyer variable g; and the arc representing the ecliptic longitude is $(\lambda - h)$ where λ is the ecliptic longitude of the body reckoned from the equinox γ_0 of epoch, and h is the longitude of N_2.

These two successive transformations lead to the following expression for $P_2(\sin \delta)$ in terms of the ecliptic longitude and latitude (λ, β) of the perturbing celestial body and the Andoyer variables g, h, L, H, G, of which the last three enter through I and J via the relations $\cos J = L/G$ and $\cos I = H/G$ presented in Eqs. (5.64):

$$P_2(\sin \delta) = R_0(3 \cos^2 J - 1) + R_1 \sin 2J + R_2 \sin^2 J, \qquad (5.76)$$

$$R_0 = \frac{1}{2}\left[\frac{1}{4}(3 \cos^2 I - 1) P_2(\sin \beta) - \frac{1}{4} \sin 2I\, P_2^1(\sin \beta)\right.$$

$$\left. \times \sin(\lambda - h) - \frac{1}{8} \sin^2 I P_2^2(\sin \beta) \cos(2\lambda - 2h)\right], \qquad (5.77)$$

$$R_1 = -\frac{3}{4} \sin 2I\, P_2(\sin \beta) \cos g$$

$$- \sum_{\epsilon=\pm1} \frac{1}{4}(1 + \epsilon \cos I)(-1 + 2\epsilon \cos I)\, P_2^1(\sin \beta) \sin(\lambda - h - \epsilon g)$$

$$- \sum_{\epsilon=\pm1} \frac{1}{8}\epsilon \sin I (1 + \epsilon \cos I)\, P_2^2(\sin \beta) \cos(2\lambda - 2h - \epsilon g), \quad (5.78)$$

$$R_2 = \frac{3}{4} \sin^2 I\, P_2(\sin \beta) \cos 2g$$

$$+ \sum_{\epsilon=\pm1} \frac{1}{4}\epsilon \sin I (1 + \epsilon \cos I)\, P_2^1(\sin \beta) \sin(\lambda - h - 2\epsilon g)$$

$$- \sum_{\epsilon=\pm1} \frac{1}{16}(1 + \epsilon \cos I)^2\, P_2^2(\sin \beta) \cos 2(\lambda - h - \epsilon g). \qquad (5.79)$$

Details of the derivation leading to this form may be found in Section 5 of Kinoshita (1977). It may be observed from Eqs. (5.77)–(5.79) that λ, β, and Andoyer's angle variables h, g appear in these expressions through functions of the form $P_2^m(\sin \beta) {\cos \atop \sin}(m\lambda - mh \pm u)$, where u takes one of the values 0 or g or

$2g$ in the different terms. Introduction of $P_2(\sin \delta)$ from (5.76)–(5.79) into (5.72) results in an expression for U in which the Earth's orientation appears solely through Andoyer's canonical variables. Consequently, the equations governing the Earth's rotational motion may be obtained by direct application of the Hamiltonian Eqs. (5.63).

It was observed in the last paragraph of Section 5.2.3 that the variations of h and I constitute the nutation–precession motions in longitude and obliquity. The spectrum of the nutations arises from that of the potential U, and knowledge of the secular component (precession) and of the amplitudes of the spectral components of nutation is of considerable value for gaining insight into the physics of the Earth (especially in the case of a non-rigid Earth). Therefore we shall seek to obtain solutions of the relevant equations in the form of spectral representations.

For this purpose, it is necessary to have a spectral representation of the potential U. It may be arrived at with the use of the spectral expansions of the quantities $(a/R)^3 P_2^m(\sin \beta) {\cos \atop \sin} m(\lambda - h)$ based on theories of the orbital motions of the Moon and the Sun. Kinoshita employed Newcomb's (1895) theory for the Sun, and Brown's (1919) theory with improvements by Eckert *et al.* (1966) for the Moon. Improved analytical theories of high accuracy for the Sun and the planets are now provided by the VSOP theory and for the Moon by ELP2000. The following expansions are taken from Kinoshita (1977):

$$-\left(\frac{a}{R}\right)^3 P_2(\sin \beta) \cos u = \sum_{\epsilon = \pm 1} \sum_s A_s^{(0)} \cos(u - \epsilon \Xi_s),$$

$$\frac{1}{3}\left(\frac{a}{R}\right)^3 P_2^1(\sin \beta) \sin(\lambda - h - \epsilon u) = \sum_{\epsilon = \pm 1} \sum_s A_s^{(1)} \cos(u - \epsilon \Xi_s), \quad (5.80)$$

$$\frac{1}{3}\left(\frac{a}{R}\right)^3 P_2^2(\sin \beta) \cos(2\lambda - 2h - \epsilon u) = \sum_{\epsilon = \pm 1} \sum_s A_s^{(2)} \cos(u - \epsilon \Xi_s),$$

where u is 0 or g or $2g$ as already mentioned.

The amplitudes $A_s^{(i)}$ fall into two classes, one pertaining to the Moon and the other to the Sun. Their values, as determined from the theories of the Moon and the Sun, are very nearly constant, though a very small correction in the form of a term linear in time needs to be taken into account for very accurate computations. Tables of values of these amplitudes (the dominant constant part and the coefficient of t in each of the amplitudes) may be found in Kinoshita (1977); they are tabulated separately for the Moon and the Sun.

The nutation term with the argument $(u - \epsilon \Xi_s)$ has the angular frequency $du/dt - \epsilon \, d\Xi_s/dt$, with $du/dt = 0$, n_g or $2n_g$, where $n_g = (C/A)\omega_z$ as given by Eq. (5.71). Ξ_s is a linear combination of five *fundamental arguments* (l, l', F, D, Ω)

called the *Delaunay arguments* (see also Sections 5.5.5 and 5.7):

$$\Xi_s = i_1 l + i_2 l' + i_3 F + i_4 D + i_5 \Omega, \qquad s = (i_1, i_2, i_3, i_4, i_5), \qquad (5.81)$$

where $i_1, i_2, \cdots$ are integers which may be positive, negative, or zero, except for i_5 which is non-negative; s is a composite index consisting of these five integers. For the definitions of the fundamental arguments and their characterization in relation to the motions of the Sun and the Moon in their Keplerian orbits in space, see Section 5.5.5.

It must be kept in mind that the arguments l, F, Ω are entirely unrelated to the Andoyer variable l, the Hamiltonian, and the Earth's rotation rate, for which these symbols are used elsewhere in this chapter; similarly the mean longitude of the Moon, for which the symbol L is used elsewhere, is unrelated to the Andoyer variable L.

It turns out, in general, that only terms with $i_5 = 0, 1,$ or 2 are of interest. Furthermore, the amplitude $A_s^{(0)} = 0$ for any term (labeled by s) for which $(i_5)_s = 1$ or 2, while $A_s^{(1)} = 0$ whenever $(i_5)_s = 0$ or 2, and $A_s^{(2)} = 0$ for $(i_5)_s = 0$ or 1. Exceptions to these rules occur, however, with extremely small values for the "prohibited" coefficients: $A_s^{(1)}$ and $A_s^{(2)}$ with $s = (0, 0, 0, 0, 0)$, and $A_s^{(0)}$ and $A_s^{(2)}$ with $s = (0, 0, 0, 0, 1)$.

In the particular case $s = (0, 0, 0, 0, 0)$, the argument $\Xi_s = 0$. The time independent term with $\Xi_s = 0$ in the spectral expansion of the perturbation term U of the Hamiltonian is responsible for the precession. The term with argument Ω, corresponding to $s = (0, 0, 0, 0, 1)$, gives rise to the largest of the nutation terms, the 18.6 year nutation.

It is of importance to note that Delaunay's Ω involves the Andoyer variable h. The reason is that the positions of the nodes of the Moon's orbital plane and of the Earth's angular momentum plane relative to the origin of longitudes on the ecliptic plane (which is the direction of the X-axis of the ecliptic reference frame) are $h_{\mathbb{C}}$ and h, respectively, and hence $\Omega = h_{\mathbb{C}} - h$. The resulting implicit dependence of Ξ_s on h has to be taken into account while developing the perturbation theory in the following sections. In particular,

$$\frac{\partial \Xi_s}{\partial h} = -(i_5)_s. \qquad (5.82)$$

We can now introduce the expressions (5.80) into Eqs. (5.76)–(5.79), to obtain the following spectral expansion:

$$U = k'\left(\frac{a}{R}\right)^3 P_2(\sin \delta) = U^{(0)} + U^{(1)} + U^{(2)}, \qquad (5.83)$$

where the three terms on the right are of the zeroth, first, and second orders in J, and involve frequencies in the low frequency, diurnal, and semidiurnal bands, respectively:

$$U^{(0)} = \frac{3k'}{2}(3\cos^2 J - 1) \sum_s B_s \cos \Xi_s,$$

$$U^{(1)} = -\frac{3k'}{2}\sin 2J \sum_{\epsilon=\pm 1}\sum_s C_s(\epsilon)\cos(g - \epsilon\Xi_s), \qquad (5.84)$$

$$U^{(2)} = \frac{3k'}{4}\sin^2 J \sum_{\epsilon=\pm 1}\sum_s D_s(\epsilon)\cos(2g - \epsilon\Xi_s),$$

wherein the coefficients B_s, C_s, D_s are functions of I:

$$B_s = -\frac{1}{6}(3\cos^2 I)A_s^{(0)} - \frac{1}{2}\sin 2I\, A_s^{(1)} - \frac{1}{4}\sin^2 I A_s^{(2)}, \qquad (5.85)$$

$$C_s = -\frac{1}{4}\sin 2I\, A_s^{(0)} + (1 + \epsilon\cos I)(-1 + 2\epsilon\cos I)A_s^{(1)}$$
$$+ \frac{1}{4}\epsilon\sin I(1 + \epsilon\cos I)A_s^{(2)}, \qquad (5.86)$$

$$D_s = -\frac{1}{2}\sin^2 I A_s^{(0)} + \epsilon\sin I(1 + c\cos I)A_s^{(1)}$$
$$- \frac{1}{4}(1 + \epsilon\cos I)^2 A_s^{(2)}. \qquad (5.87)$$

Kinoshita (1977) lists the values of the coefficients $A_s^{(i)}$ for the Moon in his Tables 1a–1c and those for the Sun in Tables 1d–1e.

Equations (5.83)–(5.87) exhibit the explicit dependence of the potential U on the Andoyer variables of the Earth rotation problem through I, J, g, and h, the last of which enters through Ξ_s as noted above Eq. (5.82); the orbital motion of the external body is represented by the Ξ_s and the coefficients $A_s^{(0)}$, $A_s^{(1)}$, $A_s^{(2)}$.

The sum over s in $U^{(0)}$ includes a term with $s = (0, 0, 0, 0, 0)$, which is a constant term, since $\Xi_s = 0$ in this case. All other terms in $U^{(0)}$ are periodic, each with its own frequency $d\,\Xi_s/dt$. In $U^{(1)}$ and $U^{(2)}$, the frequencies are $n_g \pm d\,\Xi_s/dt$ and $2n_g \pm d\,\Xi_s/dt$, respectively, with n_g given by Eq. (5.71).

Nutation of the angular momentum axis from solution of canonical equations

The form of the Hamiltonian for the rotational motion of the rigid Earth subject to the gravitational potential of the celestial body B is now

$$F = F_0 + U = \left[\frac{1}{2A}G^2 + \left(\frac{1}{2C} - \frac{1}{2A}\right)L^2\right] + U^{(0)} + U^{(1)} + U^{(2)}, \qquad (5.88)$$

with $U^{(0)}$, $U^{(1)}$, $U^{(2)}$ given by Eq. (5.84) in terms of the Andoyer variables and time. (Remember that B_s, C_s, D_s are functions of I, and that the angular momentum variables G, H, and L enter through I and J by virtue of the relations (5.64), while g appears explicitly.) The variable h lies hidden in the fifth Delaunay argument Ω which is contained in Ξ_s, as pointed out already. The remaining variable l is absent from F, and hence

$$\dot{L} = -\frac{\partial F}{\partial l} = 0, \quad \rightarrow \quad L = \text{constant}. \tag{5.89}$$

All the other five Andoyer variables are now time dependent.

Since nutations in longitude and obliquity are determined by the variations in h and I, we shall focus on these two now. The following relations obtained by differentiating $\cos I = H/G$ will be useful in expressing the canonical equations of motion for these quantities:

$$\frac{\partial I}{\partial H} = -\frac{1}{G \sin I}, \qquad \frac{\partial I}{\partial G} = \frac{\cos I}{G \sin I}. \tag{5.90}$$

Consider now the canonical equation of motion for h:

$$\dot{h} = \frac{\partial F}{\partial H} = \frac{\partial U}{\partial H} = \frac{\partial U}{\partial I}\frac{\partial I}{\partial H} = -\frac{1}{G \sin I}\frac{\partial U}{\partial I}. \tag{5.91}$$

The contributions to $\dot{h}$ from the parts $U^{(1)}$ and $U^{(2)}$ of U are negligible because they are proportional to $\sin J$ and $\sin^2 J$, respectively, where J is of $\mathcal{O}(10^{-6})$ as mentioned earlier. Taking therefore only the part $U^{(0)}$ involving B_s, we obtain

$$\dot{h} = \frac{\partial U^{(0)}}{\partial H} = -\frac{3k'}{2G \sin I}(3\cos^2 J - 1)\sum_s \frac{\partial B_s}{\partial I}\cos \Xi_s. \tag{5.92}$$

The integral of $\dot{h}$, with the integration constant ignored, is Δh, the contribution to h from the perturbation. (The constant of integration, which will be dropped, is the unperturbed part of h.) The time independent part of $\dot{h}$ gives the precession rate directly; the periodic terms in Δh are the various nutation terms in longitude.

If we are seeking solutions only to the first order in the perturbation, we can use zeroth-order expressions (i.e., solutions of the unperturbed problem) for the quantities appearing within the first-order quantity U. In that case we can take I and J to be constants in U since G, H, and L are constants in the zeroth order. (In fact we can ignore the very small value of J which is of $\mathcal{O}(10^{-6})$; so we set $J = 0$.) Similarly, g will be varying linearly with time ($g = n_g t$), and h will be a constant, as in the free case. It turns out that a first-order treatment is sufficient for the computation of nutations; but one has to make calculations to the second order for the precession, as the second-order contribution, though very small, cannot be ignored.

We proceed now to a first-order calculation of the nutation terms. Of course we must then exclude from the summation in (5.92) the constant term $s = (0, 0, 0, 0, 0)$, abbreviated as $s = 0$, which represents the precession rate; only the periodic terms ($s \neq 0$) are involved in nutation. Since it is sufficient to take the zeroth-order value 0 for J we set $(3/2)(3\cos^2 J - 1)$ equal to 3. Integration of Eq. (5.92) may then be done trivially. We obtain

$$\Delta h = -\frac{3k'}{G \sin I} \sum_{s \neq 0} \frac{\partial B_s}{\partial I} \frac{1}{N_s} \sin \Xi_s. \tag{5.93}$$

The expression for $\partial B_s / \partial I$ may be found from that for B_s given in Eq. (5.85):

$$\frac{\partial B_s}{\partial I} = \frac{1}{2} \sin 2I A_s^{(0)} - \cos 2I A_s^{(1)} - \frac{1}{4} \sin 2I A_s^{(2)}. \tag{5.94}$$

The first-order expression for Δh is obtained by substituting from the last equation into the previous one and setting I equal to its constant zeroth-order value (say I_0).

Let us go on now to the variation of I. It can be expressed in terms of variations of H and G:

$$\dot{I} = \frac{\partial I}{\partial H} \dot{H} + \frac{\partial I}{\partial G} \dot{G}, \tag{5.95}$$

wherein

$$\dot{G} = \frac{\partial F}{\partial g} \sim \mathcal{O}(J), \tag{5.96}$$

$$\dot{H} = -\frac{\partial F}{\partial h} = -\frac{\partial U}{\partial h} = -3k' \sum_{s \neq 0} (i_5)_s B_s \sin \Xi_s. \tag{5.97}$$

We have used $\partial \Xi_s / \partial h = -(i_5)_s$ from Eq. (5.82) in the last step. Introducing these results for $\dot{G}$ and $\dot{H}$ into $\dot{I}$ and taking $\partial I / \partial H = -1/(G \sin I)$ from (5.90), we integrate $\dot{I}$ over time to obtain

$$\Delta I = -\frac{3k'}{G \sin I} \sum_{s \neq 0} (i_5)_s \frac{B_s}{N_s} \cos \Xi_s. \tag{5.98}$$

Once again, the zeroth-order values may be taken for G and I here.

The expressions (5.93) and (5.98) for Δh and ΔI are the results obtained by Kinoshita. The nutation quantities that we needed to determine are now given by

$$\Delta \psi = -\Delta h, \qquad \Delta \epsilon = -\Delta I. \tag{5.99}$$

Nutations of the figure axis and rotation axis

The nutation of the figure axis differs from that of the angular momentum axis by the increments to $(h_f - h)$ and $(I_f - I)$ that are produced by the perturbation U.

We find from Eqs. (5.67) that

$$\Delta h_f = \Delta h + \Delta\left(\frac{J}{\sin I}\sin g\right), \qquad \Delta I_f = \Delta I + \Delta(J\sin g). \tag{5.100}$$

So we need to evaluate

$$\Delta\left(\frac{J}{\sin I}\sin g\right) = \frac{1}{\sin I}(\sin g\,\Delta J + J\cos g\,\Delta g - J\sin g\cot I\,\Delta I), \tag{5.101}$$

as well as

$$\Delta(J\cos g) = \cos g\,\Delta J - J\sin g\,\Delta g, \tag{5.102}$$

where the quantities Δg, ΔI, ΔJ to be considered are the increments produced by the perturbation U.

One may safely neglect terms that are of the first or higher order in J in these expressions. We have already determined ΔI, which is given by Eq. (5.98) above; it is of the zeroth order in J. Hence $J\Delta I$ is of $\mathcal{O}(J)$ and is therefore negligible. What we need to do then is to identify and evaluate the terms of $\mathcal{O}(J^0)$ present in ΔJ and $J\Delta g$.

Consider first the quantity $J\Delta g$. We observe that the rate $\dot g$ is given by the canonical equation $\dot g = \partial F/\partial G$. The increment Δg due to the perturbation part of F is evidently the time integral of $\partial U/\partial G$. Since G enters into U only through I and J, it follows that

$$\frac{\partial U}{\partial G} = \frac{\partial I}{\partial G}\frac{\partial U}{\partial I} + \frac{\partial J}{\partial G}\frac{\partial U}{\partial J} = \frac{\cot I}{G}\frac{\partial U}{\partial I} + \frac{\cot J}{G}\frac{\partial U}{\partial J}. \tag{5.103}$$

The presence of $\cot J$ which is of $\mathcal{O}(J^{-1})$ makes the second term in the above expression dominant; we can safely neglect the other term. Furthermore, within the factor $\partial U/\partial J$, the part $\partial U^{(1)}/\partial J$ is of the zeroth order while the derivatives of $U^{(0)}$ and $U^{(2)}$ are of $\mathcal{O}(J)$, as is evident from Eq. (5.84). It follows then that to the leading order in J, $\partial U/\partial G$, and hence the perturbation part of $\dot g$, are given by $(\cot J/G)\partial U^{(1)}/\partial J$. On integrating this quantity with $U^{(1)}$ taken from (5.84) and then multiplying by J, we immediately obtain $J\Delta g$:

$$J\Delta g = -3k'\frac{J\cot J\cos 2J}{G}\sum_{\epsilon=\pm1}\sum_s \frac{C_s(\epsilon)}{n_g - \epsilon N_s}\sin(g - \epsilon\Xi_s). \tag{5.104}$$

To $\mathcal{O}(J^0)$, the factor $J\cot J\cos 2J$ may be replaced by unity; the resulting expression gives $J\Delta g$ for our purposes.

Consider ΔJ next. Since $\cos J = L/G$,

$$\frac{\partial J}{\partial L} = -\frac{1}{G\sin J}, \quad\text{and}\quad \frac{\partial J}{\partial G} = \frac{\cos J}{G\sin J}, \tag{5.105}$$

and hence

$$\Delta J = \frac{\partial J}{\partial L}\Delta L + \frac{\partial J}{\partial G}\Delta G = \frac{1}{G \sin J}(-\Delta L + \cos J \, \Delta G). \tag{5.106}$$

We now need to find the terms of order J^0 in the increments ΔL and ΔG produced by the potential U. They may be obtained by integration of the canonical equations $\dot{L} = \partial F/\partial l$, and $\dot{G} = \partial F/\partial g$. Since l does not enter into the expression for F, it follows that $\dot{L} = 0$ and hence that

$$\Delta L = 0. \tag{5.107}$$

To obtain ΔG due to the perturbation U, we have to integrate $\partial U/\partial g$ with respect to time. The part $U^{(0)}$ of U does not contain g, and the part $U^{(2)}$ is of $\mathcal{O}(J^2)$ and is therefore negligible. Thus,

$$\Delta G = -\int \frac{\partial U^{(1)}}{\partial g} dt = \frac{3k'}{2} \sin 2J \sum_{\epsilon=\pm 1} \sum_s C_s(\epsilon) \frac{\cos(g - \epsilon \Xi_s)}{n_g - \epsilon N_s}. \tag{5.108}$$

Introducing the last two results into (5.106), we obtain the following explicit expression for ΔJ:

$$\Delta J = 3k' \frac{\cos^2 J}{G} \sum_{\epsilon=\pm 1} \sum_s C_s(\epsilon) \frac{\cos(g - \epsilon \Xi_s)}{n_g - \epsilon N_s}. \tag{5.109}$$

We are in a position, finally, to write down explicit expressions for $\Delta h_f - \Delta h$ and $\Delta I_f - \Delta I$. We just have to substitute for ΔJ and $J \Delta g$ from Eqs. (5.104) and (5.109) into Eqs. (5.101) and (5.102) and set $J = 0$ (its zeroth-order value) to get

$$\Delta h_f - \Delta h = \frac{3k'}{G \sin I} \sum_{\epsilon=\pm 1} \sum_{s \neq 0} \epsilon C_s(\epsilon) \sin \Xi_s, \tag{5.110}$$

$$\Delta I_f - \Delta I = \frac{3k'}{G} \sum_{\epsilon=\pm 1} \sum_{s \neq 0} C_s(\epsilon) \cos \Xi_s. \tag{5.111}$$

The Oppolzer terms for the figure axis are thus determined; those for the rotation axis are obtainable simply by multiplying by $(-e)$, see Eq. (5.68). These results are the same as in Kinoshita (1977).

Precession

The computation of precession, unlike that of nutation, has to be carried to the second order in perturbation theory (i.e., to the second order in k') because a small but non-negligible contribution arises from terms of the second order. We begin with the same expression (5.92) for $\dot{h}$ as before, but in evaluating the zero-frequency (precessional) part of this quantity, one cannot be content with using for I, J, and G

their constant initial values; one has to take into account their incremental changes ΔI, ΔJ, ΔG with time, and determine their contributions to the zero-frequency part of $\dot{h}$.

Before proceeding to do this, we take note of the rate of precession in the first order, which is simply the negative of the constant ($s = 0$) term in the expression (5.92) for $\dot{h}$, with the initial constant values assigned to I, J, and G (in particular, $J = 0$):

$$\dot{h}^{(1)} = -\frac{3k'}{G \sin I} \frac{\partial B_0}{\partial I}, \tag{5.112}$$

with k' defined as in (5.73) and with the unperturbed (constant) values for I and G. We obtain a more explicit expression for $\dot{h}^{(1)}$ on substituting from Eq. (5.85) for B_0:

$$\dot{h}^{(1)} = -\frac{3k'}{G} \left(A_0^{(0)} \cos I - A_0^{(1)} \frac{\cos 2I}{\sin I} - \frac{1}{2} A_0^{(2)} \cos I \right). \tag{5.113}$$

It is to be kept in mind that we have to sum over two such expressions, one for the Moon, and the other for the Sun. The values appropriate to the body are to be used in each case for the parameters which appear in the definition (5.73) of k' and for the coefficients $A_0^{(i)}$ (see Kinoshita, 1977).

To compute the second-order contribution to the precession rate, we have to go back to the expression (5.92) for $\dot{h}$ and calculate the contributions to the zero-frequency part of $\dot{h}$ from the increments ΔG, ΔI, ΔJ to the zeroth-order values of G, I, J. The total contribution is evidently

$$\dot{h}^{(2)} = \left(\frac{\partial \dot{h}}{\partial G} \Delta G + \frac{\partial \dot{h}}{\partial J} \Delta J + \frac{\partial \dot{h}}{\partial I} \Delta I \right)_0, \tag{5.114}$$

where the subscript 0 indicates that the zero-frequency component of the expression within parentheses is to be taken. It is understood also that the zeroth-order values are to be assigned for I, J, and G after evaluating the partial derivatives and introducing the explicit expressions for ΔG, ΔI, ΔJ.

Consider first the last term of Eq. (5.114), which is the contribution from ΔI to $\dot{h}$ of (5.92). It is seen to be

$$\frac{\partial \dot{h}}{\partial I} \Delta I = -\frac{3k'}{G} \sum_s \frac{\partial}{\partial I} \left(\frac{1}{\sin I} \frac{\partial B_s}{\partial I} \right) \Delta I \cos \Xi_s, \tag{5.115}$$

on using the fact that $J = 0$ in the zeroth order. Now, the spectral expansion (5.98) of ΔI has the form

$$\Delta I = \sum_{s'} (\Delta I)_{s'} \cos \Xi_{s'} \quad \text{with} \quad (\Delta I)_{s'} = -\frac{3k'}{G \sin I} \frac{B_{s'}}{N_{s'}} (i_5)_{s'} \tag{5.116}$$

for all $s' \neq 0$. It is evident that the product $\Delta I \cos \Xi_s$ in (5.115) has a time independent part $(1/2)\Delta I_s$, which arises from the term with $s' = s$ in ΔI.

In contrast, the time dependence of any term in the spectral expansion (5.108) of ΔG is through a factor of the type $\cos(g - \epsilon\Xi_{s'})$, while the time dependence of $\partial \dot{h}/\partial G$ is through $\cos \Xi_s$. The product $\Delta G \cos \Xi_s$ does not contain any time independent part, and so the first term in (5.114) does not yield any contribution to the second-order precession rate. The same is true of ΔJ, as is evident from its spectral expansion (5.109) of ΔJ.

Thus the sole contribution to the second-order precession rate is from the last term of (5.114); it is given by (5.115) with $\Delta I \cos \Xi_s$ replaced by its time independent part, $(1/2)\Delta I_s$:

$$\dot{h}^{(2)} = -\frac{3k'}{2G} \frac{\partial}{\partial I} \left(\frac{1}{\sin I} \frac{\partial B_s}{\partial I} \right) \Delta I_s. \tag{5.117}$$

After substituting for ΔI_s from the second of Eqs. (5.116), we rewrite the above expression as

$$\dot{h}^{(2)} = \left(\frac{3k'}{G} \right)^2 \sum_{s\neq 0} \frac{(i_5)_s X_s}{2N_s \sin I}, \qquad X_s = \frac{\partial}{\partial I} \left(\frac{1}{\sin I} \frac{\partial B_s}{\partial I} \right) B_s. \tag{5.118}$$

Now we recall the observations made following Eq. (5.81) that all terms of interest have $(i_5)_s = 0, 1,$ or 2, and further, that out of the three coefficients $A_s^{(0)}, A_s^{(1)}, A_s^{(2)}$ that are present in the expression (5.85) for B_s, only $A_s^{(1)}$ is non-vanishing in terms which have $(i_5)_s = 1$, and only $A_s^{(2)}$ is non-vanishing in the case $(i_5)_s = 2$. Terms with $(i_5)_s = 0$ drop out of the expression (5.118) for $\dot{h}^{(2)}$, of course. So the relevant terms X_s reduce to one of two types, $X_s^{(1)}$ involving $A_s^{(1)}$ and $X_s^{(2)}$ involving $A_s^{(2)}$. With the use of the expressions for B_s and $\partial B_s/\partial I$ from Eqs. (5.85) and (5.94), we see that

$$X_s^{(1)} = Y^{(1)}(I)(A_s^{(1)})^2, \quad \text{and} \quad X_s^{(2)} = Y^{(2)}(I)(A_s^{(2)})^2, \tag{5.119}$$

with

$$Y^{(1)}(I) = \frac{1}{2} \left[\frac{\partial}{\partial I} \left(\frac{\cos 2I}{\sin I} \right) \sin 2I \right] = -\frac{(2\sin^2 I + 1)\cos^2 I}{\sin I}, \tag{5.120}$$

$$Y^{(2)}(I) = \frac{1}{16} \left[\frac{\partial}{\partial I} \left(\frac{\sin 2I}{\sin I} \right) \sin^2 I \right] = -\frac{1}{8} \sin^3 I. \tag{5.121}$$

Therefore $\dot{h}^{(2)}$ reduces to

$$\dot{h}^{(2)} = \left(\frac{3k'}{G} \right)^2 \frac{1}{2\sin I} \left[Y^{(1)}(I) \sum_s \frac{1}{N_s} (A_s^{(1)})^2 + Y^{(2)}(I) \sum_s \frac{2}{N_s} (A_s^{(2)})^2 \right],$$

$$\tag{5.122}$$

wherein the first and second sums contain only terms with $(i_5)_s = 1$, and those with $(i_5)_s = 2$, respectively.

The values of $A_s^{(1)}$ and $A_s^{(2)}$ for various spectral terms s are listed in Tables 1b and 1c of Kinoshita (1977). The largest value of $A_s^{(1)}$ is 4.48718×10^{-2} rad, belonging to the 18.6 year term corresponding to $s = (0, 0, 0, 0, 1)$, for which $N_s = 2\pi \, 365.25/(-6798.38) = -0.33757$ rad/cy.

The second-order part of the precession rate, proportional to M_3, in Eq. (10.1) of Kinoshita (1977) contains the same lunar factor $(A_s^{(1)})^2$ as in (5.119): $M_3 = (3/2)[A_{(0,0,0,0,1),0}^{(1)}]_M^2$, where $A_{(0,0,0,0,1),0}^{(1)}$ is the time independent part of the coefficients of the 18.6 year term in the tesseral part of the potential U due to the Moon: see definition on p. 291 and its value (448718×10^{-7} rad) in Table 1b on p. 293 of Kinoshita.

5.3 Earth's gravitational potential on a celestial body

5.3.1 *The external gravitational potential of the Earth*

We considered in Chapter 2 the potential of an external body B at a point $\mathbf{r}$ in the Earth up to the second order in (r/d), where d is the distance of B from the geocenter, and derived from it the torque produced on the ellipsoidal Earth. We consider here the alternative approach which focuses on the gravitational potential of the Earth itself at some point $\mathbf{R}$ outside it. The purpose is to use the expression to compute the *torque due to the Earth on a celestial body* B idealized as a point mass; this torque, with its direction reversed, is the torque *due to B on the Earth*.

The gravitational potential produced at $\mathbf{R}$ by a mass element $\rho(\mathbf{r})d^3r$ of the Earth is

$$dW_E(\mathbf{R}) = -\frac{G\rho(\mathbf{r})d^3r}{|\mathbf{R} - \mathbf{r}|}. \tag{5.123}$$

(The minus sign is in conformity with the physicists' sign convention employed in this book: see Section 2.13.) The potential $W_E(\mathbf{R})$ due to the whole Earth is the integral of the above quantity over the volume of the Earth. The origin with reference to which the vectors $\mathbf{r}$ and $\mathbf{R}$ are defined is, as usual, the center of mass of the Earth (or more precisely, that of the Earth excluding the atmospheric and oceanic layers at the surface). The distance r or R from the origin is referred to as the geocentric distance.

In order to demonstrate how the detailed features of the density distribution in the Earth are reflected in the expression for the potential, it is necessary to use the full expansion of $1/|\mathbf{R} - \mathbf{r}|$ in powers of (r/R) and in terms of spherical harmonics. The expansion assumes that $\mathbf{R}$ lies outside the sphere circumscribing the Earth. i.e., that R exceeds the maximum geocentric distance of any point $\mathbf{r}$ in the Earth.

For the ellipsoidal Earth, this requirement means that R must be greater than the equatorial radius a_e.

We begin by considering the approximation to the potential up to order $(r/R)^2$, in the same manner as in Chapter 2, and demonstrate that the properties of the Earth appear in this approximate potential solely through the Earth's inertia tensor.

5.3.2 MacCullagh's theorem

Taking the expansion of $1/|\mathbf{R} - \mathbf{r}|$ up to terms of order $(r/d)^2$ where $d = |\mathbf{R}|$, we write the external potential of the Earth as

$$W_E(\mathbf{R}) = \int dW_E(\mathbf{R})$$

$$= -\frac{G}{d} \int \rho(\mathbf{r}) \left(1 + \frac{\mathbf{R} \cdot \mathbf{r}}{d^2} - \frac{r^2}{2d^2} + \frac{3(\mathbf{R} \cdot \mathbf{r})^2}{2d^4} + \cdots \right). \quad (5.124)$$

The first term yields GM_E/d, which is the potential in the spherically symmetric approximation; the second term, proportional to $\int \rho(\mathbf{r})r\,d^3r$ vanishes if the center of mass is taken as the origin, as is usually done. It is the remaining two second-degree terms that are really of interest. Their contribution to the potential may be written as

$$W_E^{(2)}(\mathbf{R}) = \frac{G}{2d^5} X_i X_j \int \rho(\mathbf{r})[r^2 \delta_{ij} - 3x_i x_j]d^3r, \quad (5.125)$$

where we have used the summation convention: any index that appears twice within a term is to be summed over. The x_i and X_i are the components of $\mathbf{r}$ and $\mathbf{R}$, respectively, in some terrestrial reference frame.

The integral in the potential (5.125) may be simply expressed in terms of the elements I_{ij} of the inertia tensor, which are defined as in Section 2.9:

$$I_{ij} = I_{ji} = \int \rho(\mathbf{r})(r^2 \delta_{ij} - x_i x_j)d^3r. \quad (5.126)$$

A corollary of this definition is that

$$I_{11} + I_{22} + I_{33} = 2 \int \rho(\mathbf{r})r^2 d^3r. \quad (5.127)$$

On writing $(r^2\delta_{ij} - 3x_i x_j)$ in the integrand of (5.125) as $[3(r^2\delta_{ij} - x_i x_j) - 2r^2\delta_{ij}]$ and using Eqs. (5.126) and (5.127), we see that the integral reduces to $3I_{ij} - I_{kk}\delta_{ij}$. Therefore the second-degree part (5.125) of the Earth's gravitational potential is given by

$$W_E^{(2)}(\mathbf{R}) = \frac{G}{2d^5}(3I_{ij} - I_{kk}\delta_{ij})X_i X_j. \quad (5.128)$$

This is MacCullagh's theorem.

The inertia tensor elements are determined purely by the density distribution and are independent of the rigidity or composition of the Earth. The theorem applies equally well for any rigid or non-rigid Earth model. Therefore if the changes in the inertia tensor elements I_{ij} that are produced by a change $\delta\rho(\mathbf{r})$ in the density function are denoted by c_{ij}, then there is an associated change in the external gravitational potential of the Earth, given by

$$dW_E^{(2)}(\mathbf{r}) = \frac{G}{2d^5}(3c_{ij} - c_{kk}\delta_{ij})X_iX_j. \tag{5.129}$$

5.3.3 Interaction of degree two potential with external body: torque produced on the Earth

The above expression for the potential (which is truncated at degree two) may be simplified by choosing the principal axes of inertia as the coordinate axes, thereby making the tensor diagonal. Then $I_{ij} \to 0$ for $i \neq j$, and the diagonal elements become the principal moments of inertia:

$$I_{11} \to A, \qquad I_{22} \to B, \qquad I_{33} \to C. \tag{5.130}$$

The reduced expression for potential is then

$$W_E^{(2)}(\mathbf{R}) = \frac{G}{2d^5}[(2A - B - C)X^2 + (2B - C - A)Y^2 + (2C - A - B)Z^2]. \tag{5.131}$$

Suppose now that $\mathbf{R}$ is the position of a celestial body of mass M_B. Then the potential $W_E^{(2)}$ exerts a force $-M_B\nabla_R W_E^{(2)}$ where the gradient is to be taken with respect to $\mathbf{R}$. The corresponding torque on the body is $\mathbf{\Gamma}^{E\to B} = -M_B\mathbf{R} \wedge \nabla_R W_E^{(2)}$. The body exerts an opposite torque, $\mathbf{\Gamma} = -\mathbf{\Gamma}^{E\to B}$, on the Earth:

$$\mathbf{\Gamma} = M_B\mathbf{R} \wedge \nabla_R W_E^{(2)}. \tag{5.132}$$

The components of this torque are easily evaluated. We obtain

$$\mathbf{\Gamma} = \frac{3GM_B}{d^5}((C - B)YZ, (A - C)ZX, (B - A)XY). \tag{5.133}$$

Here X, Y, Z are the components, relative to the TRF, of the position vector $\mathbf{R}$ of the celestial body. They are evidently functions of time; and so the above expression gives the torque components in the time domain.

The Earth's ellipticity and its deviation from axial symmetry are characterized by the parameters e and e', respectively, where

$$e = \frac{C - \bar{A}}{\bar{A}}, \qquad e' = \frac{B - A}{2\bar{A}}, \qquad \bar{A} = \frac{A + B}{2}. \tag{5.134}$$

It is known that the triaxiality parameter e' is very small – smaller than the ellipticity e by a factor of the order of 10^{-3}. So the equatorial components of the torque, Γ_1 and Γ_2, which are responsible for the wobbles and nutations, are much larger than the axial component Γ_3. Let us consider the familiar complex combination $\tilde{\Gamma} = \Gamma_1 + i\Gamma_2$. We may write it as

$$\tilde{\Gamma} = -i\frac{3GM_B}{d^5}\left((C - \bar{A})(X + iY)Z + \frac{B - A}{2}(X - iY)Z\right)$$

$$= -i\frac{3GM_B\bar{A}}{d^5}[e(X + iY)Z + e'(X - iY)Z]. \tag{5.135}$$

5.3.4 Spherical harmonics

For the derivation of the torque generated by the full potential acting on the Earth, we need to expand the potential in terms of spherical harmonics. This section is devoted to an introduction to surface spherical harmonics and their basic properties, to the extent that we need them for our purposes.

We begin by introducing the Legendre polynomials, which are basic to the definition of spherical harmonics, through the expansion of $1/|\mathbf{R} - \mathbf{r}|$ in powers of (r/d) where $d \equiv |\mathbf{R}|$ is the geocentric distance of the external point:

$$\frac{1}{|\mathbf{R} - \mathbf{r}|} = (d^2 + r^2 - 2rd\mu)^{-1/2} = \frac{1}{d}\sum_{n=0}^{\infty}\left(\frac{r}{d}\right)^n P_n(\mu), \qquad (r \le d), \tag{5.136}$$

where

$$\mu = \hat{\mathbf{R}} \cdot \hat{\mathbf{r}} = \cos w, \tag{5.137}$$

w being the angle between $\hat{\mathbf{r}}$ and $\hat{\mathbf{R}}$.

The second of the equalities in Eq. (5.136) serves to define the $P_n(\mu)$, ($n = 0, 1, 2\ldots$), from their *generating function*,

$$g(\mu; s) \equiv (1 - 2s\mu + s^2)^{-1/2} = \sum_{n=0}^{\infty} P_n(\mu)s^n, \tag{5.138}$$

where $s = r/d$. The coefficient of s^n in the expansion of the $g(\mu; s)$ is the *Legendre polynomial* $P_n(\mu)$ which is a polynomial of degree n in μ. One can see from the first two terms in the expansion, which are trivially obtained, that $P_0(\mu) = 1$ and $P_1(\mu) = \mu$. Terms of higher degree may then be calculated by employing the recurrence relation

$$(n + 1)P_{n+1} = (2n + 1)\mu P_n - nP_{n-1}, \tag{5.139}$$

Table 5.1 *Legendre polynomials, associated Legendre functions $P_n^m(c)$.*

$n =$	0	1	2	3	4
$m = 0$	1	c	$(3c^2 - 1)/2$	$(5c^3 - 3c)/2$	$(35c^4 - 30c^2 + 3)/8$
$m = 1$	$\cdots$	s	$3cs$	$(15c^2 - 3)s/2$	$(35c^3 - 15c)s/2$
$m = 2$	$\cdots$	$\cdots$	$3s^2$	$15cs^2$	$(105c^2 - 15)s^2/2$
$m = 3$	$\cdots$	$\cdots$	$\cdots$	$15s^3$	$105cs^3$
$m = 4$	$\cdots$	$\cdots$	$\cdots$	$\cdots$	$105s^4$

c stands for $\cos w$ and s for $\sin w$ in the table. The expressions for $P_n^m(s)$ are obtained by simple interchange of c and s throughout.

which is obtainable by equating the derivative of $g(\mu; s)$ with respect to s with that of the expansion and simplifying. This provides

$$P_n(\mu) = \frac{(-1)^n}{2^n n!} \frac{d^n (1 - \mu^2)^n}{d\mu^n}.$$

(5.140)

We need also the *associated Legendre functions*, which are defined in terms of P_n by the relation

$$P_n^m(\mu) = (1 - \mu^2)^{m/2} \frac{d^m P_n(\mu)}{d\mu^m}, \qquad (m = 0, 1 \ldots n).$$

(5.141)

(Note that sometimes in the literature there is a definition for the associated Legendre functions which is different by $(-1)^m$, denoted $P_n^m(\mu)$ as well. And the definition used above is often also used for $P_{nm}(\mu)$.) Using the above definition (Eq. (5.141)) and the definition of the Legendre polynomials (Eq. (5.140)), the associated Legendre functions can be expressed by:

$$P_n^m(\mu) = \frac{(-1)^n}{2^n n!} (1 - \mu^2)^{m/2} \frac{d^{n+m}(1 - \mu^2)^n}{d\mu^{n+m}}, \qquad (m = 0, 1 \ldots n).$$

(5.142)

By this definition, $P_n = P_n^0$. The P_n^m for n up to 4 are listed in Table 5.1 for ready reference. The abbreviations

$$c = \cos w = \mu, \qquad \text{and} \qquad s = \sin w = (1 - \mu^2)^{1/2},$$

(5.143)

have been used in the table.

The next step is to introduce the classical spherical harmonic expansion of $P_n(\mu)$. Each term in the expansion is a product of two spherical harmonics, one of which has the colatitude and the longitude of the Earth point $\mathbf{r}$, namely (θ, λ), as its arguments, while the arguments of the other are the colatitude and longitude variables (β, Λ) representing the direction of $\mathbf{R}$:

$$P_n(\mu) = \frac{4\pi}{2n+1} \sum_{m=-n}^{n} Y_n^{m*}(\beta, \Lambda) Y_n^m(\theta, \lambda), \tag{5.144}$$

where the Y_n^m, $(-n \leq m \leq n)$, are the surface spherical harmonics defined by

$$Y_n^m(\theta, \lambda) = N_{nm} P_n^m(\cos \theta) e^{im\lambda},$$

$$N_{nm} = (-1)^m \left(\frac{2n+1}{4\pi} \right)^{1/2} \left(\frac{(n-m)!}{(n+m)!} \right)^{1/2}, \tag{5.145}$$

for $m \geq 0$, and through

$$Y_n^m = (-1)^m Y_n^{-m*} \tag{5.146}$$

for negative m. The full set of spherical harmonics Y_n^m (also called *surface spherical harmonics*), $(n = 0, 1, 2, \ldots,$ with $-n \leq m \leq n)$ has the orthonormality property

$$\int Y_n^{m*}(\theta, \lambda) Y_q^p(\theta, \lambda) \sin \theta \, d\theta \, d\lambda = \delta_{n,q} \delta_{m,p}. \tag{5.147}$$

The equivalent orthonormality property for the associated Legendre functions is

$$\int P_n^m(\cos \theta) P_q^m(\cos \theta) \sin \theta \, d\theta = \frac{\delta_{nq}}{2\pi N_{nm}^2}. \tag{5.148}$$

Note that P_n^m and P_q^p are not orthogonal for $p \neq m$.

In view of (5.146), the term for a given negative value of m in (5.144) is the complex conjugate of the term for the corresponding positive m. Therefore (5.144) may be written also as

$$P_n(\mu) = \frac{4\pi}{2n+1} \left\{ Y_n^0(\beta, \Lambda) Y_n^0(\theta, \lambda) + \sum_{m=1}^{n} [Y_n^{m*}(\beta, \Lambda) Y_n^m(\theta, \lambda) + c.c.] \right\},$$

$$\tag{5.149}$$

where *c.c.* stands for "complex conjugate." This expression may be reduced to a form which is fully in terms of real quantities by using (5.145):

$$P_n(\mu) = P_n(\cos \beta) P_n(\cos \theta) + 2 \sum_{m=1}^{n} \frac{(n-m)!}{(n+m)!}$$

$$\times [P_n^m(\cos \beta) P_n^m(\cos \theta) (\cos m\Lambda \cos m\lambda + \sin m\Lambda \sin m\lambda)]. \tag{5.150}$$

$P_n^m(\cos \theta) \cos m\lambda$ and $P_n^m(\cos \theta) \sin m\lambda$ are real versions of the complex spherical harmonic functions introduced above.

5.3.5 Spherical harmonic expansion of the Earth's gravitational
potential; geopotential coefficients

We are now in a position to evaluate the gravitational potential $W_E(\mathbf{R})$ of the Earth itself at a point $\mathbf{R}$ outside the Earth by integrating the expression (5.123) over the volume of the Earth after substituting for $1/|\mathbf{R} - \mathbf{r}|$ from (5.136) taken together with (5.150):

$$
W_E(\mathbf{R}) = -G \int \rho(\mathbf{r}) \sum_{n=0}^{\infty} \frac{r^n}{d^{n+1}} \left\{ P_n(\cos\beta) P_n(\cos\theta) + 2 \sum_{m=1}^{n} \frac{(n-m)!}{(n+m)!} \right.
$$

$$
\left. \times \left[P_n^m(\cos\beta) P_n^m(\cos\theta) (\cos m\Lambda \, \cos m\lambda + \sin m\Lambda \, \sin m\lambda) \right] \right\} d^3r.
$$

$$(5.151)$$

The generic term in the double sum above is said to be of *spherical harmonic degree n and order m*. Integrals of the individual terms in the sum define the geopotential coefficients $C_{n,m}$ and $S_{n,m}$ to within a factor $M_E a_e^n$ which serves to make these coefficients dimensionless:

$$
M_E a_e^n C_{n,m} = (2 - \delta_{m,0}) \frac{(n-m)!}{(n+m)!} \int \rho(\mathbf{r}) r^n P_n^m(\cos\theta) \cos m\lambda \, d^3r,
$$

$$(5.152)$$

$$
M_E a_e^n S_{n,m} = (2 - \delta_{m,0}) \frac{(n-m)!}{(n+m)!} \int \rho(\mathbf{r}) r^n P_n^m(\cos\theta) \sin m\lambda \, d^3r,
$$

where M_E is the Earth's mass, and the origin of the coordinate system is taken to be coincident with the geocenter (center of mass of the Earth).

A word of caution about terminology: the term *geopotential* refers to the sum of the Earth's gravitational potential and the centrifugal potential associated with its mean diurnal rotation. It must be kept in mind that the geopotential coefficients defined above relate, despite their name, to the gravitational potential alone.

The definitions (5.152) of the $C_{n,m}$ and $S_{n,m}$ hold for both rigid and non-rigid Earth models. It may be readily seen from these definitions that $C_{0,0} = 1$ and that the $S_{n,0}$ vanish for all n. Furthermore, in the case $n = 1, r P_1^1(\cos\theta) \cos\lambda = r \sin\theta \cos\lambda = x, r P_1^1(\cos\theta) \sin\lambda = r \sin\theta \cos\lambda = y$, and $r P_1(\cos\theta) = r \cos\theta = z$. Therefore the integrals involving the $n = 1$ quantities in (5.152) are just M_E times the coordinates of the Earth's center of mass. Consequently, $C_{1,m} = S_{1,m} = 0$ for both values (0 and 1) of m if the origin of the coordinate system is at the center of mass. In practise, terrestrial coordinate frames that we use are attached to the Earth's crust, and small displacements of the center of mass relative to the origin of the coordinate system (called "geocenter motions") can take place, for example as a result of motions of the surface fluid regions (atmosphere/oceans) as a whole relative to the solid Earth; $C_{1,m}$ and $S_{1,m}$ may thus have very small time dependent values. In general, the small changes in the Earth's gravitational potential that are caused by tidal or other perturbations may be parameterized by increments to the geopotential coefficients. Redistributions of the ocean mass, in particular, are routinely represented by the associated incremental geopotential coefficients determined from ground based observations of orbits of artificial Earth satellites like LAGEOS, GeoSat, Stella, Starlette, . . . or more recently GRACE (Gravity Recovery And Climate Experiment), CHAMP (CHAllenging Minisatellite Payload) or GOCE (Gravity field and steady-state Ocean Circulation Explorer) (see Section 1.6) and Swarm (constellation of three satellites to study mainly the Earth's magnetic field).

In the following, we shall take $C_{n,m}$ and $S_{n,m}$ to refer to the unperturbed Earth unless otherwise stated; and it will be assumed that the origin of the reference frame is at the geocenter.

Returning to the expression (5.151) for the external gravitational potential of the Earth, we rewrite it with the help of Eqs. (5.152) as

$$W_E(\mathbf{R}) = -\frac{GM_E}{d} \sum_{n=0}^{\infty} \left(\frac{a_e}{d}\right)^n$$

$$\times \left[C_{n,0} P_n(\cos\beta) + \sum_{m=1}^{n} P_n^m(\cos\beta)(C_{n,m} \cos m\Lambda + S_{n,m} \sin m\Lambda) \right].$$

$$(5.153)$$

The expression (5.153) is the starting point of the treatment of rigid Earth nutations by Bretagnon *et al.* (1997).

The contributions of the individual spherical harmonic parts of the density distribution within the Earth to the gravitational potential outside the Earth are thus determined by the values of the respective geopotential coefficients. The contributions from the zonal ($m = 0$) terms, being independent of Λ, are axially symmetric. Conversely, if the Earth were axially symmetric (i.e., if $\rho(\mathbf{r})$ did not depend on λ), all non-zonal geopotential coefficients as well as all the $S_{n,0}$ would vanish, as is readily seen from the definitions (5.152). The potential of an axially symmetric Earth would thus reduce to

$$W_E(\mathbf{R}) = -\frac{GM_E}{d} \left\{ 1 + \sum_{n=2}^{\infty} \left(\frac{a_e}{d}\right)^n C_{n,0} P_n(\cos\beta) \right\}. \tag{5.154}$$

The notation J_n is often used for $(-C_{n,0})$:

$$J_n = -C_{n,0} = -\frac{1}{M_E a_e^n} \int \rho(\mathbf{r}) r^n P_n(\cos\theta) \, d^3r. \tag{5.155}$$

A measure of the equatorial bulge of the unperturbed Earth is provided by J_2, since the integral in the above equation reduces to $\frac{1}{2}(A + B) - C$ when $n = 2$. To verify this, one rewrites $r^2 P_2(\cos\theta)$ as $\frac{1}{2}(z^2 + x^2) + \frac{1}{2}(z^2 + y^2) - (x^2 + y^2)$; it becomes evident then that the value of the integral is indeed as just stated. Thus

$$J_2 = (C - \bar{A})/(M_E a_e^2). \tag{5.156}$$

The orbital motions of Earth satellites are determined by the Earth's gravitational potential, except for small perturbations by other forces (e.g., solar radiation pressure). Ground based observations of the motions of satellites such as LAGEOS, GeoSat, Stella, Starlette or others by satellite laser ranging (SLR) provide the means for precise estimation of the geopotential coefficients. Other very recent missions such as GOCE, GRACE, CHAMP defined in the first chapter also help. The estimated values are time dependent because of the contributions from the Earth's deformations due to a variety of causes – ocean tides, solid Earth tides, atmospheric pressure variations, etc. In fact, SLR observations and other missions like GOCE provide an important source of quantitative information on ocean tides as well. Geopotential coefficients for the static (unperturbed) Earth can be obtained by removal of the time dependent parts. Values thus estimated have been published in a number of tables, for the different tidal periods and for n and m up to maximum values with a precision which has been increasing over the years (see GOT00v2 (Ray, 1999), FES2004 model (Lyard *et al.*, 2006), GOT4.7 model (the 2004 version of the Goddard ocean tide model produced by Ray in 2008) and EOT08a model (Savcenko and Bosch, 2008), see also Han *et al.*, 2010, Ray and Egbert, 2012, 2013).

The values of a few geopotential coefficients of low degrees and orders, taken from JGM3 (Joint Gravity Model 3) and transformed to the normalization used here, are shown in Table 5.2. The tabulated quantities in the original tables are the "barred" geopotential coefficients $\bar{C}_{nm}$, $\bar{S}_{nm}$, which are related to the unbarred coefficients as follows:

$$\begin{pmatrix} C_{nm} \\ S_{nm} \end{pmatrix} = K_{nm} \begin{pmatrix} \bar{C}_{nm} \\ \bar{S}_{nm} \end{pmatrix}, \tag{5.157}$$

$$K_{nm} = \left(\frac{(2 - \delta_{m0})(2n + 1)(n - m)!}{(n + m)!} \right)^{1/2}. \tag{5.158}$$

Note that $K_{nm} = [4\pi(2 - \delta_{m0})]^{1/2}|N_{nm}|$, where N_{nm} is the normalization factor defined in (5.145). The barred coefficients appear when the P_{nm} in the expansions (5.153) and (5.168) of the potential are replaced by $\bar{P}_{nm} = K_{nm} P_{nm}$. The real functions $\bar{P}_{nm}(\cos\theta)\cos m\lambda$ and $\bar{P}_{nm}(\cos\theta)\sin m\lambda$ are usually referred to in the geophysics literature on potential theory as the *fully normalized spherical*

Table 5.2 *Values of JGM3 geopotential coefficients (units: 10^{-6}).*

(n, m)	$C_{n,m}$	$S_{n,m}$	$\bar{C}_{n,m}$	$\bar{S}_{n,m}$
(2, 0)	−1082.636023	0	−484.169548	0
(2, 2)	1.574 536	−0.903 868	2.439261	−1.400266
(3, 0)	2.532435	0	0.957171	0
(3, 1)	2.192 799	0.268 012	2.030137	0.248131
(3, 2)	0.309 016	−0.211 402	0.904706	−0.618923
(3, 3)	0.100 559	0.197 201	0.721145	1.414204
(4, 0)	1.619331	0	0.539777	0
(4, 1)	−0.508 725	−0.449 460	−0.536244	−0.473772
(4, 2)	0.350 670	0.662 571	0.350670	0.662571

harmonics. Their normalization is according to

$$\int \int \bar{P}_{nm}(\cos\theta)\bar{P}_{qm}(\cos\theta) \cos^2 m\lambda \, \sin\theta \, d\theta \, d\lambda = 4\pi\delta_{nq}, \tag{5.159}$$

$$\int \int \bar{P}_{nm}(\cos\theta)\bar{P}_{qm}(\cos\theta) \sin^2 m\lambda \, \sin\theta \, d\theta \, d\lambda = 4\pi\delta_{nq}(1 - \delta_{m0}). \tag{5.160}$$

5.3.6 C_{2m} *and* S_{2m}: *relation to elements of the inertia tensor*

It is useful to note that the coefficients of degree two are directly related to the elements of the inertia tensor. The relations may be established by setting $n = 2$ and $m = 0, 1, 2$ in succession in Eqs. (5.152) and rewriting the integrands in terms of Cartesian coordinates:

$$r^2 P_2(\cos\theta) = \frac{1}{2}(2z^2 - x^2 - y^2),$$

$$r^2 P_2^1(\cos\theta) \begin{pmatrix} \cos\lambda \\ \sin\lambda \end{pmatrix} = 3 \begin{pmatrix} xz \\ yz \end{pmatrix}, \tag{5.161}$$

$$r^2 P_2^2(\cos\theta) \begin{pmatrix} \cos 2\lambda \\ \sin 2\lambda \end{pmatrix} = 3 \begin{pmatrix} x^2 - y^2 \\ 2xy \end{pmatrix}.$$

The resulting integrals may then be identified with the various inertia tensor elements I_{ij} defined by Eq. (5.126). We find thus that

$$M_E a_e^2 C_{20} = -\left(I_{33} - \frac{1}{2}(I_{11} + I_{22})\right),$$

$$M_E a_e^2 C_{21} = -I_{13}, \qquad M_E a_e^2 S_{21} = -I_{23}, \tag{5.162}$$

$$M_E a_e^2 C_{22} = \frac{1}{4}(I_{22} - I_{11}), \qquad M_E a_e^2 S_{22} = -\frac{1}{2}I_{12}.$$

It is of much interest to relate the I_{ij} and the geopotential coefficients of the unperturbed Earth to the principal moments of inertia A, B, C. To do so, we have to transform the integral expressions for them, given in (2.4) and (2.5), to the principal axis frame. The commonly used TRF does not have the equatorial coordinate axes along the principal axes though the polar axes of the two frames coincide. The longitude of the direction of the axis of the first principal axis (the axis of the smaller equatorial moment of inertia A) is

$$\lambda_0 \approx -14°.9 \tag{5.163}$$

in the usual terrestrial frame (ITRF). So the transformation between the two frames is

$$x = x' \cos \lambda_0 - y' \sin \lambda_0, \quad y = x' \sin \lambda_0 + y' \cos \lambda_0, \quad z = z', \tag{5.164}$$

where the primed coordinates are in the principal axis frame. On substituting these into the definitions of the I_{ij} and noting that $I'_{11} = \int \rho(\mathbf{r})(y'^2 + z'^2)d^3r' = A$, etc., and that the integrals for the off-diagonal elements such as $I'_{12} = -\int \rho(\mathbf{r})x'y'd^3r'$ vanish as stated in Eqs. (2.7) and (2.8), we obtain

$$I_{11} = (1/2)[(A + B) - (B - A)\cos 2\lambda_0] = \bar{A}(1 - e' \cos 2\lambda_0),$$

$$I_{22} = (1/2)[(A + B) + (B - A)\cos 2\lambda_0] = \bar{A}(1 + e' \cos 2\lambda_0),$$

$$I_{12} = -(1/2)(B - A)\sin 2\lambda_0 = -\bar{A}e' \sin 2\lambda_0, \tag{5.165}$$

$$I_{12} = I_{23} = 0,$$

$$I_{33} = C,$$

where $\bar{A}$ is the mean equatorial moment of inertia and e' is the triaxiality parameter as defined in Eq. (5.134). On introducing these results into Eqs. (5.162), we get the following expressions for C_{2m} and S_{2m} in terms of the principal moments of inertia:

$$M_E a_e^2 C_{20} = -(C - \bar{A}) = -\bar{A}e,$$

$$M_E a_e^2 C_{2,2} = (1/4)(B - A)\cos 2\lambda_0 = (1/2)\bar{A}e' \cos 2\lambda_0, \tag{5.166}$$

$$M_E a_e^2 S_{2,2} = (1/4)(B - A)\sin 2\lambda_0 = (1/2)\bar{A}e' \sin 2\lambda_0,$$

and $C_{2,1} = S_{21} = 0$. Actually, small non-vanishing values are found for $C_{2,1}$ and S_{21} in tables of geopotential coefficients. This means that the intended alignment of the third axis of the ITRF (along which $\theta = 0$) with the Earth's axis of maximum moment of inertia is not perfect. We shall, however, ignore this very slight misalignment.

The general set of relations (5.162) is needed when perturbations to the inertia tensor caused, for instance, by the time dependent deformations produced by ocean tides have to be taken into account (see Section 6.8.5).

5.3.7 Torque due to an external body, in terms of geopotential coefficients

The torque $\boldsymbol{\Gamma}$ exerted by an external body B on the Earth is the negative of the torque $\boldsymbol{\Gamma}^{E\to B}$ exerted by the Earth on that body. The latter can be readily calculated, given the potential $W_E(\mathbf{R})$ of the Earth at the (time dependent) position $\mathbf{R} = (d, \beta, \Lambda)$ of the body. Thus

$$\boldsymbol{\Gamma} = -\boldsymbol{\Gamma}^{E\to B} = M_B\,(\mathbf{R} \wedge \nabla_\mathbf{R})\,W_E(\mathbf{R}), \tag{5.167}$$

where $\nabla_\mathbf{R}$ stands for the gradient with respect to $\mathbf{R}$.

In evaluating this torque, it is of some advantage to use, instead of the expression (5.153) for $W_E(\mathbf{R})$, an alternative version obtained simply by expressing $\cos m\Lambda$ and $\sin m\Lambda$ in Eq. (5.153) in terms of the complex quantities $e^{\pm im\Lambda}$:

$$W_E(\mathbf{R}) = -\frac{GM_E}{d} \sum_{n=0}^{\infty} \left(\frac{a_e}{d}\right)^n \Bigg\{ C_{n,0} P_n(\cos\beta)$$

$$+ \frac{1}{2} \sum_{m=1}^{n} P_n^m(\cos\beta)[(C_{n,m} - i\,S_{n,m})e^{im\Lambda} + (C_{n,m} + i\,S_{n,m})e^{-im\Lambda}] \Bigg\}. \tag{5.168}$$

The above expression, being a sum of mutually complex conjugate parts, is clearly real, as it should be.

To evaluate the expression (5.167) for the torque, we can use the fact that

$$(\mathbf{R} \wedge \nabla_\mathbf{R})_\pm\, Y_n^m(\beta, \Lambda) = ic_n^{\pm m}\, Y_n^{m\pm 1}(\beta, \Lambda), \tag{5.169}$$

$$(\mathbf{R} \wedge \nabla_\mathbf{R})_z\, Y_n^m(\beta, \Lambda) = im\, Y_n^m(\beta, \Lambda), \tag{5.170}$$

where

$$(\mathbf{R} \wedge \nabla_\mathbf{R})_\pm = (\mathbf{R} \wedge \nabla_\mathbf{R})_x \pm i(\mathbf{R} \wedge \nabla_\mathbf{R})_y \tag{5.171}$$

and

$$c_n^{\pm m} = [(n \mp m)(n \pm m + 1)]^{1/2}. \tag{5.172}$$

See, for instance, Hill (1954) or Abramowitz and Stegun (1964) or Mathews and Venkatesan (1976; second edition 2010) or Carrascal *et al.* (1991). Now, it turns out, by virtue of the definition of N_{nm} in Eq. (5.145), that

$$c_n^m N_{n,m+1} = -N_{nm}, \qquad c_n^{-m} N_{n,m-1} = -(n+m)(n-m+1)N_{nm}. \tag{5.173}$$

We can use the first of these relations to rewrite (5.169) with the upper sign as

$$(\mathbf{R} \wedge \nabla_{\mathbf{R}})_+ P_n^m(\cos\beta)e^{im\Lambda} = -i P_n^{m+1}(\cos\beta)e^{i(m+1)\Lambda}. \tag{5.174}$$

Again, if we take (5.169) with the lower sign and take its complex conjugate, we get

$$(\mathbf{R} \wedge \nabla_{\mathbf{R}})_+ Y_n^{m*}(\beta, \Lambda) = -i c_n^{-m} Y_n^{m-1*}(\beta, \Lambda). \tag{5.175}$$

We can reduce this, by the application of the second of the relations (5.173), to

$$(\mathbf{R} \wedge \nabla_{\mathbf{R}})_+ P_n^m(\cos\beta)e^{-im\Lambda} = i(n+m)(n-m+1)P_n^{m-1}(\cos\beta)e^{i(m-1)\Lambda}. \tag{5.176}$$

We are now in a position to evaluate the complex combination $\tilde{\Gamma} = \Gamma_x + i\Gamma_y$ of the two equatorial components of the torque. On substituting the expression (5.168) for W_E into (5.167) and making use of the results (5.174) and (5.176), we obtain

$$\tilde{\Gamma} = i \frac{GM_E M_B}{d} \sum_{n=2}^{\infty} \left(\frac{a_e}{d}\right)^n \left\{ C_{n0} P_n^1(\cos\beta)e^{i\Lambda} \right.$$

$$+ \frac{1}{2}\sum_{m=1}^{n-1}\left[(C_{n,m} - i S_{n,m})P_n^{m+1}(\cos\beta)e^{i(m+1)\Lambda} \right.$$

$$\left. - \sum_{m=1}^{n}(n+m)(n-m+1)(C_{n,m} + i S_{n,m})P_n^{m-1}(\cos\beta)e^{-i(m-1)\Lambda}\right]\right\}. \tag{5.177}$$

The first line of the above expression constitutes the contribution to the torque from the axially symmetric part of the Earth's structure which is represented by the C_{n0}.

One may re-express (5.177) in the following equivalent form by changing the summation over m to a sum over $m+1$ in the second line and to a sum over $m-1$ in the third line:

$$\tilde{\Gamma} = i \frac{GM_E M_B}{d} \sum_{n=2}^{\infty} \left(\frac{a_e}{d}\right)^n \left\{ \sum_{m=1}^{n}\frac{1}{2 - \delta_{m1}}(C_{n,m-1} - i S_{n,m-1})P_n^m(\cos\beta)e^{im\Lambda} \right.$$

$$\left. - \sum_{m=0}^{n-1}\frac{1}{2}(n+m+1)(n-m)(C_{n,m+1} + i S_{n,m+1})P_n^m(\cos\beta)e^{-im\Lambda}\right\}. \tag{5.178}$$

The real-valued equatorial components of the torque may now be obtained from either of the above expressions as

$$\Gamma_x = (1/2)(\tilde{\Gamma} + \tilde{\Gamma}^*), \qquad \Gamma_y = (1/2i)(\tilde{\Gamma} - \tilde{\Gamma}^*). \tag{5.179}$$

We have thus the equatorial components of the torque exerted on the Earth by a celestial body as functions of the time dependent position coordinates of the body.

To obtain the third component of the torque (which generates variations in the rate of axial rotation), we use Eq. (5.170), rewriting it as

$$(\mathbf{R} \wedge \nabla_{\mathbf{R}})_z \, P_n^m(\cos \beta) e^{im\Lambda} = im \, P_n^m(\cos \beta) e^{im\Lambda},$$

$$(\mathbf{R} \wedge \nabla_{\mathbf{R}})_z \, P_n^m(\cos \beta) e^{-im\Lambda} = -im \, P_n^m(\cos \beta) e^{-im\Lambda}. \tag{5.180}$$

On introducing (5.168) into (5.167) and using the above relations we obtain after a simple rearrangement of terms,

$$\Gamma_z = M_B \, (\mathbf{R} \wedge \nabla_{\mathbf{R}})_z \, W_E(\mathbf{R})$$

$$= \frac{GM_E M_B}{d} \left\{ \sum_{n=2}^{\infty} \left(\frac{a_e}{d} \right)^n \times \sum_{m=1}^{n} m \, P_n^m(\cos \beta) \, (C_{n,m} \sin m\Lambda - S_{n,m} \cos m\Lambda) \right\}. \tag{5.181}$$

Note that the C_{n0} do not figure in this expression, meaning that the axially symmetric part of the Earth's structure does not contribute to the axial component of the torque. This is to be expected for physical reasons.

The rotational motion of a rigid body is governed by Euler's equations of motion which have been presented in Section 2.22.2. The expressions obtained above for the torque components may be introduced into these equations and solutions obtained for the equatorial and axial components of the angular velocity vector (or equivalently, the wobbles and spin rate variations) after making appropriate approximations. Explicit consideration of the solution of the equations will be done after the spectral representations of the torque components are obtained.

5.4 Torque on an axially symmetric Earth

We take the celestial frame, for definiteness, to be the GCRF defined in Chapter 2. Its principal plane coincides with the mean equatorial plane of J2000. The angular position of $\mathbf{R}$ in this frame is described by the right ascension α and the declination δ, which are the counterparts, in the GCRF, of the longitude and latitude, respectively, in the terrestrial frame. (The right ascension is measured from the mean equinox of J2000 along the equator of the GCRF.) Thus the components d_X, d_Y, d_Z of $\mathbf{R}$ in the celestial frame are

$$d_X = d \cos \delta \cos \alpha, \quad d_Y = d \cos \delta \sin \alpha, \quad d_Z = d \sin \delta. \tag{5.182}$$

As we have already noted in Chapter 2, the density function of the Earth in the celestial frame is, in general, time dependent because of the variations in orientation of the Earth in space as a result of Earth rotation variations. For example, nutation causes the position of the figure axis (and hence that of the equatorial bulge) in space to keep changing. In order to keep such complications out of the picture, we shall, for the purpose of computation of the torque, assume that its rotation is about the figure axis and that this axis remains fixed in the direction of the pole of the GCRF. (In other words, we neglect the reaction of the wobble and nutation of the Earth on the external torque which produces these motions.) We also take the rotation of the Earth to be with the uniform angular velocity Ω_0, thus neglecting the spin rate variations. With these approximations, one sees that

$$\delta = \pi/2 - \beta, \qquad \alpha = \Lambda + \text{GMST}, \tag{5.183}$$

where Λ is the longitude coordinate of the vector $\mathbf{R}$ in the TRF as before, and GMST is the Greenwich Mean Sidereal Time, defined to be the time measured by the angle traversed along the mean equator of date by the Greenwich meridian at the mean sidereal rotation rate Ω_0:

$$\text{GMST} = \text{GMST}_0 + \Omega_0 t + \cdots, \qquad \text{GMST}_0 = 280°.4606\ldots, \tag{5.184}$$

where t is measured from J2000. The first of the equalities (5.183) follows from the coincidence of the principal planes of the terrestrial and celestial reference frames that results on neglecting the nutation–precession motion of the first of these planes in space. The second equality follows from the fact that α is the arc along the equator from the equinox to the meridian in which the Sun/Moon is located, which is the sum of GMST and the arc λ from the Greenwich meridian to the meridian of the celestial object.

In the following we make the further approximation of neglecting also the deviations of the Earth's structure from axial symmetry; then the density at an Earth point $\mathbf{r}$ with spherical polar coordinates (r, δ', α') in the celestial reference frame becomes independent of time and of α'; the density function has the same axially symmetric form as in the terrestrial frame despite the Earth's diurnal rotation. It is self-evident then that the Earth's potential in the celestial frame will have the same form as (5.154), though we now use the celestial coordinate δ instead of $(\pi/2 - \beta)$:

$$W_E(\mathbf{R}) = -\frac{GM_E}{d}\left(1 - \sum_{n=2}^{\infty}\left(\frac{a_e}{d}\right)^n J_n P_n(\sin\delta)\right), \tag{5.185}$$

with J_n as defined in (5.155). Since the gravitational torque $\mathbf{L}$ exerted by a celestial body B located at $\mathbf{R}$ on the Earth is the negative of the torque due to the Earth on B, we have $\mathbf{L} = M_B \mathbf{R} \wedge \nabla_{\mathbf{R}} W_E(\mathbf{R})$. The expression for this vector may be written down very simply by choosing a vector basis suited to the polar coordinates in terms of which W_E is written. Such a basis is provided by the triad of unit vectors $\hat{\mathbf{R}}, \mathbf{e}_\alpha, \mathbf{e}_\delta$ in the directions of $\mathbf{R}$, increasing δ, and increasing α, respectively; their Cartesian components in the GCRF are given by

$$\hat{\mathbf{R}} = \begin{pmatrix} \cos\delta\cos\alpha \\ \cos\delta\sin\alpha \\ \sin\delta \end{pmatrix}, \quad \mathbf{e}_\alpha = \begin{pmatrix} -\sin\alpha \\ \cos\alpha \\ 0 \end{pmatrix}, \quad \mathbf{e}_\delta = \begin{pmatrix} -\sin\delta\cos\alpha \\ -\sin\delta\sin\alpha \\ \cos\delta \end{pmatrix}. \quad (5.186)$$

We observe now that since $P_n(\sin\delta)$ depends only on δ,

$$\nabla_\mathbf{R} P_n(\sin\delta) = \mathbf{e}_\delta \frac{1}{d} \frac{d P_n(\sin\delta)}{d\delta} = \mathbf{e}_\delta \frac{1}{d} P_n^1(\sin\delta). \quad (5.187)$$

(We have used Eq. (5.141) in the last step.) Terms in $\nabla_\mathbf{R} W_E(\mathbf{R})$ that are proportional to $\nabla_\mathbf{R} d = \hat{\mathbf{R}}$ get eliminated when $\mathbf{R}\wedge$ is taken. Furthermore, $\hat{\mathbf{R}} \wedge \mathbf{e}_\delta = -\mathbf{e}_\alpha$. Hence

$$\mathbf{L} = M_B \mathbf{R} \wedge \nabla_\mathbf{R} W_E(\mathbf{R}) = -\frac{GM_B M_E}{d} \mathbf{e}_\alpha \sum_{n=2}^{\infty} \left(\frac{a_e}{d}\right)^n J_n P_n^1(\sin\delta). \quad (5.188)$$

We can make use of the components of $\mathbf{e}_\alpha$ from (5.186) to write down the complex combination of the first two components of the torque in the celestial frame:

$$\tilde{L} = L_1 + i L_2 = -i \frac{GM_B M_E}{d} \sum_{n=2}^{\infty} \left(\frac{a_e}{d}\right)^n J_n P_n^1(\sin\delta) e^{i\alpha}. \quad (5.189)$$

An observation of interest is that by virtue of the relations (5.183), the above expression for $\tilde{L}$ is simply $e^{i\,\text{GMST}}$ times the equatorial torque $\tilde{\Gamma}$ on an axially symmetric Earth as seen in the terrestrial frame, the latter being given by the first line of Eq. (5.177). This simple relation between the equatorial torques referred to the celestial and terrestrial frames holds even if the assumption of axial symmetry is dropped, as we shall see later, as long as the other approximations made above are adhered to.

Terms in the expression (5.189) with $n > 4$ are too small to be of interest even when B is the Moon, the celestial body that is nearest to the Earth. Explicit expressions for $P_n^1(\sin\delta)$ for degrees n up to 4 may be taken from Table 5.1, with the replacements $c \to \sin\delta, s \to \cos\delta$.

5.4.1 Precession–nutation of angular momentum axis

One may now determine the nutation and precession of the angular momentum axis by introducing the above expression for the torque into the equations of

Section 2.22.1, in particular, the pair of Eqs. (2.38) of Chapter 2:

$$X_H = \frac{1}{C\Omega_0} \int L_1 dt, \qquad Y_H = \frac{1}{C\Omega_0} \int L_2 dt. \qquad (5.190)$$

For numerical computations, one may take the position coordinates d, δ, and α of the Moon, Sun, and planets as functions of time from the appropriate ephemerides, and evaluation of the integrals in the equations cited above may then be carried out. This approach has been employed by Jeffreys and Vicente (1957a and b), Melchior and Georis (1968), Capitaine (1982), Bretagnon *et al.* (1997, 1998), Bretagnon (1998, 1999), and Roosbeek and Dehant (1998).

Once the nutation and precession of the angular momentum axis are computed, those of the figure and rotation axes follow on using the Oppolzer terms (see Section 2.28).

5.5 Tide generating potential (TGP)

In the foregoing sections, the determination of the gravitational torque on the Earth by a celestial body was based on the equality of this torque to the negative of the torque exerted by the Earth on the celestial body.

An alternative approach, which is more direct, starts with the gravitational potential $W^B(\mathbf{r}, t) = -GM_B/|\mathbf{R} - \mathbf{r}|$ produced by a celestial body with the time dependent position $\mathbf{R}$ relative to the geocenter at a generic point in the Earth with the geocentric position vector $\mathbf{r}$. This approach was employed in Chapter 2, but with truncation of the expansion of the potential at the second degree in (r/d), where $d = |\mathbf{R}| \gg r$. For a complete treatment of the effects of the potential, one resorts to a spherical harmonic expansion following the expansion in powers of (r/d). Before we consider the full expansion, it may be useful to see how the degree two part of the potential itself may be decomposed into parts of different orders.

5.5.1 Degree two tidal potential: Zonal, tesseral, and sectorial parts; tidal frequency bands

The degree two part of $W^B(\mathbf{r}, t)$ may be taken from Eq. (2.53). We write it for convenience in the form

$$W_2^B(\mathbf{r}, t) = -\frac{GM_B}{d^5} Q(\mathbf{r}, t), \qquad Q(\mathbf{r}, t) = -\frac{1}{2}d^2 r^2 + \frac{3}{2}(\mathbf{R} \cdot \mathbf{r})^2, \qquad (5.191)$$

where the time dependent position vector $\mathbf{R}$ of the external body B and its polar and Cartesian coordinates are written simply as (d, β, Λ) and (d_x, d_y, d_z) without using an identifying subscript B, to avoid cluttering the notation.

We can break up $Q(\mathbf{r}, t)$ into three parts identified as zonal, tesseral, and sectorial by subscripts Z, T, S, respectively:

$$Q = Q_Z + Q_T + Q_S, \qquad (5.192)$$

where

$$Q_Z = (1/4)\,(2d_z^2 - d_x^2 - d_y^2)(2z^2 - x^2 - y^2) = (1/4)(3d_z^2 - d^2)(3z^2 - r^2),$$

$$Q_T = 3(d_z d_x zx + d_y d_z yz), \qquad (5.193)$$

$$Q_S = (3/4)[(d_x^2 - d_y^2)(x^2 - y^2) + 4d_x d_y xy].$$

It is a matter of elementary algebra to verify that the coefficients of the $x_i x_j$ in the sum of the above terms agree with those in the expression (5.191) for all pairs i, j.

The motivation for the decomposition in this particular manner becomes clear on expressing each of the above parts in terms of the polar coordinates (r, θ, λ) of $\mathbf{r}$ and (d, β, Λ) of the position $\mathbf{R}$ of the celestial body:

$$Q_Z = (1/4)\,d^2(3\cos^2\beta - 1)\,r^2(3\cos^2\theta - 1)$$

$$= d^2 P_2(\cos\beta)\,r^2 P_2(\cos\theta), \qquad (5.194)$$

$$Q_T = 3\,(d^2\,\cos\beta\sin\beta)\,(r^2\cos\theta\sin\theta)\cos(\lambda - \Lambda)$$

$$= (1/3)\,d^2 P_2^1(\cos\beta)\,r^2 P_2^1(\cos\theta)\,\cos(\lambda - \Lambda), \qquad (5.195)$$

$$Q_S = (3/4)\,(d^2\,\sin^2\beta)\,(r^2\sin^2\theta)\cos 2(\lambda - \Lambda)$$

$$= (1/12)\,d^2 P_2^2(\cos\beta)\,r^2 P_2^2(\cos\theta)\cos 2(\lambda - \Lambda). \qquad (5.196)$$

The distinctive features of the three parts may be readily seen from these expressions.

The zonal potential Q_Z vanishes on parallels of latitude corresponding to $\cos^2\theta = (1/3)$, thus dividing the Earth's surface into three zones, justifying the name "zonal." It is independent of longitude. Since it is also independent of Λ, it is unaffected by the westward movement of the celestial body in the terrestrial frame. But it does vary on the annual timescale, if B is the Sun, because of the seasonal migration of the Sun from north of the equator to the south and back; if B is the Moon, which has an orbit that is inclined at about $27.5°$ to the equatorial plane and has a period of about 27.3 days for its motion around this orbit, this is the dominant period in the temporal variation of its potential at the Earth. The precession of the lunar orbit in space with a period of about 18.6 years generates a corresponding period too in the potential. The spectrum of the zonal potential is dominated by these frequencies, which are all very much below the diurnal scale. They constitute a low frequency band in the terrestrial frame.

As for the tesseral potential Q_T, it vanishes at the poles and on the equator, and its magnitude is a maximum at $\pm45°$ latitudes; it vanishes also along the meridians $\lambda = \Lambda \pm \pi/2$. It has equal values at diametrically opposite points, and changes sign on going from one point to another that is at the same latitude but is separated in longitude by $180°$. The factor $\cos(\lambda - \Lambda)$ gives it the character of a moving wave: its peak, which is evidently at $\lambda = \Lambda$, moves westwards accompanying the apparent westward movement of the body B in the sky as a consequence of the Earth's (eastward) diurnal rotation. The mean rate of change of Λ is negative and close to -1 cpsd for the Sun, Moon, and other solar system bodies.

Thus the tesseral potential consists of retrograde waves. The -1 cpsd frequency coming from the factor discussed in the last paragraph is modulated by the frequencies present in the time dependence of the factor $\sin 2\beta$, which has a predominantly semiannual character in the case of the Sun, and fortnightly in the case of the Moon. The net result is that the tesseral potential is made up of retrograde waves with nearly diurnal frequencies.

The sectorial potential vanishes at the poles and at longitudes λ differing from Λ by $\pm\pi/4$ or $\pm\pi/2$. It is largest along the equator. Considerations similar to those of the last paragraph show that Q_S too consists of retrograde waves, but since it involves $\cos 2(\lambda - \Lambda)$, its frequency spectrum is *semidiurnal*.

Thus the decomposition of the second degree potential serves to bring out the clearly distinct spatial and temporal behaviors of the three parts. Only the tesseral part generates equatorial components of the torque; so the other two parts are not relevant to the nutations and wobbles of an ellipsoidal Earth. The sectorial part is responsible for the axial component of the torque, and gives rise to spin rate variations with frequencies in the semidiurnal range.

5.5.2 Complex representation of the potential; conventions

The use of a complex representation for the tidal potential is quite common in theoretical treatments of Earth rotation variations. The degree two potentials of orders 0, 1, 2, which are given by the expressions (5.194)–(5.196) multiplied by $(-GM_B/d^5)$, serve to provide simple examples of such a representation.

Consider first the tesseral potential $W_{2,1}^B(\mathbf{r}, t) = (-GM_B/d^5)Q_T$ with Q_T as in Eq. (5.195), which is by far the most important for the excitation of nutations. It may be written as the real part of one or the other of two mutually complex conjugate expressions:

$$-\frac{3GM_B}{d^5}[(d_x - id_y)d_z(x + iy)z], \tag{5.197}$$

or

$$-\frac{3GM_B}{d^5}[(d_x + id_y)d_z(x - iy)z]. \tag{5.198}$$

One may proceed to develop the theory of tides or Earth rotation variations by taking the potential to be given by either of the above two complex expressions, leaving the real part to be taken only at the very end of the derivation; the choice between (5.197) and (5.198) is a matter of convention, at one's discretion. Not surprisingly, examples of both the choices of conventions are abundant in the literature. For instance, Wahr (1981a–d) adopts the first of the above alternatives, while Sasao *et al.* (1980) uses the second one. Much confusion can be caused as a result when comparing different papers, because complex quantities that are introduced in the course of the theoretical development, like $\tilde{\eta}$ in Section 2.19 and $\tilde{m}$ and $\tilde{\Gamma}$ in Section 2.23.5, have to be defined differently in the two cases: the expression in one case has to be the complex conjugate of the other. For instance, the definition of the complex form $\tilde{\Gamma} = \Gamma_1 + i\Gamma_2$ for the equatorial torque produced by the degree two tesseral potential in the section just referred to implied the choice of convention (5.198) for the potential, as is evident from the last paragraph of that section. If $\tilde{\Gamma}$ had been defined as $\Gamma_1 - i\Gamma_2$, that would have implied the choice of the form (5.197) for the complex potential. It will be evident that one has to be well aware of the convention used in each work if one is to avoid pitfalls in interpretation of the complex quantities.

The convention used here and elsewhere in this book corresponds to the choice of the second form (5.198) for the tesseral potential. It may be recalled that we had written this expression already in Chapter 2 (see Eq. (2.63)) as

$$- \Omega_0^2 \tilde{\phi}(t)(x - iy)z = -\frac{\Omega_0^2}{3}\tilde{\phi}(t)\, r^2\, P_2^1(\cos \theta)e^{-im\lambda}, \qquad (5.199)$$

wherein all the dependence on the mass and coordinates of the body B is contained in the dimensionless complex quantity $\tilde{\phi}$, which is used repeatedly through the rest of this book:

$$\tilde{\phi}(t) = \phi_1 + i\phi_2 = \frac{3GM_B}{\Omega_0^2 d^5}(d_x + id_y)d_z = \frac{GM_B}{\Omega_0^2 d^3}P_2^1(\cos \beta)e^{i\Lambda}. \qquad (5.200)$$

If the opposite convention were to be used, the expression corresponding to (5.199) would be $-(\Omega_0^2/3)\tilde{\phi}^*(t)(x + iy)z$, which is equal to (5.197).

Before going on to the expansion of the total tidal potential in terms of spherical harmonic functions, it may be useful to illustrate such expansions by recasting the expressions (5.194)–(5.196) in terms of these functions. We observe that Q_Z, Q_T, and Q_S involve $\cos m(\lambda - \Lambda) = (1/2)(e^{i\lambda}e^{-im\Lambda} + c.c.)$, with $m = 0, 1$, and 2, respectively. Furthermore, $P_2^m(\cos \theta)\, e^{im\lambda} = (1/N_{2m})Y_2^m(\theta, \lambda)$ by the definition (5.145) of the spherical harmonics Y_n^m. Therefore

$$P_2^m(\cos \beta)P_2^m(\cos \theta)\cos m(\lambda - \Lambda) = (1/N_{2m}^2)\,\mathrm{Re}[Y_2^{m*}(\beta, \Lambda)Y_2^m(\theta, \lambda)]. \quad (5.201)$$

The definition of N_{nm} given in Eq. (5.145) gives the values $1/(N_{2m})^2 = 4\pi/5$, $24\pi/5$, $96\pi/5$ for $m = 0, 1, 2$, respectively. We obtain the zonal, tesseral, and sectorial parts of the degree two potential in terms of spherical harmonics on substituting the above expression with the relevant value of m in the polar coordinate forms of Q_Z, Q_T, Q_S. The result is

$$W_{2m}^B(\mathbf{r}, t) = -\frac{GM_B}{d^3} \frac{(2 - \delta_{m0})4\pi}{5} \operatorname{Re}\left[r^2 Y_2^{m*}(\beta, \Lambda) Y_2^m(\theta, \lambda)\right]. \tag{5.202}$$

If the opposite of the convention adopted in this book is to be employed, the factors within the square brackets have to be replaced by their complex conjugates.

$W_{2m}^B(\mathbf{r}, t)$ is merely one term in the general expansion of the full tidal potential that is to be presented now. We have included the above derivation, nevertheless, to show explicitly how the initial expressions for the dominant degree two potentials, which are in terms of Cartesian coordinates, go over to their equivalents in terms of spherical harmonics.

5.5.3 General spherical harmonic expansion

The gravitational potential in and near the Earth due to a celestial body B located at $\mathbf{R}_B$ may be expanded in spherical harmonics by employing the expansion formulas (5.136) and (5.149) for $1/|\mathbf{R} - \mathbf{r}|$, which were used in Section 5.3.5 for expanding the Earth's own external gravitational potential. With a trivial modification in the form of the latter formula, we have:

$$W^B(\mathbf{r}, t) = -\frac{GM_B}{2d_B} \sum_{n=0}^{\infty} \frac{4\pi}{2n + 1} \left(\frac{r}{d_B}\right)^n$$

$$\times \sum_{m=0}^{n} (2 - \delta_{m0})[Y_n^{m*}(\beta_B, \Lambda_B) Y_n^m(\theta, \lambda) + \text{c.c.}]$$

$$= -\sum_{n=0}^{\infty} \sum_{m=0}^{n} r^n [F_{nm}^{B*} Y_n^m(\theta, \lambda) + F_{nm}^B Y_n^{m*}(\theta, \lambda)], \tag{5.203}$$

where $(d_B, \beta_B, \Lambda_B)$ are the geocentric distance, colatitude, and longitude of B, and

$$F_{nm}^B = GM_B \frac{2\pi(2 - \delta_{m0})}{2n + 1} \left(\frac{1}{d_B}\right)^{n+1} Y_n^m(\beta_B, \Lambda_B). \tag{5.204}$$

The spherical harmonics Y_n^m are defined as in Eqs. (5.145)–(5.147).

The generic term in the double sum above is the tidal (or tide generating) potential of spherical harmonic *degree n and order m* (or more succinctly, of spherical harmonic *type (nm)*) due to the body B. This term depends on Λ_B and

λ through the factor $e^{\pm im(\lambda - \Lambda_B)}$, which goes over into $\cos m(\lambda - \Lambda_B)$ when the real part is taken. Special cases of this type of dependence (with $m = 0, 1$, and 2) were discussed in the last section, below Eq. (5.196). As was explained there, this function represents (for any $m > 0$) a wave traveling westwards; for general m, the frequencies in its spectrum are close to m times diurnal. Thus the terms of orders $m = 1, 2, 3 \ldots$ consist of waves that move in the retrograde direction with frequencies that are diurnal, semidiurnal, terdiurnal, etc.

5.5.4 Spectral expansion of the potential

We have seen in Chapter 2 that the nutational and wobble responses to tidal forcing are dependent on the frequency of the forcing. It is important, therefore, to have a spectral expansion of the TGP. As was made clear in Chapter 2, the terms of degree less than two in the spherical harmonic expansion of W^B do not produce any tides or torques on the Earth; so the TGP due to B consists of the terms with $n \geq 2$ in (5.203). The parameters of B appear in the potential through F_{nm}^B; to obtain the total potential due to all the perturbing bodies, one simply sums over all the bodies B. The (nm) part of the total potential is then

$$W_{nm}(\mathbf{r}, t) = \sum_B W_{nm}^B(\mathbf{r}, t) = -2r^n \, \text{Re} \, [F_{nm}^* \, Y_n^m(\theta, \lambda)], \qquad (5.205)$$

$$F_{nm} = \sum_B F_{nm}^B. \qquad (5.206)$$

The dependence of W_{nm} on time arises, of course, through the coordinates of the solar system bodies, entering through F_{nm}. The orbital positions of the solar system bodies as functions of time may be found either from analytical expressions or from numerical tables that may be found in different types of ephemerides. Presentation of the main steps in the derivation of the analytical formulae, knowing the mutual gravitational interactions of the bodies, will be deferred to Section 5.11. We have already discussed in Chapter 4 the ephemerides constructed from observations of the solar system bodies. The nature of the spectrum of frequencies present in F_{nm}^* will be dealt with in the following subsection.

The spectral expansion of the potential may be inferred directly from the work of Cartwright and Taylor (1971) by noting that our F_{nm}^* goes over into $g_e c_{nm}^*(t)/2$ of those authors. Knowing the spectral expansion of the latter, we can write down that of F_{nm}^* as

$$F_{nm}^* = \sum_{\omega \geq 0} \frac{g_e H_\omega^{nm}}{2 a_e^n} e^{i(\Theta_\omega(t) - \zeta_{nm})}, \qquad (5.207)$$

where H_ω^{nm} is the amplitude, expressed as a "tide height," of the spectral component with frequency ω, and the *argument* Θ_ω determines ω through the relation

$$\Theta_\omega(t) = \omega t + \Theta_\omega(0). \tag{5.208}$$

The phase ζ_{nm} is defined by

$$\zeta_{nm} = 0 \quad \text{for } (n-m) \text{ even} \quad \text{and} \quad \zeta_{nm} = \pi/2 \quad \text{for } (n-m) \text{ odd.} \tag{5.209}$$

Since the tidal potential waves are retrograde (reflecting the westward motion of the celestial objects relative to the rotating Earth-fixed reference frame), Eq. (5.205) taken together with (5.207) implies that the frequencies ω are non-negative for all the spectral components, whatever the spherical harmonic type (nm).

The expansion (5.207) leads to

$$W_{nm}(\mathbf{r}, t) = \sum_{\omega \geq 0} W_{nm}^\omega(\mathbf{r}, t), \qquad (n \geq 2), \tag{5.210}$$

$$W_{nm}^\omega(\mathbf{r}, t) = -\frac{g_e H_\omega^{nm}}{2} \left(\frac{r}{a_e}\right)^n \left(Y_n^m(\theta, \lambda) e^{i(\Theta_\omega(t) - \zeta_{nm})}\right.$$

$$\left. + Y_n^{m*}(\theta, \lambda) e^{-i(\Theta_\omega(t) - \zeta_{nm})}\right). \tag{5.211}$$

Now, $e^{i\zeta_{nm}}$ is 1 for $(n-m)$ even and i for $(n-m)$ odd, in view of (5.209). Therefore on writing the above expression for W_{nm}^ω in real form, we obtain

$$W_{nm}^\omega(\mathbf{r}, t) = -g_e H_\omega^{nm} \left(\frac{r}{a_e}\right)^n P_n^m(\cos\theta) \cos(\Theta_\omega(t) + \lambda) \tag{5.212}$$

if $(n-m)$ is even (e.g., zonal and sectorial potentials of degree two), and

$$W_{nm}^\omega(\mathbf{r}, t) = -g_e H_\omega^{nm} \left(\frac{r}{a_e}\right)^n P_n^m(\cos\theta) \sin(\Theta_\omega(t) + \lambda) \tag{5.213}$$

if $(n-m)$ is odd, as in the case of the degree two tesseral potential. These expressions conform to the Cartwright–Taylor characterization of tidal waves.

5.5.5 Fundamental arguments of the tidal potential; frequency bands

With the neglect of very small terms that are non-linear in the time variable, the argument $\Theta_\omega(t)$ is of the form $\omega t + \alpha_\omega$ where α_ω is the initial phase. A term in the spectral expansion of $W_{nm}(\mathbf{r}, t)$ for any given n and m is identified by a set of values of the integer coefficients $n_1, n_2, \ldots, n_6$ in the expression for $\Theta_\omega(t)$ in terms of *Doodson's fundamental arguments*:

$$\Theta_\omega = n_1(\tau - \lambda) + n_2 s + n_3 h + n_4 p + n_5 N' + n_6 p_s + \cdots, \tag{5.214}$$

where $n_1 = m$. The n_i are small integers, positive, negative, or zero. (Spectral terms with large values for any of the n_i have negligibly small amplitudes and are therefore ignorable.) The fundamental arguments $s, h, p, -N', p_s$ are the mean tropic longitudes of the Moon, Sun, Moon's perigee, Moon's node, and the Sun's perigee, respectively. The fundamental angular frequencies pertaining to these five fundamental arguments, which are the time derivatives of the arguments, are all much less than 1 cpsd in magnitude, as the corresponding periods lie in the range from 27.32 days for s to about 209 years for p_s. The Doodson argument τ is such that

$$\tau + s - \lambda = \Phi \equiv 180° + \text{GMST},\tag{5.215}$$

when expressed in degrees. In view of the definition (5.184) of GMST, the period associated with $\tau + s - \lambda$ is that of the mean sidereal rotation, namely, 1 sidereal day.

The mean tropic longitudes of the planets from Mercury to Saturn have to be included as additional fundamental arguments when spectral terms of very small amplitudes are to be taken into account; the dots in (5.214) represent linear combinations of these arguments with integer coefficients $n_7, n_8, \ldots$

The frequency of the term with a given set of values for the n_i becomes more transparent if $\Theta_\omega(t)$ is rewritten as

$$\Theta_\omega(t) = m\Phi + (n_2 - n_1)s + n_3 h + n_4 p + n_5 N' + n_6 p_s + \cdots\tag{5.216}$$

In any potential of given order m, the term with zero values for $(n_2 - n_1), n_3, n_4, \ldots$ has the frequency m cpsd. If $m > 0$, the frequencies of the terms with various small positive or negative integer values for these quantities constitute a spectrum of 1 cpsd width, lying between $(m - 1/2)$ and $(m + 1/2)$ cpsd. For $m = 0$, the band of frequencies extends only from 0 to 1/2 cpsd. (Though one can cause $\Theta_\omega(t)$ to go over to $-\Theta_\omega(t)$ and thus change the sign of the frequency simply by flipping the signs of all the n_i when $m = 0$, the effect on $W_{n0}^\omega(t)$ is merely to interchange the two terms in (5.211) since Y_n^0 is real; no new spectral term is created.) The spectrum of frequencies in the TGP thus lies in *bands*, each of which is associated with a particular value of m. For given m, the spectra of potentials of different degrees n lie within the same band.

In tables listing the amplitudes H_ω^{nm} of the spectral components of the tidal potential, each component is identified by its Doodson number which is written in the form $abc.def$ wherein each symbol is an integer; a is $n_1 = m$, and the remaining symbols stand for $n_2 + 5, n_3 + 5, \ldots$ Thus, for example, the terms with Doodson numbers 145.555, 164.556, and 165.555 are the tesseral tides named O_1, S_1, K_1, having frequencies ω with the respective values $(1 - 2\,ds/dt)$ cpsd, $(1 - dh/dt + dp_s/dt)$ cpsd, and 1 cpsd; 002.000 is the solar semiannual zonal tide (S_{sa}) with the frequency $(2\,dh/dt)$ cpsd; and 255.555 is the sectorial tide M_2

with frequency $\omega = (2 - 2\,ds/dt)$ cpsd. The amplitude shown in Table 5.3 is the coefficient of $\cos(\Theta_\omega(t) + m\lambda)$ in W_{nm}/g_e at $r = a_e$ if $(n - m)$ is even, and of $\sin(\Theta_\omega(t) + m\lambda)$ if $(n - m)$ is odd.

5.5.6 *Fundamental arguments of nutations*

A different set of fundamental arguments is in use for the expansion of the arguments of the nutations and polar motions produced by the tidal potential. This set consists of the Delaunay arguments l, l', D, F, together with a fifth argument Ω; an additional argument Φ, defined by Eq. (5.215), is needed when nutations outside the low frequency band are to be considered. As already noted in Chapter 4, the argument Ω is the mean longitude of the ascending node of the Moon; l and l' are the mean anomalies of the Moon and the Sun, respectively, and are denoted in some of the literature by l_M and l_S (or l_m, l_s); F is $L - \Omega$ where L is the mean longitude of the Moon; and D is the mean elongation of the Moon from the Sun: $D = L - L'$, where L' is the mean longitude of the Sun.

The astronomy terms used in the above characterization of the fundamental arguments are defined in terms of arc lengths of the relevant great circles on a geocentric celestial sphere. One of these is the ecliptic of date, representing the intersection of the celestial sphere with the plane of the apparent orbit of the Sun relative to the Earth; and the other is the intersection of the plane of the Moon's orbit around the Earth with the same sphere. These circles are evidently projections of the Sun's and Moon's orbits on to the celestial sphere. The definitions of the fundamental arguments involve altogether six reference points lying on the two circles. One of these is common to both the circles: the ascending node N of the lunar orbit circle on the ecliptic. (It may be recalled that "nodes" are the two points at which the line of intersection of the two planes meets the celestial sphere, and that the ascending node of the Moon is the one at which the orbital motion of the Moon takes it from the "south" side of the ecliptic toward the "north" side.) Of the other five reference points, three are on the ecliptic: the equinox of date (γ); the projection, on to the celestial sphere, of the Sun's perigee π_S (i.e., the point of closest approach of the Sun to the Earth); and the projection on to the celestial sphere of the position P_S that the Sun would have if it had been moving with a constant mean angular velocity (2π) radians/year. The two remaining reference points are on the other circle: the projected position π_M of the perigee of the Moon, and the point P_M, projected on to the celestial sphere, of the position that the Moon would have if it were moving with a constant angular velocity. Ω, l', and L' are all arcs on the ecliptic: Ω is the arc γN, l' is $\pi_S P_S$, and L' is γP_S. The mean anomaly l of the Moon is the arc $\pi_M P_M$ along the Moon's orbit projection. The longitude L of the Moon is the sum of two arcs γN and $N P_M$ on the two different circles;

Table 5.3 *Values of the largest tidal potential harmonics.*

Usual symbol	Doodson number	Frequency (cpd)	Amplitude (m)
Long-period tides			
M_0, S_0	055.555	0.0000000	−0.31459
M_f	075.555	0.0732022	−0.06661
M_m	065.455	0.0362916	−0.03518
S_{sa}	057.555	0.0054758	−0.03099
	055.565	0.0001471	0.02793
	075.565	0.0733493	−0.02762
M_{tm}	085.455	0.1094938	−0.01275
M_{sm}	063.655	0.0314347	−0.00673
M_{sf}	073.555	0.0677264	−0.00584
	085.465	0.1096409	−0.00529
Diurnal tides			
K_1	165.555	1.0027379	0.36864
O_1	145.555	0.9295357	−0.26223
P_1	163.555	0.9972621	−0.12199
Q_1	135.655	0.8932441	−0.05021
	165.565	1.0028850	0.05003
	145.545	0.9293886	−0.04947
J_1	175.455	1.0390296	0.02062
M_1	155.655	0.9664463	0.02061
OO_1	185.555	1.0759401	0.01128
ρ_1	137.455	0.8981010	−0.00953
	135.645	0.8930970	−0.00947
σ_1	127.555	0.8618093	−0.00801
	155.455	0.9658274	0.00741
	165.545	1.0025908	−0.00730
	185.565	1.0760872	0.00723
π_1	162.556	0.9945243	−0.00713
$2Q_1$	125.755	0.8569524	−0.00664
ϕ_1	167.555	1.0082137	0.00525
Semidiurnal tides			
M_2	255.555	1.9322736	0.63221
S_2	273.555	2.0000000	0.29411
N_2	245.655	1.8959820	0.12105
K_2	275.555	2.0054758	0.07991
	275.565	2.0056229	0.02382
	255.545	1.9321265	−0.02359
ν_2	247.455	1.9008389	0.02299
μ_2	237.555	1.8645472	0.01933
L_2	265.455	1.9685653	−0.01787
T_2	272.556	1.9972622	0.01719
$2N_2$	235.755	1.8596903	0.01602
ϵ_2	227.655	1.8282556	0.00467
λ_2	263.655	1.9637084	−0.00466

so the mean elongation D of the Moon from the Sun is the difference $N P_M$ minus $N P_S$.

The following relations connect the Delaunay arguments and Ω to Doodson's arguments s, h, p, N', p_s:

$$l = s - p, \quad l' = h - p_s, \quad D = s - h, \quad F = s + N', \quad \Omega = -N'. \quad (5.217)$$

Conversely,

$$s = F + \Omega, \qquad h = F + \Omega - D, \qquad p = -l + F + \Omega,$$
$$N' = -\Omega, \qquad p_s = -l' + F + \Omega - D. \quad (5.218)$$

The expansion of $\Theta_\omega(t)$ in terms of the new set of fundamental arguments has the form

$$\Theta_\omega(t) = m\Phi + n_1'l + n_2'l' + n_3'D + n_4'F + n_5'\Omega + \sum_{i=6}^{14} n_i'\lambda_{i-5}. \quad (5.219)$$

The additional arguments λ_i, $(i = 1, \ldots, 8)$, stand for the mean tropic longitudes of the planets from Mercury to Saturn, and $\lambda_9 = p_a$ is the lunisolar precession. The multipliers n_i' are all integers (positive, zero, or negative).

The frequency associated with Φ is of course the same as that of GMST, which is Ω_0 (1 cpsd). The five other fundamental arguments have the following periodicities: 27.5545 days for l, 365.2596 days for l', 27.2122 days for F, 29.5306 days for D, and -6798.3835 days for Ω; consequently, the frequencies associated with these five fundamental arguments are $\ll 1$ cpsd. The band structure associated with the numerous sets of values of n_i' is the same as before, of course.

As the orbits of the celestial bodies undergo secular variations over very long timescales, the amplitudes of nutations as well as the precession rate are not strictly constant but undergo slow variations. The fundamental arguments too are not strictly linear in time: they contain small terms in T^2, T^3, etc. This means that the frequency $\omega = (d\,\Theta_\omega/dt)$ of any spectral term varies slowly with time. It has also been pointed out in some recent works that certain of the fundamental arguments come very close to being linear combinations of others, leading some to question the practical utility and rationality of retaining the full set of fundamental arguments. Some remarks on the different ways in which such questions have been dealt with in the literature may be found in Section 5.10 below.

5.5.7 Torque produced by the TGP

We turn now to evaluation of the torque exerted by the tidal potential on the Earth. Let the torque produced by W_{nm} be denoted by $\mathbf{\Gamma}^{nm}$. It is given by

$$\mathbf{\Gamma}^{nm} = -\int \rho(\mathbf{r})\,(\mathbf{r} \wedge \nabla)\,W_{nm}(\mathbf{r}, t)d^3r, \quad (5.220)$$

where $\rho(\mathbf{r})$ is, as before, the density at the position $\mathbf{r}$, and the integral is over the volume of the Earth.

The operator $(\mathbf{r} \wedge \nabla)$ acts on $r^n Y_n^m(\theta, \lambda)$ and on its complex conjugate which appears in the expression for W_{nm}. It is evident that the operator commutes with r^n, i.e., $(\mathbf{r} \wedge \nabla)(r^n Y_n^m) = r^n(\mathbf{r} \wedge \nabla)Y_n^m$. Furthermore, this operator is just i times the angular momentum operator of quantum mechanics. Knowing the action of the latter on Y_n^m (see, for instance, Mathews and Venkatesan, 1976; second edition, 2010), one finds immediately that for any m from $-n$ to $+n$,

$$(\mathbf{r} \wedge \nabla)_+ Y_n^m = i c_n^m Y_n^{m+1}, \tag{5.221}$$

$$(\mathbf{r} \wedge \nabla)_- Y_n^m = i c_n^{-m} Y_n^{m-1}, \tag{5.222}$$

$$(\mathbf{r} \wedge \nabla)_3 Y_n^m = i m Y_n^m, \tag{5.223}$$

where $(\mathbf{r} \wedge \nabla)_\pm = (\mathbf{r} \wedge \nabla)_1 \pm i(\mathbf{r} \wedge \nabla)_2$ and $c_n^{\pm m}$ is given by Eq. (5.172).

Since $(\mathbf{r} \wedge \nabla)_+$ and $(\mathbf{r} \wedge \nabla)_-$ are complex conjugates of each other, we see by taking the complex conjugates of (5.222) and (5.221) that

$$(\mathbf{r} \wedge \nabla)_+ Y_n^{m*} = -i c_n^{-m}(Y_n^{m-1})^*, \tag{5.224}$$

$$(\mathbf{r} \wedge \nabla)_- Y_n^{m*} = -i c_n^m(Y_n^{m+1})^*. \tag{5.225}$$

Now we can introduce the expression (5.205) for W_{nm} into the integral (5.220) for Γ^{nm} and evaluate the familiar complex combination $\tilde{\Gamma}^{nm}$ of the first two components with the help of Eqs. (5.221)–(5.225). We obtain

$$\tilde{\Gamma}^{nm} = i[c_n^m \, Q_{n,m+1} \, F_{nm}^* - c_n^{-m} \, Q_{n,m-1}^* \, F_{nm}] \tag{5.226}$$

for $m > 0$, and

$$\tilde{\Gamma}^{n0} = i c_n^0 \, Q_{n,1} \, 2F_{n0}, \tag{5.227}$$

where the Q_{nm} are defined by

$$Q_{nm} = \int \rho(\mathbf{r}) \, r^n \, Y_n^m(\theta, \lambda) \, d^3r. \tag{5.228}$$

Comparison of this definition with Eqs. (5.152), which define the geopotential coefficients (C_{nm}, S_{nm}), shows that

$$(2 - \delta_{m0})4\pi N_{nm} Q_{nm} = (2n + 1)M_E a_e^n(C_n^m + i S_n^m), \tag{5.229}$$

where M_E is the mass of the Earth.

The coefficients of F_{nm}^* and F_{nm} in (5.226) may now be expressed in terms of the geopotential coefficients:

$$c_n^m Q_{n,m+1} = \frac{2n+1}{8\pi} \frac{c_n^m}{N_{n,m+1}} M_E a_e^n (C_{n,m+1} + i S_{n,m+1}), \qquad (5.230)$$

$$c_n^{-m} Q_{n,m-1}^* = \frac{2n+1}{(2-\delta_{m,1})4\pi} \frac{c_n^{-m}}{N_{n,m-1}} M_E a_e^n (C_{n,m-1} - i S_{n,m-1}). \quad (5.231)$$

Now, we can see from the definitions for $N_{n,m}$ and $c_n^{\pm m}$ as given in Eqs. (5.145) and (5.172) that

$$\frac{c_n^m}{N_{n,m+1}} = -\frac{(n-m)(n+m+1)}{N_{nm}}, \qquad \frac{c_n^{-m}}{N_{n,m-1}} = -\frac{1}{N_{nm}}. \qquad (5.232)$$

On introducing Eqs. (5.230) and (5.231) taken together with (5.232) into (5.226), we obtain an expression relating the equatorial torque due to W_{nm} to the geopotential coefficients:

$$\tilde{\Gamma}^{nm} = -i \frac{2n+1}{8\pi} \frac{M_E a_e^n}{N_{nm}} \Big((n-m)(n+m+1)(C_{n,m+1} + i S_{n,m+1}) F_{nm}^*$$

$$- \frac{2}{2-\delta_{m1}} (C_{n,m-1} - i S_{n,m-1}) F_{nm} \Big) \qquad (5.233)$$

for $m > 0$, while for $m = 0$, Eq. (5.227), together with Eqs. (5.230) and (5.232), yields

$$\tilde{\Gamma}^{n0} = -i \frac{2n+1}{4\pi} \frac{M_E a_e^n}{N_{n0}} n(n+1)(C_{n1} + i S_{n1}) F_{n0}. \qquad (5.234)$$

Equations (5.204), (5.206), and their complex conjugates may now be used to substitute for F_{nm} and F_{nm}^* in the above equations to obtain explicit expressions for the equatorial torque; they involve the time dependent position coordinates of the celestial bodies, which give rise to the potential as well as the geopotential coefficients.

A general expression for the total equatorial torque on the Earth has been derived already in Section 5.3.7 by a different approach: see Eq. (5.178) for the torque due to a celestial body B. If it is summed over all the relevant bodies B, one may readily verify that the part pertaining to any particular m does coincide with (5.233) if $m > 0$ and to (5.234) if $m = 0$.

The axial component Γ_z of the torque is easy to obtain. It agrees with Eq. (5.181) of Section 5.3.7.

It is of considerable interest to observe from the derivation presented above that the two mutually complex conjugate terms in the potential (5.203) act on two quite different structural features of the Earth's density distribution to generate the two terms in the torque (5.233). The part of the potential involving $Y_n^m(\theta, \lambda)$ which is proportional to $e^{im\lambda}$ acts on the geopotential coefficients of one order higher, i.e., $(C_{n,m+1}, S_{n,m+1})$ while the complex conjugate Y_n^{m*} part of the potential (proportional to $e^{-im\lambda}$) acts on the geopotential coefficients of order $m - 1$.

For axially symmetric Earth models, the only geopotential coefficients which are non-vanishing are the C_{n0}. For any given n, the torque involving C_{n0} is $\tilde{\Gamma}^{n1}(\omega_r)$, which is generated by the Y_{n1}^* term in the expression (5.203) for the tidal potential. The complex conjugate term containing Y_{n1} cannot produce any torque when the Earth is axially symmetric. This observation provides the rationale for choosing the potential in the form Y_{21}^*, proportional to $P_{21}(\cos\theta) \, e^{-i\lambda}$, in the work of Sasao *et al.* (1980) on the principal class of low frequency nutations excited by the degree two tesseral potential.

5.5.8 Spectral representation of the torque: terrestrial frame

The spectral representation is obtained by using the expansion (5.207) of F_{nm}^*, together with its complex conjugate. The resulting expression for the spectral component of frequency ω may be written in compact form as

$$\tilde{\Gamma}_\omega^{nm}(t) = \tilde{\Gamma}^{nm}(\omega_p)e^{i(\Theta_\omega(t)-\zeta_{nm})} + \tilde{\Gamma}^{nm}(\omega_r)e^{-i(\Theta_\omega(t)-\zeta_{nm})}, \tag{5.235}$$

where $\omega_p = \omega$ and $\omega_r = -\omega$ are the frequencies of the prograde and retrograde motions, respectively, and the respective amplitudes are

$$\tilde{\Gamma}^{nm}(\omega_p) = (i\Omega_0^2\bar{A}) \, (-1)^{m+1} \, G_{nm}^{(+)} \, H_\omega^{nm} \, [(C_{n,m+1} + iS_{n,m+1})] \tag{5.236}$$

and

$$\tilde{\Gamma}^{nm}(\omega_r) = (i\Omega_0^2\bar{A}) \, (-1)^{m} \, G_{nm}^{(-)} \, H_\omega^{nm} \, [(C_{n,m-1} - iS_{n,m-1})] \tag{5.237}$$

for $m > 0$, while for $m = 0$,

$$\tilde{\Gamma}^{n0}(\omega_p) = \tilde{\Gamma}^{n0}(\omega_r) = -(i\Omega_0^2\bar{A}) \, G_{n0}^{(+)} \, H_\omega^{n0} \, (C_{n1} + iS_{n1}). \tag{5.238}$$

Here $\bar{A} \equiv (A + B)/2$ stands for the mean equatorial moment of inertia as before, and $G_{nm}^{(+)}$ and $G_{nm}^{(-)}$ are given by

$$G_{nm}^{(+)} = (n - m)(n + m + 1)G_{nm}, \qquad G_{nm}^{(-)} = \frac{2}{(2 - \delta_{m,1})}G_{nm}, \tag{5.239}$$

with

$$G_{nm} = \left(\frac{2n + 1}{4\pi} \frac{(n + m)!}{(n - m)!}\right)^{1/2} \frac{g_e M_E}{4\Omega_0^2\bar{A}}. \tag{5.240}$$

It is a remarkable fact that the torque has both prograde and retrograde parts though the tidal potential itself is purely retrograde. The phase in the first term in the complex torque $\tilde{\Gamma}_\omega^{nm}$ of (5.235) involves the time t through ωt with ω positive, and therefore this term represents a vector in prograde motion (see Section 2.18). For potentials of any order $m > 0$, the spectrum of ω consists of a band of width 1 cpsd centered at m cpsd, and hence, so does the spectrum of the prograde part of the torque. The second term of (5.235), which has $-\omega t$ in the phase, is retrograde, with a spectrum of frequencies in the band centered at $-m$ cpsd. Thus there are two distinct bands, one prograde and the other retrograde, for any $m > 0$. For $m = 0$, it may be recalled that the spectrum of ω constitutes just a half-band from 0 to 0.5 cpsd. It follows then that prograde and retrograde components of the torque together make up a low frequency spectrum consisting of a single band from -0.5 to $+0.5$ cpsd.

An examination of the expression (5.235) together with (5.236) and (5.237) for the torque produced by a potential of spherical harmonic type (n, m) is quite revealing. The prograde part of the torque is excited by the action of the potential on the structures of the Earth represented by the geopotential coefficients $(C_{n,m+1}, S_{n,m+1})$. The retrograde part originates in the action of the potential on geopotential coefficients $(C_{n,m-1}, S_{n,m-1})$, except when $m = 0$ in which case the geopotential coefficients involved are $(C_{n,1}, S_{n,1})$ in the retrograde part as well as in the prograde part. For every frequency in a prograde band, there is a corresponding frequency, with just the sign reversed, in the retrograde band. Both of them are excited by the same H_ω^{nm}; nevertheless, their amplitudes can be very different because the geopotential coefficients of type $(n, m + 1)$ responsible for the prograde component of the torque and those of type $(n, m - 1)$ responsible for the other can have very different magnitudes (see Table 5.2). In the particular case $n = 2, m = 1$, the difference is vast, by a factor of the order of 1000. In fact, $C_{2,0}$ is larger than all the other coefficients listed in that table by a factor of this order. Consequently, the wobbles of the retrograde diurnal band are by far the largest; this feature carries over to the associated nutations which have frequencies in the low frequency band, according to the kinematical relations (see Section 2.26, especially Eq. (2.105)). In the case of the $(n, 1)$ potential for any n, the expression (5.237) for the part of the torque which generates retrograde diurnal wobbles reduces to

$$\tilde{\Gamma}^{n1}(\omega_r) = (-i\Omega_0^2 \bar{A}) \, G_{n1}^{(-)} \, H_\omega^{n1} \, C_{n,0}$$

$$= i \left(\frac{2n + 1}{16\pi} n(n + 1) \right)^{1/2} g_e M_E J_n H_\omega^{n1}, \tag{5.241}$$

where the last step uses Eqs. (5.239) and (5.240) along with the fact, noted below Eq. (5.154), that $J_n = -C_{n0}$. The part of the Earth structure that is represented

Table 5.4 *Origin of nutations, wobbles, and polar motions (PM)*
in various frequency bands.

Nutations	Frequency band (cpsd)	Due to action on	By potentials	Wobbles & PM
Long period (LP)	$(-0.5, +0.5)$	$C_{n,0}$	$(n, 1)$	RD
Pro diurnal (PD)	$(+0.5, +1.5)$	$(C_{n,1}, S_{n,1})$	$(n, 0)$	LP
Retro diurnal (RD)	$(-1.5, -0.5)$	$(C_{n,1}, S_{n,1})$	$(n, 2)$	RS
Pro semidiurnal (PS)	$(+1.5, +2.5)$	$(C_{n,2}, S_{n,2})$	$(n, 1)$	PD
Retro semidiurnal (RS)	$(-2.5, -1.5)$	$(C_{n,2}, S_{n,2})$	$(n, 3)$	RT
Pro terdiurnal (PT)	$(+2.5, +3.5)$	$(C_{n,3}, S_{n,3})$	$(n, 2)$	PS

by J_3 is asymmetric about the equator, and the largest of the spectral components generated by it has an amplitude of about 0.2 mas only.

The listing of the origin of nutations and polar motions in the various frequency bands in Table 5.4 is a summary of the above considerations (see Mathews and Bretagnon, 2003).

In order to compute the nutations or polar motions in a particular band of frequencies, one starts by identifying from Table 5.4 the types of potentials and geopotential coefficients relevant to that band, then one picks out the expressions for the torques which they produce from Eqs. (5.236) to (5.238), and finally, solves the dynamical equations with these as the driving torques. For prograde semidiurnal nutations (or prograde diurnal polar motions), for instance, the table shows that the relevant potentials are of type $(n, 1)$, acting on $C_{n,2}$ and $S_{n,2}$.

Turning now to the third (axial) component of the torque, we find on using Eq. (5.223) and noting that $(\mathbf{r} \wedge \nabla) Y_n^{m*} = -im Y_n^{m*}$, that

$$\Gamma_{\omega 3}^{nm} = i \frac{g_e}{2a_e^n} H_\omega^{nm} \left[m\, Q_{n,m}\, e^{i(\Theta_\omega(t) - \zeta_{nm})} - m\, Q_{n,m}^*\, e^{-i(\Theta_\omega(t) - \zeta_{nm})} \right]. \qquad (5.242)$$

Substitution for Q_{nm} from (5.229) leads to

$$\Gamma_{\omega 3}^{nm}(t) = (i\Omega_0^2 \bar{A}) (-1)^m\, mG_{nm}\, H_\omega^{nm}$$
$$\times [(C_n^m + i S_n^m) e^{i(\Theta_\omega(t) - \zeta_{nm})} - (C_n^m - i S_n^m) e^{-i(\Theta_\omega(t) - \zeta_{nm})}]. \qquad (5.243)$$

This expression reduces to

$$2\Omega_0^2 \bar{A} (-1)^{m+1}\, mG_{nm}\, H_\omega^{nm} [C_n^m \sin \Theta_\omega(t) - S_n^m \cos \Theta_\omega(t)] \qquad (5.244)$$

if $(n - m)$ is even, and otherwise, to

$$2\Omega_0^2 \bar{A} (-1)^{m+1}\, mG_{nm}\, H_\omega^{nm} [C_n^m \cos \Theta_\omega(t) - S_n^m \sin \Theta_\omega(t)]. \qquad (5.245)$$

5.6 Wobbles and nutations excited by arbitrary potentials

5.6.1 Solution of Euler's equations

Euler's equations which govern the temporal variation of the components of the angular velocity vector of a rigid body with reference to the principal frame were presented in Section 2.22.2. We reproduce them here, expressed in terms of the wobble variables m_x, m_y and the spin rate variation quantity m_z:

$$A\dot{m}_x + (C - B)\Omega_0 m_y(1 + m_z) = \Gamma_x/\Omega_0,$$
$$B\dot{m}_y + (A - C)\Omega_0 m_x(1 + m_z) = \Gamma_y/\Omega_0, \tag{5.246}$$
$$C\dot{m}_z + (B - A)\Omega_0 m_x m_y = \Gamma_z/\Omega_0.$$

Dropping the terms that are of the second order in m_x, m_y, m_z, and dividing by $\bar{A} = (A + B)/2$, we rewrite these equations in terms of the ellipticity and triaxiality parameters e and e' defined in Eq. (5.134) above:

$$(1 - e')\dot{m}_x + (e - e')\Omega_0 m_y = \Gamma_x/\bar{A}\Omega_0,$$
$$(1 + e')\dot{m}_y - (e + e')\Omega_0 m_x = \Gamma_y/\bar{A}\Omega_0, \tag{5.247}$$
$$(1 + e)\dot{m}_z = \Gamma_z/\bar{A}\Omega_0.$$

The complex combination of the first two equations yields

$$\left(\frac{d}{dt} - ie\Omega_0\right)\tilde{m} - e'\left(\frac{d}{dt} + i\Omega_0\right)\tilde{m}^* = \frac{\tilde{\Gamma}}{\bar{A}\Omega_0}, \tag{5.248}$$

where $\tilde{m} = m_x + im_y$ as usual, and $\tilde{m}^* = m_x - im_y$. It was shown in the last section that a spectral component of frequency $\omega > 0$ of the TGP of any spherical harmonic type (nm) gives rise to a torque with both prograde and retrograde components:

$$\tilde{\Gamma}(t) = \tilde{\Gamma}_p e^{i\omega t} + \tilde{\Gamma}_r e^{-i\omega t}, \tag{5.249}$$

where the prograde and retrograde parts have frequencies ω and $-\omega$, respectively, and are indicated by subscripts p and r as usual. We have suppressed the indices (nm) which are not essential for our immediate purposes. We have also replaced the factors $e^{\pm i[\Theta_\omega(t) - \zeta_{nm}]}$ for brevity by $e^{\pm i\omega t}$; the full form may be restored when necessary.

In consequence of the structure of $\tilde{\Gamma}$ as in (5.249), it follows that $\tilde{m}$ also may be decomposed in a similar manner:

$$\tilde{m}(t) = \tilde{m}_p e^{i\omega t} + \tilde{m}_r e^{-i\omega t}, \qquad \tilde{m}^*(t) = \tilde{m}_r^* e^{i\omega t} + \tilde{m}_p^* e^{-i\omega t}. \tag{5.250}$$

Substitution of these expressions into Eq. (5.248) followed by separation into prograde and retrograde parts leads to the pair of equations

$$(\omega - e\Omega_0)\tilde{m}_p - e'(\omega + \Omega_0)\tilde{m}_r^* = \frac{\tilde{\Gamma}_p}{iA\Omega_0}, \qquad (5.251)$$

$$-(\omega + e\Omega_0)\tilde{m}_r + e'(\omega - \Omega_0)\tilde{m}_p^* = \frac{\tilde{\Gamma}_r}{iA\Omega_0}. \qquad (5.252)$$

We take the complex conjugate of the first equation to obtain

$$e'(\omega + \Omega_0)\tilde{m}_r - (\omega - e\Omega_0)\tilde{m}_p^* = \frac{\tilde{\Gamma}_p^*}{iA\Omega_0}. \qquad (5.253)$$

We now have the pair of coupled equations (5.252) and (5.253), which may be solved for the amplitudes $\tilde{m}_r$ and $\tilde{m}_p^*$; and $\tilde{m}_p$ can be recovered from the latter to complete the solution (5.250) for $\tilde{m}(t)$:

$$\tilde{m}_r = \frac{i}{A\Omega_0} \frac{(\omega - e\Omega_0)\tilde{\Gamma}_r + e'(\omega - \Omega_0)\Gamma_p^*}{(\omega^2 - e^2\Omega_0^2) - e'^2(\omega^2 - \Omega_0^2)}, \qquad (5.254)$$

$$\tilde{m}_p = -\frac{i}{A\Omega_0} \frac{(\omega + e\Omega_0)\Gamma_p + e'(\omega + \Omega_0)\tilde{\Gamma}_r^*}{(\omega^2 - e^2\Omega_0^2) - e'^2(\omega^2 - \Omega_0^2)}. \qquad (5.255)$$

One feature of these solutions stands out: the retrograde part of the wobble motion is not determined purely by the retrograde part of the torque! Thanks to the coupling due to triaxiality, the prograde torque also gives a contribution (proportional to $e'\tilde{\Gamma}_p^*$) to the retrograde motion. The reverse is also true: the retrograde torque too plays a small part in the prograde wobble. These results are valid for excitation by potentials of any type (nm).

The amplitudes $\tilde{\eta}$ of the nutations associated with the wobbles $\tilde{m}_r$ and $\tilde{m}_p$ may now be obtained by applying the kinematic relations of Section 2.26. Note that $\tilde{m}(\omega_w)$ of that section stands for the amplitude of a prograde wobble if the frequency ω_w of the wobble is positive (equal to the tidal frequency ω), and for the amplitude of a retrograde wobble if $\omega_w = -\omega$. Stated differently, $\tilde{m}_r = \tilde{m}(\omega_w)$ with $\omega_w = -\omega$, and $\tilde{m}_p = \tilde{m}(\omega_w)$ with $\omega_w = \omega$. The corresponding nutations are of frequencies $(\Omega_0 - \omega)$ and $(\Omega_0 + \omega)$, respectively. In the present example, where ω is in the retrograde diurnal band, $\Omega_0 - \omega$ is close to 2 cpsd and $\Omega_0 + \omega$ is in the low frequency band, i.e., they relate to prograde semidiurnal and low frequency nutations, respectively.

5.6.2 Wobble, nutation, polar motion

Application of the kinematic relations (2.105) of Chapter 2 to the retrograde and prograde wobbles $\tilde{m}_r$ and $\tilde{m}_p$ leads to the following expressions for the amplitudes of the corresponding nutations of the figure axis:

$$\tilde{\eta}(\Omega_0 - \omega) = -\frac{\Omega_0}{\Omega_0 - \omega}\tilde{m}_r, \qquad \tilde{m}_r = \tilde{m}(-\omega), \tag{5.256}$$

$$\tilde{\eta}(\Omega_0 + \omega) = -\frac{\Omega_0}{\Omega_0 + \omega}\tilde{m}_p, \qquad \tilde{m}_p = \tilde{m}(\omega). \tag{5.257}$$

Nutations of the figure axis which have frequencies lying in the low frequency band extending from $(-0.5\,\Omega_0)$ to $(+0.5\,\Omega_0)$ are the only ones that are equivalent to nutations of the axis of the CIP, as explained in Section 3.2.3. Considering the frequencies $(\Omega_0 - \omega)$ in (5.256) and $(\Omega_0 + \omega)$ in (5.257), it is only the former that can fall within the low frequency range, and that too only when the excitation is by potentials of type $(n, 1)$; this is because potentials of this type are the only ones for which ω lies in the diurnal tidal band extending from $0.5\Omega_0$ to $1.5\Omega_0$. In every other case, the spectrum of the nutations of the figure axis lies outside the low frequency band; these "high frequency" nutations are to be viewed as polar motions of the CIP with the same frequencies $(-\omega$ or $\omega)$ as the respective wobbles, all lying outside the retrograde diurnal band. The amplitudes of the polar motion are:

$$\tilde{p}(-\omega) = -\tilde{\eta}(\Omega_0 - \omega), \qquad \tilde{p}(\omega) = -\tilde{\eta}(\Omega_0 + \omega), \tag{5.258}$$

with the restriction that the frequencies $(-\omega)$ in the first relation are to lie outside the retrograde diurnal range.

5.6.3 Excitation by the degree two tesseral potential

It is of interest to illustrate the above with the most important case of excitation by a spectral component of the degree two tesseral potential. In this case, ω is in the diurnal frequency band centered at Ω_0, $\tilde{\Gamma}_r$ is what we had called $\Gamma^{21}(\omega_r)$ earlier, and $\tilde{\Gamma}_p$ stands for $\Gamma^{21}(\omega_p)$. We see from (5.235) and (5.237) that since $\zeta_{21} = \pi/2$, the retrograde torque is

$$\tilde{\Gamma}^{21}(\omega_r)e^{-i(\Theta_\omega(t)-\pi/2)} = i\tilde{\Gamma}^{21}(\omega_r)e^{-i\Theta_\omega(t)}, \tag{5.259}$$

$$\tilde{\Gamma}^{21}(\omega_r) = -i\Omega_0^2\bar{A}G_{21}^- H_\omega^{21}C_{20} = i\left(\frac{15}{8\pi}\right)^{1/2}\frac{g_e H_\omega^{21}}{a_e^2}\bar{A}e. \tag{5.260}$$

Here we have taken G_{21}^- from Eqs. (5.239) and (5.240), and used (5.166) to replace C_{20} by $-\bar{A}e/(M_E a_e^2)$.

The same potential acts on $(C_{22} + i S_{22})$ to produce the prograde torque given by Eqs. (5.236) and (5.239):

$$\tilde{\Gamma}^{21}(\omega_p)e^{i(\Theta_\omega(t) - \pi/2)} = -i\tilde{\Gamma}^{21}(\omega_p)e^{i\Theta_\omega(t)}, \tag{5.261}$$

$$\tilde{\Gamma}^{21}(\omega_p) = (i\Omega_0^2\bar{A})\, G_{21}^{(+)}\, H_\omega^{21}\, (C_{22} + i S_{22})$$

$$= i\left(\frac{15}{2\pi}\right)^{1/2} \frac{g_e H_\omega^{21}}{a_e^2} \bar{A}e'e^{2i\lambda_0}, \tag{5.262}$$

where we have noted from Eq. (5.166) that

$$C_{22} + i S_{22} = (\bar{A}e'e^{2i\lambda_0})/(2M_E a_e^2). \tag{5.263}$$

We see then that $(\tilde{\Gamma}_p/\tilde{\Gamma}_r) = (e'/e)e^{2i\lambda_0}$ in this example.

Considering that ω is close to Ω_0 for the excitation in the present example, and that $e' \approx 10^{-5} \approx e^2$, one sees readily that the magnitude of the second term in the numerator of (5.254) relative to the first term is about $(2e'^2/e)(\omega/\Omega_0 - 1) \approx 5 \times 10^{-8}(\omega/\Omega_0 - 1)$. So the contribution of the prograde torque to the retrograde diurnal wobbles is negligible. The reader may verify that the magnitude of the prograde wobble amplitude $\tilde{m}_p$ is close to $(2e'/e)$ times that of the corresponding retrograde $\tilde{m}_r$.

Returning to the low frequency nutations, we observe that the first factor on the right-hand side in Eq. (5.256) becomes very large for frequencies very close to Ω_0 (e.g., ± 6816 for the prograde and retrograde 18.6 year nutations in the excitations due to the degree two order one potential). This factor characterizes the so-called TOM (tilt-over-mode) resonance. (There is no such resonance in the polar motions because $(\Omega_0 \pm \omega)$ is never close to zero in $\tilde{p}(\pm\omega)$). The TOM resonance is responsible for the dominance of the low frequency part in the nutations. The extra dominance of the nutations excited by the degree two tesseral potential even among the low frequency nutations is explained by the fact that the magnitude of C_{20} is higher than that of any of the other C_{nm} and S_{nm} by a factor of the order of 1000.

5.7 Torque in the CRF and nutation arguments

We return now to the expression (5.233) for the complex combination of the equatorial components of the torque in ITRF, the terrestrial reference frame with its first axis in the Greenwich meridian, and transform it to the geocentric celestial reference frame (GCRF), the fixed equatorial celestial reference frame with its first axis in the direction of the mean equinox of J2000. As has been done in earlier sections, we shall neglect the Earth rotation variations for the purpose of this transformation. The principal planes of the two frames then coincide, and the transformation is simply a rotation through the angle (−GMST) around the common third axis to bring the first axis of the terrestrial frame into coincidence with that of the other. It follows that

$$L_1^{nm} = \Gamma_1^{nm} \cos(\text{GMST}) - \Gamma_2^{nm} \sin(\text{GMST}),$$

$$L_2^{nm} = \Gamma_1^{nm} \sin(\text{GMST}) + \Gamma_2^{nm} \cos(\text{GMST}), \qquad (5.264)$$

and hence that

$$\tilde{L}^{nm} \equiv L_1^{nm} + i L_2^{nm} = \tilde{\Gamma}^{nm} e^{i\,\text{GMST}}. \qquad (5.265)$$

Thus $\tilde{L}^{nm}$ differs from $\tilde{\Gamma}^{nm}$ by the simple factor $e^{i\,\text{GMST}}$, a fact already observed in the case of the axially symmetric Earth in Section 5.4. This relationship implies that any spectral component of $\tilde{L}^{nm}$ differs from the corresponding component of the torque $\tilde{\Gamma}^{nm}$ in the terrestrial frame only in the value of the frequency. Since $d(\text{GMST})/dt = \Omega_0$, it is evident that the frequency of any spectral component of the equatorial torque (and hence that of the nutational motion in space excited by the torque) is algebraically greater than the frequency in the terrestrial frame (both of the torque and of the wobble or polar motion excited by it) by Ω_0. That such a relation between the frequencies in the two frames exists is no surprise: it is simply a manifestation of the kinematic relations, and is shown in Table 5.4.

We now focus our attention on the spectral components of $\tilde{L}^{nm}$. Equation (5.265) taken together with Eq. (5.235) shows that the part of $\tilde{L}^{nm}$ that is produced by a spectral component of frequency ω of the tidal potential of type (nm), consists of two components,

$$\tilde{L}_+^{nm} = \tilde{\Gamma}^{nm}(\omega_p)e^{i[\text{GMST}+(\Theta_\omega - \zeta_{nm})]} \text{ and } \tilde{L}_-^{nm} = \tilde{\Gamma}^{nm}(\omega_r)e^{i[\text{GMST}-(\Theta_\omega - \zeta_{nm})]},$$

$$ \qquad (5.266)$$

which relate to the prograde and retrograde components, respectively, of the torque $\tilde{\Gamma}_\omega^{nm}$ referred to the terrestrial frame.

The two spectral components $L_\pm^{nm}$ of the torque referred to the celestial frame generate corresponding spectral components of the nutation. We denote the conventionally defined arguments of these nutations by $\Xi^\pm$. The defining relation which connects $\Xi^\pm$ to the arguments $\pm\Theta_\omega$ associated with the respective torque components in the terrestrial frame is

$$\Xi^\pm = \text{GMST} \pm (\Theta_\omega - m\pi). \qquad (5.267)$$

(It may be recalled from Eqs. (5.216) and (5.215) that the Θ_ω associated with a potential of type (nm) includes a term $m\pi$.) We can now rewrite the quantities in the exponentials in (5.266) in terms of the respective nutation arguments; then the expressions for $L_\pm^{nm}$ become

$$\tilde{L}_+^{nm} = (-1)^m \tilde{\Gamma}^{nm}(\omega_p)e^{i(\Xi^+(t)-\zeta_{nm})}, \quad \tilde{L}_-^{nm} = (-1)^m \tilde{\Gamma}^{nm}(\omega_r)e^{i(\Xi^-(t)+\zeta_{nm})}. \quad (5.268)$$

The expansion (5.219) of $\Theta_\omega(t)$ in terms of the fundamental arguments of nutation may be employed now to obtain expansions for $\Xi^\pm(t)$, in the general form

$$\Xi^\pm(t) = \text{GMST} \pm [m(\text{GMST}) + n_1'l + n_2'l' + n_3'D + n_4'F + n_5'\Omega + \cdots]. \quad (5.269)$$

Returning to the definition (5.267) of the arguments $\Xi^\pm(t)$, we observe that since $d(\text{GMST})/dt = \Omega_0$, the frequency ω_n associated with $\Xi^+(t)$ is $d\,\Xi^+(t)/dt = \Omega_0 + \omega$, where $\omega = d\Theta_\omega/dt$ stands for the frequency of the spectral component of the forcing potential as usual. So for any given order $m > 0$ of the potential, the spectrum of frequencies ω_n present in $\tilde{L}_+^{n0}$ lies in the prograde band centered at $(1 + m)\Omega_0$. Similarly, the frequencies $\omega_n = \Omega_0 - \omega$ present in $\tilde{L}_-^{n0}$ lie in the band centered at $(1 - m)\Omega_0$. This band consists of retrograde frequencies for any $m > 1$. In the case $m = 0$ where ω lies within the half-band from 0 to $\Omega_0/2$, the frequencies present in $\tilde{L}_+^{n0}$ and $\tilde{L}_-^{n0}$ constitute half-bands, which together make up the prograde diurnal band centered at Ω_0.

Consider finally the case $m = 1$. Since GMST drops out of Ξ^- when $m = 1$, the spectrum of $\tilde{L}_-^{n1}$ lies in the low frequency band centered at $\omega_n = 0$. The particular frequency $\omega_n = 0$ belongs to the spectral term with zero values for all the n_i'. This time independent term in the torque is what drives the precession. As for the other spectral terms for which at least one of the numbers $n_1', n_2', \ldots$ is nonzero, it is evident from (5.269) that along with any such term identified by a given set of values of the n_i', there exists also another term with the signs of all the n_i' reversed; the argument of the latter is the negative of the argument of the first. We denote the frequencies of these two terms, which must clearly have the same magnitude but opposite signs, by $\omega_{n\pm}$ where $\omega_{n+} > 0$ and $\omega_{n-} = -\omega_{n+}$; they characterize a pair of prograde and retrograde circular nutations. The arguments of these nutations will be denoted by Ξ_+^- and $\Xi_-^- = -\Xi_+^-$, respectively, with the subscript indicating the sign of the nutation frequency. (Thus $\omega_{n+} = d\,\Xi_+^-/dt$ and $\omega_{n-} = d\,\Xi_-^-/dt$.) The torque terms with the frequencies ω_{n+} and ω_{n-} arise from distinct components of the tidal potential, having the frequencies $\omega_+ = \omega_{n+} - \Omega_0$ and $\omega_- = -\omega_{n+} - \Omega_0$ in the terrestrial frame.

Let us consider such a pair of terms in $\tilde{L}_-^{n1}$ in the special case $n = 2$. The geopotential coefficient which gives rise to $\tilde{L}_-^{21}$ is that which is contained in $\tilde{\Gamma}^{21}(\omega_r)$, namely, C_{20}; the dominant class of low frequency nutations is generated by this torque. The two terms pertaining to frequencies ω_{n+} and ω_{n-} may be written down by making use of Eq. (5.268) along with (5.260); they are

$$\tilde{L}^{21}_{\omega_{n+}}(t) = Q_+ e^{i\Xi^-_+}, \qquad \tilde{L}^{21}_{\omega_{n-}}(t) = Q_- e^{-i\Xi^-_+}, \tag{5.270}$$

$$Q_\pm = -\left(\frac{15}{8\pi}\right)^{1/2} M_E g_e C_{20} H^{21}_{\omega_\pm}. \tag{5.271}$$

These terms generate the prograde and retrograde members of a pair of low frequency nutations, e.g., the prograde and retrograde 18.6 year nutations, or the semiannual or annual ones.

We observe now from Eq. (5.190) that the position (X_H, Y_H) of the pole of the angular momentum vector in the celestial frame is given by $(X_H + iY_H) = \int \tilde{L} dt/(C\Omega_0)$. On substituting the spectral component of $\tilde{L}$ from Eq. (5.270) we see that the corresponding spectral component of the nutation is given by

$$(X_H + iY_H)_{\omega_{n+}}(t) = -\frac{iQ_+}{C\Omega_0\omega_{n+}} e^{i\Xi^-_+},$$

$$(X_H + iY_H)_{\omega_{n-}}(t) = \frac{iQ_-}{C\Omega_0\omega_{n+}} e^{-i\Xi^-_+}, \tag{5.272}$$

so that

$$(X_H)_{\omega_{n\pm}} = (\Delta\psi_H)_{\omega_{n\pm}} \sin\epsilon_H = \frac{1}{C\Omega_0\omega_{n+}} Q_\pm \sin\Xi^-_+, \tag{5.273}$$

$$(Y_H)_{\omega_{n\pm}} = (\Delta\epsilon_H)_{\omega_{n\pm}} = \mp\frac{1}{C\Omega_0\omega_{n+}} Q_\pm \cos\Xi^-_+. \tag{5.274}$$

Of the $\pm$ and $\mp$ signs appearing in these expressions, the upper sign pertains to the prograde component of the motion and the lower one to the retrograde component. The sum of the two components represents the elliptical nutation, with its frequency ω_n taken to be either ω_{n+} or ω_{n-} according to the historical conventions regarding the signs to be assigned to the frequencies of individual elliptical nutations.

In the case of the rigid Earth, the Oppolzer terms derived in Section 2.28 enable us to calculate the nutations of the rotation and figure axes from the above results for the angular momentum axis.

The zero-frequency term mentioned earlier in this section may now be considered. It has $\tilde{L}^{21}_{\omega_n=0} = Q_0$, which leads to

$$(X_H + iY_H)_{\omega_n=0} = \frac{Q_0}{C\Omega_0} t = \frac{g_e}{a_e^2} H^{21}_{\omega=\Omega_0} H_d t. \tag{5.275}$$

This quantity, which varies linearly with t, represents precession. It is real and therefore $(Y_H)_{\omega_n=0} = 0$, meaning that the precession is in longitude only. The coefficient of t, divided by $\sin\epsilon_0$, is the rate of precession in longitude.

Expressions for the torque in terms of the amplitudes of the tidal potential, equivalent to (5.270), were obtained by Melchior and Georis (1968) and used for the computation of nutation amplitudes. Similar studies were made by Capitaine (1982) and others. More recent computations by Hartmann and Wenzel (1995a and 1995b) made use of tidal amplitudes H_ω obtained by spectral analysis of the orbital coordinates of the Moon and the Sun given in the time domain by the ephemerides DE200-LE200 in Standish (1990, 1998) and Standish *et al.* (1995). This approach had the advantage that the effects of the mutual gravitational interactions of the Moon, Sun, and the planets on the orbit of each body are automatically taken into account as they are included in the ephemerides. (If the orbits were computed from analytical treatments of the dynamics, the perturbations of the orbits by various influences would have to be considered and computed individually.) However, they encountered the problem of distinguishing between very closely spaced frequencies present in the time dependent tidal potential, and though special techniques were devised for overcoming this problem some uncertainties remained in the tidal amplitudes that resulted. These were not significant in applications to tidal phenomena other than nutations, but when computing nutation amplitudes, the small errors got amplified because of division by the frequency ω_{n+} in space (see Eqs. (5.273)), especially in the case of very long period nutations.

Roosbeek and Dehant (1998) employed series expansions of the ecliptic coordinates of the Moon, Sun, and planets (obtained from ELP2000 and VSOP87) to compute the corresponding expansions for the torque in the celestial frame; each term in the series is characterized by an argument which is a linear combination of the same fundamental arguments as in (5.269). The derivation of the torque components led to expressions in the form of similar series; and the nutation series for the angular momentum axis and the precession of the axis were then obtained by integration of the torque as in (5.275). The full treatment involved the application of corrections for small effects, which are not taken into account in the above process. See Section 5.12 for a brief explanation of the corrections.

The value of H_d to be used in making calculations of the various expressions above has to be estimated from the first-order lunisolar precession rate that is obtained from observational estimates of the precession rate by subtracting the contributions to precession from a variety of small effects, as was done by Williams (1994), see Section 5.12.

5.8 Nutation of the figure axis

Up to this point, our presentation of the treatments of nutation and precession through torque equations, whether in the CRF or in the TRF, have not been exact: we made use of either the kinematical relations (2.97) or (2.105), which are very good but not quite exact; or we made equivalent approximations as in the first paragraph of Section 5.7, where the transformation of the torque between the terrestrial and celestial frames was done with neglect of the effect of the variations in Earth orientation on the relation between the two frames.

In this section we follow closely (except for some changes in notation) the work of Bretagnon *et al.* (1997, 1998), which avoids such approximations by making use of the exact kinematical relation (3.9). These authors start out with the torque equations governing the temporal variation of the components of the angular velocity vector of a rigid body in the terrestrial reference frame ITRF, which has the true equatorial plane of date (perpendicular to the instantaneous figure axis) as its principal plane and has its coordinate axes along the Earth's principal axes. These are just the Euler equations (2.43); but we shall change the notation by adding primes to the symbols used there. Since we also have to deal with quantities referred to the ITRF (with its x-axis in the Greenwich meridian), we reserve the unprimed symbols for those. With the changed notation, we write the Euler equations as

$$A\dot{\Omega}'_x + (C - B)\Omega'_y\Omega'_z = \Gamma'_x,$$
$$B\dot{\Omega}'_y + (A - C)\Omega'_z\Omega'_x = \Gamma'_y, \qquad\qquad (5.276)$$
$$C\dot{\Omega}'_z + (B - A)\Omega'_x\Omega'_y = \Gamma'_z.$$

Expressions for the components of the external gravitational torque on the Earth in terms of the geopotential coefficients C_{nm}, S_{nm} were derived in Section 5.3.7. These coefficients are defined relative to the ITRF, and the torque components referred to the same frame are denoted by Γ_x, Γ_y, Γ_z.

To compute the nutation and precession of the figure axis (identified with the third axis of the ITRF), we need to set up and solve the equations governing the motion of the pole of this axis in space. Bretagnon *et al.* (1997) employed the exact kinematical relations derived in Section 3.2.2 to arrive at the relevant equations. The kinematical relations express the components Ω_x, Ω_y, Ω_z of the angular velocity vector relative to the ITRF in terms of the time derivatives of the Euler angles Ψ, θ, φ involved in the transformation to the ITRF from the space-fixed frame referred to the mean equinox and ecliptic of J2000. (This is the frame which has the ecliptic of the fixed epoch J2000 as its principal plane and the direction of the mean equinox of that epoch as its first coordinate axis, which is the origin of longitude in this frame.) The transformation, which involves the three Euler angles Ψ, θ, φ, is described in Section 3.2.1.

If one were to choose the terrestrial frame to be the principal axis frame (instead of the ITRF), then the Euler angle φ would have to be replaced by $\tilde{\varphi}$, with

$$\tilde{\varphi} = \varphi + \alpha, \qquad\qquad (5.277)$$

where α is the longitude of the first principal axis in the ITRF.

The temporally varying parts $\Delta\Psi = \Psi(t) - \Psi_{J2000}$ and $\Delta\theta = \theta(t) - \theta_0$ of Ψ and θ are denoted by $-\psi_A - \Delta\psi$ and $-\Delta\epsilon$. (Here $\theta_0 = -\epsilon_0 \approx -23°.5$.) ψ_A, $\Delta\psi$, and $\Delta\epsilon$ have been defined in Section 2.16 as the precession and nutation in longitude and the nutation in obliquity, respectively, relative to the fixed ecliptic. They correspond closely to the nutation–precession in longitude and obliquity, respectively, as defined classically relative to the moving ecliptic. The latter include a slow secular variation of the obliquity due to the very slow rotation of the ecliptic plane in space. The angle φ represents the axial rotation of the Earth in space:

$$\varphi = \varphi_0 + \varphi_1 t + \Delta\varphi. \tag{5.278}$$

Now, φ is the angle to the Greenwich meridian from the intersection between the ecliptic of J2000 and the true equator of date (point R in Fig. 3.1). But the Greenwich sidereal time (GST), as defined in Section 3.2.3, is measured from the true equinox of date (γ_D in the same figure); so φ is not identical to GST but is extremely close to it. Hence φ_1 differs from $d(\text{GST})/dt$ by a very small amount as a consequence of the variation of the arc $R\gamma_D$ (denoted by χ in the same figure) which is caused by the slow motion of the ecliptic plane (see Bretagnon *et al.*, 1997, where the values of φ_0 and φ_1 are given).

The kinematic relations that are relevant when the terrestrial frame employed is the principal axis frame, as here, are those obtained by making the replacements indicated in Eq. (3.10) in Eq. (3.8) within the same section. On inverting the relations thus obtained, we find that:

$$\Omega'_x = \dot\theta \cos\tilde\varphi + \dot\Psi \sin\theta \sin\tilde\varphi,$$
$$\Omega'_y = -\dot\theta \sin\tilde\varphi + \dot\Psi \sin\theta \cos\tilde\varphi, \tag{5.279}$$
$$\Omega'_z = \dot\Psi \cos\theta + \dot\varphi.$$

These equations constitute the exact kinematical relations. We transform them from the principal axis frame to a frame which has its first axis along the nodal line referred to above. This is accomplished by a rotation through $-\tilde\varphi$ around the third axis. The resulting equations are:

$$(A\dot\Omega'_x + (C - B)\Omega'_y\Omega'_z)\cos\tilde\varphi - (B\dot\Omega'_y + (A - C)\Omega'_z\Omega'_x)\sin\tilde\varphi$$
$$= \Gamma'_x \cos\tilde\varphi - \Gamma'_y \sin\tilde\varphi = \Gamma_x \cos\varphi - \Gamma_y \sin\varphi, \tag{5.280}$$
$$(A\dot\Omega'_x + (C - B)\Omega'_y\Omega'_z)\sin\tilde\varphi + (B\dot\Omega'_y + (A - C)\Omega'_z\Omega'_x)\cos\tilde\varphi$$
$$= \Gamma'_x \sin\tilde\varphi + \Gamma'_y \cos\tilde\varphi = \Gamma_x \sin\varphi + \Gamma_y \cos\varphi, \tag{5.281}$$
$$C\dot\Omega'_z + (B - A)\Omega'_x\Omega'_y = \Gamma'_z. \tag{5.282}$$

On substituting Eq. (5.279) into the left-hand members of these equations, one gets three expressions involving the Euler angles and their first and second derivatives.

We denote them by D_1, D_2, D_3.

$$D_1 = C\dot{\Psi}\dot{\varphi}\sin\theta + A\ddot{\theta} + \frac{1}{2}(C-A)\dot{\Psi}^2\sin2\theta - (B-A)F_1, \tag{5.283}$$

$$D_2 = -C\dot{\theta}\dot{\varphi} + A\ddot{\Psi}\sin\theta + (A+B-C)\dot{\Psi}\dot{\theta}\cos\theta - (B-A)F_2, \tag{5.284}$$

$$D_3 = C\ddot{\varphi} + C\ddot{\Psi}\cos\theta - C\dot{\theta}\dot{\Psi}\sin\theta - (B-A)F_3, \tag{5.285}$$

with

$$F_1 = \frac{1}{2}\ddot{\Psi}\sin2\tilde{\varphi}\sin\theta + \dot{\varphi}\dot{\Psi}\cos2\tilde{\varphi}\sin\theta - \dot{\theta}\dot{\varphi}\sin2\tilde{\varphi},$$

$$-\frac{1}{2}\ddot{\theta}(1-\cos2\tilde{\varphi}) + \frac{1}{4}\dot{\Psi}^2(1+\cos2\tilde{\varphi})\sin2\theta, \tag{5.286}$$

$$F_2 = -\frac{1}{2}\ddot{\Psi}(1+\cos2\tilde{\varphi})\sin\theta + \dot{\varphi}\dot{\Psi}\sin2\tilde{\varphi}\sin\theta + \dot{\theta}\dot{\varphi}\cos2\tilde{\varphi}$$

$$+\frac{1}{2}\ddot{\theta}\sin2\tilde{\varphi} + \frac{1}{4}\dot{\Psi}^2\sin2\tilde{\varphi}\sin2\theta, \tag{5.287}$$

$$F_3 = \frac{1}{2}\dot{\theta}^2\sin2\tilde{\varphi} - \dot{\Psi}\dot{\theta}\cos2\tilde{\varphi}\sin\theta - \frac{1}{4}\dot{\Psi}^2\sin2\tilde{\varphi}(1-\cos2\theta). \tag{5.288}$$

For the torque components appearing as the right-hand members in Eqs. (5.280)–(5.282), which we denote by L_1, L_2, L_3, explicit expressions may be obtained by substituting from Eqs. (5.177) and (5.181), noting that $\Gamma_x = \mathrm{Re}\,\tilde{\Gamma}$ and $\Gamma_y = \mathrm{Im}\,\tilde{\Gamma}$:

$$\binom{L_1}{L_2} = \frac{GM_EM_B}{d}\sum_{n=2}^{\infty}\left(\frac{a_e}{d}\right)^n\left[\sum_{m=0}^{n-1}P_n^{m+1}(\cos\beta)\right.$$

$$\times\binom{-C_{n,m}\sin[(m+1)\Lambda+\varphi] + S_{n,m}\cos[(m+1)\Lambda+\varphi]}{C_{nm}\cos[(m+1)\Lambda+\varphi] + S_{nm}\sin[(m+1)\Lambda+\varphi]}$$

$$+\sum_{m=1}^{n}(n+m)(n-m+1)P_n^{m-1}(\cos\beta)$$

$$\left.\times\binom{-C_{n,m}\sin[(m-1)\Lambda-\varphi] + S_{n,m}\cos[(m-1)\Lambda-\varphi]}{C_{nm}\cos[(m-1)\Lambda-\varphi] + S_{nm}\sin[(m-1)\Lambda-\varphi]}\right]. \tag{5.289}$$

We do not write down the expression for L_3 which is the same as Γ_z of Eq. (5.181). The variations of the Euler angles Ψ, θ, φ are now governed by the set of coupled second-order differential equations

$$D_1 = L_1, \qquad D_2 = L_2, \qquad D_3 = L_3, \tag{5.290}$$

with the D_i given by Eqs. (5.283)–(5.285) taken together with Eqs. (5.286)–(5.288), and the L_i by Eqs. (5.289) and (5.181).

In seeking to obtain the solution of Eq. (5.290), it is convenient to rearrange them, retaining on the left-hand side only the terms linear in the derivatives of the Euler angles:

$$\ddot{\theta} + \dot{\Psi}\varphi' \sin\theta_0 = \frac{1}{A}[L_1 + G_1 + (B - A)F_1], \tag{5.291}$$

$$\ddot{\Psi} \sin\theta_0 - \dot{\theta}\varphi' = \frac{1}{A}[L_2 + G_2 + (B - A)F_2], \tag{5.292}$$

$$\ddot{\varphi} = \frac{1}{C}[L_3 + G_3 + (B - A)F_3], \tag{5.293}$$

$$\varphi' = \frac{C}{A}\varphi_1, \tag{5.294}$$

$$G_1 = -C\dot{\Psi}\varphi_1(\sin\theta - \sin\theta_0) - C\dot{\Psi}\Delta\dot{\varphi}\sin\theta - \frac{1}{2}(C - A)\dot{\Psi}^2 \sin 2\theta, \tag{5.295}$$

$$G_2 = C\dot{\theta}\Delta\dot{\varphi} - A\ddot{\Psi}(\sin\theta - \sin\theta_0) - (A + B - C)\dot{\Psi}\dot{\theta}\cos\theta, \tag{5.296}$$

$$G_3 = C(\dot{\theta}\dot{\Psi}\sin\theta - \ddot{\Psi}\cos\theta). \tag{5.297}$$

The main terms on the right-hand side of Eqs. (5.291)–(5.293) are the torques L_i which excite the rotation variations. As regards the other terms, the G_i are of the second and higher orders in the small quantities $\dot{\Psi}$, $\Delta\dot{\varphi}$, and $\dot{\theta} = \Delta\dot{\theta}$, and the F_i terms include the small factor $(B - A)$, which is only about $10^{-5}A$. The equations are then quite amenable to solution by an iterative process wherein all the small terms are neglected at the start of the process.

At the outset, the first two of the equations are converted to two pairs of first-order equations by the so-called method of variation of constants. One assumes the solutions to have the form

$$\dot{\theta} = K_s \sin\varphi't + K_c \cos\varphi't, \tag{5.298}$$

$$\sin\theta_0 \dot{\Psi} = K_c \sin\varphi't - K_s \cos\varphi't. \tag{5.299}$$

If K_s and K_c were constants, the above expressions would satisfy the homogeneous equations obtained by setting the right-hand sides of (5.291) and (5.292) equal to zero. However, if we take K_c and K_s to be time dependent, and substitute these expressions into the full inhomogeneous equations, we see readily that the equations would be satisfied, provided that

$$A\dot{K}_s = [L_1 + G_1 + (B - A)F_1]\sin\varphi't - [L_2 + G_2 + (B - A)F_2]\cos\varphi't, \tag{5.300}$$

$$A\dot{K}_c = [L_1 + G_1 + (B - A)F_1]\cos\varphi't + [L_2 + G_2 + (B - A)F_2]\sin\varphi't. \tag{5.301}$$

The problem is now reduced to that of integrating these two first-order equations to determine K_s and K_c as functions of time, and then substituting them into Eqs. (5.298) and (5.299) and carrying out a direct integration of those equations to obtain Ψ and θ. For actual integration of the equations for K_s and K_c, one needs to know the time dependence of the right-hand members of the above pair of equations. The terms involving the F_i and the G_i depend on the Euler angles which are unknown to begin with; but these terms are very small and can be neglected in the initial integrations, as was noted earlier. The L_i too contain φ, which may be approximated by $\varphi_0 + \varphi_1 t$ to start with. Once this is done, the time dependence of the L_i may be determined by making use of the time dependent position coordinates of the gravitating external bodies, which are obtainable from the appropriate ephemerides (the VSOP series for the Sun and the planets, ELP for the Moon, see Chapter 4). Integration of the approximated Eqs. (5.300) and (5.301) may be done to obtain K_s and K_c, and Ψ and θ may then be obtained by integration of Eqs. (5.298) and (5.299). The newly determined Ψ and θ may be substituted into the terms on the right-hand side of Eq. (5.293), while neglecting the still-unknown $\Delta\varphi$, before performing a double integration of the equation to obtain φ. The whole set of Eqs. (5.300) and (5.301), (5.298) and (5.299), and (5.293) may then be reintegrated in the same sequence as before after introducing in the right-hand members the approximate solutions already obtained. The process may be iterated further if necessary to obtain the final converged solutions to the required degree of accuracy.

There is one element of the general procedure just outlined that was left unaddressed: the fact that one needs to input the initial conditions on K_s, K_c, Ψ, θ, φ, and $\dot\varphi$ to make the determination of these quantities complete. It is of relevance, in this context, to consider the terms linear in time, (say, $P_\Psi t$ and $P_\theta t$), in Ψ and θ; they represent the precessions in longitude and obliquity. Their contributions to $\dot\Psi$ and $\dot\theta$ are P_Ψ and P_θ, respectively; and those to $\ddot\Psi$ and $\ddot\theta$ are zero. If we denote the constant parts of L_1 and L_2 by X_1 and X_2, one sees then from Eqs. (5.291) and (5.292) that

$$C\varphi_1 P_\Psi \sin\theta_0 = X_1, \qquad -C\varphi_1 P_\theta = X_2, \qquad (5.302)$$

to the lowest order.

The values of X_1 and X_2 are to be obtained from the expressions (5.289) for L_1 and L_2. One might wonder, at first glance, whether these expressions do have any constant parts at all. It is pretty obvious, in fact, that any term therein having $m \neq 0$ cannot yield any time independent part; but the quantities $P_n^1(\cos\beta)\sin(\Lambda + \varphi)$ and $P_n^1(\cos\beta)\cos(\Lambda + \varphi)$ in the $m = 0$ terms in L_1 and L_2 would need to be scrutinized. Let us restrict our consideration to degree $n = 2$, which is the most important. Since $P_2^1(\cos\beta) = 3\cos\beta\sin\beta$ and since the terrestrial Cartesian coordinates (x, y, z) corresponding to (β, Λ) are given by the familiar expressions $x = d \cos \beta \cos \Lambda$, etc., it becomes evident that $P_2^1(\cos \beta) \sin(\Lambda + \varphi)$ in L_1 is proportional to $z(y \cos \varphi + x \sin \varphi)$. It may be readily recognized that this quantity is simply ZY, where (X, Y, Z) are the coordinates of the external body in a frame which differs from the ITRF in having its first axis in the direction of R; this frame becomes the celestial frame GCRF if we ignore the nutation–precession motions of the equatorial plane in space. Let us make this approximation and make the further idealization as in Section 2.24 that the body (taken to be the Sun) is moving with a uniform angular velocity p in a circular orbit in the ecliptic which is assumed to remain fixed. Then (X, Y, Z) become (d_X, d_Y, d_Z) with the explicit time dependence given by Eq. (2.70): both d_Y and d_Z are proportional to $\sin pt$ while d_X is proportional to $\cos pt$. It becomes evident then, in this approximation, that YZ is proportional to $\sin^2 pt = (1 - \cos 2pt)/2$, which has a time independent part. One sees consequently, on working backwards, that L_1 does have a time independent part X_1. A similar analysis of L_2 would show that it is proportional to $\cos pt \sin pt$ and hence has no constant part: $X_2 = 0$. If we had not used the approximations and simplifications made above, the results would still be qualitatively the same, though the precise values of X_1 and X_2 could not be obtained so simply.

Turning now to Eqs. (5.300) and (5.301), we observe that their integration yields

$$K_s = (K_s)_0 - \frac{X_1}{A\varphi'} \cos \varphi' t - \frac{X_2}{A\varphi'} \sin \varphi' t + \cdots, \tag{5.303}$$

$$K_c = (K_c)_0 + \frac{X_1}{A\varphi'} \sin \varphi' t - \frac{X_2}{A\varphi'} \cos \varphi' t + \cdots, \tag{5.304}$$

where only the contributions from the constant terms in the torques are explicitly shown; everything else is encompassed under the dots. Since we are interested only in the nutation–precession due to forcing, we can set $(K_s)_0 = (K_c)_0$. Then substitution into Eqs. (5.298) and (5.299) shows that

$$\dot{\theta} = -(X_2/A\varphi') + \cdots, \qquad \dot{\Psi} \sin \theta_0 = (X_1/A\varphi') + \cdots, \tag{5.305}$$

which are in agreement with the precession values shown in (5.302) since φ', as defined in (5.295), is $(C/A)\varphi_1$.

The other integration constants needed to complete the solutions for Ψ and θ are their values at J2000. They are, by definition, $\Psi_0 = 0$ and $\theta_0 = -23°.43928\ldots$ Finally, the integration constants for Eq. (5.293), namely the values of φ and $\dot{\varphi}$ at J2000, are still needed. They are

$$\varphi_0 = (\text{GMST})_0, \qquad \varphi_1 = \Omega_0 + \dot{\chi}, \tag{5.306}$$

where χ, represented by the arc from R to γ_D in Fig. 3.1, arises from the motion of the ecliptic in space. $\dot{\chi}/\Omega_0$ is extremely small, being only about 2.2×10^{-10}. As for $(GMST)_0$, it is about $280°.4606$.

Tables of geopotential coefficients obtained by analysis of satellite orbit data suffice, when taken together with an adopted value of C, to provide the values of all the parameters entering into the equations of motion. The equations are solved by an iterative process as already remarked. The final solutions obtained for Ψ and θ after the iterations converge to within a specified accuracy are expressed as sums of secular terms (including powers of t multiplied by periodic factors) and purely periodic terms, which represent nutations. The arguments of the spectral components are written as linear combinations of a set of fundamental arguments which differs from the classical Delaunay arguments. For the actual arguments used, and the reasons for the choice differing from the classical ones, see Section 5.10 below.

5.9 Numerical integration

Another way to compute the nutations is to take the evolving instantaneous positions of the celestial bodies involved from the appropriate ephemerides (e.g., DExxx/LExxx) and to compute, at each moment, the torque acting on Earth. This does not provide us with amplitudes of nutation in the spectral domain: it gives the instantaneous positions of the pole of the nutating and precessing axis over the desired time interval. This is the approach used by Kubo and Fukushima (1988). It must be mentioned that the authors, who have computed rigid Earth nutation series, have also made computations in the time domain directly from the ephemerides, in order to compare their original results with the numerical results in the time domain, to verify the precision of their computation. This has been done by Souchay and Kinoshita (1991), Souchay (1993), Bretagnon *et al.* (1997 and 1998), and Roosbeek and Dehant (1998). As the positions given by the ephemerides as functions of time contain the effects of the perturbations of the Keplerian orbits, second-order coupling, indirect planetary effects, etc., all these are taken into account at once in the numerical computations, and need not (and indeed cannot) be computed individually.

5.10 Arguments of the nutations: additional remarks

We presented in Sections 5.5.5 and 5.5.6 the classical sets of fundamental arguments employed in tidal and nutation theories, and observed that while the arguments are very nearly linear in time, they do include small non-linear terms which cause slow variations of the fundamental frequencies and hence also of the frequencies of all the spectral terms. If one prefers to work with constant frequencies, one could split the argument Ξ of a generic nutation term into a part $\alpha_0 + \omega_c t$ and the remaining part, say $\Delta\Xi$, involving the second and higher powers of time. Then one writes

$$\sin \Xi = \sin(\alpha_0 + \omega_c t) + \cos(\alpha_0 + \omega_c t)\,\Delta\Xi, \tag{5.307}$$

to the first order in $\Delta\Xi$. Thus the price to pay for switching over to the constant frequency ω_c from the varying $d\,\Xi/dt$ is the creation of an out-of-phase term (cosine as against the in-phase term involving sine) with an amplitude which has quadratic and higher degree dependences on the time variable. Similarly, if one started with $\cos \Xi$, one would get an out-of-phase (sine) term with a variable amplitude. Terms in the spectral representation which have amplitudes depending polynomially on time are usually referred to as *Poisson terms*.

A fact that has been remarked upon (see, for example, Bretagnon *et al.*, 1997, Roosbeek and Dehant, 1998) is that certain of the fundamental arguments among the set consisting of the Delaunay variables l, l', D, F, Ω, and the planetary arguments λ_i are very nearly equal to some other argument or some linear combination of the arguments. For example, the difference between λ_3 (the mean tropic longitude of the Earth) and l' is the mean longitude of the perigee of the Earth which moves with a very long period of about $111\,600$ years. So the fundamental frequencies associated with λ_3 and l' differ only by a minuscule amount, and it seems redundant to have both of them as fundamental arguments. Adopting this view, Bretagnon *et al.* dropped the Delaunay argument l' from the set of fundamental arguments, replacing it by λ_3 and adding Poisson terms as in (5.307) with amplitude $-k_2\lambda_{\text{perigee}}$ to take account of the difference $(l' - \lambda_3)$ that was left out of the argument Ξ (k_2 is the multiplier of l' in Ξ). Unlike in the general case, $\Delta\Xi$ here is linear in time since λ_{perigee} is. (Roosbeek and Dehant (1988) made the opposite choice, retaining l' and replacing λ_3 by l' while adding Poisson terms.) Bretagnon *et al.* went one step further and dropped Ω too from the set of fundamental arguments, replacing it by the combination $\lambda_3 + D - F$ which is equal to $\Omega_{\text{J2000}} + 180°$, where Ω_{J2000} is the node of longitude of the Moon referred to the equinox of J2000 (as against the equinox of date for Ω). The periods of the Ω and Ω_{J2000} are -6798.38 and -6798.48 days, respectively, so that the frequency difference between the two is extremely small. Of course, Poisson terms were added to take account of the difference $\Omega - (\lambda_3 + D - F)$ which was dropped from the argument. Though dropping one or more of the fundamental arguments results in a reduction of the total number of spectral terms in the nutation series, it is at the expense of introducing Poisson terms for many of the frequencies. A second, rather unfortunate, consequence is that since the set of spectral frequencies in the nutation series with the reduced number of fundamental arguments (e.g., the SMART series of Bretagnon and collaborators) is different from that in any series employing the full set of classical arguments, it is not possible to make a direct comparison between the two kinds of nutation series. Neither is it possible to convert the series with the smaller number of fundamental arguments to a form which has the full set of arguments (though the reverse transformation is possible).

5.11 Orbital motions of the solar system bodies

An accurate determination of the positions of the Sun, Moon, and the extraterrestrial planets relative to the Earth as functions of time is fundamental to the computation of the potential energy of interaction of the Earth with these bodies as a function of time. This is also a prerequisite to the determination of the variations in Earth rotation by calculations based on the torque approach or any other approach. In an approximation which neglects the mutual interactions of the different planets, each planet moves in an elliptical Keplerian orbit around the Sun. (It would be more precise to say that the center of mass of each planet and its moons moves in such an elliptical orbit; this will be taken as understood.) Though the planes of the different planetary orbits are distinct, their inclinations to the ecliptic are small. It is natural therefore to refer the positions of the bodies to a heliocentric coordinate system based on the ecliptic and equinox. When the mutual interactions of the planets are taken into account, the orbits are perturbed to a small extent. The time dependent positions of the planets are presented in analytical form and as numerical tables in the relevant ephemerides: the analytical ephemerides VSOP82 (Bretagnon, 1982) and its later versions, and the numerical ones of the DExxx series developed at JPL (see Sections 4.6 and 4.5).

In considering the variations in Earth rotation that result from the gravitational action of external bodies, the action of the Moon and Sun are the most important, by far. It makes sense, therefore, to begin the computation of the orbital motions by limiting consideration to the Earth–Moon–Sun system only, and developing the theory for the determination of the motions of these three bodies. This task is commonly referred to as the "main problem." The larger problem includes also the motions of all the planets other than the Earth itself; action of these planets, which is very much weaker, is to be dealt with at the second stage.

5.11.1 The main problem

The potential energies of mutual gravitational interactions of the Sun, Moon, and the Earth consist of three terms. They are $\frac{-GM_EM_S}{r_S}$, $\frac{-GM_EM_M}{r_M}$, and $\frac{-GM_MM_S}{r_{MS}}$ where r_S, r_M are the distances of the Sun and the Moon from the geocenter, and r_{SM} is the distance between the Sun and the Moon. The time dependence of the gravitational action of the Sun and the Moon on the Earth is determined fully by the first two potential energy terms and hence by the variations of r_S and r_M with time. But these variations are influenced by the Sun–Moon interaction governed by the time dependent distance r_{MS} which is nothing but $|\mathbf{r}_M - \mathbf{r}_S|$, where $\mathbf{r}_M, \mathbf{r}_S$ are the vectors from the geocenter to the Moon and the Sun, respectively. So the objective of a full treatment of the three-body problem is to determine the time dependence of these two geocentric vectors, i.e. the positions of the Moon and the Sun in their orbits, relative to the geocenter, as functions of time. This is the "main problem."

Analytical treatments of this problem start with the approximation of neglecting the Sun–Moon interaction. The apparent orbital motion of the Sun relative to Earth and the orbital motion of the Moon relative to the Earth are then independent of each other: each is a Keplerian ellipse in its own plane, with its own eccentricity and orientation in space which remain invariant. At the next step, one takes into account the effect of the Sun–Moon interaction on these two orbital motions. An immediate consequence which should be expected is that these orbits cease to be strictly elliptical, or even strictly planar, though the perturbations from the original orbits are very small. A more prominent effect is the precession of the lunar orbit around the pole of the ecliptic with a period of 18.6 years while keeping a constant angle of about 27.5° with the ecliptic plane; this precession is the result of the torque produced on the Earth–Moon system about their center of mass by the action of the Sun.

The full solution of the three-body problem of the Sun–Earth–Moon system is accomplished by perturbation methods, beginning with the two Keplerian orbits as starting solutions. The orbital motions of the Sun and the Moon relative to the Earth, thus determined, may be represented by the time dependent coordinates of the Sun and the Moon. The coordinates which are the natural choice for this purpose are those referred to the ecliptic-and-equinox reference frame with its origin at the geocenter. In analytical theories, the temporal variations of the coordinates are described by a superposition of spectral components with a rich spectrum of frequencies; these frequencies are the time derivatives of linear combinations of the fundamental arguments defined in Section 5.5.6.

The orbital motions of the Earth and Moon as obtained by the solution of the main problem are not accurate enough for the computation of Earth rotation variations to the level of accuracy of present-day observational data: the interactions that have been left out of consideration in the above treatment have non-negligible effects at that level. The first of these consists of the perturbations of orbits in the Sun–Moon–Earth system that results from the gravitational attraction by the planets. The second one concerns the fact that the Earth's gravitational potential is not strictly spherically symmetric as had been assumed in the treatment of the main problem. The Earth's ellipsoidal structure, characterized by the parameter J_2 (or by e or H_d), causes a deviation from spherical symmetry of the gravitational field, which is of sufficient magnitude to have a non-negligible effect on the orbit of the Moon. These effects are summarized below.

5.11.2 Planetary perturbations of Moon's and Earth's orbits: indirect planetary effect

The gravitational action of the planets (other than the Earth) on the Moon and on the Earth causes further perturbations of the orbital motions of these two bodies, and resultant increments to the lunar and solar torques on the Earth. These are referred to as the *indirect planetary effect*. One consequence of these perturbations is that the five main fundamental arguments l, l', F, D, Ω are no longer sufficient to characterize the periodicities present in the motions of the Sun and the Moon relative to the Earth: one needs additional arguments relating to the motions of the various planets, as noted already below Eq. (5.219) in Section 5.5.5. The perturbation of the Moon's orbit, in particular, includes a tilting of that orbit relative to the ecliptic plane by about $1.5''$, which is called *planetary tilt*. This tilt results in increments in the ecliptic longitude and latitude of the Moon, wherein terms appear with arguments which are no longer pure linear combinations of the fundamental arguments but have constant phase shifts added on. As a consequence of the phase shifts, the nutations having the arguments Ω and 2Ω (with periods of 18.6 years and 9.3 years, respectively) acquire non-negligible "out-of-phase" parts even for the rigid Earth; moreover, the obliquity of the Earth's axis develops a secular variation (obliquity rate) of -0.254 mas/year. The contribution from the planetary tilt to the precession rate is -0.056 mas/year. The numbers presented here are from Williams (1994), where a transparent treatment of this tilt effect as well as the effect of the J_2-tilt (considered immediately below) is given. This paper presents also a comparison with the results of others, for example, Kinoshita and Souchay (1990) and earlier works by Kinoshita (1975, 1977) and Woolard (1953).

5.11.3 The J_2-tilt effect

The non-central part of the Earth's gravitational potential arising from the ellipsoidal structure of the Earth represented by the J_2 parameter acts on the Moon to produce a small tilt ($0.8''$) of the plane of the lunar orbit. The effect of this tilt on the Earth's rotation variations goes under the name J_2-tilt effect. The treatment of Williams (1994) leads to a contribution of -2.630 mas/year to the precession rate, and contributions of -1.478 mas and 0.156 mas, respectively, to the 18.6 year nutations in longitude and obliquity besides smaller contributions to the 9.3 year and 13.66 day nutations.

In the Hamiltonian approach of Kinoshita–Souchay (see Souchay *et al.*, 1999), this effect is called the "spin–orbit coupling effect" because they mention that this effect is related to the perturbation induced by the Earth flattening on the orbital motion of the Moon.

5.11.4 Direct planetary torques exerted on the Earth by the planets

Our main focus in the foregoing sections has been on the time dependence of the orbital positions of the Moon and the Sun under the mutual interactions of all the solar system bodies. The effect of the planets on the lunar orbit in particular was considered in Section 5.11.2. Besides this indirect effect, planets also have a direct torquing action on the Earth. The torque due to a planet may be computed, as a function of the position coordinates of the planet concerned, as the negative of the torque due to the Earth's ellipticity on the planet. A simple derivation of the precession rate caused by this torque may be found in Williams (1994). He finds that the direct planetary contributions to the precession rate amount to 0.318 mas/year, with Venus and Jupiter accounting for all but 5% of this amount.

5.12 Additional small effects

5.12.1 Second order effects: nutation on nutation

Because the Earth's orientation changes in the CRF due to precession, obliquity variation, and nutations, the torques acting on the Earth from the Moon and the Sun are changing with time. This induces corrections on the nutations called the precession/nutations-on-the-nutations effects (it contains implicitly the effects of precession and obliquity rate on the nutations). Let us consider for instance the expression of nutation due to the lunisolar torque as given by (2.38), (2.66), and (2.107). In the arguments used to characterize the motions of the celestial objects involved in these expressions, fundamental arguments are used as stated before involving the true coordinates such as the true longitude of the Sun and the Moon. The true longitude of the celestial body for instance can be written:

$$\lambda_{\text{body}} = \lambda_{\text{mean body}} + \psi = \lambda^0_{\text{body}} + \delta\dot{\psi}t + \delta\psi, \qquad (5.308)$$

where $\psi = \psi_0 + \dot{\psi}_0 t + \delta\dot{\psi}t + \delta\psi$. This last term is usually not considered in the fundamental arguments. For that reason, all terms involving the longitude of the celestial body λ_{body} contain implicitly the effect of the change in the Earth orientation (the nutation in longitude essentially), considered to exist but to be small. For that reason, these terms can be expanded using:

$$\sin\lambda_{\text{body}} = \sin\lambda^0_{\text{body}} + \cos\lambda^0_{\text{body}}\delta\dot{\psi}t + \cos\lambda^0_{\text{body}}\delta\psi. \qquad (5.309)$$

This expansion provides additional terms in the expression of the nutations in obliquity and in longitude.

Another consequence of the change in the Earth orientation with time when computing the torques on the Earth is the change in the obliquity. Considering in Eq. (2.66), that the reference plane has changed and that consequently ϵ instead of ϵ_0 appears in the equation, one has:

$$\frac{d}{dt}(\Delta\psi_H \sin\epsilon) = \frac{L_1}{C\Omega_0}. \tag{5.310}$$

Taking $\Delta\psi_H = \Delta\psi_H^0 + \Delta\psi_H^1$, substituting $\epsilon = \epsilon_0 + \Delta\epsilon_F \approx \epsilon_0 + \Delta\epsilon_H$ (or eventually accounting for the Oppolzer terms), considering (2.66) for L_1, and neglecting second-order terms, the only remaining part (with respect to the classical first-order relation between the torque and the nutations, involving $\Delta\psi_H^0$ and $\Delta\epsilon_H$) is:

$$\frac{d}{dt}(\Delta\psi_H^1)\sin\epsilon_0 + \Delta\psi_H^0 \cos\epsilon_0 \frac{d}{dt}(\Delta\epsilon_H) = 0, \tag{5.311}$$

which enables one to compute the second-order additional part of the nutation in longitude $\Delta\psi_H^1$. The contributions to the other different axes of nutations are computed as previously using the Oppolzer terms.

5.12.2 Coupling effect

In the classical Euler equations used to compute nutations when using the torque approach, a coupling appears between the orientation of the ellipsoidal Earth and the orbital position of the Moon, both of which are changing with time. In computing its effect, the Earth orientation is first taken as constant, at its initial state, and then, in the second order, the time dependent solution obtained from the first calculation is employed to obtain the second-order correction to the Earth orientation. The resulting correction terms represent the "coupling effect" or the "J_2-tilt effect." These are accounted for as second-order terms appearing in the equations. A detailed computation of these terms is provided in Chapter 10 on the refinement of the nutation theory.

5.12.3 Relativistic effects: geodesic precession and nutations

Due to the fact that nutations are observed in a kinematical reference frame (looking at fixed directions in space) and that the computations are done in a dynamical reference frame (where Newtonian laws apply), relativistic effects must be taken into account (see Barker and O'Connell, 1970, 1975, Fukushima, 1991, Brumberg *et al.*, 1992, Soffel and Klioner, 1998, Klioner, 1998, 2003). These are called geodetic precession and nutations. The expression for the geodetic precession is

$$P_g = -3(na/c)^2 n/2(1 - e^2),$$ (5.312)

where n and a are the mean motion and the semi-major axis of the orbit of the Earth–Moon system about the Sun. (P_g is negative because the geodesic precession is in the direction opposite to that of the vector $\Delta\psi \sin\epsilon$ shown in Fig. 2.3.) The corrections to the precession rate and nutations should be added after the application of the transfer function to the rigid Earth nutation series.

5.13 Precession constant

The precession constant is a wording sometimes used for the constant precession rate determined from observation and corrected for relativistic effects (geodesic precession). But usually precession constant stands for the precession rate coming from observation and corrected for relativity effects, the motion of the ecliptic and the effects of the planets; it thus corresponds to the luni-solar precession rate and is proportional to the hydrodynamical flattening. It must be mentioned however that the precession constant as used in Laskar (1986, see Eq. (24)) is not strictly proportional to the dynamical flattening but has a correction proportional to the square of the flattening. In Laskar *et al.* (1993, see Eq. (9)), the numerical value of the dynamical flattening is derived from the precession rate at $t = 0$ and it is clearly stated that the precession constant depends on the model of precession used and the observational initial conditions for the precession and obliquity rates at the origin.

5.13.1 General motion: general precession in longitude, and obliquity rate

Exactly as for nutation, in addition to the classical luni-solar attraction producing a precessional motion of 50406.568 mas/year, there are contributions amounting to about the percent level of that last value, which must be taken into account. These contributions change the precession rate or precession constant. In the new IAU2000 precession–nutation model, secular corrections are introduced to precession changing the old value precession constant (IAU 1976), which was proportional to H_d the dynamical flattening (see Eq. (2.77) and Eq. (2.10)).

These contributions are due, in order of importance, to: (1) relativistic effects (geodesic precession); (2) J_2-tilt effect; (3) second-order terms; (4) direct planetary effect; (5) planetary-tilt effect; (6) J_4 contribution; and (7) effects of tidal attraction.

(1) The geodesic precession has been explained in the previous paragraph. It amounts to -19.194 mas/year as quoted by Williams (1994) (or -19.193 mas/year from Roosbeek and Dehant, 1998).

(2) The next important contribution to precession comes from the J_2-tilt effect described in Section 5.11.3. It amounts to -2.630 mas/year in the rate of precession in longitude and -0.254 mas/year in the obliquity rate (Williams, 1994). (A slightly different value, -0.251 mas/year, is found by Roosbeek and Dehant (1998), who combine the contribution from J_2-tilt to the precession rate in longitude with that from (5) below.)

(3) The second-order terms in the torque inducing precession and nutation are due to orientation changes of the Earth (precession and nutation) while the torque applies and are determined from products of two factors (see Eq. (2.66) for instance), each having a spectral expansion wherein the terms have a sine or cosine dependence on the argument. One finds, in the product of the two series, terms proportional to squares of sines or cosines, each of which has a zero-frequency part. These zero-frequency contributions to the torque produce a small increment to the precession rate. This amounts to -0.468 mas/year as taken from Williams (1994) (or -0.454 mas/year according to Roosbeek and Dehant (1998)).

(4) Exactly as for the action of the Moon and the Sun on the Earth, inducing precession, there exists a contribution due to the effect of the planetary direct attraction on Earth. For precession, the direct planetary contribution is mainly due to Venus and Jupiter, but also Mercury, Mars, Saturn, Uranus, and Neptune contribute. Similarly for the obliquity rate, Venus provides the main contribution, but the other planets contribute also. In total, the direct-planetary contributions to precession take the value of 0.318 mas/year, and the direct-planetary contributions to obliquity rate take the value of -0.014 mas/year (Williams, 1994) (or 0.301 mas/year and -0.021 mas/year, respectively, from Roosbeek and Dehant, 1998).

(5) The so-called planetary-tilt effect is an effect of the planets on the orbit of the Moon, tilting it. There is a direct effect which induces a tilt, but this is not the most important one. An additional tilt arises from the fact that the lunar orbit is carried along with the ecliptic which keeps tilting very slowly under the attractions of the planets on the Earth–Moon system. The planets also perturb the distances of the Moon, Earth, and Sun from each other relative to what they would be in the three-body (Moon–Earth–Sun) system. But these perturbations have little effect on the precession and obliquity rate, as observed by Williams (1994). The total planetary-tilt effect on precession amounts to -0.056 mas/year, as mentioned in that same paper. The total tilt (sum of effects (2) and (5)) amounts to -2.686 mas/year for Williams (1994) and -2.683 mas/year for Roosbeek and Dehant (1998).

(6) Exactly as for nutation, one can compute the contribution to precession from the J_4 contribution to torque, given by the C_4' term in Eq. (5.177). This amounts to 0.026 mas/year as reported by Williams (1984) (or 0.025 mas/year from Roosbeek and Dehant, 1998).

(7) The so-called "tidal effect" contribution to precession, which is not the effect of the tidal deformations of the planet Earth (considered only very recently and examined in the refinement of the theory in Chapter 10). This contribution to precession arises from tidal friction in the Earth and from the consequence of angular momentum conservation in the Earth–Moon system, i.e. the slowing down of the rotation rate of the Earth (at about a few tens of nanosecond per day or equivalently about 2.3 milliseconds per century) leading primarily to a phase-lag[1] and the gradual increase of the radius of Moon's orbit (at about 4.5 cm/year). This long-term change in the Earth–Moon distance (a very slow increase) and in the phase lag in the relative positions/orientations of the Moon with respect to the Earth induces, in turn, a corresponding long-term variation in the tidal torque on Earth, and consequently a variation of the precessional motion and a change in the obliquity rate amounting to 0.024 mas/year (Williams, 1994).

(8) Additionally, the contributions to obliquity and precession rates are not constant with time, and the derivatives of the phenomena mentioned above should be considered, i.e. small contributions to the obliquity and precession rates that are proportional to time (inducing changes in the obliquity and precession angles proportional to the square of the time) and to higher powers of time (inducing changes in the obliquity and precession angles proportional to the power of the time greater than 3). The largest of such contributions arises from the Earth's J_2 rate (about -3×10^{-9}). It amounts to $-0.140*T$ mas/year or 0.1 arcsecond/century where T is expressed in Julian centuries from the reference epoch J2000 (i.e. the time in days elapsed since 2000 January 1st 12h, divided by 36 525).

Summing all these contributions (see Williams, 1994), one gets total motions in space of 50 384.565 $+0.1655\,T$ mas/year as precession and -0.244 $-0.0004\,T$ mas/year as obliquity rate.

In addition to these contributions to precession and obliquity rates, there is an ecliptic motion in space (see Fig. 3.1) which results in a further contribution to these rates. These amount to -96.865 mas/year for the precession rate and -468.096 mas/year for obliquity rate (Williams, 1994, or -96.865 mas/year and -468.093 mas/year, respectively, from Roosbeek and Dehant, 1998). The necessity of including the former in the precession rate is clear when it is recalled that the precession in longitude, as classically defined, represents the motion of the equinox along the moving ecliptic. The total motion is called the general motion. The general precession rate amounts to 50 287.700 mas/year and the general obliquity rate to -468.340 mas/year (Williams, 1994). In fact, one scales the Moon's and Sun's

---

[1] The tidal bulges (both in the oceans and in the solid Earth) are pulled forward of the Moon because the Earth is not perfectly elastic and rotates faster than the Moon around the Earth. This so-called friction drag inducing a non-alignment between the tidal budge and the Moon, induces a retroaction of the Moon's gravitational attraction, which slows down the Earth rotation.

effects on the J_2 Earth gravitational potential in order to get this magnitude of the general precession (see next paragraph).

Very long baseline interferometry (VLBI) allows us to obtain the total motion in space, which is a combination of motions related to different sources. Consequently, the general precession and obliquity rates must be used together with computations from ephemerides in order to deduce the luni-solar precession, which is the dominant part.

It must be mentioned that the ecliptic motion effect on precession was named the planetary precession, which is not the planetary effect quoted here. The planetary effects contribute to the space motion with the direct planetary effect and the planetary tilt effect.

Also, the obliquity rate is sometimes called the precession in obliquity in the literature (Woolard, 1953).

5.13.2 Inference of the precession constant from observations of the luni-solar precession

The dynamical flattening $H_d = \frac{C-A}{C}$ can be deduced from the observation of precession in space. But this determination is not a direct determination: one starts from the precession in space as observed by VLBI, and one corrects for the tilt effects (planetary-tilt and J_2-tilt), for the second-order effect, for the J_4 effect, and the direct-planetary effect. One obtains then an estimation for the luni-solar precession of which the expression is proportional to H_d, and to two factors containing, respectively the Moon's and the Sun's masses relative to the Earth mass. Fixing these other constants in the luni-solar precession, it is possible to evaluate H_d.

It must be mentioned that the nutations are also proportional to H_d and when evaluating parameters involved in the expression, H_d could be one of them. In Chapter 7 on the transfer function, we will see that both the luni-solar precession and the nutations can be used together for evaluating H_d.

An evaluation of H_d from the dynamical flattening $e = \frac{C-A}{A}$ can be done using the relation

$$H_d = \frac{e}{1+e}. \tag{5.313}$$

The results obtained by various authors are presented in Figure 1 of Dehant and Capitaine (1997). Above, we have given the values for precession and obliquity rates of Williams (1994) or Roosbeek and Dehant (1998) for getting an order of magnitude of these phenomena, but there are many possibilities, any particular solution being related: (1) to the starting observed value; (2) to the corrections applied; and (3) to the ephemerides used. Each rigid Earth nutation theory can provide its own value for the global dynamical flattening (see Dehant *et al.*, 1999a). Nevertheless these values, based on the last observed precession, differ only at the level of a few parts in 10^{-6}. For instance, Williams (1994) gets a value of 0.0032737640, and Roosbeek and Dehant (1998), 0032737674.

The comparison between the theory and the observed precession shows that there is a difference between the dynamical flattening derived from precession, and the dynamical flattening computed from the hydrostatic equilibrium starting from the density profile of a spherical Earth (as given by seismology). This difference is at the percent level showing that the Earth is not in hydrostatic equilibrium. An Earth model which accounts for mantle tomography heterogeneities may provide a better correspondence, but the parameters involved in this modeling are not yet well constrained. What has been done by Dehant and Defraigne (1997) is the construction of an Earth model based on mass anomalies in the mantle observed by seismology (mantle tomography) and their associated computed internal boundary deformation (as done in Defraigne *et al.*, 1996), and which was consistent with the observed gravity coefficient J_2 and the observed free core nutation frequency (constraint on the degree two deformation of the core mantle boundary (CMB)).

6

Deformable Earth – Love numbers

6.1 Hydrostatic equilibrium Earth models

The structure of the Earth is pictured as that of a molten ball which has since frozen into the present structure, with what is regarded as a solid mantle enclosing the core regions. If the Earth had been a non-rotating body, its structure would have been spherically symmetric under a balance of internal forces: the force of self-gravitation which tends to pull the matter towards the Earth's center, balanced by the elastic forces that are called into play by the gravitational compression. However, the rotation about the polar axis brings with it the associated centrifugal force. When we ignore the variations in Earth rotation, which are extremely small compared to the steady part of the rotation characterized by the mean angular velocity vector Ω_0, the associated time independent centrifugal force acts on every element of matter and is directed outward from the axis of rotation along a line that is perpendicular to this axis and passes through the matter element. The effect of this additional force is to cause the symmetry of the Earth's shape and structure to become ellipsoidal instead of spherical, with the axis of the mean rotation as the axis of symmetry. The actual Earth structure is very nearly the same as the *hydrostatic equilibrium (H.E.) structure* which is what a wholly fluid body would assume under the combined action of the centrifugal force and the much larger gravitational and elastic forces. This structure is characterized by surfaces of constant density which are axially symmetric oblate ellipsoids. Other properties such as the elastic moduli and the geopotential (made up of the gravitational and centrifugal potentials) are also constant on each of these surfaces. The construction of such an ellipsoidal model is done by the application of Clairaut's theory of hydrostatic equilibrium structure (Clairaut, 1743), starting from a spherically symmetric model representing the equilibrium structure of a hypothetical non-rotating Earth. The latter is specified by giving the density and elastic moduli as functions of the radial variable r alone. The spherical model, which may be thought of as a laterally averaged version of the

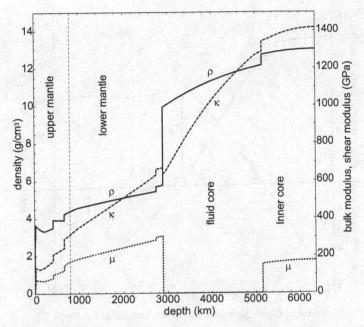

Figure 6.1 Profiles of the Earth's density, ρ, and the bulk and shear moduli, κ and μ, as derived from seismic data (ak135 model of Kennett *et al.*, 1995).

actual ellipsoidal structure, is inferred by theoretical analyses of the observations of normal modes of elastic oscillations of the Earth as well as of the traveling seismic waves; these waves and the oscillations are set off by seismic events (earthquakes) occurring in many regions of the world. The data gathering is done through the large network of seismic stations distributed over the Earth's surface. The spherically symmetric model PREM (Dziewonski and Anderson, 1981) has more or less superseded earlier models like 1066A (Gilbert and Dziewonski, 1975) and has been most widely used since. At present one finds new models based on more and more data such as the ak135 model of Kennett *et al.* (1995). Figure 6.1 displays the density profile and the bulk and shear moduli (inferred from seismic velocities) as functions of radius inside the Earth.

The relation of the ellipsoidal Earth model to the underlying spherically symmetric model will be shown in Chapter 7 while presenting one of the approaches to the theory of nutations of a deformable Earth. Figure 6.2 displays the gravity and pressure derived as a function of the radius.

It should be clear, of course, that the H. E. model includes only the $n = 0$, $m = 0$ and $n = 2$, $m = 0$ spherical harmonic components of the structure of the real Earth. It is also to be noted that observational evidence ($C_{2,0}$ from satellite tracking studies, e or H_d from precession–nutation) shows that the actual ellipticity of the Earth is about 1% higher than that of the H. E. model.

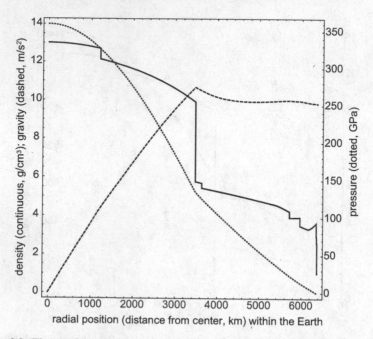

Figure 6.2 The profiles of gravity (dashed line) and pressure (dotted line), derived from the density profile (full black line, ak135 model of Kennett *et al.*, 1995). The scales for density and gravity are shown on the left vertical side (same scale – different units), and that for pressure along the right vertical side of the figure.

6.2 Perturbing potentials

The action of the tidal potentials of celestial bodies produces time dependent deformations of the Earth. The action of the same potentials on aspects of the Earth's structure that deviates from spherical symmetry (most importantly, the ellipticity) gives rise to variations in Earth rotation. These variations lead to perturbations of the centrifugal potential associated with Earth rotation, and these incremental centrifugal potentials cause further deformations which, in turn, react back on the Earth's rotation. We are concerned in this chapter with the deformations only – those produced by the direct action of the tidal potentials and those due to the centrifugal perturbations. In developing the theory of the deformations, we shall make the approximation of treating the structure of the Earth as spherically symmetric.

6.2.1 Tidal potentials

The tidal potentials were discussed in Chapter 5. We have seen there how the gravitational potential at a point $\mathbf{r} = (r, \theta, \varphi)$ in or near the Earth due to an external body B may be expanded in a series of spherical harmonics. (Note that we use φ for the longitude instead of λ as earlier, because we need the latter symbol in the present chapter to denote another quantity.) The term of a specific spherical harmonic type (nm) in the expansion is

$$W^B_{nm}(\mathbf{r}, t) = -2r^n \, \mathrm{Re} \left[F^{B*}_{nm} \, Y^m_n(\theta, \varphi) \right], \tag{6.1}$$

as in (5.205) of Chapter 5. The usual surface spherical harmonic Y^m_n and its normalization constant N_{nm} are defined in Eqs. (5.145), and F^B_{nm} by Eq. (5.206). F^B_{nm} is proportional to $Y^{m*}_n(\beta_B, \Lambda_B)$, where (θ_B, Λ_B) are the time dependent direction coordinates of the body B. One has to sum over all the solar system bodies B which contribute significantly to the potential acting on the Earth, in order to get the total tidal potential which is given by (6.1) with F^B_{nm} replaced by $F_{nm} = \sum_B F^B_{nm}$.

An alternative expression for the above potential is a generalization of the form shown in Eq. (5.199) for the special case $(nm) = (21)$ which was introduced by Sasao *et al.* (1980). Basic formulations of the type employed by these authors have been used in a number of later papers presenting significant developments in nutation theory. The expression for the potential that is used in such works is of the form

$$W_{nm}(\mathbf{r}, t) = -\frac{\Omega^2_0}{3a^{n-2}_e} r^n \mathrm{Re}[\tilde{\phi}^{nm}_e(t)\, \mathcal{Y}^m_n(\theta, \varphi)], \quad \tilde{\phi}^{nm}_e(t) = \frac{6N_{nm}a^{n-2}_e}{\Omega^2_0} F_{nm}, \tag{6.2}$$

where N_{nm} is the normalization constant of Y^m_n, and the $\mathcal{Y}^m_n$ are spherical harmonics defined by

$$\mathcal{Y}^m_n(\theta, \varphi) = P^m_n(\cos\theta)e^{-im\varphi} = \frac{1}{N_{nm}} Y^{m*}_n(\theta, \varphi), \tag{6.3}$$

with the normalization

$$\int \mathcal{Y}^{m*}_n(\theta, \varphi)\mathcal{Y}^{m'}_{n'}(\theta, \varphi) = \frac{1}{N^2_{nm}} = \left(\frac{4\pi}{2n+1} \frac{(n+m)!}{(n-m)!} \right) \delta_{nn'}\delta_{mm'}. \tag{6.4}$$

The use of spherical harmonics normalized in this manner leads to considerable simplicity in the appearance of the differential equations in the radial variable, to which the basic equations of motion of the deformation problem are reduced. Like the Y^m_n, the $\mathcal{Y}^m_n$ satisfy Laplace's equation:

$$\nabla^2 \mathcal{Y}^m_n = -\frac{n(n+1)}{r^2} \mathcal{Y}^m_n. \tag{6.5}$$

We return now to Eq. (6.2). The subscript e on $\tilde{\phi}$ in that equation is intended as a reminder that the potential is due to *external* gravitating bodies (the Moon, the Sun, and the planets). The factor $(\Omega^2_0/3a^{n-2}_e)$ on the right-hand side serves to make the time dependent factor $\tilde{\phi}^{nm}_e(t)$ dimensionless and to ensure that the expression for W_{nm} is in agreement with that of Sasao *et al.* (1980) in the special case $n = 2$, $m = 1$. It may be readily verified that the sum over B of the expression appearing within the square brackets in (6.1) is simply the complex conjugate of the quantity within the square brackets in the alternative expression (6.2). We shall find it convenient to introduce the notation $\tilde{\phi}_e^{nm}(\mathbf{r}, t)$ for the complex potential on the right-hand side of (6.2), the real part of which is $W_{nm}(\mathbf{r}, t)$:

$$\tilde{\phi}_e^{nm}(\mathbf{r}, t) = -\frac{\Omega_0^2}{3a_e^{n-2}} r^n [\tilde{\phi}_e^{nm}(t) \mathcal{Y}_n^m(\theta, \varphi)], \quad W_{nm}(\mathbf{r}, t) = \mathrm{Re}\ \tilde{\phi}_e^{nm}(\mathbf{r}, t). \quad (6.6)$$

The special case of degree two potentials (zonal, tesseral, and sectorial, corresponding to $m = 0, 1, 2$, respectively) will be of particular interest in this chapter. We write down their explicit expressions here by making use of the definition of F_{nm}^B (Eqs. (5.205) and (5.206)):

$$\tilde{\phi}_e^{20}(t) = \sum_B \frac{3GM_B}{\Omega_0^2 d_B^3} P_2(\cos \beta_B), \quad (6.7)$$

$$\tilde{\phi}_e^{21}(t) = \sum_B \frac{3GM_B}{\Omega_0^2 d_B^3} \frac{1}{3} P_2^1(\cos \beta_B)e^{i\Lambda_B} = \phi_1^{21}(t) + i\phi_2^{21}(t), \quad (6.8)$$

$$\tilde{\phi}_e^{22}(t) = \sum_B \frac{3GM_B}{\Omega_0^2 d_B^3} \frac{1}{12} P_2^2(\cos \beta_B)e^{2i\Lambda_B} = \phi_1^{22}(t) + i\phi_2^{22}(t). \quad (6.9)$$

The expression for the tesseral potential ($n = 2$, $m = 1$) is in agreement with that of $\tilde{\phi}$ in Eq. (5.200), as is to be expected.

The tesseral potential of degree two plays the dominant role in wobble, nutation, and precession, as we have observed already. The roles of the zonal and sectorial potentials in wobble and nutation are marginal, but the tidal deformations produced by them are of magnitudes comparable to the tesseral deformations.

It was noted in the last paragraph of Section 5.5.7 that in the case of potentials of order $m = 1$, only the part proportional to the complex conjugate of Y_n^1 generates a torque on an axially symmetric Earth; that is the part that is explicitly displayed in the form (6.2). Since we are primarily interested in the low frequency nutations arising from the action of the degree two tesseral potential (of order $m = 1$), we choose to use the form (6.2) for the potential.

6.2.2 Spectral representation

The spectral representation of $\tilde{\phi}_e^{nm}(t)$ is readily obtained by using the complex conjugate of the expansion of F_{nm}^* given in Eq. (5.207):

$$\tilde{\phi}_e^{nm}(t) = \frac{6N_{nm}}{\Omega_0^2} \sum_{\omega \geq 0} \frac{g_e H_\omega^{nm}}{2a_e^2} e^{-i(\Theta_\omega(t) - \zeta_{nm})}, \quad (6.10)$$

where $\Theta_\omega(t)$ is the argument of the spectral component of frequency ω of the tidal potential (see Eq. (5.216) and also (5.209) which defines ζ_{nm} to be 0 or $\pi/2$ according to whether $(n - m)$ is even or odd). It is the practise, in the notation of Sasao *et al.* (1980), to express frequencies in units of cpsd. More specifically, one writes

$$-(\Theta_\omega(t) - \zeta_{nm}) = \sigma\Omega_0 t - \alpha_\omega, \tag{6.11}$$

$$\sigma\Omega_0 = -\omega, \qquad \alpha_\omega = \Theta_\omega(0) - \zeta_{nm}. \tag{6.12}$$

The frequency σ (in cpsd) is negative since ω is positive (except that $\sigma = \omega = 0$ for the zero-frequency component present when the potential is of order $m = 0$). With this change in notation, we rewrite (6.10) as

$$\tilde{\phi}_e^{nm}(t) = \sum_\sigma \tilde{\phi}(\sigma)e^{i(\sigma\Omega_0 t - \alpha_\omega)}, \qquad \tilde{\phi}(\sigma) = \frac{3g_e N_{nm}}{\Omega_0^2 a_e^2} H_\omega^{nm}. \tag{6.13}$$

The spectral expansion of the complex potential in (6.6) is then

$$\tilde{\phi}_e^{nm}(\mathbf{r}, t) = \sum_\sigma \tilde{\phi}(\mathbf{r}, \sigma)e^{i(\sigma\Omega_0 t - \alpha_\omega)}, \tag{6.14}$$

with

$$\tilde{\phi}(\mathbf{r}, \sigma) = -K\,\tilde{\phi}(\sigma)\,r^n\,\mathcal{Y}_{nm}(\theta, \varphi), \qquad K = \frac{\Omega_0^2}{3a^{n-2}}. \tag{6.15}$$

Note that in writing the amplitude of the spectral component simply as $\tilde{\phi}(\sigma)$, we have suppressed, for convenience, the subscript and superscripts that appear on $\tilde{\phi}$ in the time domain. The fact that the amplitude pertains to the external potential of spherical harmonic type (nm) will be amply clear from the second relation in (6.13).

It was shown in Section 5.5.5 that the frequencies ω of spectral components of the tidal potential of general type (nm) lie in the range $(m - 1/2)\Omega_0 < \omega < (m + 1/2)\Omega_0$ except that if $m = 0$, the range is $0 \le (1/2)\Omega_0$. This means, in view of the relation (6.12) of σ to ω, that in the representation (6.2) of the potential with $\tilde{\phi}_e^{nm}$ given by (6.13), the frequencies σ lie in the range $-(m + 1/2) < \sigma < -(m - 1/2)$ cpsd for any $m > 0$, and $-1/2 < \sigma \le 0$ cpsd for $m = 0$. The factor $e^{i(\sigma\Omega_0 t - m\varphi)}$ that is present in the expression (6.2) on introducing (6.13) implies that for any $m > 0$, the spectral components (with $\sigma < 0$) are retrograde waves, traveling in the direction of decreasing φ.

6.2.3 Centrifugal potentials

Another source of deformations, as indicated earlier, consists of the incremental centrifugal potentials associated with the variations in Earth rotation.

As shown in Chapter 7, the presence of the FOC with ellipsoidal outer and inner boundaries (CMB and ICB) in the rotating Earth causes the vector angular velocities of rotation of the mantle, FOC, and solid inner core to differ from each other under the influence of the tidal potentials of celestial bodies. The three temporally varying angular velocities are usually denoted by $\Omega = \Omega_0 + \omega$, $\Omega_f = \Omega + \omega_f$, and $\Omega_s = \Omega + \omega_s$, respectively, where Ω_0 is the constant angular velocity vector, with components $\Omega_0(0, 0, 1)$, of the steadily rotating unperturbed Earth; the subscripts f and s refer to the *fluid* outer core and *solid* inner core, respectively. The equatorial components of ω, $(\omega + \omega_f)$, and $(\omega + \omega_s)$ represent the wobbles of the mantle, fluid outer core (FOC), and the solid inner core (SIC), respectively, while those of ω_f and ω_s represent the *differential* wobbles of the FOC and SIC relative to the mantle. These wobbles are generated by the action of the tesseral part of the degree two potential on the ellipsoidal Earth. The components of ω are written as in earlier chapters as $\Omega_0(m_x, m_y, m_z)$, and those of ω_f and ω_s will be denoted by $\Omega_0(m_{fx}, m_{fy}, m_{fz})$ and $\Omega_0(m_{sx}, m_{sy}, m_{sz})$. The complex combinations $\tilde{m}_f = m_{fx} + im_{fy}$ and $\tilde{m}_s = m_{sx} + im_{sy}$ of the equatorial components of ω_f/Ω_0 and ω_s/Ω_0 are often used to represent the two differential wobbles, like $\tilde{m} = m_x + im_y$ for the common part of the wobbles of all three regions. The axial component $\Omega_0 m_z$ of ω represents variations in the Earth's spin rate, caused almost entirely by the deforming action of the zonal part of the potential.

The centrifugal potential caused by the rotation with the angular velocity $\Omega = \Omega_0 + \omega$ is

$$\phi_c(\Omega) = -\frac{1}{2}[\Omega^2 r^2 - (\Omega \cdot \mathbf{r})^2] = \phi_c(\Omega_0) - [(\Omega_0 \cdot \omega)r^2 - (\Omega_0 \cdot \mathbf{r})(\omega \cdot \mathbf{r})], \quad (6.16)$$

to the first order in ω. Therefore the part of the incremental centrifugal potential $\phi_c(\Omega) - \phi_c(\Omega_0)$ that arises from the wobble represented by the equatorial components $(\Omega_0 m_x, \Omega_0 m_y)$ of Ω is

$$\Delta\phi_c = \Omega_0^2(m_x xz + m_y yz). \quad (6.17)$$

Since $r^2 \mathcal{Y}_2^1 = 3r^2 \cos\theta \sin\theta e^{-i\phi} = 3(z - iy)z$, we may rewrite $\Delta\phi_c$ as

$$\Delta\phi_c(\mathbf{r}, t) = \Omega_0^2 \text{Re}\,[\tilde{m}(t)(x - iy)z] = \frac{\Omega_0^2}{3}r^2\text{Re}\,[\tilde{m}(t)\mathcal{Y}_2^1]. \quad (6.18)$$

This expression has the same form as the degree two tesseral tidal potential given by Eq. (6.2) with $(nm) = (21)$, except for the replacement $\tilde{\phi}^{21} \to -\tilde{m}$; it acts throughout the Earth. The important point to note is that $\Delta\phi_c$ is a potential of spherical harmonic degree two and order one. In the spectral representation too, $\Delta\phi_c$ has $-\tilde{m}(\sigma)$ in place of the amplitude $\tilde{\phi}(\sigma)$ of the tidal potential.

Additional centrifugal potentials are operative in the FOC and SIC because of the differential wobbles $\omega_f = (\Omega_0 m_{fx}, \Omega_0 m_{fy}, 0)$ and $\omega_s = (\Omega_0 m_{sx}, \Omega_0 m_{sy}, 0)$. The additional potential (over and above $\Delta\phi_c$) in the fluid core is, to the first order in the wobbles,

$$\Delta\phi_{cf} = \phi_c(\mathbf{\Omega}_f) - \phi_c(\mathbf{\Omega}) = \frac{\Omega_0^2}{3}r^2\mathrm{Re}\,[\tilde{m}_f(t)\,\mathcal{Y}_2^1], \qquad (a_s < r < a_c), \qquad (6.19)$$

where a_c and a_s are the radii of the CMB and ICB. Similarly, the additional centrifugal potential within the SIC is

$$\Delta\phi_{cs} = \phi_c(\mathbf{\Omega}_s) - \phi_c(\mathbf{\Omega}) = \frac{\Omega_0^2}{3}r^2\mathrm{Re}\,[\tilde{m}_s(t)\,\mathcal{Y}_2^1], \qquad (0 < r < a_s). \qquad (6.20)$$

The additional potentials $\Delta\phi_{cf}$ and $\Delta\phi_{cs}$ are also of degree two tesseral character; the first one vanishes outside the FOC, and the second outside the SIC.

The deformations produced by the above three incremental centrifugal potentials play a non-trivial role in the theory of wobble and nutation, as we shall see when the torque equations for the non-rigid Earth are considered in Chapter 7.

Also of interest is the incremental centrifugal potential associated with spin rate variations represented by the third component $\Omega_0 m_3$ of $\mathbf{\Omega}$. (We are using m_3 interchangeably with m_z.) It is given by the part of the expression (6.17) that is proportional to $\omega_3 = \Omega_0 m_3$:

$$\Delta\phi_c = -m_3(t)(\mathbf{\Omega}_0 \wedge \mathbf{r}) \cdot (\mathbf{\Omega}_0 \wedge \mathbf{r}) = -\Omega_0^2 m_3(t)(x^2 + y^2). \qquad (6.21)$$

When expressed in terms of spherical harmonics, it is

$$\Delta\phi_{c3} = -\frac{2}{3}\Omega_0^2 m_3(t)\,r^2[1 - P_2(\cos\theta)]. \qquad (6.22)$$

Thus the incremental potential arising from spin rate variations has a spherically symmetric (degree zero order zero) part besides the expected degree two zonal ($m = 0$) part. The latter part has the same form as the zonal tidal potential (the special case of (6.2) for $n = 2, m = 0$), except for the replacement $\tilde{\phi}^{20} \rightarrow -2m_3$.

The deformation produced by the degree zero part is spherically symmetric. It generates equal increments to the three diagonal elements of the inertia tensor and gives rise to radial oscillations in accordance with the time dependence of m_3; but it produces no changes in the gravitational potential outside the oscillating location of the instantaneous outer boundary, as the spherically symmetric part of the external gravitational field is determined purely by the total mass of the Earth and is unaffected by the nature of the dependence of the density on the radial variable.

6.3 Deformations of a spherical Earth

The gravitational forces due to external bodies cause displacements of the elements of matter in the Earth from their positions in the equilibrium spherical Earth, and consequent deformations and perturbations of the matter distribution (characterized by the density function) in the Earth. The perturbation of the density function results in a corresponding perturbation of the Earth's own gravitational potential and this, in turn, affects the displacement field. In the case of an ellipsoidal Earth, the field of displacements contains a part which represents variations of the orientation of the Earth as a whole in space which are manifested as nutations and spin rate variations.

We represent the displacement of an element from its position $\mathbf{r}$ in the equilibrium Earth by $\mathbf{u}(\mathbf{r}; t)$ and the perturbation of the gravitational potential by $\phi_1(\mathbf{r}; t)$; the self-gravitational potential of the unperturbed Earth is $\phi_{0g}(\mathbf{r})$. The mutual influences of the displacements and perturbations of the gravitational potential cause the equations governing the spatio-temporal variations of $\mathbf{u}$ and ϕ_1 to be mutually coupled; the two fields have to be obtained by simultaneous solution of these equations. We will be concerned in the following sections with the solution for the case of a spherically symmetric Earth, which yields a displacement field that is purely deformational: it does not contain a part representing rotation variations, as the action of the gravitational potential of external bodies on a spherical Earth does not produce any torques. This restricted case is nevertheless of considerable interest because of the simplicity, in this case, of the relations connecting the field $\mathbf{u}(\mathbf{r}; t)$ and the perturbation $\phi_1(\mathbf{r}; t)$ of the potential at the Earth's surface (and more generally, at internal geocentric spherical surfaces such as the boundaries of the fluid core) to the external forcing potential. These relations involve certain constant parameters called *Love numbers*, to be introduced in the next section, which reflect the structure and properties of the Earth's interior.

Our interest in the Love numbers stems primarily from the intimate relation of one of the Love numbers to a *deformability parameter* (often referred to more succinctly as a *compliance* in an important segment of the recent literature), which appears in the torque equations for the deformable Earth; other compliances too are closely related to what may be called internal Love numbers (e.g., those relating the values of ϕ_1 at the inner and outer boundaries of the fluid core to the external potential on the same surfaces). It will be seen that the deformability properties of the Earth, to the extent that they affect nutation, enter solely through the compliance parameters. The values of these parameters as computed on the basis of a spherical Earth model do need corrections of the order of the ellipticity when the fact that the Earth is ellipsoidal and rotating is taken into account; these corrections can be evaluated separately.

A general method for the determination of $\mathbf{u}$ and ϕ_1 in the case of an ellipsoidal Earth that is perturbed by the gravitational potential of an external body is presented in Section 7.5. The field $\mathbf{u}$ of the ellipsoidal case comprises both deformational and nutational parts. Compared to the treatment of the spherical model presented below, the mathematical formulation is of much greater complexity, which is unavoidable when the Earth model is ellipsoidal. The displacement field approach is adopted for the simultaneous solution of the nutation and deformation problems.

6.4 Love numbers

The deformational responses of a non-rotating spherically symmetric Earth to the gravitational potential of celestial bodies were studied by Love (1909a and b, 1911). He showed that the response to a spherical harmonic term of the potential is especially simple. For instance, the radial component $u_r(\mathbf{r}, t)$ of the displacement $\mathbf{u}(\mathbf{r}, t)$ of the matter element which was at a position $\mathbf{r}$ in the unperturbed Earth is proportional to $W_{nm}(\mathbf{r}, t)$, with a proportionality factor which depends on r and is independent of (θ, φ); so is the perturbation ϕ_1 of the gravitational potential.

For points at the Earth's surface, i.e., for $\mathbf{r} = (a, \theta, \varphi)$, the proportionality relations are written as

$$u_r(\mathbf{r}, t) \equiv \hat{\mathbf{r}} \cdot \mathbf{u} = -\frac{h_n}{g} W_{nm}(\mathbf{r}, t), \quad \phi_1(\mathbf{r}, t) = k_n W_{nm}(\mathbf{r}, t), \quad \text{(for } r = a), \quad (6.23)$$

where g is the acceleration due to gravity at the surface, and h_n and k_n are dimensionless constants called Love numbers. Shida and Matsuyama (1912) observed that the relation between the tangential displacements $\mathbf{u}_T(\mathbf{r}, t)$ of points at the Earth's surface and the perturbing potential at the surface is only slightly less simple:

$$\mathbf{u}_T(\mathbf{r}, t) = \mathbf{u} - u_r \hat{\mathbf{r}} = -\frac{l_n}{g} \nabla W_{nm}(\mathbf{r}, t) \quad \text{(for } r = a). \quad (6.24)$$

The Love and Shida numbers are determined by the radial profiles of density and elastic parameters within the Earth, which characterize the Earth model.

The minus signs appearing on the right-hand sides of the defining relations (6.23) and (6.24) for the Love and Shida numbers h_n and l_n may look strange to readers who are accustomed to the use of the geophysicists' convention for the potential. The minus sign is rendered necessary by our switchover to the physicists' convention. No minus sign arises in the case of k_n as it relates two potentials, both of which have switched signs.

Note that according to the last two equations, neither the vector field $\mathbf{u}$ nor the scalar field ϕ_1 produced in response to the perturbing potential W_{nm} involves any spherical harmonics other than the Y_n^m that is present in W_{nm}.

In the case of an ellipsoidal Earth, the relations connecting **u** and ϕ_1 to the forcing potential are more complicated. The response to the tidal potential of given degree n (whether it be the radial or transverse displacement field or the perturbation of gravitational potential) is no longer related to the forcing potential by a simple constant factor. The Love number acquires a slight dependence on the order m, so that one has to have, for example, a k_{nm} for each (nm) rather than just k_n. Thus the values of any of the Love numbers for the degree two zonal, tesseral, and sectorial tides differ from each other, the fractional difference being of the order of the Earth's ellipticity. Furthermore, each of the Love numbers (for a given type (nm) of the forcing potential) is not strictly constant but varies slightly with latitude. The latitude dependence is generally parameterized in terms of two subsidiary Love number parameters (for each (nm)). It may be added that in the case of the ellipsoidal Earth, the acceleration due to gravity at the equatorial radius, denoted by g_e, is used instead of $g(a)$ in the defining equations of the h and l Love numbers.

6.5 Load Love numbers

We digress here to consider Love numbers of a different kind, relating to displacement fields and increments in the Earth's gravitational potential that are generated by surface loading, e.g., by atmospheric pressure variations or by ocean tides. The increments in the elements of the Earth's inertia tensor that are caused by such surface loads are of significance in nutation theory because of their contributions to the angular momentum of the rotating Earth and consequent effects on Earth rotation.

The loading mass is taken to occupy a layer of negligible thickness at the Earth's surface (the depth of the ocean and the height of the atmospheric layer being very small compared to the radius of the Earth). The distribution of the mass over the Earth's surface is then described by a density function

$$\rho(\mathbf{r}) = \delta(r - a)L(\theta, \lambda), \tag{6.25}$$

where $L(\theta, \lambda)$ is the mass of the surface layer per unit area at the angular position indicated. This function may be decomposed into a sum of spherical harmonic components:

$$L(\theta, \lambda) = \mathrm{Re}\left[\sum_{q,p} L_{qp} \mathcal{Y}_{qp}(\theta, \lambda)\right]. \tag{6.26}$$

L_{qp} is, in general, time dependent.

The gravitational potential $W_{SL}(\mathbf{r})$ of the surface mass layer at any point $\mathbf{r}$ outside the Earth may be expressed in the same form as the right-hand member of Eq. (5.151), but with the difference that $\rho(\mathbf{r})$ is now that of the surface layer as shown above. With a change of notation for the external point from $\mathbf{R}$ to $\mathbf{r} = (r, \theta, \lambda)$ and that for the integration variable from $\mathbf{r}$ to $\mathbf{r}' = (r', \theta', \lambda')$, we have:

$$W_{SL}(\mathbf{r}) = -G \int \rho(\mathbf{r}') \sum_{n=0}^{\infty} \frac{r'^n}{r^{n+1}} \left\{ P_n(\cos\theta')P_n(\cos\theta) + 2 \sum_{m=1}^{n} \frac{(n-m)!}{(n+m)!} \right.$$

$$\left. \times [P_n^m(\cos\theta')P_n^m(\cos\theta)(\cos m\lambda' \cos m\lambda + \sin m\lambda' \sin m\lambda)] \right\} d^3r'. \quad (6.27)$$

After substitution for the density function from Eqs. (6.25) and (6.26), we facilitate the integration by making use of the identity

$$P_n(\cos\theta') P_n(\cos\theta) + 2 \sum_{m=1}^{n} \frac{(n-m)!}{(n+m)!}$$

$$\times [P_n^m(\cos\theta')P_n^m(\cos\theta)(\cos m\lambda' \cos m\lambda + \sin m\lambda' \sin m\lambda)]$$

$$= \mathcal{Y}_n^0(\cos\theta')\mathcal{Y}_n^0(\cos\theta) + \sum_{m=1}^{n} \frac{(n-m)!}{(n+m)!}$$

$$\times [\mathcal{Y}_n^{m*}(\theta', \lambda')\mathcal{Y}_n^m(\theta, \lambda) + \mathcal{Y}_n^m(\theta', \lambda')\mathcal{Y}_n^{m*}(\theta, \lambda)]. \quad (6.28)$$

Evaluation of the integral in (6.27), after introducing ρ from (6.25), may now be trivially done with the aid of the orthogonality and norm of the $\mathcal{Y}_n^m$ as shown in (6.4). The result is

$$W_{SL}(\mathbf{r}) = -\sum_{n,m} \frac{4\pi Ga}{2n+1} \left(\frac{a}{r}\right)^{n+1} \text{Re}\,[L_{nm}\mathcal{Y}_n^m(\theta, \lambda)], \quad (r \geq a). \quad (6.29)$$

If we were considering some point $\mathbf{r}$ within the Earth, the factor (r'^n/r^{n+1}) in (6.27) would have to be replaced by (r^n/r'^{n+1}), and one is then led to

$$W_{SL}(\mathbf{r}) = -\sum_{n,m} \frac{4\pi Ga}{2n+1} \left(\frac{r}{a}\right)^n \text{Re}\,[L_{nm}\mathcal{Y}_n^m(\theta, \lambda)], \quad (r \leq a). \quad (6.30)$$

The deformation produced by the action of this potential on the matter within the Earth results in an incremental gravitational potential outside the Earth (just as the deformation due to the tidal potential does). The (nm) spherical harmonic component dW_{SL}^{nm} of this incremental potential has the same $(a/r)^n$ radial dependence as the corresponding term in (6.29). At the surface ($r = a$), it is a factor k_n' times the corresponding term of (6.30) or (6.29) at the same surface:

$$dW_{SL}^{nm}(\mathbf{r})|_{r=a} = \frac{4\pi Ga}{2n+1} \text{Re}\,[k_n' L_{nm}\mathcal{Y}_n^m(\theta, \lambda)]. \quad (6.31)$$

k_n' is designated as the *load Love number* of degree n for the potential. The radial displacement $u_{r(SL)}^{nm}$ associated with the loading deformation is characterized by another load Love number h'_n, and is given by the expression on the right-hand side of the above equation with k'_n replaced by $-(h'_n/g)$.

The incremental potential (6.31) induced (at any $r \geq a$) by the deformation caused by the surface loading is in addition to the potential (6.29) which arises simply from the presence of the surface mass layer. The (nm) part of the combination of these two incremental potentials is thus $(1 + k'_n)$ times the corresponding component of (6.29).

An observation that is of importance concerns the radial component of the gradient (radial derivative) of the potential W_{SL}. Differentiating (6.29) and (6.30) with respect to r, then letting $r \to a$ from outside the Earth and from within the Earth and taking the difference, one finds that

$$\left[\left(\frac{\partial W_{SL}^+}{\partial r} \right) - \left(\frac{\partial W_{SL}^-}{\partial r} \right) \right]_{r=a} = 4\pi Ga \sum_{n,m} L_{nm} \mathcal{Y}_n^m(\theta, \lambda) = 4\pi GaL(\theta, \lambda), \quad (6.32)$$

where superscripts $+$ and $-$ refer to the expressions for W_{SL} that are valid for $r \geq a$ and $r \leq a$, respectively.

6.6 Equations for displacement and gravitational fields

We turn now to the equations governing the space–time dependence of the displacement field $\mathbf{u}(\mathbf{r}; t)$ and the perturbation $\phi_1(\mathbf{r}; t)$ of the Earth's gravitational potential. The effect of an external force on the displacements of material elements is determined by the elasto-dynamical equation for a continuous medium. As formulated by Dahlen (1982), the equation in the general case of an Earth which does not necessarily have spherical symmetry and is rotating with mean angular velocity Ω_0 is, in a frame rotating with this angular velocity in space,

$$\rho \left(\frac{\partial^2 \mathbf{u}}{\partial t^2} + 2\mathbf{\Omega}_0 \wedge \frac{\partial \mathbf{u}}{\partial t} - (\nabla \cdot \mathbf{u})\nabla \phi + \nabla(\mathbf{u} \cdot \nabla \phi) + \nabla \phi_1 \right) - \nabla \cdot \mathbf{T} = -\rho \nabla \phi_e. \tag{6.33}$$

The forcing potential $\phi_e(\mathbf{r}, t)$ due to the external body is just $W_{nm}(\mathbf{r}, t)$ under another name; the potential $\phi(\mathbf{r}, t)$ is $(\phi_0 + \phi_1)$ where ϕ_0 is the gravitational-cum-centrifugal potential of the unperturbed uniformly rotating Earth and ϕ_1 is the incremental potential resulting from the perturbation of the mass distribution within the Earth by ϕ_e; and $\mathbf{T}$ is the incremental stress tensor defined by

$$\mathbf{T} = \lambda(\nabla \cdot \mathbf{u})\mathbf{I} + \mu(\nabla \mathbf{u} + (\nabla \mathbf{u})^T). \tag{6.34}$$

The elements of $\mathbf{T}$ are

$$T_{ij} = \lambda(\nabla \cdot \mathbf{u})\delta_{ij} + \mu \left(\frac{\partial u_i}{\partial x_j} + \frac{\partial u_j}{\partial x_i} \right), \tag{6.35}$$

where λ and μ are the so-called Lamé parameters which are simply related to the compressional and shear elastic moduli, and the Kronecker delta function δ_{ij} is 1 for $i = j$ and 0 otherwise. Specifically, μ is the shear modulus representing the magnitude of the shear stress needed to produce unit shear strain (which does not cause any change in volume of the material) and $\lambda = k - 2\mu/3$ where k is the Young's modulus (bulk modulus) expressing the degree of compressive stress needed to produce unit strain.

The displacement field $\mathbf{u}$ gives rise to a density perturbation ρ_1 which is the difference between the density of the matter after its displacement from $\mathbf{r}$ to $\mathbf{r} + \mathbf{u}$ and the density of the matter which was originally at $\mathbf{r} + \mathbf{u}$. The former is $\rho(1 - \nabla \cdot \mathbf{u})$ where $(\nabla \cdot \mathbf{u})$ is the dilatation associated with $\mathbf{u}$, and the latter is $\rho(1 + \mathbf{u} \cdot \nabla \rho)$. On taking the difference we get

$$\rho_1 = -\nabla \cdot (\rho \mathbf{u}). \tag{6.36}$$

The incremental gravitational potential ϕ_1 is determined by ρ_1 through the Poisson equation, which is

$$\nabla^2 \phi_1 = 4\pi G \rho_1 = -4\pi G \nabla \cdot (\rho \mathbf{u}). \tag{6.37}$$

Unlike $\mathbf{u}$, the potential ϕ_1 extends to the exterior of the Earth. The Poisson equation simplifies in the exterior region (where $\rho_0 = 0$) to

$$\nabla^2 \phi_1 = 0. \tag{6.38}$$

The solution of the pair of coupled Eqs. (6.33) and (6.37) constitutes the essence of the problem of determination of the effects of the external perturbation.

The treatment in the remainder of this chapter employs a spherically symmetric non-rotating model for the Earth, for which $\boldsymbol{\Omega}_0 = 0$. The density and the Lamé parameters are then functions of the geocentric distance r only. Recognizing this, we make the following replacements in Eqs. (6.33) and (6.34):

$$\rho(\mathbf{r}) \rightarrow \rho_0(r), \qquad \lambda(\mathbf{r}) \rightarrow \lambda_0(r), \qquad \mu(\mathbf{r}) \rightarrow \mu_0(r). \tag{6.39}$$

Furthermore, in the absence of rotation and the associated centrifugal potential, $\phi_0(\mathbf{r})$ consists of just the spherically symmetric Earth's own gravitational potential ϕ_{g0}, which also is a function of r only.

6.6.1 Boundary conditions

The quantities appearing in the above set of equations have to obey boundary conditions appropriate to the boundaries involved. With the use of the generic symbol $\hat{\mathbf{n}}$ for the unit normal to any boundary surface, the boundary conditions are:

1. Continuity of $\hat{\mathbf{n}} \cdot \mathbf{u}$, $\hat{\mathbf{n}} \cdot \mathbf{T}$, ϕ_1, and $\hat{\mathbf{n}} \cdot (\nabla\phi_1 + 4\pi G\rho\mathbf{u})$ across every boundary, with $\hat{\mathbf{n}} \cdot \mathbf{T}$ required to vanish at the outer (free) surface, assuming that there is no surface loading;

2. Continuity of $\mathbf{u}$ across any welded boundary between two solid regions;

3. Conditions at fluid–solid boundaries: inviscid fluids cannot support shear stresses. Consequently the part of $\hat{\mathbf{n}} \cdot \mathbf{T}$ which is perpendicular to $\hat{\mathbf{n}}$ vanishes on the fluid side of the boundary; so it must vanish on the solid side too, in view of the above-stated continuity condition on $\hat{\mathbf{n}} \cdot \mathbf{T}$. Similarly, if the core fluid has no viscosity, it can "glide" along the solid boundary, making it possible for the components of $\mathbf{u}$ that are tangential to the boundary to be discontinuous across the boundary.

4. Conditions at $r = 0$: except for nonzero stress at the center of the Earth, all the variables (ϕ_1, $\nabla\phi_1$ and $\mathbf{u}$) vanish at the center.

5. For the deformations caused by a surface mass distribution (see Section 6.5), the boundary conditions have to take into account the loading on the surface as well as the discontinuity of the gradient of the potential between the inner and outer sides of the thin surface layer (see Eq. (6.32)). Because of the loading at the Earth's surface, $\hat{\mathbf{n}} \cdot \mathbf{T} \cdot \hat{\mathbf{n}}$, which is the normal component of the stress, has to be $gL(\theta, \lambda)$ at the surface, $L(\theta, \lambda)$ being the surface loading mass per unit area. Additionally, $\hat{\mathbf{n}} \cdot (\nabla\phi_1 + 4\pi G\rho\mathbf{u})$ has to be discontinuous at the Earth's surface, with a jump of $4\pi GaL(\theta, \lambda)$ on going from the inner side to the outer (see Eq. (6.32)).

In the spherically symmetric case to which we are restricting ourselves, the boundaries of interest (the outer surface, the CMB and ICB, etc.) are all geocentric spheres having their normals in the radial direction, so that $\hat{\mathbf{n}} = \hat{\mathbf{r}}$.

6.6.2 Non-rotating spherical Earth: equations of motion in the frequency domain

We return now to the momentum balance Eq. (6.33) and Poisson's equation (6.37), and go on to the presentation of a method of solution to obtain $\mathbf{u}$ and ϕ_1. It is simplest to carry out the solution in the frequency domain, so we take the forcing potential $\phi_e(\mathbf{r}, t)$ in Eq. (6.33) to be the real part of a spectral component of the complex potential $\tilde{\phi}_e^{nm}(\mathbf{r}, t)$ of Eq. (6.14). It is given by

$$\tilde{\phi}(\mathbf{r}, \sigma)e^{i(\sigma\Omega_0 t - \alpha_\omega)}, \qquad \tilde{\phi}(\mathbf{r}, \sigma) = -K\,\tilde{\phi}(\sigma)\,r^n\,\mathcal{Y}_{nm}(\theta, \varphi), \qquad (6.40)$$

with $K = \Omega_0^2/(3a^{n-2})$, as noted in Eq. (6.15). It may be recalled here that the amplitude $\tilde{\phi}(\sigma)$ is related to the Cartwright-Tayler amplitude H_ω^{nm} through the second equation in (6.13).

Given the above time dependence for the driving potential, it is evident that $\mathbf{u}$ and ϕ_1 too must have the same kind of time dependence:

$$\mathbf{u}(\mathbf{r}, t) = \text{Re}\left[\mathbf{u}(\mathbf{r}; \sigma)e^{i(\sigma\Omega_0 t - \alpha_\omega)}\right], \quad \phi_1(\mathbf{r}, t) = \text{Re}\left[\phi_1(\mathbf{r}; \sigma)e^{i(\sigma\Omega_0 t - \alpha_\omega)}\right]. \quad (6.41)$$

It follows then that we may make the replacement $\partial/\partial t \to i\sigma\Omega_0$ in Eq. (6.33). Therefore the magnitudes of the first two terms, which are the inertial and Coriolis terms, go roughly as $\sigma^2\Omega_0^2\mathbf{u}$ and $2\sigma\Omega_0^2\mathbf{u}$, respectively; with $\sigma \approx -m$, they are both of order $\Omega_0^2\mathbf{u}$ (within a factor of 4) for $m = 1$ and 2, and very much smaller for $m = 0$. Compare these with other terms in the equation, e.g., $(\nabla \cdot \mathbf{u})\nabla\phi$, which may be deemed to be of $\mathcal{O}(g/a)\mathbf{u}$, where a is the radius of the Earth. Evidently the time derivative terms are very much smaller, by a factor of the order of $(\Omega_0^2 a/g)$, which closely approximates the surface ellipticity ($\epsilon \approx 1/300$). The adoption of a non-rotating Earth model implies that the second term which represents the Coriolis force due to Earth rotation is dropped; we see now that the resulting error is of $\mathcal{O}(\epsilon)$. Neglect of the first (inertial) term is problematical, however, because the process of obtaining the solutions within the fluid core region becomes more and more unstable as the frequency approaches zero. So this term has to be retained despite its being formally very small. The effects of ellipticity, which are neglected in going over to a spherically symmetric Earth model, entail errors of $\mathcal{O}(\epsilon)$, of course.

Subject to the above approximations, Eq. (6.33) becomes

$$\rho_0[-\sigma^2\Omega_0^2\mathbf{u} - (\nabla \cdot \mathbf{u})\nabla\phi_0 + \nabla(\mathbf{u} \cdot \nabla\phi_0) + \nabla\phi_1] - \nabla \cdot \mathbf{T} = -\rho_0\nabla\phi_e(\mathbf{r}, \sigma).$$
$$(6.42)$$

It will be tacitly assumed hereafter that the common factor $e^{i(\sigma\Omega_0 t - \alpha_\omega)}$ has been removed from all the quantities in the above equation, and that $\mathbf{u}$, ϕ_1, and other quantities derived from them are now functions of $\mathbf{r}$ and σ, like ϕ_e in this equation. There will then be no change in the appearance of the stress tensor (6.34) or of Poisson's equation (6.37), though it has to be kept in mind that ρ, λ, μ are to be taken as ρ_0, λ_0, μ_0, respectively.

6.7 Spherical harmonic expansions; radial functions

Arbitrary scalar fields on a geocentric spherical surface may be expanded in terms of the spherical harmonics $\mathcal{Y}_n^m$, ($n = 0, 1, 2, \ldots; m = -n, -n+1, \ldots, n$), with coefficients that are functions of the radius r of the surface. Expansion of vector fields on such a surface requires three sets of basic *vector spherical harmonic fields*. We choose them to be

$$\mathbf{R}_n^m = \hat{\mathbf{r}}\mathcal{Y}_n^m, \quad \mathbf{S}_n^m = r\nabla\mathcal{Y}_n^m, \quad \mathbf{T}_n^m = \mathbf{r} \wedge \nabla\mathcal{Y}_n^m. \quad (6.43)$$

The spherical harmonics used here are those defined in (6.3). We observe in passing that ∇ in the above expressions could well be replaced by the part $\nabla_1 = (\nabla - \hat{\mathbf{r}}\partial/\partial r)$ that is transverse to the radial direction, since $\mathcal{Y}_n^m$ does not depend on r anyway.

$\mathbf{R}_n^m$ and $\mathbf{S}_n^m$ are said to be *spheroidal* fields of spherical harmonic order and degree (nm); the first of these is a *radial* spheroidal field, it being in the radial direction everywhere, while the second is a *transverse* spheroidal field being everywhere transverse to the radial direction. $\mathbf{T}_n^m$, which is also a transverse field, is described as a *toroidal* field of degree and order (nm). The action of a component of $\mathbf{r} \wedge \nabla$ on a scalar field $f(\theta, \varphi)$ represents the effect of a rotation of the field *as a whole* about the corresponding axis; to be more specific, $-i\alpha\hat{\mathbf{n}} \cdot (\mathbf{r} \wedge \nabla) f(\theta, \varphi)$ is the difference $(f' - f)$ where $f'(\theta, \varphi)$ is the field produced by rotating f through an infinitesimal angle α about the axis $\hat{\mathbf{n}}$. It becomes evident, on setting $f(\theta, \varphi) = \mathcal{Y}_n^m(\theta, \varphi)$, that the increment to $\mathcal{Y}_n^m$ produced by the same infinitesimal rotation is $-i\alpha\hat{\mathbf{n}} \cdot \mathbf{T}_n^m$.

Let us now express the displacement field $\mathbf{u}(\mathbf{r}, \sigma)$ in terms of the vector spherical harmonics defined above, and $\phi_1(\mathbf{r}, \sigma)$ in terms of $\mathcal{Y}_n^m$:

$$\mathbf{u}(\mathbf{r}, \sigma) = (\mathbf{u})_n^m = y_1(r)\mathbf{R}_n^m + y_3(r)\mathbf{S}_n^m + y_7(r)\mathbf{T}_n^m, \qquad (6.44)$$

$$\phi_1(\mathbf{r}, \sigma) = (\phi_1)_n^m = y_5(r)\mathcal{Y}_n^m. \qquad (6.45)$$

The coefficients $y_i(r)$ of the vector/scalar spherical harmonics in the above expansions are functions of the radial variable r; the same is true of y_2, y_4, y_6, and y_8, which are to be introduced below. All of them have a weak dependence on σ which comes from the inertial term in Eq. (6.42); we leave this dependence as understood and do not display it in the notation. The y_i ought to carry the indices n, m, but we have suppressed them for brevity. It will turn out that all the y_i are nearly independent of m while being strongly dependent on n. These facts will become apparent later when the ordinary differential equations governing the y_i are derived.

For $r > a$ where a is the radius of the Earth, y_5 is proportional to $r^{-(n+1)}$ as a consequence of the simplified form (6.38) of the Poisson equation. This statement may be readily verified with the help of Eq. (6.60), which gives the expression for $\nabla^2\phi_1$ when $f(r)$ is set equal to $y_5(r)$. It follows then, on applying the continuity condition of Section 6.6.1 on ϕ_1, that

$$\phi_1 = (\phi_1)_n^m(\mathbf{r}; \sigma) = y_5(a) \left(\frac{a}{r}\right)^{n+1} \mathcal{Y}_n^m \qquad (6.46)$$

in the exterior region.

Solutions of the form (6.44) and (6.45) which involve only spherical harmonics of the same type (nm) as that of the forcing potential are valid when the Earth model is non-rotating and spherically symmetric. If either of these conditions is not met, the same potential would produce an admixture of an infinite number of parts of spherical harmonic types $(qp) \neq (nm)$ with the main part, which is still of type (nm).

As we shall see shortly, the solution of the basic equations of motion is facilitated by the introduction of three more radial functions $y_2(r)$, $y_4(r)$, $y_8(r)$ through an expansion of the vector field $(\hat{\mathbf{r}} \cdot \mathbf{T})_n^m$ in terms of the three basic vector fields:

$$(\hat{\mathbf{r}} \cdot \mathbf{T})_n^m = y_2(r)\,\mathbf{R}_n^m + y_4(r)\,\mathbf{S}_n^m + y_8(r)\,\mathbf{T}_n^m. \tag{6.47}$$

It may be kept in mind that $\mathbf{T}$ here, like $\mathbf{u}$ and ϕ_1, is in the frequency domain: it stands for $\mathbf{T}(\mathbf{r}, \sigma)$.

In addition to the seven radial functions introduced so far, one defines another function y_6 as

$$y_6(r) = \frac{dy_5(r)}{dr} + 4\pi G\rho_0\, y_1(r) + \frac{n+1}{r} y_5(r). \tag{6.48}$$

The inclusion of the last term proportional to y_5 in the definition of y_6 is a matter of convenience. If this term is dropped, as is often done, (or if its coefficient is modified), any equation involving y_6 as well as the boundary conditions on y_6 will get correspondingly modified.

The symbols y_i used here for the various radial functions are in accordance with the standard notation (see, for instance, Alterman *et al.* (1959), except for (a) the sign differences arising from our switchover to the physicists' convention for potentials from the geophysicists' convention used in Alterman *et al.* (1959), and (b) a difference in the defining expression for y_6 which includes a y_5 term in our Eq. (6.48) but not in Alterman *et al.* (1959)).

The above expansions serve to simplify the process of solving the coupled partial differential equations for $\mathbf{u}$ and ϕ_1 by reducing them to sets of first-order ordinary differential equations for the radial functions y_i. Before carrying out this reduction, we present a brief review of the basic properties of the vector spherical harmonic functions which are needed for the expansion of the various terms in the equations of motion.

6.7.1 Properties of the vector spherical harmonics

The following properties may be readily proved from the definitions (6.43) of the basic fields:

$$\mathbf{r} \cdot \mathbf{R}_n^m = r\mathcal{Y}_n^m, \qquad \mathbf{r} \cdot \mathbf{S}_n^m = 0, \qquad \mathbf{r} \cdot \mathbf{T}_n^m = 0, \tag{6.49}$$

$$\nabla \mathcal{Y}_n^m = \frac{1}{r} \mathbf{S}_n^m, \tag{6.50}$$

$$\nabla \cdot \mathbf{R}_n^m = \frac{2}{r} \mathcal{Y}_n^m, \qquad \nabla \cdot \mathbf{S}_n^m = -\frac{n(n+1)}{r} \mathcal{Y}_n^m, \qquad \nabla \cdot \mathbf{T}_n^m = 0. \tag{6.51}$$

The property (6.5) has been used in arriving at the second of the relations (6.51).

The following additional properties may be proved with a little more effort, keeping in mind the expression (6.5) for $\nabla^2 \mathcal{Y}_n^m$:

$$\nabla^2 \mathbf{R}_n^m = -\frac{2 + n(n+1)}{r^2} \mathbf{R}_n^m + \frac{2}{r^2} \mathbf{S}_n^m, \tag{6.52}$$

$$\nabla^2 \mathbf{S}_n^m = \frac{n(n+1)}{r^2} (2\mathbf{R}_n^m - \mathbf{S}_n^m), \tag{6.53}$$

$$\nabla^2 \mathbf{T}_n^m = \nabla^2 (\mathbf{r} \wedge \nabla \mathcal{Y}_n^m) = -\frac{n(n+1)}{r^2} \mathbf{T}_n^m. \tag{6.54}$$

To obtain the last of these results, one makes use of the fact that the operators ∇^2 and $\mathbf{r} \wedge \nabla$ are mutually commuting.

We also make note of the following results for completeness, though we have no immediate need for them:

$$\hat{\mathbf{r}} \wedge \mathbf{R}_n^m = 0, \qquad \hat{\mathbf{r}} \wedge \mathbf{S}_n^m = \mathbf{T}_n^m, \qquad \hat{\mathbf{r}} \wedge \mathbf{T}_n^m = -\mathbf{S}_n^m, \tag{6.55}$$

$$\nabla \wedge \mathbf{R}_n^m = -\frac{1}{r}\mathbf{T}_n^m, \quad \nabla \wedge \mathbf{S}_n^m = \frac{1}{r}\mathbf{T}_n^m, \quad \nabla \wedge \mathbf{T}_n^m = -\frac{1}{r}[n(n+1)\mathbf{R}_n^m + \mathbf{S}_n^m]. \tag{6.56}$$

The vector spherical harmonics are accompanied by factors $f(r)$ in expansions of vector fields, as in the terms $y_1(r)\mathbf{R}_n^m$, etc., in $\mathbf{u}$. It is necessary therefore to evaluate quantities like $\nabla \cdot (f\mathbf{V})$ and $\nabla^2(f\mathbf{V})$ where $\mathbf{V}$ is some vector field. Evaluation can be done by use of the vector identities

$$\nabla \cdot (f\mathbf{V}) = f\nabla \cdot \mathbf{V} + \nabla f \cdot \mathbf{V}, \quad \nabla^2(f\mathbf{V}) = f\nabla^2\mathbf{V} + 2(\nabla f \cdot \nabla)\mathbf{V} + (\nabla^2 f)\mathbf{V}. \tag{6.57}$$

Our basic vector fields being functions of the angle variables only, the application of $(\nabla f \cdot \nabla) = (df/dr)(\hat{\mathbf{r}} \cdot \nabla) = (df/dr)\,\partial/\partial r$ to these fields yields zero. Therefore the second term in the identity for $\nabla^2(f\mathbf{V})$ contributes nothing. (This is the case also if $\mathcal{Y}_n^m$ appears instead of the vector field in this identity.) Use of the Eqs. (6.51) and (6.52)–(6.54) then leads to the following results. Primes in these equations and elsewhere indicate differentiation with respect to r:

$$\nabla(f\mathcal{Y}_n^m) = f'\mathbf{R}_n^m + \frac{1}{r} f\mathbf{S}_n^m, \tag{6.58}$$

$$\nabla \cdot (f\mathbf{R}_n^m) = \left(f' + \frac{2}{r}f\right)\mathcal{Y}_n^m, \quad \nabla \cdot (f\mathbf{S}_n^m) = \frac{n(n+1)}{r}f\mathcal{Y}_n^m, \quad \nabla \cdot (f\mathbf{T}_n^m) = 0, \tag{6.59}$$

$$\nabla^2(f\mathcal{Y}_n^m) = \left(f'' + \frac{2}{r}f'\right)\mathcal{Y}_n^m - \frac{n(n+1)}{r^2}f\mathcal{Y}_n^m, \tag{6.60}$$

$$\nabla^2(f\mathbf{R}_n^m) = \left(f'' + \frac{2}{r}f'\right)\mathbf{R}_n^m + f\left(-\frac{2}{r^2}\mathbf{R}_n^m - \frac{n(n+1)}{r^2}\mathbf{R}_n^m + \frac{2}{r^2}\mathbf{S}_n^m\right), \quad (6.61)$$

$$\nabla^2(f\mathbf{S}_n^m) = \left(f'' + \frac{2}{r}f'\right)\mathbf{S}_n^m + \frac{n(n+1)}{r^2}f(2\mathbf{R}_n^m - \mathbf{S}_n^m), \quad (6.62)$$

$$\nabla^2(f\mathbf{T}_n^m) = \left(f'' + \frac{2}{r}f'\right)\mathbf{T}_n^m - \frac{n(n+1)}{r^2}f\mathbf{T}_n^m. \quad (6.63)$$

The last of the identities (6.59) is of special interest: it implies that a displacement field of the form $f(r)\mathbf{T}_n^m$ does not cause any compression or dilatation of the material, which means in turn for the case $n = 2, m = 1$, that there is no deformation associated with such a displacement; it represents just a global rotation.

As immediate applications of the above results, we write down the expansions of $\nabla \cdot \mathbf{u}_n^m$ and $\nabla^2\mathbf{u}_n^m$ which will be needed in the next section. Equations (6.59) show immediately that

$$\nabla \cdot \mathbf{u}_n^m = D\,\mathcal{Y}_n^m \quad \text{with} \quad D = \frac{dy_1}{dr} + \frac{1}{r}[2y_1 - n(n+1)y_3], \quad (6.64)$$

while (6.61)–(6.63) lead to

$$\nabla^2\mathbf{u}_n^m = \left(y_1'' + \frac{2y_1'}{r} - \frac{2y_1}{r^2} - \frac{n(n+1)y_1}{r^2} + \frac{2n(n+1)y_3}{r^2}\right)\mathbf{R}_n^m$$

$$+ \left(y_3'' + \frac{2y_3'}{r} + \frac{2y_1}{r^2} - \frac{n(n+1)y_3}{r^2}\right)\mathbf{S}_n^m$$

$$+ \left(y_7'' + \frac{2y_7'}{r} - \frac{n(n+1)y_7}{r^2}\right)\mathbf{T}_n^m. \quad (6.65)$$

6.7.2 Equations for the radial functions

We are now in a position to use the results derived above to obtain the spherical harmonic expansions of the various quantities that appear in the partial differential equations of the deformation problem, and to use the expansions to reduce those equations, namely, (6.37) and (6.42) taken together with (6.34), to a set of ordinary differential equations for the radial functions involved.

Reduction of the stress–strain relationship

To begin with, one observes from the expression (6.34) for $\mathbf{T}$ specialized to the spherically symmetric Earth, that

$$\hat{\mathbf{r}} \cdot \mathbf{T} = \lambda_0(\nabla \cdot \mathbf{u})\hat{\mathbf{r}} + \mu_0\partial_r\mathbf{u} + \frac{\mu_0}{r}(\nabla(\mathbf{r} \cdot \mathbf{u}) - \mathbf{u}). \quad (6.66)$$

The coefficients of $\mathbf{R}_n^m$, $\mathbf{S}_n^m$, and $\mathbf{T}_n^m$ in $\hat{\mathbf{r}} \cdot \mathbf{T}$ are defined to be y_2, y_4, and y_8, respectively, according to Eq. (6.47). On the other hand, the corresponding coefficients in the expansion of the right-hand side of the above equation are obtainable directly by introducing into it the expressions (6.44) and (6.64) for $\mathbf{u}$ and $\nabla \cdot \mathbf{u}$ and using the properties (6.49). On equating the expansions of the left- and right-hand sides, we obtain

$$y_2 = (\lambda_0 + 2\mu_0)\frac{dy_1}{dr} + \frac{\lambda_0}{r}[2y_1 - n(n+1)y_3] = \lambda_0 D + 2\mu_0 \frac{dy_1}{dr}, \quad (6.67)$$

$$y_4 = \mu_0 \frac{dy_3}{dr} + \frac{\mu_0}{r}(y_1 - y_3), \quad (6.68)$$

$$y_8 = \mu_0 \frac{dy_7}{dr} - \frac{\mu_0}{r}y_7. \quad (6.69)$$

Reduction of Poisson's equation

We take up next the reduction of the Poisson equation (6.37), which is rather straightforward. Since $\phi_1 = y_5 \mathcal{Y}_n^m$, we can evaluate $\nabla^2 \phi_1$ with the help of Eq. (6.60). At the same time, we can use Eqs. (6.59) to express $\nabla \cdot (\rho_0 \mathbf{u})$ in terms of the relevant radial functions and their derivatives. On substituting the resulting expressions into the two sides of the Poisson equation (6.37) we obtain the following equation for y_5:

$$y_5'' + \frac{2}{r}y_5' - \frac{n(n+1)}{r^2}y_5 = -4\pi G[\rho_0' y_1 + \rho_0 D], \quad (6.70)$$

where we have substituted for D from Eq. (6.64) in the last step. Now, differentiation of the defining expression (6.48) for y_6 yields

$$y_6' = y_5'' + 4\pi G(\rho_0 y_1)' + \frac{n+1}{r}y_5' - \frac{n+1}{r^2}y_5. \quad (6.71)$$

We obtain the following equation which contains neither y_5'' nor ρ_0' by subtracting Eq. (6.70) from Eq. (6.71);

$$y_6' = 4\pi G\rho_0[y_1' - D] + \frac{n-1}{r}y_5' + \frac{(n-1)(n+1)}{r^2}y_5. \quad (6.72)$$

We may use (6.48) once again to eliminate y_5' from this equation, and then use the expression (6.64) for D to arrive at the final form of the radial equation equivalent to Poisson's equation:

$$\frac{dy_6}{dr} = \frac{n-1}{r}y_6 - \frac{4\pi G\rho_0}{r}(n+1)(y_1 - ny_3). \quad (6.73)$$

Reduction of the equation for the displacement field **u**

The procedure for the reduction of Eq. (6.42), which governs the displacement field **u**, is similar in principle, though more complicated because of the vector nature of the equation. A common feature is that the initial reduction leads to second-order equations relating y_1'', y_3'', y_7'' to the y_i and their first derivatives; these second-order derivatives are then eliminated in favor of y_2', y_4', y_8' by using the expressions obtained by differentiating Eqs. (6.67)–(6.69). The first derivatives of the other y_i which remain at the end of these steps are finally replaced by substitution from the triplet of equations just mentioned.

The essential first step in the reduction is a spherical harmonic expansion of the individual terms in the momentum balance equation (6.42). The results (6.58)–(6.63) will be used in making the expansions.

We observe at the outset that the forcing term on the right-hand side of the equation, with ϕ_e given by Eqs. (6.14), (6.40), and (6.15), has the expansion

$$-\rho_0 \nabla \phi_e(\mathbf{r}; \sigma) = K \rho_0 \tilde{\phi} r^{n-1} (n \mathbf{R}_n^m + \mathbf{S}_n^m). \tag{6.74}$$

As for the left-hand side of the momentum balance equation (6.42), the expansions of the first four terms can be written down without much effort. The first term is

$$-\sigma^2 \Omega_0^2 \rho_0 \mathbf{u} = -\sigma^2 \Omega_0^2 \rho_0 (y_1 \mathbf{R}_n^m + y_3 \mathbf{S}_n^m + y_7 \mathbf{T}_n^m). \tag{6.75}$$

The second is

$$-\rho_0 (\nabla \cdot \mathbf{u}) \nabla \phi_0 = -\rho_0 D g_0 \mathbf{R}_n^m, \tag{6.76}$$

where $g_0 = d\phi_0/dr$ is the magnitude of the acceleration due to gravity at the radial position r in the unperturbed Earth. For the third term, we have

$$\rho_0 \nabla(\mathbf{u} \cdot \nabla \phi_0) = \rho_0 \nabla(y_1 \mathcal{Y}_n^m g_0) = \rho_0 \left[(g_0 y_1)' \mathbf{R}_n^m + \frac{g_0}{r} y_1 \mathbf{S}_n^m \right]. \tag{6.77}$$

The fourth term $\rho_0 \nabla \phi_1$ yields

$$\rho_0 \nabla \phi_1 = \rho_0 \left[y_5' \mathbf{R}_n^m + \frac{1}{r} y_5 \mathbf{S}_n^m \right]. \tag{6.78}$$

The last three are "gravitation" terms, involving either ϕ_0 or ϕ_1. The sum of these three expressions simplifies considerably on substituting for y_5' from Eq. (6.48) and using the fact that

$$g_0' = \frac{d^2 \phi_0}{dr^2} = \nabla^2 \phi_0 - \frac{2}{r} \frac{d\phi_0}{dr} = 4\pi G \rho_0 - \frac{2}{r} g_0. \tag{6.79}$$

The resulting expression for the sum is

$$\rho_0 \left[y_6 - \frac{n+1}{r} y_5 - g_0 \frac{4y_1 - n(n+1)y_3}{r} \right] \mathbf{R}_n^m + \rho_0 \left[\frac{g_0 y_1 + y_5}{r} \right] \mathbf{S}_n^m. \quad (6.80)$$

The expansion of the remaining term $\nabla \cdot \mathbf{T}$ of Eq. (6.42) calls for rather more effort. Taking $\mathbf{T}$ from (6.34), with the replacements $\rho \to \rho_0(r)$, etc., we readily see that

$$\nabla \cdot \mathbf{T} = \nabla[\lambda_0(\nabla \cdot \mathbf{u})] + \mu_0'(\hat{\mathbf{r}} \cdot \nabla)\mathbf{u} + \frac{\mu_0'}{r}[\nabla(\mathbf{r} \cdot \mathbf{u}) - \mathbf{u}] + \mu_0[\nabla^2 \mathbf{u} + \nabla(\nabla \cdot \mathbf{u})]. \quad (6.81)$$

We have used here the fact that $\nabla \mu_0 = \mu_0' \hat{\mathbf{r}}$. Spherical harmonic expansion of the individual terms in the above expression can be carried out with the help of the relevant identities from the set (6.58)–(6.63) and with the use of the results (6.64) and (6.65) obtained above. Then the first three terms of (6.81) yield

$$\nabla[\lambda_0(\nabla \cdot \mathbf{u})] = \nabla(\lambda_0 D \mathcal{Y}_n^m) = (\lambda_0 D)' \mathbf{R}_n^m + \frac{\lambda_0 D}{r} \mathbf{S}_n^m, \quad (6.82)$$

$$\mu_0'(\hat{\mathbf{r}} \cdot \nabla)\mathbf{u} = \mu_0' \frac{\partial \mathbf{u}}{\partial r} = \mu_0'(y_1' \mathbf{R}_n^m + y_3' \mathbf{S}_n^m + y_7' \mathbf{T}_n^m), \quad (6.83)$$

$$\frac{\mu_0'}{r}[\nabla(\mathbf{r} \cdot \mathbf{u}) - \mathbf{u}] = \frac{\mu_0'}{r}[\nabla(r y_1 \mathcal{Y}_n^m) - \mathbf{u}]$$

$$= \frac{\mu_0'}{r}[r y_1' \mathbf{R}_n^m + (y_1 - y_3) \mathbf{S}_n^m - y_7 \mathbf{T}_n^m]. \quad (6.84)$$

The next term is

$$\mu_0 \nabla^2 \mathbf{u} = \mu_0[(\nabla^2 \mathbf{u})_R \mathbf{R}_n^m + (\nabla^2 \mathbf{u})_S \mathbf{S}_n^m + (\nabla^2 \mathbf{u})_T \mathbf{T}_n^m], \quad (6.85)$$

where the explicit forms of the coefficients $(\nabla^2 \mathbf{u})_R$, etc., are to be taken from the expression (6.65) for $\nabla^2 \mathbf{u}$. The final term is

$$\mu_0 \nabla(\nabla \cdot \mathbf{u}) = \mu_0 \left[D' \mathbf{R}_n^m + \frac{D}{r} \mathbf{S}_n^m \right]. \quad (6.86)$$

The complete expansion of $\nabla \cdot \mathbf{T}$ is obtained in the form

$$\nabla \cdot \mathbf{T} = (\nabla \cdot \mathbf{T})_R \mathbf{R}_n^m + (\nabla \cdot \mathbf{T})_S \mathbf{S}_n^m + (\nabla \cdot \mathbf{T})_T \mathbf{T}_n^m, \quad (6.87)$$

on summing the expressions (6.82)–(6.86). Rather than writing out the result in full, we focus on the elimination of the derivatives λ_0' and μ_0' through the use of the defining Eqs. (6.67)–(6.69) for y_2, y_4, and y_8 and on the cancellation of second derivative terms that takes place when summation of the five terms is done. Take for instance, the coefficient of $\mathbf{R}_n^m$ in the sum:

$$(\nabla \cdot \mathbf{T})_R = (\lambda_0 D)' + 2\mu_0' y_1' + \mu_0(\nabla^2 \mathbf{u})_R + \mu_0 D'. \quad (6.88)$$

Each of the quantities $(\nabla^2 \mathbf{u})_R$ and D' contains a term y_1''; when these y_1'' terms are taken out of the last two terms on the right-hand side of (6.88) and combined with the first two terms, the result is $(\lambda_0 D + 2\mu_0 y_1')'$, which may be identified as y_2' (see the second form given in the defining Eq. (6.67) for y_2). Thus the coefficient of $\mathbf{R}_n^m$ in $\nabla \cdot \mathbf{T}$ is seen to be

$$(\nabla \cdot \mathbf{T})_R = y_2' + \mu_0[(\nabla^2 \mathbf{u})_R + D' - 2y_1''].\tag{6.89}$$

One can verify in a similar fashion that

$$(\nabla \cdot \mathbf{T})_S = y_4' + (\lambda_0 + \mu_0)\frac{D}{r} - \frac{\mu_0}{r}(y_1' - y_3')$$

$$+ \frac{\mu_0}{r^2}(y_1 - y_3) + \mu_0[(\nabla^2 \mathbf{u})_S - y_3''],\tag{6.90}$$

$$(\nabla \cdot \mathbf{T})_T = y_8' + \frac{\mu_0}{r}y_7' - \frac{\mu_0}{r^2}y_7 + \mu_0[(\nabla^2 \mathbf{u})_T - y_7''].\tag{6.91}$$

We have to substitute for $\nabla^2 \mathbf{u}$ from (6.65) and for D from (6.64) in order to make the above expressions explicit; when this is done, the second derivative terms cancel out within the square brackets in the three equations above. Next, one eliminates the first derivatives y_1', y_3', y_7' in favor of y_2, y_4, y_8 with the aid of Eqs. (6.67)–(6.69). The quantities (6.89)–(6.91) are thus given, respectively, by y_2', y_4', y_6' plus a host of other terms which do not involve any derivatives.

The spherical harmonic expansion of the terms in the displacement Eq. (6.42) is now complete. The next task is to pick out the coefficients of $\mathbf{R}_n^m$ from the expressions obtained for the individual terms in the derivation above, and to equate their sum to the coefficient of $\mathbf{R}_n^m$ from the forcing term; this exercise is to be repeated for the $\mathbf{S}_n^m$ and $\mathbf{T}_n^m$ terms. The expansion of the first of the terms on the left-hand side of (6.42) may be found in Eq. (6.75); the sum of the next three terms is given by (6.80) and the last term is the negative of $\nabla \cdot \mathbf{T}$ which we have just dealt with. The forcing term on the right-hand side of the equation is given by (6.74). Putting together all the expressions, one obtains the equations

$$-\sigma^2\Omega_0^2\rho_0 y_1 + \rho_0\left[y_6 - \frac{n+1}{r}y_5 - g_0\frac{4y_1 - n(n+1)y_3}{r}\right] - (\nabla \cdot \mathbf{T})_R$$

$$= K\rho_0\tilde{\phi}nr^{n-1},\tag{6.92}$$

$$-\sigma^2\Omega_0^2\rho_0 y_3 + \rho_0\frac{g_0 y_1 + y_5}{r} - (\nabla \cdot \mathbf{T})_S = K\rho_0\tilde{\phi}r^{n-1},\tag{6.93}$$

$$-\sigma^2\Omega_0^2\rho_0 y_7 - (\nabla \cdot \mathbf{T})_T = 0.\tag{6.94}$$

In view of what has been said in the lines following Eqs. (6.89)–(6.91) about the explicit form of the coefficients $(\nabla \cdot \mathbf{T})_{R,S,T}$, the last three equations express

y'_2, y'_4, y'_8, in terms of a combination of the y_i, ($i = 1, \ldots, 8$), but not any of their derivatives. One also rewrites Eqs. (6.67)–(6.69) and (6.48) as equations relating y'_1, y'_3, y'_7 and y'_5 to various y_i; and (6.73) provides the differential equation for y_6. Thus we have a full set of eight first-order ordinary differential equations for the eight radial functions $y_1, y_2, \ldots, y_8$. We shall write out the full set of equations explicitly only for the special case $n = 2$ which is by far the most important.

An examination of the terms in the differential equations for the y_i shows that the set of eight equations forms two disjoint sets. The first set of six equations for $y_1, y_2, \ldots y_6$ are mutually coupled; they do not involve y_7 or y_8. The two equations for y_7 and y_8 are also mutually coupled; they contain none of the other y_i. Furthermore, no forcing term appears in the second set of equations: the toroidal part of the displacement field, representing rotational motion, is not excited by the external potential in the case of a spherically symmetric Earth. This is a confirmation of what we already know: perturbations of the rotational motion (nutation, in particular) take place only when there is deviation from spherical symmetry.

It is worthy of note that m has not made an explicit appearance in the coefficients of the vector or scalar spherical harmonics in any of the expressions derived so far. So it would appear that the equations governing the y_i do not involve m, and that their solutions y_i are independent of m. Actually there is a subtle dependence on m through σ^2 in the inertia term because, as has been noted more than once earlier, the frequencies σ are close to $-m$. Nevertheless the effect of this term is of $\mathcal{O}(\epsilon)$ as we have observed in Section 6.6.2. Therefore, when effects of this order are neglected, one gets the same set of solutions y_i for any given degree n of the forcing potential irrespective of the order m.

In the case of Earth models that include deviations from spherical symmetry (including, in particular, the most important case of the axially symmetric ellipsoidal Earth), the spherical harmonics present in the fields $\mathbf{u}(\mathbf{r})$ and $\phi_1(\mathbf{r})$ include those of types (qp) other than the type (nm) of the forcing potential. Consequently, radial functions $y_i^{(qp)}$ associated with the components of different types (qp) have to be defined; and the spherical harmonic analysis of the momentum balance and Poisson equations leads to radial differential equations for all of them. These equations couple the radial functions of all types (qp), in general; and the equations for the spheroidal functions $y_i^{(qp)}$ with $i = 1, 2, \ldots, 6$ are no longer disjoint from the toroidal ones corresponding to $i = 7, 8$. As a result, toroidal displacements associated with rotational motions also get excited by the action of the external potential; and this results in nutation and wobble, unlike the spherically symmetric case. The theory relating to the most important case of an axially symmetric spheroidal Earth, leading to the determination of nutations as well as deformations is presented in

Chapter 7. We are concerned in this chapter with deformations only, and the effect of ellipticity on these is only of $\mathcal{O}(\epsilon)$.

6.7.3 Forcing potentials of degree two: radial equations

The radial equations pertaining to the special case $n = 2$, which emerge from the procedure detailed in the foregoing section, are as follows:

$$\frac{dy_1}{dr} = \frac{1}{\lambda_0 + 2\mu_0}\left(y_2 - \frac{\lambda_0}{r}(2y_1 - 6y_3)\right), \tag{6.95}$$

$$\frac{dy_2}{dr} = -\sigma^2\Omega_0^2\rho_0 y_1 - \frac{\rho_0 g_0}{r}(4y_1 - 6y_3) + \rho_0 y_6 - \frac{3}{r}\rho_0 y_5 + \frac{6}{r}y_4$$
$$- \frac{4}{r}\frac{\mu_0}{\lambda_0 + 2\mu_0}\left(y_2 - \frac{1}{r}(3\lambda_0 + 2\mu_0)(y_1 - 3y_3)\right) - \frac{2}{3}\Omega_0^2\tilde{\phi}(\sigma)\,\rho_0 r, \tag{6.96}$$

$$\frac{dy_3}{dr} = \frac{1}{\mu_0}y_4 + \frac{1}{r}(y_3 - y_1), \tag{6.97}$$

$$\frac{dy_4}{dr} = -\sigma^2\Omega_0^2\rho_0 y_3 - \frac{\lambda_0}{\lambda_0 + 2\mu_0}\left(\frac{1}{r}y_2 + \frac{4\mu_0}{r^2}(y_1 - 3y_3)\right)$$
$$+ \frac{\rho_0 g_0}{r}y_1 + \frac{\rho_0}{r}y_5 - \frac{3}{r}y_4 - \frac{2\mu_0}{r^2}(y_1 - 5y_3) - \frac{1}{3}\Omega_0^2\tilde{\phi}(\sigma)\,\rho_0 r, \tag{6.98}$$

$$\frac{dy_5}{dr} = y_6 - 4\pi G\rho_0 y_1 - \frac{3}{r}y_5, \tag{6.99}$$

$$\frac{dy_6}{dr} = \frac{1}{r}y_6 - 12\pi G\rho_0\frac{1}{r}(y_1 - 2y_3), \tag{6.100}$$

$$\frac{dy_7}{dr} = \frac{1}{r}y_7 + \frac{1}{\mu_0}y_8, \tag{6.101}$$

$$\frac{dy_8}{dr} = -\sigma^2\Omega^2\rho_0 y_7 - \frac{3}{r}y_8. \tag{6.102}$$

None of the equations has an explicit dependence on the order m of the forcing potential; but as noted in the penultimate paragraph of Section 6.7.2, a weak dependence on m arises from the fact that σ has to be close to $-m$. It may be observed, in confirmation of what was stated in the preceding paragraphs, that the first six of the above equations do not contain y_7 or y_8 while the last pair of equations do not involve any of the other y_i. The tidal forcing, proportional to $\tilde{\phi}$, enters only into the second and fourth equations; nevertheless it affects $y_1, y_2, \ldots, y_6$ through their mutual couplings in the six equations, while y_7, y_8 are not affected. Therefore

y_7 and y_8 may be ignored when considering the deformations of a spherical Earth due to tidal forcing.

It is self-evident that all the y_i must be proportional to the amplitude $\tilde{\phi}$ of the forcing potential. It is useful to indicate this explicitly by writing

$$y_i(r) = y_i^u(r)\tilde{\phi}(\sigma), \tag{6.103}$$

where the superscript u indicates that y_i^u is the solution for unit value of the amplitude $\tilde{\phi}$.

6.7.4 Radial equations in the case of a fluid layer

The above set of differential equations cannot be applied, as they stand, to any fluid region. The reason is that fluids have no shear resistance: the shear modulus μ_0 vanishes. One sees then, on setting $\mu_0 = 0$ in the equations obtained from (6.97) and (6.101) after multiplication by μ_0, that $y_4 = y_8 = 0$. (Alternatively, the expression (6.66) for $\hat{\mathbf{r}} \cdot \mathbf{T}$ and the definition (6.47) of the y_2, y_4, y_8 ensure that y_4 and y_8 do not exist when $\mu_0 = 0$.) Consequently Eqs. (6.97) and (6.101) drop out and the left-hand sides of Eqs. (6.98) and (6.102) become zero and hence also the complete Eq. (6.101) drops out. The five equations which remain are as follows:

$$\frac{dy_1}{dr} = \frac{1}{\lambda_0}y_2 - \frac{1}{r}(2y_1 - 6y_3), \tag{6.104}$$

$$\frac{dy_2}{dr} = -\omega^2\rho_0 y_1 - \frac{\rho_0 g_0}{r}(4y_1 - 6y_3) + \rho_0 y_6 - \frac{3}{r}\rho_0 y_5 - \frac{2}{3}\Omega_0^2\tilde{\phi}\,\rho_0 r, \tag{6.105}$$

$$0 = -\omega^2\rho_0 y_3 - \frac{1}{r}y_2 + \frac{\rho_0 g_0}{r}y_1 + \frac{\rho_0}{r}y_5 - \frac{1}{3}\Omega_0^2\tilde{\phi}\,\rho_0 r, \tag{6.106}$$

$$\frac{dy_5}{dr} = y_6 - 4\pi G\rho_0 y_1 - \frac{3}{r}y_5, \tag{6.107}$$

$$\frac{dy_6}{dr} = \frac{1}{r}y_6 - 12\pi G\rho_0\frac{1}{r}(y_1 - 2y_3). \tag{6.108}$$

Note that Eq. (6.106) is a constraint equation connecting y_1, y_2, y_3, and y_5. The remaining four equations have to be solved subject to this constraint in order to obtain these four functions and y_6.

6.7.5 Boundary conditions on the y_i

The boundary conditions stated in Section 6.6.1 translate into the following conditions on the radial functions y_i:

1. All the y_i, $(i = 1, \dots, 8)$ are to be continuous at any solid–solid welded boundary.

2. y_1 and y_2 are to be continuous across any fluid–solid boundary, while there is no continuity requirement on y_3 if the fluid is non-viscous; the shear stresses y_4 and y_8 vanish in a non-viscous fluid and hence, by the continuity condition stated in Section (6.6.1), also on the solid side of the fluid–solid boundary.

3. $y_2 = y_4 = y_6 = y_8 = 0$ at the Earth's surface, assuming that there is no surface loading.

4. All the y_i vanish at the center ($r = 0$), except that y_2 and y_4 remain nonzero in the important special case $n = 2$ (and except the geocenter motion ($n = 1$) as shown in Section 5.3.5).

5. The vanishing of y_6 at the surface (for excitation by potentials of any degree n) rests on the continuity of ϕ_1 and $\hat{\mathbf{r}} \cdot (\nabla \phi_1 + 4\pi G \rho \mathbf{u})$ at the spherical outer surface, and on the particular definition of y_6 as in (6.48). Expressed in terms of the radial functions, the quantities that are to be continuous between the inner and outer sides of the outer surface $r = a$ are y_5 and $dy_5/dr + 4\pi G \rho y_1$. The requirement on the latter may equally well be treated as a continuity condition on $y_6 = dy_5/dr + 4\pi G \rho y_1 + (n + 1)y_5/r$, since y_5 in the added term is itself continuous. Now look at the value of this quantity outside the Earth, where y_5 is proportional to $r^{-(n+1)}$, causing $dy_5/dr + (n + 1)y_5/r$ to vanish. Since ρ too vanishes in the exterior region, it follows that y_6 vanishes, which is what we wanted to prove.

   In works where the Alterman *et al.* (1959) definition of y_6 (i.e., $y_6(r) = dy_5(r)/dr + 4\pi G \rho_0 y_1(r)$) is employed, the boundary condition is $y_6(a) = -(n + 1)y_5(a)/a$.

6. The alterations in the boundary conditions necessitated by the presence of a surface layer (ocean, atmosphere) have been stated under item 5 of Section 6.6.1. As applied to the spherical harmonic components, the new conditions require that $y_2(a) = gL_{nm}$ and that there be a discontinuity in y_6 at the surface, with a jump of $4\pi GaL_{nm}$ on going from the inner side to the outer side of the thin fluid layer at the Earth's surface.

6.7.6 Integration of the radial equations

We have the set of six coupled first-order ordinary differential Eqs. (6.95)–(6.100) for $y_1, y_2, \ldots, y_6$ to be solved within the mantle and in the SIC, while the number of equations is four within the FOC. The general solutions in any region will involve as many integration constants as the number of equations, and may be taken as a linear combination, with constant coefficients, of any six independent solutions in each of the solid regions, and of four independent solutions in the fluid region. In the actual process of integration, it is convenient to start at the geocenter, $r = 0$. By treating a very small spherical region around the center as homogeneous, one can obtain a general analytical solution within that sphere (see below for the general forms of the y_i in the homogeneous case). It turns out that there are three independent solutions which diverge at the origin and are therefore not physically admissible, and three others which remain finite. Admissible solutions within the inner core are therefore arbitrary linear combinations of the three independent solutions that remain finite at the geocenter. On crossing over to the fluid core across the inner core boundary (ICB), the continuity conditions (namely that the normal components of the displacement and the stress which are represented by y_1 and y_2, as well as y_5 and y_6, be continuous across the boundary) cause the values of these four quantities on the fluid side of the ICB to be related to the three (arbitrary) integration constants on the solid side; the constraint (6.106) which relates y_3 on the fluid side to y_1, y_2, and y_5 implies that the value of y_3 on the fluid side of the ICB as well is related to the same three integration constants. The integration of the four differential equations in the fluid region can be carried outwards from the ICB up to the CMB, using the constraint equation at every stage to relate y_3 to the other four functions. (Note that y_4 is required to be zero everywhere in the fluid). There are only three independent solutions in the mantle (as in the inner core) because of the vanishing boundary conditions on y_2, y_4, and y_6 at the Earth's surface. One integrates them downwards from the surface to the CMB. Finally, on matching the solutions in the mantle and those in the outer core across the CMB, all the integration constants get fully determined, and a unique solution is obtained.

The above discussion concerns the radial functions related to the deformation produced by tidal potentials of degree two; these potentials enter the equations of motion through the terms proportional to $\tilde{\phi}(\sigma)$, which appear in the equations for y_2 and y_4. We need to consider also deformations resulting from the action of the centrifugal potentials $\Delta\phi_c$, $\Delta\phi_{cf}$, and $\Delta\phi_{cs}$ of Eqs. (6.17), (6.19), and (6.20), which are associated with the wobble $\tilde{m}$ and the differential wobbles $\tilde{m}_f$ and $\tilde{m}_s$ of the two core regions. All three of these potentials have the degree two tesseral character, just like the tidal potentials of degree two and order one. We have noted in Section 6.2.3 that the potential $\Delta\phi_c$ generated by $\tilde{m}$ differs from the tesseral tidal potential only in the replacement of $\tilde{\phi}^{21}$ by $-\tilde{m}$. Therefore the solutions $y_i(r)$ pertaining to this case are such that

$$y_i(r)/(-\tilde{m}) = y_i^u(r), \tag{6.109}$$

with the same $y_i^u(r)$ as in Eq. (6.103) pertaining to the tidally excited case.

The radial functions characterizing the deformations generated by the centrifugal potentials $\Delta\phi_{cf}$ and $\Delta\phi_{cs}$ of Section 6.2.3 due to the differential wobbles of the FOC and SIC, respectively, do not enjoy any such simple relationship to the tidal $y_i^u(r)$. The reason, of course, is that these potentials do not act throughout the Earth: the first one is non-vanishing only within the FOC, and the second one only within the SIC, though within the respective regions where they are nonzero, they differ from the tesseral tidal potential only by the replacements $\tilde{\phi} \to -\tilde{m}_f$ and $\tilde{\phi} \to -\tilde{m}_s$, respectively. In integrating the set of radial equations through the three regions, one has to set the forcing to zero outside the FOC/SIC; as a consequence, the radial dependences of these functions differ from each other and from the radial dependence of $y_i(r)$ relating to the tidal excitation. We denote by $y_{fi}(r)$ and $y_{si}(r)$ the radial functions generated by $\Delta\phi_{cf}$ and $\Delta\phi_{cs}$, respectively, and define

$$y_{fi}^u(r) = -\frac{y_{fi}(r)}{\tilde{m}_f(\sigma)}, \qquad y_{si}^u(r) = -\frac{y_{si}(r)}{\tilde{m}_s(\sigma)}. \tag{6.110}$$

The subscripts f and s are to indicate that the tesseral potential was operative only within the FOC and SIC, respectively.

6.7.7 The solutions for the homogeneous case

We present here the analytical solutions that may be obtained in the simplified case where the Earth consists of two or three homogeneous layers (with ρ, λ, μ constant within each layer), so that one may get a better feel for the nature of the solutions for the y_i.

The radial equations provided by Eq. (6.67) with D given by Eq. (6.64) for dy_1/dr, Eq. (6.92) for dy_2/dr, Eq. (6.68) for dy_3/dr, Eq. (6.93) with $(\Delta \cdot T)_s$ given by Eq. (6.90) for dy_4/dr, Eq. (6.48) for dy_5/dr, and Eq. (6.73) for dy_6/dr constitute a set of ordinary differential equations of the first order in d/dr. The coefficients of the linear combinations of the y_i are either (1) directly proportional to the properties of the material inside the Earth such as density, shear or bulk moduli, or gravity, or (2) inversely proportional to r and proportional to the properties of the material inside the Earth such as density, shear or bulk moduli, or gravity. One then considers three homogeneous layers inside the Earth, with densities ρ_m, ρ_c, and ρ_g being the mantle, outer core, and inner core densities, respectively, and ρ_M, the mean density of the Earth. For g_0 being the surface gravity, the initial gravity inside the inner core, outer core, and mantle, g_g, g_c, and g_m, can be expressed as a function of r (integrating the initial gravity inside a homogeneous layered Earth):

$$g_g = g_0 \frac{\rho_g r}{\rho_0 r_{\text{surf}}}, \tag{6.111}$$

$$g_c = g_0 \frac{\rho_g r_{\text{ICB}}^3 + \rho_c(r^3 - r_{\text{ICB}}^3)}{\rho_0 r^2 r_{\text{surf}}}, \tag{6.112}$$

$$g_m = g_0 \frac{\rho_g r_{\text{ICB}}^3 + \rho_c(r_{\text{CMB}}^3 - r_{\text{ICB}}^3) + \rho_m(r^3 - r_{\text{CMB}}^3)}{\rho_0 r^2 r_{\text{surf}}}, \tag{6.113}$$

where r_{ICB}, r_{CMB}, and r_{surf} are the radius of the inner core, outer core, and surface, respectively.

Considering that the initial gravity is thus inversely proportional to r^2 or proportional to r, the solutions of such equations (Eq. (6.67) with D given by Eq. (6.64) for dy_1/dr, Eq. (6.92) for dy_2/dr, Eq. (6.68) for dy_3/dr, Eq. (6.93) for dy_4/dr, Eq. (6.48) for dy_5/dr, and Eq. (6.73) for dy_6/dr; and Eqs. (6.95)–(6.100) for the degree two forcing potential case) for this homogeneous layered sphere are thus either proportional to r^{n-2}, r^{n-1}, r^n, r^{n+1}, $1/r^n$, $1/r^{n+1}$, $1/r^{n+2}$, or $1/r^{n+3}$. The r-dependence of the y_i turns out to be as in the following expressions:

$$y_1 = \frac{C_1}{r^n} + \frac{C_2}{r^{n+2}} + C_3 r^{n+1} + C_4 r^{n-1}, \tag{6.114}$$

$$y_2 = 2\mu_m \left(-\frac{(n^2 - 1 + 3n)}{n+1} \frac{C_1}{r^{n+1}} - (n+2)\frac{C_2}{r^{n+3}} + \frac{(n^2 - n - 3)}{n} C_3 r^n \right.$$
$$\left. + (n-1)C_4 r^{n-2} \right) - \rho_m \left(C_5 r^n + \frac{C_6}{r^{n+1}} \right)$$
$$+ \rho_m g_m \left(\frac{C_1}{r^n} + \frac{C_2}{r^{n+2}} + C_3 r^{n+1} + C_4 r^{n-1} \right), \tag{6.115}$$

$$y_3 = -\frac{(n-2)}{n(n+1)} \frac{C_1}{r^n} - \frac{1}{n+1} \frac{C_2}{r^{n+2}} + \frac{(n+3)}{n(n+1)} C_3 r^{n+1} + \frac{1}{n} C_4 r^{n-1}, \tag{6.116}$$

$$y_4 = 2\mu_m \left(\frac{n-1}{n} \frac{C_1}{r^{n+1}} + \frac{n+2}{n+1} \frac{C_2}{r^{n+3}} + \frac{n+2}{n+1} C_3 r^n + \frac{n-1}{n} C_4 r^{n-2} \right), \tag{6.117}$$

$$y_5 = C_5 r^n + \frac{C_6}{r^{n+1}}, \tag{6.118}$$

$$y_6 = nC_5 r^{n-1} - (n+1)\frac{C_6}{r^{n+2}} + 3\frac{g_0 \rho_m}{\rho_M r_{surf}} \left(\frac{C_1}{r^n} + \frac{C_2}{r^{n+2}} + C_3 r^{n+1} + C_4 r^{n-1} \right)$$
$$+ (n+1) \left(C_5 r^{n-1} + \frac{C_6}{r^{n+2}} \right), \tag{6.119}$$

where this last term must be dropped in the case of a definition of y_6 which does not involve the term $\frac{n+1}{r} y_5$ (see Eq. (6.48)) and on considering that $4\pi G\rho_M = 3g_0/r_{surf}$.

The values of the constants C_j differ from one region to the other, of course. The solutions in the core regions have to be consistent with items 2 and 4 of the boundary conditions enumerated in the last section. The coefficients C_1, C_2, and

C_6 have to vanish in the innermost region to prevent the solutions from diverging at $r = 0$.

It is necessary to point out here that the above expressions contain no reference to the forcing potential: they may therefore be ordinarily thought of as solutions of the problem of free deformational modes, e.g., free oscillations. Nevertheless, the above solutions may be made applicable to the equations of motion including the forcing potential in the following manner. Note first that the forcing potential ϕ_e in the equation of motion (6.33) may be transferred to the left-hand side and combined with ϕ_1, and the combined potential may be redefined as a new ϕ_1. The resulting equation will appear to be a homogeneous equation. The new ϕ_1 also satisfies the Poisson equation (6.37), since $\nabla^2 \phi_e = 0$; but the fact that ϕ_e is included in it shows up in the value of the new y_6 at the surface $r = a$: it will now be $-Kna^{n-1}\tilde{\phi}$ instead of zero. It is necessary that the reader be aware of this approach, as it is employed in a significant part of the literature. However, we shall continue to use elsewhere in this book our earlier definition of ϕ_1 as the perturbation of the Earth's own gravitational potential.

6.7.8 The Love numbers in terms of the radial functions

The definitions of the Love numbers h_n and k_n given in equations (6.23) relate the values, at the surface $r = a$, of the radial component u_r of $\mathbf{u}_n^m$ and of the incremental potential $(\phi_1)_n^m$, respectively, to that of ϕ_e at the same surface. We see from (6.41) taken together with (6.44) that the generic spectral components of these quantities are given, at $r = a$, by

$$u_r(\mathbf{r}; t) \equiv \hat{\mathbf{r}} \cdot \mathbf{u}(\mathbf{r}; t) = \text{Re}\,[y_1(a)\mathcal{Y}_n^m e^{i(\sigma\Omega_0 t - \alpha_\omega)}], \tag{6.120}$$

$$\phi_1(\mathbf{r}; t) = \text{Re}\,[y_5(a)\mathcal{Y}_n^m e^{i(\sigma\Omega_0 t - \alpha_\omega)}], \tag{6.121}$$

where α_ω is defined in Eq. (6.11). Furthermore,

$$\phi_e(\mathbf{r}; t) = -\frac{\Omega_0^2 a^2}{3}\text{Re}\,[\tilde{\phi}(\sigma)\mathcal{Y}_n^m e^{i(\sigma\Omega_0 t - \alpha_\omega)}] \tag{6.122}$$

at $r = a$ for any n. This is a spectral component of (6.14) with (6.15) at $r = a$. (The indices n and m on $\mathbf{u}$, ϕ_1 and ϕ_e have been suppressed in the above equations.) On substituting the above expressions into (6.23), we obtain

$$h_n = \frac{3}{\Omega_0^2 a^2 g(a)}y_1^u(a), \qquad k_n = \frac{3}{\Omega_0^2 a^2}y_5^u(a), \tag{6.123}$$

where the y_i^u are defined as in Eq. (6.103). Thus the Love numbers get determined as soon as the solutions of the radial differential equations are obtained. A quick look at the differential Eqs. (6.95)–(6.102) for the y_i makes it evident that the radial profiles of the density, gravitational potential and gravity, and the Lamé parameters determine the behavior of these radial functions, and hence also the values of the Love numbers.

6.8 The inertia tensor: contribution from deformations

Any deformation of the Earth results in a redistribution of the matter within it and an associated change in the density function from the unperturbed time independent density $\rho_0(\mathbf{r})$ to $\rho(\mathbf{r})$, which is in general a function of time. We proceed now to consider the increments c_{ij} in the elements of the inertia tensor which result from deformations.

Let us write the general expression for the elements I_{ij} of the inertia tensor of a body which has undergone deformation as (see Section 2.3)

$$I_{ij} = \int \rho(\mathbf{r}') \, (r'^2 \delta_{ij} - x_i' x_j') dV'. \tag{6.124}$$

Deformation itself means that an element of matter occupying a volume element dV at $\mathbf{r}$ in the undeformed body gets displaced to

$$\mathbf{r}' = \mathbf{r} + \mathbf{u}(\mathbf{r}) \tag{6.125}$$

and occupies a volume dV' at $\mathbf{r}'$. The use of the primed variables in Eq. (6.124) serves as a reminder that the integration is to be carried out over the instantaneous volume of the deformed body. Though the density $\rho(\mathbf{r})$ in the deformed state differs from $\rho_0(\mathbf{r})$ of the undeformed body, the mass of the element of matter which moves from dV at $\mathbf{r}$ to dV' at $\mathbf{r}'$ remains unchanged: $\rho(\mathbf{r}')dV' = \rho_0(\mathbf{r})dV$. So we can write

$$I_{ij} = \int \rho_0(\mathbf{r}) \, (r'^2 \delta_{ij} - x_i' x_j') dV, \tag{6.126}$$

with the components x_k' of $\mathbf{r}'$ given by $x_k' = x_k + u_k$. The integration is now over the undeformed volume. The inertia tensor C_{ij} of the unperturbed body is given by the same expression with $\mathbf{u}(\mathbf{r})$ set equal to zero, i.e., with the primed variables in the above integral replaced by the unprimed ones. It follows then that the perturbation $c_{ij} = I_{ij} - C_{ij}$ of the inertia tensor is given by

$$c_{ij} = \int \rho_0(\mathbf{r})[2(\mathbf{r} \cdot \mathbf{u})\delta_{ij} - x_i u_j - x_j u_i]dV. \tag{6.127}$$

6.8.1 *Deformations of general spherical harmonic type*

We are interested here in deformation-induced increments to the inertia tensors of the whole Earth and also to the inertia tensors of the individual regions of the Earth such as the mantle, the fluid core, etc. What we have so far referred to as the "body" may be identified with the whole Earth or with whichever region we are interested in. The contribution from the region of interest to the inertia tensor is given by the above integral with the volume of integration restricted to that region.

Let us take **u** in (6.127) to be a spectral component of frequency σ cpsd of the displacement field $\mathbf{u}_n^m(\mathbf{r}; t)$:

$$\mathbf{u}_n^m(\mathbf{r}; t) = \mathrm{Re}\left\{ [y_1(r)\mathbf{R}_n^m + y_3(r)\mathbf{S}_n^m] e^{i(\sigma\Omega_0 t - \alpha_\omega)} \right\}. \tag{6.128}$$

We have not included a toroidal term in $\mathbf{u}_n^m$, because the action of the potential on a spherically symmetric Earth does not excite any toroidal displacement. After substituting for **u** in (6.127) the quantity appearing within the braces in the above expression, and evaluating the integral for any i and j, the real part of the resulting quantity is to be taken to obtain c_{ij}. An immediate observation is that with **u** given by the above expression (for any $n > 0$),

$$c_{11} + c_{22} + c_{33} = 4 \int (\mathbf{u}_n^m \cdot \mathbf{r}) d^3 r = 4 \int r y_1(r) \mathcal{Y}_n^m d^3 r = 0, \tag{6.129}$$

since the angular integration of $\mathcal{Y}_n^m$ yields zero except for $n = m = 0$. Thus the trace of the inertia tensor is not altered by the perturbation.

Let us return now to the integral (6.127). It is useful, for the purpose of evaluating it, to express $\mathbf{S}_n^m$ using the last of Eqs. (6.55), as

$$\mathbf{S}_n^m = -\hat{\mathbf{r}} \wedge \mathbf{T} = -\hat{\mathbf{r}} \wedge (\mathbf{r} \wedge \nabla)\mathcal{Y}_n^m. \tag{6.130}$$

The utility of this form in the present context comes from the fact, which we have seen in Eqs. (5.221)–(5.223), that the components of $(\mathbf{r} \wedge \nabla)\mathcal{Y}_n^m$ contain only linear combinations of $\mathcal{Y}_n^{m\pm1}$ and $\mathcal{Y}_n^m$; it is particularly relevant that operation with $(\mathbf{r} \wedge \nabla)$ leaves the order n unchanged. Thus the components of the right-hand side of the foregoing equation consist of terms of the form $(x_k/r)\mathcal{Y}_n^\mu$ where μ takes one of the values $m - 1, m, m + 1$. The components of $\mathbf{R}_n^m$ are also of similar form, except that the index μ takes only the value m. Consequently each component of $\mathbf{u}_n^m$ consists of terms of the form $f(r)x_k\mathcal{Y}_n^\mu$, and the quantities $x_i\mathbf{u}_j$ and $x_j\mathbf{u}_i$, which appear in the integral expression for c_{ij}, consist of terms of the form $f(r)x_kx_l\mathcal{Y}_n^\mu$. Now, the components x, y, z of **r**, for which we have used the generic symbol x_i, are $r(\mathcal{Y}_1^1 \pm \mathcal{Y}_1^{-1})$ and $r\mathcal{Y}_1^0$, ignoring constant factors. Therefore $f(r)x_kx_l\mathcal{Y}_n^\mu$ is a linear combination of terms of the form $g(r)\mathcal{Y}_1^\beta \mathcal{Y}_1^\gamma \mathcal{Y}_n^\mu$, where each of the indices β and γ takes values from the set $(-1, 0, 1)$.

We shall now show how the above facts, taken together with the following basic properties of spherical harmonics, enable one to demonstrate that non-vanishing c_{ij} can exist only when the perturbing potential is of degree $n = 2$.

Firstly,

$$\int \mathcal{Y}_n^m \sin\theta \, d\theta d\varphi \neq 0 \quad \text{only if} \quad n = m = 0, \tag{6.131}$$

when the integration is over the whole surface of the unit sphere. This property may be seen to be a special case of the orthonormality relation (6.4): just set $n' = m' = 0$ in that relation, reducing $\mathcal{Y}_q^p$ to a constant. Secondly,

$$\mathcal{Y}_n^m(\theta, \varphi)\mathcal{Y}_q^p(\theta, \varphi) = \sum_{l=|n-q|}^{n+q} c(n, q, l; m, p)\mathcal{Y}_l^{m+p}, \tag{6.132}$$

where $c(n, q, l; m, p)$ is a numerical coefficient. Proof of this property is beyond our scope. (The reader may consult Edmonds (1960), for instance.) As a corollary, one sees from the limits of the sum in (6.132) that $\mathcal{Y}_0^0$ can be present on the right-hand side only if $q = n$ and $p = -m$.

Consider now the product $\mathcal{Y}_1^\beta \mathcal{Y}_1^\gamma \mathcal{Y}_n^\mu$ through which the angle variables (θ, φ) enter the integral in c_{ij}. According to (6.132), the highest degree of the spherical harmonics present in the expansion of the product $\mathcal{Y}_1^\beta \mathcal{Y}_1^\gamma$ is two. When further multiplication by $\mathcal{Y}_n^\mu$ is done, the same formula shows that there is no possibility of $\mathcal{Y}_0^0$ being present in the product except when the degree n is two (remembering that tidal potentials have $n \geq 2$). Therefore it becomes clear, in view of (6.131), that the angular part of the integration in the expression (6.127) for c_{ij} can yield a nonzero result only for $n = 2$.

Thus we have verified the earlier statement that a non-vanishing increment c_{ij} to the inertia tensor can arise only from deformations caused by tidal potentials of degree two. We proceed now to a consideration of the c_{ij} contributed by degree two deformations.

6.8.2 Deformations of degree two: corollary to MacCullagh's theorem

MacCullagh's theorem, presented in Section 5.3.2, relates the second degree part $W_E^{(2)}(\mathbf{r}; t)$ of the Earth's own external gravitational potential to the elements I_{ij} of the inertia tensor of the Earth (see Eq. (5.128)). The change $dW_E^{(2)}(\mathbf{r}; t)$ in this potential, which is caused by the deformation, is related to the changes $c_{ij}(t)$ in the inertia tensor in the same manner as $W_E^{(2)}$ is to the I_{ij}:

$$dW_E^{(2)}(\mathbf{r}; t) = \frac{G}{2r^5}(3c_{ij} - c_{kk}\delta_{ij})x_i x_j, \qquad (r \geq a). \tag{6.133}$$

(We have changed the notation for the position vector of the external point from $\mathbf{R}$ to $\mathbf{r}$ and for its components from X_i to x_i.) This result is valid irrespective of the cause of the deformation.

In case the deformation is produced by the action of a tidal potential W_2^B of degree two due to a celestial body B, the change in a spherical Earth's external potential that results from the deformation is $k_2 W_2^B$, where k_2 is the Love number for degree two. In fact this is how the Love number was defined. Thus

$$dW_E^{(2)}(\mathbf{r};t) = k_2 W_2^B(\mathbf{r};t) = -\frac{k_2 GM_B}{2d_B^5}[-d_B^2 r^2 + 3(\mathbf{R} \cdot \mathbf{r})^2], \qquad (6.134)$$

where the explicit expression for W_2^B is taken from Eq. (5.191), with minor changes of notation: $\mathbf{R}$ is the geocentric position vector of the celestial body (with components X_i), $d_B = |\mathbf{R}|$ is its geocentric distance, and M_B is its mass.

We may now equate the above two alternative expressions for $dW_E^{(2)}(\mathbf{r};t)$ with r set equal to a, and in particular, equate the coefficients of the various $x_i x_j$ in the two expressions. We obtain thus the following relations connecting the time dependent c_{ij} to the position coordinates of the celestial body B which gives rise to the deformations:

$$c_{11} = \frac{1}{3}q_B(d_B^2 - 3X_{B1}^2), \quad c_{23} = c_{32} = -q_B X_{B2} X_{B3},$$

$$c_{22} = \frac{1}{3}q_B(d_B^2 - 3X_{B2}^2), \quad c_{31} = c_{13} = -q_B X_{B3} X_{B1}, \qquad (6.135)$$

$$c_{33} = \frac{1}{3}q_B(d_B^2 - 3X_{B3}^2), \quad c_{12} = c_{21} = -q_B X_{B1} X_{B2},$$

where

$$q_B = k_2 M_B \left(\frac{a}{d_B}\right)^5. \qquad (6.136)$$

These relations may be combined into the single equation

$$c_{ij} = q_B \left(\frac{1}{3}d_B^2 \delta_{ij} - X_{Bi} X_{Bj}\right). \qquad (6.137)$$

Note that the above relations confirm that $c_{11} + c_{22} + c_{33} = 0$, a property that was proved in Section 6.8.1.

The above expression gives the increments c_{ij} produced by a particular celestial body B. A summation of this expression over all the relevant bodies B must be done to get the full c_{ij}.

It may now be observed that on taking c_{33} from Eqs. (6.135) and carrying out the summation over B, the resulting expression for the full c_{33} is proportional to the dimensionless quantity $\tilde{\phi}^{20}$ defined in Eq. (6.7):

$$c_{33} = -(c_{11} + c_{22}) = -\sum_B \frac{2}{3} q_B d_B^2 \, P_2(\cos \beta_B) = -\frac{2}{3} A\kappa \, \tilde{\phi}^{20}(t), \qquad (6.138)$$

where

$$\kappa = \frac{\Omega_0^2 a^5}{3GA} k_2. \qquad (6.139)$$

Similarly, we find that

$$\tilde{c}_3 \equiv c_{13} + i c_{23} = -\sum_B \frac{1}{3} q_B d_B^2 \, P_2^1(\cos \beta) e^{i\Lambda} = -A\kappa \, \tilde{\phi}^{21}(t), \qquad (6.140)$$

$$\tfrac{1}{2}(c_{11} - c_{22}) + i c_{12} = -\sum_B \frac{1}{6} q_B d_B^2 \, P_2^2(\cos \beta) e^{2i\Lambda} = -2A\kappa \, \tilde{\phi}^{22}(t), \quad (6.141)$$

where $\tilde{\phi}^{21}(t)$ and $\tilde{\phi}^{22}(t)$ are defined by Eqs. (6.8) and (6.9).

Since the $\tilde{\phi}^{2m}$ for $m = 0, 1, 2$ pertain to the zonal, tesseral, and sectorial parts, respectively, of the total degree two tidal potential, it follows that c_{33} and $(c_{11} + c_{22})$ are generated by the zonal potential, c_{13} and c_{23} by the tesseral potential, and $(c_{11} - c_{22})$ and c_{12} by the sectorial potential. It is interesting to note that on combining the expression (6.139) for κ with the second of Eqs. (6.123) by which k_2 is defined for any order m, one gets the relation

$$A\kappa\tilde{\phi}^{2m} = \frac{a^3}{G} y_5(a), \qquad \text{or} \qquad \kappa = \frac{a^3}{GA} y_5^u(a). \qquad (6.142)$$

The first form of the relation shows, in view of Eqs. (6.138), (6.140), and (6.141), that the increments in the elements of the inertia tensor of the whole of the spherical Earth due to the direct deforming actions of the respective tidal potentials are simply related to $y_5(a)$ through constant factors which are different for different orders m, though y_5 itself is very nearly independent of m, as noted earlier. The second form of the above relation highlights the fact that the compliance κ gets determined as soon as the radial equations are solved.

The relations (6.138)–(6.141) derived above are for the c_{ij} in the time domain. The same results hold good for individual spectral components: one simply makes the replacements $c_{ij} \rightarrow c_{ij}(\sigma)$, $\tilde{\phi}^{2m} \rightarrow \tilde{\phi}^{2m}(\sigma)$. The spectral expansion of $\tilde{\phi}^{nm}$ in general is given in Eq. (6.10).

6.8.3 *Contributions to c_{ij}: relation to radial functions*

In nutation theory, we need to know the increments in the inertia tensors of the whole Earth, the FOC, and the SIC due to the actions of both tidal and centrifugal potentials. The expressions for these involve, in general, the values of both y_5 and y_6 at the boundaries of such regions, as we shall show. The proof rests on the evaluation of the integral expression (6.127) for the c_{ij} in terms of the displacement field $\mathbf{u}$ produced by the perturbation, whether tidal or centrifugal.

In proceeding to deal with all these aspects, we start with the basic vector and scalar spherical harmonic fields of degree two appearing in the expansion of $\mathbf{u}$. The explicit expressions for these fields, which are special cases of Eqs. (6.3) and (6.43), are as follows:

$$\mathcal{Y}_2^0 = \frac{3}{2}\cos^2\theta - \frac{1}{2} = \frac{1}{2r^2}(3z^2 - r^2), \tag{6.143}$$

$$\mathcal{Y}_2^1 = 3\cos\theta\sin\theta e^{-i\varphi} = \frac{3}{r^2}z(x - iy), \tag{6.144}$$

$$\mathcal{Y}_2^2 = 3\sin^2\theta e^{-2i\varphi} = \frac{3}{r^2}(x - iy)^2, \tag{6.145}$$

$$\mathbf{R}_2^m = \frac{\mathcal{Y}_2^m}{r}(x, y, z). \tag{6.146}$$

Similarly $\mathbf{S}_2^m = r\nabla\mathcal{Y}_2^m$ has the components

$$\mathbf{S}_2^0 = -\frac{3z^2}{r^3}(x, y, z) + \frac{3}{r}(0, 0, z), \tag{6.147}$$

$$\mathbf{S}_2^1 = -\frac{6z(x - iy)}{r^3}(x, y, z) + \frac{3}{r}(z, -iz, x - iy), \tag{6.148}$$

$$\mathbf{S}_2^2 = -\frac{6(x - iy)^2}{r^3}(x, y, z) + \frac{6}{r}(x - iy, -i(x - iy), 0). \tag{6.149}$$

Evaluation of the integral expression (6.127) for c_{ij} may now be done for each m. We take $\mathbf{u}$ in this integral to be $\mathbf{u}_2^m(\mathbf{r}, t)$ obtained by setting $n = 2$ in Eq. (6.128): it is the spectral component associated with the frequency σ cpsd of the part of $\mathbf{u}$ that is of spherical harmonic degree two and order m. The explicit forms of the scalar and vector spherical harmonics for the relevant value of m are taken from Eqs. (6.143)–(6.149). The integration is then carried out to obtain the spectral component $c_{ij}(\sigma)$ of c_{ij}. All terms in the integrand which contain an odd power of x or y or z may be dropped because such terms yield zero on integration as a result of the symmetry of the Earth's structure. Evaluation of the integrals of the remaining terms may be done after expressing the components of the vector spherical harmonics in spherical polar coordinates. The process is tedious though elementary. As we already know from Eqs. (6.138)–(6.141), only a few of the c_{ij} are nonzero for any given m. They are:

for $m = 0$,

$$c_{33} = -2c_{11}^Z = -2c_{22}^Z = -\frac{2}{3} I \tilde{\phi}^{20}(\sigma) \cos(\sigma \Omega_0 t - \alpha_\omega) \tag{6.150}$$

and

$$c_{11}^Z = c_{22}^Z; \tag{6.151}$$

for $m = 1$,

$$c_{13} = c_{31} = -I \tilde{\phi}^{21}(\sigma) \cos(\sigma \Omega_0 t - \alpha_\omega), \tag{6.152}$$

$$c_{23} = c_{32} = -I \tilde{\phi}^{21}(\sigma) \sin(\sigma \Omega_0 t - \alpha_\omega), \tag{6.153}$$

$$\tilde{c}_3 = c_{13} + i c_{23} = -I \tilde{\phi}^{21}(\sigma) e^{i(\sigma \Omega_0 t - \alpha_\omega)}; \tag{6.154}$$

and for $m = 2$,

$$c_{11}^S = -c_{22}^S = -2I \tilde{\phi}^{22}(\sigma) \cos(\sigma \Omega_0 t - \alpha_\omega), \tag{6.155}$$

$$c_{12} = c_{21} = -2I \tilde{\phi}^{22}(\sigma) \sin(\sigma \Omega_0 t - \alpha_\omega), \tag{6.156}$$

$$c_{11}^S + i c_{12} = -2I \tilde{\phi}^{22}(\sigma) e^{i(\sigma \Omega_0 t - \alpha_\omega)}. \tag{6.157}$$

The factor I which appears in all the c_{ij}, irrespective of the order m of the perturbing potential, is

$$I = \frac{8\pi}{5} \int [y_1^u(r) + 3y_3^u(r)] r^3 \rho(r) dr, \tag{6.158}$$

where $y_1^u(r)$ and $y_3^u(r)$ have been defined in Eq. (6.103). It is solely through this factor that the deformations of the solid Earth translate into perturbations of the elements of the inertia tensor.

On carrying out the integration above from $r = 0$ to $r = a$ and introducing the resulting value of I into Eqs. (6.150) to (6.157), one gets the increments in the inertia tensor of the whole Earth. However there is no reason why one should not limit the integration to some particular region of interest. If one carried out the integration only from the radius a_s of the inner core to the radius a_c of the CMB, one would get the increments in the inertia tensor of the FOC, for which we use the notation c_{ij}^f. Similarly one would obtain c_{ij}^s pertaining to the inertia tensor of the SIC if the integration were limited to the interval 0 to a_s.

In the expressions given above for the various c_{ij}, we have indicated the parts of c_{11} and c_{22} that are generated by zonal and by sectorial potentials by the superscripts

Z and *S*, respectively. c_{11} and c_{22} are the only elements that are generated by more than one type of potential. The zonal contributions appearing in Eq. (6.150) are quite distinct from the sectorial ones in Eqs. (6.155)–(6.157): σ in the former is in the low frequency band while in the latter, σ is in the semidiurnal band. It is useful to note that because of the relation (6.150) between the zonal parts of c_{11} and c_{22} and the relation (6.155) in the sectorial case,

$$c_{11} + c_{22} = 2c_{11}^Z, \quad \text{and} \quad c_{11} - c_{22} = 2c_{11}^S. \tag{6.159}$$

It turns out that the integral in the expression (6.158) for I can be evaluated explicitly using the following result:

$$\frac{d}{dr}[r^3(5y_5 - ry_6)] = -8\pi G\rho r^3(y_1 + 3y_3). \tag{6.160}$$

To prove this equation, we carry out the differentiation in the left-hand member, and then substitute for y_6' from (6.100). The result is seen to be $r^2[5ry_5' + 12\pi Gr(y_1 - 2y_3) - 5ry_6 + 15y_5]$. Equation (6.99) may now be used to eliminate $ry_5' - y_6 + 3y_5$ from this expression; what we are left with then is just the right-hand side of Eq. (6.160). Having proved the validity of this equation, we can integrate it between any two radii a_1 and a_2 and then divide by $\tilde{\phi}$ to obtain

$$I(a_1, a_2) = \frac{1}{5G} \left\{ \left[r^3(5y_5^u - ry_6^u)\right]_{r=a_2} - \left[r^3(5y_5^u - ry_6^u)\right]_{r=a_1} \right\}. \tag{6.161}$$

With the use of this expression for I in Eqs. (6.150)–(6.157), one gets the increments in the inertia tensor elements of the region lying between the radii a_1 and a_2.

What we are primarily interested in are the incremental inertia tensors for the whole Earth, the FOC, and the SIC. $I(0, a)$, which will appear in the incremental inertia tensor elements c_{ij} of the whole Earth; $I(a_s, a_c)$ in the c_{ij}^f pertaining to the FOC alone; and $I(0, a_s)$ in the c_{ij}^s relating to the SIC alone. In the expressions for the c_{ij} of the whole Earth, in particular, we have

$$I(0, a) = \frac{a^3}{G}y_5^u(a) = A\kappa, \tag{6.162}$$

since y_5 and y_6 vanish at the center of the Earth for $n \geq 2$ and $y_6(a) = 0$ as we have seen in Section 6.7.5. We have used Eq. (6.142) in the last step.

6.8.4 The c_{ij} in terms of compliance parameters

Other compliance parameters besides κ were defined by SOS (denoting Sasao, Okubo, Saito and referring to Sasao *et al.*, 1980), and the set of compliances was expanded by Mathews *et al.* (1991a) while introducing the SIC into nutation theory. Two of these, denoted by γ and θ in the notation of the above-cited papers, characterize the deformations of the FOC and the SIC, respectively, under the direct action of the tidal potential. They are defined in terms of the $I(a_1, a_2)$ of Eq. (6.161) by the relations

$$I(a_s, a_c) = A_f \gamma, \qquad \text{and} \qquad I(0, a_s) = A_s \theta. \qquad (6.163)$$

Consider now the contributions to the incremental inertia tensors of the various regions from the deformations produced by the centrifugal potentials of Section 6.2.3. It was pointed out in Section 6.7.6 that $y_i(r)/\tilde{\phi}(\sigma)$ in the case of tidal excitation and $y_i(r)/(-\tilde{m})$ in the case of excitation by the centrifugal potential (6.18) associated with the wobble $\tilde{m}$ are equal and are denoted by $y_i^u(r)$ in both cases. So the compliances relating to this centrifugal potential are the same κ, γ, θ as in the tidal case.

It was also noted in Section 6.7.6 that in the case of the centrifugal potentials (6.19) and (6.20) associated with the differential wobbles $\tilde{m}_f$ and $\tilde{m}_s$, the radial functions $y_{fi}^u(r)$ and $y_{si}^u(r)$ per unit excitation, defined as in (6.110), are unequal to $y_i^u(r)$. The compliances ξ, β, χ characterize the deformations of the whole Earth, the FOC, and the SIC, respectively, that are generated by the differential wobble $\tilde{m}_f$. They are related to the y_{fi}^u through

$$I_f(0, a) = A\xi, \qquad I_f(a_s, a_c) = A_f\beta, \qquad I_f(0, a_s) = A_s\chi, \qquad (6.164)$$

where $I_f(a_1, a_2)$ is given by the expression (6.161) with the replacements $y_5^u \to y_{f5}^u$ and $y_6^u \to y_{f6}^u$. The compliances ζ, δ, τ characterizing the deformations produced by the differential wobble $\tilde{m}_s$ of the inner core are similarly defined:

$$I_s(0, a) = A\zeta, \qquad I_s(a_s, a_c) = A_f\delta, \qquad I_s(0, a_s) = A_s\tau, \qquad (6.165)$$

where $I_s(a_1, a_2)$ is to be obtained from (6.161) by the replacements $y_5^u \to y_{s5}^u$, $y_6^u \to y_{s6}^u$.

The nine compliance parameters $\kappa, \gamma, \ldots, \tau$ represent the extent to which the Earth and its core regions are susceptible to deformation under forcing by potentials acting in the various regions. (The last of these was denoted by ν in Mathews *et al.* (1991 a and b), but we have changed it to τ to avoid confusion with the frequency ν in subsequent chapters.)

The deformations of the greatest interest in connection with nutation theory are those produced by the degree two tesseral potential which drives precession and the low frequency nutations, and by the centrifugal effects of the accompanying wobbles. The contributions from potentials of this type are only to the elements $c_{13} = c_{31}$ and $c_{23} = c_{32}$ (or equivalently, to $\tilde{c}_3$), as observed earlier in this chapter. In applying Eq. (6.154) to the deformation of the whole Earth or of the individual regions (FOC, SIC) due to a tesseral spectral component, one has to add up the contributions from the effects of $\tilde{\phi}$, $\tilde{m}$, $\tilde{m}_f$, and $\tilde{m}_s$. The result of adding them all up, when considering the deformation of the whole Earth, is the replacement of the amplitude $-I\tilde{\phi}^{21}(\sigma)$ in Eq. (6.154) for $\tilde{c}_3$ by

$$- [I(0, a)(\tilde{\phi}^{21} - \tilde{m}) - I_f(0, a)\tilde{m}_f - I_s(0, a)\tilde{m}_s]. \tag{6.166}$$

On expressing this quantity in terms of compliances by the use of the relevant relations in Eqs. (6.162), (6.164), and (6.165), we find for a spectral component of $c_3(t)$,

$$\tilde{c}_3(t) = \tilde{c}_3(\sigma) e^{i(\sigma\Omega_0 t - \alpha_\omega)}, \qquad \tilde{c}_3(\sigma) = -A \left[\kappa(\tilde{\phi} - \tilde{m}) - \xi\tilde{m}_f - \zeta\tilde{m}_s\right]. \tag{6.167}$$

The amplitude in $\tilde{c}_3^f$ of the fluid core will become

$$-[I(a_s, a_c)(\tilde{\phi}^{21} - \tilde{m}) - I_f(a_s, a_c)\tilde{m}_f - I_s(a_s, a_c)\tilde{m}_s], \tag{6.168}$$

leading to

$$\tilde{c}_3^f(t) = \tilde{c}_3^f(\sigma) e^{i(\sigma\Omega_0 t - \alpha_\omega)}, \quad \tilde{c}_3^f(\sigma) = -A_f \left[\gamma(\tilde{\phi} - \tilde{m}) - \beta\tilde{m}_f - \delta\tilde{m}_s\right]. \tag{6.169}$$

For the SIC, the arguments of I will be $(0, a_s)$, and we obtain

$$\tilde{c}_3^s(t) = \tilde{c}_3^s(\sigma) e^{i(\sigma\Omega_0 t - \alpha_\omega)}, \quad \tilde{c}_3^s(\sigma) = -A_s \left[\theta(\tilde{\phi} - \tilde{m}) - \chi\tilde{m}_f - \tau\tilde{m}_s\right]. \tag{6.170}$$

In these equations, $\tilde{\phi}$ stands for $\tilde{\phi}^{21}(\sigma)$, and $\tilde{m}$, $\tilde{m}_f$, $\tilde{m}_s$ are all spectral amplitudes at the frequency σ cpsd. The expressions (6.167), (6.169), and (6.170) play an essential role in the derivation, in the next chapter, of the wobble equations of a deformable Earth which has FOC and SIC regions.

Consider next the deformation due to the zonal potential given by Eq. (6.2) with $n = 2$, $m = 0$. It is of primary importance in the theory of variations in the spin rate represented by m_3. The zonal potential does not produce any torque on the spherical or ellipsoidal Earth, but it does produce deformations of a zonal character. The consequent incrementing of the moment of inertia about the z-axis results in the deviation $\Omega_0 m_3$ of the axial rotation rate from the steady unperturbed value Ω_0; this deviation is needed to conserve the angular momentum about the z-axis. Associated with m_3 is the axially symmetric centrifugal potential (6.21) which contains a zonal part of degree two and a spherically symmetric part, both of amplitude $-2m_3$ as displayed in Eq. (6.22). These, in turn, affect m_3 itself to a small extent through the incremental change in the axial moment of inertia produced by the centrifugal potential (6.21).

The total zonal potential (tidal plus centrifugal) that is acting over the whole Earth has thus the amplitude $(\tilde{\phi}^{20} - 2m_3)$. A differential spin rate m_{f3} of the fluid core would produce an additional centrifugal potential in the core region which is of the same form as (6.22) except for the replacement $-2m_3 \to -2m_{f3}$. The deformations caused by the zonal tidal potential and by both the zonal and spherically symmetric parts of the centrifugal potential result in increments in the diagonal elements of the inertia tensor. A spectral component of the increments due to the zonal parts has the form (6.150) with

$$I = I(0, a)(\tilde{\phi}^{20} - 2m_3) + I_f(0, a)(-2m_{f3}),   \quad (6.171)$$

ignoring any differential spin of the SIC. Here $I(0, a) = \kappa A$ and $I_f(0, a) = \xi A$ as stated in Eqs. (6.162) and (6.164). Thus the spectral amplitudes of the increments produced by the zonal potential and the spin rate variations, are:

$$c_{33}(\sigma) = -\frac{2}{3} A[\kappa(\tilde{\phi}^{20} - 2m_3) - 2\xi m_{f3}] + c^{(00)},   \quad (6.172)$$

$$c_{11}^Z(\sigma) = c_{22}^Z(\sigma) = \frac{1}{3} A[\kappa(\tilde{\phi}^{20} - 2m_3) - 2\xi m_{f3}] + c^{(00)},   \quad (6.173)$$

with the time dependence $\cos(\sigma \Omega_0 t - \alpha_\omega)$ as in (6.150). The superscripts Z on c_{11} and c_{22} indicate that these pertain to the zonal part of the deformations.

The additional term $c^{(00)}$ in each of the above expressions comes from the deformations caused by the spherically symmetric part $-(2/3)\Omega_0^2 m_3(t)r^2$ of the centrifugal potential (6.22) and the corresponding potential associated with m_{f3} of the fluid core. Evidently, these spherically symmetric perturbing potentials can produce only purely radial displacement fields, of the form $y_1^0(r)\hat{r}$. (The superscript 0 in y_1^0 is a reminder that the deformation is of degree zero, unlike the degree two deformations that we have been considering so far.) Use of this form for **u** in the integral expression (6.127) for c_{ij} leads to equal values

$$c^{(00)} = \frac{16\pi}{3} \int \rho\, y_1^0(r) r^3 dr   \quad (6.174)$$

for the contributions to c_{11}, c_{22}, and c_{33}. For actual determination of the function $y_1^0(r)$, one has to substitute $\mathbf{u} = y_1^0(r)\hat{r}$ in the momentum balance Eq. (6.42) and in the Poisson equation (6.37) and solve the relatively simple radial equations that result. Since the forcing is by the spherically symmetric part of the centrifugal perturbations arising from m_3 and m_{f3}, $y_1^0(r)$ and hence $c^{(00)}$ must be linear combinations of terms proportional to these two quantities.

Finally, we consider the increments produced by the sectorial potential in the inertia tensor elements. Axial rotation variations induced by the sectorial potential are extremely small, being dependent on the triaxiality. So only the direct effect of the potential is relevant, and we have from (6.157),

$$c_{11}^S(\sigma) + i c_{12}(\sigma) = -2A\kappa\tilde{\phi}^{22}(\sigma)   \quad (6.175)$$

for any spectral component, with a time dependence $e^{i(\sigma \Omega_0 t - \alpha_\omega)}$, as in that equation.

Though the values of the compliances κ and ξ are independent of the order m of the forcing potential for a spherical Earth (except for a subtle dependence through the different bands in which the frequency σ lies for different values of m), ellipticity of the actual Earth, and the effect of the Coriolis term in the momentum balance equation that we had neglected, result in fractional differences of $\mathcal{O}(\epsilon)$ between the sets of values for the tesseral, zonal, and sectorial cases. They have to be taken into account in some situations where high accuracy is needed.

The increments in the inertia tensor that result from the action of the zonal and sectorial potentials do not enter into nutation theory except through the very small second-order contributions that they make to the tidal torques acting on a deformable Earth. These are dealt with in Chapter 7.

6.8.5 Deformation caused by ocean tides: contribution to the c_{ij}

Ocean tides are generated by the gravitational attraction of the water mass in the oceans by the Moon and the Sun. The sea level rises in some parts of the global oceans and falls in other parts, depending on the positions of these celestial bodies in the sky. The tide height function $\zeta(\theta, \lambda; t)$ represents the difference between the instantaneous height of the column of seawater and its mean height, as a function of the geographical location (θ, λ) and time. (The column height is taken from the ocean floor up to the ocean surface, of course.) The observed variations in the tide height function thus defined include changes due to atmospheric pressure changes, winds, etc., and these have to be accounted for while estimating the height function that is directly attributable to the gravitational effect of the Moon and the Sun. It is to be understood that it is the gravitational part that is meant when we use the term "ocean tide" in the following. The time dependence of ζ is characterized by the same spectrum as that of the tidal potential. In particular, the second degree zonal, tesseral, and sectorial potentials produce ocean tidal components with frequencies in the low frequency, diurnal, and semidiurnal bands, respectively. However, unlike the case of solid Earth tides, the spatial variation of the tide height is very different from that of the tidal potential: the ocean tide due to a potential of degree n and order m has significant admixtures of spherical harmonics of other degrees and orders. The reasons are the highly irregular shapes of the continental boundaries of the oceans and the varying bathymetry (ocean depth profile), as well as the associated ocean dynamics, which have a profound influence on the ocean response to the tidal potentials. In fact, even the degree two order one part of the ocean tide excited by a spectral component of the degree two tesseral potential has both prograde and retrograde parts, though the forcing potential itself is purely retrograde: the retrograde waves accompanying the motion of the celestial body get partially reflected because of the existence of continental boundaries and the non-uniformity of the depth of the ocean, thus creating the prograde waves.

The spatial and temporal variations in the ocean mass distribution that constitute the essence of the ocean tides result in corresponding perturbations of the pressure exerted by the water on the ocean floor, which cause deformations of the solid Earth underneath and accompanying changes in the Earth's own gravitational potential. Both the deformation of the ocean floor and the perturbation of the potential (which affects the geoid and hence the ocean surface) react on the ocean tide height. Such geophysical influences on the height of the tide raised by a given component of the external tidal potential could be dealt with without too much difficulty if the oceans were idealized as covering the whole globe and being of uniform depth. The most notable feature of the ocean tidal response to the tesseral tidal potential, irrespective of whether the ocean is treated as global or more realistically, is the existence of a resonance in the admittance function (ocean tide height per unit forcing potential). This resonance arises from the contributions of the wobble of the whole Earth and the differential wobbles of the core regions to the effective Love numbers which determine the deformations and geopotential perturbations; all of these wobbles display the NDFW resonance. The admittance function has an additional frequency dependence arising from the influence of the shape of the ocean basins (both the coastlines and the bathymetry) on the ocean response to tidal forcing. For a treatment of ocean tides excited by the tesseral potential, see, for instance, Wahr and Sasao (1981).

What we need to know in the context of Earth rotation theory are the increments to the elements in the inertia tensors of the whole Earth and of the core alone from the ocean tidal effects, and also the angular momentum contained in the ocean tidal currents. The increments to the inertia tensor give rise to corresponding increments to the Earth's external gravitational potential, which are given by Eq. (6.133); these incremental potentials may be estimated from satellite laser ranging (SLR) data or dedicated spacecraft such as GRACE, GOCE, and CHAMP (see Chapter 2). The estimates are presented usually in the form of increments to the "barred" geopotential coefficients $\bar{C}_{nm}$, $\bar{S}_{nm}$ which are dimensionless (see Section 5.3.5). The spectral representation of the increments $\Delta \bar{C}_{nm}^{OT}$, $\Delta \bar{S}_{nm}^{OT}$ is given in Chapter 6 of the IERS Conventions 2010. On combining it with the relations between the barred and unbarred coefficients which are also given in the same section, we find that

$$\Delta C_{nm}^{OT} + i \Delta S_{nm}^{OT} = Q_n \sum_{\omega(nm)} [(C_{\omega nm}^{+} + i S_{\omega nm}^{+})e^{-i\Theta_\omega} + (C_{\omega nm}^{-} - i S_{\omega nm}^{-})e^{i\Theta_\omega}],$$

(6.176)

$$Q_n = \frac{4\pi G \rho_w}{g_e} \left(\frac{1 + k_n'}{2n + 1} \right),$$

(6.177)

where ρ_w is the density of seawater, g_e is the acceleration due to gravity at the equator ($g_e = GM_E/a_e^2$, with a_e standing for the Earth's equatorial radius), k_n' is the load Love number of degree n, and Θ_ω is the argument of the spectral component of the tide with frequency ω.

As was mentioned earlier in this section, the ocean tide raised by any spectral component of the tidal potential wave (which is retrograde) has both retrograde and prograde parts. The retrograde part of $\Delta C_{nm}^{OT} + i \Delta S_{nm}^{OT}$ is the part in (6.176) containing the factor $e^{-i\Theta_\omega}$. The quantities $C_{\omega nm}^+$, $S_{\omega nm}^+$ are the coefficients of the $\cos \Theta_\omega$ and $\sin \Theta_\omega$ terms, respectively, in the spectral expansion of the retrograde part of the ocean tide height function $\zeta(\theta, \lambda)$, and $C_{\omega nm}^-$, $S_{\omega nm}^-$ are the corresponding coefficients for the prograde part.

The real and imaginary parts of the expression (6.176) are

$$\Delta C_{nm}^{OT} = Q_n \sum_{\omega(nm)} [(C_{\omega nm}^+ + C_{\omega nm}^-) \cos \Theta_\omega + (S_{\omega nm}^+ + S_{\omega nm}^-) \sin \Theta_\omega], \quad (6.178)$$

$$\Delta S_{nm}^{OT} = Q_n \sum_{\omega(nm)} [(S_{\omega nm}^+ - S_{\omega nm}^-) \cos \Theta_\omega - (C_{\omega nm}^+ - C_{\omega nm}^-) \sin \Theta_\omega]. \quad (6.179)$$

Note that since S_{nm} and any increments to it do not exist when $m = 0$,

$$C_{\omega n0}^+ = C_{\omega n0}^-, \qquad S_{\omega n0}^+ = S_{\omega n0}^-. \quad (6.180)$$

The ocean tidal increments in the inertia tensor elements may now be obtained from the special case $n = 2$ of the above by the use of the incremental version of Eqs. (5.162):

$$\Delta c_{33}^{OT} = -\frac{2}{3} M_E a_e^2 \Delta C_{20}^{OT},$$

$$\Delta \tilde{c}_3^{OT} \equiv \Delta c_{13}^{OT} + i \Delta c_{23}^{OT} = -M_E a_e^2 (\Delta C_{21}^{OT} + i \Delta S_{21}^{OT}), \quad (6.181)$$

$$(\Delta c_{11}^{OT})^S + i \Delta c_{12}^{OT} = -2 M_E a_e^2 (\Delta C_{22}^{OT} + i \Delta S_{22}^{OT}),$$

where we have employed the fact that the trace of the increments in the inertia tensor due to zonal deformations is zero and so is the sum of the increments in the two equatorial principal moments due to sectorial deformations. The superscript S on (Δc_{11}^{OT}) in the last equation indicates that this quantity arises from the action of the sectorial tides.

The spectral representation of $\tilde{c}_3^{OT}$ is of particular interest in connection with the effect of ocean tides on the wobbles of a deformable Earth. The low frequency nutations are influenced (in the first order) only by the retrograde part of the ocean tides with frequencies in the diurnal band, raised by the degree two tesseral potential. A spectral component of $\tilde{c}_3$ belonging to this part, having frequency $-\omega$, is obtained by introducing the corresponding spectral component of

$(\Delta C_{21}^{OT} + i\,\Delta S_{21}^{OT})$ from Eq. (6.176) into the second of the Eqs. (6.181):

$$\Delta\tilde{c}_3^{OT}(t) = -M_E a_e^2 Q_2 (C_{\omega 21}^+ + i\,S_{\omega 21}^+)\,e^{-i\Theta_\omega(t)}$$

$$= -\frac{4\pi\rho_w a_e^4}{5}(1+k_{21}')(C_{\omega 21}^+ + i\,S_{\omega 21}^+)\,e^{-i\Theta_\omega(t)}. \tag{6.182}$$

The amplitude of this spectral component will be defined in the same way as the amplitude of the tidal potential in Eqs. (6.10)–(6.13) in the special case $(nm) = (21)$, namely as the coefficient of $e^{-i(\Theta_\omega(t)-\zeta_{21})}$ with $\zeta_{21} = \pi/2$. Thus the amplitude is

$$\Delta\tilde{c}_3^{OT}(\sigma) = -\frac{4\pi\rho_w a_e^4}{5}(1+k_{21}')(S_{\omega 21}^+ - i\,C_{\omega 21}^+). \tag{6.183}$$

This is the increment, due to ocean tidal loading, in $\tilde{c}_3(\sigma)$ of a deformable Earth. It may be expressed as an equivalent increment to the compliance κ by using the incremental form of the relation $\tilde{c}_3 = -A\kappa\tilde{\phi}$ from (6.140):

$$\Delta\kappa^{OT}(\sigma) = \frac{4\pi\rho_w a_e^4}{5A\tilde{\phi}(\sigma)}(1+k_{21}')(S_{\omega 21}^+ - i\,C_{\omega 21}^+), \tag{6.184}$$

with $\tilde{\phi}$ standing for $\tilde{\phi}^{21}$.

It will be noted that we have replaced k_2' by k_{21}' in the above. Though the load Love numbers representing the response to surface loading of spherical harmonic type (nm) are independent of the order m for a spherical Earth, fractional differences of the order of the ellipticity arise between the k_{2m}' for different m when the ellipticity of the Earth is taken into account; therefore one needs to take note of the m dependence when high accuracy is needed. More important than this is the fact that the Love number k_{21}' here does not pertain to the static Earth, but is the effective Love number for the wobbling Earth. The resonances in the wobbles, especially the FCN resonance, cause the load Love numbers (like the body tide Love numbers) to be strongly frequency dependent in the retrograde diurnal tidal band. Therefore it has to be remembered that k_{21}' in the foregoing expressions stands for $k_{21}'(\sigma)$.

The above discussion can be repeated for the ocean tides raised by the degree two zonal and sectorial potentials, with appropriate modifications which should be reasonably obvious. The increments in the inertia tensors in these cases do not affect the low frequency nutations in the first order, but could make a very small contribution to second-order torques on a deformable Earth, as will be seen in Chapter 10.

7

Nutations of a non-rigid Earth

The term "non-rigid Earth" has been employed in the literature to refer to any Earth model other than one that is wholly solid and rigid, and we shall follow this usage. Thus, a model consisting of a rigid mantle enclosing a fluid core (whether the fluid be compressible or not) is labeled as "non-rigid," as is one that has no fluid region at all but is deformable. We deal in this chapter with axially symmetric ellipsoidal Earth models which are deformable, in general, and have a fluid core with or without a SIC.

7.1 Formulations of the theory for a non-rigid Earth

Most of the treatments of the nutation and precession of the non-rigid Earth have been based on one or other of two types of formulations, which are outlined below. Both have to take account of the fact that the external gravitational potentials which cause forced nutation and wobble produce also (tidal) deformations and that these deformations translate into changes in the inertia tensors of the Earth and its core regions which affect Earth rotation variations. Furthermore, rotation variations (whether forced or free) produce perturbations of the centrifugal potential associated with Earth rotation, and these centrifugal perturbations too cause deformation and an associated perturbation of the inertia tensors which, in turn, affects Earth rotation. The two formulations differ from each other in the way the interlinked phenomena of nutation/wobble and deformation are dealt with.

The simpler formulation focuses directly on the rotation variations which are governed by the equations of angular momentum conservation (often called the torque equations) for the whole Earth and for its core regions. A TRF is employed in this formulation. A simple example of this approach has been seen already in Chapter 2. The angular momenta of the Earth and its core regions involve their moments of inertia and the angular velocities of the mantle and the core regions. The moments of inertia are obtainable by integrations over the density function of the

257

model Earth. The angular velocities of the different regions in a wobbling/nutating Earth are unequal. In the case of a deformable Earth, the deformations produce perturbations of the inertia tensors of the various regions; these perturbations enter the expressions for the angular momenta (in the torque equations) through certain *compliance parameters* representing the deformability properties of the different regions. The values of the compliance parameters are computed, for a chosen Earth model, by use of the expressions derived in Chapter 6; these expressions are in terms of the solutions of the equations governing the Earth's deformations under tidal and centrifugal forcings.

As will be seen presently, the torque equations, which involve the compliance parameters, are expressible as a system of coupled ordinary differential equations for the components of the angular velocity vectors of the different regions. The solutions of the coupled equations, with the amplitudes of the spectral components of the forcing potential set equal to unity, are sufficient for a computation of the transfer function from the rigid Earth to the non-rigid Earth. The transfer function method described in Section 2.27 may then be employed, together with values of the rigid Earth nutation amplitudes taken from accurate tables available in the literature, to obtain the amplitudes of the nutations of the non-rigid Earth model.

This approach is the one used by e.g. Hough (1895), Sasao and Wahr (1981), Molodensky (1961), Mathews *et al.* (1991a and b, 1995a and b, 2001, 2002), de Vries and Wahr (1991), Buffet *et al.* (1993, 2002), Herring *et al.* (1991, 2002), Dehant *et al.* (2003a and b, 2005a and b), Guo *et al.* (2004), Mathews and Guo (2005), Greff-Lefftz *et al.* (2000, 2002, 2004, 2005).

In the second formulation, an example of which is outlined in Section 7.5, the central role is played by the displacement field which represents the change in position of an element of matter at an arbitrary point in the perturbed (variably rotating and deforming) Earth relative to its position in the unperturbed (steadily rotating) Earth, as seen from a CRF. Rotation and deformation are characterized by different parts of the displacement field. This approach is the one used by e.g. Dahlen (1972, 1976, 1982), Dahlen and Smith (1975), Smith (1974), Smith and Dahlen (1981), Tromp and Dahlen (1990), Wahr (1979, 1981a–c, 1982, 1988, 2005), Dehant (1986, 1987a and b, 1989, 1990, 1991), Dehant and Ducarme (1987), Dehant and Zschau (1989), Dehant and Defraigne (1997) (see also earlier works, such as those of Alterman *et al.* (1959), Jeffreys and Vicente (1957a and b), Jeffreys (1959a and b, 1978, 1979, 1983)). The dynamical equation governing this field, which is the momentum balance equation for the elements of matter in the Earth, is necessarily a partial differential equation, unlike the ordinary differential equations of the first formulation, which is global in nature. Deformations perturb the density distribution within the Earth, and consequently also the gravitational potential due to the Earth itself, both within the Earth and outside. The relation between the deformation and the perturbation of the potential is through Laplace's equation. This equation and the momentum balance equation are mutually coupled; they constitute the fundamental equations of this formulation of the Earth's rotation-deformation problem. The density and elastic parameters, as functions of position in the Earth, have to be input into the equations; they are taken from a chosen Earth model of the Earth's interior, constructed from seismological data. Model 1066A (Gilbert and Dziewonski, 1975) was adopted for this purpose in the work of Wahr (1979, 1981a–c); the more recent preliminary reference earth model (PREM) of Dziewonski and Anderson (1981) has been commonly used in later works (e.g., Dehant 1987a and b, Mathews *et al.* 1991b). Recent works are based on the ak135 Model of Kennett *et al.* (1995). The fundamental differential equations are reduced to an infinite set of coupled ordinary differential equations in the radial variable by carrying out spherical harmonic expansions of the displacement, stress, and gravitational fields. Numerical integration methods are resorted to in order to solve a suitably truncated subset of the infinite set of radial equations. From the solutions thus obtained for the displacement field, one part that represents deformation and another part that represents rotation variations are identified. The latter determines the wobbles and nutations as well as the variations in the rotation rate (length-of-day (LOD) variations).

7.2 Idealized two-layer Earth: Poincaré's theory

Poincaré gave an elegant treatment of the variations in rotation of a model Earth which is "non-rigid" only in the sense that it has a fluid core. The model consists of a *rigid* shell (mantle) with an ellipsoidal cavity (core) filled with a homogeneous, inviscid, incompressible fluid; the principal axes of the shell (which need not be ellipsoidal) were taken to be aligned with those of the fluid ellipsoid. The Poincaré treatment does not, strictly speaking, belong to either of the two types of formulations described above. It is unique in that it starts out by assuming simple expressions, linear in the position coordinates, for the components of the flow velocity field within the fluid core. The expression for the flow field, presented below, involves the components (p, q, r) of the angular velocity of rotation of the mantle in space, which is necessarily shared by the core also. The flow relative to the mantle within the core involves three other parameters (p', q', r'), which turn out to be nearly the same as, though not identical with, the components of the angular velocity of a rotation of the core as a whole *relative to the mantle*. In the notation of the earlier chapters,

$$p = \Omega_0 m_1, \quad q = \Omega_0 m_2, \quad r = \Omega_0 (1 + m_3), \qquad (7.1)$$

where Ω_0 is the mean angular velocity of Earth rotation; and to a degree of approximation that will be specified in a later section,

$$p' \approx \Omega_0 m_{f1}, \quad q' \approx \Omega_0 m_{f2}, \quad r' \approx \Omega_0 m_{f3}. \tag{7.2}$$

The dynamical equations of the flow are the torque equation for the whole body (mantle + core) and the Helmholtz vorticity equation for the flow within the core. These vector equations reduce to coupled ordinary differential equations, which may be readily solved for the above sets of parameters.

It is a remarkable fact that the equations derived by Sasao *et al.* (1980) for the much more general (and correspondingly complicated) case of a stratified compressible fluid core within a deformable mantle are essentially identical in form to the Poincaré equations, the only difference being that the angular momenta in the former case include contributions from perturbations of the inertia tensor that result from deformability. We give a detailed presentation of the Poincaré theory for this reason as well as for the purpose of gaining insights about the errors due to the approximations made, which are not readily discernible in the more general case.

7.2.1 The Poincaré flow

In considering the fluid flow within the ellipsoid

$$x^2/a^2 + y^2/b^2 + z^2/c^2 = 1, \tag{7.3}$$

Poincaré assumed the following forms for the components (u, v, w) of the flow velocity $\mathbf{v}$ at $\mathbf{r} = (x, y, z)$:

$$u = a(q'z/c - r'y/b) + (qz - ry),$$

$$v = b(r'x/a - p'z/c) + (rx - pz), \tag{7.4}$$

$$w = c(p'y/b - q'x/a) + (py - qx).$$

The terms involving (p, q, r) represent the rigid rotational motion common to the mantle and the core, and the remaining terms represent the flow within the core relative to the mantle. It may be readily verified that this differential flow is tangential to the core mantle boundary (CMB), as it should be. Furthermore, at any interior point, the flow is tangential to the scaled-down version of the boundary surface which passes through the point.

The vorticity vector associated with the flow in the core is, by definition,

$$\mathcal{V} = \text{curl}(\mathbf{v}). \tag{7.5}$$

Since **v** is linear in the position coordinates, $\mathcal{V}$ is independent of position; its components are

$$\xi = 2p + \left(\frac{b}{c} + \frac{c}{b}\right)p', \quad \eta = 2q + \left(\frac{c}{a} + \frac{a}{c}\right)q', \quad \zeta = 2r + \left(\frac{a}{b} + \frac{b}{a}\right)r'. \quad (7.6)$$

Before going on to consider the equations of motion, we note that on carrying out integrations over the ellipsoidal volume of the core only, one readily obtains the following results:

$$\rho \int x^2 dx dy dz = Ma^2/5,$$

$$\rho \int y^2 dx dy dz = Mb^2/5, \quad (7.7)$$

$$\rho \int z^2 dx dy dz = Mc^2/5,$$

where M is the total mass of the core: $M = (4\pi/3)\rho abc$. Since the principal moments of inertia of the core are $A' = \rho \int (y^2 + z^2)d^3x$, etc., it follows from the above equations that

$$A' = M(b^2 + c^2)/5, \quad B' = M(c^2 + a^2)/5, \quad C' = M(a^2 + b^2)/5. \quad (7.8)$$

We consider next the angular momentum of the core alone as well as that of the whole Earth. Both are integrals of $\rho\, \mathbf{r} \wedge \mathbf{v}$, carried out over the volume of the core alone in the first case and over the whole Earth in the second. For the x-component, for instance, the integrand is, apart from the constant factor ρ,

$$\frac{c}{b}p'y^2 + \frac{b}{c}p'z^2 + p(y^2 + z^2), \quad (7.9)$$

plus terms in xy and xz which vanish on integration. It is to be remembered, while carrying out the integration, that (p', q', r') are defined only within the core and hence that terms involving any of them must be integrated over the volume of the core only; the remaining terms are to be integrated over the core alone if the angular momentum H'_x of the core is desired, and over the whole Earth if the total H_x is wanted. We find in this fashion that

$$H_x = F'p' + Ap, \quad H_y = G'q' + Bq, \quad H_z = H'r' + Cr, \quad (7.10)$$

where

$$F' = 2Mbc/5, \quad G' = 2Mca/5, \quad H' = 2Mab/5. \quad (7.11)$$

Similarly, by restricting the integration to the core region, we obtain

$$H'_x = F'p' + A'p, \quad H'_y = G'q' + B'q, \quad H'_z = H'r' + C'r, \quad (7.12)$$

where A', B', C' are the principal moments of inertia of the core, given in (7.8).

The term Ap in the expression for H_x in (7.10) is the x-component of the angular momentum of rigid rotation of the whole Earth. It is clear then that $F'p'$ is the angular momentum h_x due to the flow in the core relative to the mantle. This quantity may be viewed as the angular momentum associated with a differential rotation (relative to the mantle) of the core as a whole, the x-component ω'_x of the angular velocity of this rotation being such that

$$A'\omega'_x = F'p'. \tag{7.13}$$

This does not mean, of course, that the motion of the fluid in the core is just a pure rotation relative to the mantle – it is not. What it does mean is that if we *define* the angular velocity vector ω' through the equation (7.13) for ω'_x and its counterparts for the other two components, i.e., if $\omega' = (\omega'_x, \omega'_y, \omega'_z)$ is defined by

$$\omega'_x = (F'/A')p', \qquad \omega'_y = (G'/B')q', \qquad \omega'_z = (H'/C')r', \tag{7.14}$$

then the *residual* flow (the Poincaré flow (7.4) minus the pure rotational flow with angular velocity ω') carries no angular momentum.

Since $F' \neq A'$, it is clear that the parameter p' is not the same as the angular velocity component ω'_x. Their ratio may be expressed in terms of the flattening of the CMB (in the y-z plane), defined as

$$\epsilon = \frac{b^2 - c^2}{b^2 + c^2}, \qquad b^2 = \frac{1 + \epsilon}{1 - \epsilon} c^2. \tag{7.15}$$

If $\epsilon \ll 1$, one can readily verify that

$$(b/c) + (c/b) = 2 + \epsilon^2 \tag{7.16}$$

to the second order in ϵ. It follows then that the fractional difference between A' and F' and hence between ω'_x and p' is given up to the same order by

$$(p'/\omega'_x) - 1 = (A'/F') - 1 = (b - c)^2/2bc = \epsilon^2/2. \tag{7.17}$$

So the second term in ξ of equation (7.6) is $2(1 + \epsilon^2)\omega'_x$.

A separation of the flow velocity relative to the mantle into a "rigid" rotation ($\omega' \wedge \mathbf{r}$) and a residual part, say $\mathbf{v}'$, is of considerable interest in the context of the more general derivation of Eq. (7.27) given in Section 7.3.2. It is a simple matter to verify that the core flow part of u may be expressed with the help of (7.11) and (7.8) as

$$a(q'z/c - r'y/b) = (\omega' \wedge \mathbf{r})_x + v'_x, \tag{7.18}$$

$$v'_x = \frac{a}{c} \frac{a^2 - c^2}{(a^2 + c^2)} q'z - \frac{a}{b} \frac{a^2 - b^2}{(a^2 + b^2)} r'y \approx \epsilon q'z, \tag{7.19}$$

where ϵ is the flattening of the boundary of the core in the x-z plane, defined analogously to (7.15), and $(a^2 - b^2)$ has been taken to be zero, it being smaller than $(a^2 - c^2)$ by some three orders of magnitude. The important point is that the core flow field is predominantly that of a rigid rotation, the residual part $\mathbf{v}'$ being smaller by at least $\mathcal{O}(\epsilon)$. This is not surprising, since it is the flattening that is responsible for the residual flow.

7.2.2 Equations of rotational motion

We go on now to the equations governing the rotational motions of the whole Earth and of the fluid core. The first of these is the equation of angular momentum balance (or torque equation) for the rotating Earth, referred to a mantle-fixed frame:

$$\frac{d\mathbf{H}}{dt} + \boldsymbol{\Omega} \wedge \mathbf{H} = \boldsymbol{\Gamma}, \qquad (7.20)$$

where $\boldsymbol{\Omega}$ is the angular velocity vector with components (p, q, r), and $\boldsymbol{\Gamma}$ is the external torque on the Earth. We obtain the following set of differential equations on substituting from (7.10) for the components of $\mathbf{H}$:

$$A\dot{p} + F'\dot{p}' + q(H'r' + Cr) - r(G'q' + Bq) = \Gamma_x,$$
$$B\dot{q} + G'\dot{q}' + r(F'p' + Ap) - p(H'r' + Cr) = \Gamma_y, \qquad (7.21)$$
$$C\dot{r} + H'\dot{r}' + p(G'q' + Bq) - q(F'p' + Ap) = \Gamma_z,$$

where the dots indicate differentiation with respect to time.

For the second of the equations of motion, Poincaré took the Helmholtz vorticity equation. This equation is constructed starting from the Navier–Stokes equation and taking the curl. For a homogeneous and non-viscous fluid, as in Poincaré's theory, the equation is

$$\frac{d\mathcal{V}}{dt} + (\mathbf{v} \cdot \nabla)\mathcal{V} = (\mathcal{V} \cdot \nabla)\mathbf{v}. \qquad (7.22)$$

In the present case, $\mathcal{V}$ is independent of position. Therefore the term $(\mathbf{v} \cdot \nabla)\mathcal{V}$ vanishes. When the resulting equation is transformed to a frame rotating with angular velocity $\boldsymbol{\Omega}$, it goes over into

$$\frac{d\mathcal{V}}{dt} + \boldsymbol{\Omega} \wedge \mathcal{V} = (\mathcal{V} \cdot \nabla)\mathbf{v}. \qquad (7.23)$$

Recalling that the components of $\mathcal{V}$ are (ξ, η, ζ) and substituting from equations (7.4) for the x-component of $\mathbf{v}$, we reduce the x-component of the above equation, after multiplication by bc, to $bc\dot{\xi} + ca\eta r' - ab\zeta q' = 0$. Then substitution for the components of $\mathcal{V}$ from equations (7.6) leads to the following equation for the parameters in the Poincaré flow:

$$2bc\dot{p} + (b^2 + c^2)\dot{p}' = -[2acq + (a^2 + c^2)q']r' + [2abr + (a^2 + b^2)r']q'.$$

$$(7.24)$$

Finally use of (7.11) and (7.8) enables us to write this equation and the corresponding equations for the y and z-components of the vorticity equation as

$$F'\dot{p} + A'\dot{p}' + (G'q + B'q')r' - (H'r + C'r')q' = 0,$$
$$G'\dot{q} + B'\dot{q}' + (H'r + C'r')p' - (F'p + A'p')r' = 0, \qquad (7.25)$$
$$H'\dot{r} + C'\dot{r}' + (F'p + A'p')q' - (G'q + B'q')p' = 0.$$

Equations (7.21) and (7.25) constitute a complete set from which the time dependence of the parameters (p, q, r) and (p', q', r') may be determined, given that of the torque Γ. The temporal variations of the rotation (angular velocity) vectors Ω and ω' may thus be obtained.

Before we proceed further, we recall that

$$\frac{p'}{\omega'_x} \approx \frac{F'}{A'} \approx \frac{q'}{\omega'_y} \approx \frac{G'}{B'} \approx \frac{r'}{\omega'_z} \approx \frac{H'}{C'} = 1 + \mathcal{O}(\epsilon^2). \qquad (7.26)$$

If the parts of $\mathcal{O}(\epsilon^2)$ in these ratios are neglected, equations (7.25) may be identified as the components of the simple equation

$$\frac{d\mathbf{H}'}{dt} - \omega' \wedge \mathbf{H}' = 0. \qquad (7.27)$$

Remarkably, Sasao *et al.* (1980) were able to show that the torque equation reduces to an equation of precisely the same form even for a stratified and compressible fluid core (see Section 7.3.2).

7.2.3 Solution of the equations: axially symmetric case

We consider next the solution of the equations (7.21) and (7.25) in the special case of axial symmetry. The dynamical ellipticity $e = [2C/(A + B) - 1]$ of the Earth is about 3×10^{-3} while the fractional difference between B and A is only about 10^{-5} which is of $\mathcal{O}(e^2)$. So one may neglect the difference $(B - A)$ to a very good approximation and consider the axially symmetric case:

$$b = a, \quad B = A, \quad B' = A', \quad G' = F', \quad H' = C'. \qquad (7.28)$$

In this case, Γ_z of the external gravitational torque vanishes. Then, on neglecting the second-order difference between F' and A', the third components of the two sets of equations (7.21) and (7.25) become

$$C\dot{r} + C'\dot{r}' + A'(pq' - p'q) = 0, \tag{7.29}$$

$$C'(\dot{r} + \dot{r}') + A'(pq' - p'q) = 0. \tag{7.30}$$

Differencing the two equations, we find that $\dot{r} = 0$; this is as it should be in the absence of a z-component to the external torque. Thus r has a constant value which must be Ω_0. Furthermore, the deviation of the Earth's rotation axis from the symmetry axis (which is our z axis) is at most about 0.4 arcseconds $\approx 2 \times 10^{-6}$ radian, primarily due to the Chandler free wobble. The deviation is represented by (p/Ω_0) and (q/Ω_0), which are thus no larger than of this order; (p'/Ω_0) and (q'/Ω_0) as well as (r'/Ω_0), all of which are nonzero only for the perturbed Earth, may be taken to be similarly small. Then products of any two of these five quantities become ignorable, and as a result, we may conclude from the pair of equations (7.29) and (7.30) that r' is a constant.

The first two of equations (7.21) now take the following simplified forms:

$$A\dot{p} + F'\dot{p}' + (C - A)\Omega_0 q - F'\Omega_0 q' = \Gamma_x, \tag{7.31}$$

$$A\dot{q} + F'\dot{q}' - (C - A)\Omega_0 p + F'\Omega_0 p' = \Gamma_y, \tag{7.32}$$

and the first two equations from (7.25) simplify to

$$F'\dot{p} + A'\dot{p}' - H'\Omega_0 q' = 0, \tag{7.33}$$

$$F'\dot{q} + A'\dot{q}' + H'\Omega_0 p' = 0. \tag{7.34}$$

Each of the above pairs of equations may be combined into a single equation involving the complex quantities

$$\tilde{P} = p + iq, \qquad \tilde{P}' = p' + iq', \tag{7.35}$$

by adding i times the second equation of the pair to the first one. One gets

$$A\frac{d\tilde{P}}{dt} + F'\frac{d\tilde{P}'}{dt} - i(C - A)\Omega_0 \tilde{P} + iF'\Omega_0 \tilde{P}' = \tilde{\Gamma}, \tag{7.36}$$

$$F'\frac{d\tilde{P}}{dt} + A'\frac{d\tilde{P}'}{dt} + iC'\Omega_0 \tilde{P}' = 0, \tag{7.37}$$

where we have used the fact that $H' = C'$ for an axially symmetric Earth. What we now have is a pair of coupled ordinary linear differential equations. In the following, we shall take $F' = A'$, neglecting the difference of order ϵ^2 between them. It may be recalled that in this approximation, $\tilde{P}'$ is the differential angular velocity ω' of the core relative to the mantle.

Consider now a spectral component of the torque $\tilde{\Gamma}$ (see Eqs. (2.49) and (2.61)) which is responsible for a wobble of angular frequency ω_w. Following the usage of Sasao *et al.* (1980), we express the frequency in units of cpsd as $\sigma = \omega_w / \Omega_0$:

$$\tilde{\Gamma} = K A \Omega_0^2 e^{i\sigma\Omega_0 t}, \qquad (7.38)$$

$\tilde{P}$ and $\tilde{P}'$ have the same kind of time dependence, and hence the above equations reduce to their frequency domain versions

$$[A\sigma - (C - A)]\tilde{P}(\sigma) + A'(\sigma + 1)\tilde{P}'(\sigma) = K A\Omega_0, \qquad (7.39)$$

$$A'\sigma \tilde{P}(\sigma) + (A'\sigma + C')]\tilde{P}'(\sigma) = 0. \qquad (7.40)$$

It may be noted that K is a dimensionless measure of the strength of the torque. It is seen later on that it is of the order of 10^{-8} for the torque on the Earth.

7.2.4 Free wobbles: eigenfrequencies

In the absence of any external forcing ($K = 0$), the determinant $\Delta(\sigma)$ of the coefficients of $\tilde{P}(\sigma)$ and $\tilde{P}'(\sigma)$ has to vanish:

$$\Delta(\sigma) \equiv A'[A(\sigma - e)(\sigma + 1 + e') - A'(\sigma + 1)\sigma] = 0, \qquad (7.41)$$

where e and e' are the dynamical ellipticities of the whole Earth and the core, respectively:

$$e \equiv C/A - 1, \qquad e' \equiv C'/A' - 1. \qquad (7.42)$$

$\Delta(\sigma)$ of equation (7.41) may be written, neglecting very small terms of the second order in the ellipticities, as

$$\Delta(\sigma) = AA'(\sigma - \sigma_1)(\sigma - \sigma_2). \qquad (7.43)$$

The roots of the determinantal equation are, correct to the first order in the ellipticities,

$$\sigma_1 = (A/A_m)e \quad \text{and} \quad \sigma_2 = -1 - (A/A_m)e', \qquad (7.44)$$

where $A_m = A - A'$. The first of these is the eigenfrequency (in units of Ω_0, i.e., in cpsd) of the Chandlerian free wobble of a non-deformable Earth with a fluid core. The second is the eigenfrequency of the nearly diurnal free wobble (NDFW), so called because the magnitude of this frequency is very close to 1 cpsd. The latter mode plays a very significant role in the forced nutations of the Earth by the resonance effect it produces in response to forcing at nearby frequencies.

7.2.5 Forced wobbles

An important special case of the forced wobbles resulting from forcing at -1 cpsd ($\tilde{\Gamma} = KA\Omega_0^2 e^{-i\Omega_0 t}$) will be dealt with first. This is the forcing due to a torque vector that is stationary in space. On setting $\sigma = -1$ in equation (7.39), one immediately gets

$$\tilde{P}(-1) = -KA\Omega_0/C, \tag{7.45}$$

independently of the existence of the core. We shall see later that the same result holds even for a deformable Earth. Poincaré, who drew attention to this property, called it *gyrostatic rigidity*. It is, understandably, not shared by $\tilde{P}'(\sigma)$ which does depend on the Earth's structure and properties. With the use of the above expression for $\tilde{P}(-1)$, Eq. (7.40) yields $\tilde{P}'(-1) = \tilde{P}(-1)/e' = -KA\Omega_0/(Ce')$.

In view of the form (7.43) of the secular determinant of the problem, one obtains immediately the following solution for the coupled equations (7.39) and (7.40) for any $\sigma \neq 1$:

$$\tilde{P}(\sigma) = \frac{K\Omega_0(\sigma + 1 + e')}{(\sigma - \sigma_1)(\sigma - \sigma_2)}, \qquad \tilde{P}'(\sigma) = \frac{-K\Omega_0\sigma}{(\sigma - \sigma_1)(\sigma - \sigma_2)}. \tag{7.46}$$

The factors $1/(\sigma - \sigma_1)$ and $1/(\sigma - \sigma_2)$ give rise to the *resonances* associated with the Chandlerian free wobble and the NDFW, respectively.

It is worth noting that the magnitude of the ratio $\tilde{P}'(\sigma)/\tilde{P}(\sigma) = -\sigma/(\sigma + 1 + e')$ is much larger than unity for the bulk of the wobbles in the retrograde diurnal band. For instance, for the wobbles of frequencies $\sigma = -1 - 1/366.25$, $-1 - 1/6817$ and $-1 + 1/13.70$ cpsd which are associated with the retrograde annual, retrograde 18.6 year, and prograde fortnightly nutations, respectively, one finds (with $e' \approx 1/250$) that the magnitudes of the ratio are about 790, 260, and 12, respectively.

7.2.6 Second-order corrections

In theoretical work aiming at high accuracy in the computation of nutations, it has been found that various second-order effects make non-ignorable contributions. Against this background, it is of interest to examine the possibility that the terms of $\mathcal{O}(\epsilon^2)$ which have been neglected in reducing the equations for the fluid core to the form (7.27) might not be insignificant. In theories such as that of Sasao *et al.* (1980), which employ more realistic (stratified, compressible) models for the fluid core, analytical expressions for the second-order terms are not available. The nature of the expressions that can be obtained from Eqs. (7.25) for the Poincaré problem should provide a good indication of what to expect in general. It would be worthwhile, therefore, to obtain the second-order terms in the present problem. We proceed now to do so.

We note first that in the axially symmetric case ($b = a$), the definition (7.15) of the flattening may be rewritten as $\epsilon = (a^2 - b^2)/(a^2 + b^2)$. Then, on making use of the definitions (7.8) and (7.11) of the parameters A', B', C', F', G', H' as well as (7.17) specialized to $b = a$ one finds that, to $\mathcal{O}(\epsilon^2)$,

$$\frac{C'}{A'} = \frac{2a^2}{a^2 + c^2} = 1 + \epsilon, \qquad \frac{F'}{A'} = \frac{G'}{A'} = \frac{2ac}{a^2 + c^2} = 1 - \frac{1}{2}\epsilon^2, \qquad (7.47)$$

$$\frac{\omega'_x}{p'} = \frac{\omega'_y}{q'} = \frac{a^2 + c^2}{2ac} = 1 + \frac{1}{2}\epsilon^2. \qquad (7.48)$$

It must be noted that the dynamical ellipticity e' of the fluid core is equal to the surface flattening ϵ of the core in the present case. This is a consequence of the definition $e' = (C'/A') - 1$ taken together with the first of equations (7.47). Therefore we replace ϵ by e' in the remainder of this section.

We introduce the above results into the first two of Eqs. (7.25), and neglect terms involving products of any two of the quantities p, q, p', q', r', as was done earlier, but retaining the terms of $\mathcal{O}(e'^2)$. The resulting equations are, after transferring all the $\mathcal{O}(e'^2)$ terms to the right-hand side,

$$\dot{\omega}_x + \dot{\omega}'_x - \Omega_0(1 + \epsilon)\omega'_y = (e'^2/2)(\dot{\omega}_x - \dot{\omega}'_x + \Omega_0\omega'_y), \qquad (7.49)$$

$$\dot{\omega}_y + \dot{\omega}'_y + \Omega_0(1 + \epsilon)\omega'_x = (e'^2/2)(\dot{\omega}_y - \dot{\omega}'_y - \Omega_0\omega'_x). \qquad (7.50)$$

The complex combination of these gives, in terms of the wobble quantities $\tilde{m} = (\omega_x + i\omega_y)/\Omega_0$ and $\tilde{m}_f = (\omega'_x + i\omega'_y)/\Omega_0$,

$$\dot{\tilde{m}} + \dot{\tilde{m}}_f + i\Omega_0(1 + e')\tilde{m}_f = (e'^2/2)(\dot{\tilde{m}} - \dot{\tilde{m}}_f - i\Omega_0\tilde{m}_f). \qquad (7.51)$$

For a spectral component having the time dependence $e^{i\sigma\Omega_0 t}$, the above equation yields

$$\sigma\tilde{m} + (1 + \sigma + e')\tilde{m}_f = (e'^2/2)[\sigma\tilde{m} - (1 + \sigma)\tilde{m}_f], \qquad (7.52)$$

wherein $\tilde{m}$ and $\tilde{m}_f$ refer to the spectral amplitudes $\tilde{m}(\sigma)$ and $\tilde{m}_f(\sigma)$.

As for the torque equation for the whole Earth, we note that Eqs. (7.36) of the axially symmetric case may be rewritten in terms of $\tilde{m}$ and $\tilde{m}_f$ defined through the relations

$$\tilde{P} = \Omega_0\tilde{m}, \qquad F'\tilde{P}' = A'\Omega_0\tilde{m}_f, \qquad (7.53)$$

where the second relation has been obtained with the aid of equation (7.14). The resulting equation is simply

$$(\dot{\tilde{m}} - ie\Omega_0\tilde{m}) + \frac{A'}{A}(\dot{\tilde{m}}_f + i\Omega_0\tilde{m}_f) = \frac{\tilde{\Gamma}}{A\Omega_0}, \qquad (7.54)$$

or in the frequency domain,

$$(\sigma - e)\tilde{m} + \frac{A'}{A}(1 + \sigma)\tilde{m}_f = \frac{\tilde{\Gamma}}{i A \Omega_0^2}. \tag{7.55}$$

Observe that there are no order e'^2 terms in this equation.

The correction to $\tilde{m}$ that is necessitated by the presence of the second-order term in (7.52) may easily be evaluated from that equation taken together with (7.55). It is found that the correction is $e'^2 X \tilde{m}$, where

$$X = (1 + \sigma)\frac{1}{2}\left(\frac{1}{1 + \sigma + e'} - \frac{A_m}{A}\frac{\sigma - e + \sigma A'/A}{(\sigma - \sigma_1)(\sigma - \sigma_2)}\right), \tag{7.56}$$

where $A_m = A - A'$ and σ_1 and σ_2 are the Chandler and FCN eigenfrequencies, respectively. The equivalent correction to the nutation amplitude $\tilde{\eta}$ is $e'^2 X \tilde{\eta}$.

The correction to $\tilde{m}$ is clearly of $\mathcal{O}(e'^2)$ if X remains of $\mathcal{O}(1)$ for all σ of interest. The magnitude of X is expected to be the largest for a value of σ which minimizes one of the factors in the denominator. With $e' \approx (1/250)$, $\sigma_2 \approx -1 - (1/433)$, and $A_m/A \approx 8/9$, one sees that the σ corresponding to the retrograde annual nutation ($\sigma = -1 - 1/366$) is the one to worry about. Actual evaluation shows that the magnitude of X is about 4 for this frequency, and is smaller for any other. So one can conclude that the correction to $\tilde{m}$ at any spectral frequency is less than about $(3 \times 10^{-5})\tilde{m}$.

7.3 Analytical treatment of more general Earth models

A brief outline of the analytical approach was presented in Section 7.1. One starts with an axially symmetric ellipsoidal Earth model consisting of a mantle region enclosing a fluid core and possibly also a SIC. All the regions are taken to be stratified and deformable (unlike the treatment in the Poincaré formalism). The first step is to formulate the torque equations for the Earth as a whole and for its core regions. These equations involve the angular momentum vectors of the different regions which are expressed in terms of their inertia tensors and angular velocities. The global deformability properties of the various regions are encapsulated in a few compliance parameters appearing in their inertia tensors. Once the torque equations are solved to obtain the angular velocities of the different regions (or equivalently, their wobbles and spin rate variations) for unit external forcing, one employs the transfer function method to obtain the amplitudes of the nutations of the non-rigid model Earth. We shall first apply this procedure to a two-layer Earth model which consists of a solid mantle enclosing a fluid core that extends all the way to the geocenter. Our first task is to develop the torque equations for the model.

7.3.1 Earth model with two layers

General equations

The obvious starting point for a treatment of the wobbles of such a model is the pair of angular momentum balance equations (torque equations), one for the Earth as a whole and the other for the core alone, with reference to a frame fixed to the mantle rotating with variable angular velocity $\mathbf{\Omega}$ which remains very close to the mean angular velocity vector $\mathbf{\Omega}_0$ in space:

$$\mathbf{\Omega} = \mathbf{\Omega}_0 + \boldsymbol{\omega}, \qquad \mathbf{\Omega}_0 = \Omega_0 \mathbf{i}_3, \tag{7.57}$$

where $\mathbf{i}_3$ is the direction of the Earth's axis of symmetry. The torque equations are, respectively,

$$\frac{d\mathbf{H}}{dt} + \mathbf{\Omega} \wedge \mathbf{H} = \mathbf{\Gamma}, \tag{7.58}$$

$$\frac{d\mathbf{H}_f}{dt} + \mathbf{\Omega} \wedge \mathbf{H}_f = \mathbf{\Gamma}_f. \tag{7.59}$$

Here $\mathbf{\Gamma}$ is the torque acting on the whole Earth, which is necessarily due to external bodies, while $\mathbf{\Gamma}_f$ is the torque on the core alone. The latter is partly of external gravitational origin, but includes also contributions from interactions of the core with the rest of the Earth (the mantle). The most important of these is the so-called "inertial torque" or "inertial coupling." Any rotational motion of the ellipsoidal core as a whole relative to the mantle about an equatorial axis (i.e., a differential wobble between the core and the mantle) is impeded, because such a rotation causes the part of the core that is outside the sphere inscribed in the ellipsoidal CMB to impinge on the corresponding parts of the boundary and produce an inertial reaction on the core itself. To compute the inertial torque which is due to this reaction, one needs to determine the dynamical pressure P in the fluid (or at least over the CMB) when the Earth is under the action of the external perturbation. This complication is avoided in an alternative form of the angular momentum equation for the core which was formulated by Sasao, Okubo, and Saito (1980), building upon and clarifying an earlier treatment due to Molodensky (1961). Their equation, which we call the SOS equation for the core, has just the same form as Poincaré Eq. (7.27) when electromagnetic or other non-pressure couplings are ignored; but unlike the latter, which was derived under the assumption that the core is uniform and incompressible, the derivation by Sasao *et al.* explicitly allowed for compressibility and axially symmetric ellipsoidal stratification of the core fluid. The inclusion of the additional torques, which are produced by electromagnetic forces generated by the differential wobble in the presence of magnetic fields crossing the CMB or by viscous drag at the boundary, introduces an additional core

mantle coupling term proportional to the differential wobble in the SOS equation
for the core. The effect of such a term on the solutions of the equations will be
considered, but derivation of the theoretical expression for the coupling strength
parameter will be deferred to Chapter 10.

7.3.2 Torque equation for the core in the SOS form

Derivation of the torque equation from the equation for flow in the core

We present now the essential steps in derivation of the Sasao *et al.* (1980) form
for the rotation equation of the core. The starting point is the momentum balance
equation governing the flow velocity of a mass element of the fluid, written with
reference to a frame fixed to and rotating with the mantle with angular velocity $\mathbf{\Omega}$.
If the fluid velocity *relative to the mantle* at $\mathbf{r}$ in this frame is $\mathbf{v}_f(\mathbf{r}, t)$, the velocity
in space is

$$\mathbf{v}_S = (\mathbf{\Omega} \wedge \mathbf{r} + \mathbf{v}_f). \tag{7.60}$$

A mass element which is at $\mathbf{r}$ at time t, moves to $\mathbf{r} + \mathbf{v}dt$ at time $(t + dt)$; so the
increment in its velocity in the interval dt is

$$\mathbf{v}_S(\mathbf{r} + \mathbf{v}_f dt, t + dt) - \mathbf{v}_S(\mathbf{r}, t) = \frac{D(\mathbf{\Omega} \wedge \mathbf{r} + \mathbf{v}_f)}{Dt} dt, \tag{7.61}$$

where the operator D/Dt, often called the substantial derivative, is given by

$$\frac{D}{Dt} = \frac{\partial}{\partial t} + (\mathbf{v}_f \cdot \nabla). \tag{7.62}$$

The acceleration as seen in the mantle-fixed frame is therefore $D\mathbf{v}_S/Dt$. To obtain
the acceleration with respect to the inertial frame, one has to add the term $\mathbf{\Omega} \wedge \mathbf{v}_S$
as usual. The Newtonian equation of motion of a fluid element of mass dm is
then

$$dm \left(\frac{D\mathbf{v}_S}{Dt} + \mathbf{\Omega} \wedge \mathbf{v}_S \right) = d\mathbf{F}, \tag{7.63}$$

where $d\mathbf{F}$ is the force acting on the element. We assume for the present that the only
forces are those arising from the fluid pressure P and a gravitational potential ϕ_g
which is the sum of the potential associated with the Earth's own mass distribution
and that due to celestial bodies:

$$d\mathbf{F} = -dV \nabla P - dm \nabla \phi_g, \tag{7.64}$$

dV being the volume occupied by the fluid element. On expanding the left-hand
part of (7.63) using (7.62) and (7.60), and dividing by dV, the equation of motion

becomes

$$\rho\,[\dot{\mathbf{v}}_f + (\mathbf{v}_f \cdot \nabla)\mathbf{v}_f + 2\boldsymbol{\Omega} \wedge \mathbf{v}_f + \dot{\boldsymbol{\Omega}} \wedge \mathbf{r} + \boldsymbol{\Omega} \wedge (\boldsymbol{\Omega} \wedge \mathbf{r})] = -\nabla P - \rho\nabla\phi_g,$$

$$(7.65)$$

wherein time derivatives are indicated by dots over the quantities concerned and ρ is the density of the fluid. This is the momentum balance equation governing the fluid flow. It may be noted that the terms $2\boldsymbol{\Omega} \wedge \mathbf{v}_f$ and $\boldsymbol{\Omega} \wedge (\boldsymbol{\Omega} \wedge \mathbf{r})$ represent the Coriolis and centrifugal accelerations, respectively.

We are more directly concerned here with the variation of the angular momentum. The angular momentum of a fluid element of mass dm is $dm\,\mathbf{r} \wedge \mathbf{v}_S$. The total angular momentum of the fluid core is therefore

$$\mathbf{H}_f = \int \rho\,\mathbf{r} \wedge \mathbf{v}_S dV = \int \rho\,\mathbf{r} \wedge (\boldsymbol{\Omega} \wedge \mathbf{r} + \mathbf{v}_f)\,dV. \qquad (7.66)$$

The integration here, as in the remainder of this section, is over the volume of the fluid core.

The reasoning employed in deriving (7.65) may be applied equally well to the angular momentum of an element of mass dm of the fluid. One simply replaces $\mathbf{v}_S$ of the earlier derivation by $\mathbf{r} \wedge \mathbf{v}_S$. One obtains then the following expression for the rate of change of the angular momentum per unit volume in the terrestrial frame:

$$\rho\,\frac{D}{Dt}(\mathbf{r} \wedge \mathbf{v}_S) = \rho\,\mathbf{r} \wedge [\dot{\mathbf{v}}_f + (\mathbf{v}_f \cdot \nabla)\mathbf{v}_f + 2\boldsymbol{\Omega} \wedge \mathbf{v}_f + \dot{\boldsymbol{\Omega}} \wedge \mathbf{r}] + \rho\boldsymbol{\Omega} \wedge (\mathbf{v}_f \wedge \mathbf{r}).$$

$$(7.67)$$

The integral of this quantity over the volume of the fluid is $d\mathbf{H}_f/dt$, the rate of change of $\mathbf{H}_f$ in the terrestrial frame. We expect that the rate of change in the inertial frame, which is

$$\frac{d\mathbf{H}_f}{dt} + \boldsymbol{\Omega} \wedge \mathbf{H}_f = \int \rho \left(\frac{D}{Dt}(\mathbf{r} \wedge \mathbf{v}_S) + \boldsymbol{\Omega} \wedge (\mathbf{r} \wedge \mathbf{v}_S) \right) dV, \qquad (7.68)$$

should be reducible to the torque expression

$$\boldsymbol{\Gamma}_f = \int \mathbf{r} \wedge (-\nabla P - \rho\nabla\phi_g)\,dV. \qquad (7.69)$$

The first step in demonstrating this is to eliminate the square bracketed terms in (7.67) by substituting for them from (7.65). The result is

$$\rho\,\frac{D}{Dt}(\mathbf{r} \wedge \mathbf{v}_S) = \mathbf{r} \wedge [(-\nabla P - \rho\nabla\phi_g) - \rho\,\boldsymbol{\Omega} \wedge (\boldsymbol{\Omega} \wedge \mathbf{r})] + \rho\boldsymbol{\Omega} \wedge (\mathbf{v}_f \wedge \mathbf{r}).$$

$$(7.70)$$

The second step is to substitute this in (7.68) and to introduce the expression (7.60) for v_S in the second term of the integrand. Then, on simplifying the combination of terms involving $\boldsymbol{\Omega}$ with the aid of vector identities like $\mathbf{A} \cdot (\mathbf{B} \wedge \mathbf{C}) = (\mathbf{A} \wedge \mathbf{B}) \cdot \mathbf{C}$ and $\mathbf{A} \wedge (\mathbf{B} \wedge \mathbf{C}) = (\mathbf{A} \cdot \mathbf{C})\mathbf{B} - (\mathbf{A} \cdot \mathbf{B})\mathbf{C}$, one finds that the integrand in (7.68) reduces to the integrand of the torque $\boldsymbol{\Gamma}_f$ acting on the core as given by the expression (7.69). We have thus verified that the torque Eq. (7.59) may be obtained starting from (7.65).

Equation for the core angular momentum in SOS form

Our objective now is to derive the equation for the angular momentum of the core in the alternative form obtained by Sasao *et al.* (1980). It has the important advantage that it does not involve either the pressure P or the potential ϕ_g, surprising as this may seem.

The derivation proceeds through a number of steps. One begins by separating the velocity field in the fluid into a part $(\omega_f \wedge \mathbf{r})$ corresponding to a "rigid" rotation of the whole core relative to the mantle with angular velocity ω_f and a residual part $\mathbf{v}$,

$$\mathbf{v}_f = \omega_f \wedge \mathbf{r} + \mathbf{v}. \tag{7.71}$$

The former part cannot be tangential to the boundary (CMB), which is ellipsoidal. The residual part $\mathbf{v}$ is needed to make the total flow tangential to the boundary; it is expected to be smaller by $\mathcal{O}(\epsilon)$ than the rigid part of the flow, i.e., of $\mathcal{O}(\omega_f r)$, where $\omega_f = |\omega_f|$,

$$\mathbf{v} \sim \mathcal{O}(\epsilon \omega_f r). \tag{7.72}$$

Not surprisingly, this expectation is borne out in the special case of the Poincaré flow at the end of 7.2.1, where explicit expressions are available.

The angular velocity ω_f of the rigid rotation is fixed by the requirement that the angular momentum of the flow $\mathbf{v}_f$ be solely due to this rotation, or equivalently, that the angular momentum in the residual flow be zero:

$$\int \rho \, \mathbf{r} \wedge \mathbf{v}_f dV = \int \rho \, \mathbf{r} \wedge (\omega_f \wedge \mathbf{r}) dV, \qquad \int \rho \, \mathbf{r} \wedge \mathbf{v} dV = 0. \tag{7.73}$$

The total angular momentum vector (7.66) of the core then reduces to

$$\mathbf{H}_f = \int \rho \, \mathbf{r} \wedge (\boldsymbol{\Omega}_f \wedge \mathbf{r}) dV, \qquad \boldsymbol{\Omega}_f = \boldsymbol{\Omega} + \omega_f. \tag{7.74}$$

which is equivalent to the relation

$$\mathbf{H}_f = [C_f] \cdot \boldsymbol{\Omega}_f, \tag{7.75}$$

wherein the $[C_f]$, the inertia tensor of the fluid core, acts upon the angular velocity vector. This relation implies that the angular momentum of the core is the same as if the whole core were rigidly rotating with angular velocity $\mathbf{\Omega}_f$ in space, even though the flow within the core is not of this nature.

The key point in the derivation of the Sasao *et al.* form is the proof that, to a sufficient degree of approximation, which will be specified later on,

$$\int \rho \mathbf{r} \wedge \mathbf{G} dV \approx 0, \tag{7.76}$$

where $\mathbf{G}$ is defined by

$$\mathbf{G} = \frac{\nabla P}{\rho} + \nabla \phi_{gc}^f. \tag{7.77}$$

ϕ_{gc}^f is the total (gravitational + centrifugal) potential in the fluid:

$$\phi_{gc}^f = \phi_g + \phi_c(\mathbf{\Omega}_f), \tag{7.78}$$

$\phi_c(\mathbf{\Omega}_f)$ being the centrifugal potential associated with the rotation of the core with angular velocity $\mathbf{\Omega}_f$,

$$\phi_c(\mathbf{\Omega}_f) = -(1/2)[(\mathbf{\Omega}_f \cdot \mathbf{\Omega}_f)r^2 - (\mathbf{\Omega}_f \cdot \mathbf{r})^2]. \tag{7.79}$$

We shall now proceed to make use of (7.76) in anticipation of its proof. We begin by using (7.77) together with (7.78) to write

$$-\nabla P - \rho \nabla \phi_g = -\rho \mathbf{G} + \rho \nabla \phi_c(\mathbf{\Omega}_f). \tag{7.80}$$

The expression on the left-hand side of this equation appears in equation (7.70); we replace it now by the quantity on the right-hand side. The relation (7.80) can be used together with (7.76) to evaluate the pressure-cum-gravitational torque. We leave this to the reader to do if this is of interest. On integrating the resulting equation, and recalling that the integral of the left-hand member is $d\mathbf{H}_f/dt$, we obtain

$$\frac{d\mathbf{H}_f}{dt} = \int \rho\{\mathbf{r} \wedge [\nabla \phi_c(\mathbf{\Omega}_f) - \mathbf{\Omega} \wedge (\mathbf{\Omega} \wedge \mathbf{r})] + \mathbf{\Omega} \wedge (\mathbf{v}_f \wedge \mathbf{r})\}dV - \int \rho \mathbf{r} \wedge \mathbf{G} dV.$$

$$\tag{7.81}$$

The striking fact about this equation is that once the term involving $\mathbf{G}$ is neglected on the basis of (7.76), neither the pressure P nor the gravitational potential ϕ_g has any role in the equation. We can now carry out simplifications of the remaining terms on the right-hand side. Firstly, we replace the last term in the integrand of the first integral in the above equation by $\mathbf{\Omega} \wedge [(\omega_f \wedge \mathbf{r}) \wedge \mathbf{r}]$ since the part $\mathbf{v}$ of $\mathbf{v}_f$ does not contribute to the integral in view of (7.73). We also write $\nabla \phi_c(\mathbf{\Omega}_f)$ explicitly as

$-[\Omega_f^2 \mathbf{r} - (\Omega_f \cdot \mathbf{r})\Omega_f]$. The use of elementary identities of vector algebra enables one then to re-express (7.81) (neglecting the **G** term) as

$$\frac{d\mathbf{H}_f}{dt} = \omega_f \wedge \mathbf{H}_f, \tag{7.82}$$

which is the Sasao *et al.* (SOS) form that we seek. As already mentioned, the Poincaré equation is expressible in exactly the same form: see (7.27).

The negligibility of $\int \rho \mathbf{r} \wedge \mathbf{G} dV$ requires evidently that it should be very much smaller than either member of the equation above. Consider for instance the first component of the right-hand part, which is $(\omega_f)_y(H_f)_z - (\omega_f)_z(H_f)_y \approx (\omega_f)_y A_f \Omega_0$. We invoked the fact that $(H_f)_z$ in the first term involves the mean angular velocity Ω_0 so that $(H_f)_z \approx C_f \Omega_0 \approx A_f \Omega_0$ while $(H_f)_y$ in the second term involves only the very much smaller $(\omega + \omega_f)_y$. Thus what we need to establish is that the magnitude of the **G** term is sufficiently small compared to $A_f \Omega_0 |\omega_f|$. It turns out, in fact that it is indeed very much smaller, by a factor of order (ϵ^2) where the flattening ϵ of the CMB is about (1/400).

Estimation of error in (7.82): recasting the momentum balance equation

We turn now to the proof that $\int \rho \mathbf{r} \wedge \mathbf{G} dV$, is of negligible magnitude, a fact which was basic to the derivation of Eq. (7.82).

We observe, to begin with, that the quantity **G** pertaining to the core of the tidally perturbed variably rotating Earth must be of the first order in the perturbation because the corresponding quantity $\mathbf{G}_0$ for the Earth in the unperturbed state, of uniform rotation with the angular velocity Ω_0 vanishes as a result of the balance of gravitational and centrifugal forces in this state:

$$\mathbf{G}_0 \equiv \frac{\nabla P_0}{\rho_0} + \nabla \phi_0 = 0, \tag{7.83}$$

with

$$\phi_0 \equiv \phi_{gc0} = \phi_{g0} + \phi_c(\Omega_0). \tag{7.84}$$

The subscript 0 refers to the unperturbed state, which is referred to as the *hydrostatic equilibrium* state. In this state, the density ρ_0 and the geopotential ϕ_{gc0} are both constant on the same set of axially symmetric ellipsoidal surfaces, the outermost of which is the outer surface of the Earth. Each of the surfaces is labeled by a constant value of a function s of the coordinates and by a flattening $\epsilon(s)$, where s is defined by

$$s = r\left(1 + \frac{2}{3}\epsilon(s)P_2(\cos\theta)\right). \tag{7.85}$$

Here r and θ are the radial and colatitude variables for points in/on the Earth, and the flattening

$$\epsilon(s) = (a(s)^2 - c(s)^2)/(a(s)^2 + c(s)^2), \tag{7.86}$$

where $a(s)$ and $c(s)$ are the semi-major and semi-minor axes of the ellipsoid in question. The value of $\epsilon(s)$ varies from about $1/300$ at the Earth's surface to about $1/400$ near the center; $\epsilon(s)$ in (7.85) is usually replaced by $\epsilon(r)$, the error introduced in s thereby being only of the second order in ϵ. Both ρ_0 and ϕ_0 may evidently be viewed as functions of s, and so one may write

$$\nabla \rho_0 = \frac{d\rho_0}{ds} \nabla s, \qquad \nabla \phi_0 = \frac{d\phi_0}{ds} \nabla s, \tag{7.87}$$

where, by virtue of (7.85),

$$\nabla s = \left(\hat{\mathbf{r}} + \frac{2s\epsilon z}{r^2} \hat{\mathbf{z}} \right). \tag{7.88}$$

Now, $\mathbf{G}$ is equal to $\mathbf{G} - \mathbf{G}_0$, since $\mathbf{G}_0 = 0$, as noted in (7.83), and so we can equate $\mathbf{G}$ to the difference between the expressions (7.77) and (7.83). Define

$$\Pi_1 = \frac{P_1}{\rho_0} + \phi_1, \tag{7.89}$$

with

$$P = P_0 + P_1, \quad \text{and} \quad \phi_{gc}^f = \phi_0 + \phi_1. \tag{7.90}$$

Then, on taking the gradient of Π_1 and using the results (7.87) we can readily verify that $\mathbf{G} = \mathbf{G} - \mathbf{G}_0$ may be expressed as

$$\mathbf{G} = \nabla \Pi_1 + R_1 \nabla s, \tag{7.91}$$

with

$$R_1 = \left(\frac{P_1}{\rho_0^2} \frac{d\rho_0}{ds} + \frac{\rho_1}{\rho_0} \frac{d\phi_0}{ds} \right). \tag{7.92}$$

We have used (7.83) to express ∇P_0 in terms of $\nabla \phi_0$ in arriving at the above expression for R_1.

With $\mathbf{G}$ given by (7.91), we can express the integral $\int \rho \mathbf{r} \wedge \mathbf{G} dV$ (which we want to show to be negligible) as

$$\int \rho \mathbf{r} \wedge \mathbf{G} dV = \int \left[\nabla \wedge (\rho_0 \Pi_1 \mathbf{r}) + \mathbf{r} \wedge \left(-\Pi_1 \frac{d\rho_0}{ds} + \rho_0 R_1 \right) \nabla s \right] dV. \tag{7.93}$$

The second part of the integral involves the factor $\mathbf{r} \wedge \nabla s$, which is of $\mathcal{O}(\epsilon r)$ in view of (7.88). The first part of the integral may be converted into a surface integral which involves the same factor $\mathbf{r} \wedge \nabla s$. By applying the identity

$$\int \nabla \wedge \mathbf{F} dV = \int_S d\mathbf{S} \wedge \mathbf{F} \tag{7.94}$$

(from vector analysis), we see that

$$\int \nabla \wedge (\rho_0 \Pi_1 \mathbf{r}) dV = \int_S d\mathbf{S} \wedge (\rho_0 \Pi_1 \mathbf{r}), \tag{7.95}$$

where the integration is over the surface (or surfaces) enclosing the volume V, and $d\mathbf{S} = \hat{\mathbf{n}} dS$, dS being an element of area on the surface. $\hat{\mathbf{n}}$ is the unit normal to this area, pointing outwards from the enclosed volume. In our case, S is the CMB, which is one of the family of surfaces $s = $ constant, and so $\hat{\mathbf{r}} \wedge \hat{\mathbf{n}} = \hat{\mathbf{r}} \wedge \nabla s / |\nabla s|$ is of $\mathcal{O}(\epsilon r)$ in view of Eq. (7.88). Since the surface integral is taken over the CMB, ϵ and r here are the flattening and the radius, respectively, of the CMB: $\epsilon = \epsilon_c \approx (1/400)$ and $r = a_c$. Note that ϵ in the interior of the core is less than ϵ_c.

The magnitudes of both Π_1 and rR_1 are of $\mathcal{O}(\Omega_0 \epsilon \omega_f r^2)$, as we shall prove shortly. We use these results meanwhile to obtain a rough estimate of the magnitude of the integral (7.93). For the first part of the integral, taken in the alternative form (7.95), the integral of the magnitude of the integrand provides an upper bound:

$$\int_S \rho_0 (\Omega_0 \epsilon \omega_f a_c^2)(\epsilon a_c) dS = 4\pi a_c^5 \rho_0 \epsilon^2 \Omega_0 \omega_f, \tag{7.96}$$

ρ_0 here being the fluid density at the CMB. The core moment of inertia, say A_{c0}, if the fluid with this density were to fill the core region, would be $(8\pi/15)a_c^5 \rho$; it is less than the moment of inertia A_c of the actual core because ρ_0 at the CMB is less than in the interior. When we express in terms of A_c the estimate provided by the right-hand side of Eq. (7.96) for the magnitude of the integral (7.95), we see that its order of magnitude is, roughly, of $\mathcal{O}(A_c \epsilon^2 \Omega_0 \omega_f)$. Arguments of a similar nature, with necessary changes, may be used to conclude that the order of magnitude of the second part of the integral in (7.93) is the same as that of the first part. Since the sum of the two parts stands for $\int \mathbf{r} \wedge \mathbf{G} dV$, our aim of estimating the order of magnitude of this quantity is accomplished; the result is that it is smaller than the magnitude of the retained terms in the Sasao *et al.* equation by $\mathcal{O}(\epsilon^2)$ and is therefore indeed negligible, given that the magnitudes of Π_1 and rR_1 are as claimed at the beginning of this paragraph. Our next task is to substantiate this claim.

For this purpose, we need to return to the momentum balance equation (7.65) and rewrite it as an equation for $\mathbf{v}$, which takes the form

$$\dot{\mathbf{v}} + 2\boldsymbol{\Omega} \wedge \mathbf{v} + \dot{\boldsymbol{\Omega}}_f \wedge \mathbf{r} + (\boldsymbol{\Omega} \wedge \boldsymbol{\omega}_f) \wedge \mathbf{r} = -\frac{\nabla P}{\rho} - \nabla \phi_{gc}^f \equiv -\mathbf{G}, \tag{7.97}$$

when terms that are of much smaller magnitude than the first two are neglected.

To arrive at this form, we use $\mathbf{v}_f = \mathbf{v} + \boldsymbol{\omega}_f$ from (7.71), and add $-\nabla\phi_c(\boldsymbol{\Omega}_f)$ to both sides of the equation. This last step carries the term $-\nabla\phi_g$ into $-\nabla\phi_{gc}^f$ and thus converts the right-hand side to $-\mathbf{G}$. As for the left-hand side, consider first the terms which do not involve $\mathbf{v}$. They are seen to be

$$((\boldsymbol{\omega}_f \wedge \mathbf{r}) \cdot \nabla)(\boldsymbol{\omega}_f \wedge \mathbf{r}) + 2\boldsymbol{\Omega} \wedge (\boldsymbol{\omega}_f \wedge \mathbf{r}) + \dot{\boldsymbol{\Omega}}_f \wedge \mathbf{r} - \boldsymbol{\Omega}_f \wedge (\boldsymbol{\Omega}_f \wedge \mathbf{r}) + \boldsymbol{\Omega} \wedge (\boldsymbol{\Omega} \wedge \mathbf{r}),$$

$$(7.98)$$

wherein the penultimate term is just the expression for $-\nabla\phi_c(\boldsymbol{\Omega}_f)$. Noting that the first term reduces to $\boldsymbol{\omega}_f \wedge (\boldsymbol{\omega}_f \wedge \mathbf{r})$ and combining it with the remaining terms, one can readily verify that (7.98) simplifies to

$$\dot{\boldsymbol{\Omega}}_f \wedge (\mathbf{r} + (\boldsymbol{\Omega} \wedge \boldsymbol{\omega}_f) \wedge \mathbf{r}).$$

$$(7.99)$$

The terms involving $\mathbf{v}$ that arise on using (7.71) for $\mathbf{v}_f$ are

$$\dot{\mathbf{v}} + 2\boldsymbol{\Omega} \wedge \mathbf{v},$$

$$(7.100)$$

with the omission of terms that are of significantly smaller magnitude than the above two. The first term $\dot{\mathbf{v}}$ is of $\mathcal{O}(\Omega_0 v)$ because the spectrum of $\mathbf{v}$ is the same as that of the tidal forcing which has frequencies in the diurnal tidal band, very close to $-\Omega_0$; the Coriolis term $2\boldsymbol{\Omega} \wedge \mathbf{r}$ is evidently of the same order. Of the three omitted terms, all arising from $(\mathbf{v}_f \cdot \nabla)\mathbf{v}_f$, the term $\mathbf{v} \cdot (\boldsymbol{\omega}_f \wedge \mathbf{r})$ is clearly of $\mathcal{O}(\omega_f v)$ and so is the term $[(\boldsymbol{\omega}_f \wedge \mathbf{r}) \cdot \nabla]\mathbf{v}$, and these are negligible since $\omega_f \equiv |\boldsymbol{\omega}_f| \ll \Omega_0$; the other omitted term $(\mathbf{v} \cdot \nabla)\mathbf{v}$ is of even smaller magnitude, being of the second order in v.

The sum of the expressions (7.100) and (7.99) yields the left-hand side of (7.97), which is thus established.

Order of magnitude of Π_1 and R_1

We can use Eq. (7.97) now to estimate the order of magnitude of Π_1 and R_1 in terms of that of v. To do this, we perform certain operations on the equation to eliminate the terms involving $\mathbf{r}$ on the left-hand side. An obvious one is to take the scalar product of the equation with $\mathbf{r}$. The result is

$$\mathbf{r} \cdot (\dot{\mathbf{v}} + 2\boldsymbol{\Omega} \wedge \mathbf{v}) = -\mathbf{r} \cdot \mathbf{G} = -r\left(\frac{\partial \Pi_1}{\partial r} + R_1\right),$$

$$(7.101)$$

on noting that $\mathbf{r} \cdot \nabla s \approx r$ in view of (7.88). The quantity on the left-hand side is evidently of $\mathcal{O}(\Omega_0 v r)$. It follows therefore that

$$r\left(\frac{\partial \Pi_1}{\partial r} + R_1\right) \sim \mathcal{O}(\Omega_0 v r).$$

$$(7.102)$$

Secondly, we perform the operation $\mathbf{r} \cdot [\nabla \wedge (\mathbf{r} \wedge \ldots)]$ on the equation. Once again, the unwanted terms on the left side drop out. Moreover, the term involving R_1 is negligible since $\mathbf{r} \wedge \nabla s$ is of $\mathcal{O}(\epsilon)$ as may be seen from (7.92). What remains after the operation may then be written as

$$i\mathbf{M} \cdot \left[\mathbf{r} \wedge \left(\frac{\partial \mathbf{v}}{\partial t} + 2\mathbf{\Omega} \wedge \mathbf{v} \right) \right] = \mathbf{M}^2 \Pi_1, \qquad (7.103)$$

where the operator $\mathbf{M}$ is proportional to the angular momentum operator of quantum mechanical theory:

$$\mathbf{M} = -i\mathbf{r} \wedge \nabla. \qquad (7.104)$$

It acts only on the angular variables θ, λ, and has the property that

$$\mathbf{M}^2 Y_n^m(\theta, \lambda) = n(n+1) Y_n^m(\theta, \lambda). \qquad (7.105)$$

Recall now that all except a very small part of the nutational motion and the associated perturbations (in the pressure, matter distribution, gravitational potential, flows in the core, etc.) is due to the degree two order one (i.e., Y_2^1) component of the potential which varies very slowly with the angular coordinates. The spatial variation of Π_1, which is determined by P_1 and ϕ_1, is expected to be almost wholly of order two, and therefore $\mathbf{M}^2 \Pi_1$ is expected to be close to $6\Pi_1$. The action of $\mathbf{M}$ on the square bracketed quantity on the left-hand side of equation (7.103) is also expected to leave its order of magnitude unchanged; it is $\mathcal{O}(\Omega_0 vr)$. Thus we conclude that

$$\Pi_1 \sim \mathcal{O}(\Omega_0 vr). \qquad (7.106)$$

The variation of Π_1 with r is expected to be smooth, and so $r(\partial \Pi_1 / \partial r)$ should also be of the same order as Π_1, i.e., of $\mathcal{O}(\Omega_0 vr)$. (It is easy to verify these results for the Poincaré flow discussed in Section 7.2.1 where an explicit expression is available for $\mathbf{v}$.) It follows then from (7.102) that $R_1 \sim \mathcal{O}(\Omega_0 v)$.

Since v is of $\mathcal{O}(\epsilon \omega_f r)$, we see finally that both Π_1 and rR_1 are of $\mathcal{O}(\Omega_0 \omega_f r^2)$. It will be recalled that this result was employed in anticipation of a proof while establishing that $\int \rho \mathbf{r} \wedge \mathbf{G} dV$ is of $\mathcal{O}(\epsilon^2)$ negligibly small.

With this, the proof of the SOS form for the torque equation of the fluid core is complete.

7.3.3 Additional coupling torque at the CMB

In the foregoing derivation of the angular momentum balance equation (7.82) for the fluid core, the forces acting on the fluid were derived from the gravitational potential and the pressure, together with additional inertial forces (centrifugal, Coriolis) which appear in the rotating reference frame employed. Sasao *et al.* (1980) considered the possibility that torques other than those generated by these forces might be acting on the fluid core through some mechanism which couples the core to the mantle. The presence of a magnetic field crossing the CMB could produce such a coupling that would tend to impede the differential wobble motion between the mantle and the FOC. The basic reason is that the differential wobble results in a motion of the conducting matter on both sides of the CMB past the magnetic field which threads through both regions, and thereby gives rise to the Lorentz force which causes a "drag" against such motion. Other possible mechanisms which can influence the relative wobble motion between the mantle and the core, such as viscosity of the core fluid and irregular structures (topography) at the CMB, have also been investigated in the literature. These will be considered in a later chapter.

Calculations of the electromagnetic coupling torque at the CMB have been made by Toomre (1974), Sasao *et al.* (1977), Buffett (1992), and Greff-Lefftz and Legros (1999a and b), and more recently by Buffett *et al.* (2002), Koot *et al.* (2010, 2011), and Koot and Dumberry (2013). All but the last of these works ignore the possibility that the flow within the FOC associated with the differential wobble may itself be significantly modified by the Lorentz force. Buffett *et al.* (2002) and Mathews *et al.* (2002) developed the theory applicable to the situation where the magnetic field at the boundary is strong enough that such modifications are not ignorable, and found that the understanding of the observational data on nutations does call for the use of such a theory. The theory is outlined in Chapter 10, where study by these authors on the role of the electromagnetic coupling in improving the fit of the nutation theory to data is also presented.

With the inclusion of a torque, say $\mathbf{\Gamma}_b$, due to electromagnetic and other boundary effects, Eq. (7.82) gets generalized to

$$\frac{d\mathbf{H}_f}{dt} - \boldsymbol{\omega}_f \wedge \mathbf{H}_f = \mathbf{\Gamma}_b. \tag{7.107}$$

The rotational motions of the Earth and its fluid core are then governed by this equation together with the torque equation (7.58) for the whole Earth, which we reproduce here for convenience:

$$\frac{d\mathbf{H}}{dt} + \boldsymbol{\Omega} \wedge \mathbf{H} = \mathbf{\Gamma}. \tag{7.108}$$

This last equation is unaffected by the boundary couplings: the torque $\mathbf{\Gamma}_b$ exerted by the mantle on the core is balanced by an equal and opposite torque $-\mathbf{\Gamma}_b$ due to the core on the mantle.

Both the electromagnetic and viscous interactions are dissipative. Therefore the torque $\mathbf{\Gamma}_b$ that they produce on the fluid core is out of phase with the external forcing torque $\mathbf{\Gamma}$; this will result in a difference in phase between any spectral component of the tidal forcing and the Earth rotation variations that it gives rise to. Secondly, the dissipative coupling $\mathbf{\Gamma}_b$ exists whenever there is relative motion between the body of the core and the mantle, which is the case even in the free wobble modes (i.e., in the absence of any external torque). Consequently it causes the free wobbles and associated nutations to be damped and thus to die away with time, though the degree of damping depends very much on the magnitude of the differential wobble between the core and the mantle in the particular free wobble mode.

7.3.4 Solution of the SOS equations

The angular momentum **H** may be written as the sum of the angular momentum of rotation of the whole Earth with angular velocity $\mathbf{\Omega}$ and the angular momentum **h** of rotation of the core relative to the mantle with the differential angular velocity ω_f; and $\mathbf{H}_f$ may also be written in similar fashion:

$$\mathbf{H} = [C] \cdot \mathbf{\Omega} + \mathbf{h}, \qquad \mathbf{H}_f = [C_f] \cdot \mathbf{\Omega} + \mathbf{h}, \qquad \mathbf{h} = [C_f] \cdot \omega_f, \qquad (7.109)$$

where $[C]$ and $[C_f]$ are the inertia tensors of the whole Earth and of the fluid core, respectively. As for Γ_b, its equatorial part is expected, from physical considerations, to be proportional $(\mathbf{i}_3 \wedge \omega_f)$. For the complex combination $\tilde{\Gamma}_b = \Gamma_{b1} + i\Gamma_{b2}$ of its equatorial components, this proportionality would translate into a relation of the form

$$\tilde{\Gamma}_b = -i K_b A_f \Omega_0 (\omega_{f1} + i\omega_{f2}) = -i K_b A_f \Omega_0^2 \tilde{m}_f, \qquad (7.110)$$

where K_b is a dimensionless coupling constant which represents the strength of the boundary coupling. The parameter K_b is complex, in general, reflecting the dissipative nature of the boundary coupling. Specific examples will be considered in due course, but it may be mentioned here that estimates of K_b obtained from a recent fit of the theoretical results to observational data (Mathews *et al.*, 2002) are about two orders of magnitude smaller than e.

Let us now write down the quantities appearing in the above equations in component form:

$$[C] = \begin{pmatrix} A + c_{11} & c_{12} & c_{13} \\ c_{21} & A + c_{22} & c_{23} \\ c_{31} & c_{32} & C + c_{33} \end{pmatrix}, \quad \mathbf{\Omega} = \begin{pmatrix} \Omega_0 m_1 \\ \Omega_0 m_2 \\ \Omega_0 (1 + m_3) \end{pmatrix}, \qquad (7.111)$$

$$[C_f] = \begin{pmatrix} A_f + c_{11}^f & c_{12}^f & c_{13}^f \\ c_{21}^f & A_f + c_{22}^f & c_{23}^f \\ c_{31}^f & c_{32}^f & C_f + c_{33}^f \end{pmatrix}, \quad \omega_f = \begin{pmatrix} \Omega_0 m_1^f \\ \Omega_0 m_2^f \\ \Omega_0 m_3^f \end{pmatrix}. \qquad (7.112)$$

The c_{ij} and c_{ij}^f arise from the deformations produced by the direct action of the tidal potential and by the centrifugal potentials associated with the rotation variations represented by the wobbles m_1, m_2 of the mantle and the differential wobble m_1^f, m_2^f of the core, as well as the axial rotation rate variations m_3, m_3^f of the two regions. We take all these perturbation quantities to be smaller than unity by several orders of magnitude. Terms which are of the second (or higher) order in these quantities will be neglected while using Eqs. (7.109) together with (7.111) and (7.112) for obtaining explicit expressions for $\mathbf{H}$, $\mathbf{H}_f$, etc. In this linear approximation, we readily obtain

$$\mathbf{h} = \begin{pmatrix} h_1 \\ h_2 \\ h_3 \end{pmatrix} = \Omega_0 \begin{pmatrix} A_f m_1^f \\ A_f m_2^f \\ C_f m_3^f \end{pmatrix}, \tag{7.113}$$

$$\mathbf{H} = \begin{pmatrix} (Am_1 + c_{1,3})\Omega_0 + h_1 \\ (Am_2 + c_{2,3})\Omega_0 + h_2 \\ (C(1+m_3) + c_{3,3})\Omega_0 + h_3 \end{pmatrix}, \quad \mathbf{H}_f = \begin{pmatrix} (A_f m_1 + c_{1,3}^f)\Omega_0 + h_1 \\ (A_f m_2 + c_{2,3}^f)\Omega_0 + h_2) \\ (C_f(1+m_3) + c_{3,3}^f)\Omega_0 + h_3 \end{pmatrix}, \tag{7.114}$$

$$\mathbf{\Omega} \wedge \mathbf{H} = \Omega_0 \begin{pmatrix} ((C-A)m_2 - c_{2,3})\Omega_0 - h_2 \\ ((A-C)m_1 + c_{1,3})\Omega_0 + h_1 \\ 0 \end{pmatrix}, \quad \omega_f \wedge \mathbf{H}_f = \Omega_0^2 C_f \begin{pmatrix} m_2^f \\ -m_1^f \\ 0 \end{pmatrix}. \tag{7.115}$$

Of the set of three equations that are obtained on introducing these expressions into the torque equation (7.108), the first two may be combined in the familiar fashion by adding i times the second to the first. The result is

$$\frac{d}{dt}(A\tilde{m} + \tilde{c}_3 + A_f \tilde{m}_f)\Omega_0 - i(Ae\tilde{m} - \tilde{c}_3 - A_f \tilde{m}_f)\Omega_0^2 = \tilde{\Gamma}(t), \tag{7.116}$$

where $\tilde{m} = m_1 + im_2$, etc., as usual, and

$$\tilde{c}_3 = c_{13} + ic_{23}, \qquad \tilde{c}_3^f = c_{13}^f + ic_{23}^f. \tag{7.117}$$

In a similar fashion, Eq. (7.82) leads to

$$\frac{d}{dt}[A_f(\tilde{m} + \tilde{m}_f) + \tilde{c}_3^f]\Omega_0 + i(1 + e_f + K_b)A_f \tilde{m}_f \Omega_0^2 = 0, \tag{7.118}$$

where K_b is the dimensionless constant representing the CMB coupling, as defined in (7.110).

Note that the dynamical ellipticities e and e_f of the whole Earth and of the fluid core appear by virtue of their definitions: $(C - A) = Ae$ and $(C_f - A_f) = A_f e_f$.

The torque $\tilde{\Gamma}$ is known from Section 2.23.5; see, in particular, Eq. (2.62), where it is expressed in the succinct form

$$\tilde{\Gamma} = -iAe\Omega_0^2\tilde{\phi}, \qquad \tilde{\phi} = \phi_1 + i\phi_2, \qquad (7.119)$$

the dimensionless quantity $\tilde{\phi}$ being

$$\tilde{\phi} = (3GM_B/d^5\Omega_0^2)\,d_z(d_x + id_y) = (3GM_B/d^3\Omega_0^2)\sin\beta\cos\beta e^{i\Lambda}. \qquad (7.120)$$

The factor $\tilde{\phi}$ in the torque, which is a function of time because of the orbital motion of the external body, comes from the degree two tesseral part of the potential of the body. This potential, which is given by Eq. (2.63), is

$$W_{B2}(\mathbf{r})|_{m=1} = -\Omega_0^2 z(\phi_1 x + \phi_2 y). \qquad (7.121)$$

It is the real part of the complex potential

$$-\Omega_0^2\tilde{\phi}\,z(x - iy). \qquad (7.122)$$

This complex potential is widely used in computations, with the understanding that the physical results are given by the real part of the resulting expressions. (It turns out that the complex conjugate part $[-\Omega_0^2\tilde{\phi}\,z(x + iy)]$, which is proportional to $Y_2^1(\theta, \lambda)$, does not contribute to the torque on an axially symmetric Earth because the torque generated by it is proportional to the triaxiality parameter. (The reader may verify this by observing that in the potential (5.205), $Y_2^1(\theta, \lambda)$ appears multiplied by F_{21}^*, and that the term containing this factor in the torque (5.233) has the triaxiality quantities C_{22} and S_{22} as coefficients.)

For proceeding further with equations (7.116) and (7.118), it is convenient to go over to the spectral domain by taking

$$\tilde{\phi} = \tilde{\phi}(\sigma)\,e^{i\sigma\Omega_0 t}, \qquad \tilde{m} = \tilde{m}(\sigma)\,e^{i\sigma\Omega_0 t}, \qquad \tilde{m}_f = \tilde{m}_f(\sigma)\,e^{i\sigma\Omega_0 t}, \qquad (7.123)$$

and similarly for $\tilde{c}_3$, $\tilde{c}_3^f$, and $\tilde{\Gamma}$, ignoring the initial phase. Expressions for $\tilde{c}_3$ and $\tilde{c}_3^f$ in terms of compliance parameters are taken from Eqs. (6.167) and (6.169), dropping the initial phase. As we are considering an Earth model without an inner core here, the terms containing $\tilde{m}_s$ in those equations are to be omitted. On substituting the resulting expressions into Eqs. (7.116) and (7.118) and introducing $\tilde{\Gamma}$ from (7.119), we are led to the following coupled algebraic equations for $\tilde{m}(\sigma)$ and $\tilde{m}_f(\sigma)$:

$$[(1 + \kappa)\sigma - (e - \kappa)]\tilde{m}(\sigma) + (1 + \sigma)(A_f/A + \xi)\tilde{m}_f(\sigma) = -[e - (1 + \sigma)\kappa]\tilde{\phi}(\sigma), \qquad (7.124)$$

$$(1 + \gamma)\sigma\tilde{m}(\sigma) + [(1 + \beta)\sigma + 1 + e_f + K_b]\tilde{m}_f(\sigma) = \gamma\sigma\tilde{\phi}(\sigma). \qquad (7.125)$$

The moments of inertia, ellipticities, compliance parameters, and the CMB coupling parameter K_b which appear in the above equations are often referred to collectively as the basic Earth parameters. More such parameters appear, of course, in the equations for a three-layer Earth.

It is useful to note, before going on to the solution of the equations for arbitrary σ, that the values of the compliances are of the same order as, or smaller than, the ellipticities e and e_f. The moment of inertia ratio A_f/A is approximately 1/9. The values of e and e_f computed for a hydrostatic equilibrium Earth based on the model PREM for the Earth's structure, and the values obtained for the compliances from solutions of the tidal deformation equations for the same Earth model, are as follows:

$$e = 0.003247, \qquad e_f = 0.002548,$$

$$\kappa = 0.001039, \qquad \gamma = 0.001965, \qquad (7.126)$$

$$\xi = 0.000222, \qquad \beta = 0.000616.$$

(Work of the past two decades comparing nutation theory with observations has necessitated small revisions of some of the values, as explained later.) In writing down the solutions below, we make the approximation of neglecting quantities that are of second order in the above Earth parameters.

7.3.5 Normal modes of the two-layer Earth

In the absence of external forcing (i.e., if $\tilde{\phi}(\sigma)$ is equal to zero), non-trivial solutions of the equations exist just for those values of σ which make the determinant of coefficients of $\tilde{m}$ and $\tilde{m}_f$ vanish. These are the eigenfrequencies of the system, and the corresponding solutions give the wobble normal modes (i.e., free wobble modes or eigenmodes). It is evident that the determinant is quadratic in σ; so there are just two normal modes. In the approximation just mentioned, the two eigenfrequencies are

$$\sigma_1 = \sigma_{\text{CW}} = \frac{A}{A_m}(e - \kappa), \qquad (7.127)$$

$$\sigma_2 = \sigma_{\text{NDFW}} \equiv -1 + \nu_2 = -1 - \frac{A}{A_m}(e_f + K_b - \beta), \qquad (7.128)$$

where ν_2 is ν_{FCN}, the frequency of the FCN mode in space, and $A_m \equiv A - A_f$ is the moment of inertia of the mantle about an equatorial axis. It may be readily verified that if either of the above values is used for σ, the determinant of coefficients does vanish, as required, on neglecting second-order quantities. The determinant itself may be seen to be

$$D(\sigma) = (A_m/A)(\sigma - \sigma_1)(\sigma - \sigma_2). \qquad (7.129)$$

It is worth observing from the expressions (7.127) and (7.128) for the eigenfrequencies that the effect of deformability is to cause the ellipticities to be replaced by reduced "effective" values in both.

The first of the two normal modes is the Chandler wobble (CW) with the frequency σ_1. It goes over into the Eulerian free wobble of frequency $\sigma_E = e$ cpsd for the rigid Earth. The deformability of the Earth, represented by the compliance κ which has a value $\approx e/3$, causes the eigenfrequency σ_1 to be reduced by about one-third. The low value ($\approx 1/400$ cpsd) of the CW frequency σ_1 places it well outside the retrograde diurnal band of the tidal excitation frequencies. Therefore this normal mode plays only a very minor role in the wobble response of the Earth in this frequency band. In the free CW mode, the differential wobble of the core is very small: $\tilde{m}_f/\tilde{m} \approx \sigma_{CW}$.

The picture is quite different in the second eigenmode, the so-called nearly diurnal free wobble mode (NDFW), in which $\tilde{m}_f/\tilde{m} \approx -(A_m/A_f)/(e_f + K_b - \beta)$. So $\tilde{m}_f$ is about 4000 times as large as $\tilde{m}$. The NDFW eigenfrequency is in the middle of the retrograde diurnal band, which gives it an important role in the forced wobbles. With the values quoted above for the Earth parameters, and with $A/A_m \approx 1.128$ one finds that $\sigma_{NDFW} \approx (-1 - 1/460)$ cpsd. The nutation mode associated with the NDFW is the FCN. Its frequency (in a space-fixed reference frame) is $\nu_{FCN} \equiv 1 + \sigma_{NDFW} \approx -1/460$ cpsd. The presence of the complex parameter K_b in σ_2 results in damping of this mode as a result of the dissipative nature of the CMB coupling torque $\mathbf{\Gamma}_b$. Incidentally, the apparent absence of any damping of the CW mode is a consequence of the very small magnitude of the differential wobble at the CW frequency.

7.3.6 *Forced wobbles*

Consider now the wobbles induced by a tidal perturbation $\tilde{\phi}(\sigma)$.

To begin with, we look at the important special case of the forcing torque component with frequency $\sigma = -1$ which corresponds to a tidal wave that is stationary as viewed from space (frequency in space $= \nu = \sigma + 1 = 0$). In this case, the $\tilde{m}_f$ term drops out of Eq. (7.124), and we see immediately that

$$\tilde{m}(-1) = \frac{e}{1+e}\tilde{\phi}(-1) = H_d\,\tilde{\phi}(-1), \qquad (7.130)$$

which is exactly the same as for a rigid Earth having the same value for e. Deformability and the presence of the core make no difference at this frequency: e is the only one of the basic Earth parameters that is relevant for the wobble at the exactly retrograde diurnal frequency $\omega_w = -\Omega_0$ or $\sigma = -1$. This property, which we encountered already in the special case of the Poincaré model, is what Poincaré (1910) characterized as *"gyrostatic rigidity"* (see Section 7.2). Using the above value of $\tilde{m}(-1)$ in Eq. (7.125), we obtain

$$\tilde{m}_f(-1) = \frac{(e - \gamma)}{(1 + e)(e_f - \beta + K_b)} \tilde{\phi}(-1). \qquad (7.131)$$

An important point to be noted is that gyrostatic rigidity does not imply equality of the wobble amplitudes of the mantle and the core at $\sigma = -1$. The ratio $\tilde{m}_f(-1)/\tilde{m}(-1)$ is seen to be about 200 if the complex coupling constant K_b is ignored. (Recent studies, considered in Chapter 10, gave estimates that are roughly of the order of 10^{-5} for the real and imaginary parts of K_b.)

Recall now the relation (2.106) between the rate of precession in longitude and the wobble amplitude $\tilde{m}$ at $\omega_w = -\Omega_0$. With a change in the argument of $\tilde{m}$ from ω_w to its equivalent σ, this relation reads:

$$\dot{\psi}_A \sin \epsilon_0 = \Omega_0 \tilde{m}(-1) = \frac{e\Omega_0}{1 + e} \tilde{\phi}(-1). \qquad (7.132)$$

Thus the precession rate of a non-rigid Earth is just the same as for a rigid Earth having the same ellipticity.

Going over now to the solution of Eqs. (7.124) and (7.125) for an arbitrary σ, we find readily that

$$\tilde{m}(\sigma) = (M_{22}y_1 - M_{12}y_2)\tilde{\phi}(\sigma)/D(\sigma), \qquad (7.133)$$

$$\tilde{m}_f(\sigma) = (M_{11}y_2 - M_{21}y_1)\tilde{\phi}(\sigma)/D(\sigma), \qquad (7.134)$$

where

$$M_{11} = (1 + \kappa)\sigma - (e - \kappa), \qquad M_{12} = (1 + \sigma)(A_f/A + \xi), \qquad (7.135)$$

$$M_{21} = (1 + \gamma)\sigma, \qquad M_{22} = (1 + \beta)\sigma + 1 + e_f + K_b, \quad (7.136)$$

$$y_1 = -(e - \kappa) + \sigma\kappa, \qquad y_2 = \gamma\sigma, \qquad (7.137)$$

and $D(\sigma) = M_{11}M_{22} - M_{12}M_{21}$ may be factorized as in (7.129).

The point of greatest interest, considering that the tidal excitation frequencies are in the retrograde diurnal band, is the presence of the NDFW eigenmode in the middle of this band. The factor $1/(\sigma - \sigma_{\text{NDFW}})$ in $1/D(\sigma)$ becomes very large when the frequency σ of the forcing is close to σ_{NDFW}; the result is what is called a "resonant enhancement" of the amplitudes of the forced wobbles at such frequencies. The most important case is that of the wobble due to the ψ_1 tide with a frequency of $-1 - 1/366.26$ cpsd; it is associated with the retrograde annual nutation of period -365.26 solar days (equivalent to 366.26 sidereal days). The amplitude of this nutation according to rigid Earth theory is about -25 mas while the observed value for the non-rigid Earth is about -33 mas. The FCN resonance is primarily responsible for the difference.

7.3.7 Transfer function; resonance formula

The ratio of the amplitude of the wobble (or nutation) of the non-rigid Earth to that of the rigid Earth, as a function of the frequency of excitation, is the transfer function, defined earlier (see e.g. Eq. (2.51)). If the Earth were rigid, the wobble amplitude would be

$$\tilde{m}_R(\sigma) = e\tilde{\phi}(\sigma)/(e - \sigma), \qquad (7.138)$$

as has been noted before. One may verify this directly from Eq. (7.124) by setting $\tilde{m}_f = 0$ (there is no core) and $\kappa = 0$. The transfer function is then

$$T(\sigma) = \frac{(M_{22}y_1 - M_{12}y_2)(e - \sigma)}{(M_{11}M_{22} - M_{21}M_{12})e} = \frac{(M_{22}y_1 - M_{12}y_2)(e - \sigma)}{e(A_m/A)(\sigma - \sigma_1)(\sigma - \sigma_2)}. \qquad (7.139)$$

This expression will suffice for computation of the amplitude of any nutation for the non-rigid Earth, given the corresponding amplitude for the rigid Earth and the values of the basic Earth parameters appearing in the M_{ij} and y_i.

However it is possible to express $T(\sigma)$ in the form of a *resonance expansion* which helps very much to bring out the importance (or otherwise) of the individual parameters in determining the nutation amplitudes of the non-rigid Earth, as we shall see.

The numerator of $T(\sigma)$ is a cubic polynomial in σ, while the denominator is quadratic. Therefore one can express $T(\sigma)$ in the form

$$T(\sigma) = R + R'(1 + \sigma) + \frac{R_1}{\sigma - \sigma_1} + \frac{R_2}{\sigma - \sigma_2}, \qquad (7.140)$$

which is the resonance formula. The last two terms in the formula represent the CW and NDFW resonances, respectively. The coefficients R, R', R_1, R_2 are to be determined by equating (7.140) to the expression (7.139). It may be readily seen by multiplying the resonance expansion by $(\sigma - \sigma_i)$, $(i = 1, 2)$, and then setting $\sigma = \sigma_i$, that

$$R_i = [(\sigma - \sigma_i)T(\sigma)]_{\sigma = \sigma_i}, \qquad (i = 1, 2). \qquad (7.141)$$

R_1 and R_2 may now be evaluated by taking $T(\sigma)$ from equation (7.139) after introducing the expressions (7.135)–(7.137) into it. One finds, after some algebraic reductions, that:

$$R_1 = -\sigma_1(1 - \sigma_1/e), \tag{7.142}$$

$$R_2 = (A_f/A_m)(1 - \gamma/e)v_2, \tag{7.143}$$

to the lowest non-vanishing order in each case, remembering that σ_1 and v_2 are of $\mathcal{O}(\epsilon)$.

Having determined R_1 and R_2, we can use the gyrostatic rigidity property, verified earlier in this section, to obtain R. This property says that $\tilde{m}(\sigma)$ and $\tilde{m}_R(\sigma)$ are equal at $\sigma = -1$, which means that $T(-1) = 1$. Consequently, we see from (7.140) that

$$R = 1 + R_1/(1 + \sigma_1) + R_2/(1 + \sigma_2) \approx 1 + R_1 + R_2/v_2 \tag{7.144}$$

to the lowest order.

Finally we note that $T(\sigma) = 0$ at $\sigma = e$ because of the factor $(\sigma - e)$ in (7.139). On using this fact in (7.140) we obtain

$$R'(1 + e) = -R - R_1/(e - \sigma_1) - R_2/(e - \sigma_2) \approx -1 + (\sigma_1/e) - R_2/v_2, \tag{7.145}$$

on substituting for R from (7.144) and R_1 from (7.142) and neglecting relatively small terms.

The coefficients R, R', R_1, R_2 determined above may of course be written in terms of the basic Earth parameters by using the expressions (7.127) and (7.128) for σ_1 and v_2.

Since the transfer function for the nutations is equal to that for the wobbles, we may use $T(\sigma)$ of (7.140) with the replacements $(1 + \sigma) \to v$ and $(\sigma - \sigma_i) \to (v - v_i)$ to multiply the rigid Earth nutation amplitude for obtaining the non-rigid Earth amplitude.

The important role that resonances can play in nutations of the non-rigid Earth can be readily gauged from the fact that the frequency $v_2 = 1 + \sigma_2$ of the FCN mode as given by Eq. (7.128) is in the midst of the low frequency nutation band. On using the PREM values shown in (7.126) for the various parameters (along with $A/A_m \approx 1.128$), one finds the value of v_2 to be about $-1/459$ cpsd, which is quite close to the frequency of the retrograde annual nutation ($-1/366$ cpsd.) Evaluation of the FCN resonance term $R_2/(v - v_2)$ with R_2 taken from Eq. (7.143) shows that it is about 0.20. The Chandler resonance term and the R' term in the transfer function are both very small; but the frequency independent "background" term is about 1.05. So the transfer function is about 1.25 at this frequency, which would make the amplitude of the retrograde annual nutation about 25% higher for the non-rigid two-layer Earth than for the rigid Earth. Estimates obtained around 1985 from VLBI observations showed that the actual amplitude is over 30% higher than for the rigid Earth. Analyzing the possible reasons for this increase on the basis of the dependence of the FCN resonance frequency on the Earth parameters, Gwinn *et al.* (1986) and Herring *et al.* (1986a and b) concluded that the explanation has to be that the dynamical ellipticity e_f of the fluid core is actually about 5% higher than for the hydrostatic equilibrium Earth. This explanation has become generally accepted, thus establishing the potential that nutation studies hold for obtaining certain types of information about the interior of the Earth. Subsequent theoretical work, aimed at obtaining the closest fit to the ever-increasing amount of observational data of increasing precision and at making high-accuracy predictions possible, has resulted in additional information being obtained about the Earth's interior. Such work is outlined later in this book.

7.4 Earth with fluid outer core and solid inner core

The presence of the SIC as the innermost region of the Earth surrounded by the FOC necessitates, in principle, a generalization of the treatment given so far. Theoretical formulations including the role of the inner core have been given by Mathews *et al.* (1991a and b), de Vries and Wahr (1991), and Dehant *et al.* (1993a and b) (see also Legros *et al.*, 1993). They ignored possible couplings of the FOC to the mantle and the SIC due to gravimetric, electromagnetic, viscous, and topographic effects at the CMB and ICB or between the layers. The following presentation is based on the first of these papers. The ignored boundary couplings are taken up for consideration in Chapter 10.

The moment of inertia of the SIC is only about 1/1400 of that of the whole Earth and about 1/150 of that of the FOC. So it might appear at first sight that the inner core should not have any significant effect on nutations. However, one expects additional normal modes to appear, and one has to consider the possibility that the frequency of a new normal mode may be very close to one of the tidal excitation frequencies, evoking thereby a considerably enhanced nutational response. In fact, recent work indicates that this does happen, and that the SIC plays a non-trivial role in the understanding of the Earth's nutations.

Because the SIC is surrounded by a fluid medium, the possibility exists for its symmetry axis i_s to move around its equilibrium direction which coincides with that of the symmetry axis i_3 of the mantle. This freedom implies that the introduction of this region into the theoretical formalism brings with it two new degrees of freedom: the misalignment

$$\mathbf{i}_s - \mathbf{i}_3 = \Omega_0 \mathbf{n}_s \tag{7.146}$$

between the two axes, and the differential wobble $\mathbf{m}_s$ between the inner core and the mantle:

$$\mathbf{\Omega}_s - \mathbf{\Omega} = \Omega_0 \mathbf{m}_s. \tag{7.147}$$

Note that the subscript s identifies quantities pertaining to the *solid* inner core.

7.4.1 Inertia tensors, angular momenta, and equations of motion

Tilt of the inner core has the effect of modifying the inner boundary of the FOC, by changing its symmetry axis from $\mathbf{i}_3$ to $\mathbf{i}_s$. A little reflection shows that the change in the inertia tensor of the FOC that is caused thereby is equal to the difference between the inertia tensors in the untilted and tilted configurations of a body that is identical to the SIC but has a uniform density ρ_f equal to that of the core fluid near the ICB. This difference may be written as $(C' - A')(\mathbf{i}_3\mathbf{i}_3 - \mathbf{i}_s\mathbf{i}_s)$ where C', A' are the principal moments of inertia of the body defined as above; the products of vectors that appear in this expression are tensor products. Similar considerations show that $(C_s - A_s)(\mathbf{i}_s\mathbf{i}_s - \mathbf{i}_3\mathbf{i}_3)$ has to be added to the inertia tensor of the untilted inner core to obtain that of the tilted one. The inertia tensor of the whole Earth is changed by the sum of the above two increments. The modified inertia tensors, which we shall continue to denote by $[C]$, $[C_f]$, and $[C_s]$, now include the tilt contributions as above; the contribution to the ij element of $[C_f]$, for instance, is

$$(C' - A')\mathbf{i}_i \cdot ((\mathbf{i}_3\mathbf{i}_3 - \mathbf{i}_s\mathbf{i}_s) \cdot \mathbf{i}_j = -(C' - A')\Omega_0[\delta_{i3}n_{sj} + n_{si}\delta_{3j}], \tag{7.148}$$

to the first order in the components of $\mathbf{n}_s$ (of which the third component is already of the second order in the first two and is hence negligible). The net effect of this contribution is therefore to add $-(C' - A')n_{s1}$ to c_{12}^f and c_{21}^f and $-(C' - A')n_{s2}$ to c_{23}^f and c_{32}^f in (7.112). The tilt contribution to $[C_s]$ (which has, in the absence of the inner core tilt, the same structure as that of $[C_f]$ except for the replacement of the subscripts/superscripts f by s), may be written down similarly; so also for the contribution to $[C]$.

The angular momentum vectors $\mathbf{H}$, $\mathbf{H}_f$, and $\mathbf{H}_s$ are now

$$\mathbf{H}_s = [C_s] \cdot \mathbf{\Omega}_s, \quad \mathbf{H}_f = [C_f] \cdot \mathbf{\Omega}_f, \quad \mathbf{H} = [C] \cdot \mathbf{\Omega} + [C_f] \cdot \boldsymbol{\omega}_f + [C_s] \cdot \boldsymbol{\omega}_s,$$

$$\tag{7.149}$$

where $\Omega_f = \Omega + \omega_f$ and $\Omega_s = \Omega + \omega_s$ are the angular velocity vectors of the FOC and the SIC. It must be kept in mind that though we continue to use the same symbols as before for the inertia tensors, they include the tilt contributions now. With this caution, we can continue to use the equations of motion (7.108) and (7.82) for the angular momenta of the whole Earth and the FOC. The equation for the SIC is again of the standard form:

$$\frac{d\mathbf{H}_s}{dt} + \Omega \wedge \mathbf{H}_s = \Gamma_s, \qquad (7.150)$$

where Γ_s is the sum of the torque due to the gravitational action of the external tidal potential and that of the mantle and the core on the body of the inner core, and the torque exerted by the fluid pressure acting on the ellipsoidal ICB. The derivation of the expression for Γ_s is rather complicated and will not be presented here. The interested reader can find it in Mathews *et al.* (1991a, see also de Vries and Wahr, 1991). It is shown there that the complex combination $\tilde{\Gamma}_s \equiv \Gamma_{s1} + i\Gamma_{s2}$ of the equatorial components of the torque is:

$$\tilde{\Gamma}_s = i\Omega_0^2 A_s e_s [-\alpha_1(\tilde{m} + \tilde{m}_f) + \alpha_2 \tilde{n}_s - \alpha_3 \tilde{\phi}] + i\Omega_0^2 \tilde{c}_3^s, \qquad (7.151)$$

where

$$\alpha_1 = \frac{C' - A'}{C_s - A_s} = \frac{A'e'}{A_s e_s}, \qquad \alpha_3 = 1 - \alpha_1,$$

$$\alpha_2 = \alpha_1 - \alpha_3 \alpha_g, \qquad \alpha_g = \frac{8\pi G}{5\Omega_0^2} \left(\int_{a_s}^a \rho_0(s)\frac{d\epsilon(s)}{ds}ds + \rho_f \epsilon_s \right). \qquad (7.152)$$

No explanation is needed regarding the meaning of α_1 and α_3. The parameter α_g determines the strength of the coupling between a tilted SIC and the rest of the Earth contained between the radius a_s of the SIC and the outer radius a of the Earth; ρ_f in the expression for α_g is the fluid density near the ICB, to which $(C' - A')$ is proportional, and ϵ_s is the geometrical flattening of the ICB; the integral term in α_g involves the geometrical flattening $\epsilon(s)$ of the ellipsoidal constant density surfaces within the equilibrium Earth, characterized by Eq. (7.85), and $\rho_0(s)$ is the density at these surfaces. For the PREM Earth model, $\alpha_1 \approx 0.946$, $\alpha_2 \approx 0.829$, $\alpha_g \approx 2.175$.

The three angular momentum balance Eqs. (7.108), (7.82), and (7.150) for the whole Earth, the FOC alone, and the SIC alone, are not sufficient to solve for the four variables $\tilde{m}, \tilde{m}_f, \tilde{m}_s$, and $\tilde{n}_s$. The needed fourth equation is provided by the kinematic relation which connects the motion of the axis $\mathbf{i}_s$ of the SIC relative to the mantle-fixed reference frame when its angular velocity of rotation relative to this frame is ω_s:

$$\frac{d\mathbf{i}_s}{dt} = \omega_s \wedge \mathbf{i}_s \approx \omega_s \wedge \mathbf{i}_3, \tag{7.153}$$

which reduces to

$$\frac{d\mathbf{n}_s}{dt} = \Omega_0(\mathbf{m}_s \wedge \mathbf{i}_3) \quad \rightarrow \quad \frac{d\tilde{n}_s}{dt} = -i\Omega_0\tilde{m}_s. \tag{7.154}$$

We can now write the equatorial components of the angular momentum balance equations in complex form as usual. The off-diagonal elements of the inertia tensors appear in the equations through $\tilde{c}_3, \tilde{c}_3^f, \tilde{c}_3^s$ and these may be replaced by expressions similar to (7.117). Additional terms $\zeta\tilde{m}_s$ and $\delta\tilde{m}_s$ have to be included in $\tilde{c}_3$ and $\tilde{c}_3^f$, respectively, to take account of the deformations due to the centrifugal effect of the differential wobble of the SIC; and we also have

$$\tilde{c}_3^s = A_s[\theta(-\tilde{\phi} + \tilde{m}) + \chi\tilde{m}_f + \tau\tilde{m}_s]. \tag{7.155}$$

The coefficients $\zeta, \delta, \theta, \chi, \tau$ in this context are all compliances. (All these parameters have values below 10^{-4}; δ and χ are negative.)

The set of four equations including (7.154) may now be written, in the frequency domain, as a matrix equation,

$$M(\sigma)x(\sigma) = \tilde{\phi}(\sigma)y(\sigma), \tag{7.156}$$

where

$$M = \begin{pmatrix} \sigma + y_1 & (1+\sigma)\left(\xi + A_f/A\right) & (1+\sigma)\left(\zeta + A_s/A\right) & (1+\sigma)\alpha_3 e_s A_s/A \\ \sigma + y_2 & (1+\sigma) + \beta\sigma + e_f & \delta\sigma & -\sigma\alpha_1 e_s A_s/A_f \\ \sigma + y_3 & \sigma\chi + \alpha_1 e_s & 1+\sigma+\sigma\tau & (1+\sigma-\alpha_2)e_s \\ 0 & 0 & 1 & \sigma \end{pmatrix}, \tag{7.157}$$

$$x = \begin{pmatrix} \tilde{m} \\ \tilde{m}_f \\ \tilde{m}_s \\ \tilde{n}_s \end{pmatrix}, \quad \text{and} \quad y = \begin{pmatrix} y_1 \\ y_2 \\ y_3 \\ y_4 \end{pmatrix} = \begin{pmatrix} (1+\sigma)\kappa - e \\ \gamma\sigma \\ \sigma\theta - \alpha_3 e_s \\ 0 \end{pmatrix}. \tag{7.158}$$

It may be kept in mind that the electromagnetic and viscous boundary couplings at the CMB and ICB have been excluded here, as mentioned in the opening paragraph of Section 7.4.

7.4.2 Normal modes

The eigenfrequencies σ_i are the solutions of the equation $\det(M) = 0$. Since this equation is quartic in σ, there are four eigenfrequencies and corresponding normal modes. One may search for solutions (like the CW of the simpler problem) with $|\sigma_i| \ll 1$ and for others, like the NDFW, with σ_i close to -1; for the latter category, one writes $\sigma_i = -1 + \nu_i$ with $|\nu_i| \ll 1$. It turns out that there are two solutions of each type, with the following frequency values:

$$\sigma_1 = \sigma_{CW} = \left(\frac{A}{A_m}\right)(e - \kappa), \tag{7.159}$$

$$\sigma_2 = \sigma_{NDFW} = -1 - \left(1 + \frac{A_f}{A_m}\right)(e_f - \beta), \tag{7.160}$$

$$\sigma_3 = \sigma_{PFCN} = -1 + \left(1 + \frac{A_s}{A_m}\right)(\alpha_2 e_s + \tau), \tag{7.161}$$

$$\sigma_4 = \sigma_{ICW} = (1 - \alpha_2)e_s = \alpha_3(1 + \alpha_g)e_s, \tag{7.162}$$

on neglecting relatively small terms.

The expression for σ_{CW} is the same as we had in the absence of the SIC; that for σ_{NDFW} does not contain K_b now since we have ignored the CMB coupling, and otherwise differs only very slightly (through the factor involving the moments of inertia) from the value in (7.143). Equations (7.161) and (7.162) give the frequencies of the two new wobble normal modes. The nutation associated with the first of these has a *positive* frequency $\nu_3 \equiv 1 + \sigma_3$, meaning that the nutational motion in space is prograde (unlike the FCN with frequency $\nu_2 \equiv 1 + \sigma_2$, which is retrograde); this is the reason for the terminology PFCN (prograde free core nutation). This mode is known alternatively as FICN (free inner core nutation), a term which emphasizes the fact that, in this mode, the inner core has a very much larger amplitude than the other regions. (Strictly speaking, the notation σ_{PFCN} is inappropriate because PFCN refers to the nutation rather than the wobble; but we follow the current practice here.) With parameter values pertaining to Earth model PREM, the FCN and PFCN periods in space turn out to be about -456 days and $+476$ days, respectively. The ICW period is about 2400 days.

7.4.3 Forced wobbles and nutations; transfer functions

The transfer function is an immediate extension of (7.140) to include the resonance terms associated with the additional normal modes of a three-layer Earth. We write it here in terms of the frequency ν of the forced nutation and the nutation eigenfrequencies ν_i:

$$T(\nu) = R + R'\nu + \sum_{i=1}^{4} \frac{R_i}{\nu - \nu_i}. \tag{7.163}$$

The two new modes (PFCN and ICW) owe their existence to the SIC, which has a very small moment of inertia, approximately 1/1400 of that of the whole Earth. Therefore it is expected, a priori, that the resonance coefficients R_3 and R_4 would be correspondingly small and that these modes are unlikely to have any significant effect on the nutations. This perception must certainly be justified for the ICW with its frequency very much outside the retrograde diurnal band. There is a chance, however, that the PFCN (or FICN) which has a frequency within this band, not far from -1 cpsd, could cause a detectable resonant enhancement of the amplitude of a forced nutation if its frequency is close enough to that of this normal mode. The work of Mathews *et al.* (2002), Buffett *et al.* (2002), and Herring *et al.* (2002), which takes account of a number of small effects not considered in the basic theory of this section and optimizes the values of a few of the Earth parameters for best fit between theory and observational data on nutations and precession, suggests that this possibility is in fact realized (see Chapter 10).

The coefficients R_1 and R_2 in the above formula may be shown to differ only insignificantly from the expressions (7.142) and (7.143) derived for the two-layer case. The expressions for R_3 and R_4 in terms of the basic Earth parameters are rather complicated; they do not reveal much useful information and will not be presented here.

The coefficients R, R', and the R_i obey sum rules that are of the same form as in Eqs. (7.144) and (7.145) except for the presence of additional terms involving R_3 and R_4. These relations (7.144) and (7.145) may then be used to express R and R' in terms of the resonance frequencies σ_i and the coefficients R_i. The result is a new form for the resonance formula:

$$T(\sigma; e) = \frac{e - \sigma}{e + 1} \left(1 + (1 + \sigma) \sum_{i=1}^{4} \frac{N_i}{\sigma - \sigma_i} \right). \qquad (7.164)$$

In the literature, another notation can be found for the parameters in the expression of the system (7.154) or (7.156) with the matrix (7.157). Instead of using the above-introduced compliances, *generalized Love numbers* have been used (see for instance, Dehant *et al.*, 2003a and b, Greff-Lefftz *et al.*, 2002), knowing that mass redistribution can be represented by a Love number of k-type and deformation by a Love number of h-type.

Love numbers express deformation or mass redistribution potential induced by a global potential or by a pressure (in that case they are denoted by a bar over).

If the mass redistribution is computed at the CMB or if the deformation is at the CMB, one indicates it with a superscript f; if it is at the surface, there is no superscript; and if it is at the ICB, one indicates it with a superscript s.

If the acting potential is induced by the mass redistribution inside the core or if the pressure is at the CMB, one applies a subscript 1. If the acting potential is induced by the mass redistribution inside the inner core or if the pressure is at the ICB, one applies a subscript 2. No superscript means acting potential or pressure at the surface.

These Love numbers are computed for a spherical Earth and are the so-called "static Love numbers" (they are not frequency dependent).

These generalized Love numbers that are used for the expression of the wobble of a three-layered Earth are:

- k used to express the mass redistribution potential induced by an external potential;
- h used to express the surface deformation induced by an external potential;
- $\bar{h}$ used to express the surface deformation induced by a surface pressure;
- k^f used to express the mass redistribution potential computed at the CMB induced by an external potential;
- h^f used to express the deformation of the CMB induced by an external potential;
- $\bar{h}^f$ used to express the deformation of the CMB induced by a surface pressure;
- k_1^f used to express the mass redistribution potential computed at the CMB induced by a potential from the core;
- h_1^f used to express the deformation of the CMB induced by a potential from the core;
- $\bar{h}_1^f$ used to express the deformation of the CMB induced by a pressure at the CMB;
- k^s used to express the mass redistribution potential computed at the ICB induced by an external potential;
- h^s used to express the deformation of the ICB induced by an external potential;
- $\bar{h}^s$ used to express the deformation of the ICB induced by a surface pressure;
- k_1^s used to express the mass redistribution potential computed at the ICB induced by a potential from the core;
- h_1^s used to express the deformation of the ICB induced by a potential from the core;
- $\bar{h}_1^s$ used to express the deformation of the ICB induced by a pressure at the CMB;
- k_2^f used to express the mass redistribution potential computed at the CMB induced by a potential from the inner core;
- h_2^f used to express the deformation of the CMB induced by a potential from the inner core;
- $\bar{h}_2^f$ used to express the deformation of the CMB induced by a pressure at the ICB;
- k_2^s used to express the mass redistribution potential computed at the ICB induced by a potential from the inner core;

- h_2^s used to express the deformation of the ICB induced by a potential from the inner core;
- $\bar{h}_2^s$ used to express the deformation of the ICB induced by a pressure at the ICB.

Note that some of these Love numbers can be very small and that the particular case of the centrifugal potential induced by the incremental rotation of the different layers inside the Earth (i.e. $\Delta\phi_{cf}$ defined in Eq. (6.19) for the core and $\Delta\phi_{cs}$ defined in Eq. (6.20) for the inner core) can also be treated by this approach.

The fact that the pressure acting on a boundary can also be expressed in terms of the centrifugal potential leads to a relation between the mass redistribution Love number and the pressure-induced deformation Love number:

$$k_1^f = (q_0/2)\bar{h}_1^f, \qquad k_1^s = (q_0/2)\bar{h}_1^s, \tag{7.165}$$

$$k_2^f = (q_0/2)\bar{h}_2^f, \qquad k_2^s = (q_0/2)\bar{h}_2^s, \tag{7.166}$$

where q_0 is the ratio between the centrifugal acceleration $\Omega^2 a_e$ at the equator (radius a_e) due to diurnal Earth rotation and gravity (g_e) at the equator (i.e. GM/a_e^2 at the first order):

$$q_0 = \Omega^2 a_e/g_e \approx \Omega^2 a_e^3/GM \approx 2\Omega^2 a_e^5/5GA. \tag{7.167}$$

These Love numbers have for instance been used in expressions of the moments of inertia $\tilde{c}_3$, $\tilde{c}_3^f$, or $\tilde{c}_3^s$.
For the Earth, this yields

$$\tilde{c}_3 = \frac{eA}{k_{fl}}\left(k\tilde{m} + k_1\tilde{m}^f + k_2\tilde{m}^s\right) - \frac{ik}{k_{fl}}\frac{\tilde{\Gamma}}{\Omega^2}, \tag{7.168}$$

where k_{fl} is the so-called secular Love number or fluid Love number obtained for the fluid limit (i.e., in the limit of resistance to translational displacement vanishing for the whole Earth) of the model computing k Love number,

$$k_{fl} = \frac{3GeA}{\Omega^2 a_e^5}. \tag{7.169}$$

Using Eq. (2.62) giving $\tilde{\Gamma} = -iA\Omega^2 e\tilde{\phi}$, Eq. (7.168) can be written

$$\tilde{c}_3 = \frac{eA}{k_{fl}}\left(k(\tilde{m} - \tilde{\phi}) + k_1\tilde{m}^f + k_2\tilde{m}^s\right). \tag{7.170}$$

On using Eqs. (7.165) and (7.166), the part that does not involve the external potential can be re-written as $A_f(q_0/2)(\bar{h}\tilde{m} + \bar{h}_1\tilde{m}^f + \bar{h}_2\tilde{m}^s)$.

For the outer core, this yields

$$\tilde{c}_3^f = \frac{eA_f}{k_{fl}}\left(k^f(\tilde{m} - \tilde{\phi}) + k_1^f\tilde{m}^f + k_2^f\tilde{m}^s\right), \tag{7.171}$$

of which the part that does not involve the external potential can be re-written as $A_f(q_0/2)(\bar{h}^f \tilde{m} + \bar{h}_1^f \tilde{m}^f + \bar{h}_2^f \tilde{m}^s)$.

For the inner core, we have

$$\tilde{c}_3^s = \frac{eA_s}{k_{fl}} \left(k^s(\tilde{m} - \tilde{\phi}) + k_1^s \tilde{m}^f + k_2^s \tilde{m}^s \right), \tag{7.172}$$

of which the part that does not involve the external potential can be re-written using Eqs. (7.165) and (7.166):

$$A_s \frac{q_0}{2} \left(\bar{h}^s \tilde{m} + \bar{h}_1^s \tilde{m}^f + \bar{h}_2^s \tilde{m}^s \right).$$

The last three equations are to be compared with Eqs. (6.167), (6.169), and (6.170), respectively.

One immediately sees the relations connecting the compliances $\kappa, \xi, \zeta, \gamma, \beta, \delta$, θ, χ, and τ with the generalized Love numbers:

$$\kappa = \frac{ek}{k_{fl}}, \qquad \xi = \frac{ek_1}{k_{fl}}, \qquad\qquad \zeta = \frac{ek_2}{k_{fl}}, \tag{7.173}$$

$$\gamma = \frac{ek^f}{k_{fl}}, \qquad \beta = \frac{ek_1^f}{k_{fl}} = \frac{q_0}{2}\bar{h}_1^f, \qquad \delta = \frac{ek_2^f}{k_{fl}} = \frac{q_0}{2}\bar{h}_2^f, \tag{7.174}$$

$$\theta = \frac{ek^s}{k_{fl}}, \qquad \chi = \frac{ek_1^s}{k_{fl}} = \frac{q_0}{2}\bar{h}_1^s, \qquad \tau = \frac{ek_2^s}{k_{fl}} = \frac{q_0}{2}h_2^s. \tag{7.175}$$

7.5 Displacement field approach

The theoretical formulations of wobble and nutation in the present chapter have been based essentially on torque equations for the whole Earth and for its core regions, which could be solved analytically or semi-analytically.

The other approach to be considered now consists in solving the equations which govern the spatial and temporal variations of the field of displacements of mass elements in the Earth that are produced by gravitational perturbations. The field consists of a part which describes rotational motion of the globe as a whole, and a part which represents deformations. It is the former which is of direct interest in connection with Earth rotation variations, but the equations for the two parts are mutually coupled and have to be solved simultaneously, starting with a detailed model of the Earth's interior. The reader may refer to Smith's (1974) works setting up the basic formalism (with reference to Goldstein, 1950, Hill, 1954, Jeffreys, 1958, 1959a and b, Jeffreys and Vicente, 1957a and b, Backus, 1967, Longuet-Higgins, 1968, Burridge, 1969, Dahlen, 1968, 1972, 1973, 1974, 1976, 1982, Dahlen and Smith, 1975) which was further developed by Wahr (1979, 1981, 1982, 1988, 2005), Wahr and Sasao (1981), Wahr and Bergen (1986), Wahr *et al.* (1981), Dehant (1986, 1987, 1989, 1990, 1991, 1993), Dehant and Ducarme (1987), Dehant and Zschau (1989), Dehant and Wahr (1991), Dehant and Defraigne (1997), Dehant *et al.* (1999), Huang (1999), Huang *et al.* (2001, 2004, 2011), and more recently by Trinh (2013).

We first present the dynamical equations for the displacement field in a continuous medium, subjected to external forcing, and for the associated perturbation of the Earth's gravitational potential. We explain then how to pass from these second-order partial differential equations to ordinary differential equations of the first order in the radial variable by means of spherical harmonic expansions and passage to the frequency domain. The equations are also used as in the analytical case to compute the Earth's normal modes. As in the previous chapters, we also try to use a notation that sticks to the one used by the group of authors mentioned above. Concerning the sign convention for the potential, we use the physicists' convention as usual (see Section 2.13).

7.5.1 *Vectorial equation for the deformation of the Earth*

The reference system in which we shall work is one which rotates with a constant angular velocity,

$$\mathbf{\Omega} = \Omega \hat{\mathbf{e}}_z \tag{7.176}$$

in space. The rotation speed Ω is the Earth's mean rotation rate: $\Omega = 2\pi$ rad/day $= 7.292115 \times 10^{-5}$ rad/s. This reference frame would become a reference frame tied to the Earth if the Earth rotation variations were neglected.

The input parameters

The parameters that will be used in our equations are the physical properties of an ellipsoidal Earth. They can be expressed as functions of the radial distance from the geocenter and the colatitude. They are considered to be constant on ellipsoidal surfaces. This is the case for the density $\rho(r, \theta)$, the shear modulus $\mu(r, \theta)$, expressing the resistance to a shear force, and the bulk modulus $k(r, \theta)$, expressing the resistance to compression. These last two parameters are equivalent to another set of two other parameters called the Lamé parameters ($\mu(r, \theta)$ and $\lambda(r, \theta)$). The second Lamé parameter $\lambda(r, \theta)$ is related to $\mu(r, \theta)$ and $k(r, \theta)$ by $\lambda(r, \theta) = k(r, \theta) - \frac{2}{3}\mu(r, \theta)$.

The initial gravity potential $\Phi_{\text{in}}(r, \theta)$ is also constant on an ellipsoid as the Earth is considered as initially in hydrostatic equilibrium. This means that the gravity potential at the initial state of the Earth (prior to the deformation) is the sum of the gravitational potential related to the self-gravitation $W_{\text{in}}(r, \theta)$ and the centrifugal potential $\psi(r, \theta)$ related to the uniform rotation of the Earth:

$$\Phi_{\text{in}}(r, \theta) = W_{\text{in}}(r, \theta) + \psi(r, \theta), \tag{7.177}$$

where

$$W_{in}(\mathbf{r}) = W_{in}(r, \theta) = -G \int_{V_{in}} \frac{\rho(\mathbf{r}')}{|\mathbf{r} - \mathbf{r}'|} \, dV_{in}, \tag{7.178}$$

$$\psi(\mathbf{r}) = \psi(r, \theta) = -\frac{1}{2} \left((\mathbf{\Omega} \wedge \mathbf{r}) \cdot (\mathbf{\Omega} \wedge \mathbf{r}) \right), \tag{7.179}$$

so that

$$\nabla \psi(\mathbf{r}) = \mathbf{\Omega} \wedge (\mathbf{\Omega} \wedge \mathbf{r}), \tag{7.180}$$

where V_{in} is the initial volume of the Earth and $\rho(\mathbf{r}')$, the initial density. We note here again that we use the physicists' convention for expressing the force of initial gravity as

$$\mathbf{g}_{in} = -\nabla W_{in}(\mathbf{r}) - \nabla \psi(\mathbf{r}). \tag{7.181}$$

For an ellipsoidal Earth with flattening $\epsilon(r)$, we can write the parameters appearing in the deformation equation as:

$$\rho(r, \theta) = \rho_0(r) + \rho_2(r) P_2(\cos(\theta)), \tag{7.182}$$

$$\mu(r, \theta) = \mu_0(r) + \mu_2(r) P_2(\cos(\theta)), \tag{7.183}$$

$$\lambda(r, \theta) = \lambda_0(r) + \lambda_2(r) P_2(\cos(\theta)), \tag{7.184}$$

$$\Phi_{in}(r, \theta) = \Phi_0(r) + \Phi_2(r) P_2(\cos(\theta)). \tag{7.185}$$

It is easy to get the value of $\rho_2(r)$, as the mean density on the sphere corresponding to the Earth without rotation is the same constant value which remains on the ellipsoid ($\rho(r, \theta) = \rho_0(r_0)$). On the ellipsoid one can consider that the density in each point $\rho(r, \theta)$ is the density of the sphere passing at that point $\rho_0(r)$ plus a correction related to flattening and which can be computed from a Taylor expansion. Knowing that

$$r = r_0 - \frac{2}{3}\epsilon(r_0) r_0 P_2(\cos(\theta)), \tag{7.186}$$

one obtains:

$$\rho(r, \theta) = \rho_0(r_0) - \frac{2}{3}\epsilon(r_0) r_0 \frac{d\rho_0}{dr} P_2(\cos(\theta)) + \rho_2(r) P_2(\cos(\theta)) = \rho_0(r_0) \tag{7.187}$$

and consequently

$$\rho_2(r) = \frac{2}{3}\epsilon(r_0) r_0 \frac{d\rho_0}{dr}. \tag{7.188}$$

We have similarly

$$\mu_2(r) = \frac{2}{3}\epsilon(r_0)r_0\frac{d\mu_0}{dr}, \tag{7.189}$$

$$\lambda_2(r) = \frac{2}{3}\epsilon(r_0)r_0\frac{d\lambda_0}{dr}, \tag{7.190}$$

$$\Phi_2(r) = \frac{2}{3}\epsilon(r_0)r_0\frac{d\Phi_0}{dr}. \tag{7.191}$$

Note that Φ_0 and Φ_2 can be determined from the initial gravitational potential and the rotation of the Earth and that the flattening profile inside the Earth can be determined from Clairaut theory (see Appendix B). The gravitational parts of Φ_2 and Φ_0 are denoted by W_2 and W_0, respectively:

$$W_{\text{in}}(r, \theta) = W_0(r) + W_2(r)P_2(\cos(\theta)). \tag{7.192}$$

The numerical values for the spherical Earth mean density and Lamé parameters are determined from seismology as done by Dziewonski and Anderson (1981) (the PREM model, preliminary reference earth model, see Fig. 6.1 and 6.2) or other models such as the model ak135 of Kennett *et al.* (1995).

The variables

The variables used in the deformation equation must be related to the difference between the new position vector resulting from the deformation induced by the forcing and the initial position vector of an element of matter. We denote this difference of vectors $s(r, t)$. This displacement is the prime variable of the problem as knowing it at each time and each initial position in the Earth enables one to know the deformation of the whole Earth at each time.

The second set of variables is the components of the stress tensor $T(r, t)$, as given in Eq. (6.34) for the elastic Earth.

The next variable of the problem is the mass redistribution potential $\Phi_1^E(r, t)$ induced by the forcing. This comes from the fact that the total gravitational potential after the application of the forcing $W_{def}(r)$ is given by:

$$W_{def}(\mathbf{r}, t) = W_{def}(r, \theta, t) = -G \int_{V_{def}} \frac{\rho'(\mathbf{r}')}{|\mathbf{r} - \mathbf{r}'|} dV_{def} = W_{\text{in}}(r, \theta) + \Phi_1^E(\mathbf{r}, t),$$

$$\tag{7.193}$$

where V_{def} is the volume of the deformed Earth, $\mathbf{r}'$ are the new positions inside the deformed Earth, ρ' is the new Earth density. $\rho'(\mathbf{r}')$ is equal to $\rho(\mathbf{r}') + \rho_1^E(\mathbf{r}')$, the initial density plus a perturbation named ρ_1^E where we keep the same notation as in Chapter 6, except for the superscript E stemming from the notation used by

Smith (1974) and subsequent authors. As indicated there, the displacement field **s** gives rise to a density perturbation $\rho_1^E = -\nabla \cdot (\rho \mathbf{s})$ as may be seen from Eq. (6.36); it comes from the difference between the density of the matter after its displacement from $\mathbf{r}$ to $\mathbf{r}' = \mathbf{r} + \mathbf{s}$ and the density of the matter which was originally at $\mathbf{r} + \mathbf{s}$.

The forcing

The forcing by a tidal potential is written in the equation of motion as $-\nabla \Phi^{ext}(\mathbf{r}, t)$. A new variable is introduced later, which is the sum of the external potential and the mass redistribution potential:

$$\Phi(\mathbf{r}, t) = \Phi_1^E(\mathbf{r}, t) + \Phi^{ext}(\mathbf{r}, t). \tag{7.194}$$

Vectorial equations

In what follows, we suppress the dependence on $\mathbf{r}$ and t for economy in writing.

The first equation is the *Poisson equation*:

The Laplacian of the mass redistribution potential is related, as explained in Eq. (6.37), to the new density distribution after the deformation:

$$\nabla^2 \Phi_1^E = 4\pi G \rho_1^E = -4\pi G \nabla \cdot (\rho \mathbf{s}). \tag{7.195}$$

Considering the expression of the initial density (Eq. (7.182) with Eq. (7.188)), this yields:

$$\nabla^2 \Phi_1^E = -4\pi G \nabla \cdot \left[(\rho_0 + \rho_2 P_2) \mathbf{s} \right]. \tag{7.196}$$

It must be noted, as seen in Eq. (6.38), that the tidal potential obeys the Laplace equation:

$$\nabla^2 \Phi^{ext} = 0. \tag{7.197}$$

So the Poisson equation can be written for the total gravitational potential including the external tidal potential:

$$\nabla^2 \Phi = -4\pi G \nabla \cdot \left[(\rho_0 + \rho_2 P_2) \mathbf{s} \right]. \tag{7.198}$$

Partial differential equations

The *constitutive equation* enables one to relate the stress to the strain. For an elastic Earth, it is the generalization of Hooke's law for a spring:

$$\mathbf{T} = (\lambda_0 + \lambda_2 P_2)(\nabla \cdot \mathbf{s}) \mathbf{I} + (\mu_0 + \mu_2 P_2)\left(\nabla \mathbf{s} + (\nabla \mathbf{s})^T\right), \tag{7.199}$$

where ^{T} indicates that we must take the transpose of the matrix as before. We then consider *the equation of motion* relating the deformation to the forcing:

$$\nabla \cdot \mathbf{T} = (\rho_0 + \rho_2 P_2)(\eta_1 + \eta_2 + \eta_3), \tag{7.200}$$

where η_i are defined by:

$$\eta_1 = \frac{d^2\mathbf{s}}{dt^2} + \nabla\Phi - (\nabla\cdot\mathbf{s})\,\nabla W_0 + \nabla(\mathbf{s}\cdot\nabla W_0), \qquad (7.201)$$

$$\eta_2 = 2\mathbf{\Omega}\wedge\frac{d\mathbf{s}}{dt}, \qquad (7.202)$$

$$\eta_3 = -(\nabla\cdot\mathbf{s})\nabla(W_2 P_2) + \nabla(\mathbf{s}\cdot\nabla(W_2 P_2)). \qquad (7.203)$$

These equations are valid in all the layers of the Earth where the density, the Lamé parameters, and the potential are continuous.

Boundary conditions

The vectorial equations given above are partial differential equations. The unique solution of this system of equations can only be obtained by solving for the integration constant using the boundary conditions, exactly as done for a spherical Earth in Chapter 6 (see Section 6.7, and in particular the homogeneous case in Subsection 6.7.7). These boundaries must be expressed for each layer of the Earth (when we have a discontinuity in the rheological properties of the Earth) and at the surface. The first boundary condition is the continuity of the normal stress (projection of the stress tensor along the normal to the surface). As the surfaces delimiting the layers and the external surface are ellipsoidal, the normal is the normal to the ellipsoid. Let us first compute this normal. For an ellipsoidal surface $r = a(1 - \frac{2}{3}\epsilon(r)P_2)$, the normal can be expressed by:

$$\hat{\mathbf{n}}_0 = \left(1 + \frac{d(h_2(r)P_2)}{dr}\right)\hat{\mathbf{e}}_r - \nabla(h_2(r)P_2), \qquad (7.204)$$

where

$$h_2(r) = -\frac{2}{3}\epsilon(r)r. \qquad (7.205)$$

The projection of the stress tensor $\mathbf{T}$ along the normal can be written:

$$\left[\mathbf{T}\cdot\hat{\mathbf{e}}_r + h_2 P_2\frac{d}{dr}\mathbf{T}\cdot\hat{\mathbf{e}}_r + \mathbf{T}\cdot\left(\frac{d(h_2 P_2)}{dr}\hat{\mathbf{e}}_r - \nabla(h_2 P_2)\right)\right]_{\text{int}}^{\text{ext}} = \mathbf{0}, \qquad (7.206)$$

where $[\ ]_{\text{int}}^{\text{ext}}$ expresses that one takes the value on one side of the boundary minus the value on the other side of the boundary.

The second boundary condition expresses that the displacement along the normal to the layer surfaces and the external surfaces is continuous:

$$\left[\mathbf{s}\cdot\hat{\mathbf{e}}_r + \mathbf{s}\cdot\frac{d(h_2(r)P_2)}{dr}\hat{\mathbf{e}}_r - \mathbf{s}\cdot\nabla(h_2(r)P_2)\right]_{\text{int}}^{\text{ext}} = 0. \qquad (7.207)$$

The third boundary condition imposes the continuity of the mass redistribution potential or of the total additional potential due to tides:

$$[\Phi]_{\text{int}}^{\text{ext}} = 0. \tag{7.208}$$

The fourth boundary condition must be imposed in order to be able to apply the Poisson equation:

$$\left[\frac{d\Phi}{dr} + 4\pi G \left(\rho_0 + \rho_2 P_2 \right) \mathbf{s} \cdot \hat{\mathbf{e}}_r + \frac{d\left(h_2 P_2 \right)}{dr} \left(\frac{d\Phi}{dr} + 4\pi G \left(\rho_0 + \rho_2 P_2 \right) \mathbf{s} \cdot \hat{\mathbf{e}}_r \right) \right.$$

$$\left. - \mathbf{s} \cdot \nabla \left(h_2 P_2 \right) \left(\frac{d\Phi}{dr} + 4\pi G \left(\rho_0 + \rho_2 P_2 \right) \mathbf{s} \cdot \hat{\mathbf{e}}_r \right) \right]_{\text{int}}^{\text{ext}} = 0. \tag{7.209}$$

It must be noted that in the boundary conditions (7.206)–(7.209), the variables $\mathbf{T}$, $\mathbf{s}$, and Φ must be expressed on the ellipsoid as the value on the mean sphere at the point plus a correction due to the flattening of the surface.

Preliminary definitions of the GSH and of the basis vectors that will be used

The canonical basis that is used to express the vector fields given above so that one can get scalar equations and boundary conditions is a complex basis, related to the classical spherical polar coordinate basis by:

$$\hat{\mathbf{e}}_- = \frac{1}{\sqrt{2}} \left(\hat{\mathbf{e}}_\theta - i \hat{\mathbf{e}}_\lambda \right),$$

$$\hat{\mathbf{e}}_0 = \hat{\mathbf{e}}_r, \tag{7.210}$$

$$\hat{\mathbf{e}}_+ = -\frac{1}{\sqrt{2}} \left(\hat{\mathbf{e}}_\theta + i \hat{\mathbf{e}}_\lambda \right).$$

In this basis, the identity tensor or metric tensor is written:

$$\mathbf{I} = \begin{pmatrix} 0 & 0 & -1 \\ 0 & 1 & 0 \\ -1 & 0 & 0 \end{pmatrix}. \tag{7.211}$$

A vector $\mathbf{s}$ can be written as:

$$\mathbf{s} = s^- \hat{\mathbf{e}}_- + s^0 \hat{\mathbf{e}}_0 + s^+ \hat{\mathbf{e}}_+ = \begin{pmatrix} s^- \\ s^0 \\ s^+ \end{pmatrix}. \tag{7.212}$$

Similarly for a vector $\mathbf{u}$:

$$\mathbf{u} = u^- \hat{\mathbf{e}}_- + u^0 \hat{\mathbf{e}}_0 + u^+ \hat{\mathbf{e}}_+ = \begin{pmatrix} u^- \\ u^0 \\ u^+ \end{pmatrix}. \tag{7.213}$$

Their scalar product is:

$$\mathbf{s} \cdot \mathbf{u} = -s^- u^- + s^0 u^0 - s^+ u^+. \tag{7.214}$$

As seen in Eq. (6.44) with Eq. (6.43), one can decompose any vector field using three scalars. The displacement field $\mathbf{s}(\mathbf{r})$, for example, may thus be decomposed in terms of scalars P, Q, and R:

$$\mathbf{s}(\mathbf{r}) = P(\mathbf{r})\hat{\mathbf{e}}_r + \nabla_H Q(\mathbf{r}) + \hat{\mathbf{e}}_r \wedge \nabla_H R(\mathbf{r}), \tag{7.215}$$

with $\nabla_H = \frac{d}{d\theta}\hat{\mathbf{e}}_\theta + \frac{1}{\sin\theta}\frac{d}{d\lambda}\hat{\mathbf{e}}_\lambda$ (denoted ∇_1 in the previous chapter). This is the classical way of expressing a vector field. The first part is the radial part, the second is called the spheroidal part and the last is called the toroidal part. s^-, s^0, and s^+ are uniquely related to P, Q, and R, which is easy to show when developments in spherical harmonics are used.

We introduce now the generalized spherical harmonics functions (GSH) D_{mn}^ℓ; they are functions of the colatitude θ, of the longitude λ, and of three numbers ℓ, m, and n, which are three integers so that $\ell \in [0, +\infty]$, $m \in [-\ell, +\ell]$, and $n \in \{-1, 0, +1\}$. ℓ is called the harmonics degree and m, the harmonics order. Hence,

$$D_{mn}^\ell (\theta, \lambda) = (-1)^{m+n} Y_{\ell m}^n (\theta, \lambda) = (-1)^{m+n} P_\ell^{mn}(\mu)e^{im\lambda}, \tag{7.216}$$

where $Y_{\ell m}^n$, called the generalized spherical harmonic function, and P_ℓ^{mn} (with $\mu = \cos(\theta)$) called the generalized Legendre function are both known in the literature:

$$P_\ell^{mn}(\mu) = \frac{(-1)^{\ell-n}}{2^\ell(\ell-n)!} \sqrt{\frac{(\ell-n)!(\ell+m)!}{(\ell+n)!(\ell-m)!}} \sqrt{\frac{1}{(1-\mu)^{m-n}(1+\mu)^{m+n}}}$$

$$\times \frac{d^{\ell-m}}{d\mu^{\ell-m}} \left((1-\mu)^{\ell-n}(1+\mu)^{\ell+n}\right), \tag{7.217}$$

or

$$P_\ell^{-mn}(\mu) = \frac{(-1)^{\ell-n}}{2^\ell(\ell-n)!} \sqrt{\frac{(\ell-n)!(\ell-m)!}{(\ell+n)!(\ell+m)!}} \sqrt{(1-\mu)^{m+n}(1+\mu)^{m-n}}$$

$$\times \frac{d^{\ell+m}}{d\mu^{\ell+m}} \left((1-\mu)^{\ell-n}(1+\mu)^{\ell+n}\right), \tag{7.218}$$

with

$$P_\ell^{mn}(\mu) = (-1)^{m+n} P_\ell^{-m-n}(\mu). \tag{7.219}$$

And for $n = 0$,

$$P_\ell^{m0}(\mu) = \frac{(-1)^\ell}{2^\ell(\ell)!}\sqrt{\frac{(\ell+m)!}{(\ell-m)!}}\sqrt{\frac{1}{(1-\mu^2)^m}}\frac{d^{\ell-m}}{d\mu^{\ell-m}}\left((1-\mu^2)^\ell\right), \qquad (7.220)$$

or as this formula is valid for any m, positive or negative:

$$P_\ell^{-m0}(\mu) = \frac{(-1)^\ell}{2^\ell(\ell)!}\sqrt{\frac{(\ell-m)!}{(\ell+m)!}}\sqrt{(1-\mu^2)^m}\frac{d^{\ell+m}}{d\mu^{\ell+m}}\left((1-\mu^2)^\ell\right) = (-1)^m P_\ell^{m0}.$$
$$(7.221)$$

Recalling the definition of the associated Legendre polynomials (also called associated Legendre function or ordinary associated Legendre function) given in Eq. (5.142) for positive m:

$$P_\ell^m(\mu) = \frac{(-1)^\ell\sqrt{(1-\mu^2)^m}}{2^\ell\ell!}\frac{d^{\ell+m}}{d\mu^{\ell+m}}\left((1-\mu^2)^\ell\right), \qquad (7.222)$$

one sees immediately that

$$P_\ell^{-m0}(\mu) = \sqrt{\frac{(\ell-m)!}{(\ell+m)!}}\,P_\ell^m(\mu). \qquad (7.223)$$

Or, knowing that $P_\ell^{-m}(\mu) = \frac{(\ell-m)!}{(\ell+m)!}(-1)^m P_\ell^m(\mu)$, this yields

$$P_\ell^{-m0}(\mu) = (-1)^m\sqrt{\frac{(\ell+m)!}{(\ell-m)!}}\,P_\ell^{-m}(\mu), \qquad (7.224)$$

and on using in addition the relation $P_\ell^{-m0}(\mu) = (-1)^m P_\ell^{m0}$ stated above,

$$P_\ell^{m0}(\mu) = (-1)^m\sqrt{\frac{(\ell-m)!}{(\ell+m)!}}\,P_\ell^m(\mu). \qquad (7.225)$$

So, more generally, for any m, be it positive or negative, the associated Legendre polynomials $P_\ell^m(\mu)$ can be related to the generalized Legendre functions $P_\ell^{m0}(\mu)$ by:

$$P_\ell^{m0}(\mu) = (-1)^m\sqrt{\frac{(\ell-m)!}{(\ell+m)!}}\,P_\ell^m(\mu). \qquad (7.226)$$

Here we remind the reader that the definition used for the associated Legendre polynomials is the one with the convention used in quantum mechanics where there is usually a $(-1)^m$ difference with respect to the geodesy convention. This does not change the definition of surface spherical harmonics. On using the definition introduced in Eq. (5.145) for the associated Legendre functions:

$$Y_\ell^m(\theta, \lambda) = (-1)^m \sqrt{\frac{2\ell + 1}{4\pi}} \sqrt{\frac{(\ell - m)!}{(\ell + m)!}} P_\ell^m(\mu) e^{im\lambda}, \qquad (7.227)$$

and using Eqs. (7.216), (7.226), and (7.227) one obtains:

$$Y_{\ell m}^0(\theta, \lambda) = \sqrt{\frac{4\pi}{2\ell + 1}} Y_\ell^m(\theta, \lambda). \qquad (7.228)$$

Or, using (7.216) for $n = 0$,

$$D_{m0}^\ell(\theta, \lambda) = (-1)^m \sqrt{\frac{4\pi}{2\ell + 1}} Y_\ell^m(\theta, \lambda). \qquad (7.229)$$

The normalization of the GSH is provided by the integral:

$$\int_0^{2\pi} \int_0^\pi D_{mn}^\ell(\theta, \lambda) D_{m'n'}^{\ell'}(\theta, \lambda) \sin(\theta) d\theta d\lambda = \frac{4\pi}{2\ell + 1} \delta_\ell^{\ell'} \delta_m^{m'} \delta_n^{n'}, \qquad (7.230)$$

while the normalization of the surface spherical harmonics is obtainable from the integral:

$$\int_0^{2\pi} \int_0^\pi Y_\ell^m(\theta, \lambda) Y_{\ell'}^{m'}(\theta, \lambda) \sin(\theta) d\theta d\lambda = \delta_\ell^{\ell'} \delta_m^{m'}. \qquad (7.231)$$

The product of two GSH is given by:

$$D_{m_1 n_1}^{\ell_1} D_{m_2 n_2}^{\ell_2} = \sum_{\ell = |\ell_2 - \ell_1|}^{\ell = |\ell_2 + \ell_1|} \begin{vmatrix} \ell & \ell_1 & \ell_2 \\ n & n_1 & n_2 \\ m & m_1 & m_2 \end{vmatrix} D_{mn}^\ell, \qquad (7.232)$$

where $n = n_1 + n_2$ and $m = m_1 + m_2$ and the "determinant" coefficient in the sum is not a determinant but rather a coefficient (sometimes called the J-square coefficient) defined from the Wigner coefficients:

$$\begin{vmatrix} \ell & \ell_1 & \ell_2 \\ n & n_1 & n_2 \\ m & m_1 & m_2 \end{vmatrix} = (-1)^{m+n}(2\ell + 1) \begin{pmatrix} \ell & \ell_1 & \ell_2 \\ -n & n_1 & n_2 \end{pmatrix} \begin{pmatrix} \ell & \ell_1 & \ell_2 \\ -m & m_1 & m_2 \end{pmatrix}, \qquad (7.233)$$

where the 3j-Wigner symbols are defined by:

$$\begin{pmatrix} \ell & \ell_1 & \ell_2 \\ n & n_1 & n_2 \end{pmatrix} = (-1)^{\ell-\ell_1-n_2} \sqrt{\frac{(\ell+\ell_1-\ell_2)!(\ell-\ell_1+\ell_2)!(-\ell+\ell_1+\ell_2)!}{(\ell+\ell_1+\ell_2+1)!}}$$

$$\times \sqrt{((\ell+n)!(\ell-n)!(\ell_1+n_1)!(\ell_1-n_1)!(\ell_2+n_2)!(\ell_2-n_2)!}$$

$$\times \sum_k \frac{(-1)^k}{k!(\ell+\ell_1-\ell_2-k)!(\ell-n-k)!(\ell_1+n_1-k)!(\ell_2-\ell_1+n+k)!(\ell_2-\ell-n_1+k)!}.$$

$$(7.234)$$

The sum over k contains all the integers such that the arguments in the factorials are not negative.

We also note here some properties of the 3j-Wigner symbols that will be used later on:

$$\begin{vmatrix} \ell & \ell_1 & \ell_2 \\ n & n_1 & n_2 \\ m & m_1 & m_2 \end{vmatrix} = (-1)^{(\ell+\ell_1+\ell_2)} \begin{vmatrix} \ell & \ell_1 & \ell_2 \\ -n & -n_1 & -n_2 \\ m & m_1 & m_2 \end{vmatrix}, \qquad (7.235)$$

$$\begin{vmatrix} \ell & \ell_1 & \ell_2 \\ n & n_1 & n_2 \\ m & m_1 & m_2 \end{vmatrix} = (-1)^{(\ell+\ell_1+\ell_2)} \begin{vmatrix} \ell & \ell_1 & \ell_2 \\ n & n_1 & n_2 \\ -m & -m_1 & -m_2 \end{vmatrix}, \qquad (7.236)$$

$$\begin{vmatrix} \ell & \ell_1 & \ell_2 \\ n & n_1 & n_2 \\ m & 0 & m \end{vmatrix} = (-1)^{(\ell+\ell_1+\ell_2)} \begin{vmatrix} \ell & \ell_1 & \ell_2 \\ -n & -n_1 & -n_2 \\ m & 0 & m \end{vmatrix}, \qquad (7.237)$$

$$\begin{vmatrix} \ell & \ell_1 & \ell_2 \\ n & n_1 & n_2 \\ 0 & 0 & 0 \end{vmatrix} = 0 \text{ if } (\ell+\ell_1+\ell_2) \text{ is odd}, \qquad (7.238)$$

$$\begin{vmatrix} \ell & \ell_1 & \ell_2 \\ 0 & 0 & 0 \\ m & m_1 & m_2 \end{vmatrix} = 0 \text{ if } (\ell+\ell_1+\ell_2) \text{ is odd}, \qquad (7.239)$$

and

$$\begin{vmatrix} \ell & \ell_1 & \ell_2 \\ 0 & 0 & 0 \\ m & 0 & m \end{vmatrix} = 0 \text{ if } (\ell+\ell_1+\ell_2) \text{ is odd}. \qquad (7.240)$$

The ellipticity terms in the Poisson equation (Eq. (7.198)), the constitutive equation (Eq. (7.199)), and the motion equations (Eqs. (7.200)–(7.203)) of this section involve multiplication by $P_2(\cos\theta)$. Since this function is proportional to Y_2^0 and hence to D_{00}^2, it is worthwhile to make a special mention of Eq. (7.232) with

$\ell_1 = 2, n_1 = m_1 = 0$ and to identify the values of ℓ_2 for which the respective terms on the right-hand side of (7.232) are non-vanishing. The relevance of this will become apparent in the next few paragraphs:

$$\begin{vmatrix} \ell & 2 & \ell \\ n & 0 & n_2 \\ 0 & 0 & 0 \end{vmatrix} \neq 0, \qquad \begin{vmatrix} \ell & 2 & \ell \pm 2 \\ n & 0 & n_2 \\ 0 & 0 & 0 \end{vmatrix} \neq 0,$$

$$\begin{vmatrix} \ell & 2 & \ell \\ 0 & 0 & 0 \\ m & 0 & m_2 \end{vmatrix} \neq 0, \qquad \begin{vmatrix} \ell & 2 & \ell \pm 2 \\ 0 & 0 & 0 \\ m & 0 & m_2 \end{vmatrix} \neq 0,$$

$$\begin{vmatrix} \ell & 2 & \ell \\ 0 & 0 & 0 \\ m & 0 & m \end{vmatrix} \neq 0, \qquad \begin{vmatrix} \ell & 2 & \ell \pm 2 \\ 0 & 0 & 0 \\ m & 0 & m \end{vmatrix} \neq 0, \qquad (7.241)$$

$$\begin{vmatrix} \ell & 2 & \ell \pm 1 \\ n & 0 & n_2 \\ 0 & 0 & 0 \end{vmatrix} = \begin{vmatrix} \ell & 2 & \ell \pm 1 \\ 0 & 0 & 0 \\ m & 0 & m_2 \end{vmatrix} = \begin{vmatrix} \ell & 2 & \ell \pm 1 \\ 0 & 0 & 0 \\ m & 0 & m \end{vmatrix} = 0. \qquad (7.242)$$

The GSH constitute a functional basis on which we can decompose any component of a tensor, of a vector, and of a scalar. Any of these components can be written as the sum of products between a function of the radius and a GSH function of the colatitude θ and the longitude λ.

For a tensor such as the stress tensor, we have:

$$T_{\alpha_1}^{\alpha_2}(r, \theta, \lambda) = \sum_{\ell=0}^{\infty} \sum_{m=-\ell}^{\ell} T_{\ell\alpha_1}^{m\alpha_2}(r) D_{mn}^{\ell}(\theta, \lambda) \text{ with } n = \alpha_1 + \alpha_2. \qquad (7.243)$$

For a vector such as the deformation vector **s** given by equation (7.212):

$$\mathbf{s}(r, \theta, \lambda) = s^- \hat{\mathbf{e}}_- + s^0 \hat{\mathbf{e}}_0 + s^+ \hat{\mathbf{e}}_+ = \begin{pmatrix} \sum_{\ell=0}^{\infty} \sum_{m=-\ell}^{\ell} S_{\ell}^{m-}(r) D_{m-}^{\ell}(\theta, \lambda) \\ \sum_{\ell=0}^{\infty} \sum_{m=-\ell}^{\ell} S_{\ell}^{m0}(r) D_{m0}^{\ell}(\theta, \lambda) \\ \sum_{\ell=0}^{\infty} \sum_{m=-\ell}^{\ell} S_{\ell}^{m+}(r) D_{m+}^{\ell}(\theta, \lambda) \end{pmatrix}.$$

$$(7.244)$$

For a scalar such as the mass redistribution and the external potential Φ:

$$\Phi(r, \theta, \lambda) = \sum_{\ell=0}^{\infty} \sum_{m=-\ell}^{\ell} \Phi_{\ell}^{m}(r) D_{m0}^{\ell}(\theta, \lambda). \qquad (7.245)$$

Making use of the well-known spherical harmonic expansions of scalar functions to expand the potential functions P, Q, and R in equation (7.215), one has:

$$\mathbf{s}(\mathbf{r}) = \left(\sum_{\ell=0}^{\infty}\sum_{m=-\ell}^{\ell} P_\ell^m(r)Y_\ell^m(\theta,\lambda)\right)\hat{\mathbf{e}}_r + \nabla_H\left(\sum_{\ell=0}^{\infty}\sum_{m=-\ell}^{\ell} Q_\ell^m(r)Y_\ell^m(\theta,\lambda)\right)$$

$$+ \hat{\mathbf{e}}_r \wedge \nabla_H\left(\sum_{\ell=0}^{\infty}\sum_{m=-\ell}^{\ell} R_\ell^m(r)Y_\ell^m(\theta,\lambda)\right), \tag{7.246}$$

$$= \sum_{\ell=0}^{\infty}\sum_{m=-\ell}^{\ell}\left[P_\ell^m(r)Y_\ell^m(\theta,\lambda)\hat{\mathbf{e}}_r \right.$$

$$+ Q_\ell^m(r)\left(\frac{d}{d\theta}Y_\ell^m(\theta,\lambda)\hat{\mathbf{e}}_\theta + \frac{1}{\sin(\theta)}\frac{d}{d\lambda}Y_\ell^m(\theta,\lambda)\hat{\mathbf{e}}_\lambda\right)$$

$$\left. + R_\ell^m(r)\left(\frac{1}{\sin(\theta)}\frac{d}{d\lambda}Y_\ell^m(\theta,\lambda)\hat{\mathbf{e}}_\theta - \frac{d}{d\theta}Y_\ell^m(\theta,\lambda)\hat{\mathbf{e}}_\lambda\right)\right]. \tag{7.247}$$

In order to make the link between Eqs. (7.244) and (7.247) for $\mathbf{s}$, we must consider Eq. (7.229) and the following expressions:

$$D_{m+}^\ell + D_{m-}^\ell = (-1)^m\sqrt{\frac{4\pi}{2\ell+1}}\frac{2}{\sqrt{\ell(\ell+1)}}\frac{-i}{\sin(\theta)}\frac{dY_\ell^m}{d\lambda}, \tag{7.248}$$

$$D_{m+}^\ell - D_{m-}^\ell = (-1)^m\sqrt{\frac{4\pi}{2\ell+1}}\frac{2}{\sqrt{\ell(\ell+1)}}\frac{dY_\ell^m}{d\theta}. \tag{7.249}$$

These relations (see Edmonds (1960) for the properties of the GSH) are then used to compute the expression (7.247) in terms of the D_{m0}^ℓ, D_{m+}^ℓ, and D_{m-}^ℓ:

$$Q_\ell^m(r)\left(\frac{d}{d\theta}Y_\ell^m(\theta,\lambda)\hat{\mathbf{e}}_\theta + \frac{1}{\sin(\theta)}\frac{d}{d\lambda}Y_\ell^m(\theta,\lambda)\hat{\mathbf{e}}_\lambda\right)$$

$$= Q_\ell^m(r)\frac{L_0^\ell}{\sqrt{2}}\sqrt{\frac{2\ell+1}{4\pi}}(-1)^m\left[\left(D_{m+}^\ell - D_{m-}^\ell\right)\hat{\mathbf{e}}_\theta + i\left(D_{m+}^\ell + D_{m-}^\ell\right)\hat{\mathbf{e}}_\lambda\right], \tag{7.250}$$

$$R_\ell^m(r)\left(\frac{1}{\sin(\theta)}\frac{d}{d\lambda}Y_\ell^m(\theta,\lambda)\hat{\mathbf{e}}_\theta - \frac{d}{d\theta}Y_\ell^m(\theta,\lambda)\hat{\mathbf{e}}_\lambda\right)$$

$$= R_\ell^m(r)\frac{L_0^\ell}{\sqrt{2}}\sqrt{\frac{2\ell+1}{4\pi}}(-1)^m\left[i\left(D_{m+}^\ell + D_{m-}^\ell\right)\hat{\mathbf{e}}_\theta - \left(D_{m+}^\ell - D_{m-}^\ell\right)\hat{\mathbf{e}}_\lambda\right], \tag{7.251}$$

where

$$L_0^\ell = \sqrt{\frac{\ell(\ell+1)}{2}}. \tag{7.252}$$

The sum of these two terms represents the non-radial components of the deformation vector **s**. By equating the non-radial part of Eq. (7.244) to these terms, one can show that the amplitude of the spheroidal term Q_ℓ^m is proportional to $S_\ell^{m+} + S_\ell^{m-}$, and that the amplitude of the toroidal term R_ℓ^m is proportional to $S_\ell^{m+} - S_\ell^{m-}$.

From now on, we shall omit to write the dependence on the radius r to simplify the writing.

Derivative of the GSH

The derivative in one of the three directions $\hat{\mathbf{e}}_-$, $\hat{\mathbf{e}}_0$, or $\hat{\mathbf{e}}_+$ of a component (with respect to one component α_3 where $\alpha_3 = 0, -1,$ or 1) of a tensor as given in (7.243) is written:

$$T_{\alpha_1}^{\alpha_2\,'\alpha_3} = \sum_{\ell=0}^{\infty} \sum_{m=-\ell}^{\ell} T_{\ell\alpha_1}^{m\alpha_2\,'\alpha_3} D_{mn}^\ell(\theta, \lambda) \ \text{ with } \ n = \sum_{i=1}^{3} \alpha_i, \tag{7.253}$$

where

$$T_{\ell\alpha_1}^{m\alpha_2\,'-} = \frac{-1}{r} \left(L_{(\alpha_1+\alpha_2)}^\ell T_{\ell\alpha_1}^{m\alpha_2} + T_{\ell\alpha_1}^{m0} \delta_{\alpha_2}^+ + T_{\ell 0}^{m\alpha_2} \delta_{\alpha_1}^+ + T_{\ell\alpha_1}^{m-} \delta_{\alpha_2}^0 + T_{\ell-}^{m\alpha_2} \delta_{\alpha_1}^0 \right),$$

$$T_{\ell\alpha_1}^{m\alpha_2\,'0} = \frac{d}{dr} T_{\ell\alpha_1}^{m\alpha_2}, \tag{7.254}$$

$$T_{\ell\alpha_1}^{m\alpha_2\,'+} = \frac{-1}{r} \left(L_{(\alpha_1+\alpha_2+1)}^\ell T_{\ell\alpha_1}^{m\alpha_2} + T_{\ell\alpha_1}^{m0} \delta_{\alpha_2}^- + T_{\ell 0}^{m\alpha_2} \delta_{\alpha_1}^- + T_{\ell\alpha_1}^{m+} \delta_{\alpha_2}^0 + T_{\ell+}^{m\alpha_2} \delta_{\alpha_1}^0 \right),$$

and where

$$L_n^\ell = \sqrt{\frac{(\ell - n + 1)(\ell + n)}{2}}; \tag{7.255}$$

the derivative of a component of a vector as given in (7.244) with respect to one component α_1 is:

$$s^{n\,'\alpha_1} = \sum_{\ell=0}^{\infty} \sum_{m=-\ell}^{\ell} S_\ell^{mn\,'\alpha_1} D_{m(n+\alpha_1)}^\ell, \tag{7.256}$$

where

$$S_\ell^{mn\,'-} = \frac{-1}{r} \left(L_n^\ell S_\ell^{mn} + S_\ell^{m0} \delta_n^+ + S_\ell^{m-} \delta_n^0 \right),$$

$$S_\ell^{mn\,'0} = \frac{d}{dr} S_\ell^{mn}, \tag{7.257}$$

$$S_\ell^{mn\,'+} = \frac{-1}{r} \left(L_{(n+1)}^\ell S_\ell^{mn} + S_\ell^{m0} \delta_n^- + S_\ell^{m+} \delta_n^0 \right);$$

the derivative of a scalar as given in (7.245) with respect to one component α_1 is:

$$\Phi^{'\alpha_1} = \sum_{\ell=0}^{\infty} \sum_{m=-\ell}^{\ell} \Phi_\ell^{m\,'\alpha_1} D_{m\alpha_1}^\ell, \qquad (7.258)$$

where

$$\Phi_\ell^{m\,'-} = \frac{-L_0^\ell}{r} \Phi_\ell^m,$$

$$\Phi_\ell^{m\,'0} = \frac{d}{dr} \Phi_\ell^m, \qquad (7.259)$$

$$\Phi_\ell^{m\,'+} = \frac{-L_0^\ell}{r} \Phi_\ell^m,$$

on noting from Eq. (7.255) that $L_1^\ell = L_0^\ell$.

The gradient of a scalar developed in GSH can thus be written as:

$$\nabla\Phi = \sum_{\ell=0}^{\infty} \sum_{m=-\ell}^{\ell} \begin{bmatrix} -\frac{L_0^\ell}{r} \Phi_\ell^m D_{m-}^\ell \\ \frac{d\Phi_\ell^m}{dr} D_{m0}^\ell \\ -\frac{L_0^\ell}{r} \Phi_\ell^m D_{m+}^\ell \end{bmatrix}. \qquad (7.260)$$

The divergence of a vector becomes:

$$\nabla \cdot \mathbf{s} = \sum_{\ell=0}^{\infty} \sum_{m=-\ell}^{\ell} \left[\frac{dS_\ell^{m0}}{dr} + \frac{L_0^\ell}{r} \left(S_\ell^{m+} + S_\ell^{m-} \right) + \frac{2}{r} S_\ell^{m0} \right] D_{m0}^\ell. \qquad (7.261)$$

The divergence of a tensor can thus be written as:

$$\nabla \cdot \mathbf{T} = \sum_{\ell=0}^{\infty} \sum_{m=-\ell}^{\ell} \begin{bmatrix} \left(\left(\frac{d}{dr} + \frac{3}{r}\right) T_{\ell 0}^{m-} + \frac{L_0^\ell}{r} T_{\ell+}^{m-} + \frac{L_2^\ell}{r} T_{\ell-}^{m-} \right) D_{m-}^\ell \\ \left(\left(\frac{d}{dr} + \frac{2}{r}\right) T_{\ell 0}^{m0} + \frac{L_0^\ell}{r} \left(T_{\ell+}^{m0} + T_{\ell-}^{m0} \right) + \frac{2}{r} T_{\ell+}^{m-} \right) D_{m0}^\ell \\ \left(\left(\frac{d}{dr} + \frac{3}{r}\right) T_{\ell 0}^{m+} + \frac{L_0^\ell}{r} T_{\ell-}^{m+} + \frac{L_2^\ell}{r} T_{\ell+}^{m+} \right) D_{m+}^\ell \end{bmatrix}, \qquad (7.262)$$

where L_0^ℓ is given by Eq. (7.252) and

$$L_2^\ell = \sqrt{\frac{(\ell-1)(\ell+2)}{2}}. \qquad (7.263)$$

The gradient of a vector can thus be written as:

$$\nabla \mathbf{s} = \sum_{\ell=0}^{\infty} \sum_{m=-\ell}^{\ell} \begin{bmatrix} \frac{-L_2^\ell}{r} S_\ell^{m-} D_{m-2}^\ell & \frac{-1}{r} \left(L_0^\ell S_\ell^{m0} + S_\ell^{m-} \right) D_{m-}^\ell & \frac{-1}{r} \left(L_0^\ell S_\ell^{m+} + S_\ell^{m0} \right) D_{m0}^\ell \\ \frac{dS_\ell^{m-}}{dr} D_{m-}^\ell & \frac{dS_\ell^{m0}}{dr} D_{m0}^\ell & \frac{dS_\ell^{m+}}{dr} D_{m+}^\ell \\ \frac{-1}{r} \left(L_0^\ell S_\ell^{m-} + S_\ell^{m0} \right) D_{m0}^\ell & \frac{-1}{r} \left(L_0^\ell S_\ell^{m0} + S_\ell^{m+} \right) D_{m+}^\ell & \frac{-L_2^\ell}{r} S_\ell^{m+} D_{m+2}^\ell \end{bmatrix}. \qquad (7.264)$$

The Laplacian of a scalar is also written as:

$$\nabla^2 \Phi = \nabla \cdot \nabla \Phi = \sum_{\ell=0}^{\infty} \sum_{m=-\ell}^{\ell} \left(\frac{d^2}{dr^2} + \frac{2}{r} \frac{d}{dr} - \frac{2L_0^{\ell\,2}}{r^2} \right) \Phi_\ell^m D_{m0}^\ell. \qquad (7.265)$$

Time derivative

Computation of tides and nutations are performed by using an external tidal potential. This potential may be decomposed into a large number of periodic components with a discrete spectrum of frequencies, as noted in earlier chapters. This allows us to work component per component and to work in the frequency domain. Consequently, the time derivative of one component f can be written $i\Omega\sigma f$ where $\Omega\sigma$ is the frequency in the terrestrial frame (σ is as usual, the frequency expressed in cycles per day). We then use the functional basis that has been introduced above and project on that basis. The scalar equations that are derived from vectorial equations after this procedure are written for one frequency $\Omega\sigma$ and one degree and order of GSH. Typically the nutations are obtained from the displacement corresponding to a rotation around an axis in the equator, in response to a tidal potential of degree two order one; the LOD variations are obtained from the displacement corresponding to a rotation around the z-axis, in response to a tidal potential of degree two and order zero; the long-period tides are obtained from the deformational response to a tidal potential of degree two order zero; the diurnal tides are obtained from the deformation in response to a tidal potential of degree two and order one; the semidiurnal tides are obtained from the deformation in response to a tidal potential of degree two and order two.

New variables

Starting from the definition (7.212) for the vector **s**, we can define three new variables given by:

$$U_\ell^m = S_\ell^{m0}, \qquad (7.266)$$

$$V_\ell^m = S_\ell^{m+} + S_\ell^{m-}, \qquad (7.267)$$

$$W_\ell^m = S_\ell^{m+} - S_\ell^{m-}. \qquad (7.268)$$

Starting from the definition (7.243) for the tensor **T**, let us consider now $\mathbf{T} \cdot \hat{\mathbf{e}}_r$, describing the stress (force per unit area) applied on a surface element of a geocentric sphere (which is normal to $\hat{\mathbf{e}}_r$) with the following form:

$$\mathbf{T} \cdot \hat{\mathbf{e}}_r = \sum_{\ell=0}^{\infty} \sum_{m=-\ell}^{\ell} \begin{pmatrix} T_{\ell 0}^{m-} D_{m-}^\ell \\ T_{\ell 0}^{m0} D_{m0}^\ell \\ T_{\ell 0}^{m+} D_{m+}^\ell \end{pmatrix}. \qquad (7.269)$$

The radial component of this vector is called the "radial stress" and is given by $T_{\ell 0}^{m0}$, and the "tangential stresses" are given by $T_{\ell 0}^{m+}$ and $T_{\ell 0}^{m-}$. One then defines the radial and spheroidal and toroidal tangential stresses by:

$$P_\ell^m = T_{\ell 0}^{m0}, \tag{7.270}$$

$$Q_\ell^m = T_{\ell 0}^{m+} + T_{\ell 0}^{m-}, \tag{7.271}$$

$$R_\ell^m = T_{\ell 0}^{m+} - T_{\ell 0}^{m-}. \tag{7.272}$$

Note that the symbols P_ℓ^m, Q_ℓ^m, R_ℓ^m that were defined in Eqs. (7.215) and (7.247) as components of the displacement field **s** are not the same, but while this could be confusing, we decided to use these notations as they are found in the literature.

In order to decrease the order of the radial derivative in the Poisson equation, we shall also introduce a new variable called g_1^E, related to the radial derivative of Φ_1^E:

$$g_{1\ell}^{Em} = \frac{d\Phi_{1\ell}^{Em}}{dr} + 4\pi G\rho_0 U_\ell^m. \tag{7.273}$$

Scalar equations

We are now ready to write the scalar equations. Using Eq. (7.273), the Poisson equation (Eq. (7.198)) can be split into two equations of the first order in d/dr:

$$\frac{d\Phi_{1\ell}^{Em}}{dr} = g_{1\ell}^{Em} - 4\pi G\rho_0 U_\ell^m, \tag{7.274}$$

$$\frac{dg_{1\ell}^{Em}}{dr} = \frac{\ell(\ell+1)}{r^2}\Phi_{1\ell}^{Em} - \frac{2}{r}g_{1\ell}^{Em} - \frac{L_0^\ell}{r}4\pi G\rho_0 U_\ell^m$$

$$- 4\pi G \sum_{\ell'=|\ell-2|}^{|\ell+2|} \begin{vmatrix} \ell & 2 & \ell' \\ 1 & 0 & 1 \\ m & 0 & m \end{vmatrix} \frac{L_0^\ell}{r}\rho_2 \begin{Bmatrix} V_{\ell'}^m \\ W_{\ell'}^m \end{Bmatrix} - 4\pi G \sum_{\ell'=|\ell-2|}^{|\ell+2|} \begin{vmatrix} \ell & 2 & \ell' \\ 0 & 0 & 0 \\ m & 0 & m \end{vmatrix}$$

$$\times \left[\frac{d\rho_2}{dr}U_{\ell'}^m + \frac{2}{r}\rho_2 U_{\ell'}^m + \rho_2 \left(\frac{1}{\beta_0}P_{\ell'}^m - \frac{\lambda_0}{\beta_0 r}\left(L_0^{\ell'}V_{\ell'}^m + 2U_{\ell'}^m\right)\right)\right]. \tag{7.275}$$

Using all the material given before, the motion equation (Eqs. (7.200)–(7.203)) is projected on the three basis vectors ($\hat{e}_-, \hat{e}_0, \hat{e}_+$). One then takes the radial component (component along $\hat{e}_0$), the sum and the difference of the two other components (along $\hat{e}_-$ and $\hat{e}_+$), and employs the new variables defined in

Eqs. (7.266)–(7.272). We thus obtain the following three scalar equations:

$$\frac{dU_\ell^m}{dr} = \frac{1}{\beta_0} P_\ell^m - \frac{\lambda_0}{\beta_0 r} \left(L_0^\ell V_\ell^m + 2U_\ell^m \right)$$

$$- \sum_{\ell'=|\ell-2|}^{|\ell+2|} \begin{vmatrix} \ell & 2 & \ell' \\ 0 & 0 & 0 \\ m & 0 & m \end{vmatrix} \left[\frac{\beta_2}{\beta_0^2} P_{\ell'}^m - 2\frac{\lambda_0\mu_2 - \lambda_2\mu_0}{\beta_0^2 r} \left(L_0^{\ell'} V_{\ell'}^m + 2U_{\ell'}^m \right) \right], \quad (7.276)$$

$$\frac{dV_\ell^m}{dr} = \frac{1}{\mu_0} Q_\ell^m + \frac{1}{r} \left(V_\ell^m + 2L_0^\ell U_\ell^m \right) - \frac{\mu_2}{\mu_0^2} \sum_{\ell'=|\ell-2|}^{|\ell+2|} \begin{vmatrix} \ell & 2 & \ell' \\ 1 & 0 & 1 \\ m & 0 & m \end{vmatrix} \left\{ \begin{matrix} Q_{\ell'}^m \\ R_{\ell'}^m \end{matrix} \right\}, \quad (7.277)$$

$$\frac{dW_\ell^m}{dr} = \frac{1}{\mu_0} R_\ell^m + \frac{1}{r} W_\ell^m - \frac{\mu_2}{\mu_0^2} \sum_{\ell'=|\ell-2|}^{|\ell+2|} \begin{vmatrix} \ell & 2 & \ell' \\ 1 & 0 & 1 \\ m & 0 & m \end{vmatrix} \left\{ \begin{matrix} R_{\ell'}^m \\ Q_{\ell'}^m \end{matrix} \right\}. \quad (7.278)$$

Similarly, the constitutive equation (7.199) yields:

$$\frac{dP_\ell^m}{dr} = -\frac{2}{r} P_\ell^m - \frac{L_0^\ell}{r} Q_\ell^m + \frac{2}{r} \left[\lambda_0 X_\ell^m + \frac{\mu_0}{r} \left(L_0^\ell V_\ell^m + 2U_\ell^m \right) \right]$$

$$+ \rho_0 \left(\eta_{1\ell}^{Um} + \eta_{2\ell}^{Um} + \eta_{3\ell}^{Um} \right) + \sum_{\ell'=|\ell-2|}^{|\ell+2|} \begin{vmatrix} \ell & 2 & \ell' \\ 0 & 0 & 0 \\ m & 0 & m \end{vmatrix}$$

$$\times \left[\rho_2 \left(\eta_{1\ell'}^{Um} + \eta_{2\ell'}^{Um} + \eta_{3\ell'}^{Um} \right) + \frac{2}{r} \left(\lambda_2 X_\ell^m + \frac{\mu_2}{r} \left(L_0^{\ell'} V_{\ell'}^m + 2U_{\ell'}^m \right) \right) \right], \quad (7.279)$$

$$\frac{dQ_\ell^m}{dr} = -\frac{3}{r} Q_\ell^m - \frac{2L_0^\ell L_2^\ell}{r^2} \mu_0 V_\ell^m + \frac{2L_0^\ell}{r} \left[\lambda_0 X_\ell^m + \frac{\mu_0}{r} \left(L_0^\ell V_\ell^m + 2U_\ell^m \right) \right]$$

$$+ \rho_0 \left(\eta_{1\ell}^{Vm} + \eta_{2\ell}^{Vm} + \eta_{3\ell}^{Vm} \right)$$

$$+ \sum_{\ell'=|\ell-2|}^{|\ell+2|} \begin{vmatrix} \ell & 2 & \ell' \\ 1 & 0 & 1 \\ m & 0 & m \end{vmatrix} \rho_2 \left\{ \begin{matrix} \eta_{1\ell'}^{Vm} + \eta_{2\ell'}^{Vm} + \eta_{3\ell'}^{Vm} \\ \eta_{1\ell'}^{Wm} + \eta_{2\ell'}^{Wm} + \eta_{3\ell'}^{Wm} \end{matrix} \right\}$$

$$+ \frac{2L_0^\ell}{r} \sum_{\ell'=|\ell-2|}^{|\ell+2|} \begin{vmatrix} \ell & 2 & \ell' \\ 0 & 0 & 0 \\ m & 0 & m \end{vmatrix} \left[\lambda_2 X_{\ell'}^m + \frac{\mu_2}{r} \left(L_0^{\ell'} V_{\ell'}^m + 2U_{\ell'}^m \right) \right]$$

$$+ \frac{2L_0^\ell}{r^2} \mu_0 \sum_{\ell'=|\ell-2|}^{|\ell+2|} \begin{vmatrix} \ell & 2 & \ell' \\ 2 & 0 & 2 \\ m & 0 & m \end{vmatrix} L_2^{\ell'} \left\{ \begin{matrix} V_{\ell'}^m \\ W_{\ell'}^m \end{matrix} \right\}, \quad (7.280)$$

$$\frac{dR_\ell^m}{dr} = -\frac{3}{r}R_\ell^m + \frac{2L_0^\ell L_2^\ell}{r^2}\mu_0 W_\ell^m + \rho_0\left(\eta_{1\ell}^{Wm} + \eta_{2\ell}^{Wm} + \eta_{3\ell}^{Wm}\right)$$

$$+ \sum_{\ell'=|\ell-2|}^{|\ell+2|} \begin{vmatrix} \ell & 2 & \ell' \\ 1 & 0 & 1 \\ m & 0 & m \end{vmatrix} \rho_2 \left\{ \begin{matrix} \eta_{1\ell'}^{Wm} + \eta_{2\ell'}^{Wm} + \eta_{3\ell'}^{Wm} \\ \eta_{1\ell'}^{Vm} + \eta_{2\ell'}^{Vm} + \eta_{3\ell'}^{Vm} \end{matrix} \right\}$$

$$+ \frac{2L_0^\ell}{r^2}\mu_0 \sum_{\ell'=|\ell-2|}^{|\ell+2|} \begin{vmatrix} \ell & 2 & \ell' \\ 2 & 0 & 2 \\ m & 0 & m \end{vmatrix} L_2^{\ell'} \left\{ \begin{matrix} W_{\ell'}^m \\ V_{\ell'}^m \end{matrix} \right\}. \tag{7.281}$$

The braces $\{\ \}$ mean that one must take the upper component if $|\ell - \ell'|$ is even, and the lower component if $|\ell - \ell'|$ is odd. One writes $\beta_0 = \lambda_0 + 2\mu_0$ and $\beta_2 = \lambda_2 + 2\mu_2$. One also uses, as in Smith (1974), the following definitions for X_ℓ^m, $\eta_{i\ell'}^{Um}$, $\eta_{i\ell'}^{Vm}$, and $\eta_{i\ell'}^{Wm}$:

$$X_\ell^m = \frac{dU_\ell^m}{dr} + \frac{1}{r}\left(L_0^\ell V_\ell^m + 2U_\ell^m\right), \tag{7.282}$$

$$\eta_{1\ell}^{Um} = -\Omega^2\sigma^2 U_\ell^m + g_{1\ell}^{Em} - \frac{2(g+\tilde{g})}{r}U_\ell^m - \frac{L_0^\ell}{r}\tilde{g}V_\ell^m, \tag{7.283}$$

$$\eta_{1\ell}^{Vm} = -\Omega^2\sigma^2 V_\ell^m - \frac{L_0^\ell}{r}\left(\Phi_{1\ell}^{Em} + \tilde{g}U_\ell^m\right), \tag{7.284}$$

$$\eta_{1\ell}^{Wm} = -\Omega^2\sigma^2 W_\ell^m, \tag{7.285}$$

$$\eta_{2\ell}^{Um} = 2\sigma\Omega^2 \sum_{\ell'=|\ell-1|}^{|\ell+1|} \begin{vmatrix} \ell & 2 & \ell' \\ 0 & 1 & -1 \\ m & 0 & m \end{vmatrix} \left\{ \begin{matrix} V_{\ell'}^m \\ -W_{\ell'}^m \end{matrix} \right\}, \tag{7.286}$$

$$\eta_{2\ell}^{Vm} = -2\sigma\Omega^2 \sum_{\ell'=|\ell-1|}^{|\ell+1|} \left[\begin{vmatrix} \ell & 2 & \ell' \\ -1 & -1 & 0 \\ m & 0 & m \end{vmatrix} \left\{ \begin{matrix} 2U_{\ell'}^m \\ 0 \end{matrix} \right\} + \begin{vmatrix} \ell & 2 & \ell' \\ -1 & 0 & -1 \\ m & 0 & m \end{vmatrix} \left\{ \begin{matrix} V_{\ell'}^m \\ -W_{\ell'}^m \end{matrix} \right\} \right], \tag{7.287}$$

$$\eta_{2\ell}^{Wm} = 2\sigma\Omega^2 \sum_{\ell'=|\ell-1|}^{|\ell+1|} \left[\begin{vmatrix} \ell & 2 & \ell' \\ -1 & -1 & 0 \\ \dot{m} & 0 & m \end{vmatrix} \left\{ \begin{matrix} 0 \\ 2U_{\ell'}^m \end{matrix} \right\} + \begin{vmatrix} \ell & 2 & \ell' \\ -1 & 0 & -1 \\ m & 0 & m \end{vmatrix} \left\{ \begin{matrix} -W_{\ell'}^m \\ V_{\ell'}^m \end{matrix} \right\} \right], \tag{7.288}$$

$$\eta_{3\ell}^{Um} = \sum_{\ell'=|\ell-2|}^{|\ell+2|} \begin{vmatrix} \ell & 2 & \ell' \\ 0 & 0 & 0 \\ m & 0 & m \end{vmatrix} \left[\frac{d\Phi_2}{dr}\left(\frac{dU_{\ell'}^m}{dr} - X_{\ell'}^m + \frac{d^2\Phi_2}{dr^2}\right) \right]$$

$$+ \sqrt{3} \sum_{\ell'=|\ell-2|}^{|\ell+2|} \begin{vmatrix} \ell & 2 & \ell' \\ 0 & 1 & -1 \\ m & 0 & m \end{vmatrix} \left(\frac{1}{r}\frac{d\Phi_2}{dr} - \frac{\Phi_2}{r^2} + \frac{\Phi_2}{r}\frac{d}{dr}\right) \left\{ \begin{matrix} V_{\ell'}^m \\ -W_{\ell'}^m \end{matrix} \right\}, \tag{7.289}$$

$$
\eta_{3\ell}^{Vm} = -\frac{2L_0^\ell}{r}\frac{d\Phi_2}{dr} \sum_{\ell'=|\ell-2|}^{|\ell+2|}
\begin{vmatrix} \ell & 2 & \ell' \\ 0 & 0 & 0 \\ m & 0 & m \end{vmatrix} U_{\ell'}^m
$$

$$
-\sqrt{3}\frac{\Phi_2}{r} \sum_{\ell'=|\ell-2|}^{|\ell+2|}
\begin{vmatrix} \ell & 2 & \ell' \\ 0 & 1 & -1 \\ m & 0 & m \end{vmatrix}
\left\{ \begin{matrix} V_{\ell'}^m \\ -W_{\ell'}^m \end{matrix} \right\}
$$

$$
+2\sqrt{3}\frac{\Phi_2}{r} \sum_{\ell'=|\ell-2|}^{|\ell+2|}
\begin{vmatrix} \ell & 2 & \ell' \\ 1 & 1 & 0 \\ m & 0 & m \end{vmatrix}
\left\{ \begin{matrix} X_{\ell'}^m \\ 0 \end{matrix} \right\},
\tag{7.290}
$$

$$
\eta_{3\ell}^{Wm} = 2\sqrt{3}\frac{\Phi_2}{r} \sum_{\ell'=|\ell-2|}^{|\ell+2|}
\begin{vmatrix} \ell & 2 & \ell' \\ 1 & 1 & 0 \\ m & 0 & m \end{vmatrix}
\left\{ \begin{matrix} 0 \\ X_{\ell'}^m \end{matrix} \right\},
\tag{7.291}
$$

where $\tilde{g} = \frac{dW_0}{dr}$ and $g = 2\pi G\rho_0 r + \frac{r}{2}\frac{d^2W_0}{dr^2}$, W_0 being the spherically symmetric part of the gravitational potential of the undeformed ellipsoidal Earth (see Eq. (7.192)).

Solutions for the scalar equations

In order to find the solutions of the above equations, one defines a vector $\overline{\sigma}_\ell^m$ with the spheroidal variables U_ℓ^m, P_ℓ^m, $\Phi_{1\ell}^{Em}$, $g_{1\ell}^{Em}$, V_ℓ^m, Q_ℓ^m as components:

$$
\overline{\sigma}_\ell^m = \begin{pmatrix} U_\ell^m \\ P_\ell^m \\ \Phi_{1\ell}^{Em} \\ g_{1\ell}^{Em} \\ V_\ell^m \\ Q_\ell^m \end{pmatrix},
\tag{7.292}
$$

and a vector $\overline{\tau}_\ell^m$ with the toroidal variables W_ℓ^m, R_ℓ^m as components:

$$
\overline{\tau}_\ell^m = \begin{pmatrix} W_\ell^m \\ R_\ell^m \end{pmatrix}.
\tag{7.293}
$$

We use σ_ℓ^m and τ_ℓ^m to denote the spheroidal and toroidal parts of the deformation vectors. They can be written in terms of the basis $(\hat{e}_0, \hat{e}_+, \hat{e}_-)$ as:

$$
\sigma_\ell^m = U_\ell^m D_\ell^{m0}\hat{e}_0 + \frac{1}{2}V_\ell^m \left(D_\ell^{m+}\hat{e}_+ + D_\ell^{m-}\hat{e}_-\right),
\tag{7.294}
$$

$$
\tau_\ell^m = \frac{1}{2}W_\ell^m \left(D_\ell^{m+}\hat{e}_+ - D_\ell^{m-}\hat{e}_-\right).
\tag{7.295}
$$

And the total displacement field **s** can be written:

$$\mathbf{s} = \sum_{\ell=0}^{+\infty} \sum_{m=-\ell}^{\ell} \left(\sigma_\ell^m + \tau_\ell^m \right). \tag{7.296}$$

Now, a careful inspection of the equations (7.273)–(7.291) reveals that they connect any spheroidal displacement σ_ℓ^m with toroidal displacements of the same order m but of degrees $\ell \pm 1, \ell \pm 3, \ldots$ subject to the degree being $\geq m$, and similarly for the toroidal displacements. Thus we have, for any m, two independent infinite sets of fields, with the members of each set interconnected by the set of equations referred to above. For a fixed m:

$$\mathbf{s} = \tau_{|m|}^m + \sigma_{|m|+1}^m + \tau_{|m|+2}^m + \sigma_{|m|+3}^m + \cdots \tag{7.297}$$

or

$$\mathbf{s} = \sigma_{|m|}^m + \tau_{|m|+1}^m + \sigma_{|m|+2}^m + \tau_{|m|+3}^m + \cdots \tag{7.298}$$

Scalar expressions of the boundary conditions

We require continuity of the normal stress at any boundary between two layers within the Earth (Eq. (17.206)). This condition calls for continuity of the following quantities relating to radial and tangential scalar fields appearing in the stress tensor:

$$P_\ell^m + h_2 \sum_{\ell'=|\ell-2|}^{|\ell+2|} \left[\begin{vmatrix} \ell & 2 & \ell' \\ 0 & 0 & 0 \\ m & 0 & m \end{vmatrix} \frac{d}{dr} P_{\ell'}^m - \sqrt{3} \begin{vmatrix} \ell & 2 & \ell' \\ 0 & 1 & -1 \\ m & 0 & m \end{vmatrix} \left\{ \begin{matrix} Q_{\ell'}^m \\ -R_{\ell'}^m \end{matrix} \right\} \right], \tag{7.299}$$

and

$$Q_\ell^m + h_2 \sum_{\ell'=|\ell-2|}^{|\ell+2|} \begin{vmatrix} \ell & 2 & \ell' \\ 1 & 0 & 1 \\ m & 0 & m \end{vmatrix} \frac{d}{dr} \left\{ \begin{matrix} Q_{\ell'}^m \\ R_{\ell'}^m \end{matrix} \right\}$$

$$+ h_2 \frac{\sqrt{3}}{r} \sum_{\ell'=|\ell-2|}^{|\ell+2|} \begin{vmatrix} \ell & 2 & \ell' \\ 1 & 1 & 0 \\ m & 0 & m \end{vmatrix} \left\{ \begin{matrix} \lambda_0 X_{\ell'}^m + \frac{\mu_0}{r} \left(L_0^{\ell'} V_{\ell'}^m + 2U_{\ell'}^m \right) \\ 0 \end{matrix} \right\}$$

$$+ h_2 \frac{\sqrt{3}}{r^2} \sum_{\ell'=|\ell-2|}^{|\ell+2|} \begin{vmatrix} \ell & 2 & \ell' \\ 1 & -1 & 2 \\ m & 0 & m \end{vmatrix} L_2^{\ell'} \mu_0 \left\{ \begin{matrix} V_{\ell'}^m \\ -W_{\ell'}^m \end{matrix} \right\}, \tag{7.300}$$

and

$$R_\ell^m + h_2 \sum_{\ell'=|\ell-2|}^{|\ell+2|} \begin{vmatrix} \ell & 2 & \ell' \\ 1 & 0 & 1 \\ m & 0 & m \end{vmatrix} \frac{d}{dr} \left\{ \begin{matrix} R_{\ell'}^m \\ Q_{\ell'}^m \end{matrix} \right\}$$

$$+ h_2 \frac{\sqrt{3}}{r} \sum_{\ell'=|\ell-2|}^{|\ell+2|} \begin{vmatrix} \ell & 2 & \ell' \\ 1 & 1 & 0 \\ m & 0 & m \end{vmatrix} \left\{ \begin{matrix} 0 \\ \lambda_0 X_{\ell'}^m + \frac{\mu_0}{r} \left(L_0^{\ell'} V_{\ell'}^m + 2U_{\ell'}^m \right) \end{matrix} \right\}$$

$$+ h_2 \frac{\sqrt{3}}{r^2} \sum_{\ell'=|\ell-2|}^{|\ell+2|} \begin{vmatrix} \ell & 2 & \ell' \\ 1 & -1 & 2 \\ m & 0 & m \end{vmatrix} L_2^{\ell'} \mu_0 \left\{ \begin{matrix} W_{\ell'}^m \\ V_{\ell'}^m \end{matrix} \right\}, \tag{7.301}$$

where the last terms in the last three expressions above can be dropped in the solid–solid boundary case at the first order in h_2 because of the continuity conditions Eqs. (7.300), (7.301), (7.304), and (7.305).

Continuity of the normal displacement at fluid–solid boundaries requires vanishing of:

$$U_\ell^m + h_2 \sum_{\ell'=|\ell-2|}^{|\ell+2|} \left[\begin{vmatrix} \ell & 2 & \ell' \\ 0 & 0 & 0 \\ m & 0 & m \end{vmatrix} \frac{d}{dr} U_{\ell'}^m - \frac{\sqrt{3}}{r} \begin{vmatrix} \ell & 2 & \ell' \\ 0 & 1 & -1 \\ m & 0 & m \end{vmatrix} \left\{ \begin{matrix} V_{\ell'}^m \\ -W_{\ell'}^m \end{matrix} \right\} \right]. \tag{7.302}$$

At all solid–solid boundaries, we need the continuity of all the displacement field components:

$$U_\ell^m + h_2 \sum_{\ell'=|\ell-2|}^{|\ell+2|} \begin{vmatrix} \ell & 2 & \ell' \\ 0 & 0 & 0 \\ m & 0 & m \end{vmatrix} \frac{d}{dr} U_{\ell'}^m, \tag{7.303}$$

$$V_\ell^m + h_2 \sum_{\ell'=|\ell-2|}^{|\ell+2|} \begin{vmatrix} \ell & 2 & \ell' \\ 1 & 0 & 1 \\ m & 0 & m \end{vmatrix} \left\{ \begin{matrix} \frac{d}{dr} V_{\ell'}^m \\ \frac{d}{dr} W_{\ell'}^m \end{matrix} \right\}, \tag{7.304}$$

$$W_\ell^m + h_2 \sum_{\ell'=|\ell-2|}^{|\ell+2|} \begin{vmatrix} \ell & 2 & \ell' \\ 1 & 0 & 1 \\ m & 0 & m \end{vmatrix} \left\{ \begin{matrix} \frac{d}{dr} W_{\ell'}^m \\ \frac{d}{dr} V_{\ell'}^m \end{matrix} \right\}. \tag{7.305}$$

The continuity of the mass redistribution potential and Eqs. (7.208) and (7.209) require the continuity conditions on $g_{1\ell}^{Em}$ and $\Phi_{1\ell}^{Em}$:

$$\Phi_{1\ell}^{Em} + h_2 \sum_{\ell'=|\ell-2|}^{|\ell+2|} \begin{vmatrix} \ell & 2 & \ell' \\ 0 & 0 & 0 \\ m & 0 & m \end{vmatrix} \frac{d}{dr} \Phi_{1\ell}^{Em}, \tag{7.306}$$

and

$$
g_{1\ell}^{Em} + h_2 \sum_{\ell'=|\ell-2|}^{|\ell+2|} \left[\begin{vmatrix} \ell & 2 & \ell' \\ 0 & 0 & 0 \\ m & 0 & m \end{vmatrix} 4\pi G \left(-\frac{L_0^{\ell'}\rho_0}{r} V_{\ell'}^m - \frac{d\rho_0}{dr} U_{\ell'}^m \right) \right.
$$

$$
\left. - \begin{vmatrix} \ell & 2 & \ell' \\ 0 & 1 & -1 \\ m & 0 & m \end{vmatrix} 4\pi G \rho_0 \frac{\sqrt{3}}{r} \left\{ \begin{matrix} V_{\ell'}^m \\ -W_{\ell'}^m \end{matrix} \right\} \right].
\tag{7.307}
$$

By "continuity" of the various quantities considered above at some boundary, we mean of course that their values on one side of the boundary ($|_{int}$) should not differ from their respective values on the other side ($|_{ext}$) of the boundary, i.e. $|_{ext} - |_{int} = 0$ (here $|_{ext}$ does not mean at the exterior of the Earth (which is a particular case) but outside the boundary in the direction of the exterior of the Earth).

Examining the boundary conditions (7.299)–(7.301), (7.303)–(7.307) for solid–solid boundaries, we see that they are of the form:

$$
(1 + \mathcal{H})\mathcal{X}|_{int}^{ext} = (1 + \mathcal{H})\mathcal{X}|_{ext} - (1 + \mathcal{H})\mathcal{X}|_{int} = 0,
\tag{7.308}
$$

where $\mathcal{X}$ stands for any of the variables and $\mathcal{H}$ is proportional to h_2. We have already noted this on mentioning above that some terms can be dropped here. The case of solid–fluid boundaries will be examined later.

Resolution of the equations

We started from the set of vectorial equations:

- the Poisson equation (Eq. (7.198)),
- the constitutive equation (Eq. (7.199)), and
- the equations of motion (Eqs. (7.200)–(7.203)),

which were partial differential equations. These equations are transformed into a set of scalar ordinary differential equations in d/dr:

- the Poisson equation (Eq. (7.274) and Eq. (7.275)),
- the constitutive equation (Eq. (7.276), Eq. (7.277), Eq. (7.278)), and
- the equations of motion (Eq. (7.279), Eq. (7.280), Eq. (7.281)).

For a forcing at degree ℓ and order m, the set of scalar ordinary differential equations that must be considered involves $\overline{\sigma}_\ell^m$, $\overline{\tau}_{\ell+1}^m$, $\overline{\tau}_{\ell-1}^m$, $\overline{\sigma}_{\ell+2}^m$, $\overline{\sigma}_{\ell-2}^m$, $\overline{\tau}_{\ell+3}^m$, $\overline{\tau}_{\ell-3}^m$, ... on the one hand (the so-called "spheroidal solutions" because the main term is $\overline{\sigma}_\ell^m$) and $\overline{\tau}_\ell^m$, $\overline{\sigma}_{\ell+1}^m$, $\overline{\sigma}_{\ell-1}^m$, $\overline{\tau}_{\ell+2}^m$, $\overline{\tau}_{\ell-2}^m$, $\overline{\sigma}_{\ell+3}^m$, $\overline{\sigma}_{\ell-3}^m$, ... on the other hand (the so-called "toroidal solutions" because the main term is $\overline{\tau}_\ell^m$). On examining the ordinary differential equations, one sees that the whole set of equations can be subdivided into two sets of infinite numbers of equations, one involving the spheroidal solutions and the other involving the toroidal solutions. (We recall here that this separation was also the case for the spherical Earth, with independent solutions $\overline{\sigma}_\ell^m$ and $\overline{\tau}_\ell^m$ (see Chapter 6).) The two systems of equations each constitute an infinite set of equations that can only be solved if truncation is done. The systems can be truncated at the first order in the small quantities such as h_2. Here we concentrate on the "spheroidal" set of equations (involving $\overline{\sigma}_\ell^m$, $\overline{\tau}_{\ell+1}^m$, $\overline{\tau}_{\ell-1}^m$, $\overline{\sigma}_{\ell+2}^m$, $\overline{\sigma}_{\ell-2}^m$, $\overline{\tau}_{\ell+3}^m$, $\overline{\tau}_{\ell-3}^m$, ...) as this is the only one involving the forcing.

We denote $A_{\sigma\ell'}^m$ the spherical Earth matrix spheroidal operator in response to a degree ℓ' and order m (this matrix is defined and built from the coefficients of the y_i of Alterman *et al.* (1959) provided in Section 6.7.3, Eqs. (6.95)–(6.100)), and $A_{\tau\ell'}^m$ the spherical Earth matrix toroidal operator in response to a degree ℓ' and order m (this matrix is designed and built from the coefficients of the y_i of Alterman *et al.* (1959) provided in Section 6.7.3, Eqs. (6.101)–(6.102)). We denote ε (subscripts and superscripts ℓ' and m have been omitted for writing simplicity) the quantities of the first order in the flattening. Using this notation, the system of equations in response to a forcing ϕ^{ext} of degree ℓ and order m can be written:

$$
\begin{pmatrix}
\cdot & & & & & & & & \\
& \cdot & & & & & & & \\
\varepsilon & \varepsilon & A_{\sigma(\ell-2)}^m + \varepsilon & \varepsilon & \varepsilon & 0 & 0 & 0\ 0 \\
0 & \varepsilon & \varepsilon & A_{\tau(\ell-1)}^m + \varepsilon & \varepsilon & \varepsilon & 0 & 0\ 0 \\
0 & 0 & \varepsilon & \varepsilon & A_{\sigma\ell}^m + \varepsilon & \varepsilon & \varepsilon & 0\ 0 \\
0 & 0 & 0 & \varepsilon & \varepsilon & A_{\tau(\ell+1)}^m + \varepsilon & \varepsilon & \varepsilon\ 0 \\
0 & 0 & 0 & 0 & \varepsilon & \varepsilon & A_{\sigma(\ell+2)}^m + \varepsilon & \varepsilon\ \varepsilon \\
& & & & & & & & \cdot
\end{pmatrix}
\begin{pmatrix}
\cdot \\ \cdot \\ \overline{\sigma}_{\ell-2}^m \\ \overline{\tau}_{\ell-1}^m \\ \overline{\sigma}_\ell^m \\ \overline{\tau}_{\ell+1}^m \\ \overline{\sigma}_{\ell+2}^m \\ \cdot \\ \cdot
\end{pmatrix}
=
\begin{pmatrix}
0 \\ 0 \\ 0 \\ 0 \\ f(\phi^{ext}) \\ 0 \\ 0 \\ 0 \\ 0
\end{pmatrix}.
$$

$$(7.309)$$

One needs then to consider only the first order in the flattening (or h_2). Knowing that $\bar{\sigma}_{\ell+2}^m$ and $\bar{\sigma}_{\ell-2}^m$ are of the order of ε of $\bar{\sigma}_\ell^m$ and similarly for $\overline{\tau}_{\ell+3}^m$ and $\overline{\tau}_{\ell-3}^m$, and that $\bar{\sigma}_{\ell+4}^m$ and $\bar{\sigma}_{\ell-4}^m$ are of the order of ε of $\bar{\sigma}_{\ell+2}^m$ and $\bar{\sigma}_{\ell-2}^m$, respectively, and similarly for $\overline{\tau}_{\ell+3}^m$ and $\overline{\tau}_{\ell-3}^m$, we can truncate the system to:

$$
\begin{pmatrix}
A_{\sigma(\ell-2)}^m & \varepsilon & \varepsilon & 0 & 0 \\
\varepsilon & A_{\tau(\ell-1)}^m + \varepsilon & \varepsilon & \varepsilon & 0 \\
\varepsilon & \varepsilon & A_{\sigma\ell}^m + \varepsilon & \varepsilon & \varepsilon \\
0 & \varepsilon & \varepsilon & A_{\tau(\ell+1)}^m + \varepsilon & \varepsilon \\
0 & 0 & \varepsilon & \varepsilon & A_{\sigma(\ell+2)}^m
\end{pmatrix}
\begin{pmatrix}
\overline{\sigma}_{\ell-2}^m \\ \overline{\tau}_{\ell-1}^m \\ \overline{\sigma}_\ell^m \\ \overline{\tau}_{\ell+1}^m \\ \overline{\sigma}_{\ell+2}^m
\end{pmatrix}
=
\begin{pmatrix}
0 \\ 0 \\ f(\phi^{ext}) \\ 0 \\ 0
\end{pmatrix}.
$$

$$(7.310)$$

Noting that $\overline{\sigma}_{\ell-2}^m$ and $\overline{\sigma}_{\ell+2}^m$ are also small quantities, their product with ε is of second order and the first and last columns change as well:

$$
\begin{pmatrix}
A_{\sigma(\ell-2)}^m & \varepsilon & \varepsilon & 0 & 0 \\
0 & A_{\tau(\ell-1)}^m + \varepsilon & \varepsilon & \varepsilon & 0 \\
0 & \varepsilon & A_{\sigma\ell}^m + \varepsilon & \varepsilon & 0 \\
0 & \varepsilon & \varepsilon & A_{\tau(\ell+1)}^m + \varepsilon & 0 \\
0 & 0 & \varepsilon & \varepsilon & A_{\sigma(\ell+2)}^m
\end{pmatrix}
\begin{pmatrix}
\overline{\sigma}_{\ell-2}^m \\
\overline{\tau}_{\ell-1}^m \\
\overline{\sigma}_{\ell}^m \\
\overline{\tau}_{\ell+1}^m \\
\overline{\sigma}_{\ell+2}^m
\end{pmatrix}
=
\begin{pmatrix}
0 \\
0 \\
f(\phi^{ext}) \\
0 \\
0
\end{pmatrix}.
$$

(7.311)

The second, third, and fourth equations of this super-truncated system can be solved separately, and the first and the last of these equations can be solved, introducing the just-obtained solution for $\overline{\tau}_{\ell-1}^m$, $\overline{\sigma}_{\ell}^m$, and $\overline{\tau}_{\ell+1}^m$. The super-truncated system yields:

$$
\begin{pmatrix}
A_{\tau(\ell-1)}^m + \varepsilon & \varepsilon & \varepsilon \\
\varepsilon & A_{\sigma\ell}^m + \varepsilon & \varepsilon \\
\varepsilon & \varepsilon & A_{\tau(\ell+1)}^m + \varepsilon
\end{pmatrix}
\begin{pmatrix}
\overline{\tau}_{\ell-1}^m \\
\overline{\sigma}_{\ell}^m \\
\overline{\tau}_{\ell+1}^m
\end{pmatrix}
=
\begin{pmatrix}
0 \\
f(\phi^{ext}) \\
0
\end{pmatrix}.
$$
(7.312)

The system involves six spheroidal variables in $\overline{\sigma}_{\ell}^m$ and 2×2 toroidal variables in $\overline{\tau}_{\ell-1}^m$ and $\overline{\tau}_{\ell+1}^m$. The vector of variables in this system is denoted $\mathcal{Y}_{\ell}^m$:

$$
\mathcal{Y}_{\ell}^m =
\begin{pmatrix}
\overline{\tau}_{\ell-1}^m \\
\overline{\sigma}_{\ell}^m \\
\overline{\tau}_{\ell+1}^m
\end{pmatrix}.
$$
(7.313)

The final system of equations that must be solved is the system (7.276), (7.279), (7.274), (7.275), (7.277), (7.280), for the spheroidal variables, and

$$
\frac{dW_{\ell\pm1}^m}{dr} = \frac{1}{\mu_0} R_{\ell\pm1}^m + \frac{1}{1} r W_{\ell\pm1}^m
$$

$$
- \frac{\mu_2}{\mu_0} \sum_{\ell'=|\ell\pm2|,|\ell\pm1|,\ or\ \ell}
\begin{vmatrix}
\ell\pm1 & 2 & \ell' \\
1 & 0 & 1 \\
m & 0 & m
\end{vmatrix}
\left\{
\begin{matrix}
R_{\ell'}^m \\
Q_{\ell'}^m
\end{matrix}
\right\},
$$
(7.314)

and

$$
\frac{dR_{\ell\pm1}^m}{dr} = -\frac{3}{r} R_{\ell\pm1}^m + \frac{2 L_0^{\ell\pm1} L_2^{\ell\pm1}}{r} \mu_0 W_{\ell\pm1}^m + \rho_0 \left(\eta_{1\ell\pm1}^{Wm} + \eta_{2\ell\pm1}^{Wm} + \eta_{3\ell\pm1}^{Wm} \right)
$$

$$
+ \sum_{\ell'=|\ell\pm2|,|\ell\pm1|,\ or\ \ell}
\begin{vmatrix}
\ell\pm1 & 2 & \ell' \\
1 & 0 & 1 \\
m & 0 & m
\end{vmatrix}
\rho_2
\left\{
\begin{matrix}
\eta_{1\ell'}^{Wm} + \eta_{2\ell'}^{Wm} + \eta_{3\ell'}^{Wm} \\
\eta_{1\ell'}^{Vm} + \eta_{2\ell'}^{Vm} + \eta_{3\ell'}^{Vm}
\end{matrix}
\right\}
$$

$$
+ \frac{2 L_0^{\ell}}{r} \mu_0 \sum_{\ell'=|\ell\pm2|,|\ell\pm1|,\ or\ \ell}
\begin{vmatrix}
\ell & 2 & \ell' \\
2 & 0 & 2 \\
m & 0 & m
\end{vmatrix}
L_2^{\ell'}
\left\{
\begin{matrix}
W_{\ell'}^m \\
V_{\ell'}^m
\end{matrix}
\right\},
$$
(7.315)

for the toroidal variables. On examining the forms of $A_{\sigma\ell}{}^m$ and $A_{\tau(\ell\pm1)}{}^m$ we see that they consist of d/dr plus linear combinations of the variables and terms that are linear combinations of the variables. The system (7.312) can thus be written

$$\frac{d}{dr}\mathcal{Y}_\ell^m + \mathcal{A}_\ell^m \mathcal{Y}_\ell^m = \begin{pmatrix} 0 \\ f(\phi^{ext}) \\ 0 \end{pmatrix}. \tag{7.316}$$

This system of ten ordinary differential equations is valid for the solid parts of the Earth. It is reduced for the liquid part of the Earth as the stress components vanish in a fluid, as will be seen in the next paragraph. After integration, one gets the mass redistribution inside the Earth, as well as the total displacement field at the surface of the Earth, which includes the toroidal displacement components from which nutations can be obtained.

Solutions in the fluid part of the Earth

Exactly as in the case of a spherical Earth, the equations in the fluid part of the Earth (outer core) are different than in the solid parts (inner core and mantle). The shear modulus μ_0 (and μ_2) are equal to zero in the FOC. The equations that must be used in the core are thus different, but they are not reproduced here as they can be retrieved exactly in the same way as for the spherical Earth case (see Section 6.7.4): Q_ℓ^m and $R_{\ell\pm1}^m$ are equal to zero in the fluid; the velocities V_ℓ^m and $W_{\ell\pm1}^m$ in the fluid can be retrieved from the other variables as in the spherical Earth case for V_ℓ^m and from putting $R_{\ell\pm1}^m = 0$ in Eq. (7.315) for $W_{\ell\pm1}^m$. The fluid equivalent to system (7.316) is the same as in the spherical Earth case and can be written

$$\frac{d}{dr}\mathcal{Y}_{\text{fl }\ell}{}^m + \mathcal{A}_{\text{fl }\ell}{}^m \mathcal{Y}_{\text{fl }\ell}{}^m = f_{\text{fl}}(\phi^{\text{ext}}), \tag{7.317}$$

where

$$\mathcal{Y}_{\text{fl }\ell}{}^m = \begin{pmatrix} U_\ell^m \\ P_\ell^m \\ \Phi_{1\ell}^{Em} \\ g_{1\ell}^{Em} \end{pmatrix}. \tag{7.318}$$

Using PREM to get the solutions

The systems of Eqs. (7.316) and (7.317) are sets of ordinary differential equations to which the original partial differential equations have been reduced, with the help of GSH expansions and after truncation to the first order in the ellipticity. In the fluid case, we have obtained a system of four ordinary differential equations for four variables, and in the solid case we have obtained a system of ten ordinary differential equations for ten variables (six spheroidal, of the same degree ℓ as the degree of the tidal forcing, four toroidal, two of them being of degree $\ell + 1$ and the two others of degree $\ell - 1$). These systems must be integrated inside the Earth (often from the center up to the surface).

The properties of the Earth, such as the mean density at a certain radius (density profile inside the Earth), the mean bulk and shear moduli profiles are needed to solve the set of ten or four equations. They may be taken from the seismic PREM model (Dziewonski and Anderson, 1981) or any other model providing seismic profiles of the rheological properties such as the ak135 model (Kennett *et al.*, 1995).

The initial gravity (or equivalently gravitational potential) profile and the flattening profile are obtained from solving Clairaut's equations for an ellipsoidal Earth (see Appendix B).

Combining mass redistribution potential and external tidal potential

Note that the equations of motion are non-homogeneous equations in the presence of the external potential (i.e. there is a right-hand side in the system shown in (7.312)), and that the boundary conditions (continuity on both sides $|_{int}$ and $|_{ext}$, i.e. $|_{int} - |_{ext} = 0$) are homogeneous everywhere inside the Earth as well as at the surface of the Earth (i.e. the external potential does not enter the boundary condition expressions). However, as mentioned previously (see paragraph on "The forcing" in Section 7.5.1, Eq. (7.194)), it is possible to reverse the situation by combining the external potential with the mass redistribution potential in a new variable:

$$\Phi_\ell^m(r) = \Phi_{1\ell}^{Em}(r) + \Phi_\ell^{\text{ext } m}(r), \tag{7.319}$$

and similarly

$$G_{1\ell}^{Em}(r) = g_{1\ell}^{Em}(r) + \frac{d}{dr}\Phi_\ell^{\text{ext } m}(r). \tag{7.320}$$

On using these new variables $\Phi_\ell^m(r)$ and $G_{1\ell}^{Em}$ instead of $\Phi_{1\ell}^{Em}$ and $g_{1\ell}^{Em}$, the ten (or four in the fluid case) ordinary differential equations then become homogeneous, while the surface boundary conditions involve the external potential explicitly (see e.g. Wahr, 1979, 1981a–d, Dehant, 1986).

On replacing $\Phi_{1\ell}^{Em}$ and $g_{1\ell}^{Em}$ by Φ_ℓ^m and $G_{1\ell}^{Em}$ in $\mathcal{Y}_\ell^m$, the system (7.316) can be written

$$\frac{d}{dr}\mathcal{Y}_\ell^m + \mathcal{A}_\ell^m \mathcal{Y}_\ell^m = 0, \tag{7.321}$$

where $\mathcal{Y}_\ell^m$ now has another meaning (we keep this notation $\mathcal{Y}$ in what follows).

On replacing $\Phi_{1\ell}^{Em}$ and $g_{1\ell}^{Em}$ by Φ_{ℓ}^m and $G_{1\ell}^{Em}$ in $\mathcal{Y}_{\text{fl}\,\ell}^{\,m}$, the system (7.317) can be written

$$\frac{d}{dr}\mathcal{Y}_{\text{fl}\,\ell}^{\,m} + \mathcal{A}_{\text{fl}\,\ell}^{\,m}\mathcal{Y}_{\text{fl}\,\ell}^{\,m} = 0, \tag{7.322}$$

where again $\mathcal{Y}_{\text{fl}\,\ell}^{\,m}$ now has another meaning (we again keep this meaning for the notation $\mathcal{Y}_{\text{fl}}$ from now on).

Integration of the super-truncated system

These systems (Eqs. (7.321) and (7.322)) can be solved numerically (using classical ways of solving ordinary differential equations) in all the layers of the Earth using profiles for the rheological properties and the boundary conditions inside the Earth and at the surface. One (usually) starts from the solutions for a very small homogeneous spheroid at the very center (we can consider the solutions as provided in Chapter 6) and propagates them up to the surface.

Step one: at the very center

In principle one can thus start with ten independent solutions and integrate them, applying the boundary conditions. We take with Smith (1974), Wahr (1979), or Dehant (1986) the choice of integrating the "homogeneous" equations (with the change of variable including the external potential as explained above) starting from the center up to the surface where the boundary conditions involve the external potential. The boundary conditions at the very center express that only finite solutions for a homogeneous spheroid must be retained (i.e. five independent solutions are eliminated, similarly to Eqs. (6.114)–(6.119) where C_1, C_2, and C_6 were taken equal to zero, for a homogeneous sphere). Solutions for the incompressible case or compressible case are provided by Dahlen and Tromp (1998), Sabadini and Vermeersen (2004), and Denis (1993).

Step two: inside the inner core

We then "propagate" these five independent solutions using the ordinary differential equations for a solid (the super-truncated system (7.321)) and the continuity conditions of the variables (7.308) inside the inner core.

The boundary conditions that must be applied for the variables in $\mathcal{Y}$ at solid–solid boundaries have been mentioned in Eqs. (7.299), (7.300), (7.303), (7.304), (7.306), and (7.307), and in Eqs. (7.301) and (7.305) taking them for $\ell \pm 1$. On using their form (7.308), we find

$$(1 + \mathcal{H}|_{\text{ext}})\mathcal{Y}|_{\text{ext}} - (1 + \mathcal{H}|_{\text{int}})\mathcal{Y}|_{\text{int}} = 0 \;\; \text{or} = f(\Phi^{\text{ext}}) \;\text{at the Earth's surface.} \tag{7.323}$$

Considering that $\mathcal{H}$ is proportional to small quantities, they can be re-written

$$\mathcal{Y}|_{\text{ext}} = (1 + \mathcal{H}|_{\text{int}} - \mathcal{H}|_{\text{ext}})\mathcal{Y}|_{\text{int}} \text{ or } + f(\Phi^{\text{ext}}) \text{ at the Earth's surface}, \quad (7.324)$$

which shows that the variables on one side of a boundary can easily be obtained from their values on the other side on applying these conditions.

Step three: at the ICB

At the ICB, the continuity of the tangential stresses (Q_ℓ^m, $R_{\ell+1}^m$, and $R_{\ell-1}^m$), which must be zero on the fluid side, imposes the requirement that these quantities must vanish on the solid side also on the boundary, which reduces the number of independent solutions by three. But the free-slip conditions for the tangential displacements allow us to consider their values in the fluid part completely independent of their values in the solid part. Their values in the core are determined from the values of the other variables in the fluid. We now may go on and propagate the two independent solutions in the fluid core.

Step four: at the CMB

At the CMB, we again have the free-slip conditions adding three solutions accounting for discontinuity in V_ℓ^m, $W_{\ell+1}^m$, and $W_{\ell-1}^m$. So we are back with five independent solutions in the mantle.

We now detail how to use the boundary conditions at the CMB for passing the two independent solutions built in the fluid core into the mantle. This needs a particular treatment as the form (7.324) cannot be used due to vanishing variables in the fluid and, to a minor extent, to the fact that the toroidal displacements appearing in the core in the ellipsoidal Earth case may be quite large (resonance effects).

At the solid–fluid boundary, we have two independent solutions as mentioned above and four continuity conditions (Eqs. (7.302), (7.299), (7.306), and (7.307)) which involve the tangential displacement fields of degree ℓ and $\ell \pm 1$. We have as well three continuity conditions (Eqs. (7.300), (7.301) taken for the degree $\ell - 1$, and (7.301) taken for the degree $\ell + 1$). These are the most complicated boundary conditions:

$$(1 + \mathcal{H}'|_{\text{mantle}})\mathcal{Y}|_{\text{mantle}} = (1 + \mathcal{H}'|_{\text{fluid core}})\mathcal{Y}|_{\text{fluid core}}, \quad (7.325)$$

where

- $\mathcal{H}'|_{\text{mantle}}$ is a matrix of seven lines and ten columns (the three continuity conditions of the tangential displacements Eqs. (7.304) and (7.305) for the degree $\ell \pm 1$ have disappeared);
- $\mathcal{Y}|_{\text{mantle}}$ is the classical vector with all the variables of the super-truncated system of equations (Eq. (7.313));

- on considering that Q_ℓ^m and $R_{\ell\pm1}^m$ are equal to zero in the fluid part, $\mathcal{Y}|_{\text{fluid core}}$ is a vector of seven variables and $\mathcal{H}'|_{\text{fluid core}}$ is a matrix of seven lines and seven columns.

Four of the equations in (7.325) are dealing mainly with the continuity conditions for the variables provided in $\mathcal{Y}_{\text{fl}}$ and three of the equations in (7.325) are the continuity of the stress components Q_ℓ^m and $R_{\ell\pm1}^m$, which are equal to zero in the fluid part (we call them the "vanishing components"). One immediately sees the complexity involved in this boundary, which is in addition reinforced by the discontinuity in V_ℓ^m and $W_{\ell\pm1}^m$. As explained in Smith (1974) this needs a special treatment. A form like (7.324) cannot be derived. Indeed, for the boundary conditions involving the vanishing components, the parts proportional to h_2 are not smaller than the variables themselves: Q_ℓ^m and $R_{\ell\pm1}^m$ are zero in the core and are small in the mantle and the continuity conditions on the CMB for the vanishing variables cannot be expressed by replacing $h_2 Q_\ell^m|_{\text{mantle}}$ and $h_2 R_{\ell\pm1}^m|_{\text{mantle}}$ by $h_2 Q_\ell^m|_{\text{fluid}}$ and $h_2 R_{\ell\pm1}^m|_{\text{fluid}}$ as done in solid–solid boundaries. We need to proceed by successive approximations.

We exploit the fact that for a fluid–solid boundary as the CMB, all "nonvanishing components" of $\mathcal{H}|_{\text{fluid}}\mathcal{Y}|_{\text{fluid}}$ are small compared to the variables in $\mathcal{Y}|_{\text{fluid}}$. This allows us to use a form like (7.324) for the four non-vanishing conditions and to get first approximations for the variables $U_\ell^m|_{\text{mantle}}$, $P_\ell^m|_{\text{mantle}}$, $\Phi_\ell^m|_{\text{mantle}}$, and $G_{1\ell}^{Em}|_{\text{mantle}}$. We then use these values in the conditions of the vanishing components, on using also $Q_\ell^m|_{\text{fluid}} = R_{\ell\pm1}^m|_{\text{fluid}} = 0$ and the relation between $V_\ell^m|_{\text{fluid}}$ and $W_{\ell\pm1}^m|_{\text{fluid}}$ and the other four variables $U_\ell^m|_{\text{fluid}}$, $P_\ell^m|_{\text{fluid}}$, $\Phi_\ell^m|_{\text{fluid}}$ and $G_\ell^m|_{\text{fluid}}$. We first get values for $Q_\ell^m|_{\text{mantle}}$ and $R_{\ell\pm1}^m|_{\text{mantle}}$. We substitute these values (indicated by a superscript (1)) in the boundary conditions (7.325):

$$\mathcal{Y}^{(2)}|_{\text{mantle}} = (1 + \mathcal{H}'|_{\text{fluid core}})\mathcal{Y}|_{\text{fluid core}} - \mathcal{H}'|_{\text{mantle}}\mathcal{Y}^{(1)}|_{\text{mantle}} \qquad (7.326)$$

and re-iterate as necessary.

Step five: in the mantle up to the surface
We now propagate inside the mantle the two so-obtained independent solutions in the mantle plus the three coming from the discontinuity conditions at the CMB.

Five boundary conditions at the Earth's surface ensure the uniqueness of the solution. This approach is similar to the approach used for the spherical Earth (see Section 6.7.7).

7.5.2 *Using the solutions to get the nutations*

The response to an external tesseral potential of degree two and order one includes the displacements σ_2^1, τ_1^1, and τ_3^1. The last part of the displacement field is very small, but τ_1^1 can be quite large as it represents the nutation (a rotation around an axis in the equator). Indeed, let us consider Eq. (7.295) in the case of $\ell = 1$ and $m = 1$. Replacing D_1^{1+} and D_1^{1-} using their definitions one obtains

$$\tau_1^1 = \frac{1}{2} W_1^1 \left(\frac{1}{2}(1 + \cos\theta)\hat{\mathbf{e}}_+ - \frac{1}{2}(1 - \cos\theta)\hat{\mathbf{e}}_- \right) e^{i\lambda}. \tag{7.327}$$

On replacing $\hat{\mathbf{e}}_-$ and $\hat{\mathbf{e}}_+$ using Eqs. (7.210) by $\frac{1}{\sqrt{2}}(\hat{\mathbf{e}}_\theta - i\hat{\mathbf{e}}_\lambda)$ and $-\frac{1}{\sqrt{2}}(\hat{\mathbf{e}}_\theta + i\hat{\mathbf{e}}_\lambda)$, respectively, one obtains:

$$\tau_1^1 = -\frac{1}{2\sqrt{2}} W_1^1 (\hat{\mathbf{e}}_\theta + i\cos\theta\hat{\mathbf{e}}_\lambda) e^{i\lambda}. \tag{7.328}$$

Since the Cartesian components of $\hat{\mathbf{e}}_\theta$ and $\hat{\mathbf{e}}_\lambda$ are

$$\hat{\mathbf{e}}_\theta = \cos\theta\cos\lambda\hat{\mathbf{e}}_x + \cos\theta\sin\lambda\hat{\mathbf{e}}_y - \sin\theta\hat{\mathbf{e}}_z \tag{7.329}$$

and

$$\hat{\mathbf{e}}_\lambda = -\sin\lambda\hat{\mathbf{e}}_x + \cos\lambda\hat{\mathbf{e}}_y, \tag{7.330}$$

we find that the components of τ_1^1 are

$$
\begin{aligned}
\tau_1^1 &= -\frac{1}{2\sqrt{2}} W_1^1 \left[(\cos\theta\cos\lambda\hat{\mathbf{e}}_x + \cos\theta\sin\lambda\hat{\mathbf{e}}_y - \sin\theta\hat{\mathbf{e}}_z) \right. \\
&\quad \left. + i\cos\theta(-\sin\lambda\hat{\mathbf{e}}_x + \cos\lambda\hat{\mathbf{e}}_y) \right] e^{i\lambda} \\
&= -\frac{1}{2\sqrt{2}} W_1^1 \begin{pmatrix} \cos(\theta) \\ i\cos(\theta) \\ -\cos(\lambda)\sin(\theta) - i\sin(\lambda)\sin(\theta) \end{pmatrix},
\end{aligned}
\tag{7.331}
$$

while

$$
\begin{aligned}
\tau_1^1 &= \frac{-1}{2\sqrt{2}} W_1^1 \left((-i)(\hat{\mathbf{e}}_x + i\hat{\mathbf{e}}_y) \wedge \hat{\mathbf{e}}_r \right) \\
&= -\frac{1}{2\sqrt{2}} W_1^1 \begin{pmatrix} \cos(\theta) \\ i\cos(\theta) \\ -\cos(\lambda)\sin(\theta) - i\sin(\lambda)\sin(\theta) \end{pmatrix},
\end{aligned}
\tag{7.332}
$$

from which we see that τ_1^1 represents a rotation around an axis in the equator (in the plane $(\hat{\mathbf{e}}_x, \hat{\mathbf{e}}_y)$), i.e. a nutation.

Although not related to nutation, we now consider for completeness the response to an external zonal potential of order two and degree zero; it includes the displacements σ_2^0, τ_1^0, and τ_3^0. The last part of the displacement is very small, but τ_1^0 can be quite large as it represents the rotation variations induced by the zonal tidal potential (a rotation around the $\hat{\mathbf{e}}_z$-axis). Indeed, let us consider Eq. (7.295) in the case of $\ell = 1$ and $m = 0$. Replacing D_1^{0+} and D_1^{0-}, one obtains similarly to the above:

$$\tau_1^0 = \frac{1}{2} W_1^0 \left(i \sin \theta \hat{\mathbf{e}}_\lambda \right). \tag{7.333}$$

On replacing $\hat{\mathbf{e}}_\lambda$ using Eq. (7.330), we find τ_1^0 to be

$$\tau_1^0 = \frac{i}{2} W_1^0 \left(\hat{\mathbf{e}}_z \wedge \hat{\mathbf{e}}_r \right), \tag{7.334}$$

from which we see that τ_1^0 represents a rotation around the z-axis.

Coming back to Eq. (7.332), we now want to link the amplitude of τ_1^1 to a non-rigid Earth nutation theory. We recall that the reference frame in which we compute the solutions to get the mass redistribution and displacement field (including nutation for a degree two order one external potential) in response to an external forcing is a frame tied to a uniformly rotating Earth. The frame is tied to the equator and the mean figure axis of the Earth, also called the Tisserand axis.

In order to get the nutations of the other axis, specific computations are necessary, as explained in full detail in Wahr (1979): (1) for the instantaneous figure axis, we have to account for the tidal deformations changing the moment of inertia of the Earth; (2) for the rotation vector, we have to account for the displacement field changing the rotation/position of each point inside each layer of the Earth (inner core, outer core, and mantle) and thus changing the mean position (over the whole deformed ellipsoidal volume of the layer considered) of the rotation vector as well (for this computation it is necessary to invoke the minimum rotation condition); and (3) for the angular momentum axis, we have to consider the instantaneous inertia tensor affected by the mass redistribution and the instantaneous rotation vector.

7.5.3 Non-rigid Earth models: transfer function, Bratio, and convolution

It may be recalled that the concept of the transfer function was introduced in Section 2.27. In the following, we shall denote the frequency of a spectral component of the nutation in cpsd by ν, while the notation used in Chapter 2 was $\omega_n \equiv \nu \Omega_0$, for the frequency in rad/unit time.

The amplitude $\tilde{\eta}(\nu)$ of a circular nutation produced by a spectral component of the tidal potential is proportional to the amplitude of that spectral component, whether the Earth model employed be a rigid one or non-rigid. Therefore the ratio

$$T(\nu) = \frac{\tilde{\eta}_{NR}(\nu)}{\tilde{\eta}_R(\nu)} \tag{7.335}$$

of the nutation amplitude of a non-rigid Earth model to that of the rigid Earth for a given frequency v is independent of the amplitude of the perturbing potential (see Eq. (2.109)); it is determined solely by the structure and properties of the non-rigid Earth. The *transfer function* $T(v)$ is called the "Bratio" by Wahr (1979, 1981a–d); it is defined *only for circular nutations*. Once it is computed for a chosen non-rigid Earth interior model for any frequency v of the excitation, the non-rigid Earth amplitude $\tilde{\eta}_{NR}(v)$ is obtained simply as $\tilde{\eta}_{NR}(v) = T(v)\tilde{\eta}_R(v)$. The practical advantage of this procedure is that the results available in the literature on the amplitudes of rigid Earth nutations may be employed directly to get the amplitudes for the non-rigid Earth without having to worry about the various kinds of contributions to the tidal potential (e.g., from the Sun's influence on the orbit of the Moon, see Chapter 5), which are systematically taken into account in the rigid Earth calculations. It is sufficient to compute the nutation amplitude of the chosen non-rigid Earth model per unit amplitude of the potential, and use the known expression for the corresponding amplitude of the rigid Earth. The sole Earth parameter on which the rigid Earth amplitudes depend is the dynamical ellipticity e. In constructing a rigid Earth nutation theory, the value of e (or rather, the related quantity $H_d = e/(1+e)$) is inferred from observational data on the rate of precession of the real Earth. In constructing the non-rigid Earth transfer function, the dynamical ellipticity e (or H_d) corresponds to the value obtained by computing the hydrostatic equilibrium and the corresponding geometrical flattening (Clairaut's theory, see Appendix B). In a precise computation of the nutations this must be taken into account (see Chapter 10).

A circular nutation with $v > 0$ is prograde according to our convention, and one with $v < 0$ is retrograde. A pair of circular nutations, one with frequency $v = v_p > 0$ and the other with frequency $v = -v_p$, together constitute an elliptical nutation which is usually represented in terms of nutations in longitude and obliquity with a common frequency. The way to go from the circular motions to the elliptical motions has been detailed in Sections 2.18 and 2.19.

8

Anelasticity

When an *elastic* body is subjected to some force, its deformational response is instantaneous; if the forcing is periodic, the response has the same period and the same phase as the forcing.

In some types of materials, in addition to the instantaneous elastic response, there is a small additional delayed response which keeps changing with time. The behavior of *anelastic* and *viscoelastic* materials is of this kind. The response of such materials to periodic forcing has, of course, the same periodicity, but the amplitude of the response is slightly larger than in the absence of anelasticity, and the phase lags behind that of the forcing, with a consequent dissipation of the energy of the oscillations in the material. Both these effects are dependent on the forcing frequency.

In terms of material properties, *elastic* materials strain instantaneously when a stress is applied and quickly return to their original state once the stress is removed; *viscous* materials resist shear and also undergo strain when a stress is applied but do not return instantaneously to their original state once the stress is removed; *viscoelastic* materials have elements of both of these properties and exhibit a time-dependent strain in response to stress; *anelastic* materials represent a subset of viscoelastic materials: they ultimately recover fully after removal of a transient stress.

Some properties of viscoelastic/anelastic materials can be put forward, which summarize the above statements: (1) if the stress is held constant, the strain increases with time (*creep*); (2) if the strain is held constant, the stress decreases with time (*relaxation*); if cyclic stress is applied, *hysteresis* (a phase lag) occurs, leading to a dissipation of mechanical energy. Anelastic dissipation results also in damping of the free oscillations of anelastic bodies.

The Earth's mantle behaves anelastically in responding to forcing over a wide range of frequencies (e.g. seismic frequencies, free oscillation frequencies, tidal frequencies). The fact that seismic waves die out with the passage of time after an earthquake or that the free oscillations initiated in the Earth after an earthquake also die out after some time are consequences of the dissipation of energy. Models of the Earth's structure and properties that are constructed from seismic travel times and normal modes include information about the energy dissipation. It has been inferred from such data that the lower mantle (below a depth of 670 km from the surface) is more dissipative than the upper mantle. This observation is reflected in models of mantle anelasticity.

8.1 Complex rheological parameters for the anelastic Earth

Mantle anelasticity causes a small frequency-dependent phase lag in the Earth's response to periodic forcing, besides altering the magnitude of the response. The shear and bulk moduli, which are the parameters that characterize the stresses needed to produce unit shear or compression, become complex and frequency dependent in the case of anelastic materials.

Mathematically, when considering the Earth response $r(t)$ to a periodic forcing $f(t) = Fe^{i\sigma t}$ (with F real, > 0) at a particular frequency, the elastic response would be $r(t) = Re^{i\sigma t}$, while the anelastic response will be altered by a change in the amplitude as a function of time and by the presence of a phase lag:

$$r(t) = \tilde{R}(t)e^{i\sigma t} = R(t)e^{i\beta}e^{i\sigma t} = R(t)e^{i(\sigma t + \beta)} = R(t)e^{i\sigma(t + \frac{\beta}{\sigma})}. \quad (8.1)$$

In order that the phase shift of Eq. (8.1) relative to the forcing $f(t)$ be a lag, it is necessary that

$$\frac{\beta(\sigma)}{\sigma} < 0 \quad (8.2)$$

and hence that

$$\frac{\Im(\tilde{R})}{\sigma} < 0. \quad (8.3)$$

Thus the sign of $\Im(\tilde{R})$ must be opposite to that of the frequency σ.

Generally speaking, for an elastic body, the stress–strain relation is linear ($\tau = k\epsilon$) and k is constant with time. For a viscoelastic body and for unidirectional stress and strain, the stress–strain relationship in the time domain can be written as a linear function in the deformation ϵ and its time derivatives, and in the stress τ and its time derivatives:

$$a_0\epsilon + a_1\dot{\epsilon} + a_2\ddot{\epsilon} + \cdots = b_0\tau + b_1\dot{\tau} + b_2\ddot{\tau} + \cdots \quad (8.4)$$

On going over into the frequency domain, this relation yields:

$$a_0\epsilon(\sigma) + a_1 i\sigma\epsilon(\sigma) - a_2\sigma^2\epsilon(\sigma) + \cdots = b_0\tau(\sigma) + b_1 i\sigma\tau(\sigma) - b_2\sigma^2\tau(\sigma) + \cdots \tag{8.5}$$

or equivalently

$$\tau(\sigma) = k(\sigma)\,\epsilon(\sigma), \tag{8.6}$$

where $k(\sigma) = \Re(k(\sigma)) + i(-\sigma/|\sigma|)\Im(k(\sigma))$. The advantage of going over to the frequency domain is immediately apparent, as one obtains the same relation as the 1D elastic Hooke's law involving just one elastic constant, though now it involves a complex and frequency-dependent parameter. In the anelastic 1D case and in the time domain, the strain due to a particular stress is considered as not instantaneous; there is a retardation (a phase lag) due to the anelastic property of the material. In order to represent this, one uses a convolution with an elementary function, which consists of the real part, $\Re(k(\sigma))$, changing slightly with respect to the elastic case, and a strain retardation function ($\Im(k(\sigma))$), both typical for the timescale concerned (depending on the frequency). The imaginary part has been multiplied by $(-\sigma/|\sigma|)$ in order to ensure that the imaginary part leads to a phase lag.

This can be generalized for a 3D case such as for the deformation of a sphere, using a stress tensor and a strain tensor. For the 3D elastic case, as explained in Chapter 6, one uses Hooke's law (see e.g. the constitutive equation (6.34)). It is reproduced here for convenience:

$$\mathbf{T} = \lambda(\nabla \cdot \mathbf{u})\mathbf{I} + \mu(\nabla\mathbf{u} + (\nabla\mathbf{u})^T) \text{ with } \lambda = k - \frac{2\mu}{3}. \tag{8.7}$$

In the anelastic case, one can thus use the same Hooke's law but with complex and frequency-dependent rheological properties. The stress–strain relationship within the Earth then becomes complex in the frequency domain, involving complex and frequency dependent bulk and shear moduli ($k(\sigma)$, $\mu(\sigma)$, respectively).

Usually, the phase lags and damping (or equivalently the complex values of the rheological parameters) are observed at seismic frequencies and extrapolated to the other frequencies, such as those in the tidal frequency range, using rheological models including a particular frequency dependence.

The work of Sailor and Dziewonski (1978), Anderson and Minster (1979), and Sipkin and Jordan (1980) is significant in this connection. Variants of these have been treated in detail by Wahr and Bergen (1986); they conclude with the following power law variation of the real and imaginary parts of the shear modulus with excitation frequency σ:

$$\frac{\Re(\mu) + i(-\sigma/|\sigma|)\Im(\mu)}{\mu_0} - 1 = \frac{\left(\frac{2}{\pi}\ln\left(\frac{\sigma}{\sigma_0}\right) + i\right)}{Q_0}, \tag{8.8}$$

where $\mu_0(r)$ is the real part of the shear modulus of the anelastic Earth at radius r at the reference frequency σ_0 in the seismic frequency range, and $Q_0(r)$ is the quality factor of the anelastic Earth at radius r at the same reference frequency σ_0 $(Q_0 = (-\sigma_0/|\sigma_0|)\Re(\mu_0)/\Im(\mu_0))$. Again, the imaginary part has been multiplied by $(-\sigma/|\sigma|)$ in order to ensure that the imaginary part leads to a phase lag.

It is not sufficient, in general, to take Q to be not frequency dependent. Typically, the dependence, inferred from seismological studies, is of a power-law nature:

$$Q(\sigma) = Q_0(\sigma/\sigma_0)^\alpha, \tag{8.9}$$

where α is the index of the assumed power law variation of the quality factor Q of mantle dissipation. The variation of μ with σ has then been shown (Anderson and Minster, 1979) to be given by

$$\mu(\sigma) = \mu_0[1 + cot(\alpha\pi/2)[1 - (\sigma_0/\sigma)^\alpha] + i(\sigma_0/\sigma)^\alpha(1/Q_0)]. \tag{8.10}$$

The value to be assigned to α (which has to lie in the range 0 to 0.5) is estimated from seismic data.

When $\alpha = 0$, the limit $\alpha \to 0$ has to be taken for the real and imaginary part of μ; the above pair of equations then goes over into

$$Q(\sigma) = Q(\sigma_0) - \frac{2}{\pi} \ln\left(\frac{\sigma}{\sigma_0}\right), \tag{8.11}$$

$$\Re(\mu) = Q(\sigma)(-\sigma/|\sigma|)\Im(\mu) = \frac{|\mu(\sigma_0)|\sqrt{Q^2(\sigma_0) + 1}}{Q^2(\sigma) + 1}Q(\sigma), \tag{8.12}$$

which are the expressions proposed by Liu *et al.* (1976).

It may be recalled that the Q of resonances associated with normal modes is usually defined as $Q = \Re(\sigma_{mode})/(2\Im(\sigma_{mode}))$ in the literature. The discrepancy of a factor of 2 in the definition of Q in the present case must (see Eq. (8.12)) be kept in mind.

Contributions from anelasticity to nutations are mediated by the small changes arising therefrom in the values of the compliance parameters κ, γ, etc. To calculate the increments to the compliances arising from anelasticity, one needs to determine the changes produced in the solutions to the deformation equations on making an incremental change in the rigidity modulus μ, and then compute the corresponding increments in the integrals in terms of which the values of the compliances are expressed: Eqs. (6.162) and (6.163) for κ and γ. (The effects of anelasticity-induced increments in other compliances are negligible.) Then one has to go to the equations of rotational motion of the Earth and its inner regions, based on the angular momentum approach, viz., Eqs. (7.124)–(7.125) or Eqs. (7.156)–(7.158) for two-layer/three-layer Earth models. From these one calculates the increments in the wobble amplitudes (and hence in the nutation amplitudes), given the a-priori values of all the parameters involved. There are no firmly established values for the reference frequency σ_0 and α in the anelasticity model; an attempt to optimize these parameters in the process of fitting theoretical results to nutation estimates from observations was made by Mathews *et al.* (2002), starting with plausible values, namely 200 s for the reference period $(2\pi/\sigma_0)$, and 0.15 for α. The best fit was obtained when the imaginary part of the increment to μ was raised by a factor of 1.09 while the real part was left unchanged, as discussed in Chapter 10.

We have used the term "response" above in a very general sense, encompassing both primary and secondary responses. The deformation caused directly by a constituent of the TGP is a primary response; the deformations due to the periodic perturbation of the centrifugal potential caused by the wobbles excited by the TGP are a primary response to the centrifugal perturbation but are a secondary response to the TGP. The compliances appearing in nutation theory are primary deformational response parameters for the TGP or the centrifugal perturbations; the Love numbers for deformation when including resonance effects are not, since they include secondary effects due to the wobbles. The wobbles of the mantle and core regions themselves depend in a complicated way on the primary compliance parameters and so do the nutations. They are to be classed then as secondary. The concept of a phase lag in the response of a dissipative Earth to the driving potential is clearly applicable to primary responses, but the phases of the secondary response of a dissipative Earth to the driving potential are determined by their relationship to the primary response parameters and need not always appear as a lag. This is especially true in the neighborhood of resonances.

The explanations above and the considerations below are restricted to primary responses.

Consider now the application of the relation (8.1) to the compliances, which become complex and frequency dependent and are denoted for instance by $\tilde{\kappa}(\sigma)$. The perturbation of the inertia tensor as well as that of the gravitational potential at the surface of the Earth, due to the deformation caused by either the direct action of the tidal potential or by the centrifugal perturbation accompanying the wobble of the whole Earth, is directly proportional to $\tilde{\kappa}$ (see Sections 6.3 and 6.8.2). In the dynamical perturbation of the forced nutations, the tidal forcing frequencies σ are retrograde diurnal in the terrestrial frame, i.e. $\sigma_{\text{nutation}} < 0$. The lag condition (8.3) requires that $\Im(\tilde{\kappa}(\sigma_{\text{nutation}})) > 0$.

Note that the sign conventions of the TGP are of importance in the representation of the effects on a dissipative Earth. As seen previously (see e.g. Section 5.5.4), the TGPs used for the tides are given in the literature as proportional to the real part of $Y_l^m(\theta, \lambda) \, e^{i(\omega t + \chi'(\omega))}$ where $(Y, +) = Y_l^m(\theta, \lambda) = N_l^m P_n^m(\cos\theta)e^{im\lambda}$ with $P_n^m(\cos\theta)$ being the associated Legendre function and the $+$ sign referring to the $e^{+i\omega t}$ time dependence; whereas in nutation theory formulations following the Sasao *et al.* (1980) conventions (see e.g. Sections 2.13 and 2.21), one takes the tidal potential as the real part of the complex conjugate of the above, written in the form $(Y^*, +) = Y_l^{m*}(\theta, \lambda)e^{i(\sigma t + \chi(\omega))}$, where $*$ expresses the complex conjugate and where the $+$ sign refers to the $e^{+i\sigma t}$ time dependence (though σ itself is negative, equal to $-\omega$, and $\chi(\sigma) = -\chi'(\omega)$). The signs of ω and σ are determined by the sense of motion of the potential wave. For a retrograde motion, the surfaces of constant phase must move westward with a negative σ in the time dependence $(\sigma t - m\lambda)$ in the $(Y^*, +)$-case, and with a positive ω in the time dependence $(\omega t + m\lambda)$ in the $(Y, +)$-case. Equivalently, for a prograde motion, the surfaces of constant phase must move eastward with a positive σ in the time dependence $(\sigma t + m\lambda)$ in the $(Y^*, +)$-case, and with a negative ω in the time dependence $(\omega t - m\lambda)$ in the $(Y, +)$-case. The responses to these forcings can be written as $R = R^{in} + iR^{out}$ in the first case and $R = R^{in} - iR^{out}$ in the second case. The imaginary parts of the Earth response to tidal potential may thus switch sign from one representation to the other.

Let us now consider the example of the Chandler wobble (CW). The Earth response to a periodic excitation at the CW period will be perturbed by mantle anelasticity. The resonance frequency will be altered as the material behavior is different when considering anelasticity. As the deformations of the Earth are slightly increased by mantle anelasticity, the frequency of the CW (involving minus the compliance κ) decreases (σ_{CW} is proportional to $(e - \kappa)$, see Eq. (7.127)), and consequently the period will increase. The resonance period for the CW increases by about 9 days (see e.g. Dehant, 1987a and b, Groten *et al.*, 1991, Gross, 2005a and b). Note that here the frequency of the CW is prograde (positive). Consequently, the imaginary part of the Earth's response to a forcing near that frequency (such as atmospheric or oceanic forcing) would be of the opposite sign. The lag condition (8.3) requires that $\Im(\tilde{\kappa}(\sigma_{CW})) < 0$.

The CW resonance is important even when considering tides and nutations. For the potential change or the perturbation of the inertia tensor due to the centrifugal deformation accompanying the free CW (prograde frequency), on the other hand, one must have $\Im(\tilde{\kappa}(\sigma_{CW})) < 0$ and thus the imaginary part of the complex frequency of the CW must be positive. However, if one considers the response of the Earth in the nutation frequency band (retrograde diurnal forcing), which also involves the CW resonance, this involves $\Im(\tilde{\kappa}(\sigma_{nutation})) > 0$. Thus the imaginary part of the CW complex frequency must be negative.

These observations have the interesting consequence, apparently unsuspected at first, that the CW free oscillation with positive frequency must have a positive sign for the imaginary part of its complex eigenfrequency, while the CW frequency that enters the resonance formula for the nutations, determined by the response to forcing at retrograde diurnal frequencies, must have a negative imaginary part.

Both CW frequencies have the same form but the complex compliance appearing in the expression has to be used with the appropriate frequency dependence. The quality factor of the CW free oscillation is thus positive and that of the CW in the nutation expansion is negative.

Similarly, for the FCN free mode, the decreasing amplitude again requires that the $\Im(\sigma_{FCN}) > 0$. As seen in Eq. (7.128), this frequency σ_{FCN} is equal to $(-1 - (A/A^m)(e_f - \tilde{\beta}))\Omega_0$, which leads to the requirement that $\Im(\tilde{\beta}) > 0$. This last condition is in agreement with the lag condition as the frequency in this case is retrograde diurnal (negative). In this case too, the quality factor Q is definitely positive.

8.2 Anelasticity contributions to nutation

8.2.1 *Numerical integration*

As explained above, one considers mantle anelasticity by working in the frequency domain (using a spectral representation of the forcing and of the response, and working with one spectral component at a time) and using a new stress–strain relationship specific for each frequency (see Eq. (8.7)). The consideration of a complex form for the shear and bulk moduli implies a complex form for the strain and for the stress tensors.

In the framework of a numerical integration method for computing nutations, as explained in Section 7.5, note that all variables have become complex and frequency dependent. This doubles the number of equations of the system to be integrated, as derived from the motion equation, the Poisson equation, and the stress–strain relationship, and doubles the boundary conditions used when integrating from the center up to the surface (see Section 7.5.1 for the ellipsoidal Earth case, and Section 6.7 for the spherical Earth case). Details on this method are not given here as it consists in simply doubling the number of equations and variables (separating real and imaginary parts) relative to when complex variables are employed. This has been done by Dehant (1986, 1987a and b), see also Dehant and Ducarme (1987).

8.2.2 *Perturbation method*

Considering, in accordance with Wahr and Bergen (1986), that mantle anelasticity effects are small, it is feasible to treat these effects by a perturbation method. Suppose that we perturb the shear modulus in the mantle from the elastic value μ_0 to $\mu_0 + \delta\mu$, where $\delta\mu$ is complex and frequency dependent. This induces a perturbation in the Earth response to the tidal forcing as well as in the normal mode frequencies. In their approach, Wahr and Bergen have taken different values for $\delta\mu$ in the upper mantle (um) and in the lower mantle (lm). Let $\delta(um) = \delta\mu(um)/\mu_0(um)$ and $\delta(lm) = \delta\mu(lm)/\mu_0(lm)$. With the modified values of μ introduced into the equations, the perturbation to the normal mode frequency ω_{nm} can be computed. $(\delta\omega_{nm}/\omega_{nm})$ may be considered to be a linear function of $\delta(lm)$ and $\delta(um)$. The coefficients can be computed from the classical vectorial equations [the Poisson equation (Eq. (7.198)), the constitutive equation (Eq. (7.199)), and the motion equation (Eqs. (7.200)–(7.203))] or their scalar forms, scalar ordinary differential equations in d/dr [the Poisson equation (Eq. (7.274) and Eq. (7.275)), the constitutive equation (Eq. (7.276), Eq. (7.277), Eq. (7.278)), and the motion equation (Eq. (7.279), Eq. (7.280), Eq. (7.281))]. Similarly, for nutation amplitudes $\eta(\sigma_{\text{nutation}})$ derived from the solutions of these equations (see Section 7.5.2), one can compute the relative variations $(\delta\eta(\sigma_{\text{nutation}})/\eta(\sigma_{\text{nutation}}))$ as a linear function of $\delta(lm)$ and $\delta(um)$. Wahr and Bergen (1986) have shown that the most important contribution comes from the shear modulus variation of the lower mantle, the other being one order of magnitude less.

9

Ocean and atmospheric corrections

The pressure exerted on the Earth's crust by the atmosphere and the oceans is spatially non-uniform and undergoes significant temporal variations; it produces a variable torque on the Earth, which causes perturbations in Earth rotation. Additional perturbations take place as a result of the transfer of angular momentum from the winds and ocean currents to the solid Earth and vice versa. Both these effects must be taken into account when seeking to achieve high accuracy in computations of nutation, polar motion, and LOD variations.

The temporal and spatial variations in the atmospheric conditions are caused by the thermal forcing of the atmosphere by the variations in temperature of the continental and oceanic surfaces due to the diurnal and seasonal variations of the incident solar radiation, together with direct absorption by the atmosphere and secondary effects like evaporation from the oceans and cloud coverage. The bulk of the atmospheric (and oceanic) variations are non-periodic in nature; but the variations include periodic changes, referred to as atmospheric tides, which have been observed (Chapman and Lindzen, 1970). The largest of these tides have periods of one half of a solar day and one solar day, for readily understandable reasons. The retrograde diurnal part of these and other components of the tides with frequencies in the retrograde diurnal band evidently influence the low frequency nutations. The non-periodic part of the atmospheric variations contributes a continuum part to the spectrum of Earth rotation variations. The atmosphere is subject to gravitational forcing too, but this is much too weak to be of any consequence.

Gravitational forcing of the oceans, on the other hand, is far from ignorable: it produces ocean tides at the various tidal frequencies, which make a significant contribution to nutation amplitudes. Periodic forcing of the oceans by the atmosphere and periodic thermal effects also are of relevance.

This chapter aims at presenting the theoretical approaches to the treatment of atmospheric and oceanic effects on nutation. The global dynamics of the ocean and the atmosphere is outside the scope of this book but our computations are based on the fact that the behavior of these geophysical fluids is presently relatively well modeled at global scales.

Both the ocean and atmospheric effects on nutation and wobble are the consequences of angular momentum transfer from the atmosphere and the oceans to the solid Earth. The transfer of angular momentum takes place through torques acting on the solid Earth.

It must be mentioned that there is actually no external torque associated with thermal forcing. In order to treat this case, we just consider the present state of the atmosphere and compute the balance of angular momentum of the solid Earth–atmosphere system.

9.1 Effects of the fluid layers on Earth rotation

We develop here two different approaches to computation of the effects of the fluid layers on Earth rotation:

(1) The angular momentum approach, wherein the solid Earth and the fluid layers at the surface are considered to constitute a closed system, which has to have a constant angular momentum. It takes advantage of the consequent fact that variations in the angular momentum of the solid Earth must be equal and opposite to those of the fluid layers.
(2) The torque approach based on computation of the torques exerted by the surface fluid layers on the solid Earth.

In both approaches, the solid Earth and the fluid layers are considered as isolated as a first approximation, i.e. we consider the states of the oceans and atmosphere as they are, without explicitly computing the external gravitational forcing acting on these fluid layers. In practice, we ignore the atmosphere and ocean effects when we compute the gravitational torque on the solid Earth and we ignore the gravitational torque when computing the effects of the oceans and atmosphere on the solid Earth. Finally we sum the two contributions.

At the diurnal timescale, the inertia tensor of the solid Earth which enters into computations of the normal modes and forced wobbles should contain all the contributions from the deformational effects induced by the fluid layers on the solid Earth. This is accomplished by computing the inertia increments associated with the deformations that must be added to the relevant compliance parameters. Here we can again consider a transfer function approach with respect to the atmosphere and ocean changes so that we only have to compute them once using a unit forcing and convolve them with the appropriate forcing depending on the actual states of the oceans and atmosphere.

As stated earlier, we are considering the solid Earth together with the surface fluid layers as a closed system. So the total angular momentum of the system is constant:

$$\left(\frac{d\mathbf{H}}{dt}\right)_S = \left(\frac{d\mathbf{H}_{\text{solid Earth}}}{dt}\right)_S + \left(\frac{d\mathbf{H}_{\text{Fluid}}}{dt}\right)_S = 0, \qquad (9.1)$$

where $\left(\frac{d}{dt}\right)_S$ indicates that the time derivative is taken in a space-fixed reference frame. So the variations in angular momentum of the solid Earth, which are reflected in variations in Earth rotation, can be computed using

$$\left(\frac{d\mathbf{H}_{\text{solid Earth}}}{dt}\right)_S = -\left(\frac{d\mathbf{H}_{\text{Fluid}}}{dt}\right)_S, \qquad (9.2)$$

$$\left(\frac{d\mathbf{H}_{\text{Fluid}}}{dt}\right)_S = \mathbf{\Omega} \wedge \mathbf{H}_{\text{Fluid}} + \frac{d\mathbf{H}_{\text{Fluid}}}{dt}, \qquad (9.3)$$

where the quantities in the last expression are referred to the TRF. Hence, precise knowledge of the evolution of the fluid angular momentum in the terrestrial frame enables us to determine the contribution of the fluid layers to the temporal variation of the direction of the angular momentum axis in space, i.e., to the nutation of the angular momentum axis.

In applying this approach, one uses general circulation models (based on well-established dynamics of the atmosphere and its interaction with the solid Earth or ocean as well as on data assimilation) to compute $\mathbf{H}_{\text{Fluid}}$ in the terrestrial frame as a time dependent function. There is no need on our part to go into the dynamics of the atmosphere. Consequently we do not get any information as to how, why, and where the changes in the atmosphere occur. For this reason, another approach has been introduced based on a direct computation of the interaction torque between the solid Earth and the fluids.

In the torque approach, the torque exerted by the fluid layer on the solid Earth is studied, and the angular momentum variations of the solid Earth under the action of this "external torque" are computed. The variations are evidently governed by the equation

$$\left(\frac{d\mathbf{H}_{\text{solid Earth}}}{dt}\right)_S = \mathbf{\Gamma}_{\text{Fluid}\rightarrow\text{Earth}}. \qquad (9.4)$$

The two approaches are related by the angular momentum budget equation of the fluid, expressing that the time derivative of the angular momentum of the fluid $((\frac{d\mathbf{H}_{\text{Fluid}}}{dt})_S)$ is equal to the total torque exerted by the Earth on the fluid ($\mathbf{\Gamma}_{\text{Earth}\rightarrow\text{Fluid}}$):

$$\left(\frac{d\mathbf{H}_{\text{Fluid}}}{dt}\right)_S = \mathbf{\Gamma}_{\text{Earth}\rightarrow\text{Fluid}}, \qquad (9.5)$$

and by the action–reaction between the fluid and the solid Earth, the torque exerted by the solid Earth on the fluid being the opposite of the torque exerted by the fluid on the solid Earth:

$$\left(\frac{d\mathbf{H}_{\text{solid Earth}}}{dt}\right)_S = \mathbf{\Gamma}_{\text{Fluid}\to\text{Earth}} = -\mathbf{\Gamma}_{\text{Earth}\to\text{Fluid}}. \tag{9.6}$$

So the two approaches are in principle perfectly equivalent, and should be equivalent numerically and analytically. Nevertheless, some studies (see e.g. de Viron and Dehant, 1999, 2003, de Viron *et al.*, 1999, 2000, 2001a–c, 2002a–c, 2004, 2005a and b) have shown problems in the computation of the torques compared to the angular momentum changes when using models based on assimilated data for the atmosphere at the diurnal timescale in the terrestrial frame, which is exactly the frequency band of interest for nutation. Indeed, for the atmosphere, the torque approach and the angular momentum approach give one order of magnitude difference for the evaluation of the retrograde diurnal term, the torques providing an atmospheric effect on the nutations about one order of magnitude larger than does the AAM (see Dehant *et al.*, 1996, Gegout *et al.*, 1998, Bizouard *et al.*, 1998, de Viron *et al.*, 2001b, and Marcus *et al.*, 2004). The verification of the angular momentum budget equation for the fluid (Eq. (9.5)) is a good way to verify the precision of the fluid models. This consists in evaluating the changes in the angular momentum of the fluid induced by the pressure and winds/currents on the one hand, and computing the torques acting at the boundaries between the fluid and the solid Earth, on the other hand. If these values are equivalent, one has "closure" of the angular momentum budget equation.

De Viron *et al.* (2005), Lott *et al.* (2008), Dickey *et al.* (2010, see also Marcus *et al.*, 2010, 2011, 2012) have examined the closure of the angular momentum budget equation using a general circulation model without assimilating any data. They show that the budget can indeed be closed but the amplitudes of some contributions such as the change of the pressure part of the angular momentum due to the rotation of the reference frame and the ellipsoidal part of the torque (pressure torque on the ellipsoid) are very large and almost the same but occurring on opposite sides of the budget equation, thus nearly canceling each other. This makes the evaluation of all the components very difficult, which may lead to order of magnitude differences at the end of the computation.

9.1.1 Angular momentum approach

Consider a fluid element $\rho_{\text{Fluid}}dV$ at position **r** or (x, y, z) within the volume V_{Fluid} of the fluid layer at the Earth's surface, moving with velocity **u** in an inertial frame. Since the fluid rotates along with the solid Earth with angular velocity **Ω**, **u** may be written as

$$\mathbf{u} = \mathbf{v} + \boldsymbol{\Omega} \wedge \mathbf{r}, \tag{9.7}$$

where $\mathbf{v}$ is the relative velocity of the fluid element with respect to the rotating Earth, and $\boldsymbol{\Omega} \wedge \mathbf{r}$ is the velocity due to the rotation. The angular momentum of the fluid in the inertial frame is evidently given by

$$\mathbf{H}_{\text{Fluid}} = \int_{V_{\text{Fluid}}} \mathbf{r} \wedge \rho_{\text{Fluid}} \mathbf{u} \, dV = \int_{V_{\text{Fluid}}} \mathbf{r} \wedge \rho_{\text{Fluid}} \left(\mathbf{v} + \boldsymbol{\Omega} \wedge \mathbf{r} \right) dV. \tag{9.8}$$

For the purpose of computing $\mathbf{H}_{\text{Fluid}}$ it is sufficient to take

$$\boldsymbol{\Omega} = \Omega \hat{\mathbf{e}}_z,$$

the contribution from Earth rotation variations being negligibly small. Then we can see readily that the contribution to the angular momentum from the part $\boldsymbol{\Omega} \wedge \mathbf{r}$ of the velocity which relates to the rotation of the atmosphere as a whole in space is, in terms of its Cartesian components,

$$\int_{V_{\text{Fluid}}} \rho_{\text{Fluid}} \begin{pmatrix} -xz \\ -yz \\ x^2 + y^2 \end{pmatrix} \Omega \, dV = \begin{pmatrix} c_{13}^{\text{Fluid}} \\ c_{23}^{\text{Fluid}} \\ c_{33}^{\text{Fluid}} \end{pmatrix} \Omega, \tag{9.9}$$

where the c_{ij}^{Fluid} are elements of the inertia tensor of the fluid region. (Recall that $c_{ij} = \int \rho(\mathbf{r})(r^2 \delta_{ij} - x_i x_j) dV$, see Eq. (6.124) with I_{ij} corresponding to c_{ij}.)

In the literature on the atmospheric angular momentum, this term is often referred to as the *matter term*, while the term involving $\mathbf{v}$ in the expression (9.8) for $\mathbf{H}_{\text{Fluid}}$ is called the *wind term*. The latter represents the angular momentum of the atmosphere relative to the rotating Earth.

In the integral expressions above, the volume V_{Fluid} of the atmosphere extends over the whole surface of the Earth in a thin layer from the surface upwards. ("Thin" here means in comparison with the radius of the Earth.) It is convenient, therefore, to use spherical polar coordinates for the position vector $\mathbf{r}$ in carrying out the integrations. We then write

$$\rho_{\text{Fluid}} \, dV = \rho_{\text{Fluid}} \, r^2 dr \sin \theta \, d\theta \, d\lambda = -(dp/g) \, r^2 \sin \theta \, d\theta \, d\lambda \tag{9.10}$$

by making use of the hydrostatic relation $dp = -g\rho_{\text{Fluid}} dr$, where g is the acceleration due to gravity, and dp is the difference between the pressures at r and $r + dr$. This step has the advantage that unlike the density ρ_{Fluid}, which is far from being a readily observable function of position in the atmosphere (or ocean), measurements of the vertical profiles of pressure and wind velocity are routinely carried out at weather stations spread around the globe. Measurements of $\mathbf{v}$ at $\mathbf{r}$ are provided by sounding or GCM and are given in spherical components v_r, v_θ, v_λ along the radially outward, southward and eastward directions (θ and λ being the colatitude and longitude).

In order to write the integrals in the matter term and the wind term in $\mathbf{H}_{\text{Fluid}}$ in terms of the above quantities, we use the familiar expressions

$$\mathbf{r} = \begin{pmatrix} r \sin \theta \cos \lambda \\ r \sin \theta \sin \lambda \\ r \cos \theta \end{pmatrix} \tag{9.11}$$

for the Cartesian components of $\mathbf{r}$, as well as the components of $\mathbf{v}$ expressed as

$$\begin{pmatrix} v_x \\ v_y \\ v_z \end{pmatrix} = \begin{pmatrix} v_r \sin \theta \cos \lambda + v_\theta \cos \theta \cos \lambda - v_\lambda \sin \lambda \\ v_r \sin \theta \sin \lambda + v_\theta \cos \theta \sin \lambda + v_\lambda \cos \lambda \\ v_r \cos \theta - v_\theta \sin \theta \end{pmatrix}. \tag{9.12}$$

We substitute these now into the integrals for the matter term and the wind term, given by (9.9) and the part involving $\mathbf{r} \wedge \mathbf{v}$ in (9.8), respectively. We also use the relation (9.10) to eliminate ρ_{Fluid} from the integrals, and furthermore treat g and r as constants in the integrals (setting $r = a$, $g = g(a)$) in view of the thinness of the atmospheric (or oceanic) layer.

The resulting expressions are:

$$\mathbf{H}_{\text{Fluid}}^{\text{matter}} = \frac{a^4}{g} \Omega \iint P_s \sin^2 \theta \begin{pmatrix} \cos \theta \cos \lambda \\ \cos \theta \sin \lambda \\ -\sin \theta \end{pmatrix} d\theta \, d\lambda, \tag{9.13}$$

$$\mathbf{H}_{\text{Fluid}}^{\text{wind}} = \frac{a^3}{g} \iiint \sin \theta \begin{pmatrix} v_\theta \sin \lambda + v_\lambda \cos \theta \cos \lambda \\ -v_\theta \cos \lambda + v_\lambda \cos \theta \sin \lambda \\ -v_\lambda \sin \theta \end{pmatrix} dp \, d\theta \, d\lambda, \tag{9.14}$$

$$\mathbf{H}_{\text{Fluid}} = \mathbf{H}_{\text{Fluid}}^{\text{matter}} + \mathbf{H}_{\text{Fluid}}^{\text{wind}} \tag{9.15}$$

where the dependence of v_r, v_θ, v_λ on r is converted into a dependence on p when the integration over r is replaced by one on p. P_s appearing in Eq. (9.13) is the surface pressure: $P_s = \int_{\text{atmospheric layer}} dp$.

The above two parts of the angular momentum vector are numerically evaluated using the global data set of the surface pressure P_s, the vertical profile of the pressure p, and the vertical profiles of the wind velocity components (v_θ, v_λ), compiled from observations and GCM made simultaneously at meteorological stations world-wide at stipulated times (on the hour and half-hour, GMT). The wind velocity is viewed as a function of the pressure p rather than of r. The series of values of the two integrals for all the times of observation constitute the time series of the matter and motion terms of the atmospheric angular momentum (see e.g. Salstein *et al.*, 1993, 2001).

The x- and y-components of the angular momentum are the ones which are relevant to nutations and polar motions. In particular, the part of their spectra which belongs to the retrograde diurnal band is what matters for the low frequency nutations. Spectral analysis of the above-mentioned time series shows that about 90% of the angular momentum in this part of the spectrum is contributed by the motion term and the remaining 10% by the matter term (Bizouard *et al.*, 1998). The motion and matter terms contribute roughly equally to the prograde diurnal part of the spectrum, while their contributions outside both bands are about 10% and 90%, respectively (Chao and Au, 1991, Salstein *et al.*, 1993, Bell, 1994).

For the z-component of the angular momentum, the motion and matter parts contribute roughly equal and opposite amounts, thus nearly canceling each other's contribution in the diurnal band, leaving a remainder of only 5% to 10% of either contribution within this band (Bizouard *et al.*, 1998, Lott *et al.*, 2008). On the contrary, the wind term is much more important (Hide and Dickey, 1992, de Viron *et al.*, 2002a–c) outside the diurnal band.

9.1.2 Torque approach

We consider now the torque $\boldsymbol{\Gamma}$ exerted by the atmosphere/oceans on the solid Earth. It is the negative of the torque exerted by the solid Earth on the fluid layer, the latter being equal to the rate of change of the angular momentum of the fluid. Hence

$$\boldsymbol{\Gamma} = \boldsymbol{\Gamma}_{\text{Fluid}\rightarrow\text{Earth}} = -\boldsymbol{\Gamma}_{\text{Earth}\rightarrow\text{Fluid}} = -\left(\frac{d\mathbf{H}_{\text{Fluid}}}{dt}\right)_S, \qquad (9.16)$$

with

$$\mathbf{H}_{\text{Fluid}} = \int_{V_{\text{Fluid}}} \rho_{\text{Fluid}} \mathbf{r} \wedge \mathbf{u}\, dV. \qquad (9.17)$$

Since $\rho_{\text{Fluid}}\, dV$ is the mass of a fluid element, it hardly changes with time. So the rate of change of the angular momentum integral may be written as

$$\left(\frac{d\mathbf{H}_{\text{Fluid}}}{dt}\right)_S = \int_{V_{\text{Fluid}}} \rho_{\text{Fluid}} \left[\left(\frac{d}{dt}\right)_S (\mathbf{r} \wedge \mathbf{u})\right] dV = \int_{V_{\text{Fluid}}} \rho_{\text{Fluid}} \left[\mathbf{r} \wedge \left(\frac{d\mathbf{u}}{dt}\right)_S\right] dV,$$
$$(9.18)$$

noting that $(d\mathbf{r}/dt)_S \wedge \mathbf{u} = \mathbf{u} \wedge \mathbf{u} = 0$. Now, the Newtonian equation of motion for the fluid is

$$\rho_{\text{Fluid}} \left(\frac{d\mathbf{u}}{dt}\right)_S = -\rho_{\text{Fluid}} \nabla\Phi - \nabla P + \mathbf{F}, \qquad (9.19)$$

where Φ is the gravitational potential of the Earth itself (with the classical physicists' convention), P is the pressure, and $\mathbf{F}$ is the acceleration due to any other mechanism, such as viscosity, at any point in the atmosphere.

On making use of this equation, the preceding equation becomes

$$\left(\frac{d\mathbf{H}_{\text{Fluid}}}{dt}\right)_S = \int_{V_{\text{Fluid}}} \mathbf{r} \wedge (-\rho_{\text{Fluid}} \nabla \Phi - \nabla P + \mathbf{F}) \, dV. \tag{9.20}$$

The three terms represent the three types of torque exerted on the fluid: gravitational torque due to the Earth's own gravitation, the pressure torque, and the torque due to other forces, respectively. The corresponding torques exerted by the fluid on the solid Earth, which we denote by $\boldsymbol{\Gamma}^G$, $\boldsymbol{\Gamma}^P$, and $\boldsymbol{\Gamma}^F$, are obtained by changing the signs of the respective terms:

$$\boldsymbol{\Gamma}^G = \int_{V_{\text{Fluid}}} \mathbf{r} \wedge (\rho_{\text{Fluid}} \nabla \Phi) \, dV, \tag{9.21}$$

$$\boldsymbol{\Gamma}^P = \int_{V_{\text{Fluid}}} \mathbf{r} \wedge \nabla P \, dV, \tag{9.22}$$

$$\boldsymbol{\Gamma}^F = -\int_{V_{\text{Fluid}}} \mathbf{r} \wedge \mathbf{F} \, dV. \tag{9.23}$$

The torque $\boldsymbol{\Gamma}^F$ is the friction torque related to the wind/currents in the case of the atmosphere/oceans.

The pressure torque

The pressure torque (9.22) may be expressed as an integral over the surface of the Earth, by using the following identity, which is a variant of Stokes' theorem:

$$\int_V \nabla \wedge \mathbf{A} \, dV = \int_S \hat{\mathbf{n}} \wedge \mathbf{A} \, dS, \tag{9.24}$$

where dS is an element of the surface S enclosing the volume V, and $\hat{\mathbf{n}}$ is the outward-pointing unit normal to this surface element. On using this identity, the pressure torque on the solid Earth becomes

$$\boldsymbol{\Gamma}^P = \int_{V_{\text{Fluid}}} \mathbf{r} \wedge \nabla P \, dV = -\int_{V_{\text{Fluid}}} \nabla \wedge (\mathbf{r}P) \, dV = -\int_{S_{\text{Fluid}}} \hat{\mathbf{n}} \wedge (\mathbf{r}P) \, dS.$$
$$\tag{9.25}$$

Note that in the case of the atmospheric layer, which is on the outer side of the solid Earth, the normal $\hat{\mathbf{n}}$ points towards the interior of the Earth. The surface S is just the surface of the solid Earth, the "outer" surface at the top of the atmosphere being irrelevant because of vanishing of the pressure there; thus P is really the surface pressure P_s. The shape includes the local topography or orography (shape of mountain surfaces); the resulting deviations of the normal from that due to the overall ellipsoidal shape have to be taken into account in computations of the pressure torque. (If the pressure on one side of a mountain is different from that on the opposite side, it is understandable that the difference would affect Earth rotation.) Similar considerations apply in the computation of the pressure on the ocean bottom. In the case of the ocean, the surface includes also the free surface at the top, where the atmospheric pressure operates.

In considering the atmospheric pressure torque $\mathbf{\Gamma}^P$ further, we replace $\hat{\mathbf{n}}$ as defined above by $-\hat{\mathbf{n}}_{SE}$ where $\hat{\mathbf{n}}_{SE}$ is the local outward unit normal to the surface of the solid Earth (SE). The last expression in (9.25) may then be written as

$$\mathbf{\Gamma}^P = -\int_S P_s \,(\mathbf{r} \wedge \hat{\mathbf{n}}_{SE}) \, dS. \tag{9.26}$$

Note that $\mathbf{r}$ in the integral is on the surface S; so $\mathbf{r}$ is a function of θ and λ which describes the topography. The vector $\hat{\mathbf{n}}_{SE} \, dS$ represents the extent and orientation of an element of area on the surface; it may evidently be written as $d\mathbf{r}_\theta \wedge d\mathbf{r}_\lambda$ where $d\mathbf{r}_\theta$ and $d\mathbf{r}_\lambda$ are infinitesimal vectors *on the surface S* in the directions towards increasing θ and λ. Hence

$$\hat{\mathbf{n}}_{SE} \, dS = \left(\frac{d\mathbf{r}}{d\theta}\right)_S \wedge \left(\frac{d\mathbf{r}}{d\lambda}\right)_S \, d\theta \, d\lambda. \tag{9.27}$$

We find then, on using the expressions for the Cartesian components of $\mathbf{r}$ in spherical polar coordinates, that

$$(\mathbf{r} \wedge \hat{\mathbf{n}}_{SE}) \, dS = \left[\left(\mathbf{r} \cdot \frac{d\mathbf{r}}{d\lambda}\right) \frac{d\mathbf{r}}{d\theta} - \left(\mathbf{r} \cdot \frac{d\mathbf{r}}{d\theta}\right) \frac{d\mathbf{r}}{d\lambda}\right]_S d\theta \, d\lambda$$

$$= r^2 \begin{pmatrix} \frac{dr}{d\theta} \sin\theta \sin\lambda + \frac{dr}{d\lambda} \cos\theta \cos\lambda \\ -\frac{dr}{d\theta} \sin\theta \cos\lambda + \frac{dr}{d\lambda} \cos\theta \sin\lambda \\ -\frac{dr}{d\lambda} \sin\theta \end{pmatrix} d\theta \, d\lambda. \tag{9.28}$$

Substituting this in the torque expression (9.26) and setting $r^2 = a^2$ in view of the thin layer approximation, we obtain

$$\mathbf{\Gamma}^P = -a^2 \int_S P_s \begin{pmatrix} \frac{dr}{d\theta} \sin\theta \sin\lambda + \frac{dr}{d\lambda} \cos\theta \cos\lambda \\ -\frac{dr}{d\theta} \sin\theta \cos\lambda + \frac{dr}{d\lambda} \cos\theta \sin\lambda \\ -\frac{dr}{d\lambda} \sin\theta \end{pmatrix} d\theta \, d\lambda. \tag{9.29}$$

An integration by parts may be done to go from partial derivatives of the radial (or height) variable in the integral to partial derivatives of the pressure. The x-component of the torque contains the following terms that must still be integrated over λ and θ, respectively:

$$\int P_s(\theta, \lambda) \frac{dr}{d\lambda} \cos \theta \cos \lambda \, d\lambda = (r P_s \cos \theta \cos \lambda)_{\lambda=0}^{\lambda=2\pi}$$
$$- \int \left[r \frac{dP_s}{d\lambda} \cos \theta \cos \lambda - r P_s \cos \theta \sin \lambda \right] d\lambda,$$

$$\int P_s(\theta, \lambda) \frac{dr}{d\theta} \sin \theta \sin \lambda \, d\theta = (r P_s \sin \theta \sin \lambda)_{\theta=0}^{\theta=\pi}$$
$$- \int \left[r \frac{dP_s}{d\theta} \sin \theta \sin \lambda + r P_s \cos \theta \sin \lambda \right] d\theta.$$

The first terms in the right-hand sides in these two equations are equal to zero and the last parts of the integrals cancel each other when integrating over λ and θ, respectively, and carrying out the further integrations over θ and λ, respectively, in the two equations. Combining what remains of the above two integrals, we get for the first component of (9.29) the expression

$$\Gamma_x^P = \iint r^3 \left(\frac{dP_s}{d\lambda} \cos \theta \cos \lambda + \frac{dP_s}{d\theta} \sin \theta \sin \lambda \right) d\theta \, d\lambda. \tag{9.30}$$

Similarly for the y-component we get:

$$\Gamma_y^P = \iint r^3 \left(\frac{dP_s}{d\lambda} \cos \theta \sin \lambda - \frac{dP_s}{d\theta} \sin \theta \cos \lambda \right) d\theta \, d\lambda. \tag{9.31}$$

The integral over λ in the third component of (9.29) yields

$$\int P_s(\theta, \lambda) \frac{dr}{d\lambda} \sin \theta \, d\lambda = (r P_s \sin \theta)_{\lambda=0}^{\lambda=2\pi} - \int r \frac{dP_s}{d\lambda} \sin \theta \, d\lambda, \tag{9.32}$$

and hence we obtain

$$\Gamma_z^P = - \iint r^3 \frac{dP_s}{d\lambda} \sin \theta \, d\theta \, d\lambda. \tag{9.33}$$

The torque vector due to the fluid pressure acting over the whole surface of the Earth is therefore given by the following integral:

$$\boldsymbol{\Gamma}^P = \iint r^3 \left(\frac{1}{\sin \theta} \left(\frac{dP_s}{d\lambda} \right)_s \begin{pmatrix} \cos \lambda \cos \theta \\ \sin \lambda \cos \theta \\ -\sin \theta \end{pmatrix} + \left(\frac{dP_s}{d\theta} \right)_s \begin{pmatrix} \sin \lambda \\ -\cos \lambda \\ 0 \end{pmatrix} \right) \sin \theta \, d\theta \, d\lambda, \tag{9.34}$$

or equivalently starting from Eq. (9.29) and using the relation between r and the topography given in spherical harmonics ($r = a(1 + h_2(\theta, \lambda) + h(\theta, \lambda)) = a(1 + h'(\theta, \lambda))$, where $h(\theta, \lambda)$ is with respect to the ellipsoid and $h'(\theta, \lambda)$ with respect to the spherical Earth):

$$\mathbf{\Gamma}^P = -a^3 \iint P_s \left(\frac{dh'(\theta, \lambda)}{d\lambda} \begin{pmatrix} \cos\lambda\cos\theta \\ \sin\lambda\cos\theta \\ -\sin\theta \end{pmatrix} + \frac{dh'(\theta, \lambda)}{d\theta} \begin{pmatrix} \sin\theta\sin\lambda \\ -\sin\theta\cos\lambda \\ 0 \end{pmatrix} \right) d\theta \, d\lambda.$$

(9.35)

The shape described by the height $h(\theta, \lambda)$ or $h'(\theta, \lambda)$ and the pressure $P_s(\theta, \lambda)$ are either developed in spherical harmonics, or are expressed by their numerical values on a system of grid points over the Earth's surface; the latter provides the more precise description.

It is interesting to note that the equivalent relations (9.34) and (9.35) make it possible to compute from the gradient of the topography instead of the gradient of the pressure, which is less known. However, the expression (9.34) will be used later. It is also important to keep in mind while performing the computations that the topography $h(\theta, \lambda)$ or $h'(\theta, \lambda)$ is the same as the one considered in the GCM model used for the pressure grid or spherical harmonics. Additionally, the gradients of the topography can be computed once and for all and used in all computations later on as they do not vary with time (at short timescale such as the diurnal timescale). Moreover, the pressure is not very well measured at the positions where we have a strong gradient. This is a problem for the reliability of the pressure grids.

Pressure torque on the Earth's bulge (related to $h_2(\theta, \lambda)$)

For an axially symmetric Earth, the pressure on the ellipsoidal surface makes no contribution to the axial component of the torque. On the other hand, the major part of the equatorial components of the torque arises from the action of the pressure on the ellipsoidal shape; only the part of $P_s(\theta, \lambda)$ that is of spherical harmonic degree two and order one contributes to the equatorial torque. On replacing the normal to the topography $\mathbf{n}_{SE}$ by the normal to the ellipsoid $\mathbf{n}_{SE}^0$ in Eq. (9.26) and on replacing h' by h_2 in Eq. (9.35), one obtains:

$$\mathbf{\Gamma}_{\text{pressure}\to\text{ellipsoid}} = -\int_S P_s \left(\mathbf{r} \wedge \hat{\mathbf{n}}_{SE}^0\right) dS$$

(9.36)

$$= -a^3 \iint P_s \left(\frac{dh_2(\theta, \lambda)}{d\lambda} \begin{pmatrix} \cos\lambda\cos\theta \\ \sin\lambda\cos\theta \\ -\sin\theta \end{pmatrix} \right.$$

$$\left. + \frac{dh_2(\theta, \lambda)}{d\theta} \begin{pmatrix} \sin\theta\sin\lambda \\ -\sin\theta\cos\lambda \\ 0 \end{pmatrix} \right) d\theta \, d\lambda.$$

(9.37)

As shown below, about one half of this torque is cancelled out by the gravitational torque associated with the so-called form factor J_2 of the Earth or equivalently the degree two order zero coefficient of the gravitational potential.

Local pressure torque (related to $h(\theta, \lambda)$)

This part of the torque is associated with the differential action of the fluid pressure on different sides of elevated (or depressed) areas of the Earth's surface. It is computed from the fluid pressure and the spatial derivative of the orography over the whole surface.

The gravitational torque

The torque due to the gravitational action of the mass elements of the fluid on those of the solid Earth is given by Eq. (9.21). It is convenient to introduce here the normalized dimensionless potential $\tilde{\Phi}$ defined by

$$\Phi = -\frac{GM}{a}\tilde{\Phi} = -ga\tilde{\Phi}, \tag{9.38}$$

where g is the acceleration due to gravity at the surface of a spherical Earth of mean radius a. The gravitational torque on the solid Earth may now be written as

$$\mathbf{\Gamma}^G = \int_{V_{\text{Fluid}}} \rho_{\text{Fluid}} \left(\mathbf{r} \wedge \nabla \Phi\right) dV = -\int_{V_{\text{Fluid}}} ga\, \rho_{\text{Fluid}} \left(\mathbf{r} \wedge \nabla \tilde{\Phi}\right) dV. \tag{9.39}$$

In a spherical basis, the gradient operator is:

$$\nabla = \hat{\mathbf{e}}_r \frac{d}{dr} + \hat{\mathbf{e}}_\theta \frac{1}{r}\frac{d}{d\theta} + \hat{\mathbf{e}}_\lambda \frac{1}{r \sin\theta}\frac{d}{d\lambda}, \tag{9.40}$$

where $\hat{\mathbf{e}}_r, \hat{\mathbf{e}}_\theta, \hat{\mathbf{e}}_\lambda$ are unit vectors in the radial, southward, and eastward directions. Since $\mathbf{r} = \hat{\mathbf{e}}_r r$, we can now re-express the torque as

$$\mathbf{\Gamma}^G = \int_{V_{\text{Fluid}}} ga\rho_{\text{Fluid}} \left(\frac{d\tilde{\Phi}}{d\theta}\hat{\mathbf{e}}_\lambda - \frac{1}{\sin\theta}\frac{d\tilde{\Phi}}{d\lambda}\hat{\mathbf{e}}_\theta\right) r^2\, dr \sin\theta\, d\theta\, d\lambda. \tag{9.41}$$

Then we can carry out the integration in the radial direction after converting it into an integration over pressure by using the fact that $g\rho_{\text{Fluid}}dr = -dp$:

$$\int ga\, \rho_{\text{Fluid}} r^2 dr = -a^3 \int dp = -P_s, \tag{9.42}$$

where we have set $r = a$ in the thin-layer approximation. Thus we obtain

$$\mathbf{\Gamma}^G = -a^3 \iint P_s(\theta, \lambda) \left(-\frac{1}{\sin\theta} \frac{d\tilde{\Phi}}{d\lambda} \hat{\mathbf{e}}_\theta + \frac{d\tilde{\Phi}}{d\theta} \hat{\mathbf{e}}_\lambda \right) \sin\theta \, d\theta \, d\lambda. \quad (9.43)$$

Since the Cartesian components of $\hat{\mathbf{e}}_\theta$ and $\hat{\mathbf{e}}_\lambda$ are $(\cos\theta\cos\lambda, \cos\theta\sin\lambda, -\sin\theta)$ and $(-\sin\lambda, \cos\lambda, 0)$, respectively (see Eqs. (7.329) and (7.330)), we find that the components of $\mathbf{\Gamma}^G$ are

$$\Gamma_x^G = -\frac{a^2}{g} \iint P_s(\theta, \lambda) \left(\frac{d\Phi}{d\lambda} \cos\theta\cos\lambda + \frac{d\Phi}{d\theta} \sin\theta\sin\lambda \right) d\theta \, d\lambda, \quad (9.44)$$

$$\Gamma_y^G = -\frac{a^2}{g} \iint P_s(\theta, \lambda) \left(\frac{d\Phi}{d\lambda} \cos\theta\sin\lambda - \frac{d\Phi}{d\theta} \sin\theta\cos\lambda \right) d\theta \, d\lambda, \quad (9.45)$$

$$\Gamma_z^G = \frac{a^2}{g} \iint P_s(\theta, \lambda) \frac{d\Phi}{d\lambda} \sin\theta \, d\theta \, d\lambda. \quad (9.46)$$

The above expressions for the components of $\mathbf{\Gamma}^G$ have exactly the same forms as the components of $\mathbf{\Gamma}^P$ in Eq. (9.29), the only difference being that the radial position $r(\theta, \lambda)$ of the solid Earth in the latter has been replaced by (Φ/g). Therefore if we take $\mathbf{r}$ now to represent a generic point on the surface of constant (Φ/g), we can see from the same kind of derivation which equated (9.29) to (9.26) that $\mathbf{\Gamma}^G$ with the above components is nothing but an explicit form of

$$\mathbf{\Gamma}^G = \int_S P_s \left(\mathbf{r} \wedge \hat{\mathbf{n}}_{\text{grav}} \right) dS, \quad (9.47)$$

where $\hat{\mathbf{n}}_{\text{grav}}$ is the normal to the surface of constant $-(\Phi/g)$ at the Earth's surface and pointing inward. It is to be noted that Φ is the gravitational potential of the Earth, which differs from the geopotential: the latter includes also the centrifugal potential due to the mean Earth rotation.

An integration by parts, similar to that which was done in the case of the pressure torque, enables us to express $\mathbf{\Gamma}^G$ in terms of the gradient of the pressure rather than the gradient of the potential. We obtain, similarly as for Eqs. (9.30), (9.31), and (9.33):

$$\Gamma_x^G = \frac{a^2}{g} \iint \Phi \left(\frac{dP_s}{d\lambda} \cos\theta\cos\lambda + \frac{dP_s}{d\theta} \sin\theta\sin\lambda \right) d\theta \, d\lambda, \quad (9.48)$$

$$\Gamma_y^G = \frac{a^2}{g} \iint \Phi \left(\frac{dP_s}{d\lambda} \cos\theta\sin\lambda - \frac{dP_s}{d\theta} \sin\theta\cos\lambda \right) d\theta \, d\lambda, \quad (9.49)$$

$$\Gamma_z^G = -\frac{a^2}{g} \iint \Phi \frac{dP_s}{d\lambda} \sin\theta \, d\theta \, d\lambda. \quad (9.50)$$

The total gravitational torque between the fluid and the solid Earth is thus expressed either by:

$$\boldsymbol{\Gamma}^G = -\frac{a^2}{g} \iint P_s \left(\frac{d\Phi}{d\lambda} \begin{pmatrix} \cos\lambda\cos\theta \\ \sin\lambda\cos\theta \\ -\sin\theta \end{pmatrix} + \frac{d\Phi}{d\theta} \begin{pmatrix} \sin\lambda\sin\theta \\ -\cos\lambda\sin\theta \\ 0 \end{pmatrix} \right) d\theta\, d\lambda,$$

$$(9.51)$$

or by:

$$\boldsymbol{\Gamma}^G = \frac{a^2}{g} \iint \Phi \left(\frac{dP_s}{d\lambda} \begin{pmatrix} \cos\lambda\cos\theta \\ \sin\lambda\cos\theta \\ -\sin\theta \end{pmatrix} + \frac{dP_s}{d\theta} \begin{pmatrix} \sin\lambda\sin\theta \\ -\cos\lambda\sin\theta \\ 0 \end{pmatrix} \right) d\theta\, d\lambda.$$

$$(9.52)$$

As was the case for the pressure torque (see Eqs. (9.36) and (9.37)), it is useful to divide the equatorial components of $\boldsymbol{\Gamma}^G$ into two parts: one part associated with the effect of the Earth's form factor (i.e. J_2, the degree two order zero of the gravitational potential of the Earth, related to the dynamical ellipticity) and the other associated with the local variations of the gravitational potential (the other spherical harmonics components). It has been shown (see for instance de Viron *et al.*, 1999) that the second part of the gravitational torque has a quite negligible contribution to the torque at all frequencies.

On replacing the normal to the geoid $\mathbf{n}_{\mathrm{grav}}$ by the normal to the ellipsoidal surface of constant gravitational potential of degree two order zero $\mathbf{n}^0_{\mathrm{grav}}$ in Eq. (9.47), one obtains:

$$\boldsymbol{\Gamma}_{\mathrm{gravitation}\to J_2} = \int_S P_s \left(\mathbf{r} \wedge \hat{\mathbf{n}}^0_{\mathrm{grav}} \right) dS.$$

$$(9.53)$$

The ellipsoidal torque

Let us consider the sum of the gravitational torque related to the Earth's form factor (Eq. (9.53)) and the pressure torque related to the Earth's ellipticity (Eq. (9.36)) (i.e., the combined torque in the axially symmetric ellipsoidal approximation):

$$\boldsymbol{\Gamma}^{\mathcal{E}} = \boldsymbol{\Gamma}_{\mathrm{pressure}\to\mathrm{ellipsoid}} + \boldsymbol{\Gamma}_{\mathrm{gravitation}\to J_2}.$$

$$(9.54)$$

We have denoted the normals to the ellipsoidal solid Earth and to the ellipsoidal surface of constant gravitational potential of degree two order zero by $\hat{\mathbf{n}}^0_{\mathrm{SE}}$ and $\hat{\mathbf{n}}^0_{\mathrm{grav}}$. Then we see on combining the torques from Eqs. (9.26) and (9.47) that

$$\boldsymbol{\Gamma}^{\mathcal{E}} = -\int_S P_s \left(\mathbf{r} \wedge \left(\hat{\mathbf{n}}^0_{\mathrm{SE}} - \hat{\mathbf{n}}^0_{\mathrm{grav}} \right) \right) dS.$$

$$(9.55)$$

The Earth has a structure that is very close to hydrostatic equilibrium under the balance of the self-gravitational force and the centrifugal force associated with its diurnal rotation. According to Clairaut's theory of the hydrostatic equilibrium structure (see Appendix B), the equidensity surfaces and surfaces of equal geopotential coincide and form a continuum of concentric ellipsoids. The flattening $\epsilon(r)$ of the ellipsoid of mean radius r increases gradually as one goes outwards from the center. The surfaces of constant gravitational potential form a different set of ellipsoids, with the ellipsoid of mean radius r characterized by the form factor $J_2(r)$ of the region inside it. From Clairaut's theory we see that the two normals at the Earth's surface are given (see Eq. (7.204)) in terms of these quantities by

$$\hat{\mathbf{n}}_{SE}^0 = \left[1 + \frac{2}{3}\left(\epsilon(r) + r\frac{d\epsilon}{dr}\right)P_{20}(\cos\theta)\right]_{r=a}\hat{\mathbf{e}}_r - 2\epsilon(a)\hat{\mathbf{e}}_\theta\cos\theta\sin\theta \quad (9.56)$$

and

$$\hat{\mathbf{n}}_{grav}^0 = \left[1 + \left(J_2(r) + r\frac{dJ_2}{dr}\right)P_{20}(\cos\theta)\right]_{r=a}\hat{\mathbf{e}}_r - 3J_2(a)\hat{\mathbf{e}}_\theta\cos\theta\sin\theta, \quad (9.57)$$

where $J_2(a)$ is the familiar J_2. Hence

$$\hat{\mathbf{n}}_q^0 = \hat{\mathbf{n}}_{SE}^0 - \hat{\mathbf{n}}_{grav}^0 = \left[\frac{2}{3}\left(1 + r\frac{d}{dr}\right)\left(\epsilon(r) - \frac{3}{2}J_2\right)P_{20}(\cos(\theta))\right]_{r=a}\hat{\mathbf{e}}_r$$

$$- 2\cos(\theta)\sin(\theta)\left(\epsilon(a) - \frac{3}{2}J_2(a)\right)\hat{\mathbf{e}}_\theta. \quad (9.58)$$

Then we see on combining the torques from Eqs. (9.26) and (9.47) for the ellipsoidal case (see Eq. (9.55)) that

$$\Gamma^{\mathcal{E}} = -\int_S P_s\left(\mathbf{r}\wedge\hat{\mathbf{n}}_q^0\right)dS. \quad (9.59)$$

We now show that the torque related to the difference $\hat{\mathbf{n}}_q^0$ between the two normals is the same as the torque related to the centrifugal effect. Indeed, Clairaut's theory (see Appendix B) shows also that

$$\epsilon - \frac{3}{2}J_2 = \frac{q}{2}, \quad (9.60)$$

where $\epsilon = \epsilon(a)$ and q is the ratio of the centrifugal and gravitational accelerations at the equator:

$$q = \frac{\Omega^2 a}{(GM/a^2)} = \frac{\Omega^2 a^3}{GM} = \frac{\Omega^2 a}{g}. \quad (9.61)$$

Thus, from Eq. (9.58)

$$\hat{\mathbf{n}}_q^0 = \frac{1}{3}\left[q + r\frac{dq}{dr}\right]_{r=a} P_{20}(\cos\theta)\hat{\mathbf{e}}_r - q\hat{\mathbf{e}}_\theta \cos\theta \sin\theta, \qquad (9.62)$$

where q must evidently be a function of r: $q(r) = \Omega^2 r^3/GM$ (which implies that $r(dq/dr) = 3q$, and the square bracketed quantity above becomes $4q$). If we now consider that the potential in the torque (9.39) is the centrifugal potential instead of the gravitational potential, we can write the torque related to the centrifugal potential in the same form as (9.43):

$$\mathbf{\Gamma}_\psi = -a^3 \iint P_s(\theta, \lambda)\left(-\frac{d\tilde{\psi}}{d\lambda}\begin{pmatrix}\cos\lambda\cos\theta \\ \sin\lambda\cos\theta \\ -\sin\theta\end{pmatrix} + \frac{d\tilde{\psi}}{d\theta}\begin{pmatrix}-\sin\lambda\sin\theta \\ \cos\lambda\sin\theta \\ 0\end{pmatrix}\right) d\theta\, d\lambda,$$

$$(9.63)$$

where $\tilde{\psi}$ is the normalized centrifugal potential given by

$$\tilde{\psi} = \frac{(\mathbf{\Omega}\wedge\mathbf{r})\cdot(\mathbf{\Omega}\wedge\mathbf{r})}{\Omega^2 a^2}. \qquad (9.64)$$

On the other hand, considering the centrifugal potential and expressing the normal to the constant centrifugal potential surface, one gets the same form as Eq. (9.59) with $\hat{\mathbf{n}}_q^0$ being the normal to the equipotential surfaces related to the centrifugal potential. Indeed, we first note that the centrifugal potential does not depend on λ and consequently the first term in (9.63) is zero and only the second term remains:

$$\mathbf{\Gamma}_\psi = a^3 \iint P_s(\theta, \lambda)\frac{d\tilde{\psi}}{d\theta}\begin{pmatrix}\sin\lambda\sin\theta \\ -\cos\lambda\sin\theta \\ 0\end{pmatrix} d\theta\, d\lambda. \qquad (9.65)$$

Just as we had $\nabla_H\tilde{\Phi}$ related to $\hat{\mathbf{n}}_{grav}$ earlier, we now have $\nabla_H\tilde{\psi}$ or equivalently $\nabla_H\tilde{\psi}/ag$ related to $\hat{\mathbf{n}}_q^0$. And consequently we have, as in equation (9.47):

$$\mathbf{\Gamma}_\psi = \int_S P_s\left(\mathbf{r}\wedge\hat{\mathbf{n}}_q^0\right) dS = -\mathbf{\Gamma}^\varepsilon. \qquad (9.66)$$

More specifically, the ellipsoidal parts of the pressure torque and the gravitational torque provide a zero contribution to the axial (z-component) torque and partially compensate one another for the equatorial component (x- and y-components).

Link between the torque on the ellipsoid and the matter term of the angular momentum

Let us now express the torques on the ellipsoid. The pressure torque on the ellipticity of the Earth is reduced to Eq. (9.35) with

$$h' = h_2 = -\frac{2}{3}\epsilon a \left(\frac{3\cos^2\theta - 1}{2} \right) \tag{9.67}$$

and the pressure developed in spherical harmonics. Using the orthogonality relation of the spherical harmonics, the topographic torque on the Earth's equatorial bulge is given by (see also, for instance, Bizouard, 1995):

$$\mathbf{\Gamma}_\epsilon = \mathbf{\Gamma}_{\text{pressure}\rightarrow\text{ellipsoid}} = \frac{12\pi}{5}a^3 \begin{pmatrix} -\frac{2}{3}\epsilon\tilde{p}_{21} \\ +\frac{2}{3}\epsilon p_{21} \\ 0 \end{pmatrix}, \tag{9.68}$$

where p_{21} and $\tilde{p}_{21}$ are the degree two order one components of the pressure written as $P_s = 3p_{21}\cos\theta\sin\theta\cos\lambda + 3\tilde{p}_{21}\cos\theta\sin\theta\sin\lambda$.

The ellipsoidal part of the gravitational torque is computed similarly from the expression of the gravitational torque for the degree two order zero part of the gravitational potential; it is given by

$$\mathbf{\Gamma}_{J_2} = \mathbf{\Gamma}_{\text{gravitation}\rightarrow J_2} = -\frac{12\pi}{5}a^3 \begin{pmatrix} C_{20}\tilde{p}_{21} \\ -C_{20}p_{21} \\ 0 \end{pmatrix}, \tag{9.69}$$

where $C_{20} = -J_2 = -1.08 \times 10^{-3}$. The sum of the two torques is thus given by

$$\mathbf{\Gamma}^{\mathcal{E}} = \frac{12\pi}{5}a^3 \begin{pmatrix} -\frac{2}{3}\mathcal{E}\tilde{p}_{21} \\ +\frac{2}{3}\mathcal{E}p_{21} \\ 0 \end{pmatrix}, \tag{9.70}$$

where $\mathcal{E} = \epsilon - \frac{3}{2}J_2 = 0.514\,\epsilon$. The ellipsoidal part of the gravitational torque thus compensates about one half the ellipsoidal part of the pressure torque. Note that from the Clairaut theory, J_2 and ϵ are related to one another by Eq. (9.60). We see immediately that $\mathcal{E} = q/2$. Substituting these expressions in Eq. (9.70), we get:

$$\mathbf{\Gamma}^{\mathcal{E}} = -\frac{4\pi}{5}a^3 q \begin{pmatrix} \tilde{p}_{21} \\ -p_{21} \\ 0 \end{pmatrix} = -\frac{4\pi}{5}\frac{\Omega^2 a^4}{g} \begin{pmatrix} \tilde{p}_{21} \\ -p_{21} \\ 0 \end{pmatrix}. \tag{9.71}$$

Considering the expression of the matter part of the angular momentum given in (9.13), we can compute $\mathbf{\Omega} \wedge \mathbf{H}_{\text{matter}}$. Considering the development of the pressure in spherical harmonics, it is easy to show that it implies

$$\mathbf{\Omega} \wedge \mathbf{H}_{\text{matter}} = -\frac{4\pi}{5}\frac{\Omega^2 a^4}{g}\begin{pmatrix} \tilde{p}_{21} \\ -p_{21} \\ 0 \end{pmatrix}. \qquad (9.72)$$

Considering that $q = \Omega^2 a/g$, one obtains equivalently:

$$\mathbf{\Omega} \wedge \mathbf{H}_{\text{matter}} = -\frac{4\pi}{5}a^3 q \begin{pmatrix} \tilde{p}_{21} \\ -p_{21} \\ 0 \end{pmatrix}. \qquad (9.73)$$

Noting that Eqs. (9.71) and (9.72) have the same right-hand sides, this shows, as computed by Bell (1994), that:

$$\mathbf{\Gamma}^{\mathcal{E}} = \mathbf{\Gamma}_{\text{pressure} \rightarrow \text{ellipsoid}} + \mathbf{\Gamma}_{\text{gravitation} \rightarrow J_2} = \mathbf{\Gamma}_{\epsilon} + \mathbf{\Gamma}_{J_2} = \mathbf{\Omega} \wedge \mathbf{H}_{\text{matter}}. \qquad (9.74)$$

This equation expresses that the ellipsoidal torque is exactly equal to the effect of the rotation of the matter term of the angular momentum change. If the ellipsoidal torque were equal to zero, the flattening of the geoid would cancel the flattening of the geometrical surface, and it is also obvious in that case that the matter part has no rotation either.

Topography used in the general circulation models

Usually the topography used in the general circulation models of the atmosphere or the oceans is described by the geopotential height with respect to the geoid as a function of position on the surface defined as the topography, rather than by the shape of the surface with respect to a mean radius. We call it the *geotopography* in order to differentiate it from the local topography or orography describing the geometrical shape of the Earth's surface with respect to its mean radius. In these models, one uses a *local* coordinate system having its third axis in the direction opposite to the local direction of gravity, i.e., along the vertical as defined by the direction of the gradient of the geopotential (the sum of the gravitational and centrifugal potentials). The reason for this is related to employing a change of coordinates used in GCM so that all the gravitational effects are on the vertical component and there is no contribution in the horizontal components. In such a coordinate system, only the z-component of the gravitational force on a mass element of the fluid is non-zero.

The mountain torque

After removing the torques acting on the ellipsoidal part of the Earth's structure, the remaining parts of the pressure and gravitational torques are related in the same way as in Eq. (9.55). What remains from the total pressure and gravitational torque when the ellipsoidal torque has been removed is

$$\mathbf{\Gamma}^G + \mathbf{\Gamma}^P - \mathbf{\Gamma}^{\mathcal{E}} = -\int_S P_s \, \mathbf{r} \wedge \left(\hat{\mathbf{n}}_{SE}^1 - \hat{\mathbf{n}}_{grav}^1\right) dS. \tag{9.75}$$

As mentioned at the end of Section 9.1.2, the remaining part of the gravitational torque relative to the J_2-part (the term $\mathbf{\Gamma}_{gravitation \to J_2}$) has a negligible contribution to the gravitation torque ($\mathbf{\Gamma}_{gravitation \to Earth}$ or $\mathbf{\Gamma}^G$) at all frequencies. This means that the term in $\hat{\mathbf{n}}_{grav}^1$ in the above equation can be considered as negligible when computing all torques induced by the atmosphere and ocean acting on the solid Earth. Given this fact, when using a topography given in geopotential height of the surface, as mentioned in the previous paragraph, this geotopography can be given with respect to the Clairaut ellipsoid instead of the geoid. The topographic torque is then simply the same as the part containing the term in n_{SE}^1 in the above equation. The torque of Eq. (9.75) simply involves the normal to the geotopography denoted $n_{topography}$. This is the reason why it is called the "topographic" torque or the "mountain" torque:

$$\mathbf{\Gamma}^{\text{Mountain}} = -\int_S P_s \left(\mathbf{r} \wedge \hat{\mathbf{n}}_{topography}\right) dS. \tag{9.76}$$

This mountain torque and the ellipsoidal torque all together can be related to the sum of the gravitational and pressure torque by:

$$\mathbf{\Gamma}^G + \mathbf{\Gamma}^P = \mathbf{\Gamma}^{\mathcal{E}} + \mathbf{\Gamma}^{\text{Mountain}}, \tag{9.77}$$

or

$$\mathbf{\Gamma}^G + \mathbf{\Gamma}^P = -\int_S P_s \left(\mathbf{r} \wedge \hat{\mathbf{n}}_q\right) dS - \int_S P_s \left(\mathbf{r} \wedge \hat{\mathbf{n}}_{topography}\right) dS, \tag{9.78}$$

$$\mathbf{\Gamma}^G + \mathbf{\Gamma}^P = -\int_S P_s \left(\mathbf{r} \wedge \left(\hat{\mathbf{n}}_{SE}^0 - \hat{\mathbf{n}}_{grav}^0\right)\right) dS - \int_S P_s \left(\mathbf{r} \wedge \left(\hat{\mathbf{n}}_{SE}^1 - \hat{\mathbf{n}}_{grav}^1\right)\right) dS. \tag{9.79}$$

The friction torque

The friction torque is associated with the local friction forces due to the relative motion of the fluid with respect to the Earth (for instance, the surface wind in the case of the atmosphere). The drag or friction force at any point on the surface is the wind/current velocity times a drag coefficient: $\mathbf{F} = f \, \mathbf{v}$; the drag coefficient f has a complicated dependence on shape, inclination, and surface roughness of the body relative to which the fluid flows, as well as on the flow conditions. The torque is computed by an integration over the Earth's surface, a sum over all small surface elements considered. The integrand involved in this integration is proportional to the friction drag at each point of interaction on the surface. The zonal friction drag, associated with the zonal winds/currents affects the three Cartesian components of the torque, and the meridional friction drag, only the equatorial Cartesian components of the torque. The mathematical explanation of this can be found by computing the Cartesian components of the torque (last part of Eq. (9.20)) from the zonal and meridional components of the friction drag, exactly as was done previously, replacing **F** by f **v** and considering a similar treatment as applied for Eq. (9.8) where **v** was expressed in spherical coordinates to provide Eq. (9.14).

Atmospheric models provide grid values for the friction drag coefficient as well as for the wind velocity, from which the surface friction force can be computed (**F**) and can be substituted in the torque equation, which can be integrated. Atmospheric winds are usually stronger and more coherent over the oceans; but the friction coefficient is higher between air and land than between air and water.

9.2 Links between torque and angular momentum

The major advantage of using the angular momentum method for computing the effect of the atmosphere on polar motion or nutations is the fact that the angular momentum is a classical output of atmospheric analysis systems, and that several angular momentum data sets are available routinely. Those data sets may be found at the site (http://euler.jpl.nasa.gov/sba/) of the Special Bureau for the Atmosphere which is a part of the Center for Global Geophysical Fluids of the IERS (Salstein *et al.*, 1993).

The torque approach may nevertheless be used as a complementary method as it facilitates physical understanding. It provides us with the knowledge as to how and where the angular momentum is transferred from the fluid layers to the solid Earth. We give here a few examples:

- In the study of de Viron *et al.* (2002a and b), it is shown that, at the annual timescale, the most important atmospheric contribution to polar motion stems from the highest topographies which enter the torque computation. Despite the ellipsoidal torque, the contributions to the x- and y-components are mainly due to large departure of the actual surface of the Earth from the idealized ellipsoidal shape, especially from large topography gradients such as those found over the Himalayas. Additionally, large pressure anomalies such as those found over the North Pacific Ocean may provide an important contribution. For the axial component, de Viron *et al.* have shown that there is a partial cancellation of the mountain torque over Asia and North America. The seasonal effect of the Asian Monsoon is visible in the axial component.
- At the diurnal timescale (see de Viron *et al.*, 2001b), one can expect important contributions from the mountain torque in Asia due to the Himalayas and in

South America due to the Andes. Antarctica is also characterized by strong topographic gradients and contributes significantly to the x- and y-components of the mountain torque. The largest contribution to the friction torque is also from the same regions, as the winds are strong there (see de Viron *et al.*, 2001b).

- In another study, de Viron *et al.* (2001c) examined the behavior of the atmosphere during an ENSO cycle, and in particular when in 1989 there was a strong episode of cool weather (La Niña event) in the North Atlantic and Arctic. This event could be studied in terms of torques on the different regions and could explain the minimum of the atmospheric angular momentum during that period. The mountain torques over North America, South America, and Europe were partially compensated by the mountain torque over Asia. The La Niña event was causing strong surface pressure over the Rockies and the Andes. The stronger-than-normal Monsoon due to this event also implied a high pressure effect along the eastern edge of the Tibetan plateau. These must be looked at in a global analysis over the different continents, as done in the paper of de Viron *et al.* (2001c).

- Lambert *et al.* (2006) have examined the short-term contributions (few days or weeks) to polar motion during a period when the CW and the annual wobble were nearly canceling each other's effects (from November 2005 to February 2006). Here again, major pressure/depression events on the continents and on the oceans are the major contributors to the torque. They noted especially the strong influence of a depression over Northern Europe in phase with similar events over North America. They have indeed studied "loops" in polar motion; they have identified the unbalanced low pressure above Europe contributing to these loops as well as similar-scale pressure anomalies in diametrically opposite areas.

Up to now, the angular momentum approach seems to yield numerical results for the effects of atmosphere on nutation that are closer to the observed nutation residuals than the results obtained using the torque approach (Brzezinski *et al.*, 2002, de Viron *et al.*, 2004). The numerical performance of the angular momentum approach is better than that of the torque approach, as the computations of the torques are delicate, requiring precise knowledge of the fluid pressure over each Earth's surface element, precise knowledge of the topography itself, as well as precise knowledge of the fluid velocities and drag over each surface element of the Earth. In addition, the torques are usually built from contributions from the different areas of the Earth's surface and these contributions are often cancelling each other. Different global circulation models (GCM) of the atmosphere lead to different results, as shown in Yseboodt *et al.* (2002).

As the amplitudes of the periodic components are not stable in time, it is better to estimate the atmospheric effects on nutation in the time domain (see Yseboodt *et al.*, 2002). But the atmospheric GCMs used for these computations are developed specifically for weather forecasting purposes; they are not really suited for the estimation of atmospheric variations at diurnal frequencies. With use of the recent gravimetry satellites such as CHAMP, GRACE, and GOCE, better constraints on the geophysical fluid models will be obtained, as these satellites are sensitive to the mass anomalies below them.

The dynamics of the atmosphere at diurnal timescales is dominated by the thermal effects on the atmosphere above the continents due to the cycle of day-time heating and night-time cooling of the land mass; the temperature changes of the ocean surface are much less pronounced because of the greater heat capacity of water and the cooling caused by evaporation. The cyclic temperature changes produced in the lower atmosphere are rather small over the oceans and much higher over the continents. As seen from the Navier-Stokes Eq. (7.65), the Coriolis term ($2\Omega \wedge \mathbf{v}$) and the inertial acceleration ($d\mathbf{v}/dt$) are of the same order of magnitude in the diurnal frequency band. The latter is thus important in the equation of motion and consequently important for the dynamics of the atmosphere. For semi-diurnal and higher frequencies, the inertial acceleration dominates over the Coriolis term, and the winds are primarily driven by the pressure gradient. At the other extreme, for frequencies much lower than diurnal, geostrophy (balance of the Coriolis force and the pressure gradient) provides a good approximation.

On the diurnal timescale, there are much larger pressure values and pressure changes over continents than over oceans. The local temperatures induced by thermal forcing (depending on the insolation) have larger changes between day and night above continents than above oceans.

During the day, when the temperature increases near the surface (this is more important above continents), the lower atmosphere is heated and expands. Equal density surfaces in the atmosphere are thus higher with respect to the Earth's surface where the Sun is shining than during the night. This implies motions of air masses in the upper atmosphere from the location where it is day to the location where it is night, increasing the ground pressure where it is night. The pressure increases thus at the surface where the Sun is not heating and an antisymmetric wind pattern with respect to the upper atmosphere is created at the Earth's surface. This is the so-called Hadley circulation. There are thus diurnal winds and pressure changes created with the successive day and night heating and cooling. These motions observed on a global Earth map show some symmetry, with divergence and convergence fields; this results in cancellations when computing the total torque which generates Earth rotation variations and wobbles.

The remaining contributions are quite small in the diurnal timescale. This is the reason why, for nutation, the computation of the effects of the atmosphere on the solid Earth is not easy. Although the diurnal thermal effect on the atmosphere is very important, the contribution to the torque is small. (However, when precise nutation theory is considered, this torque leads to a contribution of about 0.1 mas to the prograde annual nutation, and must not, therefore, be ignored.)

As mentioned before, Lott *et al.* (2008, see also de Viron, 2008) have examined the closure of the angular momentum budget equation using a theoretical general circulation model with no assimilation of observational data. This work shows how difficult it is to close the budget. De Viron *et al.* (2005a and b) have examined the order of magnitude of the different contributions. Their results are presented in what follows. Let us recall that the ellipsoidal torque is exactly equal to the effect of the Earth's rotation on the matter part of the angular momentum change (Bell, 1994, see also Eq. (9.74) of this book):

$$\Gamma^{\mathcal{E}} = \Omega \wedge \mathbf{H}_{\text{matter}}. \tag{9.80}$$

This allows us to simplify the angular momentum balance equation in space. Starting from Eqs. (9.2)–(9.6) for the atmosphere, and Eq. (9.15)

$$\left(\frac{d\mathbf{H}_{\text{Fluid}}}{dt}\right)_S - \frac{d\mathbf{H}_{\text{matter}}}{dt} + \Omega \wedge \mathbf{H}_{\text{matter}} + \frac{d\mathbf{H}_{\text{motion}}}{dt} + \Omega \wedge \mathbf{H}_{\text{motion}}$$
$$= \Gamma_{\text{Earth}\rightarrow\text{Fluid}}, \tag{9.81}$$

and using the expressions (9.77) and (9.79) we can write the total torque or angular momentum change as

$$\left(\frac{d\mathbf{H}_{\text{Fluid}}}{dt}\right)_S = \Gamma^G + \Gamma^P + \Gamma^F = \Gamma^{\mathcal{E}} + \Gamma^{\text{Mountain}} + \Gamma^F. \tag{9.82}$$

We then substitute Eq. (9.80) in Eq. (9.82) and use Eq. (9.81), giving

$$\frac{d\mathbf{H}_{\text{matter}}}{dt} + \frac{d\mathbf{H}_{\text{motion}}}{dt} + \Omega \wedge \mathbf{H}_{\text{motion}} = \Gamma^{\text{Mountain}} + \Gamma^F = \Gamma^{\text{local}}, \tag{9.83}$$

which also defines Γ^{local}, or equivalently in the diurnal frequency range (and considering first-order quantities):

$$-i\Omega\tilde{H}_{\text{matter}} + i\sigma'\tilde{H}_{\text{motion}} = \tilde{\Gamma}^{\text{Mountain}} + \tilde{\Gamma}^F = \tilde{\Gamma}^{\text{local}}, \tag{9.84}$$

where $\tilde{}$ expresses the complex sum of the first two components, where the frequency in the terrestrial frame is considered as retrograde diurnal ($\sigma \approx -\Omega$) and where the frequency in the celestial frame is denoted $\sigma' = \sigma + \Omega$. Also, (9.80) yields in the diurnal timescale:

$$\tilde{\Gamma}^{\mathcal{E}} = i\Omega\tilde{H}_{\text{matter}}. \tag{9.85}$$

Substitution of this equation into Eq. (9.84) yields:

$$-1 + \frac{\sigma'}{\Omega} \frac{\tilde{H}_{\text{motion}}}{\tilde{H}_{\text{matter}}} = \frac{\tilde{\Gamma}^{\text{local}}}{\tilde{\Gamma}^{\mathcal{E}}}. \tag{9.86}$$

As $\frac{\sigma'}{\Omega}$ is usually small for the important nutation frequencies, we have:

$$-\mathbf{\Gamma}^{\mathcal{E}} \approx \mathbf{\Gamma}^{\text{local}}. \tag{9.87}$$

It is important to note that the local torque should almost cancel the ellipsoidal torque. This situation is difficult to handle when dealing with real data. If the ellipsoidal torque were to be small as well, one would have to consider the term in $\frac{\tilde{H}_{\text{motion}}}{\tilde{H}_{\text{matter}}}$ and in that case, as seen previously, $\tilde{H}_{\text{matter}}$ would be small (see Eq. (9.80)), which ensures that the term proportional to $\tilde{H}_{\text{motion}}$ might be high enough to compensate the local torque.

For the x- and y-components and outside the diurnal frequency band, the mountain torque and the friction torque have the same order of magnitude: the mountain torque is about 50% of the local torque, and the friction torque is about 50% as well (de Viron *et al.*, 2001b and 2002a and b). For the x- and y-components and within the retrograde diurnal frequency band, the friction torque is smaller than the mountain torque (de Viron *et al.*, 2001b).

For the z-component and outside the diurnal frequency band, the mountain torque and the friction torque have the same order of magnitude: the mountain torque is larger than the friction torque (de Viron *et al.*, 2001b, 2002a and b), and also the friction torque is about 50% (de Viron *et al.*, 2001c and 2002a and b). For the z-component and within the diurnal frequency band, the friction torque is smaller than the mountain torque (de Viron *et al.*, 2001b).

9.3 Effect of a fluid acting on the solid Earth

In considering the equations governing the contributions from atmospheric angular momentum (AAM) variations to the variations in Earth rotation, it is traditional and convenient to define the dimensionless vector

$$\chi_{\text{Fluid}} = \frac{1}{eA\Omega} \mathbf{H}_{\text{Fluid}} \tag{9.88}$$

to represent the AAM. AAM variations translate into variations of χ_{Fluid}, which perturb the Earth's rotation. So the components of χ_{Fluid} are referred to as the *atmospheric excitation functions*. Each of these quantities is made up of a "matter term" and a "motion term." Oceanic excitation functions may be similarly defined.

9.3.1 Rigid Earth

We start from the Euler equation (Eq. (2.47)) for the wobble of a rigid Earth without atmosphere and without ocean, as explained in Chapter 2 and repeated here for convenience:

$$\frac{d\tilde{m}}{dt} - ie\Omega\tilde{m} = \frac{\tilde{\Gamma}^{\text{external}}}{A\Omega}, \tag{9.89}$$

where $\tilde{m}$ is the complex sum of the first two components of the wobble ($\tilde{m} = m_x + im_y$), $\tilde{\Gamma}^{\text{external}}$ is the complex sum of the first two components of the external torque acting on the solid Earth. We then consider the existence of a fluid layer above the surface of this simple Earth. An atmospheric layer undergoes thermal effects, and an oceanic layer suffers tidal perturbations, with time dependence similar to the solid Earth's response to the tidal gravitational forcing, and we compute the Earth's response to the fluids in the same frequency bands as in the case of the solid Earth's response to external gravitational forcing. In a linear approximation, we may compute separately the Earth's responses to external tidal forcing and to the behavior of the fluid layer (which is dealt with in this chapter), and add them together. To a first approximation, we consider that the system is only composed of a one-layer solid Earth under the fluid layer which may be either the atmosphere or the ocean. The system Earth–ocean or Earth–atmosphere in this computation is considered as isolated and the variations of the total angular momentum of the system Earth–ocean or Earth–atmosphere is equal to zero. This approach enables us to compute the wobble of the solid Earth from the atmosphere and ocean excitation functions. The response of the ocean to the atmosphere and these ocean effects on Earth are not considered here; they are dealt with in Section 9.4. Consider now the solid Earth–atmosphere case. As mentioned previously, one can work in two ways, either considering the system Earth–atmosphere without external torque (in that case, the atmosphere must be considered in the left-hand side of Eq. (9.89) and thus additional moment of inertia and relative angular momentum must be considered), or the atmosphere can be considered as acting on the solid Earth (entering into the torque on the right-hand side of the equation). We may then replace the external torque acting on the solid Earth by the torque $\Gamma_{\text{Fluid} \rightarrow \text{Earth}} = -\Gamma_{\text{Earth} \rightarrow \text{Fluid}}$ from Eq. (9.6). And as the torque on the surface fluid layer is expressed as the time derivative in space of the angular momentum of the fluid, we have:

$$\frac{d\tilde{m}}{dt} - ie\Omega\tilde{m} = -\frac{1}{A\Omega}\left(\frac{d\tilde{H}_{\text{Fluid}}}{dt}\right)_S. \tag{9.90}$$

Since $\tilde{H}_{\text{Fluid}} = eA\Omega\tilde{\chi}_{\text{Fluid}}$ by Eq. (9.88), the above equation becomes

$$\frac{d\tilde{m}}{dt} - ie\Omega\tilde{m} = -e\left(\frac{d\tilde{\chi}_{\text{Fluid}}}{dt}\right)_S = -ie\Omega\tilde{\chi}_{\text{Fluid}} - e\frac{d\tilde{\chi}_{\text{Fluid}}}{dt}, \tag{9.91}$$

or

$$\tilde{m} + \frac{i}{e\Omega} \frac{d\tilde{m}}{dt} = \tilde{\chi}_{\text{Fluid}} - \frac{i}{\Omega} \frac{d\tilde{\chi}_{\text{Fluid}}}{dt}. \tag{9.92}$$

The equivalent equation in the frequency domain is

$$\left(1 - \frac{\omega_w}{e\Omega}\right) \tilde{m}(\omega_w) = \left(1 + \frac{\omega_w}{\Omega}\right) \tilde{\chi}_{\text{Fluid}}(\omega_w), \tag{9.93}$$

where ω_w is the angular frequency of the wobble. It may be recalled that $e\Omega$ is the Eulerian wobble frequency of the rigid Earth, which is often denoted by σ_E in the literature. The previous equation then simplifies thus:

$$\tilde{m}(\omega_w) = \frac{\sigma_E}{\sigma_E - \omega_w} \left(1 + \frac{\omega_w}{\Omega}\right) \tilde{\chi}_{\text{Fluid}}(\omega_w). \tag{9.94}$$

Another way to obtain the same equation is to consider first, as in the case of the existence of a fluid core, the relative angular momentum of the fluid $\mathbf{h}^{\text{Fluid}}$ given by the motion term of the angular momentum for the atmosphere and the current term for the ocean ($\mathbf{h}^{\text{Fluid}} = \mathbf{H}_{\text{Fluid}}^{\text{motion}}$), and second, the induced changes in the moment of inertia due to the mass changes in the fluid (given by the matter part of the fluid angular momentum). The changes in the mass distribution in the atmosphere induce a change in the moment of inertia $c_{13}^{\text{Fluid}} + i c_{23}^{\text{Fluid}}$ and consequently there is a contribution $i\Omega\tilde{c}^{\text{Fluid}}$ in the angular momentum of the whole system. $\tilde{c}^{\text{Fluid}} = c_{13}^{\text{Fluid}} + i c_{23}^{\text{Fluid}}$ can be computed from the mass repartition in the atmosphere and it is easy to show that it is proportional to the matter part of the angular momentum of the fluid:

$$\Omega\tilde{c}^{\text{Fluid}} = \tilde{H}_{\text{Fluid}}^{\text{matter}} = e\Omega A \tilde{\chi}^{\text{matter}}, \tag{9.95}$$

where $\tilde{H}_{\text{Fluid}}^{\text{matter}}$ and $\tilde{\chi}^{\text{matter}}$ are the complex sum of the first two components of the matter parts of the angular momentum $\mathbf{H}_{\text{Fluid}}^{\text{matter}}$ and of the dimensionless angular momentum χ^{matter}, defined in (9.88). The additional fluid part related to the relative angular momentum in the angular momentum conservation equations is a term in $\mathbf{\Omega} \wedge \mathbf{H}^{\text{Fluid}} + \frac{d}{dt}\mathbf{H}^{\text{Fluid}}$, which, for the complex sum of the first two components, gives: $i\Omega\tilde{h}^{\text{Fluid}} + \frac{d}{dt}\tilde{h}^{\text{Fluid}}$. Similarly, the additional fluid part related to the inertia tensor changes due to the atmosphere in the complex sum of the first two components of the angular momentum conservation equations is a term in $i\Omega^2\tilde{c}^{\text{Fluid}} + \Omega\frac{d}{dt}\tilde{c}^{\text{Fluid}}$. In the frequency domain, these two terms show the same frequency dependence as Eq. (9.94).

Equation (9.93) for the wobble $\tilde{m}$ may be converted into an equation for the corresponding polar motion $\tilde{p}$ of the CIP (or equivalently, that of the pole of the CRF, see Gross, 1992), which is what is observationally estimated for wobble (or polar motion) frequencies outside the retrograde diurnal band. The relation between

$\tilde{m}$ and $\tilde{p}$ may be seen by combining Eqs. (2.93) and (2.105):

$$\tilde{p}(\omega_w) = \frac{\Omega}{\Omega + \omega_w} \tilde{m}(\omega_w),$$ (9.96)

where ω_w is the angular frequency of the wobble. Equation (9.93) may now be written as the following equation for $\tilde{p}$:

$$\left(1 - \frac{\omega_w}{e\Omega}\right) \tilde{p}(\omega_w) = \tilde{\chi}_{\text{Fluid}}(\omega_w).$$ (9.97)

The equivalent equation in the time domain is

$$\tilde{p}(t) + \frac{i}{e\Omega} \frac{d\tilde{p}(t)}{dt} = \tilde{\chi}_{\text{Fluid}}(t).$$ (9.98)

9.3.2 Deformable solid Earth

When considering a deformable solid Earth, we must take account of the increments to the elements of the inertia tensor from the deformational effects of the fluid loading besides those due to the incremental centrifugal potential resulting from rotation variations. The loading by the fluid is of course through the pressure acting on the Earth's surface, and is therefore related to the matter part of the angular momentum of the fluid. The angular momentum balance equation for the whole Earth (including the fluid layer) is given by Eq. (9.1) with

$$\mathbf{H} = \mathbf{H}_{\text{solid Earth}} + \mathbf{H}_{\text{Fluid}}, \qquad \left(\frac{d\mathbf{H}}{dt}\right)_S = \mathbf{\Omega} \wedge \mathbf{H} + \frac{d\mathbf{H}}{dt} = 0,$$ (9.99)

and

$$\mathbf{H} = [C] \cdot \mathbf{\Omega} + \mathbf{h}^{\text{Fluid}},$$ (9.100)

where $[C]$ is the inertia tensor of the whole system Earth, i.e., the solid Earth part plus the surface fluid part. Let us write down explicitly the quantities appearing in the above equations in component form:

$$[C] = \begin{pmatrix} A + c_{11} & c_{12} & c_{13} \\ c_{21} & A + c_{22} & c_{23} \\ c_{31} & c_{32} & C + c_{33} \end{pmatrix}, \qquad \mathbf{\Omega} = \begin{pmatrix} \Omega_0 m_1 \\ \Omega_0 m_2 \\ \Omega_0(1 + m_3) \end{pmatrix},$$ (9.101)

$$c_{ij} = c_{ij}^{\text{Earth}} + c_{ij}^{\text{Fluid}}.$$ (9.102)

The c_{ij}^{Fluid} arise from the inertia of the fluid layer, and the c_{ij}^{Earth} are produced by the deformation of the solid Earth induced by the fluid mass loading on top, as well as from the centrifugal effect of the wobbles. We take all these perturbation quantities to be smaller than A or C by several orders of magnitude as usual. The only change with respect to the angular momentum equation of the previous section (for a rigid Earth) is the occurrence of the c_{ij}^{Earth}. As explained in Chapter 6 (see Eqs. (6.152)–(6.154)), we only need the c_{i3}^{Earth} and c_{i3}^{Fluid} for the nutation forcing. The part of the solid Earth deformation arising from the fluid loading $c_{i3}^{Earth \leftarrow Fluid}$ can be expressed as proportional to the loading using a compliance expressing the Earth's inertia changes due to a unit loading potential at the Earth surface, exactly as done for the Earth's inertia changes due to a unit tidal potential (see Section 6.8.4). Using the same kind of link between Love numbers and compliances as expressed before (see Sections 6.3, 6.4, and 6.5) one can also get the link between the incremental moment of inertia and the loading (see Eq. (7.170)) using the load Love numbers such as $\hat{k}$ as done in Sections 6.5 and 7.4.3. The angular momentum of the whole system Earth + fluid layer can be written exactly as done in Chapter 7 (see Eq. (7.114)):

$$\mathbf{H} = \begin{pmatrix} (Am_1 + c_{13})\Omega_0 + h_1 \\ (Am_2 + c_{23})\Omega_0 + h_2 \\ (C(1 + m_3) + c_{13})\Omega_0 + h_3 \end{pmatrix}, \tag{9.103}$$

$$\mathbf{\Omega} \wedge \mathbf{H} = \Omega_0 \begin{pmatrix} ((C - A)m_2 - c_{2,3})\Omega_0 - h_2 \\ ((A - C)m_1 + c_{1,3})\Omega_0 + h_1 \\ 0 \end{pmatrix}. \tag{9.104}$$

Consider the system solid Earth + geophysical fluids to be isolated:

$$\left(\frac{d\mathbf{H}}{dt}\right)_S = \mathbf{\Omega} \wedge \mathbf{H} + \frac{d\mathbf{H}}{dt} = 0.$$

By substituting the equations above, evaluating at the first order in the small quantities, and taking the complex sum of the first two equations as usual, we get:

$$\frac{d}{dt}\left(A\Omega\tilde{m} + \Omega\tilde{c}_3^{Earth} + \Omega\tilde{c}_3^{Fluid} + \tilde{h}\right)$$

$$- i\Omega^2\left(Ae\left(1 - \frac{k}{k_{fl}}\right)\tilde{m} - \tilde{c}_3^{Earth \leftarrow Fluid} - \tilde{c}_3^{Fluid}\right) + i\Omega\tilde{h} = 0, \tag{9.105}$$

where $\tilde{m} = m_1 + im_2$, etc., as usual, and

$$\tilde{c}_3^{Earth} = c_{13}^{Earth} + ic_{23}^{Earth}, \quad \tilde{c}_3^{Earth \leftarrow Fluid} = c_{13}^{Earth \leftarrow Fluid} + ic_{23}^{Earth \leftarrow Fluid}, \tag{9.106}$$

$$\tilde{c}_3^{Fluid} = c_{13}^{Fluid} + ic_{23}^{Fluid}. \tag{9.107}$$

Here $\tilde{c}_3^{Earth \leftarrow Fluid}$ represents the solid Earth deformation induced by loading by the fluid; it is related to the matter term of the fluid by

$$\Omega\tilde{c}^{Earth \leftarrow Fluid} = \hat{k}\Omega\tilde{c}^{Fluid} = \hat{k}e\Omega A\tilde{\chi}^{matter}, \tag{9.108}$$

where $\hat{k}$ is the load Love number expressing the mass redistribution potential induced by a load. Noting that

$$\tilde{h} = e\Omega A \tilde{\chi}^{\text{motion}} \quad \text{and} \quad \Omega \tilde{c}^{\text{Fluid}} = e\Omega A \tilde{\chi}^{\text{matter}}, \tag{9.109}$$

and that $(C - A) = Ae$, we obtain:

$$\frac{d}{dt}\left(A\tilde{m} + (1 + \hat{k})eA\tilde{\chi}^{\text{matter}} + eA\tilde{\chi}^{\text{motion}}\right)$$

$$- i\Omega\left(Ae(1 - \frac{k}{k_{fl}})\tilde{m} - (1 + \hat{k})eA\tilde{\chi}^{\text{matter}} - eA\tilde{\chi}^{\text{motion}}\right) = 0, \tag{9.110}$$

where k is the gravitational Love number of the solid Earth (see Eq. (7.168) and also Eq. (7.170)) and k_{fl} is the fluid Love number (see Eq. (7.169)), as already introduced in Chapter 7.

The above equation with the χ terms dropped is the equation for a deformable Earth without atmosphere or ocean:

$$\frac{d\tilde{m}}{dt} - i\sigma'_{\text{CW}}\tilde{m} = 0, \tag{9.111}$$

where σ'_{CW} is the Chandler wobble frequency for a deformable Earth without a fluid core: $\sigma'_{\text{CW}} = \Omega e(1 - \frac{k}{k_{fl}}) = \sigma_E(1 - \frac{k}{k_{fl}})$ or $\sigma'_{\text{CW}} = \Omega(e - \kappa)$ where κ is the compliance introduced in the same chapter.

Equation (9.110) may be reorganized and written in the succinct form

$$\tilde{m} + i\frac{1}{\sigma'_{\text{CW}}}\frac{d\tilde{m}}{dt} = \tilde{\chi}_{\text{effective}} - i\frac{1}{\Omega}\frac{d\tilde{\chi}_{\text{effective}}}{dt}, \tag{9.112}$$

by defining the "effective" χ-function

$$\chi_{\text{effective}} = \chi_{\text{effective}}^{\text{matter}} + \chi_{\text{effective}}^{\text{motion}}, \tag{9.113}$$

with:

$$\tilde{\chi}_{\text{effective}}^{\text{matter}} = \frac{1 + \hat{k}}{\left(1 - \frac{k}{k_{fl}}\right)}\tilde{\chi}^{\text{matter}}, \qquad \tilde{\chi}_{\text{effective}}^{\text{motion}} = \frac{1}{\left(1 - \frac{k}{k_{fl}}\right)}\tilde{\chi}^{\text{motion}}. \tag{9.114}$$

With the numerical values $k_f = 0.945$, $k = 0.3$, and $\hat{k} = -0.3$, we find that

$$\tilde{\chi}_{\text{effective}}^{\text{matter}} \approx 1.03\tilde{\chi}^{\text{matter}}, \qquad \tilde{\chi}_{\text{effective}}^{\text{motion}} \approx 1.46\tilde{\chi}^{\text{motion}}. \tag{9.115}$$

The numerical values used by Barnes *et al.* (1983), i.e. $k_f = 0.945$, $k = 0.285$, and $\hat{k} = -0.3$, gave the factors 1.00 and 1.43 (instead of 1.03 and 1.46) in the above relations.

The frequency domain version of Eq. (9.112) is

$$\left(1 - \frac{\omega_w}{\sigma'_{CW}}\right) \tilde{m}(\omega_w) = \left(1 + \frac{\omega_w}{\Omega}\right) \tilde{\chi}_{effective}(\omega_w). \tag{9.116}$$

The corresponding result for the fluid effects on polar motion may be obtained by making use of Eq. (9.96):

$$\left(1 - \frac{\omega_w}{\sigma'_{CW}}\right) \tilde{p}(\omega_w) = \tilde{\chi}_{effective}(\omega_w). \tag{9.117}$$

The equivalent relation in the time domain is evidently

$$\tilde{p} + i \frac{1}{\sigma'_{CW}} \frac{d\tilde{p}}{dt} = \tilde{\chi}_{effective}. \tag{9.118}$$

9.4 Non-tidal ocean effects on the Earth rotation

The oceans, like the atmosphere, perturb the Earth rotation from very long time-scales to sub-diurnal ones, at frequencies that could be similar to those of the atmosphere. While the solid Earth tides are rather easy to compute and the frequencies involved are simply those present in the tidal potential (and those of the rotational normal modes), the ocean tides have, additionally, resonances associated with the bathymetry of the ocean basins. These resonances, as well as the ocean tides, have to be determined by numerical solutions of the Laplace tidal equations (LTE, Laplace, 1775, 1776, Lamb, 1932). LTE are obtained from momentum and mass conservation equations written in rotating coordinates for a homogeneous fluid shell surrounding a nearly spherical planet and having a gravitationally stabilized free surface. The Coriolis acceleration associated with the horizontal component of the Earth's rotation and the vertical component of the particle acceleration are neglected. Ocean tidal models are better determined when the information provided by satellite altimetry (e.g. TOPEX/POSEIDON) is used to constrain the solution of the Laplace equations. So the determination of the short-period ocean tides becomes quite good. For the long-period tides, the ocean response to tidal forcing may be considered to be that of an equilibrium ocean, and is then easy to compute. The ocean modeling community has greatly improved the quality of the ocean tide models and the data are available for geodetic use (see the website of the IERS Special Bureau for the Ocean, http://euler.jpl.nasa.gov/sbo/). Unfortunately, no operational data set is available yet for the non-tidal ocean contributions to wobbles or nutations on the diurnal timescale, though research papers can be found on the subject. The non-tidal effects at the timescale of nutation and polar motion relate mainly to the oceanic response to the atmospheric forcing. This contribution to nutation contains large quasi-periodic components at the same periods as the tides related to the Sun, as well as other components and a continuum part. The atmospheric pressure pushes on the oceanic surface, and the friction associated with the wind stress also influences the motion in the ocean. The response of the ocean to such forcing must be computed using ocean general circulation models under the relevant approximations. The commonly used types of approximations consist in straightforward hypotheses describing the possible extremes of behavior of the ocean, such as (1) that the ocean is completely "rigid" and transmits the local surface pressure faithfully to the bottom, or (2) that the ocean transmits only the globally averaged pressure; but for a more realistic computation of the response of the ocean to atmospheric forcing, we must take account of the dynamics of the oceans which could play an important role.

Inverted barometric ocean

The IBO hypothesis is one of the hypotheses concerning the ocean response to atmospheric forcing. It consists in neglecting friction and assuming that the ocean has sufficient time to readjust to changes in atmospheric pressure over the global oceans, so that at a particular depth, the pressure is the same everywhere, which means that only the mean global pressure changes over the ocean are transmitted to the bottom of the ocean. This hypothesis may be considered as valid at long timescales, but at the diurnal timescale relevant for nutation, this is far from valid.

Non-inverted barometric ocean

According to the NIBO hypothesis, the atmospheric pressure which acts on the surface of the ocean gets transmitted as such to the bottom of the ocean without any modification by the intervening column of ocean water.

IBO models driven by wind forcing

Intermediate between the above two hypotheses, we find in the literature, models convenient for timescales longer than 20 days (for which the IBO-hypothesis is not too far from reality), assuming the IBO hypothesis together with ocean dynamics driven by the wind stress. The ocean angular momentum from such a model has been used to compute the oceanic effect on the LOD and on polar motion.

The ocean response to a pressure forcing provided as a grid system evolving over time has been computed by integration of the equation of motion, for instance, by Ponte (1997), and the ocean currents and the water height fields thus obtained have been used to compute either the oceanic torque on the Earth (see for instance de Viron *et al.*, 2001a), or the oceanic angular momentum (see Gross, 2000).

Marcus *et al.* (1998), for instance, have computed the effect on LOD from two global circulation models for the ocean (the multi-layer ocean model from the Geophysical Fluid Dynamics Laboratory (Bryan, 1969), and from a model derived from the Miami Isopycnal Coordinate Ocean Model (Bleck *et al.*, 1989)) and have shown that the results explain an important part of the residuals between the observed LOD and the part caused by the atmosphere.

Concerning polar motion, the effect of ocean angular momentum has been exhibited by Ponte *et al.* (1998, see also e.g. Nastula and Ponte, 1999) who computed the oceanic excitation from an ocean model of constant density forced by surface wind stresses. Gross (2000) has also shown from another ocean model that taking account of the oceanic contribution to polar motion, especially at the CW period, helps to increase the coherence between the observed and computed temporal variations to a great extent. Ponte and Stammer (1999 and 2000) have shown that the amplitude contribution from the ocean to polar motion is of the right order of magnitude.

General circulation models for the ocean

In computing the ocean effects on short-period polar motion and nutations it is important to use a dynamical ocean in response to the atmospheric forcing, as was done by de Viron *et al.* (2004) and Brzezinski *et al.* (2004). It is shown in these papers that the observed residuals at the one-solar-day frequency in the nutation (i.e. the prograde diurnal frequency in space) can be explained (both in-phase and out-of-phase components) by the atmosphere excitation and the ocean response.

10

Refinements of non-rigid nutation

Most of the essential elements of the theory of nutation of the non-rigid Earth have been presented in Chapter 7. However a number of relatively small effects have to be taken into account before one can expect to have a theoretical framework that can yield numerical results that match the observational data on nutation and precession to approximately the same level of accuracy as the precision of the observations themselves.

The celebrated work of John Wahr (1979, 1981a–d) on the effects of the tidal potential on a rotating, ellipsoidal, elastic, and oceanless Earth led to numerical results for the coefficients of nutation for such an Earth model, as well as for the gravimetric factor and the various Love number parameters which represent various aspects of the deformational effects produced by the tidal potential. His numerical values for the coefficients of the 106 most important spectral components of the nutation were adopted as the IAU 1980 nutation series. The presence of the oceans on the actual Earth (and in particular, the effects of the ocean tides raised by the tidal potential) necessitates corrections to the results related to the above idealized model. The theory of the effects of the ocean tides (OT) raised by the tesseral potential (see Chapter 9), as developed by Sasao and Wahr (1981), was employed by Wahr and Sasao (1981) to obtain numerical results for the corrections. Studies on the influence of mantle anelasticity on Earth rotation variations (Wahr and Bergen, 1986, Dehant, 1987a and b) followed (see Chapter 8). The contributions to nutation amplitudes from both of these effects are not in phase with the tidal potential component responsible for the nutation and the contributions to certain amplitudes are rather large, of the order of a milliarcsecond. The influence of CMB topography was considered subsequently by Wu and Wahr (1997). Also, Buffett (1992) considered the effect on nutations due to the geomagnetic field passing from the fluid core into the mantle region and beyond, and showed that the electromagnetic torque generated by this field, when the core and the mantle are in relative motion, can affect some nutation amplitudes (both in-phase and out-of-phase) to the extent of a few hundred μas. Electromagnetic couplings, both at the CMB and at the ICB, played a significant role in the work of Mathews *et al.* (2002) and Buffett *et al.* (2002), which led to the nutation series adopted by the IAU as IAU2000A. Effects of non-hydrostatic CMB topography and core dynamics on Earth rotation has also been considered by Wu and Wahr (1997) and later on by Dehant *et al.* (2012a and b, 2013). In addition to small improvements on the rigid Earth nutation theory (such as better scaling related to the dynamical flattening of the Earth), another effect of relativistic origin, generally known as geodetic precession and nutation, has been routinely taken account of in rigid Earth nutation theories of the 1990s, and has its reflection in theories for the non-rigid Earth also.

Finally, there has been a growing realization that currently available Earth models cannot provide sufficiently accurate values for some of the Earth parameters (e.g., e, e_f) that are critical for nutation theory, and that they have little or nothing to say about some other parameters such as the strengths of the electromagnetic couplings at the CMB and the ICB. It has been mentioned earlier in this book that the value of e used in constructing Wahr's nutation series was not obtained from any Earth model but was inferred from the observed precession rate, and that Gwinn *et al.* (1986) argued persuasively that the observational estimate of the amplitude of the retrograde annual nutation implied a higher value for e_f than that which one obtains from Earth model computations. Thus the need to obtain a certain amount of feedback from observational data on precession and nutation for the formulation of a nutation–precession theory that accurately represents the observations has become gradually accepted, and a process of fitting theory and observations has become incorporated into recent theoretical efforts (see e.g. Koot *et al.*, 2008, 2010, Koot and Dumberry, 2011, 2013).

We begin now with a brief presentation of the various effects enumerated above and then go on to an account of the essentials of the work of Mathews *et al.* (2002), which incorporated all the effects studied till then into a theoretical formulation. This work then optimized the values of some of the Earth parameters in the theory by a least squares fit of the nutation amplitudes and precession rate computed from the theory to the values estimated from a long VLBI (very long baseline interferometry) data set.

10.1 Effect of a thermal conductive layer at the top of the core

Recently Pozzo *et al.* (2012) have demonstrated from laboratory experiments that the thermal conductivity of liquid iron under the conditions in the Earth's core is several times higher than previously estimated. As proposed by Buffett (2012), this increases the heat to be carried by conduction in this layer. As a consequence, there is less heat to drive convection in the core and thus a decrease in electrical resistance. In the induction equation for the magnetic field balance there is thus more generation than loss of magnetic field. In reassessment of how the Earth's magnetic field has been generated and maintained over time, there is thus no need to consider a large initial magnetic field.

10.2 CMB/ICB coupling effects on nutation

As we have seen in Chapter 7, the wobbles of the mantle, the FOC and the SIC differ from one another. The principal mechanism for the mutual coupling of the motions of the FOC and the mantle was the inertial coupling, determined by the ellipticity of the fluid core (see Section 7.3.1). Other mechanisms which can cause additional couplings were mentioned in Section 7.3.3: electromagnetic coupling produced by the effects of the motions, associated with the differential wobble, of the conducting matter on both sides of the CMB relative to geomagnetic field lines crossing the CMB; and the coupling caused by the viscous drag of the core fluid flowing past the boundary. In each case the coupling is reflected in a torque exerted by the mantle on the fluid core and an equal and opposite torque exerted by the fluid core on the mantle. As we have noted in Section 7.3.3, the complex combination of the equatorial components of either of these additional torques *on the fluid core* has the form

$$\tilde{\Gamma}^{\text{CMB}} = -i K^{\text{CMB}} A_f \Omega_0^2 \tilde{m}_f,$$ (10.1)

similar to Eq. (7.110), where $\tilde{m}_f$ is the differential wobble between the two regions. This torque would appear in the angular momentum balance equation for the fluid core, as was observed by Sasao *et al.* (1980). The coupling constant K^{CMB} representing the strength of the coupling is complex because of the dissipative nature of the electromagnetic / viscous effects. The value of the constant is determined by the specifics of the interaction (whether electromagnetic or viscous or both) between the two regions. Similar couplings exist between the FOC and the SIC, and lead to equatorial torques of the form

$$\tilde{\Gamma}^{\text{ICB}} = -i K^{\text{ICB}} A_s \Omega_0^2 (\tilde{m}_s - \tilde{m}_f),$$ (10.2)

acting on the SIC; it appears as a term in the equation for the inner core. An opposite torque $-\tilde{\Gamma}^{\text{ICB}}$ acts on the FOC of course (in addition to CMB torque as in Eq. (10.1) above). Expressions for the boundary coupling torques due to electromagnetic interactions and viscous coupling are derived later in Sections 10.3 and 10.4, respectively.

A different kind of mechanism that has been considered in the literature for coupling between the motions of the mantle and fluid core is that due to the "topography" (deviations from a smooth ellipsoidal shape) of the CMB surface (treated in Section 10.5).

10.2.1 Effect of CMB and ICB couplings on the nutations

The torques exerted by different layers of the Earth on one another do not affect the angular momentum balance equation for the whole Earth: only torques of external origin enter into this equation. So the angular momentum equation for the whole Earth is still given by Eq. (7.108). For the FOC, one has Eq. (7.107), with $\Gamma_b = \Gamma^{CMB} - \Gamma^{ICB}$; and for the SIC, one has Eq. (7.150) with the ICB torque Γ^{ICB} added to Γ_s. It was shown in Section 7.4 that these vector equations lead to the matrix equation (7.156)–(7.158) for the wobble quantities $\tilde{m}, \tilde{m}_f, \tilde{m}_s$ and $\tilde{n}_s$ of a three-layer Earth in the frequency domain, if the boundary coupling torques are ignored. It may be readily verified that the effect of introducing the boundary torques consists in simple modifications of a few of the elements of the matrix M of Eq. (7.157):

$$M_{22} \rightarrow M_{22} + K^{CMB} + (A_s/A_f)K^{ICB}, \qquad M_{23} \rightarrow M_{23} - (A_s/A_f)K^{ICB}$$

$$M_{32} \rightarrow M_{32} - K^{ICB}, \qquad M_{33} \rightarrow M_{33} + K^{ICB}. \tag{10.3}$$

All other elements of M remain unchanged.

How will these modifications affect the expressions (7.159)–(7.162) for the frequencies of the normal modes? The simplest way of inferring the modified frequencies is by observing how the parameter combinations appearing in the expressions just referred to are related to the matrix elements of M. One notes that the combination $(e_f - e')$ to which the frequency $(\sigma_2 + 1)$ of the FCN mode in the celestial frame (see Eq. (7.160)) is proportional, is in fact the value of M_{22} at $\sigma = -1$, and also that $(\alpha_2 e_s + \tau)$ in the frequency $(1 + \sigma_3)$ in the celestial frame of the FICN mode is proportional to the value of $(-M_{33} - M_{34})$ at $\sigma = -1$. Applying these observations to the matrix modified as in (10.3), we infer that the eigenvalues σ_2 and σ_3 in the presence of the CMB and ICB couplings are

$$\sigma_2 = -1 + \nu_2 = -1 - \left(1 + \frac{A_f}{A_m}\right)\left(e_f - e' + K^{CMB} + \frac{A_s}{A_f}K^{ICB}\right), \tag{10.4}$$

$$\sigma_3 = -1 + \nu_3 = -1 + \left(1 + \frac{A_s}{A_m}\right)(\alpha_2 e_s + \tau - K^{ICB}). \tag{10.5}$$

The other two eigenvalues (the frequencies σ_1 of the CW and σ_4 of the inner core wobble (ICW)) are unaffected in the first order by the introduction of the boundary couplings.

The various basic Earth parameters, including the boundary coupling strengths K^{CMB} and K^{ICB}, influence the nutation amplitudes through the dependence of the transfer function $T(\sigma)$ on the respective parameters. These coupling strengths influence the transfer function most strongly through their roles in the FCN and FICN resonance terms, and in particular through their contributions to the normal mode frequencies σ_2 and σ_3 given by the expressions above. (The other two eigenfrequencies are essentially independent of the coupling strengths, as just noted, and so are the corresponding resonance coefficients R_1 and R_4.) The expressions for the coefficients R_2 and R_3 in terms of the Earth parameters are rather complicated and not very transparent. But a good idea of the nature of the dependence of R_2 on the Earth parameters may be obtained from the simpler case of the two-layer Earth, by using the expression (7.143) for R_2: $R_2 = (A_f/A_m)(1 - \gamma/e)\nu_2$ where $\nu_2 \equiv 1 + \sigma_2$ with σ_2 given now by Eq. (10.4) above. An approximate expression for R_3 may be worked out for the three-layer Earth, but it is too complicated to give much insight. Suffice it to say that one may expect the amplitudes of nutations with frequencies close to the FCN or the FICN eigenfrequency to be affected most by the introduction of the couplings K^{CMB} and K^{ICB}. The finding of Mathews *et al.* (2002) is that the contributions from the CMB and ICB couplings to the real (in-phase) and imaginary (out-of-phase) parts of the retrograde annual and retrograde 18.6 year nutation amplitudes are all quite significant, of the order of 0.5 mas. An outline of their work is presented later in this chapter.

10.3 Electromagnetic coupling

The electromagnetic coupling between the core and the mantle had been investigated in the past mainly for explaining the LOD variations on decadal timescales. It is one of the major causes of such variations. Investigations in the context of nutation theory were done by Buffett (1992), Buffett *et al.* (2002), Mathews *et al.* (2002), and more recently by Rochester and Crossley (2009) and Koot *et al.* (2008, 2010, 2011, see also Koot and Dumberry, 2011, 2013). The first two of these works, in particular, developed the theory of the relevant physical processes in the differentially wobbling fluid and solid layers of the Earth in the presence of the ambient quasi-static geomagnetic field $\mathbf{B}_0(\mathbf{r})$, and determined the dependence of the coupling constant K^{CMB} on $\mathbf{B}_0$ and on properties of the Earth, the crucial one being the electrical conductivity σ_e of the material within boundary layers on both sides of the CMB. Buffett *et al.* (2002) extended the treatment also to the constant K^{ICB} pertaining to the coupling between the wobbles of the FOC and the SIC. Mathews *et al.* (2002) estimated the values of both these coupling constants by confronting the results from the nutation theory incorporating the coupling terms with the observational estimates of nutation amplitudes; and they went on to make inferences regarding the magnetic fields in both the boundary regions. Koot *et al.* (2008, 2010, 2011, see also Koot and Dumberry, 2011, 2013) did that as well.

The differential wobble motions of adjacent fluid and solid layers at the nearly diurnal tidal frequencies result in oscillatory motions of the conducting matter of the fluid through the magnetic field **B** (which is dominated by the main field $\mathbf{B}_0$). These motions generate electric currents at the same set of frequencies, which in turn cause increments to the magnetic field. At the same time, the current-carrying matter is subject to the Lorentz force exerted by the magnetic field. The torque associated with the Lorentz force tends to oppose the relative wobble motion between the outer core and the mantle, on the one hand, and the SIC on the other. These are what are referred to as the electromagnetic couplings at the CMB and ICB.

We denote the perturbations of the magnetic and velocity fields due to the electromagnetic interaction process outlined above as **b** and **v**, respectively:

$$\mathbf{B} = \mathbf{B}_0 + \mathbf{b}, \qquad \mathbf{V} = \mathbf{V}_0 + \mathbf{v}, \tag{10.6}$$

where $\mathbf{B}_0$ and $\mathbf{V}_0$ are the ambient fields present when the electromagnetic perturbations are ignored. Actually, the flow field in the fluid core in the absence of electromagnetic and other couplings between the mantle and the core is essentially that associated with a "rigid" rotation of the core as a whole with angular velocity $\mathbf{\Omega}_f = \mathbf{\Omega} + \omega_f$ in space, $\mathbf{\Omega}$ being the angular velocity of the mantle and $\omega_f = \Omega_0 \mathbf{m}_f$ that of the differential rotation of the fluid core relative to the mantle. (There is in reality a residual flow in the layers very close to the inner and outer boundaries of the fluid core on account of the ellipsoidal shapes of these boundaries. This part of the flow is small, however, and it can be shown, given the nature of this residual flow, that it has no impact on the perturbations due to the electromagnetic interaction, which are what we are interested in here.) In this situation, the special choice of a reference frame as one that is *attached to the rigidly rotating core* would make $\mathbf{V}_0 = 0$ and $\mathbf{v} = 0$. When the coupling interactions are turned on, **v** becomes non-zero. Such a reference frame was employed by Buffett *et al.* (2002), and we follow this choice. As regards $\mathbf{B}_0$, we take it to be time independent. In doing so, we are ignoring the variability inherent in the geodynamo process which generates this field. The justification for this neglect becomes understandable when it is recalled that the electromagnetic effects on nutations that are sought to be quantified are only those at the specific frequencies that correspond to the frequencies present in the retrograde diurnal spectrum of the tidal potential: there is no reason why the geomagnetic field, the variations of which are over long timescales of decades and much longer, should involve any frequencies common to the tidal spectrum. So the temporal variations of $\mathbf{B}_0$ are essentially irrelevant to the nutation problem and may be ignored. The same argument holds good for the geodynamo-related flows appearing as part of the velocity field.

10.3.1 Basic equations; assumptions and approximations

The discussion of the foregoing sections implies that the mathematical formulation of the electromagnetic coupling must involve two basic equations, one governing the perturbation of the magnetic field by the flow field, and the other governing the modification of the flow field by the magnetic field. In applying these equations to our present problem, the electrical conductivity of the matter on both sides of each of the boundaries of the fluid core plays a prominent role, especially by permitting certain approximations, which make it possible to simplify the equations to forms which can be solved analytically.

The material of the core of the Earth is almost wholly iron, in molten form in the fluid core and as solid in the inner core. It has very high electrical conductivity σ_e, for which the typical estimate is $\sigma_e = 5 \times 10^5$ Sm^{-1} (e.g., Braginsky and Roberts, 1995); the conductivity of the inner core is considered to be about the same. The bulk of the mantle is made up of non-conducting materials, but it is believed that a conducting layer of high conductivity is present at the bottom of the mantle (see e.g. Jeanloz, 1990, 1993, Poirier, 1993, Manga and Jeanloz, 1996, Lay *et al.*, 1998, Garnero *et al.*, 1998). The presence of such a conducting layer (without which no electromagnetic interaction with the core is possible) has been invoked to explain the decadal-scale LOD variations in Earth rotation (see e.g. Roden, 1963, Stix and Roberts, 1984, Stewart *et al.*, 1995, Jault and Le Mouël, 1993, Holme R., 1998a and b, 2000). Though the value of the conductivity and the thickness of this boundary layer have not been established independently with any certainty, the requirements of consistency of the results from the theories of LOD variations and nutations with observational data strongly suggest a conductivity of roughly the same order as that of the fluid core. Experiments also demonstrated that post-perovskite (which is a phase of MgSiO$_3$ only stable at high pressure and temperature conditions corresponding to the lowermost 150 km of the mantle, the D'' layer) exhibits electrical and thermal conductivity much higher than that of perovskite (Ohta *et al.*, 2008, 2010, 2012). Buffett *et al.* (2002) have assumed σ_e of the mantle layer to be the same as that of the fluid core, referred to above. Given this value, it turns out, as will be seen below, that the perturbation $\mathbf{b}(\mathbf{r})$ having spectral frequencies in the diurnal range suffers a rapid exponential decay with distance from the fluid–solid boundary over a length scale of just a couple of hundred meters, which is very small compared to the 3480 km radius of the CMB. The boundary layer in the inner core at the ICB (of radius 1220 km) is also of the same dimension. Within the FOC too, a similar decay with distance from either boundary takes place, but the thickness of the boundary layer, unlike in the solid regions, is influenced by the fluid flow perturbation which, in turn, is affected by the Coriolis force as well as the ambient magnetic field (or more precisely, by its radial component). The result is that the boundary layer thickness varies with the angular position over the boundary surface, as is seen when the outline of the theoretical development is presented below. Nevertheless, the decay takes place, on the whole, on the same small distance scale as in the solid regions. Thus the interactions which give rise to the electromagnetic couplings of the fluid core to the mantle and the SIC take place within these thin boundary layers on both sides of the CMB and the ICB. It is the approximations based on this perception that have been exploited by Buffett *et al.* (2002) to accomplish the substantial simplification of the basic equations referred to above (Eqs. (7.156)–(7.158) together with the above Eq. (10.3)). For instance, the decay in the radial direction within a penetration depth of just a few hundred meters would imply that the radial derivatives of the perturbations $\mathbf{b(r)}$ and $\mathbf{v(r)}$ are very large compared to their tangential derivatives which represent variations on length scales of the order of the radius of the boundary; neglect of the tangential derivatives while developing the theory is therefore a legitimate approximation under the above scenario. Similarly, the assumption that the electrical conductivity σ_e is constant within the thin boundary layers (with different values, in general, for the fluid and solid sides) appears not unreasonable, even though one cannot be certain that there is no lateral heterogeneity within the mantle boundary layer. It will be seen later that the evidence from the role of the electromagnetic coupling in the fitting of the predictions of the theory outlined below to the observational data on nutations appears to justify the assumptions and approximations made.

We turn now to formulation of the two equations relevant to the electromagnetic interaction. The first one is Maxwell's *induction equation* (Faraday's law) which states that

$$\frac{\partial \mathbf{B}}{\partial t} = -\nabla \wedge \mathbf{E}, \tag{10.7}$$

where $\mathbf{E}$ is the electric field. The electrical current density $\mathbf{J}$ generated by $\mathbf{E}$ in a material of conductivity σ_e is $\mathbf{J} = \sigma_e \mathbf{E}$. In case the material is in motion relative to the reference frame in which the electromagnetic fields are described, the current density gets modified to $\mathbf{J} = \sigma_e(\mathbf{E} + \mathbf{v} \wedge \mathbf{B})$, where $\mathbf{v}$ is the velocity of the conducting matter relative to the frame of reference. On the other hand, another of Maxwell's equations (Ampere's law) states that $\mathbf{J}$ is determined by $\mathbf{B}$ through the relation $\mathbf{J} = \nabla \wedge \mathbf{H} = \nabla \wedge (\mathbf{B}/\mu)$, where μ is the magnetic permeability. Combining the above two expressions for $\mathbf{J}$ and making a simple rearrangement of terms, we obtain the following expression for **E** in a conducting fluid:

$$\mathbf{E} = \frac{1}{\sigma_e}(\nabla \wedge \mathbf{H}) - \mathbf{v} \wedge \mathbf{B} = \eta(\nabla \wedge \mathbf{B}) - \mathbf{v} \wedge \mathbf{B}, \tag{10.8}$$

where η, the *magnetic diffusivity*, is defined by

$$\eta = (\sigma_e \mu_0)^{-1}. \tag{10.9}$$

Here we have set μ equal to the value $\mu_0 = 4\pi \times 10^{-7}$ for a non-magnetic medium, which is indeed what we have. The diffusivity is treated as a constant since σ_e has been assumed to be so. When σ_e is taken to have the widely accepted value of about 5×10^5 Sm^{-1} estimated for the core fluid, one finds that $\eta \approx 1.6$.

Now we substitute for **E** in Eq. (10.7) from the above relation and note, while evaluating its curl, that $\nabla \wedge (\nabla \wedge \mathbf{B}) = -\nabla^2 \mathbf{B}$ obtained using the vector identity $\nabla \wedge (\nabla \wedge \mathbf{A}) = \nabla(\nabla \cdot \mathbf{A}) - \nabla^2 \mathbf{A}$ and since $\nabla \cdot \mathbf{B} = 0$. We then consider the first order in **b** and **v** and obtain the induction equation appropriate to the conditions of our problem:

$$\frac{\partial \mathbf{b}}{\partial t} = \nabla \wedge (\mathbf{v} \wedge \mathbf{B}_0) + \eta \nabla^2 \mathbf{b} = B_r \frac{\partial \mathbf{v}}{\partial r} + \eta \nabla^2 \mathbf{b}, \tag{10.10}$$

wherein B_r is an abbreviation for the radial component $(\mathbf{B}_0)_r$ of $\mathbf{B}_0$.

In carrying out the reduction to the last expression in the induction equation above, we have gone through a number of steps, including making a couple of approximations. The first step is to expand $\nabla \wedge (\mathbf{v} \wedge \mathbf{B}_0)$ as $(\mathbf{B}_0 \cdot \nabla)\mathbf{v} - (\mathbf{v} \cdot \nabla)\mathbf{B}_0 + \mathbf{v}(\nabla \cdot \mathbf{B}_0) - \mathbf{B}_0(\nabla \cdot \mathbf{v})$ (using the vector identity $\nabla \wedge (\mathbf{A} \wedge \mathbf{B}) = (\mathbf{B} \cdot \nabla)\mathbf{A} - (\mathbf{A} \cdot \nabla)\mathbf{B} + \mathbf{A}(\nabla \cdot \mathbf{B}) - \mathbf{B}(\nabla \cdot \mathbf{A}))$. The second is to set

$$\nabla \cdot \mathbf{B}_0 = 0 \quad \text{and} \quad \nabla \cdot \mathbf{v} = 0. \tag{10.11}$$

The first of these equations is a statement of the solenoidal nature of magnetic fields, and the second equation means that the core fluid is incompressible (a good enough approximation in the present problem). The third step is to neglect $(\mathbf{v} \cdot \nabla)\mathbf{B}_0$ as being very much smaller than $(\mathbf{B}_0 \cdot \nabla)\mathbf{v}$; and the final step is to replace the last quantity by $(B_r \partial \mathbf{v}/\partial r)$. This step simply recognizes that the radial derivative of **v** is very large compared to the tangential derivatives because of the rapid decay of **v** within a span of only a few hundred meters in the radial direction while there is no such decay in the transverse directions. As for the variation of $\mathbf{B}_0$, it is visualized as taking place in all directions over large distance scales appropriate to the geodynamo mechanism; the resultant small magnitudes of the spatial derivatives of $\mathbf{B}_0$ justify the neglect of the term $(\mathbf{v} \cdot \nabla)\mathbf{B}_0$ in the penultimate step above. The form (10.10) of the induction equation is therefore appropriate to our problem. It is the first of our basic equations, and is valid in the fluid core as well as in the mantle and the SIC. However, the value of η is determined by the nature of the material in the particular region.

The second basic equation, which we now turn to, is the one governing the flow in the core fluid relative to our chosen reference frame, which has the time dependent angular velocity $\Omega_f = \Omega + \omega_f$, Ω being the angular velocity of the mantle in space. The equation is

$$\rho\left(\frac{\partial \mathbf{v}}{dt} + 2\Omega_f \wedge \mathbf{v} + \dot{\Omega}_f \wedge \mathbf{r} + \Omega_f \wedge (\Omega_f \wedge \mathbf{r})\right)$$
$$= -\nabla P - \rho\nabla\phi + \mathbf{J} \wedge \mathbf{B}, \tag{10.12}$$

wherein the last term represents the Lorentz force due to $\mathbf{B}$ acting on the matter carrying the current density $\mathbf{J}$. Now, $\mathbf{J}$ itself is related to $\mathbf{B}$ through $\mathbf{J} = \nabla \wedge (\mu^{-1}\mathbf{B})$, as already noted, with $\mu = \mu_0$ in the present case. Hence the Lorentz force may be expressed purely in terms of $\mathbf{B}$ as

$$\mathbf{J} \wedge \mathbf{B} = \frac{1}{\mu_0}(\nabla \wedge \mathbf{B}) \wedge \mathbf{B} = \frac{1}{\mu_0}\left((\mathbf{B} \cdot \nabla)\mathbf{B} - \frac{1}{2}\nabla(\mathbf{B} \cdot \mathbf{B})\right). \tag{10.13}$$

Suppose now that the Lorentz force were "switched off." Then $\mathbf{v}$ would become zero as already mentioned above. That would affect the pressure in Eq. (10.12), which would change, say, to $P - P'$. Also, $\mathbf{B}$ would get replaced by $\mathbf{B}_0$. However, ρ and ϕ would remain unaltered if the fluid is incompressible ($\nabla \cdot \mathbf{v} = 0$), as we assume; and the other terms involving neither $\mathbf{v}$ nor $\mathbf{b}$ would of course remain unchanged. We make the above-mentioned changes in (10.12) after substituting for $\mathbf{J} \wedge \mathbf{B}$ from Eq. (10.13), and then subtract the resulting reduced equation from the original equation. Thus we obtain the following perturbation equation for $\mathbf{v}$:

$$\rho\left(\frac{\partial \mathbf{v}}{dt} + 2\Omega_0 \wedge \mathbf{v}\right)$$
$$= -\nabla P' + \frac{1}{\mu_0}[(\mathbf{B}_0 \cdot \nabla)\mathbf{b} + (\mathbf{b} \cdot \nabla)\mathbf{B}_0] - \frac{1}{2\mu_0}\nabla(\mathbf{B} \cdot \mathbf{B} - \mathbf{B}_0 \cdot \mathbf{B}_0), \tag{10.14}$$

wherein we have made the approximation of replacing Ω_f by $\Omega_0 = \Omega_0\hat{\mathbf{z}}$. Now, the term $(\mathbf{B}_0 \cdot \nabla)\mathbf{b}$ in the above equation is much larger in magnitude than $(\mathbf{b} \cdot \nabla)\mathbf{B}_0$. The reason is that the radial derivative of $\mathbf{b}$ in the former is very large on account of the rapid decay of $\mathbf{b}$ in the radial direction, which we have referred to earlier, while the variation of $\mathbf{B}_0$ is slow, being over much larger distance scales. The term $(\mathbf{b} \cdot \nabla)\mathbf{B}_0$ may therefore be neglected. Finally, we take the curl of Eq. (10.14) in order to eliminate the term involving the pressure perturbation P'; this operation removes also the last term on the right-hand side. The equation which results is

$$\rho\left(\frac{\partial}{\partial t}(\nabla \wedge \mathbf{v}) - 2(\mathbf{\Omega}_0 \cdot \nabla)\mathbf{v}\right) = \frac{1}{\mu_0}\nabla \wedge (\mathbf{B}_0 \cdot \nabla)\mathbf{b}. \tag{10.15}$$

The mutually coupled equations (10.10) and (10.15) are to be solved simultaneously for the induced magnetic field $\mathbf{b}$ and the velocity field $\mathbf{v}$ within the fluid core.

Consider now the velocity fields in the mantle and the SIC. The velocity in each of these solid regions relative to our core-fixed reference frame is that associated with the differential wobble motion between that region and the fluid core; this is the motion which drives the electromagnetic interaction between the two sides of the fluid–solid boundary. In the case of the mantle, the differential wobble is characterized by the vector $-\mathbf{m}_f$ where $\mathbf{m}_f$ has the components $(m_{fx}, m_{fy}, 0)$; and the differential wobble of the SIC is $\mathbf{m}_s - \mathbf{m}_f$ where $\mathbf{m}_s$ has the components $(m_{sx}, m_{sy}, 0)$. To minimize duplication, we introduce the symbol $\Delta\mathbf{m}$ as a common notation for the two. The velocity fields associated with the two differential wobbles will also be represented by a common symbol $\mathbf{v}_S$ where the subscript S indicates that the velocity is that within a solid region. Of course, when a specific one of the two solid regions is being considered, the value pertaining to it is to be chosen. The angular velocity associated with $\Delta\mathbf{m}$ is $\Omega_0\Delta\mathbf{m}$, and the associated velocity at $\mathbf{r}$ in either solid region is

$$\mathbf{v}_S(\mathbf{r}, t) = \Omega_0\Delta\mathbf{m}(t) \wedge \mathbf{r}. \tag{10.16}$$

(Here we are ignoring the small effects of deformability, which are negligible in the present context.) On introducing this value for $\mathbf{v}$ in the induction Eq. (10.10), it reduces to

$$\frac{\partial \mathbf{b}_S}{\partial t} = \Omega_0 B_r \Delta\mathbf{m} \wedge \hat{\mathbf{r}} + \eta_S \nabla^2 \mathbf{b}_S \tag{10.17}$$

in the mantle/solid inner core.

The solution of this equation gives the general form of $\mathbf{b}_S$ in the solid regions, while $\mathbf{b}$ and $\mathbf{v}$ within the fluid core are to be determined by simultaneous solution of Eqs. (10.10) and (10.15). The solutions become unique when the appropriate boundary conditions are applied.

The quantities that we need to calculate, once the solutions are obtained, are the torques acting on the FOC and the SIC as a result of the electromagnetic interactions taking place within the boundary layers at the CMB and the ICB. Now, the electromagnetic torque acting on a volume of matter may be expressed, following Rochester (1960), as the surface integral

$$\mathbf{\Gamma} = \int_\Sigma \frac{1}{\mu_0}(\mathbf{r} \wedge \mathbf{b})(\mathbf{B}_0 \cdot \mathbf{n})\, d\Sigma, \tag{10.18}$$

where Σ denotes the boundary of volume of interest, and **n** is the *outward* unit normal to Σ. In the case of the FOC, which is bounded on the outer side by the CMB and on the inner side by the ICB, the sum of the integrals over these two surfaces has to be taken. It is sufficient, for the purposes of the present problem, to consider the two boundary surfaces to be spherical. Then $\mathbf{n} = \hat{\mathbf{r}}$ at the CMB and $\mathbf{n} = -\hat{\mathbf{r}}$ at the ICB. Therefore the sum of the integrals over the two surfaces becomes

$$\boldsymbol{\Gamma} = \int_{\text{CMB}} \frac{1}{\mu_0} (\mathbf{r} \wedge \mathbf{b}) B_r \, d\Sigma - \int_{\text{ICB}} \frac{1}{\mu_0} (\mathbf{r} \wedge \mathbf{b}) B_r \, d\Sigma \equiv \boldsymbol{\Gamma}^{\text{CMB}} - \boldsymbol{\Gamma}^{\text{ICB}}, \quad (10.19)$$

where B_r is the radial component of $\mathbf{B}_0$, and $\boldsymbol{\Gamma}^{\text{CMB}}$ and $\boldsymbol{\Gamma}^{\text{ICB}}$ are computed on the FOC and ICB, respectively. The nutation problem involves only the equatorial components of $\boldsymbol{\Gamma}$, of course.

10.3.2 Simplified forms of the induction and flow equations

We observe now that for the calculation of the electromagnetic torque (10.18) on the fluid core, it is sufficient to know the components of **b** that are transverse to the radial direction. It is most appropriate, therefore, to resolve the vector **b** into radial and transverse spherical components b_r, b_θ, b_φ (and similarly for **v**), and to express Eqs. (10.10) and (10.15) as equations for these components. It turns out, happily, that the equations for the transverse components do not involve the radial one. They are:

$$\frac{\partial b_\theta}{\partial t} = B_r \frac{\partial v_\theta}{\partial r} + \eta \frac{\partial^2 b_\theta}{\partial r^2}, \quad (10.20)$$

$$\frac{\partial b_\varphi}{\partial t} = B_r \frac{\partial v_\varphi}{\partial r} + \eta \frac{\partial^2 b_\varphi}{\partial r^2}, \quad (10.21)$$

from the induction Eq. (10.10), and

$$-\frac{\partial}{\partial t} \left(\frac{\partial v_\varphi}{\partial r} \right) - f \frac{\partial v_\theta}{\partial r} = -\left(\frac{B_r}{\rho \mu_0} \right) \frac{\partial^2 b_\varphi}{\partial r^2}, \quad (10.22)$$

$$\frac{\partial}{\partial t} \left(\frac{\partial v_\theta}{\partial r} \right) - f \frac{\partial v_\varphi}{\partial r} = \left(\frac{B_r}{\rho \mu_0} \right) \frac{\partial^2 b_\theta}{\partial r^2}, \quad (10.23)$$

from the fluid flow equation (10.15). The factor f appearing in these equations is defined by

$$f = 2\Omega_0 \cos \theta. \quad (10.24)$$

We introduce now the following complex combinations of the transverse components:

$$b_\pm = b_\theta \pm i b_\varphi, \qquad v_\pm = v_\theta \pm i v_\varphi. \tag{10.25}$$

The pairs of equations (10.20) and (10.21), and (10.22) and (10.23) may now be recast as

$$\frac{\partial b_\pm(\mathbf{r}, t)}{\partial t} = B_r \frac{\partial v_\pm(\mathbf{r}, t)}{\partial r} + \eta \frac{\partial^2 b_\pm(\mathbf{r}, t)}{\partial r^2}, \tag{10.26}$$

$$\frac{\partial}{\partial t} \left(\frac{\partial v_\pm(\mathbf{r}, t)}{\partial r} \right) \pm i f \frac{\partial v_\pm(\mathbf{r}, t)}{\partial r} = \left(\frac{B_r}{\rho \mu_0} \right) \frac{\partial^2 b_\pm(\mathbf{r}, t)}{\partial r^2}. \tag{10.27}$$

The induction Eq. (10.26) applies to both solid and fluid regions, while the flow equation is relevant only to the fluid core. We recall now that the velocity $\mathbf{v}_S$ in the solid regions is already known, being given by Eq. (10.16). Its spherical components $(v_{Sr}, v_{S\theta}, v_{S\varphi})$ are $\Omega_0 r (0, \Delta m_\varphi, -\Delta m_\theta)$. Hence

$$v_{S\pm}(\mathbf{r}, t) = \mp i \Omega_0 r \Delta m_\pm(t), \tag{10.28}$$

where $\Delta m_\pm = \Delta m_\theta \pm i \Delta m_\varphi$. Therefore one has only the induction equation to solve in the solid regions, while the two coupled Eqs. (10.26) and (10.27) have to be solved within the fluid core. The magnetic field and velocity variables pertaining to the fluid region are identified hereafter by a subscript f referring to the word "fluid," while the corresponding label for quantities in the solid region will be S (as already employed in $\mathbf{v}_S$). The perturbations $b_{f\pm}(\mathbf{r}, t)$ and $v_{f\pm}(\mathbf{r}, t)$ in the fluid region and $b_{S\pm}(\mathbf{r}, t)$ in the solid region are excited by, and are to be determined in terms of, the wobble-related $v_{S\pm}(\mathbf{r}, t)$.

Equation (10.26) would reduce to the diffusion equation if the velocity term were absent. This reduced equation may first be solved to gain a general idea of the variation of b_+ with r. Considering a spectral component with the time dependence $e^{i\omega t}$, one can see readily that the r dependence of the approximate b_+ which obeys the diffusion equation is of the form $e^{\lambda r}$ where $\lambda^2 = (i\omega/\eta)$ so that $\lambda = \pm (1 + i s_\omega)(|\omega|/2\eta)^{1/2}$, where $s_\omega = \omega/|\omega|$. The imaginary part of λ reflects a wave-like character of propagation of the disturbance $\mathbf{b}$ in the radial direction while the real part represents the rate of exponential growth or decay of $\mathbf{b}$ with increasing r. In our case, where b_+ is generated by the relative wobble motion between the fluid and solid regions across their common boundary, it should be decaying as one moves away from the boundary, and the sign of λ has to be chosen to ensure this behavior. Since $|\omega| \approx \Omega_0 = 7.29 \times 10^{-5}$ s^{-1}, an estimate of the decay constant (which is the magnitude of $\Re(\lambda)$) may be obtained by taking $\eta = 1.6$ (see below Eq. (10.9)). One finds that the decay constant λ is $\approx 5 \times 10^{-3}$ m^{-1}, indicating a penetration depth of about 200 m from the boundary for b_+. Though this estimate is obtained using the diffusion equation it does provide a useful indication that the electromagnetic perturbations do not extend beyond a boundary layer which is of very small thickness compared to the radius of either boundary of the fluid core. The results of the full treatment given in the following section are consistent with this picture.

Before going on to the solution of Eqs. (10.26) and (10.27), we need to examine the nature of the dependence of $v_{S\pm}(\mathbf{r}, t)$ on the radial and other variables and to consider its implications for the structure of $b_{f\pm}(\mathbf{r}, t)$ and $v_{f\pm}(\mathbf{r}, t)$.

We begin by considering the differential wobbles of the mantle and of the solid inner core relative to the FOC. It is convenient to refer to these as the differential wobbles of the two regions at the respective solid–fluid boundaries, namely the CMB and the ICB. They are:

$$\Delta\tilde{m}(t) = -\tilde{m}_f(t) \text{ at the CMB}, \quad \Delta\tilde{m}(t) = \tilde{m}_s(t) - \tilde{m}_f(t) \text{ at the ICB}. \quad (10.29)$$

For a spectral component of frequency ω,

$$\Delta\tilde{m}(t) = \Delta\tilde{m}_0\, e^{i\omega t}, \quad (10.30)$$

where $\Delta\tilde{m}_0$ is the amplitude of the differential wobble:

$$\Delta\tilde{m}_0 = -\tilde{m}_{f0} \quad \text{at the CMB}, \quad \Delta\tilde{m}_0 = (\tilde{m}_s - \tilde{m}_f)_0 \quad \text{at the ICB}. \quad (10.31)$$

We are interested here in the electromagnetic coupling effects on low frequency nutations; the wobbles associated with them are retrograde, with nearly diurnal frequencies. It follows that ω in the above expressions has to be negative, with $|\omega|$ close to Ω_0. In earlier chapters, the symbol ω was reserved for the frequency of the tidal potential component, which is positive. The departure made here from this convention is for conformity with the usage in the work of Buffett *et al.* (2002) so that comparison with the detailed treatment there may be facilitated.

The real and imaginary parts of the $\Delta\tilde{m}(t)$ are Δm_x and Δm_y, respectively. We write them, along with Δm_z, as

$$\Delta m_x = \Delta\tilde{m}_0\cos\omega t, \quad \Delta m_y = \Delta\tilde{m}_0\sin\omega t, \quad \Delta m_z = 0. \quad (10.32)$$

(Strictly speaking, $\tilde{m}_{f0}$ and $\tilde{m}_{s0}$ are complex, in general. One should have had $\Delta m_x = -|\tilde{m}_{f0}|(\cos\omega t + \alpha)$ at the CMB, for instance, where α is the phase of the complex amplitude $\tilde{m}_{f0}$. But this phase plays no role in the following derivation and may be ignored. It may be verified that the final results are unaffected by our simplification.)

Transformation of the components of $\Delta\mathbf{m}$ to spherical coordinates (such as is done for $\mathbf{r}$ as in Eq. (9.11) of Chapter 9), followed by the use of (10.32) yields

$$\Delta m_r = \Delta m_x \sin\theta \cos\varphi + \Delta m_y \sin\theta \sin\varphi = \Delta\tilde{m}_0 \sin\theta \cos(\omega t - \varphi),$$

$$\Delta m_\theta = \Delta m_x \cos\theta \cos\varphi + \Delta m_y \cos\theta \sin\varphi = \Delta\tilde{m}_0 \cos\theta \cos(\omega t - \varphi), \quad (10.33)$$

$$\Delta m_\varphi = -\Delta m_x \sin\varphi + \Delta m_y \cos\varphi = \Delta\tilde{m}_0 \sin(\omega t - \varphi).$$

The time dependence of the complex combinations $\Delta m_\pm$ follows directly from these expressions. Then the use of Eq. (10.28), which relates the quantities $v_{S\pm}(\mathbf{r}, t)$ pertaining to the velocity in the solid regions to $\Delta m_\pm$, leads to the following structure for the $v_{S\pm}$:

$$v_{S+}(\mathbf{r}; t) = v_{S+}^+(\mathbf{r})(1 + \cos\theta)e^{i(\omega t - \varphi)} + v_{S+}^-(\mathbf{r})(1 - \cos\theta)e^{-i(\omega t - \varphi)}, \quad (10.34)$$

$$v_{S-}(\mathbf{r}; t) = v_{S-}^+(\mathbf{r})(1 - \cos\theta)e^{i(\omega t - \varphi)} + v_{S-}^-(\mathbf{r})(1 + \cos\theta)e^{-i(\omega t - \varphi)}, \quad (10.35)$$

where the superscripts $+$ and $-$ in $v_{S+}^\pm(\mathbf{r})$ and $v_{S-}^\pm(\mathbf{r})$ indicate the sign preceding ω in the exponent, and

$$v_{S\pm}^+(\mathbf{r}) = -ir\Omega_0\Delta\tilde{m}_0/2, \qquad v_{S\pm}^-(\mathbf{r}) = -v_{S\pm}^+(\mathbf{r}). \quad (10.36)$$

Now, the perturbations $\mathbf{b}$ and $\mathbf{v}$ are both forced by the differential wobble motion represented by $\mathbf{v}_S$, and are linearly dependent on it; therefore both b_+ and v_+ should have the same general structure as v_{S+} in (10.34) and similarly for b_- and v_- and v_{S-} in (10.35). Thus we can write

$$b_+(\mathbf{r}; t) = b_+^+(\mathbf{r})(1 + \cos\theta)e^{i(\omega t - \varphi)} + b_+^-(\mathbf{r})(1 - \cos\theta)e^{-i(\omega t - \varphi)}, \quad (10.37)$$

$$b_-(\mathbf{r}; t) = b_-^+(\mathbf{r})(1 - \cos\theta)e^{i(\omega t - \varphi)} + b_-^-(\mathbf{r})(1 + \cos\theta)e^{-i(\omega t - \varphi)}, \quad (10.38)$$

with similar expressions for $v_\pm(\mathbf{r}; t)$.

Since $b_-(\mathbf{r}, t)$ and $v_-(\mathbf{r}, t)$ are the complex conjugates of $b_+(\mathbf{r}, t)$ and $v_+(\mathbf{r}, t)$, respectively, an examination of the above equations shows that

$$b_-^+(\mathbf{r}) = (b_+^-(\mathbf{r}))^*, \;\; b_-^-(\mathbf{r}) = (b_+^+(\mathbf{r}))^*, \quad v_-^+(\mathbf{r}) = (v_+^-(\mathbf{r}))^*, \;\; v_-^-(\mathbf{r}) = (v_+^+(\mathbf{r}))^*.$$

$$(10.39)$$

Therefore in seeking the solutions of the radial equations (10.26) and (10.27), we can confine our attention to the quantities $b_\pm^+(\mathbf{r})$ and $v_\pm^+(\mathbf{r})$ associated with the time dependent factor $e^{i\omega t}$. (ω is negative as mentioned above; we will need to use this fact during the derivation below.) In the following, we shall suppress the superscript $+$ to reduce the congestion of indices; thus we use the abbreviated notation

$$b_\pm(\mathbf{r}) \quad \text{for} \quad b_\pm^+(\mathbf{r}), \qquad \text{and} \qquad v_\pm(\mathbf{r}) \quad \text{for} \quad v_\pm^+(\mathbf{r}). \quad (10.40)$$

Only when any quantity with a superscript $-$ is to be referred to will the superscript be explicitly shown.

10.3.3 Solution of the equations

What we need to accomplish by solving Eqs. (10.26) and (10.27) is to obtain expressions for $b_{f\pm}(\mathbf{r})$, $v_{f\pm}(\mathbf{r})$ and $b_{s\pm}(\mathbf{r})$ within the boundary layers on the two sides of each of the boundaries of the fluid core (CMB and ICB) in terms of the velocity $v_{S\pm}$ associated with the differential wobble between the fluid core and the mantle/inner core, which gives rise to the perturbations.

The derivation of the expressions proceeds in three stages: finding the solutions for $b_{f\pm}(\mathbf{r})$ and $v_{S\pm}(\mathbf{r})$ in the fluid layer; determining $b_{s\pm}(\mathbf{r})$ in the boundary layer on the solid side in terms of the known $v_{s\pm}(\mathbf{r})$; and using a boundary condition to relate the functions in the fluid layer to $b_{s\pm}(\mathbf{r})$ and hence to $v_{S\pm}(\mathbf{r})$.

Fluid layer

Given the structures shown in Eqs. (10.37) and (10.38) for $b_{\pm}(\mathbf{r})$ and the similar structures for $v_{\pm}(\mathbf{r})$, one can see readily that Eqs. (10.26) and (10.27) lead to the following reduced forms for the induction and fluid flow equations:

$$(i\omega - \eta_f \, \partial_r^2) \, b_{f\pm}(\mathbf{r}) = B_r \, \partial_r v_{f\pm}(\mathbf{r}), \tag{10.41}$$

$$i(\omega \pm f) \, \partial_r v_{f\pm}(\mathbf{r}) = \left(\frac{B_r}{\rho\mu_0}\right) \partial_r^2 b_{f\pm}(\mathbf{r}), \tag{10.42}$$

wherein the abbreviation ∂_r has been introduced for the differential operator $\partial/\partial r$. On eliminating $\partial_r v_{f\pm}$ between the two equations, we obtain

$$\left(\frac{B_r^2}{\mu_0\rho} + i(\omega \pm f)\eta_f\right) \partial_r^2 b_{f\pm}(\mathbf{r}) + \omega(\omega \pm f) b_{f\pm}(\mathbf{r}) = 0. \tag{10.43}$$

This equation has the solution

$$b_{f\pm}(\mathbf{r}) = b_{f\pm}(\mathbf{r}_b) \, e^{\lambda_{f\pm}(r - r_b)}, \tag{10.44}$$

where

$$\lambda_{f\pm}^2 = \frac{-\omega(\omega \pm f)}{B_r^2/(\mu_0\rho) + i(\omega \pm f)\eta_f} = \frac{i\omega/\eta_f}{1 - iB_r^2/[\mu_0\rho(\omega \pm f)\eta_f]}. \tag{10.45}$$

The subscript b is used here to indicate a boundary (CMB or ICB, as the case may be) between the fluid and solid regions; thus $\mathbf{r}_b$ is a point on such a boundary. We rewrite the above expression in the compact form

$$\lambda_{f\pm}^2 = \frac{i\omega/\eta_f}{1 - i\epsilon_\pm x_\pm}, \quad \text{with} \quad x_\pm = \frac{B_r^2}{\mu\rho\eta_f|\omega \pm f|}, \quad \epsilon_\pm = \frac{\omega \pm f}{|\omega \pm f|}. \tag{10.46}$$

It may be verified then that with $\omega = -|\omega|$ and with $y_\pm$ defined as

$$y_\pm = (1 + x_\pm^2)^{1/2}, \tag{10.47}$$

$\lambda_{f\pm}$ is given by

$$\lambda_{f\pm} = \left(\frac{2|\omega|}{\eta_f}\right)^{1/2} \frac{\epsilon_b}{(y_\pm + \epsilon_\pm x_\pm)^{1/2} + i(y_\pm - \epsilon_\pm x_\pm)^{1/2}}, \tag{10.48}$$

where ϵ_b, which can take one of the values $+1$ or -1, is to be defined so as to ensure that the variation of $e^{\lambda_{f\pm}(r-r_b)}$ is an exponential decay as one moves into the interior of the fluid region adjacent to the boundary surface (CMB or ICB) that is being considered. Since $(r - r_b)$ is negative within the fluid boundary layer adjacent to the CMB and positive in the layer next to the inner core, the real part of $\lambda_{f\pm}$ has to be positive in the former and negative in the latter; hence ϵ_b has to be taken as

$$\epsilon_b = +1 \quad \text{for the CMB}, \qquad \epsilon_b = -1 \quad \text{for the ICB}. \tag{10.49}$$

Before proceeding further with the derivation we call attention to the important fact that $\lambda_{f\pm}$ is not a constant: it depends on the angle θ through $f = 2\Omega_0 \cos\theta$, and in general also on φ through B_r, which varies with angular position on the boundary surface.

When B_r is small enough such that $x_\pm \ll 1$, we have the "weak field" limit in which $\lambda_{f\pm}^2$ goes over to a constant value $(i\omega/\eta_f)$, similar to the expression (10.55) found below for λ_S^2 of the solid region. The deviation from the weak field limit, characterized by the parameter $x_\pm$, is the result of the interaction between the induced currents and the fluid flow within the boundary layer in the fluid.

The dependence of the velocity $v_{f\pm}(\mathbf{r})$ on r must evidently be of the same nature as that of $b_{f\pm}(\mathbf{r})$ in view of the interrelation of the two quantities through Eqs. (10.26) and (10.27). The latter equation leads therefore to the relation

$$i(\omega \pm f)v_{f\pm}(\mathbf{r}) = \frac{B_r}{\rho\mu_0}\lambda_{f\pm}b_{f\pm}(\mathbf{r}). \tag{10.50}$$

Now, the first of the expressions (10.46) for $\lambda_{f\pm}^2$ may be rewritten as

$$(B_r^2/\rho\mu_0)\lambda_{f\pm}^2 = -(\omega \pm f)(i\eta_f \lambda_{f\pm}^2 + \omega). \tag{10.51}$$

This relation enables us to bring the foregoing equation to the following useful form:

$$B_r v_{f\pm}(\mathbf{r}) = (i\omega/\lambda_{f\pm} - \eta_f\lambda_{f\pm})b_{f\pm}(\mathbf{r}). \tag{10.52}$$

Solid layer: Within the boundary layer in the solid region adjacent to the CMB in the mantle or to the ICB in the inner core, we need to solve just the induction Eq. (10.26). It is known already that $v_{S\pm}$ is given by the expression (10.36); it is proportional to r and therefore $\partial v_{S\pm}/\partial r$ is a constant, equal to $v_{S\pm}(r_b)/r_b$. Thus the induction equation in the solid region for a spectral component of $b_{S\pm}(\mathbf{r}, t)$ with the time dependence $e^{i\omega t}$ becomes

$$(\eta_S \partial_r^2 - i\omega) b_{S\pm}(\mathbf{r}) = -B_r v_{S\pm}(\mathbf{r}_b)/r_b. \tag{10.53}$$

This is an inhomogeneous linear differential equation, with the solution

$$b_{S\pm}(\mathbf{r}) = P_\pm e^{\lambda_S(r-r_b)} - i\frac{B_r v_{S\pm}(r_b)}{\omega r_b}, \tag{10.54}$$

where $P_\pm$ is independent of r (it being assumed as before that the radial variation of B_r within the boundary layer is negligible), and

$$\lambda_S^2 = \frac{i\omega}{\eta_S} = -\frac{i|\omega|}{\eta_S}, \qquad \lambda_S = \epsilon_b(i-1)\left(\frac{|\omega|}{2\eta_S}\right)^{1/2}. \tag{10.55}$$

The factor ϵ_b is given by Eq. (10.49); it ensures that $b_{S\pm}(\mathbf{r})$ decays exponentially as the point $\mathbf{r}$ moves away from the boundary in the radial direction towards the interior of the solid region.

The constant $P_\pm$ may now be written in terms of the value of b_{f+} at the boundary by taking $r = r_b$ in (10.54) and taking note that continuity of the magnetic induction implies that $b_{S\pm}(\mathbf{r}_b) = b_{f\pm}(\mathbf{r}_b)$:

$$P_\pm = b_{f\pm}(\mathbf{r}_b) + i\frac{B_r v_{S\pm}(r_b)}{\omega r_b}. \tag{10.56}$$

On using this result, Eq. (10.54) yields $b_{S\pm}(\mathbf{r})$ in terms of the boundary value $b_{f\pm}(\mathbf{r}_b)$.

Boundary condition

The next step is to make the connection between the solutions found for the fluid and solid regions. This can be done by using the boundary condition that is provided by the first integral of Eq. (10.26) with respect to r from a point just inside a boundary of the fluid core to the adjacent point on the solid side. Since the range of integration is infinitesimal and the induction $b_\pm$ is continuous across the boundary (i.e., $b_{f\pm}(\mathbf{r}_b) = b_{S\pm}(\mathbf{r}_b)$), the result of the integration is the relation

$$B_r[v_{S\pm}(\mathbf{r}_b) - v_{f\pm}(\mathbf{r}_b)] = -\left(\eta_S\frac{\partial b_{S+}}{\partial r} - \eta_f\frac{\partial b_{f+}}{\partial r}\right)_{r=r_b}. \tag{10.57}$$

The quantity on the right-hand side reduces to $\eta_f\lambda_{f\pm}b_{f\pm}(\mathbf{r}_b) - \eta_S\lambda_{S\pm}P_\pm$. Then, on substituting for $P_\pm$ from (10.56) and using Eq. (10.52) to express $v_{f\pm}(\mathbf{r}_b)$ in terms of $b_{f\pm}(\mathbf{r}_b)$, we obtain

$$\left(1 + \frac{i\eta_S\lambda_S}{\omega r_b}\right) B_r v_{S\pm}(\mathbf{r}_b) = \left(\frac{i\omega}{\lambda_{f\pm}} - \eta_S\lambda_S\right) b_{f\pm}(\mathbf{r}_b). \tag{10.58}$$

Rough estimates considered in the penultimate paragraph of Section 10.3.2 for the parameters η and λ of the core fluid were $1.6\,\mathrm{m^2s^{-1}}$ and $5 \times 10^{-3}\,\mathrm{m^{-1}}$, respectively. If the value of η_S and hence also that of λ_{S+} are of similar orders of magnitude, the second term within parentheses on the left-hand side of the equation will be roughly of the order of 10^{-4}. Therefore this term may be ignored, thereby reducing the factor $(1 + i\eta_S\lambda_S/\omega r_b)$ to unity and reducing the left-hand member to just $B_r v_{S\pm}(\mathbf{r}_b)$. It is this reduced form that appears in Buffett *et al.* (2002).

The factor multiplying $b_{f\pm}(\mathbf{r}_b)$ in Eq. (10.58) may be expressed as a function of the frequency and the various parameters of the problem by substituting for $\lambda_{f\pm}$ and λ_S from equations (10.48) and (10.55). After some algebraic manipulations we obtain

$$(i\omega/\lambda_{f\pm} - \eta_S\lambda_S) = \epsilon_b \left(\frac{|\omega|\eta_f}{2}\right)^{1/2}(G_\pm - iH_\pm), \qquad (10.59)$$

where

$$G_\pm = (y_\pm - \epsilon_\pm x_\pm)^{1/2} + (\eta_S/\eta_f)^{1/2}, \quad H_\pm = (y_\pm + \epsilon_\pm x_\pm)^{1/2} + (\eta_S/\eta_f)^{1/2}.$$

$$(10.60)$$

Equation (10.58), reduced in the manner described above and with $v_{S\pm}(\mathbf{r}_b) \equiv v_{S\pm}^+(\mathbf{r}_b)$ taken from (10.36), yields finally the following expression for $b_{f\pm}(\mathbf{r}_b)$:

$$b_{f\pm}(\mathbf{r}_b) = \epsilon_b \left(\frac{1}{2|\omega|\eta_f}\right)^{1/2} \frac{B_r r_b \Omega_0 \Delta\tilde{m}_0}{H_\pm + iG_\pm} = \epsilon_b \left(\frac{1}{2|\omega|\bar{\eta}}\right)^{1/2} \frac{B_r r_b \Omega_0 \Delta\tilde{m}_0}{4} sF_\pm,$$

$$(10.61)$$

where

$$\bar{\eta}^{1/2} = \frac{1}{2}(\eta_f^{1/2} + \eta_S^{1/2}), \quad \text{and} \quad F_\pm = \frac{4(\bar{\eta}/\eta_f)^{1/2}}{H_\pm + iG_\pm}. \qquad (10.62)$$

The introduction of the quantities $F_\pm$ is done for ease of comparison with Buffett *et al.* (2002). Note that they are complex, and that

$$\Re(F_\pm) > 0 \qquad \text{and} \qquad \Im(F_\pm) < 0. \qquad (10.63)$$

Equation (10.44) taken together with (10.61) provides the solution for $b_{f\pm}(\mathbf{r}) \equiv b_{f\pm}^+(\mathbf{r})$ that we were seeking. Expressions for $b_{f\pm}^-$ are then obtainable trivially from the relations (10.39): $b_{f\pm}^- = b_{f\mp}^{+*}$. The integral (10.18) for the torque on the fluid core does not involve the velocity field, and therefore we do not need the expressions for $v_{f\pm}^+(\mathbf{r})$ or $v_{f\pm}^-(\mathbf{r})$.

It may be verified that in the weak field limit, $x_\pm \ll 1$,

$$G_\pm \to 1 + (\eta_s/\eta_f)^{1/2}, \quad H_\pm \to 1 + (\eta_s/\eta_f)^{1/2}, \quad F_\pm \to 1 - i. \qquad (10.64)$$

Then $b_{\pm}(\mathbf{r}_b)$ is directly proportional to the radial component B_r of the ambient magnetic field. For stronger fields, an additional complicated dependence on B_r arises through $x_{\pm}$ and $y_{\pm}$ in terms of which $G_{\pm}$ and $H_{\pm}$ are defined.

10.3.4 Torques and coupling constants

We are now in a position to consider evaluation of the integral (10.18) for the electromagnetic coupling torques at the CMB and at the ICB. Its integrand is proportional to the vector $\mathbf{r} \wedge \mathbf{b}(\mathbf{r}, t) \equiv \mathbf{T}(\mathbf{r}, t)$, say. The spherical components of $\mathbf{T}$ are

$$T_r = 0, \qquad T_\theta = rb_\varphi, \qquad T_\varphi = -rb_\theta. \qquad (10.65)$$

Transformation to the Cartesian basis yields $T_x = T_\theta \cos\theta \cos\varphi - T_\varphi \sin\varphi$ and $T_y = T_\theta \cos\theta \sin\varphi + T_\varphi \cos\varphi$. Thus the complex combination $\tilde{T} = T_x + iT_y$ which appears in the usual complex sum of the equatorial components of the torque becomes

$$\tilde{T}(\mathbf{r}, t) = (T_\theta \cos\theta + iT_\varphi)e^{i\varphi} = \frac{1}{2}[(1 + \cos\theta)T_+ - (1 - \cos\theta)T_-]e^{i\varphi}, \qquad (10.66)$$

where

$$T_{\pm}(\mathbf{r}, t) = T_\theta \pm iT_\varphi = \mp irb_{\pm}(\mathbf{r}, t), \qquad (10.67)$$

in view of the values noted above for T_φ and T_θ. Use of the forms (10.37) and (10.38) for $b_{\pm}(\mathbf{r}, t)$ leads to the expression

$$\tilde{T}(\mathbf{r}, t) = \frac{ir}{2}[(1 + \cos\theta)^2 b_+^+(\mathbf{r}) + (1 - \cos\theta)^2 b_-^+(\mathbf{r})]e^{i\omega t}$$
$$+ ir[\sin^2\theta(b_+^-(\mathbf{r}) + b_-^-(\mathbf{r}))]e^{-l(\omega t - 2\varphi)}, \qquad (10.68)$$

for a spectral component of $\tilde{T}(\mathbf{r}, t)$. The combination $\tilde{\Gamma} = \Gamma_1 + i\Gamma_2$ of the equatorial components of the electromagnetic torque (10.18) on the fluid core by the mantle (or inner core) is now given by

$$\tilde{\Gamma}_b(\mathbf{r}; t) = \frac{1}{\mu_0} \int_\Sigma \tilde{T}(\mathbf{r}, t) B_r \, d\Sigma, \qquad (10.69)$$

where $\tilde{T}$ is to be taken from (10.68), Σ is either the CMB or the ICB and accordingly, $\tilde{\Gamma}_b$ is either $\tilde{\Gamma}^{CMB}$ or $\tilde{\Gamma}^{ICB}$. The integrand in (10.69) is proportional to B_r^2 in the weak field limit wherein $b_{\pm}(\mathbf{r})$ is simply proportional to B_r, while it has a more complicated dependence on B_r^2 in the strong field case.

The ambient field $\mathbf{B}_0$ generated by the geodynamo is not expected to have any simple dependence on the position vector; thus B_r on the boundaries will have, in general, a complicated dependence on θ and φ. Evaluation of the integral (10.69) will then have to be done by numerical methods even if the functional form of $B_r(\theta, \varphi)$ were known. Actually there is no way of making direct measurements of $B_r(\theta, \varphi)$ on the CMB or ICB, especially the latter. However, information about the distribution of the energy content of the magnetic field at the CMB among its spherical harmonic components has been inferred by downward continuation to the CMB from measurements of the field made along its orbit by the Earth satellite MagSat. The information is limited to degrees $n \leq 12$ (see Langel and Estes, 1982). The reason is that with increasing degree of the harmonics, the amplitudes of the harmonics decrease ever more rapidly with the height at which measurements are made, so that detection at satellite altitudes becomes no longer possible beyond some n. Suffice it to say that, for the evaluation of the torque integral, one has to do with some model of the structure of the magnetic field over each of the two boundaries. As for the magnetic field at the ICB, it cannot be observed from the Earth's surface or from outside by electromagnetic means because of the intervening highly conducting FOC. Available models of the field in this region come from computations based on geodynamo theory (see, for example, Glatzmaier and Roberts, 1995, 1996, Kuang and Bloxham, 1997, 1999, Olson and Christensen, 2002, Christensen and Olson, 2003, Christensen and Tilgner, 2004, Christensen *et al.*, 2001, 2009, 2010, 2012, Gubbins *et al.*, 2008, Amit and Christensen, 2008, Wicht and Christensen, 2010, Christensen, 2011, Lhuillier *et al.*, 2011, Stelzer and Jackson, 2013), occasionally derived from observation (see, for example, Kuang *et al.*, 2009, Jackson and Livermore, 2009, Chulliat and Olsen, 2010, Fournier *et al.*, 2010, Aubert J., 2013).

We observe that if the ambient magnetic field B_r is taken to be axially symmetric, i.e., to have no dependence on φ, then the part of $\tilde{T}$ that is proportional to $e^{-i(\omega t - 2\varphi)}$ will not contribute to the torque integral. This is the case with the two-parameter model employed for B_r by Buffett *et al.* (2002):

$$B_r = B_p \cos \theta + B_u, \qquad (10.70)$$

where B_p (also denoted $B_{r \text{ dipole}}$) and B_u (also denoted $B_{r \text{ uniform}}$) are constants. The "dipole" term $B_p \cos \theta$ represents the dominant long wavelength (degree one) part of the field and the other term, called the "uniform field," is intended to represent the combined effect of all shorter wavelength parts (of spherical harmonic degrees ≥ 2). Alternative considerations of a more complete field have been studied by Deleplace and Cardin (2006) or Buffett and Christensen (2007, which authors also propose considering the dynamics of the buoyant fluid in the core, see Section 10.7) or Koot and Dumberry (2013). With axial symmetry for the magnetic field, the torque integral becomes

$$\tilde{\Gamma}(\mathbf{r}, t) = \frac{i\pi r_b^3}{\mu_0} e^{i\omega t} \int B_r[(1 + \cos\theta)^2 b_+^+(\mathbf{r}) + (1 - \cos\theta)^2 b_-^+(\mathbf{r})] \sin\theta \, d\theta.$$

(10.71)

Finally, substituting for $b_\pm^+(\mathbf{r}_b) \equiv b_\pm(\mathbf{r}_b)$ from (10.61), we obtain

$$\tilde{\Gamma}_b = \epsilon_b \frac{i\pi r_b^4 \Omega_0}{4\mu_0} \left(\frac{1}{2\bar{\eta}|\omega|}\right)^{1/2} I_b \, \Delta\tilde{m}_0 \, e^{i\omega t},$$

(10.72)

$$I_b = \int B_r^2 [(1 + \cos\theta)^2 F_+(B_r) + (1 - \cos\theta)^2 F_-(B_r)] \sin\theta d\theta.$$ (10.73)

The amplitude of the torque $\tilde{\Gamma}_b$ is proportional to the differential wobble amplitude $\Delta\tilde{m}_0$ at the boundary b, as expected.

The last two equations may be used to determine the torques at the CMB and the ICB. One takes $\Delta\tilde{m}_0$ at the two boundaries from Eqs. (10.31). Evaluation of I_b has to be done for each boundary with the use of an axially symmetric model $B_r(r_b, \theta)$ for the spatial variation of the magnetic field over that boundary. The defining relations (10.1) and (10.2) enable us now to express the CMB and ICB coupling constants as $K^{\text{CMB}} = \tilde{\Gamma}^{\text{CMB}}/(-iA_f\Omega_0^2\Delta\tilde{m})$ and $K^{\text{ICB}} = \tilde{\Gamma}^{\text{ICB}}/(-iA_s\Omega_0^2\Delta\tilde{m})$, where $\Delta\tilde{m} = -\tilde{m}_{f0} e^{i\omega t}$ at the CMB and $\Delta\tilde{m} = (\tilde{m}_{s0} - \tilde{m}_{f0}) e^{i\omega t}$ at the ICB. Thus we obtain, on using the expression (10.72),

$$K^{\text{CMB}} = \frac{\pi a_c^4}{4\mu A_f \Omega_0} \left(\frac{1}{2\bar{\eta}|\omega|}\right)^{1/2} I_{\text{CMB}}, \quad K^{\text{ICB}} = \frac{\pi a_s^4}{4\mu A_s \Omega_0} \left(\frac{1}{2\bar{\eta}|\omega|}\right)^{1/2} I_{\text{ICB}},$$

(10.74)

wherein the radius a_c of the whole core and a_s of the inner core have been substituted for r_b at the CMB and ICB, respectively. Since F_+ and F_-, which are defined in Eq. (10.62), are complex, I_b is complex and so are the CMB and ICB coupling constants.

It should be noted that the real parts of K^{CMB} and K^{ICB} are both positive and that their imaginary parts are both negative, in view of the signs of the $\Re(F_\pm)$ and $\Im(F_\pm)$ as noted in Eqs. (10.63). In the weak field limit where $F_\pm = (1 - i)$, the magnitudes of the real and imaginary parts are equal, and the integral I_b simplifies to give the result obtained by Toomre (1974), Sasao *et al.* (1977), Buffett (1992), Greff-Lefftz and Legros (1999a and b):

$$I_b = 2(1 - i) \int B_r^2 (1 + \cos^2\theta) \sin\theta d\theta.$$

(10.75)

It may be evaluated explicitly if the model (10.70) is used for B_r at the boundary b. When the field becomes strong enough to alter the flow field in the fluid boundary layer significantly, one finds that $|\Im(I_b)| > |\Re(I_b)|$.

The out-of-phase contributions to nutations are large as seen from the above equation (of the same order for the weak field approximation).

The presence of the magnetic field at the CMB is expected to strengthen the coupling between the mantle and the core. Since increased flattening also increases the (inertial) coupling between the mantle and the core, the magnetic field acts in the same way as the flattening of the CMB, i.e. its presence decreases the FCN period in space, and consequently moves its period further away from the one-sidereal-day cycle in the terrestrial frame.

For the FICN, the effect of the magnetic coupling is in the opposite direction than for the FCN; while there is a decrease of the FCN period in an inertial reference frame, there is an increase of the FICN period.

A shorter FCN period in the inertial frame brings it closer to the retrograde annual period, increasing the influence of the FCN resonance on the retrograde annual nutation; similarly, a larger FICN period in space brings it closer to the prograde 18.6 year period and further away from the prograde annual nutation, increasing the influence of the FICN resonance on the prograde 18.6 year nutation and decreasing it on the prograde annual nutation. The importance of the electromagnetic couplings then becomes evident.

The analytical formulation of this section corresponds to the presently adopted theory of nutation, MHB2000 of Mathews *et al.* (2002). It is based on nutation and wobble computation using combinations of basic Earth parameters that govern the Earth's response to gravitational tidal forcing by heavenly bodies. The values of these basic parameters have been estimated from observations as explained in Section 11.2. They are presented in Table 10.1 together with other parameters determined from the Earth's interior model PREM.

10.4 Viscous and electromagnetic coupling

The viscous coupling arises if the outer core fluid is viscous. Viscosity would prevent the FOC from slipping freely past the mantle/SIC at the CMB/ICB. This coupling was believed to be very small (e.g. Wahr *et al.*, 1981), but Brito *et al.* (2004, see also Davies and Whaler, 1997) for instance have shown from laboratory experiments on a viscous rotating fluid that the presence of turbulence in a flow may enhance the effective viscosity. In addition, even such small effects may become important when considering the resonance effects induced by the FCN

Table 10.1 *Values of basic Earth parameters determined from observation or from computation for MHB2000.*

Parameter name	Nomenclature	Values
Dynamical flattening of the Earth	e	0.0032845479
For $H_d = \frac{e}{1+e}$	H_d	0.0032737949
Dynamical flattening of the core	e_f	0.0026456
Dynamical flattening of the inner core	e_s	0.
Mass of the Earth [kg]	M	5.9732 10^{24}
Mass of the core [kg]	M_f	1.9395 10^{24}
Mass of the inner core [kg]	M_s	9.8416 10^{22}
Radius of the Earth [km]	a	6371
Radius of the core [km]	a_f	3484.3
Radius of the inner core [km]	a_s	1229.5
Mean moment of inertia of the Earth [kg m^3]	A	8.0177 10^{37}
Mean moment of inertia of the mantle [kg m^3]	A_f	7.1006 10^{37}
Mean moment of inertia of the core [kg m^3]	A_f	9.1100 10^{36}
Mean moment of inertia of the inner core [kg m^3]	A_s	6.1589 10^{34}
Compliance	κ	0.0010340
Compliance	γ	0.0019662
Coupling constant at the CMB	$\Im K_{CMB}$	−0.0000185
Coupling constant at the CMB	$\Re K_{CMB}$	0.00111
Coupling constant at the ICB	$\Im K_{ICB}$	−0.00078

and the FICN. An easy way to incorporate viscosity is to consider, as for the electromagnetic effect of the nutation, the viscous torque in the set of Liouville equations for angular momentum budget. The torque itself is computed from the viscous force, which enters in the Navier–Stokes equation:

$$\underbrace{\rho \frac{\partial \mathbf{v}}{dt}}_{\text{acceleration}} + \underbrace{\rho (\mathbf{v} \cdot \nabla \mathbf{v})}_{\text{advection}} + \underbrace{2\rho (\mathbf{\Omega} \wedge \mathbf{v})}_{\text{Coriolis}} = \underbrace{-\nabla P'}_{\text{pressure}} + \underbrace{\rho' \mathbf{g}}_{\text{buoyancy}} + \underbrace{\mathbf{J} \wedge \mathbf{B}}_{\text{Lorentz}} + \underbrace{\eta' \nabla^2 \mathbf{v}}_{\text{viscosity}},$$

(10.76)

where ρ is the mean fluid density and ρ' the perturbation in fluid density, P' is the perturbation in the fluid pressure, $\mathbf{v}$ is the fluid velocity at wobble frequency, and η' is the absolute or dynamical viscosity related to the kinematic viscosity ν by $\nu\rho = \eta'$. In the context of geodynamo studies, temperature effects are considered in the above equation. In addition, it is often considered, as in this equation, that the perturbation in fluid density only appears in the buoyancy term (Boussinesq approximation). The continuity of the velocity at the CMB with respect to the mantle velocity must be imposed now and the continuity with the inner core velocity at the other boundary of the core at the ICB as well. Core–mantle interactions that include viscosity have been considered for instance by Rochester (1969, 1975) in the frame of Earth rotation. In the literature one also finds recent works mixing electromagnetic and viscous couplings between the core and the mantle in that frame (see Guo *et al.*, 2004, Mathews and Guo, 2005, Deleplace and Cardin, 2006, Buffett and Christensen, 2007, Buffett, 2012, King and Buffett, 2013) and even works including inner core viscosity (see Greff-Lefftz *et al.*, 2000, 2002, Koot *et al.*, 2010, Koot and Dumberry, 2011). Some of these works are presented in the next sections of this chapter. We concentrate here on the work of Mathews and Guo (2005), considering that the core is viscous together with the existence of electromagnetic coupling. These authors solve the induction equation for the perturbation in the magnetic field together with the motion equation as developed in the previous section and consider new boundary conditions for obtaining a new expression of the torques.

Their new solution for the induced magnetic field is, as before:

$$b_{b\pm}(\mathbf{r}, t) = b_{b\pm}^{(+)}(1 \pm \cos\theta)e^{i(\omega t - \varphi)} + b_{b\pm}^{(-)}(1 \mp \cos\theta)e^{-i(\omega t - \varphi)}. \quad (10.77)$$

But the expressions for the coefficients $b_{b\pm}^{\pm}$ involved are now functions of the viscosity ν too. These new solutions have been substituted in Eq. (10.71), in order to provide new expressions of the electromagnetic torque and the viscous torque:

$$\tilde{\Gamma}_b^{em} = e^{i\omega t}\frac{\pi a_b^4 \Delta\tilde{\omega}}{2\mu} \int_\theta B_r^2 \left[(1 + \cos\theta)^2 (F_{1+} + F_{2+})\right.$$

$$\left. + (1 - \cos\theta)^2 (F_{1-} + F_{2-})\right] \sin\theta d\theta, \quad (10.78)$$

where

$$F_{1\pm} = \frac{-\left(\eta_f\lambda_{2\pm} - \eta_s\lambda_s\right)}{F_\pm}, \quad (10.79)$$

$$F_{2\pm} = \frac{\left(\eta_f\lambda_{1\pm} - \eta_s\lambda_s\right)}{F_\pm}, \quad (10.80)$$

$$F_\pm = \left(\eta_f\lambda_{1\pm} - \eta_s\lambda_s\right)\left(\frac{i\omega}{\lambda_{2\pm}} - \eta_f\lambda_{2\pm}\right)$$

$$- \left(\eta_f\lambda_{2\pm} - \eta_s\lambda_s\right)\left(\frac{i\omega}{\lambda_{1\pm}} - \eta_f\lambda_{1\pm}\right), \quad (10.81)$$

where $\lambda_{1\pm}$ and $\lambda_{2\pm}$ are given by

$$\lambda_{1\pm}^2 = \frac{1}{2\eta_f v}\left\{\frac{B_r^2}{\rho\mu} + i(\omega \pm f)\eta_f + i\omega v \right.$$

$$\left. - \sqrt{\left[\frac{B_r^2}{\rho\mu} + i(\omega \pm f)\eta_f + i\omega v\right]^2 + 4\eta_f v\omega(\omega \pm f)}\right\}, \quad (10.82)$$

$$\lambda_{2\pm}^2 = \frac{1}{2\eta_f v}\left\{\frac{B_r^2}{\rho\mu} + i(\omega \pm f)\eta_f + i\omega v \right.$$

$$\left. + \sqrt{\left[\frac{B_r^2}{\rho\mu} + i(\omega \pm f)\eta_f + i\omega v\right]^2 + 4\eta_f v\omega(\omega \pm f)}\right\}. \quad (10.83)$$

The limit of $F_{1\pm}$ for viscosity $v \to 0$ is $F_{0\pm}$, and the limit of $F_{2\pm}$ is zero, with $F_{0\pm}$ given by

$$F_{0\pm} = \frac{1}{\left(\frac{i\omega}{\lambda_{1\pm}} - \eta_s\lambda_s\right)}. \quad (10.84)$$

The limit of $\lambda_{1\pm}$ for no viscosity is $\lambda_{0\pm}$, and the limit of $\lambda_{2\pm}$ is infinity, with $\lambda_{f\pm}$ given by Eq. (10.45).

As the term $\eta'\nabla^2\mathbf{v}$ in the Navier–Stokes equation arises from the divergence of the stress tensor $\mathbf{T} = \eta'\left(\nabla\mathbf{v}(\mathbf{r}, t) + [\nabla\mathbf{v}(\mathbf{r}, t)]^T\right)$ in the incompressible fluid case ($\nabla \cdot \mathbf{v} = 0$), the viscous torque can be expressed in terms of the stress on the surface $\mathbf{n} \cdot \mathbf{T}$ where $\mathbf{n}$ stands for the outward unit normal to the surface S (see e.g. Guo and Ning, 2002). The torque can thus be expressed in terms of the spatial rate of change of the fluid velocity at the core boundaries as:

$$\boldsymbol{\Gamma}_b^{\text{viscous}} = \int_S \rho v\, \mathbf{r} \wedge \mathbf{n} \cdot \{\nabla\mathbf{v}(\mathbf{r}, t) + [\nabla\mathbf{v}(\mathbf{r}, t)]^T\}_{r=a_b}\, dS, \quad (10.85)$$

where v is the viscosity of the fluid core as before. We now express the velocity field in terms of the spherical coordinates and consider that ρv is constant on the boundary surface. Considering in addition that, at the first order, the spatial derivatives can be approximated by the radial derivatives, the torque can be written as

$$\boldsymbol{\Gamma}_b^{\text{viscous}} = \rho v \int_S \left(r\frac{dv_\theta(\mathbf{r}, t)}{dr}\hat{\mathbf{e}}_\varphi - r\frac{dv_\varphi(\mathbf{r}, t)}{dr}\hat{\mathbf{e}}_\theta\right)_{r=a_b} dS, \quad (10.86)$$

or equivalently

$$\Gamma_{b\pm}^{\text{viscous}} = \pm i\rho v \int_S \left(r\frac{dv_\pm(\mathbf{r}, t)}{dr}\right)_{r=a_b} dS. \quad (10.87)$$

The complex sum of the first two components of the viscous torque thus takes the following form:

$$\tilde{\Gamma}_b^{\text{viscous}} = \frac{1}{2}e^{i\omega t}\int_\theta ia_b\rho v\left[(1+\cos\theta)^2\frac{\partial v_+^{(+)}}{\partial r} + (1-\cos\theta)^2\frac{\partial v_-^{(+)}}{\partial r}\right]_{r=a_b}\sin\theta\,d\theta,$$

(10.88)

which is of the same form as the electromagnetic torque given by Eq. (10.71). We now note that, as for the solutions for the incremental magnetic field, the solutions for the velocity from the motion equation and the induction equation coupled together are of the same form as the solutions for the velocity in the case without viscosity:

$$v_{b\pm}(\mathbf{r}, t) = -\frac{ir\Delta\tilde{\omega}}{2}\left[(1\pm\cos\theta)e^{i(\omega t-\varphi)} - (1\mp\cos\theta)e^{-i(\omega t-\varphi)}\right] \quad (10.89)$$

$$= v_{b\pm}^{(+)}(1\pm\cos\theta)e^{i(\omega t-\varphi)} + v_{b\pm}^{(-)}(1\mp\cos\theta)e^{-i(\omega t-\varphi)}. \quad (10.90)$$

Similarly as for the electromagnetic torque, the viscous torque then takes the form:

$$\tilde{\Gamma}_b^{\text{viscous}} = e^{i\omega t}\int\frac{\pi a_b^4\rho v\Delta\tilde{\omega}}{2}$$
$$\times \left\{(1+\cos\theta)^2\left[i\omega\,(F_{1+}+F_{2+}) - \eta_f\left(F_{1+}\lambda_{1+}^2 + F_{2+}\lambda_{2+}^2\right)\right]\right.$$
$$\left. + (1-\cos\theta)^2\left[i\omega\,(F_{1-}+F_{2-}) - \eta_f\left(F_{1-}\lambda_{1-}^2 + F_{2-}\lambda_{2-}^2\right)\right]\right\}\sin\theta\,d\theta.$$

(10.91)

The sum of the electromagnetic torque given by Eq. (10.71) and the viscous torque given by Eq. (10.91) can be written in the same form as Eqs. (10.1) for the CMB and (10.2) for the ICB, with

$$K^{\text{ICB}} = \frac{\pi a_s^4}{2i\Omega_0 A_s}\int_0^\pi\left(\frac{1}{\mu}I_{\text{em}}^{\text{ICB}} + \rho v I_{\text{vis}}^{\text{ICB}}\right)\sin\theta\,d\theta, \quad (10.92)$$

$$K^{\text{CMB}} = \frac{\pi a_s^4}{2i\Omega_0 A_f}\int_0^\pi\left(\frac{1}{\mu}I_{\text{em}}^{\text{CMB}} + \rho v I_{\text{vis}}^{\text{CMB}}\right)\sin\theta\,d\theta, \quad (10.93)$$

with

$$I_{vis}^b = (1+\cos\theta)^2\left[i\omega\,(F_{1+}+F_{2+}) - \eta_f\left(F_{1+}\lambda_{1+}^2 + F_{2+}\lambda_{2+}^2\right)\right]$$
$$+ (1-\cos\theta)^2\left[i\omega\,(F_{1-}+F_{2-}) - \eta_f\left(F_{1-}\lambda_{1-}^2 + F_{2-}\lambda_{2-}^2\right)\right] \quad (10.94)$$

and

$$I_{em}^b = B_r^2\left[(1+\cos\theta)^2\,(F_{1+}+F_{2+}) + (1-\cos\theta)^2\,(F_{1-}+F_{2-})\right]. \quad (10.95)$$

Note that using series of nutation observations (see Chapter 11), it is possible to interpret them in terms of a visco-magnetic coupling on using the above formulas.

The lowermost mantle electrical conductivity (in the 200 m layer at the base of the mantle) is unfortunately unknown. Considering in addition that (1) the outer core electrical conductivity, known from laboratory experiments is at the level of 5×10^5 Sm^{-1} (Stacey and Anderson, 2001), (2) the lowermost mantle electrical conductivity has to be lower than that of the core, it is possible to get a range of values: Koot *et al.* (2010) have considered e.g. 10 Sm^{-1}, 5×10^4 Sm^{-1}, 5×10^5 Sm^{-1}. One also further considers that the total root mean square (RMS) of the radial magnetic field at the CMB can be extrapolated downward from the surface magnetic field measurements. Concerning the viscosity of the outer core fluid close to the CMB, again a range of values may be considered between a molecular viscosity at the level of 10^{-6} m^{2}s^{-1} as determined from laboratory experiments and ab-initio computations and an eddy viscosity at the level of 10^{-4} m^{2}s^{-1} (Buffett and Christensen, 2007). With these parameters, Koot *et al.* (2010) have determined the possibilities for the RMS of the radial uniform magnetic field as a function of the core viscosity. They found that for electromagnetic coupling only (no viscosity), the RMS of the radial magnetic field at the CMB must be 0.7 mT or higher. The consideration of a viscous coupling allows for lower values of the magnetic field at the CMB. However, assuming that to be the case, the outer core viscosity is of the order of a few 10^{-2} m^{2}s^{-1}, which is unrealistic. Koot *et al.* (2010) conclude then that the viscosity of the core does not help to explain the large RMS of the magnetic field that is necessary (if other torques, such as the topographic torque, are ignored).

The estimations of Koot *et al.* (2010) concerning the magnetic field at the ICB lead to values around 6–7 mT and to a kinematic viscosity of the fluid core close to the ICB that is way too large, orders of magnitude larger than what is expected from laboratory measurements and ab-initio computations.

Deleplace and Cardin (2006) have also shown that with their visco-magnetic model at the CMB using the "observed" geomagnetic field (surface magnetic field values extrapolated downward for the degrees and orders corresponding to $\ell \leq$ 13 and decreasing power-law for the unobserved spherical harmonic component contributions), viscous dissipative effects are necessary to explain the nutation data. Their viscosity is at the upper bound of the one of Mathews and Guo (2005), which is quite large. This viscosity value could be an effective viscosity inside a boundary layer near the core rather than a global property of the fluid core: this large apparent viscosity can be the result of the turbulent transfer of energy of the flow within the core from the small scale to the large scale. As stated in the paper of Deleplace and Cardin (2006), Davies and Whaler (1997) indeed show that the effective transport of momentum in the core may be generated by convective motions associated with the dynamo process or by surfacic flows such as topographic winds, unstable motions of the boundary layers, or chemical/compositional fluxes.

10.5 Topographic coupling

The so-called topographic coupling is the coupling induced by the fluid pressure of the liquid core acting on the bumpy boundaries, in addition to the classical ellipsoidal topography. The topographic torque is a way to transfer angular momentum between the outer core and the inner core and between the outer core and the mantle. The ICB is considered to be nearly in hydrostatic equilibrium as the mass anomalies in the inner core or in the mantle would only induce very small departures from the hydrostatic equilibrium, of the order of a maximum 200 m as shown in Defraigne *et al.* (1996). The situation is different for the CMB where a topography of the order of a few kilometers is expected (in addition to the large hydrostatic bulge of the boundary) as demonstrated by (1) computations of loading on the CMB from the mass anomalies in the mantle (see e.g. Hager *et al.*, 1985, Defraigne *et al.*, 1996, Moucha *et al.*, 2007, Simmons *et al.*, 2006, 2009), as well as (2) computations from seismic observations (see e.g. Morelli and Dziewonski, 1987, Woodhouse and Dziewonski, 1989, Murphy *et al.*, 1997, Sylvander *et al.*, 1997, Sze and van der Hilst, 2003, Koelemeijer *et al.*, 2012, Soldati *et al.*, 2012, 2013). While these observations have presently large error bars (at the km level) they definitely show the possibility to have bumps and valleys on the interface between the core and the mantle. In particular, the degree two order zero contribution with respect to the hydrostatic value is interesting in the frame of nutation theory as it induces large effects on particular nutations, as shown by Herring *et al.* (1986a and b), i.e. a 500 m increase of the equatorial radius with respect to the polar radius is sufficient to explain the large retrograde annual nutation residual when comparing the VLBI observation with respect to a model based on a hydrostatic ellipsoidal Earth.

This 500 m increase with respect to the 9 km difference between the equatorial radius and the polar radius corresponds to a few percent (5%). The effects of non-hydrostatic topography at the CMB on nutations were often ignored, being considered as of the second order. However, even if the uncertainties on the departure from hydrostatic equilibrium are large, it is a potential explanation for the discrepancy between the observed nutation and a nutation theory built on hydrostatic equilibrium. The degree two order zero contribution (increase of the equatorial radius of the core with respect to its polar radius) has been considered in recent adopted nutation theory (e.g. the MHB2000 model of Mathews *et al.*, 2002). This is particularly important for the FCN period and therefore for the nutations that are in the vicinity of the FCN resonance. In MHB2000 the increase of the equatorial radius of the core with respect to its polar radius has been determined to be about 350 m (3.8% instead of 5%) on considering other coupling mechanisms and the electromagnetic coupling in particular.

In parallel, Wu and Wahr (1997) have made numerical computations of the effects of topographic coupling at the CMB on Earth rotation, with the objective of getting an estimate of the possible contributions to LOD and nutations. In order to obtain an order of magnitude for such effects we may consider a simple Earth model consisting of just a mantle and a liquid core, both homogeneous, and consider that the CMB has topographic features on top of its hydrostatic flattening. The torque acting at the CMB must be amended with two additional parts: one torque related to the fluid pressure acting on the non-hydrostatic boundary and one torque associated with the external forcing potential acting on the mass anomalies at the boundary. Wu and Wahr (1997) have reconsidered the solution of the equation of fluid motion. They have taken the velocity to be the sum of a part very close to a rigid rotation of the fluid (what one has for a Poincaré fluid, see Section 7.2.1) plus a remaining second-order part for which they solve the angular momentum balance equations. This additional velocity is expressed in terms of a scalar potential that has a form involving coefficients demonstrated to be proportional to the topography coefficients when boundary continuity conditions are used, in front of Legendre polynomials or associated Legendre polynomials, together with a complex exponential dependence with the frequency. The associated dynamical pressure from which the topographic torque on the CMB is computed shows that we do not only have an important incremental global core rotation contribution with respect to the initial mean rotation of the Earth, but also an incremental global mantle rotation contribution (as for the hydrostatic case). The topography being developed in spherical harmonics, all the degrees and orders of the topography may be considered as potential contributors to the torque.

Wu and Wahr (1997) have computed the topographic torque contributions to the retrograde annual nutation (suspected to be the most influenced nutation as it is close to the FCN). They have shown that some of the topography components have large contributions at the level of a few tenths of a milliarcsecond. From their work, we also see that the CMB topography effects on the nutations are proportional to the square of the amplitude of the topography. Small deviations from the hydrostatic equilibrium would thus remain very small contributions. However, Wu and Wahr have shown that there are large contributions to nutations for some of the topography degrees and orders.

Dehant *et al.* (2012a and b, 2013) have used the same approach and have considered a semi-analytical method in order to identify the physical origin of the enhancement of the contributions of some of the topography features. They conclude that this can be due to the amplitudes of the coefficients of the topography themselves or to the existence of resonances with inertial waves in the core. These computations are very interesting and still need to consider the existence of an inner core in the Earth model, as the velocity field in the fluid core, and the frequencies of the inertial modes in particular, can vary when adding an inner core. It is thus premature to conclude that the degree and order found by Wu and Wahr to amplify the nutation are definitive. But it is certainly true that the topography effects of nutation may cause changes in the nutations at the observable level.

10.6 Viscosity of the inner core

Before everything else, we should mention that the viscosity of a fluid is different from that of a solid, in which case one usually uses the term viscoelasticity. Viscous materials resist shear flow when a stress is applied. Elastic materials stretch almost instantaneously when pulled and quickly return to their original state once the stress is removed. Viscoelastic materials exhibit time dependent strain. The elastic materials, as previously mentioned (see Chapter 8), can be viewed as akin to elastic springs and the viscoelastic materials can be modeled as a combination of springs and dashpots. Viscosity is defined as being equal to the ratio of shear stress/shear rate and can be thought of as resistance of the material to flow. If viscosity increases, the strain rate for the same stress will decrease.

It is easy to account for inner core viscosity as it is only a means to consider changes in the stress–strain relationship when the numerical integration approach is used (see Chapter 8) or to change the compliances in the Liouville equations when the angular momentum approach is used. Using this last approach, Greff-Lefftz *et al.* (2002) have evaluated that the viscosity of the inner has a quite small contribution to nutations (which is not the case for the contributions from the viscosity of the outer core, as seen in Section 10.4). Here we are not speaking about the viscous torque at the CMB or ICB due to viscosity of the liquid part but the viscous behavior (or viscoelastic behavior) of the inner core. If the inner core material has a high viscosity it will be more resistant to flow and thus strain more like an elastic material.

In 1997, Buffett published a paper on evaluating the viscosity of the inner core from the consideration of seismic data for the inner core and a possible differential rotation of the inner core (which is now controversial). The inferred rotation rate constrains the viscosity of the inner core to be less than 10^{16} Pa s or greater than 10^{20} Pa s, as two different dynamical regimes are expected to be possible. Studies based on experiments (e.g. Frost and Ashby, 1983) reduce further the range of the effective viscosity of the inner core to be from 10^{12} Pa s up to 10^{17} Pa s.

For a viscosity of the inner core $> 10^{15}$ Pa s, the inner core has an elastic behavior and the nutation amplitudes and phases do not depend on the viscosity of the inner core. For mid-value of the viscosity, the situation is different.

In general, viscous behavior of an otherwise solid inner core could affect nutations primarily through its effect on the frequency of the FICN eigenmode.

The frequency of this mode in space is $\nu_3 = 1 + \sigma_3$, with σ_3 given by Eq. (10.5), wherein τ is the compliance parameter representing the deformational response of the inner core to rotational perturbations of the inner core; its value is determined by Eq. (6.165) taken together with preceding equations (Eq. (6.164)). (For the parameter α_2, see Eqs. 7.152.) It is important to note that in view of the values of the parameters involved, as may be seen from the para below Eqs. 7.152, the eigenfrequency ν_3 of the FICN is of the order of e_s, which is very low, $\mathcal{O}(10^{-3})$. Viscosity in the inner core modifies the value of τ and also makes it complex. It is possible that in some range of values of the viscosity, the real part of ν_3 becomes close to one of the very low frequencies in the tidal spectrum. The result would be a resonant enhancement of the nutation component at that frequency, leading to detectable effects.

For the other extreme of the viscosity, a very low viscosity, the inner core behaves closer to a fluid and again the amplitudes and phases of the nutations do not depend on the viscosity of the inner core (we are back to the case without FICN).

Observations of Earth's nutations allow thus for insights into the physical properties of the inner core. Estimations of the FICN eigenfrequency are important in this context. The frequency of the FICN is controlled by the strength of the coupling acting at the ICB and by the ability of the inner core to deform under the action of centrifugal and gravitational forces. Attenuation of the FICN reflects energy dissipated by electromagnetic and viscous friction at the ICB as well as through viscous relaxation of the inner core. Similarly, the damping of the mode reflects the energy dissipated both through the coupling at the ICB and through the effects of viscosity in inner core deformation. Estimations of the ICB coupling strength and dissipation have been obtained previously from nutation observations by assuming a purely elastic inner core (Mathews *et al.*, 2002). By using the latest series of nutation observations, Koot *et al.* (2010) have shown that it is possible to explain the observed frequency and damping of the FICN by a combination of electromagnetic coupling at the ICB and viscoelastic deformation of the inner core. The data have imposed a strong constraint on the viscosity of the inner core, which has to be in the range 2–7×10^{14} Pa s, which is in very good agreement with seismic observations of shear wave attenuation in the inner core as published in the PREM interior model of the Earth (Dziewonsky and Anderson, 1981).

10.7 Effect of stratification in the core

An alternative mechanism inducing changes in the nutation theory with respect to what is expected from observations has been proposed by Buffett (2010a). It involves the presence of fluid stratification at the top of the core, which causes the fluid close to the core–mantle boundary to be trapped by the effects of topography.

Such a strong stratification has recently been proposed on the basis of chemical interactions between the core and the mantle by Ozawa *et al.* (2009).

Stratification has been considered to have little influence on the flow induced by a tidal potential in the absence of boundary topography because the motion is very nearly parallel to constant density surfaces (close to Poincaré fluid, see Section 7.2). Such a flow causes little change ρ' in density and consequently, the buoyancy forces are expected to be weak (the term $\rho'g$ in Eq. (10.76) is expected to be small). On the other hand, when a boundary topography is considered, the flow near the boundary introduces a vertical component of motion. (The presence of stratification suppresses flow over boundary topography, trapping fluid in the boundary region.) These additional vertical components of the motion in a fluid initially stratified and incompressible disturb the density field and an additional buoyancy force arises in proportion to the strength of the stratification. By using expressions for the velocity field in such a model, Buffett (2010a) has demonstrated that there is a strong dependence of the dissipation on the magnitude of the topography, which is poorly known at the present time. The presence of stratification in the liquid core and/or topography would induce effects on nutation that could be above the observation precision and compatible with these observations. Buffett (2010a) demonstrated that nutation observations may thus support the presence of stratification as well as that of an additional topography with respect to the core flattening (a strong stratification with a small topography at the level of a few meters or a larger topography at the level of more than 100 meters and a small stratification).

Chemical interactions between the core and the mantle are capable of producing the required stratification. As proposed by Buffett and Seagle (2010), the physical model for the existence of a chemical layer at the top of the core involves the addition of light elements to the top of the core. These are brought there by chemical interactions with the mantle (transfer of O and Si). The reason why these elements remain there must be found in thermal stratification due to a superadiabatic[1] heat flow at the CMB.

10.8 Discussion on precession constant

The precession constant (see Sections 1.4 and 3.1.1 for instance, as well as Eq. (2.77)) is proportional to $H_d = \frac{C-A}{C}$, which can be deduced from the precession and nutation observations and from the non-luni-solar

---

[1] A superadiabatic temperature gradient is related to an excess with respect to the adiabatic temperature gradient which is related to a corresponding pressure gradient by $dT/dr = (1 - 1/\gamma)(T/P)(dP/dr)$, where T and P are temperature and pressure, dT/dr and dP/dr temperature and pressure gradients, respectively, and $\gamma = C_P/C_V$ with C_P and C_V the specific heat at constant pressure and constant temperature, respectively (properties of the material).

contribution to precession (see Section 5.13). It is related to the difference of the moments of inertia of the whole Earth or the so-called dynamical flattening of the Earth e ($H_d = \frac{e}{1+e}$, as seen in Eq. (5.313)). This parameter H_d also enters into the principal nutation amplitudes. A fitting of this parameter on the precession and nutation observations has therefore been done by Mathews *et al.* (2002). A readjustment of the precession constant is discussed in Capitaine *et al.* (2003a–d). In their paper, they discuss precession models consistent with IAU2000A precession–nutation (i.e. MHB2000, provided by Mathews *et al.*, 2002) and provide a range of expressions that implement them. They develop new expressions for the motion of the ecliptic with respect to the space-fixed ecliptic of J2000 using the developments from Simon *et al.* (1994) and Williams (1994), and with improved constants fitted to the most recent numerical planetary ephemerides. The final precession model, designated P03, is the replacement for the precession component of IAU2000A.

The transfer function for nutation has been defined in Eq. (2.109) as the ratio between the amplitudes of the non-rigid Earth and rigid Earth nutations as a function of frequency in the celestial frame, or equivalently as the ratio of the non-rigid Earth wobble amplitude to the rigid Earth wobble amplitude for frequencies in the terrestrial frame. The transfer function for nutation can thus be determined from a theory providing the wobble expressions as functions of the frequency for the non-rigid Earth and for the rigid Earth as in Sections 7.3.7, 7.4.3, and 7.5.3. As seen from the expression for the transfer function provided there (Eq. (7.139) or its new form Eq. (7.164)), it has a dependence on the dynamical flattening. If the values used for the dynamical flattening are the same for both models, we have:

$$\tilde{\eta}(\sigma; e) = T(\sigma; e)\tilde{\eta}(\sigma; e). \tag{10.96}$$

But this is not often the case. We use e for the dynamical flattening used for computing the transfer function and e_{rig} for the value related to the precession constant used for building the rigid Earth nutation theory. In order to be coherent, the rigid Earth nutation or equivalently the transfer function must be scaled according to the value of e used for the non-rigid Earth:

$$\tilde{\eta}(\sigma; e) = T(\sigma; e|e_{\text{rig}})\tilde{\eta}(\sigma; e_{\text{rig}}). \tag{10.97}$$

As the factor that appears in the amplitudes of the rigid Earth nutations is $H_{d\,\text{rigid}} = e_{\text{rig}}/(1 + e_{\text{rig}})$ and as the amplitudes of the rigid Earth nutation at σ are resonant at the e_{rig} frequency with a factor $e_{\text{rig}}/(e_{\text{rig}} - \sigma)$ as seen from Eq. (7.138), the transfer function must be scaled by the value of H_d corresponding to e, i.e., by $\frac{e_{\text{rig}}/(1+e_{\text{rig}})}{e/(1+e)}$, and the resonance factor by $\frac{e/(e-\sigma)}{e_{\mathrm{rig}}/(e_{\mathrm{rig}}-\sigma)}$:

$$T(\sigma;e|e_{\mathrm{rig}}) = \left(\frac{e_{\mathrm{rig}} - \sigma}{1 + e_{\mathrm{rig}}}\right)\left(\frac{1+e}{e-\sigma}\right)\left(1 + (1+\sigma)\sum_{i=1}^{4}\frac{N_i}{\sigma - \sigma_i}\right). \qquad (10.98)$$

By defining N_0 as $\frac{e/(e-\sigma)}{e_{\mathrm{rig}}/(e_{\mathrm{rig}}-\sigma)}$, we have:

$$T(\sigma;e|e_{\mathrm{rig}}) = \left(\frac{e_{\mathrm{rig}} - \sigma}{1 + e_{\mathrm{rig}}}\right)N_0\left(1 + (1+\sigma)\sum_{i=1}^{4}\frac{N_i}{\sigma - \sigma_i}\right), \qquad (10.99)$$

with N_i for $i = 0, \ldots, 4$ and σ_i containing the basic Earth parameters. The transfer function (10.99) is thus expressed as an explicit analytic function of σ and of the basic Earth parameters that are then fitted to the observations.

The nutation for the non-rigid Earth is then computed from a convolution of this new transfer function $T(\sigma;e|e_{\mathrm{rig}})$ on replacing the basic Earth parameters by their values, with the rigid Earth nutation theory using $H_{d\,\mathrm{rigid}}$.

10.9 Triaxiality

The effect of triaxiality of the Earth has been neglected in the Liouville equations as well as in the numerical integration for computing nutations. In the latter approach, triaxiality in the equilibrium state of the Earth adds a lot of complexity due to coupling between variables of different degrees and orders. However, an attempt to solve the equations of motion in a triaxial planet was published with the aim of determining the libration of planetary satellites and for Mercury (see Trinh *et al.*, 2011, 2013, Van Hoolst *et al.*, 2013). In the angular momentum approach to the computation of wobbles and nutations, triaxiality enters through the difference $A - B$ between the two equatorial principal moments as in the Poincaré theory (see Section 7.2). The triaxiality parameter for the Earth is defined by:

$$e' = \frac{A - B}{\bar{A}}. \qquad (10.100)$$

The generalization of the dynamical flattening in the triaxial case is:

$$e = \frac{C - \bar{A}}{\bar{A}} \quad \text{with} \quad \bar{A} = \frac{A + B}{2}. \qquad (10.101)$$

Let us consider a solid triaxial Earth model, as in the Poincaré theory (see Eqs. (7.58) and (7.82)). The torque equation

$$\frac{d\mathbf{H}}{dt} + \mathbf{\Omega} \wedge \mathbf{H} = \mathbf{\Gamma}, \qquad (10.102)$$

may be written as a pair of equations for the spectral components, of frequency σ, of the first two components $(\Omega m_1, \Omega m_2)$ of the wobble vector ω of the whole Earth:

$$i A \Omega^2 \sigma m_1 + \Omega^2 (C - B) m_2 = \Gamma_1, \tag{10.103}$$

$$i B \Omega^2 \sigma m_2 - \Omega^2 (C - A) m_1 = \Gamma_2. \tag{10.104}$$

In writing down these equations, we have neglected the off-diagonal elements of the Earth inertia tensors as well as the incremental angular momentum components related to the core, as we concentrate on expressing the triaxiality effects on the wobble.

On taking the complex sum of Eqs. (10.103) and (10.104), on substituting the definitions of e and e' given in Eqs. (10.100) and (10.101), and on substituting $\tilde{m} = m_1 + i m_2$, $\tilde{m}^* = m_1 - i m_2$, and $\tilde{\Gamma} = \Gamma_1 + i\Gamma_2$, we obtain

$$\sigma \tilde{m} - \tilde{m} e + (1 + \sigma) \frac{e'}{2} \tilde{m}^* = \frac{-i\tilde{\Gamma}}{\Omega^2 \tilde{A}}. \tag{10.105}$$

The expressions for the normal mode frequencies have been derived from these equations by Zharkov and Molodensky (1996), Yoder and Standish (1997), Gonzalez and Getino (1997), as well as Van Hoolst and Dehant (2002). The Chandler wobble frequency for a triaxial solid Earth has been found to be:

$$\sigma_{\text{CW}} = \sqrt{\frac{(C - A)(C - B)}{AB}} \approx \sqrt{e^2 - \frac{e'^2}{4}}. \tag{10.106}$$

Since e'/e is of $\mathcal{O}(10^{-3})$ the fractional change in σ_{CW} that is produced by triaxiality is of the order of $(1/2)(e'/e)^2 \approx 10^{-6}$. The eigenfrequencies are to be found by solving the pair of equations, Eq. (10.105) with $\tilde{\Gamma}$ set equal to zero, and its complex conjugate.

10.10 Second-order effects

While computing, in Chapter 7, the nutation and wobble of an Earth with three layers using the angular momentum approach, we have neglected terms that are of the second or higher orders in small quantities: the density perturbations produced by deformations induced by tidal forces on the torque itself, second-order couplings of the wobble components among themselves and with the increments c_{ij} to the inertia tensor arising from the deformations produced by the tidal potential, etc. To take account of the second-order effects, one has to consider, in the equations of motion, terms such as $m_i m_j$ and $c_{ij} m_k$. The spectral component of any such term at some spectral frequency σ present in the external torque will of course involve products of amplitudes of the two factors at different frequencies; these component frequencies, say σ_1 and σ_2 with $\sigma_1 + \sigma_2 = \sigma$, could come from the zonal and sectorial parts of the tidal potential, besides the tesseral part.

With the high precision of current observational data on nutations and the ever-increasing length of the nutation series, uncertainties in the amplitudes estimated from the data for many of the spectral components of the nutation have gone below 10 μas, though the uncertainties may be some tens of μas for some nutations of very long periods. Given this scenario, and anticipating further reduction in the uncertainty of observational estimates, it is appropriate to aim at accuracies at the 1 μas level in the results of theoretical computations of nutation amplitudes. It is necessary, in this context, to investigate the role of the second-order terms in the theory of low frequency nutations. (High frequency nutations and equivalent polar motions have such small amplitudes even in the first order that second-order contributions to these are totally ignorable.)

10.10.1 Wobble equations, to the second order

We reproduce here for convenience the basic equations of motion (see e.g. Eq. (7.108)), namely, the Liouville equation

$$\frac{d\mathbf{H}}{dt} + \mathbf{\Omega} \wedge \mathbf{H} = \mathbf{\Gamma} \tag{10.107}$$

and the SOS equation (7.82):

$$\frac{d\mathbf{H}_f}{dt} = \boldsymbol{\omega}_f \wedge \mathbf{H}_f, \tag{10.108}$$

where

$$\mathbf{H} = [C] \cdot \mathbf{\Omega} + [C_f] \cdot \boldsymbol{\omega}_f, \qquad \mathbf{H}_f = [C_f] \cdot (\mathbf{\Omega} + \boldsymbol{\omega}_f). \tag{10.109}$$

In proceeding to investigate the second-order terms in these equations it will be sufficient to work with a two-layer Earth model: contributions from the SIC are very small even in the first order, and would be entirely negligible in the second order.

Consider now the first two components of Eq. (10.107), which govern the wobble motions that are of direct interest to us. Their complex combination yields the following equation for $\tilde{H} = H_1 + iH_2$:

$$\frac{d\tilde{H}}{dt} - i\tilde{\Omega}H_3 + i\tilde{H}\Omega_3 = \tilde{\Gamma}, \tag{10.110}$$

where $\tilde{\Gamma} = \Gamma_1 + i\Gamma_2$. To identify the terms of different orders in the small quantities in the equation, we separate $\tilde{H}$ and H_3 into parts of different orders:

$$\tilde{H} = \tilde{H}^{(1)} + \tilde{H}^{(2)}, \qquad H_3 = H_3^{(0)} + H_3^{(1)} + H_3^{(2)}, \tag{10.111}$$

where the superscripts 0, 1, 2 label quantities of the zeroth, first, and second orders in the small quantities. Note also that

$$\tilde{\Omega} = \Omega_0 \tilde{m}, \qquad \Omega_3 = \Omega_3^{(0)} + \Omega_3^{(1)} = \Omega_0 + \Omega_0 m_3, \qquad (10.112)$$

and that the torque too may be separated into first- and second-order parts $\boldsymbol{\Gamma}^{(1)}$ or $\tilde{\Gamma}^{(1)}$ and $\boldsymbol{\Gamma}^{(2)}$ or $\tilde{\Gamma}^{(2)}$. Substituting these into Eq. (10.110), we rewrite it as

$$\frac{d\tilde{H}^{(1)}}{dt} - i\tilde{\Omega} H_3^{(0)} + i\Omega_3^{(0)} \tilde{H}^{(1)} = \tilde{\Gamma}^{(1)} + \tilde{\Delta}^{(2)} \qquad (10.113)$$

by moving all the second-order terms to the right-hand side and incorporating them along with the second-order torque $\tilde{\Gamma}^{(2)}$ into $\tilde{\Delta}^{(2)}$:

$$\tilde{\Delta}^{(2)} = \tilde{\Gamma}^{(2)} - \left(\frac{d\tilde{H}^{(2)}}{dt} + i\Omega_3^{(0)} \tilde{H}^{(2)} + i\Omega_3^{(1)} \tilde{H}^{(1)} - i\tilde{\Omega} H_3^{(1)} \right). \qquad (10.114)$$

It is worth noting that in the frequency domain, the two terms containing $\tilde{H}^{(2)}$ become $i\Omega_0(\sigma + 1)\tilde{H}^{(2)}(\sigma)$. Now, the factor $(1 + \sigma)$ multiplying the second-order quantity $\tilde{H}^{(2)}(\sigma)$ is itself very small for the frequencies pertaining to the large nutation components. So we can drop these two terms. Furthermore, $H_3^{(1)} = 0$, as shown in Section 10.10.3 below. Thus $\tilde{\Delta}^{(2)}$ of Eq. (10.114) gets reduced to $\tilde{\Gamma}^{(2)} - i\Omega_3^{(1)} \tilde{H}^{(1)}$, and Eq. (10.113) acquires the highly reduced form:

$$\frac{d\tilde{H}^{(1)}}{dt} - iC\Omega_0^2 \tilde{m} + i\Omega_0 \tilde{H}^{(1)} = \tilde{\Gamma}^{(1)} + \tilde{\Gamma}^{(2)} - i\Omega_0 m_3 \tilde{H}^{(1)}, \qquad (10.115)$$

where

$$\tilde{H}^{(1)} = (A\tilde{m} + \tilde{c}_3 + A_f \tilde{m}_f)\Omega_0. \qquad (10.116)$$

The equation governing the wobbles is thus reduced to Eq. (10.115) taken together with (10.116) and the expressions for the first- and second-order torques that are also seen below (see Subsection 10.10.2, Eqs. (10.137) and (10.138)).

Next, we need to make a similar reduction of the SOS Eq. (10.108). The wobble part of the equation goes over into

$$\frac{d\tilde{H}_f}{dt} - i\omega_{f3} \tilde{H}_f + i\tilde{\omega}_f H_{f3} = 0. \qquad (10.117)$$

Splitting $\tilde{H}_f$ and H_{f3} into parts up to the second order in the small quantities, and observing that $H_{f3}^{(0)} = C_f \Omega_0$, we write the above equation, grouping together all the second-order terms into a quantity $\Delta_f^{(2)}$, as

$$\frac{d\tilde{H}_f^{(1)}}{dt} + iC_f\Omega_0^2\tilde{m}_f = \Delta_f^{(2)}, \tag{10.118}$$

$$\Delta_f^{(2)} = -\frac{d\tilde{H}_f^{(2)}}{dt} + i\Omega_0 m_{f3}\tilde{H}_f^{(1)} - i\Omega_0\tilde{m}_f H_{f3}^{(1)}. \tag{10.119}$$

Expressions for $\tilde{H}_f^{(1)}$, $\tilde{H}_f^{(2)}$, and $H_{f3}^{(1)}$ are obtained from the components of $\mathbf{H}_f$ by substituting for $[\mathbf{C}_f]$, $\mathbf{\Omega}$ and $\boldsymbol{\omega}_f$ from Eqs. (7.111) and (7.112):

$$\tilde{H}_f^{(1)} = [A_f(\tilde{m} + \tilde{m}_f) + \tilde{c}_3^f]\Omega_0, \tag{10.120}$$

$$\tilde{H}_f^{(2)} = [c_{11}^{fZ}(\tilde{m} + \tilde{m}_f) + (c_{11}^{fS} + ic_{12}^f)(\tilde{m}^* + \tilde{m}_f^*) + \tilde{c}_3^f(m_3 + m_{f3})]\Omega_0, \tag{10.121}$$

$$H_{f3}^{(1)} = [C_f(m_3 + m_{f3}) + c_{33}^f]\Omega_0 = 0, \tag{10.122}$$

remembering that $\tilde{c}_3^f = c_{13}^f + ic_{23}^f, c_{11}^f = c_{11}^{fZ} + c_{11}^{fS}, c_{22}^f = c_{22}^{fZ} + c_{22}^{fS}, c_{11}^{fZ} = c_{22}^{fZ},$ $c_{11}^{fS} = -c_{22}^{fS}$, and $c_{33}^f = -2c_{11}^{fZ} = -2c_{22}^{fZ}$, which are analogous to Eqs. (6.150)–(6.157) but referring now to the fluid core. The vanishing of $H_{f3}^{(1)}$ is by virtue of Eq. (10.151) below (see Subsection 10.10.3).

The equation of wobble motion of the core is thus reduced to (10.118) taken together with (10.119)–(10.122).

Incremental inertia tensor elements are noticeably present in the reduced forms of both the Liouville equation and the SOS equation. They are affected by mantle anelasticity, and also include contributions from deformations caused by ocean tides.

10.10.2 Torques to the second order

We now evaluate the torque $\tilde{\Gamma}$ produced by the degree two tidal potential W_2^B to the second order in the small quantities. It is convenient to use the expressions for the zonal, tesseral, and sectorial parts of this potential in terms of the dimensionless quantities $\tilde{\phi}^{2m}$ as given in Eqs. (6.7)–(6.9). After transformation to Cartesian coordinates, they are:

$$W_{20}^B = -\frac{\Omega_0^2}{3}\tilde{\phi}^{20}\left(\frac{3}{2}z^2 - \frac{1}{2}r^2\right), \tag{10.123}$$

$$W_{21}^B = -\Omega_0^2\,\mathcal{Re}\,[\tilde{\phi}^{21}(zx - izy)] = -\Omega_0^2(\phi_1^{21}zx + \phi_2^{21}zy), \tag{10.124}$$

$$W_{22}^B = -\Omega_0^2\,\mathcal{Re}\,[\tilde{\phi}^{22}(x^2 - y^2 - 2ixy)] = -\Omega_0^2[\phi_1^{22}(x^2 - y^2) + \phi_2^{22}2xy]. \tag{10.125}$$

The torque due to W_{2m}^B is $- \int \rho(\mathbf{r})(\mathbf{r} \wedge \nabla)W_{2m}^B(\mathbf{r})d^3r$. Evaluation of this integral was done for the case $m = 1$ in Section 2.23.5. Similarly, we compute the gradient of the potential and the cross-product with this gradient. We then compute the integral over the deformed volume, noting the following relations (where the arrow means that it will provide the following contribution to the torque):

$$y^2 - z^2 = (y^2 + x^2)(z^2 + x^2) \quad \leftarrow \quad C - B + c_{33} - c_{22}, \quad (10.126)$$

$$z^2 - x^2 = (z^2 + y^2)(x^2 + y^2) \quad \leftarrow \quad A - C + c_{11} - c_{33}, \quad (10.127)$$

$$x^2 - y^2 = (x^2 + z^2)(y^2 + z^2) \quad \leftarrow \quad B - A + c_{22} - c_{11}, \quad (10.128)$$

$$yz \quad \leftarrow \quad -c_{23} = -c_{32}, \quad (10.129)$$

$$xz \quad \leftarrow \quad -c_{13} = -c_{31}, \quad (10.130)$$

$$xy \quad \leftarrow \quad -c_{12} = -c_{21}. \quad (10.131)$$

We denote again $\tilde{\phi}^{21} = \phi_1^T + i\phi_2^T$, $\tilde{\phi}^{21*} = \phi_1^T - i\phi_2^T$, $\tilde{c}_3 = c_{13}^T + ic_{23}^T$, and $\tilde{c}_3^* = c_{13}^T - ic_{23}^T$, indicating zonal, tesseral, or sectorial by ^{Z}, ^{T}, or ^{S}.

In doing so, contributions c_{ij} to the inertia tensor from tidal deformations are ignored. Now that these are to be taken into account, the integrals appearing in the torque have to be evaluated using formulae such as

$$\int \rho(\mathbf{r})(x^2 + y^2)d^3r = C + c_{33}, \qquad \int \rho(\mathbf{r})xyd^3r = -c_{12}, \quad (10.132)$$

which include the contributions from the c_{ij}. The torques due to the three potentials are obtained in a straightforward manner, and we obtain the following results for the complex combination of the first two components of the torque in each case:

$$\tilde{\Gamma}^{20}(t) = i\Omega_0^2 \tilde{\phi}^{20}\tilde{c}_3, \quad (10.133)$$

$$\tilde{\Gamma}^{21}(t) = -i\Omega_0^2 \left[(eA + c_{33} - c_{11}^Z) \tilde{\phi}^{21} - \left(c_{11}^S + ic_{12} \right) \tilde{\phi}^{21*} \right] \quad (10.134)$$

$$\tilde{\Gamma}^{22}(t) = -2i\Omega_0^2 \tilde{c}_3^* \tilde{\phi}^{22}, \quad (10.135)$$

where c_{11}^Z and c_{11}^S are the contributions to c_{11} from the zonal and sectorial parts, respectively, of the tidal potential:

$$c_{11}^Z = \frac{1}{2}(c_{11} + c_{22}) = -\frac{1}{2}c_{33}, \qquad c_{11}^S = \frac{1}{2}(c_{11} - c_{22}), \quad (10.136)$$

as noted in Chapter 6, Eqs. (6.150) and (6.159), as well as from the relations $c_{11} = c_{11}^Z + c_{11}^S$ and $c_{22} = c_{22}^Z + c_{22}^S$. The asterisks in the above equations denote complex conjugation, as usual. Note that we have set $B = A$, as triaxiality is of no relevance to the consideration of effects on low frequency nutations.

The only first-order term in the above expressions, is $-ieA\Omega_0^2\tilde{\phi}^{21}$ in the torque produced by the tesseral potential, as expected. All the remaining terms in the three torques contain products of two time-dependent factors. The first- and second-order parts of the total torque are:

$$\tilde{\Gamma}^{(1)}(t) = -iAe\Omega_0^2 \tilde{\phi}^{21}, \tag{10.137}$$

$$\tilde{\Gamma}^{(2)}(t) = i\Omega_0^2 \left[\left(\tilde{c}_3 \tilde{\phi}^{20} - \frac{3}{2} c_{33} \tilde{\phi}^{21} \right) + \left(c_{11}^S + ic_{12} \right) \tilde{\phi}^{21*} - 2\tilde{c}_3^* \tilde{\phi}^{22} \right], \tag{10.138}$$

where $\tilde{c}_3 = c_{13} + ic_{23}$ as before.

The spectra of $\tilde{\phi}^{20}$ and the inertia tensor increments c_{33} and c_{11}^Z produced by the zonal potential are in the low frequency band; those of $\tilde{\phi}^{21}$ and $\tilde{c}_3$ are in the retrograde diurnal band while their complex conjugates have spectra in the prograde diurnal band; and the spectra of $\tilde{\phi}^{22}$, c_{11}^S, and c_{12} are in the retrograde semidiurnal band. From the combination of frequencies which takes place when two factors are multiplied, it becomes evident that the spectrum of every one of the terms in $\tilde{\Gamma}^{(2)}$, like that of $\tilde{\Gamma}^{(1)}$, lies in the retrograde diurnal band; this is precisely the band that is of interest to us.

How does the magnitude of the second-order terms compare with that of the first-order torque? If we consider, for instance, the 18.6 year retrograde term in the nutation, which has the largest amplitude of about 8000 mas, the relevant spectral component of the second-order torque would have to be about 10^{-7} times the corresponding first-order torque to contribute about 1 μas to this nutation term. So second-order terms of a magnitude of 10^{-7} or higher relative to the first-order torque are the ones that are of interest.

Consider, for instance, the magnitude of the first term in the expression (10.138) for $\tilde{\Gamma}^{(2)}$ relative to $\tilde{\Gamma}^{(1)}$ of Eq. (10.137). Since the largest spectral components of the $\tilde{\phi}^{2m}$ for all m are about 10^{-5} to within a factor 3 or so, the ratio of this term to $\tilde{\Gamma}^{(1)}$ may be taken to be of the same order of magnitude as $\tilde{c}_3/(Ae)$. As seen in Eq. (6.167),

$$\tilde{c}_3 = -A[\kappa(\tilde{\phi}^{21} - \tilde{m}) - \xi \tilde{m}_f], \tag{10.139}$$

when the inner core is ignored as is done here. The leading term is the one proportional to $\tilde{\phi}^{21}$. So the magnitude of $\tilde{c}_3/(Ae)$ is dominated by $(\kappa/e)\tilde{\phi}^{21}$ which is about one-third that of $\tilde{\phi}^{21}$ and hence, of the order of 10^{-6}; this is not negligible in view of the considerations of the previous paragraph. Since $\tilde{m}$ is close to $e\tilde{\phi}^{21}$, its contribution is smaller by a factor e and is therefore negligible. However the $\tilde{m}_f$ term may not be summarily dismissed on similar grounds: because of the huge FCN resonance effect on $\tilde{m}_f$, the amplitudes of the spectral terms of $\tilde{m}_f$ which are associated with the precession and the 18.6 year nutations are roughly 200 times the corresponding amplitudes of $\tilde{m}$, so that $\tilde{m}_f$ is about 0.6 $\tilde{\phi}^{21}$ at these frequencies. Therefore, though ξ is only about one-fifth of κ, the contribution of the $(\xi/e)\tilde{m}_f$

term in $\tilde{c}_3$ could be better than 10% of that of the leading term, which is not negligible.

A similar analysis of the magnitudes of three other terms in the second-order torque may be made using the expressions for c_{33} and $(c_{11}^S + c_{12})$ given in Eqs. (6.172) and (6.175), namely,

$$c_{33} = -\frac{2}{3}A[\kappa(\tilde{\phi}^{20} - 2m_3) - 2\xi m_{3f}] + c_{33}^{00}, \qquad (10.140)$$

$$c_{11}^S + ic_{12} = -2A\kappa\tilde{\phi}^{22}. \qquad (10.141)$$

Let us consider now the sum of all the terms in the second-order torque, retaining just the dominant part involving the relevant ϕ^{2m} in each of the inertia tensor elements or their combinations. We obtain

$$\tilde{\Gamma}^{(2)}(t) \approx i\Omega_0^2[-A\kappa\tilde{\phi}^{21}\tilde{\phi}^{20} + A\kappa\tilde{\phi}^{20}\tilde{\phi}^{21} - 2A\kappa\tilde{\phi}^{22}\tilde{\phi}^{21*} + 2A\kappa\tilde{\phi}^{21*}\tilde{\phi}^{22}]. \qquad (10.142)$$

Thus it appears that the second-order torque vanishes in the leading approximation.

However, we need to remember that the symbol κ that we have employed generically in the foregoing derivation does not have the same value in equations (10.139)–(10.141). In fact, when considering the case of an ellipsoidal, anelastic Earth with oceans, the incremental inertia tensor elements include contributions from ellipticity and anelasticity which differ from one to another of the three cases. In addition, there are the ocean tidal contributions which are strongly frequency dependent within each band, partly because of the nature of the dynamics of the oceans and also because of the resonances associated with the normal modes of the wobbles. So we now identify the type of forcing involved by adding the superscripts $(2m)$ with $m = 0, 1$, or 2 as appropriate, to the compliances; we also add a tilde symbol to indicate that the compliances are complex, the anelasticity and the ocean tide effects being out of phase with the forcing torque. If we now go back to Eq. (10.138) and take only the dominant part of the inertia tensor increment appearing in each term as before, the expression (10.142) gets modified to

$$\tilde{\Gamma}^{(2)}(t) \approx -i\Omega_0^2 A[(\tilde{\kappa}^{21}\tilde{\phi}^{21})\tilde{\phi}^{20} - (\tilde{\kappa}^{20}\tilde{\phi}^{20})\tilde{\phi}^{21} + 2(\tilde{\kappa}^{22}\tilde{\phi}^{22})\tilde{\phi}^{21*} - 2(\tilde{\kappa}^{21*}\tilde{\phi}^{21*})\tilde{\phi}^{22}]. \qquad (10.143)$$

This expression is still not quite explicit because the implication of the "frequency dependence of the compliances" is not clearly spelt out. It means actually that $\tilde{\kappa}^{2m}\tilde{\phi}^{2m}$ is merely an abbreviation for the following expression wherein each spectral component of $\tilde{\phi}^{2m}(t)$ appears multiplied by the value $\tilde{\kappa}^{21}(\sigma)$ of the compliance for the same frequency:

$$\tilde{\kappa}^{2m}\tilde{\phi}^{2m} \rightarrow \sum_{\sigma} \tilde{\kappa}^{2m}(\sigma)\tilde{\phi}^{2m}(\sigma)e^{i(\sigma\Omega_0 t - \alpha_\omega)}. \tag{10.144}$$

The exponential factor is simply another way of writing $e^{-i(\Theta_\omega t - \zeta_{2m})}$, as in Eq. (6.11). Those equations imply that $\alpha_\omega = \Theta_\omega(0) - \zeta_{nm}$. Thus,

$$\tilde{\kappa}^{21}\tilde{\phi}^{21}\tilde{\phi}^{20} \rightarrow \sum_{\sigma'} [\,\tilde{\kappa}^{21}(\sigma')\tilde{\phi}^{21}(\sigma')\tilde{\phi}^{20}(\sigma - \sigma')\,]e^{i(\sigma\Omega_0 t - \alpha_\omega)}, \tag{10.145}$$

where $\alpha_\omega = \alpha_{\omega'} + \alpha_{\omega-\omega'}$. Spectral expansions of the other terms in (10.143) may be written down similarly. On combining them we obtain the following expression for the coefficient of $e^{i(\sigma\Omega_0 t - \alpha_\omega)}$ in the spectral component of frequency $\sigma\Omega_0$ in $\tilde{\mathbf{\Gamma}}^{(2)}(t)$:

$$\tilde{\mathbf{\Gamma}}^{(2)}(\sigma) \approx \sum_{\sigma'} [\tilde{\kappa}^{21}(\sigma') - \tilde{\kappa}^{20}(\sigma - \sigma')]\tilde{\phi}^{21}(\sigma')\tilde{\phi}^{20}(\sigma - \sigma')$$

$$+ \sum_{\sigma'} 2\,[\tilde{\kappa}^{22}(\sigma - \sigma') - \tilde{\kappa}^{21*}(\sigma')]\tilde{\phi}^{21}(\sigma')\tilde{\phi}^{22}(\sigma - \sigma'), \tag{10.146}$$

on noting that the spectral amplitudes $\tilde{\phi}^{21}(\sigma')$ are all real. The phase α_ω is unique for a given frequency ω (or equivalently, σ in cpsd); in the second part of the above expression, $\alpha_{\omega-\omega'} - \alpha_{\omega'} = \alpha_\omega$, the minus sign of the second term being on account of the occurrence of $\tilde{\phi}^{21*}$ rather than $\tilde{\phi}^{21}$ in the last two terms of (10.143).

Since Eq. (10.146) contains differences of the compliances as factors, it might ordinarily be expected to be of very small magnitude, of the same order as, or smaller than, that of terms containing the product of a $\tilde{\phi}$ and $\tilde{m}$ or $\tilde{m}_f$. But the presence of an ocean tidal contribution makes it necessary to check the actual magnitudes before deciding whether (10.146) can be neglected.

Finally, one has to consider the contributions to $\tilde{\mathbf{\Gamma}}^{(2)}$ from the terms other than those containing $\tilde{\phi}^{21}$ and $\tilde{\phi}^{20}$ in (10.139) and (10.140). As already remarked, only the role of the FCN resonance in the former and the CW resonance in the latter could cause the resulting contributions to nutations and precession to be not entirely negligible.

10.10.3 The axial components H_3 and H_{f3}

Before going on to introduce the above expressions for the second-order terms in the equatorial torque into the equatorial components of angular momentum balance equations, we need to consider the third component of the torques exerted by the three potentials of orders 0, 1, and 2:

$$\Gamma_3^{20} = 0, \qquad \Gamma_3^{21} = -\frac{1}{2}i\Omega_0^2(\tilde{c}_3\tilde{\phi}^{21*} - \tilde{c}_3^*\tilde{\phi}^{21}),$$

$$\Gamma_3^{22} = 2\Omega_0^2[c_{12}(\tilde{\phi}^{22} + \tilde{\phi}^{22*}) + ic_{11}^S(\tilde{\phi}^{22} - \tilde{\phi}^{22*})]. \tag{10.147}$$

These are seen to be purely of the second order. Therefore they cause variations of H_3 only in the second order. Such variations are of no relevance to the nutation problem: we have seen above (see Subsection 10.10.1) that the equatorial components of the torque Eqs. (10.107) and (10.108) do not involve H_3 beyond the first order. The first-order quantities $H_3^{(1)}$ and $H_{f3}^{(1)}$ do, however, occur in the wobble equations taken to the second order, and therefore we consider first the third (axial) components of the torque equations.

From the Liouville equation,

$$\frac{dH_3}{dt} + \Omega_1 H_2 - \Omega_2 H_1 = \Gamma_3, \tag{10.148}$$

$$H_3 = H_3^{(0)} + H_3^{(1)}, \quad H_3^{(0)} = C\Omega_0, \quad H_3^{(1)} = (Cm_3 + c_{33} + C_f m_{f3})\Omega_0. \tag{10.149}$$

The components of **H** here and further below are again obtained by using $[C]$, $[C_f]$, $\mathbf{\Omega}$, and ω_f from Eqs. (7.111) and (7.112).

To the first order, Eq. (10.148) becomes $dH_3^{(1)}/dt = 0$. Hence we have in the frequency domain

$$Cm_3 + c_{33} + C_f m_{f3} = 0 \tag{10.150}$$

for all spectral components of non-zero frequency σ. The third component of the SOS equation similarly leads to $dH_{f3}^{(1)}/dt = 0$, and hence to

$$C_f(m_3 + m_{f3}) + c_{33}^f = 0. \tag{10.151}$$

This equation shows clearly that $m_3 + m_{f3} \neq 0$. The last two equations yield the following expressions for m_3 and m_{f3} in the frequency domain:

$$C_m m_3 = c_{33}^f - c_{33}, \quad \text{where} \quad C_m = (C - C_f), \tag{10.152}$$

$$C_f m_{f3} = \frac{1}{C_m}(C_f c_{33} - C c_{33}^f). \tag{10.153}$$

When these results are used, one finds from (10.149) that

$$H_3^{(1)} = 0. \tag{10.154}$$

As for the third component H_{f3} of the angular momentum of the fluid core, we find by using the expressions (7.111) and (7.112) for $[C_f]$, $\boldsymbol{\Omega}$, and ω_f that $H_{f3}^{(0)} = C_f \Omega_0$ and that

$$H_{f3}^{(1)} = \left[C_f (m_3 + m_{f3}) + c_{33}^f \right] \Omega_0 = 0. \qquad (10.155)$$

The vanishing of $H_{f3}^{(1)}$ is by virtue of Eq. (10.151) above.

11

Comparison observation-theory

11.1 Empirical models for nutations (such as IERS96)

In order to be very close to the nutation observation, one possibility is to consider a model with a very simple form and of which the parameters can be fitted on the observation. In that case, ignoring the a-priori corrections that are explained in the next paragraph, the parameters cannot be interpreted in terms of the physics of the Earth's interior. This empirical model cannot be considered as a geophysical model.

11.2 Estimation of basic Earth parameters from VLBI

We have seen in Chapter 5 that the rigid Earth nutations are quite well determined, with the exception of a scaling factor for the global Earth flattening. The precession and low frequency nutations due to the action of the degree two tesseral gravitational potential of the solar system bodies at the first order in the potential may be calculated to high accuracy as shown in that chapter.

Other contributions to Earth rotation with frequencies outside the low frequency band, in the celestial frame are produced by the other components of the potential. They have been computed in Chapter 10 and are outside of the scope of a fit to VLBI (Very Long Baseline Interferometry) data as they contribute to polar motions of the CIP and are not considered as nutations of the CIP. However, we must consider (1) the contributions of the order of 0.1 mas or smaller to a couple of low frequency nutations from the degree three tidal potential acting on J_3, and about 25 μas/year to precession from the degree four potential acting on J_4 (see Chapter 5). We must also consider (2) contributions of the second order in the perturbing potential, which arise from the action of each of the three parts (zonal, tesseral, and sectorial) of the potential on the increment to the density function that is produced by the tidal deformation induced by these three parts. They have been evaluated in Chapter 10 where it was shown that the effect on precession–nutation is practically nil because of mutual cancellations.

These effects, as well as the geodesic precession and nutation and the contribution from atmospheric pressure tides generated by solar thermal effects must be taken out of the observations before fitting any parameter for the nutation model by comparing the observations and the results of computations.

Furthermore, the ocean tide effects on the nutation must be estimated. The model in use in the computations performed for the model adopted by the IAU and the IUGG (MHB2000, Mathews *et al.*, 2002) is a model based on a three-layered Earth with coupling mechanisms at the CMB and the ICB. We will concentrate on that model. Other models can also be fitted on the observations but we will not deal with them in this book. Note that it is difficult to fit global parameters on data when a model based on deformation field computation is considered. Only very rough estimates of basic Earth parameters used in this theory can be obtained from geophysical data. Herring *et al.* (1991, 2002) and Mathews *et al.* (1991a and b, 2002) proceeded, instead, to carry out a least squares procedure for estimating the values of those of the parameters in their nutation theory to which the residuals of nutation amplitudes after the fitting were most sensitive. Note that, as explained below, one can only compare VLBI observations with models based on the development performed in the previous chapters of this book, after applying a number of different corrections. This requires iterations for the convergence of the fit as well as for consistency between the Earth's response to the ocean model and the nutational response to the tidal forcing.

Prior to the just-mentioned least squares determination of some parameters, a change in the dynamical flattening of the Earth had already been identified as a potential adjustment for a better matching between the theory and the observation. Wahr (1981a–d) has indeed noted from observational data on the precession rate, that the Earth dynamical flattening has to be higher than the hydrostatic equilibrium value by over 1%.

The second parameter considered is the ellipticity e_f of the core; this choice was based on the observation that the retrograde annual nutation is strongly influenced by the free core nutation (FCN) resonance as the FCN eigenfrequency is quite close to the retrograde annual. A rather small adjustment of e_f can change the eigenfrequency enough to eliminate the large residual of about 2 mas that was found on the retrograde annual nutation when considering a hydrostatic Earth nutation theory. It must be mentioned though that this is not the only mechanism by which the 2 mas residuals on the out-of-phase part of the retrograde annual nutation component can be explained. We show below how one can determine e_f in conjunction with other coupling phenomena.

Concerning the real and imaginary parts of the coupling constant K^{CMB} at the CMB and K^{ICB} at the ICB, no inputs are available from Earth models to provide reasonably stringent bounds on them, which is particularly the case for the mean squared magnetic field at the ICB for instance. Consequently, these coupling constants are further parameters used in the fit. On observing the combination of geophysical parameters appearing in these coupling constants and on considering additional information from theories such as that on visco-magnetic torque at the CMB and ICB, it is possible to get better constraints on the geophysical parameters. This has been done in the work of Mathews *et al.* (2002), Buffett *et al.* (2002), Herring *et al.* (2002), as well as some earlier work of these authors, or subsequent works by Guo *et al.* (2004), Mathews and Guo (2005), Koot *et al.* (2008, 2010, 2011), Koot and Dumberry (2011, 2013).

Assuming that the coupling constants at the CMB and ICB are related to electromagnetic coupling only at the core boundaries and that some model for the structure and strengths of the core magnetic fields crossing the CMB can be built, it is possible to reduce the number of parameters by one (the real part of the coupling constant K^{CMB} at the CMB is expressed as a function of the imaginary part). Confronting nutation theory (complemented by functional relationship between geophysical parameters) with observational data provides a means of obtaining estimates with relatively small uncertainties for such parameters, and therefore of better understanding the physics of the Earth interior.

The real part of K^{CMB} appears in combination with e_f, and the real part of K^{ICB} appears in combination with e_s. However, it is believed that the ICB is close to hydrostatic equilibrium, so that e_s can be fixed to its hydrostatic value. One can thus consider that the real and imaginary parts of the coupling constant K^{ICB} at the ICB determine the magnetic field down there and that, using a theoretical relation between the magnetic field contributions to the real and imaginary parts of K^{CMB}, has the real and imaginary parts of the coupling constant K^{CMB} at the CMB determine e_f as well as the magnetic field at the CMB.

Further candidates for estimation are the compliance parameters κ and γ appearing in the theory, which have uncertainties, primarily due to the contributions from anelasticity.

Other non-trivial contributions to nutations come from ocean tide effects. They have a very strong frequency dependence as they depend on the compliance parameters (frequency dependent body tide and load Love numbers) themselves, depending on resonances involving the basic Earth parameters. These effects are a function of the ocean tide admittance (i.e., the tide height per unit strength of the TGP component that excites the tide) that has large variations (1) with frequency across the diurnal tidal band, associated with the FCN resonance of the Earth, and (2) with the ocean dynamics factors unconnected with the FCN. The inclusion of ocean tide effects is thus not easy and is described by an empirical formula representing the frequency dependence of the admittance.

The values for some of the basic Earth parameters have been provided in Table 10.1 in Section 10.4.

11.3 Consideration of the ocean tide effects on nutation

Ocean tide effects on nutations are considered (1) through the changes in the inertia tensors of the Earth, the outer core, and the inner core due to loading at the Earth's surface, and (2) through the contribution from ocean tide angular momentum to the global Earth angular momentum. The tidal changes in the inertia tensors of the whole Earth, the fluid core, and the deformable inner core have been considered by Sasao and Wahr (1981) and explained in Chapter 9:

$$\tilde{c}_3 = -A\{\kappa(\tilde{\phi} - \tilde{m}) - \xi \tilde{m}_f\} + \tilde{c}_3^O, \tag{11.1}$$

$$\tilde{c}_3^f = -A_f\{\gamma(\tilde{\phi} - \tilde{m}) - \beta \tilde{m}_f\} + \tilde{c}_3^{fO}, \tag{11.2}$$

wherein $\tilde{c}_3$ and $\tilde{c}_3^f$ stand for the complex combination $(c_{13} + ic_{23})$ and $(c_{13}^f + ic_{23}^f)$ of elements of the inertia tensor of the whole Earth and the fluid core as in Chapter 7, and where the superscript O identifies the contributions from ocean loading. According to Sasao and Wahr (1981),

$$\tilde{c}_3^O = -A(\tau - \chi)\phi_L, \qquad \text{and} \qquad \tilde{c}_3^{fO} = A_f \eta \phi_L, \tag{11.3}$$

where ϕ_L is a potential, in dimensionless units, representing the amplitude of the surface load. The symbols τ, χ, and η are those of Wahr and Sasao (1981), and are unrelated to the same symbols used elsewhere in this book. Here, $\tau = (\Omega_0^2 a^5/3GA)$, and $(-\chi/\tau)$ is simply the load Love number k' with the contributions from wobbles turned off. Thus $\tilde{c}_3^O$ is made up of two parts: the direct contribution to the inertia tensor from the ocean tide mass, and the contribution due to load-induced deformation.

The increments to the compliances κ and γ due to ocean loading (OL) are thus given by

$$\Delta\kappa^{OL} \equiv -\tilde{c}_3^O/[A(\tilde{\phi} - \tilde{m})] \quad \text{and} \quad \Delta\gamma^{OL} \equiv -\tilde{c}_3^{fO}/[A_f(\tilde{\phi} - \tilde{m})]. \tag{11.4}$$

The effect of the inner core being very small, it is neglected in this computation.

Observational estimates for the angular momentum are available only for a few of the largest tides such as K_1, P_1, O_1, and Q_1 in the diurnal band, see for instance Chao *et al.* (1996). From these angular momentum estimations, one can deduce the contributions to the compliance $\Delta\kappa^{OL}$ for the respective tides. Since $\tilde{c}_3^{fO} = (-A_f/A)[\eta/(\tau - \chi)]\tilde{c}_3^O$ according to (11.3), one can deduce the contribution to the compliance $\Delta\gamma^{OL}$. The values thus obtained for $\Delta\gamma^{OL}$ and $\Delta\kappa^{OL}$ exhibit considerable variation from one tide to another. This variation is a reflection of the well-known frequency dependence of the ocean tide admittance (amplitude of the degree two order one part of the ocean tide per unit amplitude of the exciting TGP component). In order to obtain values for other frequencies, one needs a formula expressing the frequency dependence in general. Mathews *et al.* (2002) have estimated an empirical formula, as in Wahr and Sasao (1981), separating the ocean loading admittance into a product of two factors: (1) a frequency dependence due to the FCN resonance f_{FCN}, and (2) a complex ocean dynamics factor f_{OD}:

$$\tilde{c}_3^O / (A\tilde{\phi}) = f_{\text{FCN}} \, f_{\text{OD}}. \tag{11.5}$$

Values of f_{FCN} have been given for a set of 11 tidal components by Wahr and Sasao (1981). Mathews *et al.* (2002) have found that these can be well fitted by the expression

$$f_{\text{FCN}}(\sigma) = c_0 + c_1 A_e(\sigma), \tag{11.6}$$

with, as shown by Dahlen (1976),

$$A_e(\sigma) = \frac{1 + k(\sigma) - h(\sigma)}{1 - (3\rho_w/5\rho_E)(1 + k'(\sigma) - h'(\sigma))}, \tag{11.7}$$

where k and h are body tide Love numbers for excitation by a degree two order one potential, k' and l' are the corresponding load Love numbers for the solid Earth, and ρ_w, ρ_E are the density of seawater and the mean density of the Earth, respectively. $A_e(\sigma)$ represents the dynamical response of the ocean including the effects of the ocean bottom deformations. To determine $A_e(\sigma)$ as a function of frequency, one can use the classical expression of the body tide Love numbers as linear combinations of wobble admittances, and their counterparts for the load Love numbers.

The behavior of the factor $f_{\text{OD}}(\sigma)$ representing ocean dynamic effects is expected to be smooth for low spherical harmonic orders (see, for instance Wahr and Sasao, 1981) and therefore, one can assume linear forms

$$f_{\text{OD}}^R(\sigma) = d_0^R + d_1^R \sigma, \tag{11.8}$$

$$f_{\text{OD}}^I(\sigma) = d_0^I + d_1^I \sigma, \tag{11.9}$$

for the real ^{R} and imaginary ^{I} parts of this factor. The coefficients in these expressions are iteratively fitted together with the other basic geophysical parameters in the least squares fits.

The next concern is about the role of the angular momentum $\tilde{h}$ carried by the ocean tidal currents. As shown by Wahr and Sasao (1981), this angular momentum enters into the dynamical equation simply through the replacement of $\tilde{c}_3^O$ by $(\tilde{c}_3^O + \tilde{h}/\Omega_0)$. The $\tilde{h}$ term, in effect, adds a further increment $\Delta\kappa^{OC}$ to κ while leaving all other compliances unaffected:

$$\Delta\kappa^{OC} \equiv -(\tilde{h}/\Omega_0)/[A(\tilde{\phi} - \tilde{m})]. \tag{11.10}$$

To derive an empirical formula from which this increment may be evaluated for any tidal constituent, one adopts essentially the same procedure as was described above for ocean loading. The only difference is that we now use the Chao *et al.* (1996) values of $\tilde{h}$ instead of those of $\tilde{c}_3^O \Omega_0$. The f_{OD} factor is a function of other parameters to be fitted while the FCN remains fixed at each iteration.

In the least squares fit performed by Mathews *et al.* (2002), it was found that the fit is optimal with a scale factor of about 0.8 for the current admittance function. Due to the present uncertainties and the numerical difficulties of computing the ocean current angular momentum from ocean tide maps, one can consider that this scaling might not be unreasonable.

At this point, the dynamical equations including the above effects and electromagnetic couplings are fully determined, given the values assigned to the basic Earth parameters, provided one knows the ocean tide admittances (or the iteration for getting them). An iterative process is thus part of the overall procedure that is employed.

11.4 Inclusion of anelasticity effects

Mantle anelasticity causes a small phase lag in the Earth's response to periodic forcing and alters the magnitude of the response as seen in Chapter 8. The effect is different for different frequencies; the higher the frequency, the smaller the effect. Anelasticity, as seen in Chapter 8, can be introduced by considering that the shear and bulk moduli in the mantle are complex and frequency dependent. The variations in the shear and bulk moduli depend on the anelasticity model chosen, as explained in Chapter 8. The compliances appearing in nutation theory are computed initially for an elastic Earth model, such as PREM of Dziewonski and Anderson (1981), by integrating the equations of tidal deformation. Then mantle anelasticity is introduced by taking the compliances in these equations to be complex and frequency dependent. In MHB2000, a power law dependence, proportional to $(\omega/\omega_0)^\alpha$, has been adopted, with $\alpha = 0.15$ and the reference period of 200 seconds ($\omega_0 = 2\pi/200$), as the default options. After computing the increments to the displacement fields from increments to the real and imaginary parts of the shear modulus, the corresponding increments to the compliances in the diurnal band are computed as in Sasao *et al.* (1980). Thereafter the contributions from anelasticity to the nutation amplitudes are readily obtained. For the construction of MHB2000, Mathews *et al.* (2002) have considered the effect of switching to other values for the compliance parameters and noted that the effect is just to change their real and imaginary parts by appropriate factors. They have then explored how the quality of the fit is altered on varying the scale factors from unity and found that it degrades rapidly as either of the parameters deviates significantly from unity. The fit was found to be best for a scale factor equal to 1 for the real part and 0.96

for the imaginary part. This marginal scaling has been incorporated into the model MHB2000.

11.5 Nutation using estimated parameters (MHB2000)

In constructing the nutation model MHB2000 on which the current IAU model IAU2000A is based, Mathews *et al.* (2002) employed a least squares procedure to obtain the best fit of the results of their theory to estimates of nutation and precession (see Herring *et al.*, 2002) from a long VLBI data set which was then up-to-date. The least squares process involves repeated adjustments and optimization of the values of a few Earth parameters, namely, the ellipticities e, e_f, the compliances κ, γ, the imaginary part of K^{CMB}, and the real and imaginary parts of K^{ICB}, while retaining values based on the PREM Earth model for all the other parameters. The process was iterative in view of the non-linear dependence of the nutation amplitudes on the parameters being optimized (large number of iterations).

Because ocean effect modeling also involves Love numbers and resonance parameters, within each step of the iterative process another iterative loop was necessary. Additionally, the ocean responses are only known observationally for a very small number of tidal frequencies and the responses at other frequencies must be interpolated, a process that involves additional parameters to be estimated.

The final results are a set of "best fit" values for the basic Earth parameters and a theoretical nutation series based on these parameters. Mathews *et al.* (2002) have proposed such a model called MHB2000, which has been adopted by the IAU and the IUGG.

The best-fit values of e and e_f are higher than their hydrostatic equilibrium values; these are interpreted as effects of mantle convection, as mentioned earlier. The excess (non-hydrostatic) part of e_f corresponds to an extra difference of about 350 m between the equatorial and polar radii of the CMB (it had been estimated to be 500 m when the electromagnetic coupling was ignored). The estimates obtained for K^{CMB} and K^{ICB} imply, if they are interpreted as being related to electromagnetic coupling only, high root-mean-square (rms) values for the magnetic fields at the CMB and ICB. The values are higher than suggested by geodynamo models, and also, in the case of the CMB, higher than the extrapolated values from satellite observations of magnetic fields along the satellite's orbit. No independent observations are available for the fields at the ICB. The higher estimate for the rms radial field strength at the CMB could be considered, at least in part, as a reflection of the short spatial wavelength contributions that are not detectable at satellite heights. Another explanation is that the coupling constants might include the effect of topographic coupling (Wu and Wahr, 1997) or the effect of viscous coupling at the boundaries of the fluid core. This last coupling mechanism has been considered by Mathews and Guo (2005) and Deleplace and Cardin (2006).

The estimates made by Mathews *et al.* (2002) for the strengths of magnetic fields at the CMB and the ICB were thus based on the hypothesis that the torques generated by differential wobbles between the fluid core and the mantle/inner core are only due (in addition to the inertial coupling related to the boundary flattenings) to the magnetic fields crossing the boundaries. As seen in Chapter 10, viscous drag may, however, contribute significantly to the torques. Examination of both electromagnetism and viscosity effects for the interpretation of the coupling constants obtained from the observation was thus performed as well (see Mathews and Guo, 2005, Koot *et al.*, 2010, 2011, Koot and Dumberry, 2011, 2013). These interesting works have placed new constraints on the basic Earth parameters and new interpretations have been invoked.

It is interesting to note that the use of recent observations does not change the CMB coupling constant K^{CMB} much, leaving its rather stable value for geophysical interpretation. As also discussed in Chapter 10, topographic torque at the CMB could also be involved in the interpretation of the coupling constant. However the values of the ICB coupling constant are significantly different. This can be explained by the increasing length and increasing precision of the VLBI series used as well as by the use of different inversion strategies (least squares versus Bayesian approaches). The interpretation concerning the new values at the ICB implies that a combination of both electromagnetic and viscous couplings is required.

The recently obtained values of the coupling constants place bounds on the viscosities at the core-mantle and inner core boundaries. These bounds not only rule out the extremely high viscosity called for in recent studies involving the "effective viscosity" related to turbulence effects, but are in agreement also with rheology estimations coming from the seismic normal mode studies. If viscosity in the inner core is considered, a strong constraint on its value is found, which has to be in the range 2–7×10^{14} Pa s and which is perfectly in agreement with viscoelastic rheology models for the inner core from seismic normal modes observations. In that case, the RMS strength of the radial magnetic field at the ICB has to be between 4.5 and 6.7 mT, see Section 10.4.

11.6 Free core nutation model

The free core nutation eigenmode not only induces a resonance in the forced nutation, but can also be excited at an observable level as seen from the VLBI data. In the remaining residuals between the observation and the theory, one has an important contribution at the FCN period at the level of a couple of tenths of mas. This is due to the excitation of a free nutation at the FCN period. The excitation mechanism is believed to be due to the atmosphere and is difficult to predict. But the amplitude and phase (or cosine and sine components), changing with time, can be estimated from the observation, as has been done by Herring *et al.* (2002). The estimations are not performed on the whole data set but rather on piecewise parts of it. Their estimations of the amplitudes and phases are determined on a two-year basis; Lambert and Dehant (2007) used a ten year sliding window displaced by one year (chosen to ensure that the free motion and the annual oscillation are decorrelated). Also Krásná *et al.* (2013) used a sliding window over two years and displaced by one year. The reason for such a time evolution of the FCN is that the excitation function is not stable in time. Scientists are currently investigating the possibility of using the atmospheric angular momentum function for deducing the FCN free nutation amplitude as explained in Chapter 9. Other scientists (e.g. Malkin, 2013a and b) have come up with the idea of explaining the FCN amplitude and phase variations as effects of magnetic jerks.

11.7 Present-day precision

The present-day precision of the nutation theory is at the level of milliarcseconds in the time domain. In the frequency domain, the differences are at the level of a tenth of a microsecond, except for a couple of nutations of which the period is greater than 400 days: the 9.3 year nutation and the 18.6 year nutation. The differences may be due to the observational precision at these periods. The long-term nutations are indeed not very well determined from a 20 year data set, and furthermore, imperfections in the stability of the VLBI network could also have been the cause of changes in the nutation amplitude of several tens of a microsecond (Feissel-Vernier *et al.*, 2005). This can be understood by examining the poor geometry of the VLBI network at the beginning of the VLBI observations. The observation strategies used at present avoid such contamination.

11.8 Future decimal research

Concerning observation, one should further consider the effects of the departures from stability of the ground station networks as well as those of the radio sources. Furthermore, new methods/strategies for evaluating the amplitudes and phases of the nutation components from VLBI observations must be considered, which should account for the time-dependence of the error in the observation (accounting for the fact that the precision of individual VLBI observation sessions is becoming better and better). The consideration of decontamination from Earth rotation, as proposed by the new definition of precession and nutation (see Chapter 12) using the non-rotating origin (NRO), is a development already in use.

Concerning the rigid Earth nutation theory, the precision in the effects that have been accounted for in the computation is high enough, but any physical phenomenon that has not been considered yet might produce an effect. We do not identify any such phenomena at the moment.

Concerning the non-rigid Earth transfer function, the present theory is precise enough if basic Earth parameters are estimated from the data. We see future improvement either in the procedure of evaluating these parameters from the improving data (for instance using a Bayesian approach), or in the theory considered to interpret these coefficients (for instance, in the theory used to interpret the observation of the coupling constants at the CMB and ICB).

The numerical integration method for computing the nutations and the tides does not provide results competing with the MHB2000 model. Nevertheless, attempts to model electromagnetic coupling and viscous coupling at the CMB and ICB have been made by introducing boundary layer mechanisms and by solving the induction equation for the relation between the induced magnetic field and the velocity field (Huang *et al.*, 2004, 2011). In particular, the topographic coupling should be studied as well, and should be incorporated in both the semi-analytical approach of the MHB2000-type of modeling and in the numerical integration method. This last approach will necessitate further developments in spherical harmonics or even a finite element procedure.

An approach in which all second-order effects coupling rigid Earth and non-rigid Earth nutation theories are considered simultaneously can also be envisaged in the future. This is, for instance, the case of the approach used by Getino and Ferrandiz (1991, 1994, 1995, 1999, 2000, 2001), Getino *et al.* (2001), Escapa *et al.* (2001), and Ferrandiz *et al.* (2004), all using the Hamiltonian approach.

Further evaluation of the free mode contributions to nutations must be undertaken in order to be able to better predict the nutations.

12

Conventions

This chapter has been written in order to explain to the reader the new concepts, in use from 1 January, 2003, for the precise definition of precession and nutation and other aspects of Earth rotation. More details, if desired, may be found in the references mentioned in the text. Additionally, the International Earth rotation and Reference system Service (IERS) has published the *Conventions* (IERS Conventions 2003, 2010) in which the new paradigm is given, as well as *Technical notes* (IERS TN 29) in which one finds articles on various aspects of the new paradigm by Capitaine, Capitaine *et al.*, Fukushima, Guinot, Gontier, Kovalevsky, McCarthy and Capitaine, Seidelmann, and Wallace. Furthermore, other articles can also be found useful, such as those of Capitaine (1986), Capitaine and Guinot (1988), Capitaine *et al.*, 2003a–d, 2005a and b, and Kaplan, 2003, 2005a, and b.

12.1 Definition of equator

As seen in Chapter 2 and Chapter 3 (see also Appendix C), the ecliptic is the plane of the orbit of the Earth–Moon barycenter around the Sun. It is the plane perpendicular to the mean orbital angular momentum vector of the Earth–Moon barycenter. It is a conceptual plane of which the realization depends on a specified epoch and a specified date. When not otherwise specified, the term "ecliptic" consists in the fixed ecliptic of J2000 and is referred to the celestial frame realized by coordinates of the radio sources of the ICRF by a set of fixed angles. The actual ecliptic plane undergoes a very slow motion (a very slow rotation mainly) in space as it is affected by planetary precession in longitude and obliquity.

Note that the ecliptic of the "inertial" system (space-fixed as defined in Section 2.22.1) is not the orbital plane of the Earth–Moon barycenter around the Sun, as the angular momentum vector of a particle orbiting a central mass in a moving (rotating) plane is not normal to that plane.[1] Previously (before the existence

---

[1] The ecliptic is the plane perpendicular to the mean heliocentric orbital angular momentum vector of the Earth–Moon barycenter in the BCRS, as mentioned in the IAU 2006 Resolution B1.

of precise ephemerides) the ecliptic was defined in a "rotating" sense as it was the plane of the Earth–Moon barycenter motion. In fact, the actual Earth–Moon barycenter follows a very complex orbit, so that any plane that truly contains it, wobbles all over the place with many frequencies.

The mean orbital angular momentum vector of the Earth–Moon barycenter around the Sun is not independent of the motion of the orbital plane. For the mean ecliptic in the inertial sense, Standish (e.g., 1998a–d) took a 6000-year ephemeris, computed the node and obliquity of the instantaneous normal plane for many thousands of points over these 6000 years, and then fitted them with time polynomials and Fourier series of about 60 terms. The constant and linear rates of the polynomials define the "mean" node and obliquity and their rates for the inertial ecliptic.

A fixed ecliptic is the ecliptic of a given ephemeris at an adopted epoch (as J2000). Such a fixed ecliptic has a specified obliquity and crosses the ICRF (see Section 3.1.3) equator at a specified offset from the ICRF origin.

As seen in Chapter 2 and Chapter 3, the Earth's equator used in wobble or nutation determination theories is the plane perpendicular to the figure axis of the Earth passing through the center of the Earth. In practice, the equator is a conceptual plane of which the realization depends on a specified epoch and a specified date; when this is not specified, it consists in the true equator of date and is the plane formed by the x- and y-axes of the terrestrial frame realized by the coordinates of points of the ITRF.

The term "equator" is also used to designate any conceptual plane that is perpendicular to the direction of any specified pole. In particular, the celestial intermediate equator is perpendicular to the direction of the celestial intermediate pole (CIP) at some epoch as is used in this chapter. It is synonymous with the instantaneous equator or true equator of date, or equator of the CIP.

The celestial equator is the projection on to the celestial sphere of the Earth's equator.

The instantaneous position in space of the equator resulting from the secular precession is the *mean* equator of date, while the *true* equator of date is the instantaneous position of the equator which includes the effect of nutation too. The true/mean equator of epoch (e.g., J2000) represents the position occupied by the true/mean equator at that fixed epoch, as explained in Section 2.3 and in Section 3.1.1.

12.2 Definition of equinox

The equinoxes are the two points on the celestial sphere at which the ecliptic intersects the celestial equator.

An equinox is also the time at which the Sun passes through either of these intersection points, i.e., when the ecliptic longitude of the Sun is 0° or 180°. That equinox at which the Sun crosses from the south of the equator to the north is called the vernal equinox, or the vernal point, or the spring equinox.

The equinoxes are defined with the adjective *true* or *mean* referring to the true or mean equator, and with the complement *of date* or *of epoch* if one refers to the ecliptic of date or the ecliptic of epoch as explained in Sections 2.3 and in Section 3.1.1. In summary, the true equator of date is the plane corresponding to the instantaneous positions of the celestial equator of date (it undergoes the motion due to precession and nutation); and the true equinox of date is the intersection between the true equator of date and the ecliptic of date; the mean equator and equinox of date refer to the same quantities but determined by ignoring small variations of short period (nutation) in the motions of the celestial equator (they undergo motion due to precession only). The mean/true equators and equinoxes of epoch refer to the same quantities as above for a particular epoch. It is usually referred to J2000.0, at 12 hours UT1, 1 January, 2000. UT1 is defined as not being affected by precession, nutation, and polar motion.

As specified above and in Appendix C, associated with the concepts of the ecliptic, there is a rotating dynamical equinox and an inertial dynamical equinox. We understand the equinox of a particular date or of epoch (of J2000) to be the inertial dynamical mean equinox, defined to be the intersection between the mean equator and the ecliptic, both of that date or epoch (see Section 3.1.3).

12.3 Definition of CIP related to precession and nutations

12.3.1 Link connecting GCRS–BCRS–ICRS–ICRF

The international celestial reference system (ICRS) is the idealized barycentric coordinate system to which celestial positions are referred. It is kinematically non-rotating with respect to the ensemble of distant extragalactic objects as seen in Section 3.1. See as well the report of the IAU WG on nomenclature for fundamental astronomy (NFA, 2006a–g). It has no intrinsic orientation but was aligned close to the mean equator and dynamical equinox of J2000.0 for continuity with previous fundamental reference systems. Its orientation is independent of epoch, ecliptic, or equator and is realized by a list of adopted coordinates of extragalactic sources.

The barycentric celestial reference system (BCRS) is a system of barycentric space–time coordinates assumed to be oriented according to the ICRS axes. See as well the report of the IAU WG on NFA (2006a–g).

The geocentric celestial reference system (GCRS) is a system of geocentric spatial coordinates defined such that the transformation between BCRS and GCRS spatial coordinates contains no rotation component, so that GCRS is kinematically non-rotating with respect to BCRS, as was seen in Section 3.1, as well as in the report of the IAU WG on NFA.

The international celestial reference frame (ICRF) is a set of extragalactic objects whose adopted positions and uncertainties realize the ICRS axes and give the uncertainties of the axes. It is also the name of the radio catalog whose set of 212 defining sources is currently the most accurate realization of the ICRS. Note that the orientation of the ICRF catalog was carried over from earlier IERS radio catalogs and was within the errors of the standard astrometric and dynamic frames at the time of adoption. Successive revisions of the ICRF are intended to minimize rotation from its original orientation. Other realizations of the ICRS have specific names (e.g. Hipparcos celestial reference frame). See the report of the IAU WG on NFA, as well as the next paragraph.

12.3.2 Realization of the ICRS

As explained in Sections 3.1.3 and 3.1.6 and Section 4.2, the principal plane of the ICRF was defined using the coordinates of the radio sources employed for VLBI observations, determined in agreement with the definition of the ICRS, so that it realizes the idealized solar-system-barycentric coordinate system to which celestial positions are referred. It has no intrinsic orientation but was aligned to have its principal plane close to the mean equator and its first principal axis close to the inertial dynamical equinox of J2000 for continuity with previous fundamental reference systems.

This does not mean that the present-day realization of the ICRS will be the same as when it was adopted: tiny deviations occur due to additional accumulation of VLBI data. In fact, offsets between the CIP at J2000 and the Z-axis of the ICRF have been determined. The present-day coordinates of the pole must take account of such offsets.

These constant offsets have been determined from VLBI as being equal to 17.2 mas and 5.1 mas, as determined by Feissel and Mignard (1998), and to 16.6 mas and 6.8 mas (determined by Herring, see IERS Conventions 2010 (Petit and Luzum, 2011) or 2000 (McCarthy *et al.*, 2003)) as adopted by the IAU in 2000, and from LLR of 17.7 mas and 5 mas (Chapront *et al.*, 2002). In the IERS Conventions 2010 which are the most up-to-date, one finds -16.617 mas for X and -6.951 mas for Y.

12.3.3 Transformation between celestial and terrestrial frames

As explained in Sections 2.12, 2.15, and 2.25 and Section 3.2.3, the CIP is a conceptual pole (see Capitaine, 2000, Capitaine *et al.*, 2000); its motions in the ICRF with periods greater than 2 days are defined to constitute the precessional and nutational motions of the Earth. They correspond to the precession–nutation of the ITRF pole with the same range of periods. The position of the CIP is represented by two coordinates (X, Y) in the ICRF, which includes a constant offset from the pole of the ICRF besides the lunisolar precession and nutation in longitude and obliquity.

Reduction of observational data requires the use of the transformation between the ICRF and the ITRF; it involves a succession of elementary rotations (represented by matrices) which may be chosen in different ways. They may, for instance, be rotations through the classical Euler angles (Ψ, Θ, φ) see Fig. 3.2.

The following equation gives the classical representation of the way to pass from the TRF to the CRF, in terms of the Euler angles:

$$[TRF] = R_3(\varphi(t))R_1(\Theta(t))R_3(\Psi(t))[CRF].$$ (12.1)

Here R_1, R_2, and R_3 are rotations around the x-axis, the y-axis and the z-axis, respectively (see Mueller, 1981, or also Appendix A, Eqs. (A.2)–(A.5)) and the successive rotations are about the new orientations of the axes after the earlier rotations.

An alternative expression already partly explained in Chapter 3 is in terms of another series of three rotations: (1) a rotation which carries the pole of the CRF into the intermediate pole CIP; (2) a rotation around the CIP axis, representing the axial rotation of the Earth; and (3) a rotation which brings the CIP pole into coincidence with the pole of the ITRF. These successive rotations are represented by one of two forms given by the equations below:

$$[TRF] = W'(t)^{-1}R_3(GST)PN'^{-1}(t)[CRF],$$ (12.2)

$$[TRF] = W^{-1}(t)R_3(\theta(t))Q^{-1}(t)[CRF],$$ (12.3)

where the first form has been presented in Chapter 3 (Eq. (3.11)), and the second involves the non-rotating origins and the Earth rotation angle (ERA). GST is the Greenwich sidereal time (also called the Greenwich apparent sidereal time, GAST) of date, i.e. the angle on the CIP equator (or true equator of date) between the true equinox of date (γ_D) and the prime meridian, which is the origin of longitude in the terrestrial frame (direction of Ox), and is very close to the terrestrial intermediate origin (TIO) meridian (the difference, on the equator, is only a small quantity s' that is defined later, see Eqs. (12.25)–(12.27)). GST is related to the Greenwich mean sidereal time (GMST) by (see also Fig. 12.8):

$$GST = GMST + [\alpha(\gamma_A)]_p,$$ (12.4)

where γ_A is the mean equinox of date, $\alpha(\gamma_A)$ is its right ascension relative to the true equinox and equator of date, and the subscript p stands for "periodic part." (GMST is said to be GST corrected for the "equation of the equinox" (EE in Fig. 12.8)).

The transformation (12.2) involves an intermediate frame of which the (x, y)-plane is the CIP equator but with axis directions related to the true equinox of date, while (12.3) is with reference to the same intermediate frame based on the CIP as previously but with axis directions involving the NRO (adopted by the IAU in 2000 and the IUGG in 2003).

Note that in the IERS Conventions 2010 (Petit and Luzum, 2011) and hereinafter, the inverses of the last two forms of the transformation are used:

$$[CRF] = PN'(t)R_3(-GST)W'(t)[TRF], \tag{12.5}$$

$$[CRF] = Q(t)R_3(-\theta(t))W(t)[TRF], \tag{12.6}$$

wherein the obvious relation $R_i^{-1}(\alpha) = R_i(-\alpha)$ has been made use of, PN' and $W'(t)$ introduced in Chapter 3 (Eqs. (3.11), (3.12), and (3.14)), and of $Q(t)$ and $W(t)$ defined in the "non-rotating origin" approach below.

Note that the matrix PN'^{-1} in the transformation (12.2) from the CRF to the TRF represents the precession P followed by the nutation N, and is sometimes denoted by PN:

$$PN(t) = N(t)P(t) = PN'(t)^{-1}, \quad PN'(t) = P(t)^{-1}N(t)^{-1} = PN(t)^{-1}, \tag{12.7}$$

and

$$W'(t) = R_2(x_p)R_1(y_p). \tag{12.8}$$

Sometimes one finds in the literature $W'(t) = R_1(y_p)R_2(x_p)$, as x_p and y_p are considered as small quantities. However, in the present day, when considering all second-order effects and the existence of s', we use Eq. 12.8 only, as in the IERS Conventions.

12.4 Classical representations of the precession–nutation matrix

The precession–nutation matrix $PN(t)$ carries one from the CRF ($OXYZ$-frame) to the intermediate frame referred to the CIP equator and having its x-axis toward the true equinox of date (γ_D). It can be expressed in many different ways:

- Firstly, following Lieske *et al.* 1997, we can use the angles ζ_A, z_A, and θ_A as defined in Fig. 3.5 of Chapter 3, for going from the mean equinox of epoch (γ_0) to the mean equinox of date (γ_A), and the angle of nutation ($\Delta\psi$) in longitude and in obliquity ($\Delta\epsilon$) for going from γ_A to the true equinox of date (γ_D). Lieske *et al.* (1977) have used this representation together with the angles ζ_A, θ_A, and z_A; the angle ζ_A is along the mean equator of epoch, from the mean equinox of epoch (γ_0) to the meridian passing through the poles P_0 and P_D of both mean equators of epoch and of date; the angle θ_A is the angle of rotation about OQ between the mean equator of epoch and the mean equator of date; the angle z_A is the angle of rotation about OP_D along the mean equator of date, which takes one from the mean equinox of date (γ_A) to the true equinox of date (γ_D); the dashed line is the line of fixed longitude in the geocentric celestial reference frame (GCRF), which passes through P_D, or equivalently the perpendicular to the mean equator of epoch and to the mean equator of date passing through P_0 and P_D; F_0 and F_D are the intersections of the dashed curve with the mean equators of epoch and date, respectively; the arc lengths from P_0 to F_0 and P_D to F_D are $\pi/2$. We recall that P_0 is the pole of the GCRF and P_D is the Earth's pole of date. The dashed line is the line of fixed longitude in the GCRF, which passes through P_D. F_0 and F_D are the intersections of the dashed curve with the mean equators of epoch and date, respectively. The arc lengths from P_0 to F_0 and P_D to F_D are evidently $\pi/2$. The representation of the precession–nutation matrix in terms of these quantities is now as follows:

$$PN(t) = R_1(-\epsilon_A - \Delta\epsilon)\, R_3(-\Delta\psi)\, R_1(\epsilon_A)\, P(t), \tag{12.9}$$

$$P(t) = R_3(-z_A)\, R_2(\theta_A)\, R_3(-\zeta_A), \tag{12.10}$$

where, as presented in Fig. 3.5, $(90° - \zeta_A)$ is the angle on the mean equator of epoch from the mean equinox of epoch (γ_0) to the node of this equator on the mean equator of date (denoted by Q in Fig. 3.5), θ_A is the angle between the mean equators of epoch and date, and $(90 + z_A)$ is the angle on the mean equator of date from Q to the mean equinox of date (γ_A), ϵ_A is the obliquity of the mean equator of date on the ecliptic of date, and $\Delta\epsilon$ and $\Delta\psi$ are the nutations in obliquity and longitude (see Woolard, 1953, Lieske *et al.*, 1977, Lieske, 1979, Capitaine, 1990).

• Alternatively, we can use the angles Π_A, π_A, and $-\Lambda_A$ as defined in Fig. 3.1 for going from the mean equinox of epoch (γ_0) to the mean equinox of date (γ_A), and $\Delta\psi$ and $\Delta\epsilon$ for going from γ_A to the true equinox of date (γ_D). The transformation is given in terms of these angles by:

$$PN(t) = R_1(-\epsilon_A - \Delta\epsilon)\, R_3(-\Delta\psi)\, R_1(\epsilon_A)P(t), \tag{12.11}$$

$$P(t) = R_1(-\epsilon_A)R_3(-\Lambda_A)\, R_1(\pi_A)\, R_3(\Pi_A)\, R_1(\epsilon_0), \tag{12.12}$$

where ϵ_0 is the obliquity of the mean equator of epoch on the ecliptic of epoch, Π_A is the angle on the ecliptic of epoch from γ_0 to the intersection between the ecliptic of epoch and the ecliptic of date (denoted by N in Fig. 3.1), π_A is the angle between the ecliptic of epoch and the ecliptic of date at N, and Λ_A is the angle on the ecliptic of date from N to the node γ_A of this ecliptic on the mean equator of date. The difference between Λ_A and Π_A is the general precession p_A (see Lieske *et al.*, 1977, Bretagnon *et al.*, 1997).

• As yet another alternative, we can use the angles ψ_A, ω_A, and χ_A as defined in Fig. 3.1, for going from the mean equinox of epoch (γ_0) to the mean equinox of date (γ_A), and use $\Delta\psi$ and $\Delta\epsilon$ as before to go from the mean equinox of date (γ_A) to the true equinox of date (γ_D). The resulting representation is as follows:

$$PN(t) = R_1(-\epsilon_A - \Delta\epsilon) \, R_3(-\Delta\psi) \, R_1(\epsilon_A) \, P(t), \qquad (12.13)$$

$$P(t) = R_3(\chi_A) \, R_1(-\omega_A) \, R_3(-\psi_A) \, R_1(\epsilon_0), \qquad (12.14)$$

where ψ_A is the so-called luni-solar precession, i.e. the angle on the ecliptic of epoch from γ_0 to the node of the mean equator of date on the ecliptic of epoch (denoted γ_1 in Fig. 3.1), ω_A is the obliquity of this equator on the ecliptic of epoch, χ_A is the angle on the mean equator of date from γ_1 to the mean equinox of date γ_A (see Lieske *et al.*, 1977, Capitaine, 1990, Bretagnon *et al.*, 1997).

It is not difficult to relate the angles appearing in the different representations by using the relevant spherical triangle formulas. Table 12.1 summarizes the definitions of all these angles.

12.5 Definition of CIO (non-rotating origin)

UT1, the time based on Earth rotation (as defined before 2003, see Section 2.7), is the Greenwich apparent sidereal time (GAST, or simply GST). GST is the arc length measured along the CEP equator of date from the equinox of date to the prime meridian, which marks the Ox-axis that serves as the origin of longitude. If the axis of rotation (and hence the equator perpendicular to it) had fixed orientations in space and if the ecliptic remained stationary, GST thus defined would give the angle of Earth rotation exactly.

However, the equatorial plane is moving in space because of precession and nutation, and this gives rise to small motions of the equinox along the equator, which cause small changes to the length of the arc which defines GST, though they are not caused by axial rotation around the pole of the equator. GST is therefore no longer an exact measure of the angle of axial rotation. It is contaminated by the precession and nutation motions. To obtain a rigorous measure of time based on Earth rotation, it is necessary, therefore, to define an origin on the equator, which has no motion along the instantaneous equator, or differently stated, has no rotation around the pole of the equator. Such an origin is referred to as the non-rotating origin (NRO). The NRO was introduced by Guinot (1979) and considered in detail by Capitaine *et al.* (1986, see also Capitaine, 1990).

According to the conventions adopted by the IAU in 2000 and the IUGG in 2003, the axial rotation of the Earth is to be measured around the CIP; so the equator of the CIP is the relevant equator for the NRO which is to replace the equinox in the definition of time. To distinguish the newly defined NRO from other "origins" which had earlier been referred to by the same name (NRO), the nomenclature CEO

Table 12.1 *Definition of all the angles related to precession, obliquity, equinox.*

Notation	Name	Angle between the two planes		At their intersection point	Notation of intersection point
ϵ_0	mean obliquity of epoch	ecliptic$|_{J2000}$	mean equator$|_{J2000}$	γ_0	mean equinox of epoch
ϵ_A	mean obliquity of date	ecliptic$|_{date}$	mean equator$|_{date}$	γ_A	mean equinox of date
ϵ_D, ϵ	true obliquity of date	ecliptic$|_{date}$	true equator$|_{date}$	γ_D	true equinox of date
ω_A	obliquity at γ_1	ecliptic$|_{J2000}$	mean equator$|_{date}$	γ_1	
Θ	obliquity Euler angle	true equator$|_{date}$	ecliptic$|_{J2000}$	R	
Π_A		ecliptic$|_{J2000}$	ecliptic$|_{date}$	N	
θ_A		mean equator$|_{J2000}$	mean equator$|_{date}$	Q	
$\Delta\epsilon$	nutation in obliquity at γ_D	$\Delta\epsilon = \epsilon_D - \epsilon_A$		γ_D	

Notation	Name	Plane*	Point$_1$	Point$_2$
ζ_A		mean equator$|_{J2000}$	F_0	γ_0
z_A		mean equator$|_{date}$	γ_A	F_D
Π_A		ecliptic$|_{J2000}$	γ_0	N
Λ_A		ecliptic$|_{date}$	N	γ_A
p_A	general precession	$p_A = \Lambda_A - \Pi_A$		
ψ_A	luni-solar precession	ecliptic$|_{J2000}$	γ_0	γ_1
χ_A	planetary precession	mean equator$|_{date}$	γ_1	γ_A
Ψ	longitude Euler angle	ecliptic$|_{J2000}$	γ_0	R
$\Delta\psi$	nutation in longitude	ecliptic$|_{date}$	γ_D	γ_A

* For each of the angles of the first column of this table, we mention the plane (third column) where the angle is measured and the ranging on that plane of this angle, from point 1 (fourth column) to point 2 (fifth column).

(celestial ephemeris origin) has been adopted for the NRO associated with the CIP, by IAU2000 resolutions of 2000. From 2003, the name CIO (celestial intermediate origin), has been adopted by the IAU for the NRO associated with the CIP in the celestial frame. It is usually denoted by σ.

If n̂ be the unit vector from the geocenter in the direction of the CIP, and if **W** be the angular velocity of motion of the CIO, the defining condition on the NRO as stated by Capitaine *et al.* (1986, see also Capitaine, 1990), namely that it should have no rotation around the CIP, translates into

$$\mathbf{W} \cdot \hat{\mathbf{n}} = 0. \qquad (12.15)$$

Note that in Capitaine *et al.* (1986) the notation $\mathbf{\Omega}$ was used instead of **W**; we have preferred to use the latter in order to avoid the confusion with the Earth's angular velocity for which the symbol $\mathbf{\Omega}$ was used previously. An equivalent way of stating the condition is in terms of the velocity of the CIO: that the velocity, say $\dot{\mathbf{x}}$, should be parallel to the instantaneous direction of the CIP: $\dot{\mathbf{x}} = k\hat{\mathbf{n}}$. Now, as the position **x** of the CIO is always on the equator of the vector n̂ representing the direction of the CIP, it follows that $\mathbf{x} \cdot \hat{\mathbf{n}} = 0$ at all times, and hence that its derivative also vanishes. Consequently $\mathbf{x} \cdot \dot{\hat{\mathbf{n}}} = -\hat{\mathbf{n}} \cdot \dot{\mathbf{x}} = -k$. So the defining condition for the CIO has the alternative form introduced by Kaplan (2003):

$$\dot{\mathbf{x}} = -(\mathbf{x} \cdot \dot{\hat{\mathbf{n}}})\hat{\mathbf{n}}. \qquad (12.16)$$

The equivalence of the two conditions (12.15) and (12.16) may be readily seen on observing that $\dot{\mathbf{x}}$ is due to the angular velocity **W**, so that $\dot{\mathbf{x}} = \mathbf{W} \wedge \mathbf{x}$. Thus $\dot{\mathbf{x}} \cdot \mathbf{W} = 0$, and hence $\hat{\mathbf{n}} \cdot \mathbf{W} = 0$ by virtue of (12.16). It is thus established that (12.16) implies (12.15). The converse too may be readily proved.

The equator of the CIP moves in the TRF too as a consequence of polar motion. Therefore the node of the prime meridian on the equator of the CIP has small motions along the latter, and needs to be replaced by a non-rotating origin for the purpose of time measurement. This NRO is called the terrestrial intermediate origin or TIO presently. It had been called for a while (2003–2006) the terrestrial ephemeris origin or TEO following the IAU2000 resolutions. Since then, the IAU has adopted new resolutions in which the nomenclature has changed and the names CIO and TIO have been adopted.

Figure 12.1 illustrates how the CIO $\sigma(t_1)$, $\sigma(t_2)$, $\sigma(t_3)$, ... moves as the CIP moves through successive positions $\mathrm{CIP}(t_1)$, $\mathrm{CIP}(t_2)$, $\mathrm{CIP}(t_3)$, ... As shown in the figure, we start from the $\mathrm{CIP}(t_1)$ and the CIO $\sigma(t_1)$; at the time t_2, for t_2 being the instant immediately after t_1, the CIP has moved to $\mathrm{CIP}(t_2)$ and the CIO to $\sigma(t_2)$. As seen on the figure, the circle passing through $\mathrm{CIP}(t_1)$ and $\mathrm{CIP}(t_2)$ is perpendicular to the equator of the $\mathrm{CIP}(t_1)$ and to the equator of the $\mathrm{CIP}(t_2)$. We then consider the time t_3, for t_3 being the instant immediately after t_2, and draw the circle passing through $\mathrm{CIP}(t_2)$ and $\mathrm{CIP}(t_3)$ and perpendicular to the equator of the $\mathrm{CIP}(t_2)$ and to the equator of the $\mathrm{CIP}(t_3)$.

The initial condition for the position of $\sigma(t_1)$ is arbitrary on the equator of $\mathrm{CIP}(t_1)$. The IAU has thus decided for continuity with earlier conventions at the time of the changeover, 1 January 2003, imposed by the IAU resolutions.

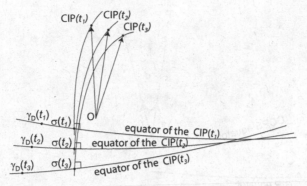

Figure 12.1 Shows how σ, the CIO, moves as the CIP goes through successive positions.

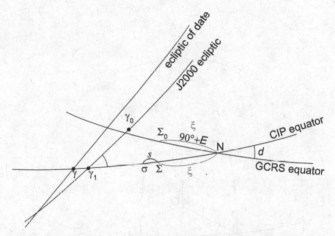

Figure 12.2 Link between Σ and Σ_0, and between σ and Σ (see Capitaine, 1990).

We consider now how σ may be located on the equator of the CIP of date, given the location Σ_0 of the CIO of the epoch J2000 on the CIP equator of epoch J2000 (or, as explained in Section 12.3.1, conceptually the GCRS equator, noted as such in Fig. 12.2 and the following). With reference to Fig. 12.2, let N be the ascending node of the CIP equator of date on the mean equator of epoch J2000 (which defines the principal plane of the GCRF if the epoch is J2000). Let us now refer to Fig. 12.3, where the pole of the CIP equator of date and the pole of the GCRF plane of epoch are marked by P and P_0, respectively. $\hat{n}$ and $\hat{n}_0$ are unit vectors pointing towards these poles. The unit vector $\hat{I}$ is in the direction of the node N; it corresponds to the unit vector along the intersection of the CIP and CGRS equatorial planes. Now with the help of those two figures (Fig. 12.2 and Fig. 12.3), we see that the point Σ_0 may be carried over to σ by the following sequence of rotations: a rotation through an angle ξ $(= 90° + E)$ around $\hat{n}_0$ which brings Σ_0 into coincidence with N, then a rotation about I through an angle d to take the principal plane into the equatorial

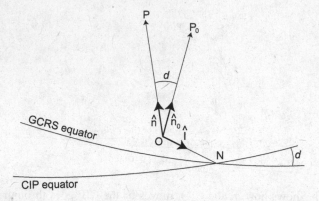

Figure 12.3 The unit vectors $\hat{\mathbf{l}}$, $\hat{\mathbf{n}}$ and $\hat{\mathbf{n}}_0$. (Figure adapted from Capitaine *et al.*, 1986.)

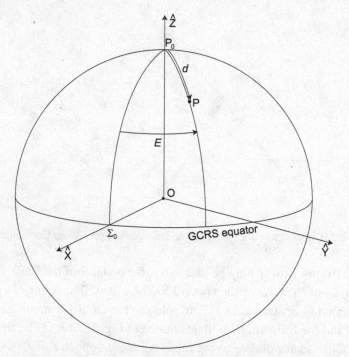

Figure 12.4 Celestial sphere showing the CIP P in the GCRS, Σ_0, and the angles d and E. (Figure based on Capitaine *et al.*, 1986.)

plane of the CIP, and finally a rotation through an angle $(-\xi - s)$ to bring the point from N over to σ. The angles d and E are presented in Fig. 12.4.

Thus

$$s = \sigma \mathrm{N} - \Sigma_0 \mathrm{N} \qquad (12.17)$$

by construction.

The point σ moves as time varies, and its angular velocity $\mathbf{W}$ can be written as a sum of angular velocities relating to the three rotations:

$$\mathbf{W} = \dot{\xi}\,\hat{\mathbf{n}}_0 + \dot{d}\,\mathbf{1} - (\dot{\xi} + \dot{s})\,\hat{\mathbf{n}}. \tag{12.18}$$

The defining condition (12.15) then becomes

$$(\cos d - 1)\,\dot{\xi} - \dot{s} = 0. \tag{12.19}$$

Hence

$$s = \int_{t_0}^{t} (\cos d - 1)\,\dot{\xi}\,dt + s_0. \tag{12.20}$$

The constant s_0 which is the value of s at the initial epoch is given by

$$s_0 = \sigma_0 N_0 - \Sigma_0 N_0, \tag{12.21}$$

where N_0 is the node of the CIP equator of initial epoch on the principal plane of the GCRF, and σ_0 is the CIO of the initial epoch.

The expression (12.20) for s may be rewritten directly in terms of Cartesian coordinates (X, Y, Z) of the CIP in the GCRF. They are evidently related to E and d (see Fig. 12.4) by

$$X = \cos E \sin d,$$

$$Y = \sin E \sin d, \tag{12.22}$$

$$Z = \cos d.$$

It is a simple matter to verify then that $(Y\dot{X} - X\dot{Y}) = -\dot{E}\sin^2 d = -\dot{\xi}\sin^2 d$. Since $1 + Z = 1 + \cos d$, one sees immediately that Eq. (12.20) has the alternative form

$$s = \int_{t_0}^{t} \frac{Y\dot{X} - X\dot{Y}}{1 + Z}\,dt + s_0. \tag{12.23}$$

The numerator of the integrand is evidently the Z component of $\hat{\mathbf{n}} \wedge \hat{\mathbf{n}}$, and $Z = \hat{\mathbf{n}} \cdot \hat{\mathbf{n}}_0$. Thus we have yet another form for s:

$$s = \int_{t_0}^{t} \frac{(\dot{\hat{\mathbf{n}}} \wedge \hat{\mathbf{n}}) \cdot \hat{\mathbf{n}}_0}{1 + (\hat{\mathbf{n}} \cdot \hat{\mathbf{n}}_0)}\,dt + s_0. \tag{12.24}$$

We move on now to the NRO in the terrestrial reference frame, the TIO, denoted by ϖ. In this case the relevant reference plane is that of the true equator of date, and a point Π_0 on this equator is chosen. The node of the equator of the CIP of date on the above equator is denoted by M (see Fig. 12.5). The two planes are very close, however.

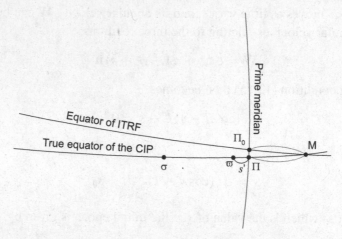

Figure 12.5 The link between ϖ, Π_0, and Π.

With Π on the equator of the CIP chosen so that the arcs $\Pi_0 M$ and ΠM are equal, one defines s' by

$$s' = \varpi M - \Pi M, \tag{12.25}$$

ϖ being the TIO on the CIP equator. By arguments similar to those used in the case of the CIO, one sees that s' is given by

$$s' = -\frac{1}{2} \int_{t_0}^{t} (\dot{x}_p y_p - \dot{y}_p x_p) dt, \tag{12.26}$$

where $(x_p, -y_p)$ are the coordinates of the CIP in the ITRF. (As noted in Section 2.14.1, the ITRS has an orientation maintained in continuity with past international agreements (BIH orientation) and with a first alignment close to the mean equator of 1900 and the Greenwich meridian, and the ITRF is realized by a set of instantaneous coordinates and velocities of reference points distributed on the topographic surface of the Earth and taking account of plate tectonics.) The arbitrary additive constant that could be present in s' is set equal to zero by choosing ϖ_0 at the initial epoch to be such that the $\varpi_0 - M_0$ and $\Pi_0 - M_0$ are equal. It may be noted that the presence of the minus sign in (12.26), unlike (12.23), is due to the fact that the position of the CIP in the TRF is $(x_p, -y_p)$ while it is (X, Y) in the CRF.

The concept of NRO also applies to the TRF, using the same kinds of formula for obtaining s' with the angle g being the angle between the CIP and the pole of the TRF of date and the angle $\xi' = 90° + F$ being the angle on the equator of date between the meridian passing through the pole of the TRF and the origin of longitude (prime meridian of the TRF), and the meridian passing through the

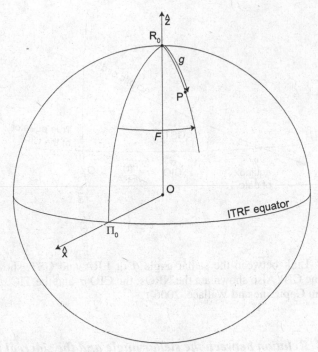

Figure 12.6 Celestial sphere showing the CIP P, Π_0, and the angles g and F, R_0 being the pole of the ITRS equator. (Figure based on Capitaine *et al.*, 1986.)

instantaneous CIP and the TIO (prime meridian of the TIO) (see Fig. 12.5 and Fig. 12.6):

$$s' = \varpi M - \bar{\Pi}_0 M = \int_{t_0}^{t} (\cos g - 1)\dot{F} \, dt. \qquad (12.27)$$

Again, the integration constant may be chosen so that it is very close to the prime meridian of the initial epoch. s' is very small in practice.

12.6 Stellar angle

The stellar angle, denoted by θ or ERA, is the angle of rotation around the CIP of date that is represented by the arc from the CIO (σ) to the TIO (ϖ) on the equator of the CIP (see Fig. 12.7).

The stellar angle is the accumulated rotation of the Earth in space about an axis with an orientation which is varying both in space and in the terrestrial frame. One denotes this angle with the nomenclature *stellar angle* or *Earth rotation angle* (ERA). For references, see Guinot (1979), Capitaine *et al.* (1986), Capitaine (1990).

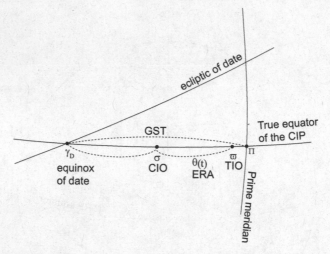

Figure 12.7 Link between the stellar angle θ or ERA and GST, shown on the equator of the CIP. Also shown are the NROs: the CIO σ and the TIO ϖ. (Figure adapted from Capitaine and Wallace, 2006.)

12.6.1 *Relation between the stellar angle and the sidereal time*

How is θ related to the sidereal rotation angle GST? In Fig. 12.7, GST is the arc $\gamma_D \Pi$ on the equator of the CIP, while θ is the arc $\sigma \varpi$. Since the arc length from ϖ to Π is s', and that from γ_D to σ is the right ascension $\alpha(\sigma)$ of σ, we have

$$\text{GST} = \alpha(\sigma) + \theta + s'. \tag{12.28}$$

Combining the angle s with the position of Σ with respect to γ_D, one obtains:

$$\text{GST} = \alpha(\Sigma) - s + \theta + s'. \tag{12.29}$$

Or equivalently:

$$\theta = \text{GST} + s - s' - \alpha(\Sigma). \tag{12.30}$$

On combining Eq. (12.28) with the relation (12.4) between GST and GMST, we have

$$\theta = \text{GST} - \alpha(\sigma) - s' = \text{GMST} + [\alpha(\gamma_A)]_p - \alpha(\sigma) - s'. \tag{12.31}$$

We can further decompose $\alpha(\sigma)$ in terms of $\alpha(\gamma_A) + A\sigma$, as shown in Fig. 12.8, or equivalently: $[\alpha(\gamma_A)]_s + [\alpha(\gamma_A)]_p + A\sigma$, where A is determined on the true equator of the CIP from the projection of the origin of the right ascension, i.e. the mean equinox of J2000 along the equator of the GCRF. Equation (12.31)

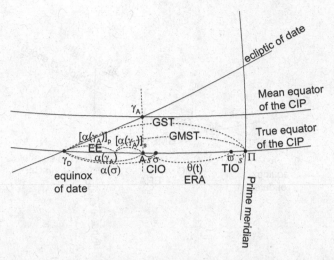

Figure 12.8 Link between the stellar angle, GMST, and GST shown on the equator of the CIP. γ_A is the mean equinox of date, and $\alpha(\gamma_A) = \gamma_D A$, is its right ascension; it can be decomposed into a periodic part $[\alpha(\gamma_A)]_p$ (also denoted EE for equation of the equinoxes), and a secular part $[\alpha(\gamma_A)]_s$. (Figure adapted from Capitaine and Wallace, 2006.)

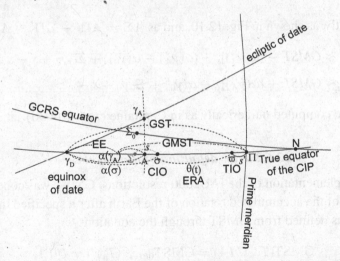

Figure 12.9 Link between the stellar angle, GMST, and GST shown on the equator of the CIP. γ_A is the mean equinox of date, and $\alpha(\gamma_A) = \gamma_D A$, its right ascension, and EE, the equation of the equinoxes. This figure shows the link between Σ_0 and Σ, and between σ and Σ (the quantity s). (Figure adapted from Capitaine and Wallace, 2006, and Kaplan, 2005a.)

becomes

$$\theta = \text{GMST} - [\alpha(\gamma_A)]_s - A\sigma - s', \tag{12.32}$$

or equivalently, as shown in Fig. 12.9:

$$\theta = \text{GMST} - [\alpha(\gamma_A)]_s - A\Sigma + s - s', \tag{12.33}$$

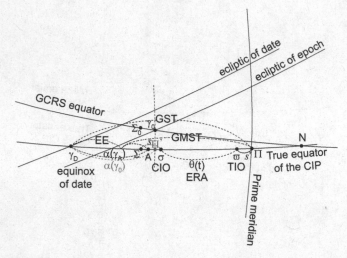

Figure 12.10 Link between the stellar angle, GMST, and GST shown on the equator of the CIP. γ_A is the mean equinox of date, and $\alpha(\gamma_A) = \gamma_D A$, its right ascension, and EE, the equation of the equinoxes. This figure shows the mean equator of J2000 and thereon, γ_0 and Σ_0. $\alpha(\gamma_0) = \gamma_D H$. σ, ϖ, s, s', and Σ are also shown on the figure. (Figure adapted from Capitaine and Wallace, 2006.)

or equivalently, as shown in Fig. 12.10, and as $A\Sigma = A\Pi' - \Sigma\Pi' \approx A\Pi' - \Sigma_0\gamma_0$:

$$\theta \approx GMST - [\alpha(\gamma_A)]_s + (\alpha(\gamma_A) - \alpha(\gamma_0)) + \Sigma_0\gamma_0 + s - s', \quad (12.34)$$

$$\theta \approx GMST + [\alpha(\gamma_A)]_p - \alpha(\gamma_0) + \Sigma_0\gamma_0 + s - s', \quad (12.35)$$

which can be computed numerically as in Capitaine *et al.* (2003d).

12.6.2 New definition of universal time

Until the implementation of the IAU2000 resolutions, GMST was conceived of as the measure of the accumulated rotation of the Earth after a specified initial epoch, and $UT1$ was defined from GMST through the equation

$$GMST(T_u, UT1) = GMST_{0hUT1}(T_u) + r\,UT1, \quad (12.36)$$

which relates the increase of GMST during day T_u to the "time of day" (UT1). $GMST_{0hUT1}(T_u)$ is given by a formula which is not strictly linear in T_u (see IERS Conventions 1996, Capitaine *et al.*, 1986, 2000, 2003d, 2004). The factor r is the number of sidereal days per solar day, namely 1.00273790935.

It is recognized now that the stellar angle (not GMST or GST) is the true measure of the accumulated Earth rotation. The new definition of UT1 is in terms of θ:

$$\theta = \theta_0 + k(UT1 - UT1_0), \quad (12.37)$$

where k represents the factor to pass from solar days to sidereal days. Recognizing that the rate of change of θ differs slightly from that of the old UT1 (because of the slightly contaminated definition of UT1 based on GST previously), the scale factor k contains a scale factor close to unity so that a day of the new UT1 remains close to the mean solar day. The value of r is 1.00273781191. This does not ensure the continuity of LOD variations. There is indeed an expected discontinuity in UT1 rate, unavoidable due to the improved models and the fixed relationship between ERA and UT1. The value of θ_0 (0.7790572732640) is chosen to ensure continuity of the new UT1 with the old one at the epoch of switchover from the old to the new (2003 January 1.0, Julian Day 2452640.5). In practice, one has adjusted the constants in the definitions in order to ensure the continuity of UT1 on 2003 January 1.0:

$$\theta = 2\pi[0.7790572732640$$

$$+ 1.00273781191(\text{Julian UT1 date} - 2451545.0)/36525]. \qquad (12.38)$$

The new definition of UT1 based on the NROs provides numerical values that can be compared with the definition of UT1 based on GST. The chosen numerical constants and the definitions used have an effect on the determination of UT1 less than a few hundreds of microseconds over the next century. Capitaine *et al.* (2003d, see also Capitaine and Gontier, 1993, Capitaine *et al.*, 2005b) have shown that the differences between the two are very small, at the level of 60 μas over 40 years (periodic terms), but can be larger as time elapses (due to quadratic terms), at the level of 800 μas over 200 years.

12.7 Link between ICRF and the ITRF involving the NRO

The different forms used for the transformation connecting the CRF and the TRF were presented in Section 12.3.3. We reproduce two of them, namely, Eqs. (12.2) and (12.3), to facilitate clarification of the relations between the sets of matrices appearing in the alternative forms:

$$[TRF] = W'^{-1} R_3(GST) PN'^{-1} [CRF], \qquad (12.39)$$

$$[TRF] = W^{-1} R_3(\theta) Q^{-1} [CRF]. \qquad (12.40)$$

In view of the relation (12.30) between GST and θ,

$$R_3(\theta) = R_3(-s')R_3(GST)R_3(s)R_3(-\alpha(\Sigma)). \qquad (12.41)$$

Or equivalently,

$$R_3(GST) = R_3(s')R_3(\theta)R_3(-s)R_3(\alpha(\Sigma)). \qquad (12.42)$$

The matrix $R_3(s')$ will be used within the matrix $W^{-1}(t)$ for going from the CRF to the TRF, and the matrix $R_3(-s')$ will be used within the matrix $W(t)$ for going from the TRF to the CRF. The matrix $R_3(-s)$ will be used within the matrix $Q^{-1}(t) = Q'(t)$ for going from the CRF to the TRF, and the matrix $R_3(s)$ will be used to multiply the matrix $Q(t)$ for going from the TRF to the CRF. It follows therefore that

$$W^{-1} = W'^{-1} R_3(s') \quad \text{and} \quad Q^{-1} = Q' = R_3(-s) R_3(\alpha(\Sigma)) PN'^{-1}. \quad (12.43)$$

The matrix $R_3(\alpha(\Sigma)) PN'^{-1}$ and W'^{-1} are functions of the coordinates of the CIP in the CRF and TRF, respectively, i.e., of (X, Y) in the former case and of $(x_p, -y_p)$ in the latter.

$PN'^{-1} = PN$ is a set of rotations that brings the CRF into the intermediate frame related to the CIP with the x-axis in the direction of γ_D and the z-axis in the direction of the CIP; $R_3(\alpha(\Sigma))$ is a rotation around the CIP that then takes γ_D into Σ and $R_3(-s)$ is a rotation around the CIP that takes Σ into the NRO σ.

W' is a set of rotations that brings the third axis of the TRF into the intermediate frame related to the CIP; $R_3(-s')$ is a rotation around the CIP that takes the prime meridian into the NRO ϖ.

12.7.1 Rotation matrix representation Q' or Q for CIP precession–nutation

Consider the intermediate coordinate frame F_2 obtained by rotating the axes of the frame F_1 about an axis x_k of this system; this is for instance the case when one considers rotation of the axes of the CRF about an axis in the (X, Y) plane so as to bring the Z-axis into the unit vector (X, Y, Z) in the direction of the CIP. F_1 is taken over into the intermediate frame F_2 by a rotation R_k around the chosen axis, represented by the matrix $R_k(x_k)$, where k stands for the axis x_k around which we rotate. As seen from Appendix A, any rotation can be expressed using a series of such rotations or equivalently by using the J_i symbols (see Eqs. (A.26) and (A.35)) to go over to a pole position at (X, Y, Z). The rotation which takes the $\hat{e}_3$-axis of the F_1 to the unit vector (X, Y, Z) giving the direction of the $\hat{e}_3$-axis of F_2 in F_1 frame, or more specifically, the rotation Q' which takes the Z-axis of the CRF to the unit vector (X, Y, Z) giving the direction of the CIP (see Eq. (A.27) of Appendix A) is

$$e^{i\boldsymbol{\alpha}\cdot\mathbf{J}} = Q', \qquad \boldsymbol{\alpha} = (Y, -X, 0). \quad (12.44)$$

On using

$$\sin\alpha = (X^2 + Y^2)^{1/2} \quad (12.45)$$

and

$$\cos \alpha = Z \qquad (12.46)$$

in Eq. (A.26) of Appendix A, with the abbreviation $q = 1/(1 + Z)$ nearly equal to $1/2$, one finds

$$Q' = \begin{pmatrix} 1 - \dfrac{X^2}{2} & \dfrac{-XY}{2} & X \\[2ex] \dfrac{-XY}{2} & 1 - \dfrac{Y^2}{2} & Y \\[2ex] -X & -Y & 1 - \dfrac{(X^2 + Y^2)}{2} \end{pmatrix}, \qquad (12.47)$$

which is what is shown in Eq. (A.43) equivalent to:

$$Q' = Q^{-1} = R_1(Y)R_2(-X)R_3\left(\frac{-XY}{2}\right). \qquad (12.48)$$

The inverse of the above matrices, namely $Q = Q'^{-1}$, may be obtained simply by making the replacements $(X, Y) \rightarrow (-X, -Y)$:

$$Q = Q'^{-1} = R_1(-Y)R_2(X)R_3\left(\frac{-XY}{2}\right) \qquad (12.49)$$

and

$$Q = \begin{pmatrix} 1 - \dfrac{X^2}{2} & \dfrac{-XY}{2} & -X \\[2ex] \dfrac{-XY}{2} & 1 - \dfrac{Y^2}{2} & -Y \\[2ex] X & Y & 1 - \dfrac{(X^2 + Y^2)}{2} \end{pmatrix}. \qquad (12.50)$$

Note also that Q can be obtained from applying the inverse successive rotations:

$$Q = Q'^{-1} = R_3\left(\frac{XY}{2}\right) R_2(X)R_1(-Y). \qquad (12.51)$$

12.7.2 Rotation matrix representation Q for CIP rotation angles

Knowing that the coordinates of the CIP in the GCRF are given by Eq. (12.22) with E and d as shown in Fig. 12.4, Q becomes

$$Q = R_3(-E)R_2(-d)R_3(E)R_3(s). \qquad (12.52)$$

On expressing the first three of the matrices appearing in (12.52) in terms of E and d, we find the product $R_3(-E)R_2(-d)R_3(E)$ to be

$$\begin{pmatrix} 1 + \cos^2 E \ (\cos d - 1) & (\cos d - 1) \ \sin E \cos E & -\sin d \cos E \\ (\cos d - 1) \ \sin E \cos E & 1 + \sin^2 E \ (\cos d - 1) & -\sin d \sin E \\ \sin d \cos E & \sin d \sin E & \cos d \end{pmatrix}, \quad (12.53)$$

or equivalently using Eq. (12.22) and $q = 1/(1 + \cos d)$:

$$R_3(-E)R_2(-d)R_3(E) = \begin{pmatrix} 1 - qX^2 & -qXY & -X \\ -qXY & 1 - qY^2 & -Y \\ X & Y & 1 - q(X^2 + Y^2) \end{pmatrix}, \quad (12.54)$$

which can be written as:

$$R_1(-Y)R_2(X)R_3\left(\frac{-XY}{2}\right). \quad (12.55)$$

12.7.3 The W' or W transformation for CIP polar motion effects

Similarly, applying this approach to the transformation between the TRF and the intermediate frame related to the CIP, knowing that the coordinates of the CIP pole in the TRF are $(x_p, -y_p)$, we obtain

$$W' = R_1(-y_p)R_2(-x_p)R_3\left(\frac{x_p y_p}{2}\right)$$

$$= \begin{pmatrix} 1 - \dfrac{x_p^2}{2} & \dfrac{x_p y_p}{2} & x_p \\ \dfrac{x_p y_p}{2} & 1 - \dfrac{y_p^2}{2} & -y_p \\ -x_p & y_p & 1 - \dfrac{x_p^2}{2} - \dfrac{y_p^2}{2} \end{pmatrix}. \quad (12.56)$$

The inverses of the above matrices, namely $W = W'^{-1}$, may be obtained simply by making the replacements $(x_p, -y_p) \to (-x_p, y_p)$,

$$W = W'^{-1} = R_1(y_p)R_2(x_p)R_3\left(\frac{x_p y_p}{2}\right) \quad (12.57)$$

and

$$W = \begin{pmatrix} 1 - \dfrac{x_p^2}{2} & \dfrac{x_p y_p}{2} & -x_p \\ \dfrac{x_p y_p}{2} & 1 - \dfrac{y_p^2}{2} & y_p \\ x_p & -y_p & 1 - \dfrac{(x_p^2 + y_p^2)}{2} \end{pmatrix}. \quad (12.58)$$

12.8 Alternative derivation of the Earth rotation angle

The ERA defined in Section 12.6 and evaluated using the locations of the non-rotating origins (CIO and TIO) as determined in Section 12.5 is, conceptually, the accumulated angle of the Earth's rotation about the time-varying axis in the direction of the CIP. In other words, ERA between a stipulated epoch t_0 (which we may take to be J2000) and any other desired epoch t is the time integral of the component of the angular velocity vector along the direction $\hat{n}$ of the CIP:

$$\text{ERA} = \int_{t_0}^{t} \mathbf{W}(t') \cdot \hat{n}(t') dt'. \tag{12.59}$$

It is an elementary fact that once we know the time dependence of the transformation matrix T, which relates the components of a vector in the TRF to those in the CRF, the angular velocity $\mathbf{W}$ can immediately be determined. The components (W_1, W_2, W_3) of $\mathbf{W}$ in the TRF are in fact the elements of the antisymmetric matrix

$$M = -\frac{dT}{dt} T^{-1}, \tag{12.60}$$

with

$$M_{32} = -M_{23} = W_1, \quad M_{13} = -M_{31} = W_2, \quad M_{21} = -M_{12} = W_3. \tag{12.61}$$

We proceed to verify this now.

Firstly, the antisymmetry of the above matrix follows from the fact that T is necessarily an orthogonal matrix since it represents a rotation: $T\tilde{T} = I$ where $\tilde{T}$ is the transpose of T and I stands for the unit matrix. On differentiating this identity, we obtain

$$\frac{dT}{dt}\tilde{T} + T\frac{d\tilde{T}}{dt} = 0. \tag{12.62}$$

Since the orthogonality relation means that $\tilde{T} = T^{-1}$, we can rewrite the above equation as

$$\frac{dT}{dt}T^{-1} + \tilde{T}^{-1}\frac{d\tilde{T}}{dt} = 0 \quad \rightarrow \quad M = -\tilde{M}, \tag{12.63}$$

proving that M is antisymmetric. Secondly, to show that M does really represent the angular velocity vector, let us consider some vector $\mathbf{V}$ with its components represented by a three-component column $\mathbf{V}_T$ in the TRF and $\mathbf{V}_C$ in the CRF. Then the relations between the two vectors may be written, by virtue of the definition of the transformation matrix T, as $T\mathbf{V}_C = \mathbf{V}_T$. On differentiating this equation, we have

$$T\frac{d\mathbf{V}_C}{dt} = \frac{d\mathbf{V}_T}{dt} - \frac{dT}{dt}\mathbf{V}_C = \frac{d\mathbf{V}_T}{dt} + M\mathbf{V}_T, \tag{12.64}$$

where we have used the definition (12.60) in the last step. This equation is the equivalent of the familiar classical vectorial relation

$$\left(\frac{d\mathbf{V}}{dt}\right)_C = \left(\frac{d\mathbf{V}}{dt} + \mathbf{W} \wedge \mathbf{V}\right)_T, \tag{12.65}$$

when the identification of the elements of M with the components of $\mathbf{W}$ is made as in Eq. (12.61). The relation between a rotation matrix and a rotation vector can also be seen from Appendix A. Consider, for example, the simplified situation where there is no wobble or nutation so that, at time t, the TRF appears rotated relative to the CRF through an angle $\chi(t)$ about the common third axis. Then

$$T \rightarrow R_3(\chi) = \begin{pmatrix} \cos\chi & \sin\chi & 0 \\ -\sin\chi & \cos\chi & 0 \\ 0 & 0 & 1 \end{pmatrix}. \tag{12.66}$$

It is easily seen that the only non-zero elements of M in this case are $M_{21} = -M_{12} = \dot{\chi}$. When M, which appears in (12.64), is applied to a vector with components (V_x, V_y, V_z), the result is the vector $\dot{\chi}(-V_y, V_x, 0)$. These are just the components of the vector $\mathbf{W} \wedge \mathbf{V}$, since $\mathbf{W}$ is $(0, 0, \dot{\chi})$ in the present example. Thus the equivalence referred to above is verified in this particular case. Similar proofs can be given for angular velocity vectors parallel to the second or third axes, and the general case follows on taking a linear combination of all three.

Let us now compute M in the general case when nutation–precession and wobble are also present. The transformation T may then be taken to be of the same form as in Eq. (12.2) or (12.3). Replacing GST by a symbol χ for convenience, and noting that $PN'^{-1} = PN$ (by virtue of Eq. (12.7)), we have

$$[TRF] = W'^{-1}(x_p, y_p)\, R_3(\chi)\, PN(X, Y)\, [CRF]. \tag{12.67}$$

The matrix M is now

$$M = -\frac{dT}{dt}T^{-1} = -\left(\frac{dW'^{-1}}{dt}W'\right) - W'^{-1}\left(\frac{dR}{dt}R^{-1}\right)W'$$

$$- (W'^{-1}R)\left(\frac{d(PN)}{dt}(PN)^{-1}\right)(W'^{-1}R)^{-1}. \tag{12.68}$$

This is a *kinematic relation* connecting the terrestrial components of the angular velocity vector to the Earth orientation parameters.

On evaluating the right-hand member of Eq. (12.68), and identifying its matrix elements with those of M as given by (12.61), one obtains the following expressions for the angular velocity components:

$$W_1 = L_{23} + L_{12}x_p - (1/2)(L_{23}x_p - L_{31}y_p)x_p - \dot{y}_p - (1/2)(x_p\dot{x}_p + y_p\dot{y}_p)y_p,$$

(12.69)

$$W_2 = L_{31} - L_{12}y_p + (1/2)(L_{23}x_p - L_{31}y_p)y_p - \dot{x}_p - (1/2)(x_p\dot{x}_p + y_p\dot{y}_p)x_p,$$

(12.70)

$$W_3 = L_{12}[1 - (1/2)(x_p^2 + y_p^2)] - (L_{23}x_p - L_{31}y_p) - (1/2)(y_p\dot{x}_p - x_p\dot{y}_p),$$

(12.71)

where

$$L_{23} = -L_{32} = (\dot{X} + \Delta_X)\sin\chi - (\dot{Y} + \Delta_Y)\cos\chi, \qquad (12.72)$$

$$L_{31} = -L_{13} = (\dot{X} + \Delta_X)\cos\chi + (\dot{Y} + \Delta_Y)\sin\chi, \qquad (12.73)$$

$$L_{12} = -L_{21} = \frac{d\chi}{dt} + \delta_3, \qquad (12.74)$$

with

$$\Delta_X = \frac{X(X\dot{X} + Y\dot{Y})}{(1+Z)}, \quad \Delta_Y = \frac{Y(X\dot{X} + Y\dot{Y})}{(1+Z)}, \quad \delta_3 = \frac{(Y\dot{X} - X\dot{Y})}{(1+Z)}. \quad (12.75)$$

Now that we know the components of $\mathbf{W}$, calculation of the projection of this vector on to the direction $\hat{\mathbf{n}}$ of the CIP is straightforward. The components of $\hat{\mathbf{n}}$ are given by the coordinates of the CIP in the TRF, which are $x_p, -y_p, (1 - x_p^2 - y_p^2)^{1/2}$. Evaluation of $\mathbf{W} \cdot \hat{\mathbf{n}}$ may then be done readily using Eqs. (12.69)–(12.71). One finds that a number of cancellations take place, leading to the end result:

$$\mathbf{W} \cdot \hat{\mathbf{n}} = \frac{d\chi}{dt} + \delta_3 + \frac{1}{2}\left(y_p\dot{x}_p - x_p\dot{y}_p\right). \qquad (12.76)$$

Integration of the above equation from an epoch t_0 up to t yields

$$\int_{t_0}^{t} \mathbf{W}(t') \cdot \hat{\mathbf{n}}(t')dt' = \chi(t) - \chi_0 + \int_{t_0}^{t} \delta_3(t')dt'$$

$$+ \frac{1}{2}\int_{t_0}^{t} \left(y_p(t')\dot{x}_p(t') - x_p(t')\dot{y}_p(t')\right)dt', \quad (12.77)$$

wherein the dots indicate differentiation with respect to t', of course. On introducing the expression given in Eq. (12.75) for δ_3,

$$\int_{t_0}^{t} \mathbf{W}(t') \cdot \hat{\mathbf{n}}(t')dt' = \chi(t) - \chi_0 + \int_{t_0}^{t} \frac{(Y(t')\dot{X}(t') - X(t')\dot{Y}(t'))}{1 + Z(t')}dt'$$

$$+ \frac{1}{2}\int_{t_0}^{t} [y_p(t')\dot{x}_p(t') - x_p(t')\dot{y}_p(t')]dt'. \qquad (12.78)$$

One recognizes readily that the first of the two integrals on the right-hand side of (12.78) is the quantity $s - s_0$ of Eq. (12.23) and that the negative of the second integral is s' of (12.26). It may also be recalled that we introduced χ, while writing down Eq. (12.67), merely as a convenient symbol for GST, related to GMST by Eq. (12.4). These identities enable us to rewrite the foregoing equation as

$$\int_{t_0}^{t} \mathbf{W}(t') \cdot \hat{\mathbf{n}}(t') dt' = \text{GMST} + s - s' - (s_0 + \text{GMST}_0)$$

$$+ [\alpha(\gamma_A)(t)]_p - [\alpha(\gamma_A)(t_0)]_p$$

$$= \text{GMST} + s - s' + [\alpha(\gamma_A)(t)]_p - \alpha(\gamma_0) + \Sigma_0 \gamma_0$$

$$- \left(s_0 + \text{GMST}_0 + [\alpha(\gamma_A)(t_0)]_p - \alpha(\gamma_0) + \Sigma_0 \gamma_0 \right). \quad (12.79)$$

Since s_0 in (12.23) is essentially an arbitrary constant, we can choose it to simplify Eq. (12.79). Then the right-hand side becomes identical with the stellar angle θ defined in Eq. (12.35). We see thus that θ is nothing but the ERA defined by the integral of the component of the angular velocity vector in the direction of the CIP. The approach employed here follows Mathews (2002); it avoids the geometrical intricacies of the definition of a moving origin (CIO) on the moving equator of the CIP.

12.9 Link between ICRF and the ITRF involving CIP and NRO

Now we have in hand the final expression for going from the CRF to the TRF in the frame of the NRO approach, substituting Eqs. (12.48) and (12.57), in Eq. (12.3):

$$[TRF] = R_1(-y_p)R_2(-x_p)R_3\left(\frac{x_p y_p}{2}\right) R_3(\theta(t))R_3\left(\frac{XY}{2}\right)$$

$$\times R_1(-Y)R_2(X)[CRF]. \quad (12.80)$$

Substituting Eq. (12.30) in Eq. (12.80), we obtain:

$$[TRF] = R_1(-y_p)R_2(-x_p)R_3\left(\frac{x_p y_p}{2} + \text{GST} + s - s' - \alpha(\Sigma) + \frac{XY}{2}\right)$$

$$\times R_1(-Y)R_2(X)[CRF]. \quad (12.81)$$

Or equivalently,

$$[TRF] = R_1(-y_p)R_2(-x_p)R_3\left(\frac{x_p y_p}{2} - s'\right) R_3(\text{GST})$$

$$\times R_3\left(\frac{XY}{2} + s - \alpha(\Sigma)\right) R_1(-Y)R_2(X)[CRF]. \quad (12.82)$$

Comparing with Eq. (12.2), we find:

$$W'(t)^{-1} = R_1(-y_p)R_2(-x_p)R_3\left(\frac{x_p y_p}{2} - s'\right), \tag{12.83}$$

$$PN'^{-1}(t) = R_3\left(\frac{XY}{2} + s - \alpha(\Sigma)\right)R_1(-Y)R_2(X). \tag{12.84}$$

12.9.1 Definition of X and Y as a function of precession and nutation

When a precession–nutation model is defined, the coordinates of the CIP X and Y in the ICRF can be determined. This relation must involve the offset (ξ_0 and η_0) between the mean celestial pole and the ICRF pole as determined by VLBI, as well as the difference between the origin of right ascension used in the ephemerides from which the precession–nutation is computed and the ICRF origin ($d\alpha_0$), and the coordinates ($\bar{X}, \bar{Y}$) of the CIP as determined from the precession–nutation model:

$$X = \bar{X} + \xi_0 - d\alpha_0\bar{Y}, \tag{12.85}$$

$$Y = \bar{Y} + \eta_0 + d\alpha_0\bar{X}. \tag{12.86}$$

The terms in $d\alpha_0$ do not introduce any shift in the right ascensions of the stars, but rather re-orient the classical underlying model of precession–nutation, which used the equinox as the reference direction.

The terms in ξ_0 and η_0 are there to take account of the shift of the pole for a better realization of the ICRF pole with respect to the one constructed at the time of the ICRF realization (Ma *et al.*, 1998, 2009).

12.9.2 Offset of the mean celestial pole

The precession and nutation observations by VLBI are expressed in the form of long-term variations in the Earth's orientation with respect to the ICRF, or equivalently motions of the CIP in space. These can be expressed as the sum of a constant offset of the CIP with respect to the ICRF, precession in longitude (times $\sin\epsilon$) and obliquity, and nutation in longitude (times $\sin\epsilon$) and in obliquity, themselves expressed as a model plus celestial pole offsets ($\delta\psi, \delta\epsilon$) as a function of time, in longitude and in obliquity.

The constant offset of the CIP with respect to the ICRF has been determined by Herring (see IERS Conventions 2010, Petit and Luzum, 2011, Capitaine *et al.*, 2003a–d) as:

$$\delta\psi_0 = -41.775 \text{ mas}, \quad \delta\psi_0 \sin\epsilon_0 = \xi_0 = -16.617 \text{ mas},$$

$$\delta\epsilon_0 = \eta_0 = -6.819 \text{ mas}. \tag{12.87}$$

These values were adopted by the IAU2000, but other values can be found in the literature. Using a combination of lunar laser ranging (LLR) and VLBI, Folkner *et al.* (1994) have determined these values to be −17.2 mas and −5.1 mas, as also mentioned in Feissel and Mignard (1998) and in the IERS Conventions 1996 (McCarthy, 1996). These numbers represent the difference between the coordinates of the realization of the mean pole of the mean equator of J2000 and the pole of the ICRF. They have been determined by Chapront *et al.* (2002) as well from LLR data over 1972–2001, together with the precession rates and the obliquity at J2000, using the new DE405 ephemerides (Standish, 1998a–d) and DE406 ephemerides (Standish, 2004). The values obtained by these authors are: −17.7 mas and −5.4 mas.

12.9.3 Origin of right ascension

As the precession–nutation model uses ephemerides of the relative positions of the Earth with respect to the Sun, of the Moon with respect to the Earth, and of the other planets with respect to the Earth, the origin for the angles such as the ecliptic longitudes of the Moon or any other angle (see Chapter 3) is a realization of the mean inertial equinox of J2000. This point is not the origin of the ICRF, which is in principle realized as close as possible to the mean equinox of J2000. These two points differ also from the present-day realization of the mean equinox of J2000. It must be mentioned that the mean equinox of epoch to be considered is not the rotational dynamical mean equinox of J2000 (Standish, 1981) corresponding to computation with respect to a rotating ecliptic, as used in the past, but the inertial dynamical mean equinox of J2000, corresponding to a computation with respect to an inertial mean ecliptic to which the recent numerical or analytical solutions refer (see Standish, 1981). The difference between the rotational inertial mean equinox of J2000 and the rotational dynamical mean equinox of J2000 is 93.66 mas. Full details are given in Appendix C.

Again as found in Feissel and Mignard (1998), using a combination of LLR and VLBI and the DE200 JPL ephemerides (Standish, 1982a and b), Folkner *et al.* (1994) have determined a difference of −78 mas between the origin of the ICRF and the mean dynamically rotating equinox of J2000. Using more recent LLR data, Chapront *et al.* (2002) have determined the relative positions, with respect to the inertial ecliptic of J2000 as determined in the ephemerides, of the mean inertial equinox of J2000, γ_{J2000} (JPL), as realized for the DE405 JPL ephemerides, secondly of the mean equinox of J2000, γ_{J2000} (LLR) from the recent LLR data, and thirdly of the IRCF mean equinox of J2000, γ_{J2000} (ICRF). They also determined the right ascension origin O(JPL) as realized by JPL for the DExxx kinds of ephemerides, the right ascension origin O(LLR) as realized from the recent LLR

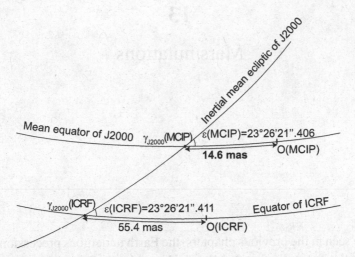

Figure 12.11 Representation of the different realizations of the equinox of J2000 and the origin of right ascension for the JPL ephemerides (DE405), with reference to the ICRF and to the mean equator linked to the mean CIP of J2000, and using LLR and VLBI observations (from Chapront *et al.*, 2002, Capitaine *et al.*, 2003a–d, and Chapront and Francou, 2006).

data, and the right ascension origin O(ICRF) in the ICRF (see Fig. 12.11). The relative angles between those points are

$$\psi_1 = \gamma_{J2000}(\text{LLR})\gamma_{J2000}(\text{ICRF}) = 44.5 \text{ mas},$$

$$\psi_2 = \gamma_{J2000}(\text{LLR})\gamma_{J2000}(\text{JPL}) = 6.4 \text{ mas},$$

$$\phi_{\text{LLR}} = \gamma_{J2000}(\text{LLR})\text{O}(\text{LLR}) = -14.6 \text{ mas},$$

$$\phi_{\text{JPL}} = \gamma_{J2000}(\text{JPL})\text{O}(\text{JPL}) = -50.3 \text{ mas},$$

$$\phi_{\text{ICRF}} = \gamma_{J2000}(\text{ICRF})\text{O}(\text{ICRF}) = -55.4 \text{ mas},$$

$$\text{O}(\text{ICRF})\text{O}(\text{JPL}) = 0.7 \text{ mas},$$

all very close to one another but not perfectly aligned.

13

Mars nutations

As we have seen in the previous chapters, the Earth undergoes precession, nutation, and wobble due to gravitational pulls of other objects of the solar system (such as the Sun and the Moon) because of its rotation, its equatorial bulge, and the 23.5° obliquity of its orbit. However, the Earth is not the only planet having precessional and nutational motions. Mars in particular is also rapidly rotating, has an equatorial bulge, and its mean orientation is tilted with respect to the ecliptic (the obliquity of Mars is about 25°). Therefore Mars also changes its orientation in space.

13.1 Rotation and obliquity of Mars

Mars is at present in the situation where the obliquity is around 25°. According to long-term studies of Mars' orbital and orientation evolution theory (Laskar *et al.*, 2002, 2004), the obliquity of Mars fluctuated between about 15 and 35° over the last million years and has a chaotic behavior on long timescales. The present situation is thus very interesting as it is incidentally very close to the situation of the Earth.

One solar day of Mars is around 24h 40min, thus corresponding to a rapid rotation very similar to that of the Earth. In the first approximation, Mars is in hydrostatic equilibrium and, due to its rotation, has an equatorial bulge; the equatorial radius is about 9 km larger than the polar radius, less (as Mars is smaller than Earth) than the 21 km of difference between the Earth's equatorial radius and polar radius.

Mars being a terrestrial planet (a rocky planet composed primarily of silicate rocks or metals) its interior is very similar to the interior of the Earth, with a mantle and a core believed to be liquid as deduced from first estimations of the tidal Love number k measured by spacecraft orbiting around Mars at present (Konopliv *et al.*, 2006, 2011). We do not know whether there is an inner core inside the liquid core of Mars as there is for the Earth. Considering the present-day constraints coming from the observation of the tides and moments of inertia, as

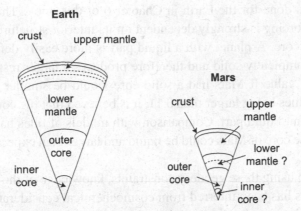

Figure 13.1 Comparison between the interiors of Mars and the Earth.

well as considering the thermo-chemical evolution of the different reservoirs inside Mars, despite the absence of a magnetic field, leaves us with the conviction that the core is completely liquid. However, this must be confirmed by measurements (by very long baseline interferometry (VLBI), see Section 13.5) and it is not excluded in the possible interior models of Mars that one would have a solid inner core (Rivoldini *et al.*, 2011). Figure 13.1 compares the gross properties of the Earth and Mars schematically.

Only a few observations are available for Mars' global properties; among the most prominent are the average density and the mean moment of inertia deduced from gravity measurements of spacecraft orbiting around the planet. The average value of the density of Mars is 3935 kg/m^3; it is about 37% smaller than that of the Earth, as expected in view of the dimension of the planet (compression within planet interiors increases their densities; the mean density of large planets is greater than that of small ones). The values of the mass and moment of inertia require a density increasing with depth and the presence of a dense core. The size, state, and composition of the core are, however, weakly constrained by these values. The precise Doppler and ranging measurements of radio links between the spacecraft and the Earth ground stations (the NASA Deep Space Network (DSN) or the European Space Agency (ESA) tracking stations), have allowed us also to measure the effect of the tidal deformations of the planet Mars on the orbiter trajectories. Yoder *et al.* (2003), Konopliv *et al.* (2006, 2011), and Marty *et al.* (2009) have determined the k_2 Love number and inferred that the core is liquid at present. The tides of Mars are calculated by numerically integrating the ordinary differential equations describing the deformations and external potential variations (see Van Hoolst *et al.*, 2003) by using the depth-dependent profiles of density and rheological properties such as the profiles of Sohl and Spohn (1997) or Rivoldini *et al.* (2011), as done for the Earth in Chapter 6 of this book. The response of a planet to tidal forcing is strongly dependent on its interior structure, especially on the state of the core. A planet with a liquid part is more easily deformable than a planet that is completely solid and therefore produces a larger response to a tidal forcing. The k_2 value if Mars had a solid core would be smaller than 0.08. The measured k_2 values being larger than 0.11, it is believed that the core of Mars contains an important liquid part. Comparison with models of tides has allowed us to conclude that the core of Mars could be liquid and larger than expected in the range 1500–1800 km.

In addition to using these geodesy constraints, knowledge of the chemical composition of Mars has been inferred from cosmochemical considerations. On using meteorites widely agreed to have originated from Mars and to be representative of the Martian mantle, it has been deduced that Mars is generally slightly different from the Earth in terms of mineral content. Sohl and Spohn (1997) have constructed a model for the interior of Mars. Rivoldini *et al.* (2011) have refined that model using the k_2 constraint.

Some analyses have also indicated that most of the sulfur has segregated to the core, enriching its light element fraction. This large fraction of sulfur in the core results in a significant depression of the melting temperature of the core material aggregate compared to pure iron and implies that models based on the above bulk composition are most likely to have a molten outer core. Interior models of Mars (and in particular those of Rivoldini *et al.*, 2011) rely also on laboratory measurements at high pressure.

This is the only way to build interior models for Mars while waiting for the first precise seismic observation from the InSIGHT mission (Interior Exploration using Seismic Investigations, Geodesy and Heat Transport), a NASA mission to be launched in 2016 that will place a single geophysical lander on Mars to study its deep interior.

13.2 Mars normal modes

Like the Earth, Mars has normal modes, including global rotational normal modes such as the free core nutation (FCN) and the Chandler wobble (CW; see Section 2.20). The existence of an equatorial bulge is enough for the CW mode to appear if it is excited. The existence of a flattened liquid core ensures the possibility of exciting the FCN. These modes can be computed using numerical integration inside the planet exactly as done for Earth (see Section 7.5), or using the classical Liouville equation for a two-layered planet (see Section 7.3.1).

The frequencies of the normal modes are deduced from their expressions obtained in Chapter 7 (see Eqs. (7.127) and (7.128)) and using parameter values

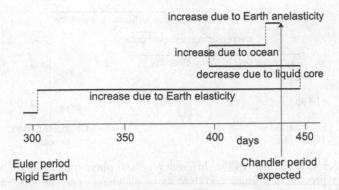

Figure 13.2 Contribution of the different physical phenomena to the CW period of the Earth (see Dehant *et al.*, 2000b, Van Hoolst *et al.*, 2000b) starting from the Earth's non-hydrostatic flattening. The period is expressed in the terrestrial frame.

appropriate for Mars for the basic geophysical parameters such as e, e_f, etc:

$$\sigma_{CW} = \frac{A}{A_m}(e - \kappa), \qquad (13.1)$$

$$\sigma_{FCN} = -1 - \frac{A}{A_m}(e_f + K_{CMB} - \beta), \qquad (13.2)$$

where κ, β are compliances that must be computed from an interior model for Mars as explained above; and where as usual, A, A_m, A_f are the moments of inertia of the whole planet, its mantle and its fluid core, e and e_f are the dynamical flattenings of the whole planet and of the core, respectively, K_{CMB} is a parameter characterizing the strength of the coupling between the core and the outer solid region. With values inferred or estimated from various kinds of studies regarding the flattening of Mars, its interior structure and its rheological parameters (see the legend of Fig. 13.3) one obtains a CW period around 200 days in a frame tied to the planet and a period around 250 days in space. Figure 13.2 and Fig. 13.3 show the different contributions to the CW periods of the Earth and Mars. The figures show the contributions of mantle deformations and liquid core, as well as of mantle anelasticity. For the Earth, observations have indicated that there are contributions from the Earth non-hydrostatic shape; this can be expected for Mars as well. Indeed the presence of the large volcanoes province called Tharsis near Mars' equator may largely change the hydrostatic shape of the planet and therewith change its dynamical flattening and CW period.

As seen from the above equation, the value of the FCN period depends on A/A_m, reflecting the percentage of core and mantle contributions to the moment of inertia, i.e. related to the dimension and mean densities of the core and the mantle. The FCN period depends also on the core flattening e_f and to a minor extent on the

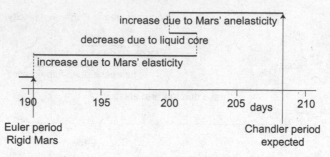

Figure 13.3 Contribution of the different physical phenomena to the CW period of Mars expressed in a frame tied to Mars (see Dehant *et al.*, 2000b, Van Hoolst *et al.*, 2000b) for a particular Mars interior model and starting from the CW period for the non-hydrostatic flattening (5.382×10^{-3}, which is 7% larger than the hydrostatic value 5.005×10^{-3}) as derived from the form factor $J_2 = (C - A)/(MR^2) = 1960.454 \times 10^{-6}$, degree-2 coefficient of the observed gravity field determined from spacecraft. Mars' elasticity depends on the Love number k value. For $k = 0.071$, elasticity lengthens the CW period by almost 6% or 11.2 days to 201.8 days. In the case of a fluid core, the Love number k increases to 0.099 (and therefore the CW period increases by 4.8 days) and the ratio of the global to the mantle moments of inertia A/A_m (only 1.031) appears in the expression for the CW frequency (and therefore the CW period decreases by 6.3 days), which induces a total decrease of the CW period of 1.5 days, down to 200.3 days. Mantle anelasticity for a quality factor Q = 50 and for a frequency-to-the-power-α model with $\alpha = 0.15$ induces an increase of 7 days.

deformation (via the compliance β) of the CMB and on the coupling mechanisms (via the coupling constant K_{CMB}) at the CMB, and e_f is also related to the dimension of the core. One immediately sees that the FCN period, if observed, would provide interesting constraints on the dimension and structure of the core. Note that this normal mode only appears in the response of Mars to an external forcing if the core is liquid. The observation of a resonance in the nutation spectrum near the FCN would immediately confirm that the core is liquid and would provide interesting interior properties of Mars. Figure 13.4 shows the nutation transfer function for a solid and liquid core, and Fig. 13.5 shows the nutation transfer function for different possible values of the core radius.

13.3 Precession and nutation of Mars

The gravitational attraction of the Sun on the ellipsoidal (flattening of the order 1/900) rotating planet Mars tilted in space with respect to its orbital plane (close to the ecliptic) induces a torque that tends to turn the equator towards the ecliptic, exactly as on Earth. As Mars is rotating, it reacts as a gyroscope and its rotation axis moves with a relative large amplitude in space around the perpendicular to the ecliptic, it "precesses" in space.

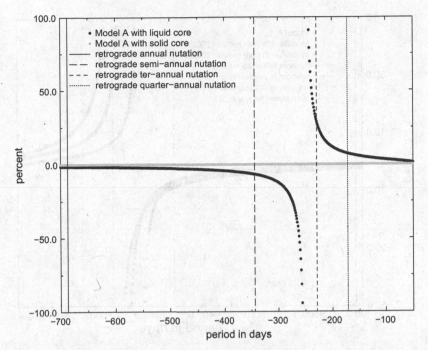

Figure 13.4 Effects of the interior properties of Mars on the nutation induced by the external gravitational forcing of the Sun, in percent with respect to the rigid Earth case. The grey and black curves correspond to Mars' interior models with solid and liquid cores, respectively. Model A is a reference to a model developed by Sohl and Spohn in 1997, using the known mass, radius, and moment of inertia of Mars. The resonance seen in the black curve is induced by the existence of an FCN when the core is liquid. The period is expressed in the celestial frame. The vertical lines indicate the periods of the forced nutations. Reprinted from Dehant *et al.* (2000a), with permission from Elsevier.

On to this precessional motion with a period of about 171 000 years is superimposed a series of periodic variations of the orientation of the rotation axis called nutations, exactly as on Earth. The most important nutation has an amplitude of about 500 milliarcsecond (mas) corresponding to a displacement at the surface of Mars of a few meters (1 mas corresponds to 1.6 cm at the surface of Mars). The nutational variations in Mars' orientation in space are in the direction along the precessional motion and perpendicular to it, nutation in longitude $\Delta\psi$ and in obliquity $\Delta\epsilon$, respectively.

13.3.1 Precession and nutation of rigid Mars

The main nutations have periods of 1 Martian year or semi-annual or ter-annual or quarter-annual, as for the Earth ignoring the influence of the Moon. Using

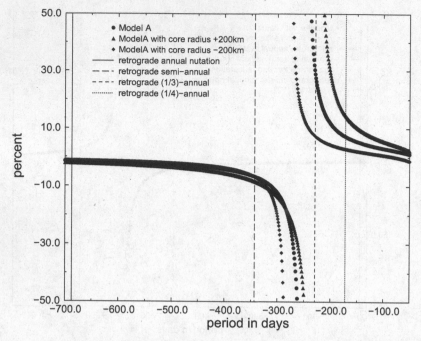

Figure 13.5 Effects of the interior properties of Mars on the nutation induced by the external gravitational forcing of the Sun, for different Mars interior models with a liquid core and different dimensions for the core. Model A is a reference to a model developed by Sohl and Spohn in 1997, using the known mass, radius, and moment of inertia of Mars. The FCN resonance induced by the existence of a liquid core is shifted for different core dimensions; it is shifted to the left for a smaller core than its nominal dimension (1468 km), and to the right for a larger core. The period is expressed in the celestial frame. The vertical lines indicate the periods of the forced nutations. Reprinted from Dehant *et al.* (2000a), with permission from Elsevier.

similar approaches as those used for Earth it is possible to compute these nutations considering that Mars is rigid (see Roosbeek, 2000, Bouquillon and Souchay, 1999). As shown in Chapter 2 with a very simple case of a planet orbiting around the Sun, the main nutation is the semi-annual nutation (see Sections 2.24 and 2.24.2).

Mars has two little moons, Phobos and Deimos, which could in principle also induce nutations. However, these moons are rather small, at the level of 10–20 km diameter; they do not contaminate significantly the nutational variations of Mars' orientation.

13.3.2 *Precession and nutation of non-rigid Mars*

The nutations of Mars are influenced by the interior of the planet as for the Earth. In particular, the presence of an ellipsoidal liquid core induces a resonance in the nutation, called the FCN (see Section 13.2). This resonance has a period close to the ter-annual retrograde nutation for a nominal core (such as 1500 km, the core dimension considered in Sohl and Spohn, 1997). The difference between the liquid-core case and the solid core case is shown in percent with respect to the rigid Earth case in Fig. 13.4.

As seen in Fig. 13.5, for a core larger than the nominal case, the FCN period can be very close to the ter-annual nutation, which induces a large amplification in its amplitude. For a smaller core, the FCN period increases except if an inner core is formed inside the liquid core (which can be the case if the mantle is relatively cold), and if an FICN (free inner core nutation) resonance is forming. In that case, the FCN period decreases when the core size decreases. For Mars, nothing is known about the existence of an inner core. There is nevertheless some support for the absence of an inner core from the study of thermal evolution, using knowledge of the remnant magnetic field and the thermal consequences of the absence of plate tectonics. A large inner core has an effect on the nutations that could be detected, via the fact that the amplification of the largest prograde semi-annual nutation due to a liquid core (to the FCN) would be canceled, while the other nutations would be amplified by the FCN (Van Hoolst *et al.*, 2000a, Dehant *et al.*, 2003a and b, Defraigne *et al.*, 2003). Failure to detect the amplification of the semi-annual nutation with a high precision geodetic experiment (like Lander Radio-science, LaRa, Dehant *et al.*, 2009, 2011, Le Maistre *et al.*, 2012, see also Section 13.5) in its more precise configuration, together with the detection of a liquid core from the retrograde band of the nutations and from the k_2 Love number, could then be interpreted as evidence for a large inner core.

13.4 Length-of-day variations and polar motion of Mars

Mars has a thin atmosphere, about 100 times less dense (at the level of 10 mbar) than that of the Earth. But the changes in the atmospheric pressure are quite large (several mbars). Indeed the CO_2 in the atmosphere evaporates with the highest temperatures on Mars and condenses into CO_2 ice at the lowest temperatures. About one fourth of the atmosphere of Mars is participating in the sublimation and condensation process of the CO_2 in the atmosphere and ice caps. Mars' atmospheric CO_2 seasonal cycle thus induces mass exchange between the polar caps and the atmosphere, which in turn induces seasonal variations of the gravitational potential as well as of the rotation rate of the planet exactly as for the Earth. Mars' rotation around its axis is thus not uniform due to angular momentum exchange between the solid planet and the atmosphere. There is a variation in the LOD which is at the seasonal timescale. A lander or a rover at the surface of Mars on the equator undergoes huge changes of position, up to 15 meters peak-to-peak, over a Martian year (about two Earth years).

This phenomenon can be computed from general circulation models (GCM) of the atmosphere. It represents the global seasonal changes in the atmosphere. Exactly as for the Earth (see Chapter 9), the angular momentum is transferred to the solid planet by three kinds of coupling mechanisms: (1) the pressure torque related to the atmospheric pressure on the topography; (2) the gravitational torque related to the mass anomalies inside Mars and in the atmosphere; and (3) the friction torque related to the wind friction on the Martian surface (see Karatekin *et al.*, 2011). Exactly as for the Earth (see Chapter 9), the angular momentum of the atmosphere consists of two parts, the matter term related to the rigid rotation of the atmosphere with the solid Mars and directly involving the surface pressure all over the surface, and the motion term related to the relative angular momentum of the atmosphere and directly involving the winds in the atmosphere.

Present-day GCMs allow computing the seasonal changes induced in the rotation of Mars as well as in the low degree gravity coefficients. But there are still unexplained differences between the computation and the observation coming from orbiters. Additionally inter-seasonal changes are expected in the time variable data of the low-degree gravity coefficients due to dust storm contributions (Mars undergoes global dust storms).

Polar motion is the motion of the planet around its rotation axis or equivalently the motion of the mean rotation axis around the figure axis in a frame tied to Mars. It is induced by the atmosphere and related to the angular momentum exchange between the atmosphere and the solid planet as for the LOD variations. One fourth of the atmosphere is participating in the sublimation and condensation phenomenon of the CO_2 in the ice caps. The components of polar motion are thus at the seasonal timescale. Additionally, as the rotation axis is not necessarily coincident with the figure axis, there might be a wobble of Mars. This is the so-called CW already mentioned in Section 13.2 above, which is at a period of about 200 days (van Hoolst *et al.*, 2000a, 2000b). As the noisy atmospheric behavior is not purely harmonic, the CW contribution to polar motion might be at the level of a couple of meters, depending on the dissipation within the planet (the quality factor of Mars). The polar motion amplitude of Mars could be a factor of ten (when looking at the polar motion in a plane perpendicular to the geographical pole it is seen in a square of about 2 m) lower than that of the Earth (in a square of about 20 m).

13.5 Rotation measurements of Mars

Mars geodesy observations that are presently used for determining Mars' rotation and interior structure are radio signals in the UHF, S-band, X-band, or Ka-band. The measurements used for reaching the geodesy objectives are the Doppler shifts and ranging times as measured at the DSN stations of NASA or the ESTRACK ESA tracking stations of ESA. The Doppler shifts are measurements of the shifts in the frequency induced by the relative position between the transmitter and receiver. These shifts are measured over a certain period of time called the integration time. The ranging measurements consist in time intervals between the signal sent to the spacecraft and received at the ground station. The geometry, or equivalently the orientation of the line-of-sight, changes during a mission. From these measurements, one is able to reconstruct the motion of a spacecraft around Mars or of a lander or a rover with respect to the Earth, which provides information on the orbit of the spacecraft and its changes in space and time (precise orbit positioning). One is also able to derive information on gravitational and non-gravitational perturbations of the spacecraft trajectory, on the acceleration undergone by the spacecraft at a particular space and time (on targets), and on the Mars orientation and rotation parameters (the so-called MOP) mainly determined by landers.

The observations must be processed with a least square approach in order to deduce the information on the MOPs or on the orbit of a satellite for instance. A precise orbit determination (POD) procedure using dedicated software is necessary to perform an iterative least squares procedure on the tracking data in order to adjust parameters related to the data themselves and to the dynamical model of the spacecraft (orbiter or lander) motion.

The rotation of Mars beneath a spacecraft orbiting around it, or the rotation, orientation, and polar motion of Mars as seen from a lander or a rover may be deduced from radio science data. Knowing almost perfectly from VLBI where the Earth is in space, the observations from Earth make possible the determination of Mars' rotation and orientation in space. The uniform rotation of the planet must be taken out of the Doppler and the known parts of the orientation of Mars as well. The residuals are then analyzed in terms of the physics of Mars' interior.

In 2016, a mission to Mars will carry a seismometer to the surface together with a heat flow measurement device and a radioscience transponder. This mission, called InSIGHT, a NASA Discovery Program mission that will place a single geophysical lander on Mars, will provide more information on the interior of Mars and on its orientation as well. Other future missions are envisaged, such as a network mission of geophysical measurements (including radioscience) on Mars.

Appendix A

Rotation representation

A.1 Rotation and infinitesimal rotations of a vector

A rotation of a vector $\mathbf{r}$ around an axis can be written by using a rotation matrix R:

$$\mathbf{r}' = R\mathbf{r}. \tag{A.1}$$

It can be decomposed into a series of rotations using the three fundamental rotations about the x-, y-, and z-axes denoted R_1, R_2, and R_3, respectively:

$$R_1(\alpha) = \begin{pmatrix} 1 & 0 & 0 \\ 0 & \cos\alpha & \sin\alpha \\ 0 & -\sin\alpha & \cos\alpha \end{pmatrix},$$

$$R_2(\beta) = \begin{pmatrix} \cos\beta & 0 & -\sin\beta \\ 0 & 1 & 0 \\ \sin\beta & 0 & \cos\beta \end{pmatrix}, \tag{A.2}$$

$$R_3(\gamma) = \begin{pmatrix} \cos\gamma & \sin\gamma & 0 \\ -\sin\gamma & \cos\gamma & 0 \\ 0 & 0 & 1 \end{pmatrix}.$$

For small values of the angles α, or β, or γ, these matrices may be written as

$$R_1(\alpha) \approx \begin{pmatrix} 1 & 0 & 0 \\ 0 & 1+\dfrac{\alpha^2}{2} & \alpha \\ 0 & -\alpha & 1+\dfrac{\alpha^2}{2} \end{pmatrix} \approx \begin{pmatrix} 1 & 0 & 0 \\ 0 & 1 & \alpha \\ 0 & -\alpha & 1 \end{pmatrix} = I + M_1(\alpha), \tag{A.3}$$

$$R_2(\beta) \approx \begin{pmatrix} 1+\dfrac{\beta^2}{2} & 0 & -\beta \\ 0 & 1 & 0 \\ \beta & 0 & 1+\dfrac{\beta^2}{2} \end{pmatrix} \approx \begin{pmatrix} 1 & 0 & -\beta \\ 0 & 1 & 0 \\ \beta & 0 & 1 \end{pmatrix} = I + M_2(\beta), \tag{A.4}$$

$$R_3(\gamma) \approx \begin{pmatrix} 1+\dfrac{\gamma^2}{2} & \gamma & 0 \\ -\gamma & 1+\dfrac{\gamma^2}{2} & 0 \\ 0 & 0 & 1 \end{pmatrix} \approx \begin{pmatrix} 1 & \gamma & 0 \\ -\gamma & 1 & 0 \\ 0 & 0 & 1 \end{pmatrix} = I + M_3(\gamma), \tag{A.5}$$

where I is the unit matrix:

$$I = \begin{pmatrix} 1 & 0 & 0 \\ 0 & 1 & 0 \\ 0 & 0 & 1 \end{pmatrix}, \tag{A.6}$$

and $M_1(\alpha)$, $M_2(\beta)$, and $M_3(\gamma)$ are given by

$$M_1(\alpha) = \begin{pmatrix} 0 & 0 & 0 \\ 0 & 0 & \alpha \\ 0 & -\alpha & 0 \end{pmatrix}, \tag{A.7}$$

$$M_2(\beta) = \begin{pmatrix} 0 & 0 & -\beta \\ 0 & 0 & 0 \\ \beta & 0 & 0 \end{pmatrix}, \tag{A.8}$$

$$M_3(\gamma) = \begin{pmatrix} 0 & \gamma & 0 \\ -\gamma & 0 & 0 \\ 0 & 0 & 0 \end{pmatrix}. \tag{A.9}$$

For an infinitesimal rotation, the displacement of a vector $\mathbf{r}$ can be expressed as a combination of the three infinitesimal rotations and infinitesimal rotation matrices; it can be written as

$$\mathbf{r}' = \mathbf{r} + M\mathbf{r} = (I + M)\mathbf{r}, \tag{A.10}$$

where the infinitesimal rotation matrix M is given by $M(\alpha, \beta, \gamma) = M_1(\alpha) + M_2(\beta) + M_3(\gamma)$, with

$$M(\alpha, \beta, \gamma) = \begin{pmatrix} 0 & \gamma & -\beta \\ -\gamma & 0 & \alpha \\ \beta & -\alpha & 0 \end{pmatrix}. \tag{A.11}$$

A.2 Rotation vector and rotation matrix

Let us consider a rotation around a particular direction, represented by the vector of rotation:

$$\boldsymbol{\omega} = \begin{pmatrix} \omega_1 \\ \omega_2 \\ \omega_3 \end{pmatrix} = \omega \mathbf{n}. \tag{A.12}$$

The rotation induced by this vector on a point at $\mathbf{r}$ is:

$$\boldsymbol{\omega} \wedge \mathbf{r} = \begin{pmatrix} \omega_1 \\ \omega_2 \\ \omega_3 \end{pmatrix} \wedge \begin{pmatrix} x \\ y \\ z \end{pmatrix}. \tag{A.13}$$

Let us now consider one component after the other and express it in terms of a rotation generator and infinitesimal rotation matrix:

- The rotation induced by the first component of the rotation vector, referring to Eq. A.7, is:

$$\omega_1 \hat{\mathbf{e}}_1 \wedge \mathbf{r} = \begin{pmatrix} \omega_1 \\ 0 \\ 0 \end{pmatrix} \wedge \begin{pmatrix} x \\ y \\ z \end{pmatrix} = \begin{pmatrix} 0 \\ -\omega_1 z \\ \omega_1 y \end{pmatrix}. \tag{A.14}$$

This expression can be written in the first order in terms of the following infinitesimal rotation matrix $M_1(-\omega_1)$ as $M_1(-\omega_1)\mathbf{r}$:

$$M_1(-\omega_1) = \begin{pmatrix} 0 & 0 & 0 \\ 0 & 0 & -\omega_1 \\ 0 & \omega_1 & 0 \end{pmatrix}. \tag{A.15}$$

- Consider now the second component:

$$\omega_2\hat{\mathbf{e}}_2 \wedge \mathbf{r} = \begin{pmatrix} 0 \\ \omega_2 \\ 0 \end{pmatrix} \wedge \begin{pmatrix} x \\ y \\ z \end{pmatrix} = \begin{pmatrix} \omega_2 z \\ 0 \\ -\omega_2 x \end{pmatrix}. \tag{A.16}$$

- Consider now the third component:

$$\omega_3\hat{\mathbf{e}}_3 \wedge \mathbf{r} = \begin{pmatrix} 0 \\ 0 \\ \omega_3 \end{pmatrix} \wedge \begin{pmatrix} x \\ y \\ z \end{pmatrix} = \begin{pmatrix} -\omega_3 y \\ \omega_3 x \\ 0 \end{pmatrix}. \tag{A.17}$$

To the first order, the results of the rotations about the second and third axes, $M_2(-\omega_2)$ and $M_3(-\omega_3)$, may similarly be expressed as $M_2(-\omega_2)\mathbf{r}$ and $M_3(-\omega_3)\mathbf{r}$, respectively, where

$$M_2(-\omega_2) = \begin{pmatrix} 0 & 0 & \omega_2 \\ 0 & 0 & 0 \\ -\omega_2 & 0 & 0 \end{pmatrix}, \tag{A.18}$$

and

$$M_3(-\omega_3) = \begin{pmatrix} 0 & -\omega_3 & 0 \\ \omega_3 & 0 & 0 \\ 0 & 0 & 0 \end{pmatrix}. \tag{A.19}$$

The rotation can thus also be expressed using an infinitesimal rotation matrix $M(-\omega_1, -\omega_2, -\omega_3) = M_1(-\omega_1) + M_2(-\omega_2) + M_3(-\omega_3)$, with

$$M(-\omega_1, -\omega_2, -\omega_3) = \begin{pmatrix} 0 & -\omega_3 & \omega_2 \\ \omega_3 & 0 & -\omega_1 \\ -\omega_2 & \omega_1 & 0 \end{pmatrix}, \tag{A.20}$$

with

$$\begin{pmatrix} \omega_1 \\ \omega_2 \\ \omega_3 \end{pmatrix} \wedge \begin{pmatrix} x \\ y \\ z \end{pmatrix} = M(-\omega_1, -\omega_2, -\omega_3) \begin{pmatrix} x \\ y \\ z \end{pmatrix}. \tag{A.21}$$

A.3 Rotation from change of pole position

Another way of expressing rotations is to use three rotation angles around different unit directions in order to bring the vector $\mathbf{r}$ to another position $\mathbf{r}'$. This is performed by using the notation of quantum mechanics to express a rotation. We use the symbols J_i for matrices defined by

$$J_1 = \begin{pmatrix} 0 & 0 & 0 \\ 0 & 0 & -i \\ 0 & i & 0 \end{pmatrix}, \quad J_2 = \begin{pmatrix} 0 & 0 & i \\ 0 & 0 & 0 \\ -i & 0 & 0 \end{pmatrix}, \quad J_3 = \begin{pmatrix} 0 & -i & 0 \\ i & 0 & 0 \\ 0 & 0 & 0 \end{pmatrix}. \tag{A.22}$$

We note that these matrices obey the commutation rules $[J_i, J_j] = i\epsilon_{ijk}J_k$ where ϵ_{ijk} are defined by:

$$\epsilon_{ijk} = \begin{cases} +1 & \text{if } (i, j, k) \text{ is } (1, 2, 3), (3, 1, 2) \text{ or } (2, 3, 1), \\ -1 & \text{if } (i, j, k) \text{ is } (1, 3, 2), (3, 2, 1) \text{ or } (2, 1, 3), \\ 0 & \text{if } i = j \text{ or } j = k \text{ or } k = i, \end{cases} \tag{A.23}$$

or

$$\epsilon_{ijk} = \frac{(i - j)(j - k)(k - i)}{2}. \tag{A.24}$$

Any rotation matrix operating on the three components of a vector can be written as $e^{i(\mathbf{J} \cdot \boldsymbol{\alpha})}$ where the vector $\boldsymbol{\alpha}$ gives the orientation of the axis of rotation as well as the *angle of rotation*, or equivalently $e^{i\,\alpha(\mathbf{J} \cdot \mathbf{n})}$ where $\mathbf{n}$ is the unit vector in the direction of $\boldsymbol{\alpha}$ and $\alpha = |\boldsymbol{\alpha}|$ is the angle of rotation about $\mathbf{n}$.

Now, it may be verified easily from the definitions of the J_i that with any unit vector $\mathbf{n}$,

$$(\mathbf{J} \cdot \mathbf{n})^3 = \mathbf{J} \cdot \mathbf{n}. \tag{A.25}$$

Using this property, we can write the rotation generator, for an unspecified value of α, as

$$e^{i\alpha(\mathbf{J} \cdot \mathbf{n})} = I + i\sin\alpha(\mathbf{J} \cdot \mathbf{n}) + (\cos\alpha - 1)(\mathbf{J} \cdot \mathbf{n})^2. \tag{A.26}$$

We now consider a rotation which takes the z-axis of a reference frame to the unit vector (X, Y, Z) giving the direction of the z-axis of another reference frame in the first one. We shall show that the rotation and the rotation axis can be written

$$e^{i\boldsymbol{\alpha} \cdot \mathbf{J}}, \qquad \boldsymbol{\omega} = (Y, -X, 0). \tag{A.27}$$

On using $\mathbf{n} = (Y, -X, 0)/(X^2 + Y^2)^{1/2}$ in Eq. (A.26) we find

$$e^{i\alpha(\mathbf{J} \cdot \mathbf{n})} = I + \frac{\sin\alpha}{\sqrt{X^2 + Y^2}} i(J_1 Y - J_2 X) - \frac{1 - \cos\alpha}{X^2 + Y^2}(J_1 Y - J_2 X)^2. \tag{A.28}$$

The two matrices appearing in this expression are

$$i(J_1 Y - J_2 X) = \begin{pmatrix} 0 & 0 & X \\ 0 & 0 & Y \\ -X & -Y & 0 \end{pmatrix} \tag{A.29}$$

and

$$(J_1 Y - J_2 X)^2 = \begin{pmatrix} X^2 & XY & 0 \\ XY & Y^2 & 0 \\ 0 & 0 & X^2 + Y^2 \end{pmatrix}. \tag{A.30}$$

On introducing these into the previous equation, it becomes evident that if the operation of $e^{i\alpha(\mathbf{J} \cdot \mathbf{n})}$ on the Z-axis vector $(0, 0, 1)$ is to yield (X, Y, Z), it is necessary that

$$\sin\alpha = (X^2 + Y^2)^{1/2}. \tag{A.31}$$

Then

$$\frac{1 - \cos\alpha}{X^2 + Y^2} = \frac{1}{1 + \cos\alpha} = \frac{1}{1 + (1 - X^2 - Y^2)^{1/2}} = \frac{1}{1 + Z}. \tag{A.32}$$

With the abbreviation $q = 1/(1 + Z)$, we can now write the final expression for the rotation generator $e^{i\alpha(\mathbf{J}\cdot\mathbf{n})}$ as

$$e^{i\alpha(\mathbf{J}\cdot\mathbf{n})} = \begin{pmatrix} 1 - qX^2 & -qXY & X \\ -qXY & 1 - qY^2 & Y \\ -X & -Y & 1 - q(X^2 + Y^2) \end{pmatrix}. \tag{A.33}$$

If the position of the new frame (X, Y, Z) is close to $(0, 0, 1)$ with $Z \approx 1$, this yields $q \approx 1/2$ at the first order and thus

$$e^{i\alpha(\mathbf{J}\cdot\mathbf{n})} \approx \begin{pmatrix} 1 - \dfrac{X^2}{2} & \dfrac{-XY}{2} & X \\[2mm] \dfrac{-XY}{2} & 1 - \dfrac{Y^2}{2} & Y \\[2mm] -X & -Y & 1 - \dfrac{X^2 + Y^2}{2} \end{pmatrix}. \tag{A.34}$$

With X and Y being small as well,

$$e^{i\alpha(\mathbf{J}\cdot\mathbf{n})} \approx \begin{pmatrix} 1 & \dfrac{-XY}{2} & X \\[2mm] \dfrac{-XY}{2} & 1 & Y \\[2mm] -X & -Y & 1 \end{pmatrix}. \tag{A.35}$$

A.4 Change in pole position

Now we make the link between the two ways of expressing the rotation. The infinitesimal rotation matrix M_1 given by Eq. (A.15) can be written in the quantum mechanics notation as a rotation of small amplitude ω_1 around the $\hat{\mathbf{e}}_1$ direction on using Eq. (A.26),

$$e^{-i\omega_1(\mathbf{J}\cdot\hat{\mathbf{e}}_1)} = \begin{pmatrix} 1 & 0 & 0 \\[1mm] 0 & 1 - \dfrac{\omega_1^2}{2} & -\omega_1 \\[2mm] 0 & \omega_1 & 1 - \dfrac{\omega_1^2}{2} \end{pmatrix}, \tag{A.36}$$

and for ω_1 being very small as

$$e^{-i\omega_1(\mathbf{J}\cdot\hat{\mathbf{e}}_1)} = R_1(-\omega_1) \approx \begin{pmatrix} 1 & 0 & 0 \\ 0 & 1 & 0 \\ 0 & 0 & 1 \end{pmatrix} + \begin{pmatrix} 0 & 0 & 0 \\ 0 & 0 & -\omega_1 \\ 0 & \omega_1 & 0 \end{pmatrix} = I + M_1(-\omega_1). \tag{A.37}$$

The infinitesimal rotation matrix M_2 given by Eq. (A.18) can be written in the quantum mechanics notation as a rotation of amplitude ω_2 around the $\hat{\mathbf{e}}_2$ direction on using Eq. (A.26),

$$e^{-i\omega_2(\mathbf{J}\cdot\hat{\mathbf{e}}_2)} = \begin{pmatrix} 1 - \dfrac{\omega_2^2}{2} & 0 & \omega_2 \\[2mm] 0 & 1 & 0 \\[2mm] -\omega_2 & 0 & 1 - \dfrac{\omega_2^2}{2} \end{pmatrix}, \tag{A.38}$$

and for ω_2 being very small as

$$e^{-i\omega_2(\mathbf{J}\cdot\hat{\mathbf{e}}_2)} = R_2(-\omega_2) \approx \begin{pmatrix} 1 & 0 & 0 \\ 0 & 1 & 0 \\ 0 & 0 & 1 \end{pmatrix} + \begin{pmatrix} 0 & 0 & \omega_2 \\ 0 & 0 & 0 \\ -\omega_2 & 0 & 0 \end{pmatrix} = I + M_2(-\omega_2). \quad \text{(A.39)}$$

The infinitesimal rotation matrix M_3 given by Eq. (A.19) can be written in the quantum mechanics notation as a rotation of amplitude ω_3 around the $\hat{\mathbf{e}}_3$ direction on using Eq. (A.26),

$$e^{-i\omega_3(\mathbf{J}\cdot\hat{\mathbf{e}}_3)} = \begin{pmatrix} 1 - \dfrac{\omega_3^2}{2} & -\omega_3 & 0 \\[2mm] \omega_3 & 1 - \dfrac{\omega_3^2}{2} & 0 \\[2mm] 0 & 0 & 1 \end{pmatrix}, \quad \text{(A.40)}$$

and for ω_3 being very small as

$$e^{-i\omega_3(\mathbf{J}\cdot\hat{\mathbf{e}}_3)} = R_3(-\omega_3) \approx \begin{pmatrix} 1 & 0 & 0 \\ 0 & 1 & 0 \\ 0 & 0 & 1 \end{pmatrix} + \begin{pmatrix} 0 & -\omega_3 & 0 \\ \omega_3 & 0 & 0 \\ 0 & 0 & 0 \end{pmatrix} = I + M_3(-\omega_3). \quad \text{(A.41)}$$

Therefore to the first order we have

$$e^{-i\omega_1(\mathbf{J}\cdot\hat{\mathbf{e}}_1)} \, e^{-i\omega_2(\mathbf{J}\cdot\hat{\mathbf{e}}_2)} \, e^{-i\omega_3(\mathbf{J}\cdot\hat{\mathbf{e}}_3)} = R_1(-\omega_1) \, R_2(-\omega_2) \, R_3(-\omega_3)$$

$$\approx I + M(-\omega_1, -\omega_2, -\omega_3) = I + M_1(-\omega_1) + M_2(-\omega_2) + M_3(-\omega_3). \quad \text{(A.42)}$$

On comparing Eq. (A.35) with the above Eqs. (A.36), (A.38), and (A.41) (no need to consider Eq. (A.38) as $\omega_3 = XY/2$ is already very small compared to X and Y), we immediately see that $\omega_1 = -Y$, $\omega_2 = X$, and $\omega_3 = XY/2$, and thus

$$\begin{pmatrix} 1 - \dfrac{X^2}{2} & \dfrac{-XY}{2} & X \\[3mm] \dfrac{-XY}{2} & 1 - \dfrac{Y^2}{2} & Y \\[3mm] -X & -Y & 1 - \dfrac{X^2 + Y^2}{2} \end{pmatrix} = R_1(Y)R_2(-X)R_3\left(\dfrac{-XY}{2}\right). \quad \text{(A.43)}$$

Appendix B

Clairaut theory

B.1 Expression of the Earth's gravitational potential

We start from the general definition of the gravitational potential of the Earth at a position $\mathbf{r}$ outside the Earth, at a distance r, colatitude θ, and longitude λ. In a reference frame with its origin at the Earth's center of mass, the expression for the gravitational potential (in the context of Clairaut's theory) is:

$$W = -\frac{GM}{r} \left\{ 1 + \sum_{n=2}^{\infty} \sum_{m=0}^{n} \left(\frac{a_e}{r} \right)^n \left(C_n^m \cos m\lambda + S_n^m \sin m\lambda \right) P_n^m(\cos\theta) \right\}, \quad \text{(B.1)}$$

where a_e is the equatorial radius of the Earth, the $P_n^m(\cos\theta)$ are the associated Legendre functions, and

$$C_n^0 = -J_n. \quad \text{(B.2)}$$

The terms for $n = 2$ in this expansion are by far the largest in magnitude when r is much larger than the Earth's dimensions. So we begin by dropping all terms with $n > 2$. In what follows we shall choose our geocentric reference frame to have its axes along the axes of principal moments of inertia.

- The degree zero part of the potential is:

$$W_0 = -\frac{G}{r} \int_V \rho(x', y', z') \, dx' \, dy' \, dz' \quad \text{(B.3)}$$

or

$$W_0 = -\frac{GM_E}{r}, \quad \text{(B.4)}$$

where M_E is the Earth's mass.
- The part of degree 1 is:

$$W_1 = -\frac{G}{r^2} \int_V \rho(x', y', z') r' \cos\psi \, dx' \, dy' \, dz'. \quad \text{(B.5)}$$

The angle ψ is between the directions of a point outside the Earth (at (x, y, z)) and of the mass element inside the Earth dm (at (x', y', z')); the cosine $\cos\psi$ may be expressed

471

as the scalar product between the two position-vectors by:

$$rr' \cos \psi = xx' + yy' + zz'. \tag{B.6}$$

We then write W_1:

$$W_1 = -\frac{G}{r^3} \left[x \int_V \rho(x', y', z') x' \, dx' \, dy' \, dz' + y \int_V \rho(x', y', z') y' \, dx' \, dy' \, dz' \right.$$

$$\left. + z \int_V \rho(x', y', z') z' \, dx' dy' \, dz' \right]. \tag{B.7}$$

When choosing the frame with its origin at the center of mass:

$$\int_V x' \rho(x', y', z') \, dx' \, dy' \, dz' = 0,$$

$$\int_V y' \rho(x', y', z') \, dx' \, dy' \, dz' = 0, \tag{B.8}$$

$$\int_V z' \rho(x', y', z') \, dx' \, dy' \, dz' = 0,$$

one finds that

$$W_1 = 0. \tag{B.9}$$

- At degree two:

$$W_2 = -\frac{G}{2r^3} \int_V \rho(x', y', z') r'^2 (3 \cos^2 \psi - 1) \, dx' \, dy' \, dz'. \tag{B.10}$$

We use the relation (B.6) to transform (B.10) to:

$$W_2 = -\frac{G}{2r^3} \int_V \rho(x', y', z') \left(\frac{3}{r^2} (xx' + yy' + zz')^2 - (x'^2 + y'^2 + z'^2) \right) dx' \, dy' \, dz'. \tag{B.11}$$

Considering that the reference frame is tied to the principal momentum axes, we get:

$$A = \int_V \rho(x', y', z')(y'^2 + z'^2) \, dV, \tag{B.12}$$

$$B = \int_V \rho(x', y', z')(x'^2 + z'^2) \, dV, \tag{B.13}$$

$$C = \int_V \rho(x', y', z')(x'^2 + y'^2) \, dV, \tag{B.14}$$

$$\int_V \rho(x', y', z') xy \, dV = 0, \tag{B.15}$$

$$\int_V \rho(x', y', z') yz \, dV = 0, \tag{B.16}$$

$$\int_V \rho(x', y', z') xz \, dV = 0. \tag{B.17}$$

On using the relations (B.15), (B.16), and (B.17), Eq. (B.11) for W_2 becomes:

$$W_2 = -\frac{G}{2r^3} \int_V \rho(x', y', z') \left(\frac{3(x^2 x'^2 + y^2 y'^2 + z^2 z'^2)}{r^2} - (x'^2 + y'^2 + z'^2) \right) dx' \, dy' \, dz'.$$
(B.18)

As $x^2 + y^2 + z^2 = r^2$,

$$\frac{x^2}{r^2} = 1 - \frac{y^2}{r^2} - \frac{z^2}{r^2},$$
(B.19)

$$\frac{y^2}{r^2} = 1 - \frac{x^2}{r^2} - \frac{z^2}{r^2},$$
(B.20)

$$\frac{z^2}{r^2} = 1 - \frac{x^2}{r^2} - \frac{y^2}{r^2}.$$
(B.21)

We then substitute Eq. (B.19), (B.20), and (B.21) in Eq. (B.18):

$$W_2 = -\frac{G}{2r^3} \int_V \rho(x', y', z') \left(2(x'^2 + y'^2 + z'^2) - 3\frac{x^2}{r^2}(y'^2 + z'^2) \right.$$

$$\left. - 3\frac{y^2}{r^2}(x'^2 + z'^2) - 3\frac{z^2}{r^2}(x'^2 + y'^2) \right) dx' \, dy' \, dz', \quad \text{(B.22)}$$

and substitute the relations (B.12), (B.13), and (B.14) in Eq. (B.22):

$$W_2 = -\frac{G}{2r^3} \left(A + B + C - 3\frac{x^2}{r^2}A - 3\frac{y^2}{r^2}B - 3\frac{z^2}{r^2}C \right).$$
(B.23)

When considering that the Earth is a rotating ellipsoid with

$$A = B,$$
(B.24)

W_2 can be written as

$$W_2 = -\frac{G}{2r^3} \left(2A + C - 3A\frac{(x^2 + y^2)}{r^2} - 3C\frac{z^2}{r^2} \right).$$
(B.25)

On using the relations between (x, y, z) and its spherical coordinates (r, θ, λ) (i.e. $x = r \sin \theta \cos \lambda$, $y = r \sin \theta \sin \lambda$, and $z = r \cos \theta$), W_2 in Eq. (B.25) can be written as

$$W_2 = -\frac{G}{2r^3} \left(2A + C - 3A \sin^2 \theta - 3C \cos^2 \theta \right)$$

$$= -\frac{G}{2r^3} \left(C - A - 3(C - A) \cos^2 \theta \right),$$
(B.26)

and thus

$$W_2 = \frac{GM_E}{r}\left(\frac{a_e}{r}\right)^2 \frac{C-A}{M_E a_e^2}\frac{3\cos^2\theta - 1}{2},\tag{B.27}$$

where a_e is the Earth's equatorial radius. This yields

$$W_2 = \frac{GM_E}{r}\left(\frac{a_e}{r}\right)^2 \frac{C-A}{M_E a_e^2}P_2(\cos\theta).\tag{B.28}$$

The degree two potential can thus be written in terms of the Legendre polynomial of degree two.

- The total potential in the case of a reference system tied to the principal inertia axes for an Earth with a rotational symmetry, may be written using (B.4), (B.9), and (B.28), and generalizing the expression (B.28):

$$W = -\frac{GM_E}{r}\left(1 - \sum_{n=2}^{\infty}\left(\frac{a_e}{r}\right)^n J_n P_n(\cos\theta)\right),\tag{B.29}$$

with

$$J_2 = \frac{C-A}{M_E a_e^2},\tag{B.30}$$

called the dynamical form factor of the Earth, a fundamental constant in astronomy and geodesy.

- When the reference system is a general reference system not related to the principal moments of inertia, and when the Earth has no rotational symmetry, a dependence in terms of the longitude λ must be introduced:

$$W = -\frac{GM_E}{r}\left\{1 + \sum_{n=2}^{\infty}\sum_{m=0}^{n}\left(\frac{a_e}{r}\right)^n \left(C_n^m \cos m\lambda + S_n^m \sin m\lambda\right)P_n^m(\cos\theta)\right\},\tag{B.31}$$

with $P_n^m(\cos\theta)$, the generalized Legendre polynomials, and

$$C_n^0 = -J_n.\tag{B.32}$$

B.2 Equipotential surface at the first order – Clairaut's ellipsoid

The fundamental hypothesis of Clairaut's (1713–1765) theory is that the Earth is in hydrostatic equilibrium, that the surface of the Earth is an equipotential corresponding to the mean sea level, and that the Earth is uniformly rotating about the polar axis.

We then have to obtain the expression for the gravity from the total potential U which is the sum of the geopotential W and the centrifugal potential:

$$U = W - \frac{1}{2}\Omega^2 r^2 \sin^2\theta.\tag{B.33}$$

On using for W the second-degree expression W_2 of Eq. (B.28) together with Eq. (B.30), we obtain

$$U_2(\mathbf{r}) = -\frac{GM}{r}\left(1 - \left(\frac{a}{r}\right)^2 J_2 \frac{3\cos^2\theta - 1}{2}\right) - \frac{1}{2}\Omega^2 r^2 \sin^2\theta. \qquad (B.34)$$

An equipotential surface corresponds to $U_2 = $ constant. Since the surface of the Earth is equipotential, the expression Eq. (B.34) must have equal values at any equatorial point $(r = a, \theta = \pi/2)$ and at the poles $(r = c, \theta = 0$ or $\pi)$. The value at the equator is

$$U_2(r = a, \theta = \pi/2) = -\frac{GM}{a}\left(1 + \frac{J_2}{2}\right) - \frac{\Omega^2 a^2}{2}. \qquad (B.35)$$

At the pole $r = c, \theta = 0$ or π

$$U_2(r = c, \theta = 0 \text{ or } \pi) = -\frac{GM}{c}\left(1 - \left(\frac{a}{c}\right)^2 J_2\right). \qquad (B.36)$$

The above two expressions have to be equal. We find thus that

$$\frac{GM}{a}\left(1 + \frac{J_2}{2}\right) + \frac{\Omega^2 a^2}{2} = \frac{GM}{c}\left(1 - \left(\frac{a}{c}\right)^2 J_2\right). \qquad (B.37)$$

The flattening of the Earth, with a value of $\approx (1/298.25)$ is defined by

$$\epsilon = \frac{a - c}{a}. \qquad (B.38)$$

Or equivalently,

$$c = a(1 - \epsilon). \qquad (B.39)$$

On using the Taylor expansion in the small quantity ϵ, this goes over to

$$\frac{1}{1+\epsilon} = 1 - \epsilon + \epsilon^2 + \cdots, \qquad (B.40)$$

and to the first order

$$c = \frac{a}{1+\epsilon}, \qquad (B.41)$$

which can be substituted in Eq. (B.37) to obtain to the first order in the small quantities such as ϵ and J_2:

$$\frac{GM}{c}\left(1 - \left(\frac{a}{c}\right)^2 J_2\right) = \frac{GM}{a}(1 + \epsilon) - \frac{GM}{a}J_2(1 + \epsilon)^3. \qquad (B.42)$$

On introducing Eq. (B.41) in Eq. (B.42) and considering that Eqs. (B.35) and (B.42) represent the same constant potential, this yields

$$\frac{\Omega^2 a^2}{2} = \frac{GM}{a}\left(\epsilon - \frac{3}{2}J_2\right).$$

(B.43)

We obtain the relation between the surface flattening ϵ, the dynamical form factor of the Earth J_2 and a coefficient q,

$$\epsilon = \frac{q}{2} + \frac{3}{2}J_2,$$

(B.44)

where q is the ratio between the centrifugal force and gravitational accelerations at the equator on a sphere of radius a,

$$q = \frac{\Omega^2 a}{\frac{GM}{a^2}} = \frac{\Omega^2 a^3}{GM}.$$

(B.45)

We note that for a homogeneous Earth, $M = \frac{4}{3}\pi a^3 \overline{\rho}$, and thus $q = (3\Omega^2)/(4\pi G\overline{\rho})$, which does not depend on the Earth's radius any more.

We now consider the surface of the Earth which is equipotential, so that $U_2(\mathbf{r})$ at any point of the surface has the same value as at the equator:

$$U_2(\mathbf{r}) = U_2\left(r = a, \theta = \frac{\pi}{2}\right).$$

(B.46)

On substituting from Eqs. (B.34) and (B.35) and then using the value of q from Eq. (B.45) as well as the relation (B.44) connecting q, ϵ, and J_2, we obtain the explicit form of (B.46)) as

$$\frac{GM}{r}\left(1 - J_2\frac{3\cos^2\theta - 1}{2} + \frac{q}{2}\sin^2\theta\right) = \frac{GM}{a}\left(1 + \frac{J_2}{2} + \frac{q}{2}\right).$$

(B.47)

On grouping the terms in r on the left-hand side of the equation and the terms in a on the right-hand side of the equation, this yields:

$$r\left(1 + \frac{J_2}{2} + \frac{q}{2}\right) = a\left(1 - J_2\frac{3\cos^2\theta - 1}{2} + \frac{q}{2}\sin^2\theta\right).$$

(B.48)

We solve for (r/a) now, neglecting the second-order terms (containing J_2^2, q^2, and qJ_2):

$$\frac{r}{a} = 1 - \frac{1}{2}J_2(3\cos^2\theta - 1) + \frac{q}{2}\sin^2\theta - \frac{1}{2}J_2 - \frac{q}{2}$$

(B.49)

or equivalently

$$r = a\left[1 - \left(\frac{3}{2}J_2 + \frac{q}{2}\right)\cos^2\theta\right].$$

(B.50)

On using Eq. (B.44), we arrive at an equation expressing the position of a point on the ellipsoid of equatorial radius a and flattening ϵ:

$$r = a(1 - \epsilon \cos^2 \theta). \tag{B.51}$$

B.3 Gravity on the surface of Clairaut's ellipsoid

The gravity g can be obtained by taking the gradient of the potential along the normal

$$g = \frac{\partial U}{\partial r} \cos \alpha - \frac{1}{r} \frac{\partial U}{\partial \theta} \sin \alpha, \tag{B.52}$$

where α is the deviation angle of the vertical with respect to the radial direction toward the Earth's center. Since α is extremely small, g is very well approximated by $-\partial U/\partial r$. On using the expression (B.34) of U_2 for U, differentiating, we obtain

$$g = \frac{\partial U_2}{\partial r} = \frac{GM}{r^2} - 3\frac{GM}{r^4} a^2 J_2 \frac{(3 \cos^2 \theta - 1)}{2} - \Omega^2 r \sin^2 \theta \tag{B.53}$$

or equivalently

$$g = \frac{GM}{r^2} \left[1 - 3\left(\frac{a}{r}\right)^2 J_2 \frac{(3 \cos^2 \theta - 1)}{2} - q\left(\frac{r}{a}\right)^3 \sin^2 \theta \right]. \tag{B.54}$$

In the terms containing the ratio $\left(\frac{a}{r}\right)^2$ or $\left(\frac{r}{a}\right)^3$, to a first approximation in the small quantities we can consider $r = a$. However, in the term Gm/r^2, r must be replaced by its expression (B.50):

$$g = \frac{GM}{a^2} \frac{\left(1 - 3 J_2 \frac{(3 \cos^2 \theta - 1)}{2} - q \sin^2 \theta\right)}{\left(1 - \left(\frac{3}{2} J_2 + \frac{q}{2}\right) \cos^2 \theta\right)^2}. \tag{B.55}$$

Developing to the first order in J_2 and q, we obtain

$$g = \frac{GM}{a^2} \left(1 - \frac{3}{2} J_2(3 \cos^2 \theta - 1) + 3 J_2 \cos^2 \theta - q \sin^2 \theta + q \cos^2 \theta\right) \tag{B.56}$$

or equivalently

$$g = \frac{GM}{a^2} \left(1 + \frac{3}{2} J_2 - q + \cos^2 \theta (2q - \frac{3}{2} J_2)\right). \tag{B.57}$$

As at the equator ($\theta = 90°$), we have

$$g_E = \frac{GM}{a^2} \left(1 + \frac{3}{2} J_2 - q\right), \tag{B.58}$$

and at the pole ($\theta = 0°$) we have

$$g_P = \frac{GM}{a^2} (1 + q). \tag{B.59}$$

We can deduce that

$$\frac{g_P}{g_E} = \frac{1 + q}{1 + \frac{3}{2} J_2 - q} = 1 + 2q - \frac{3}{2} J_2 \tag{B.60}$$

or equivalently

$$\frac{g_P - g_E}{g_E} = 2q - \frac{3}{2}J_2 = \beta. \tag{B.61}$$

On substituting Eq. (B.58) and (B.61) in Eq. (B.57) we obtain:

$$g = g_E(1 + \beta \cos^2 \theta). \tag{B.62}$$

This equation gives a succinct expression for gravity in Clairaut's theory.

Appendix C

Definitions of equinoxes

C.1 Definition of ecliptic and orbital plane

The ecliptic, loosely speaking, is the plane in which the Earth–Moon barycenter (abbreviated as EMB hereafter) moves in space. The orbital plane itself does not remain fixed in space but rotates slowly about an axis in the instantaneous position of the plane. Therefore, the orbit does not quite close on itself at the end of one period; the orbit is a tight spiral. Nevertheless, the deviation from a closed orbit in one period is very small, and one continues to talk of the orbital plane for convenience.

C.2 Definition of "inertial" and "rotating" equinox

The "inertial" and "rotating" definitions of the equinox rest upon the definitions of the corresponding ecliptics, as given by Standish (1981). The two equinoxes are determined by the intersections of the respective ecliptics with the mean equator of date.

The *ecliptic in the inertial sense* is defined to be the plane perpendicular to the angular momentum vector

$$\mathbf{L} = \mathbf{r} \wedge \frac{d\mathbf{r}}{dt} \tag{C.1}$$

of the orbital motion of the EMB in inertial space.

The velocity $d\mathbf{r}/dt$, which is in an inertial reference frame, may be written as

$$\frac{d\mathbf{r}}{dt} = \left(\frac{d\mathbf{r}}{dt}\right)_{\text{rot}} + \mathbf{b} \wedge \mathbf{r}, \tag{C.2}$$

where $\mathbf{b}$ is the angular velocity of rotation of the orbital plane in space, and the subscript rot indicates that the velocity is referred to a frame tied to the rotating orbital plane "of date."

The *ecliptic in the rotating sense* is defined to be perpendicular to the vector

$$\mathbf{L}' = \mathbf{r} \wedge \left(\frac{d\mathbf{r}}{dt}\right)_{\text{rot}}. \tag{C.3}$$

The relation between the vectors $\mathbf{L}$ and $\mathbf{L}'$ is evidently

$$\mathbf{L} = \mathbf{L}' + \mathbf{r} \wedge (\mathbf{b} \wedge \mathbf{r}). \tag{C.4}$$

The difference between $\mathbf{L}$ and $\mathbf{L}'$ is the contribution to the angular momentum from the rotation of the orbital plane in space; $\mathbf{L}'$ is the angular momentum of motion in the orbital plane of date on taking it to be instantaneously non-rotating (i.e., having $\mathbf{b} = 0$).

C.3 Expression in inertial frame

It may be instructive to see the above equations expressed in terms of components in an inertial reference frame (IRF) with its axes aligned with those of the ICRS, but with its origin at the center of the orbit of the EMB. Assume, for simplicity, that the motion of the EMB is along a circular orbit of unit radius, with uniform angular speed. The inclination of the orbit to the principal plane of the IRF is the obliquity ϵ, which varies with time.

1. Let the x-axis of the IRF, say $\hat{\mathbf{x}}$, be along the intersection between this principal plane and the orbital plane. Then the position vector $\mathbf{r}$ of the EMB in the IRF has coordinates

$$x = \cos \omega t,$$
$$y = \cos \epsilon \sin \omega t, \tag{C.5}$$
$$z = \sin \epsilon \sin \omega t,$$

where ω is the angular frequency of orbital motion of the EMB.

2. The orbital plane is known to be rotating slowly about an axis lying in the plane. The magnitude of the angular velocity vector $\mathbf{b}$ is $\dot{\pi}_A$ and the direction of $\mathbf{b}$ is almost opposite to that of the vernal equinox, i.e., close to $(-\hat{\mathbf{x}})$. Let us, for simplicity, make the approximation that the direction of $\mathbf{b}$ coincides with that of $-\hat{\mathbf{x}}$, i.e., that $\mathbf{b} = -\hat{\mathbf{x}}\dot{\pi}$. Then

$$\mathbf{b} \approx \dot{\epsilon}\,\hat{\mathbf{x}}, \tag{C.6}$$

with $\dot{\epsilon} = -\dot{\pi}_A$, noting that as π_A increases, ϵ decreases.

With the obliquity ϵ varying slowly, the orbit is no longer a strictly closed curve, and it does not, strictly speaking, lie in a plane, as already noted.

The velocity components (with reference to the IRF) of the motion in the moving ecliptic are

$$\dot{x} = -\omega \sin \omega t,$$
$$\dot{y} = \omega \cos \epsilon \cos \omega t - \dot{\epsilon} \sin \epsilon \sin \omega t, \tag{C.7}$$
$$\dot{z} = \omega \sin \epsilon \cos \omega t + \dot{\epsilon} \cos \epsilon \sin \omega t.$$

3. Consider now the angular momentum $\mathbf{L} = \mathbf{r} \wedge \dot{\mathbf{r}}$ of the motion. On using (C.5) and (C.7), its components turn out to be

$$L_x = \dot{\epsilon} \sin^2 \omega t,$$
$$L_y = -\omega \sin \epsilon - \dot{\epsilon} \cos \epsilon \sin \omega t \cos \omega t, \tag{C.8}$$
$$L_z = \omega \cos \epsilon - \dot{\epsilon} \sin \epsilon \sin \omega t \cos \omega t,$$

or, in vector form,

$$\mathbf{L} = \mathbf{L}' + \mathbf{r} \wedge (\mathbf{b} \wedge \mathbf{r}), \tag{C.9}$$

when the approximation introduced earlier is used for $\mathbf{b}$. This is the same as Eq. (C.4). The approximation (C.6) is of course not needed for its validity.

4. By construction, L_x, L_y, L_z are the components, with respect to an *inertial reference frame*, of the instantaneous angular momentum $\mathbf{L}$ of motion in the *rotating orbital plane*.

The part $\mathbf{L}'$ does not involve $\dot{\epsilon}$, and yet it is time dependent because it involves the obliquity ϵ *of date*; this reinforces the characterization of $\mathbf{L}'$, below Eq. (C.4), as the angular momentum of motion in an *instantaneously non-rotating* orbital plane. The difference $(\mathbf{L} - \mathbf{L}')$ may be visualized as arising from the fact that the EMB moves over from one side of the instantaneously non-rotating plane to the other in the course of one orbital period, as a consequence of the slow rotation of the orbital plane.

5. The equinoxes: Since the equinox lies in the ecliptic, the vector from the center of the orbit in the direction of the equinox is, by definition, perpendicular to the relevant angular momentum vector; the equinox lies also in the principal plane of the IRF. The unit vector in the direction of the equinox in the *inertial* sense (which is perpendicular to $\mathbf{L}$ of (C.8)) is then readily seen to have the components

$$\left(1, \frac{\dot{\epsilon}\sin^2\omega t}{\omega\sin\epsilon}, 0\right) \rightarrow \left(1, \frac{\dot{\epsilon}}{2\omega\sin\epsilon}, 0\right) \tag{C.10}$$

in the IRF, to the lowest order in $\dot{\epsilon}$. (The second form is the average of the first over the orbital period.) The equinox in the *rotational* sense is in the direction $(1, 0, 0)$, perpendicular to the angular momentum $\mathbf{L}'$; it is in the direction $\hat{\mathbf{x}}$. These results, though subject to the approximation (C.6), serve to illustrate the distinction between the two kinds of equinox.

6. It should be observed that Eq. (C.4) is the same as the equation at the top of the second column of the first page of the Standish (1981) paper: just replace $\mathbf{L}$ by $\mathbf{h}$, and $\mathbf{L}'$ by $\mathbf{h}'$. Note also that the mean of $\mathbf{L}$ (wherein the periodic terms in (C.8), which have frequency 2ω, are averaged out) has components

$$\bar{L}_x = (1/2)\dot{\epsilon},$$
$$\bar{L}_y = \omega\sin\epsilon, \tag{C.11}$$
$$\bar{L}_z = \omega\cos\epsilon,$$

while the components of $\mathbf{L}'$ are $(0, -\omega\sin\epsilon, \omega\cos\epsilon)$. So the former has constant x-axis bias relative to the latter. Specifically,

$$\bar{\mathbf{L}} = \bar{\mathbf{L}}' + (1/2)\mathbf{b} \quad \rightarrow \quad \bar{\mathbf{h}} = \bar{\mathbf{h}}' + (1/2)\mathbf{b}. \tag{C.12}$$

7. The use of a more realistic elliptical orbit or of a rotation axis $\hat{\mathbf{b}}$ different from $-\hat{\mathbf{x}}$ would make no material difference to the above picture, though the use of the correct $\mathbf{b}$ is essential for recovering the precession of the equinox.

Bibliography

Abramowitz M. and Stegun I., 1964, 1972, *"Handbook of Mathematical Functions."* Dover Publications, USA, 1046 pp.

Altamimi Z., Sillard P., and Boucher C., 2002, "ITRF2000: a new release of the International Terrestrial Reference Frame for Earth science applications." *J. Geophys. Res. (Solid Earth)*, **107**(*B10*), pp ETG 2-1, CiteID 2214, DOI: 10.1029/2001JB000561.

Altamimi Z., Collilieux X., Legrand J., Garayt B., and Boucher C., 2007, "ITRF2005: a new release of the International Terrestrial Reference Frame based on time series of station positions and Earth Orientation Parameters." *J. Geophys. Res. (Solid Earth)*, **112**(*B9*), CiteID B09401, DOI: 10.1029/2007JB004949.

Altamimi Z., Métivier L., and Collilieux X., 2011, "ITRF2008: an improved solution of the international terrestrial reference frame." *J. Geodesy*, **85**(*8*), 457–473, DOI: 10.1007/s00190-011-0444-4.

Altamimi Z., Métivier L., and Collilieux X., 2012, "TRF2008 plate motion model." *J. Geophys. Res. (Solid Earth)*, **117**(*B7*), CiteID B07402, DOI: 10.1029/2011JB008930.

Alterman Z., Jarosch H., and Pekeris C.L., 1959, "Oscillations of the Earth." *Proc. Roy. Soc. London A*, **252**, 80–95.

Amit H. and Christensen U.R., 2008, "Accounting for magnetic diffusion in core flow inversions from geomagnetic secular variation." *Geophys. J. Int.*, **175**(*3*), 913–924, DOI: 10.1111/j.1365-246X.2008.03948.x.

Anderson D.L. and Minster J.B., 1979, "The frequency dependence of Q in the Earth and implications for mantle rheology and Chandler wobble." *Geophys. J. Int.*, **58**(*2*), 431–440, DOI: 10.1111/j.1365-246X.1979.tb01033.x.

Andoyer M.H., 1911, "Les Formulaes de la Précessiòn d'Apres S. Newcomb." *Bull. Astron. Soc.*, **28**, 66–76.

Andoyer M.H., 1923, *"Cours de Mécanique Céleste."* Vol. 1, Gauthier-Villar, Paris.

Aoki S., Kinoshita H., Guinot B., *et al.*, 1982, "The new definition of universal time." *Astr. Astrophys.*, **105**(*2*), 359–361.

Aoki S. and Kinoshita H., 1983a, "Note on the relation between the equinox and Guinot's non-rotating origin." *Celest. Mech.*, **29**, 335–360, DOI: 10.1007/BF01228528.

Aoki S. and Kinoshita H., 1983b, "The definition of the ecliptic." *Celest. Mech.*, **31**, 329–338, DOI: 10.1007/BF01230290.

Aoki S., 1988, "Relation between the celestial reference system and the terrestrial reference system of a rigid Earth." *Celest. Mech.*, **42**(*1–4*), 309–354.

Argus D.F., Gordon R.G., Heflin M.B., *et al.*, 2010, "The angular velocities of the plates and the velocity of Earth's centre from space geodesy." *Geophys. J. Int.*, **180**(*3*), 913–960, DOI: 10.1111/j.1365-246X.2009.04463.x.

Arias E.F., Feissel M., and Lestrade J.-F., 1988a, "An extragalactic celestial reference frame consistent with the BIH Terrestrial System (1987)." BIH Annual Rep. for 1987, pp D-113–D-121.

Arias E.F., Lestrade J.-F., and Feissel M., 1988b, "Comparison of VLBI celestial reference frames." *Astr. Astrophys.*, **199**, 357–363.

Arias E.F., Charlot P., Feissel M., and Lestrade J.-F., 1995, "The extragalactic reference system of the International Earth Rotation Service, ICRS." *Astr. Astrophys.*, **303**, 604–608.

Artz T., Bernhard L., Nothnagel A., Steigenberger P., and Tesmer S., 2012, "Methodology for the combination of sub-daily Earth rotation from GPS and VLBI observations." *J. Geodesy*, **86**(*3*), 221–239, DOI: 10.1007/s00190-011-0512-9.

Atkinson d'Escourt R. and Sadler D.H., 1951, "On the use of mean sidereal time." *Mon. Not. Roy. Astron. Soc.*, **111**, 619–623.

Atkinson d'Escourt R., 1973, "On the dynamical variation of latitude and time." *Astron. J.*, **78**, 147–161, DOI: 10.1086/111391.

Atkinson d'Escourt R., 1975, "On the Earth's axes of rotation and figure." *Mon. Not. Roy. Astron. Soc.*, **171**, 381–386.

Aubert J., 2013, "Flow throughout the Earth's core inverted from geomagnetic observations and numerical dynamo models." *Geophys. J. Int.*, **192**(*2*), 537–556, DOI: 10.1093/gji/ggs051.

Backus G.E., 1967, "Converting vector and tensor equations to scalar equations in spherical coordinates." *Geophys. J. R. Astron. Soc.*, **13**, 71–101.

Ball R.H., Kahle A.B., and Vestine E.H., 1969, "Determination of surface motions of the Earth's core." *J. Geophys. Res.*, **74**(*14*), 3659–3680, DOI: 10.1029/JA074i014p03659.

Barker B.M. and O'Connell R.F., 1970, "Derivation of the equations of motion of a gyroscope from the quantum theory of gravitation." *Phys. Rev. D*, **2**(*8*), 1428–1435, DOI: 10.1103/PhysRevD.2.1428.

Barker B.M. and O'Connell R.F., 1975, "Gravitational two-body problem with arbitrary masses, spins, and quadrupole moments." *Phys. Rev. D*, **12**(*2*), 329–335, DOI: 10.1103/PhysRevD.12.329.

Barnes R.T.H., Hide R., White A.A., and Wilson C.A., 1983, "Atmospheric angular momentum fluctuation, length-of-day changes and polar motion." *Proc. Roy. Soc. London, A*, **387**, 31–73.

Behrend D., 2013, "Data handling within the International VLBI service." *Data Science Journal*, **12**, WDS81–WDS84, DOI 10.2481/dsj.WDS-011.

Bell M.J., 1994, "Oscillations in the equatorial components of the atmosphere's angular momentum and torques on the Earth's bulge." *Quarterly J. Roy. Meteorol. Soc.*, **120**(*515*), 195–213.

Bellanger E., Gibert D., and Le Moul J.L., 2002, "A geomagnetic triggering of Chandler wobble phase jumps?" *Geophys. Res. Letters*, **29**(*7*), 28-1–28-4, CiteID: 1124, DOI: 10.1029/2001GL014253.

Benjamin D., Wahr J.M., Ray R.D., Egbert G.D., and Desai S.D., 2006, "Constraints on mantle anelasticity from geodetic observations, and implications for the J2 anomaly." *Geophys. J. Int.*, **165**(*1*), 3–16, DOI: 10.1111/j.1365-246X.2006.02915.x.

Beutler G., 2005, "*Methods of Celestial Mechanics*: Volume I: *Physical, Mathematical, and Numerical Principles*." Springer Verlag, Berlin, Heidelberg, New York.

Bizouard Ch., 1995, "Modélisation astrométrique et géophysique de la rotation de la Terre." Ph.D. thesis, Observatoire de Paris.

Bizouard C., Brzezinski A., and Petrov S., 1998, "Diurnal atmospheric forcing and temporal variations of the nutation amplitudes." *J. Geodesy*, **72**, 561–577.

Bizouard C. and Lambert S., 2002, "Lunisolar torque on the atmosphere and Earth's rotation." *Planet. Space Sci.*, **50**(*3*), 323–333.

Bizouard C., Remus F., Lambert S.B., Seoane L., and Gambis D., 2011, "The Earth's variable Chandler wobble." *Astr. Astrophys.*, **526**, id.A106, DOI: 10.1051/0004-6361/201015894.

Bizouard C. and Zotov L., 2013, "Asymmetric effects on Earth's polar motion." *Celest. Mech. Dynam. Astr.*, **116**(*2*), 195–212, DOI: 10.1007/s10569-013-9483-x.

Bleck R., Hanson H.P., Hu D.M., and Kraus E.B., 1989, "Mixed layer-thermocline interaction in a three-dimensional isopycnic coordinate model." *J. Phys. Ocean*, **19C**, 1417–1439.

Bleck R., Rooth C, Hu D.M., and Smith L.T., 1992, "Salinity-driven thermocline transients in a wind-forced and thermohaline-forced isopycnic coordinate model of the North-Atlantic." *J. Phys. Oceanogr.*, **22**, *12*, 1486–1505.

Bloxham J., 1998, "Dynamics of angular momentum in the Earth's core." *Annual Rev. Earth Planet. Sci.*, **26**, 501–517.

Boehm J., Schuh H., and Weber R., 2002, "Tropospheric zenith path delays derived from GPS used for the determination of VLBI station heights." In: Proc. IVS 2002 General Meeting, Tsukuba, Japan, February 4–7 2002, Eds. N.R. Vandenberg and K.D. Baver, pp 219–222.

Boehm J. and Schuh H., 2004a, "Vienna mapping functions in VLBI analyses." *Geophys. Res. Letters*, **31**(*1*), CiteID: L01603, DOI: 10.1029/2003GL018984.

Boehm J. and Schuh H., 2004b, "Tropospheric parameters over two decades determined by VLBI as a contribution to climatological studies." *Artificial Satellites – J. Planet. Geodesy*, **39**(*2*), special issue Proc. Seminar '*Earth rotation and satellite geodesy from astrometry to GNSS*', Part II: *Earth rotation, geodynamics, and dynamics of the satellite motion*, Warsaw, September 2003, pp 121–128.

Boehm J. and Schuh H., 2007, "Troposphere gradients from the ECMWF in VLBI analysis." *J. Geodesy*, **81**(*6–8*), DOI: 10.1007/s00190-007-0144-2, 403–408.

Boehm J., Hobiger T., Ichikawa R., *et al.*, 2010, "Asymmetric tropospheric delays from numerical weather models for UT1 determination from VLBI intensive sessions on the baseline Wettzell–Tsukuba." *J. Geodesy*, **84**(*5*), 319–325, DOI: 10.1007/s00190-010-0370-x.

Boucher C., Altamimi Z., and Willis P., 1988, "Relation between BTS87, WGS84 and GPS activities." BIH Annual Report 1987, p. D-131.

Boucher C. and Altamimi Z., 1989a, "The initial IERS Terrestrial Reference Frame 1989." IERS Technical Notes No. 1.

Boucher C. and Altamimi Z., 1989b, "Evaluation of the realizations of the terrestrial reference system done by the BIH and IERS (1984–1988)." IERS Technical Notes No. 4.

Boucher C. and Altamimi Z., 1989c, "ITRF89 and other realizations of the IERS terrestrial reference system for 1989." IERS Technical Notes No. 6.

Boucher C. and Altamimi Z., 1991, "ITRF90 and other realizations of the IERS terrestrial reference system for 1990." IERS Technical Notes No. 9.

Boucher C., Altamimi Z., and Duhem L., 1994, "Results and analysis of the ITRF93." IERS Technical Notes No. 18.

Boucher C., Altamimi Z., Feissel M., and Sillard P., 1996a, "Results and analysis of the ITRF94." IERS Technical Notes No. 20.

Boucher C., Altamimi Z., and Sillard P., 1996b, "Results and analysis of the ITRF96." IERS Technical Notes No. 24.

Boucher C., Altamimi Z., and Sillard P., 1999, "The 1997 international terrestrial reference frame (ITRF97)." IERS Technical Notes No. 27.

Boucher C., Altamimi Z., Sillard P., and Feissel-Vernier M., 2004, "The ITRF2000." IERS Technical Notes No. 31.

Bouman J., Floberghagen R., and Rummel R., 2013, "More than 50 years of progress in satellite gravimetry." EOS, AGU publication, DOI: 10.1002/2013EO310001, pp 269–270.

Bouquillon S., and Souchay J., 1999, "Precise modeling of the precession-nutation of Mars." *Astr. Astrophys.*, **345**, 282–297.

Bourda G. and Capitaine N., 2004, "Precession, nutation, and space geodetic determination of the Earth's variable gravity field." *Astr. Astrophys.*, **428**, 691–702, DOI: 10.1051/0004-6361:20041533.

Bourda G., Charlot P., and Le Campion J.-F., 2008, "Astrometric suitability of optically-bright ICRF sources for the alignment with the future Gaia celestial reference frame." *Astr. Astrophys.*, **490**(*1*), 403–408, DOI: 10.1051/0004-6361:200810667.

Bourda G., Charlot P., Porcas R.W., and Garrington S.T., 2010, "VLBI observations of optically-bright extragalactic radio sources for the alignment of the radio frame with the future Gaia frame. I. Source detection." *Astr. Astrophys.*, **520**, ID. A113, 8 pp, DOI: 10.1051/0004-6361/201014248.

Bradley J., 1748, "A letter to the right honourable George Earl of Macclesfield concerning an apparent motion observed in some of the fixed stars." *Phil. Trans. Roy. Soc.*, **45**, London.

Braginsky S.I. and Roberts P.H., 1995, "Equations governing convection in Earth's core and the geodynamo." *J. Geophys. Astrophys. Fluid Dynam.*, **79**, 1–97.

Bretagnon P., 1981, "Construction of a theory of the outer planets through an iterative method." *Astr. Astrophys.*, **101**(*3*), 342–349.

Bretagnon P. and Chapront J., 1981, "A note on the numerical expressions for precession calculations." *Astr. Astrophys.*, **103**(*1*), 103–107.

Bretagnon P., 1982, "Théorie du mouvement de l'ensemble des planètes. Solution VSOP82." *Astr. Astrophys.*, **114**, 278–288.

Bretagnon P. and Francou G., 1988, "Planetary theories in rectangular and spherical variables. VSOP87 solutions." *Astr. Astrophys.*, **202**, 309–315.

Bretagnon P., Rocher P., and Simon J.L., 1997, "Theory of the rotation of the rigid Earth." *Astr. Astrophys.*, **319**, 305–317.

Bretagnon P., Francou G., Rocher P., and Simon J.-L., 1998, "SMART97: a new solution for the rotation of the rigid Earth." *Astr. Astrophys.*, **329**, 329–338.

Bretagnon P., 1998, "Proposals for a new solution of the precession-nutation." In: Proc. *Journées Systèmes de Référence Spatio-temporels 1997*, Prague, Czech Rep., September 1997, Eds. J. Vondrák and N. Capitaine, pp 61–64.

Bretagnon P., 1999, "The planetary theories and the precession of the ecliptic." XXIIIrd General Assembly of IAU, Kyoto, invited paper JD 3 on 'Precession-nutation and astronomical constants for the dawn of the 21st century'.

Bretagnon P., Rocher P., and Simon J.L., 2000, "Non-rigid Earth rotation solution." In: *Towards Models and Constants for Sub-microarcsecond Astrometry.*, IAU

Colloquium 180, Eds. K. Johnson, D. McCarthy, B. Luzum, G. Kaplan, Washington, USA, pp 230.

Bretagnon P., Rocher P., and Simon J.L., 2001, "Towards the construction of a new precession–nutation theory of non-rigid Earth." *Celest. Mech. Dynam. Astr.*, **80**, 177–184.

Bretagnon P., Fienga A., and Simon J.-L., 2003, "Expressions for precession consistent with the IAU 2000A model. Considerations about the ecliptic and the Earth Orientation Parameters." *Astr. Astrophys.*, **400**, 785–790, DOI: 10.1051/0004-6361:20021912.

Briers R., Dehant V., and Leroy O., 1996, "The influence of a density stratification and of an incompressibility modulus on the spectrum of modes of oscillation." *Geophys. J. Int.*, **27**(*3*), 588–594.

Brito D., Aurnou J., and Cardin P., 2004, "Turbulent viscosity measurements relevant to planetary core mantle dynamics." *Phys. Earth Planet. Inter.*, **141**, 3–8.

Brown E.W., 1919, "*Tables of the Motion of the Moon.*" Yale University Press, New Haven.

Brumberg V.A., 1991, "*Essential Relativistic Celestial Mechanics.*" Adam Hilger Publ., 263 pp.

Brumberg V.A., Bretagnon P., and Francou G., 1992, "Analytical algorithms of relativistic reduction of astronomical observations." *Journées Systèmes de Référence Spatio-temporels 1991*, Paris, France, June 1991, Ed. N. Capitaine, pp 141–148.

Brumberg V.A. and Ivanova T.V., 2007, "Precession/nutation solution consistent with the general planetary theory." *Celest. Mech. Dynam. Astr.*, **97**(*3*), 189–210, DOI: 10.1007/s10569-006-9060-7.

Bryan K., 1969, "A numerical method for the study of the circulation of the world ocean." (Reprinted from the *Journal of Computational Physics*, **4**, 347–376, 1969), *J. Comput. Phys.*, **135**(*2*), 154–169.

Brzezinski A., 1992, "Polar motion excitation by variations of the effective angular momentum function: considerations concerning deconvolution problem." *Manuscripta Geodaetica*, **17**, 3–20.

Brzezinski A. and Capitaine N., 1993, "The use of the precise observations of the celestial ephemeris pole in the analysis of geophysical excitation of Earth rotation." *J. Geophys. Res.*, **98**(*B4*), 6667–6675.

Brzezinski A., 1994, "Polar motion excitation by variations of the effective angular momentum function, II: extended-model." *Manuscripta Geodaetica*, **19**, 157–171.

Brzezinski A., 1995, "On the interpretation of maximum entropy power spectrum and cross-power spectrum in Earth rotation investigations." *Manuscripta Geodaetica*, **20**, 248–264.

Brzezinski A., 2000, "Review of the Chandler wobble and its excitation." Workshop 'Forcing of polar motion in the Chandler frequency band: a contribution to understanding interannual climate variations', April 21–23, 2004, Luxembourg, *Cahiers du Centre Européen de Géodynamique et de Séismologie*, Vol. 24, Eds. H.P. Plag, B. Chao, R. Gross, and T. Van Dam, pp 109–120.

Brzezinski A., Bizouard C., and Petrov S., 2002, "Influence of the atmosphere on Earth rotation: what new can be learned from recent atmospheric angular momentum estimates?" *Surv. Geophysics*, **23**, 33–69.

Brzezinski A., 2003, "Oceanic excitation of polar motion and nutation: an overview." In: Proceedings of the IERS Workshop on *Combination Research and Global Geophysical Fluids*, Bavarian Academy of Sciences, Munich, Germany, 18–21 November 2002, Eds. B. Richter, W. Schwegmann, and W.R. Dick, International Earth Rotation and Reference Systems Service (IERS), IERS Technical Note, No.

30, Frankfurt am Main, Germany: Verlag des Bundesamtes für Kartographie und Geodäsie, pp 144–149.

Brzezinski A., Ponte R.M., and Ali A.H., 2004, "Nontidal oceanic excitation of nutation and diurnal/semidiurnal polar motion revisited." *J. Geophys. Res.*, **109**(*B11*), CiteID B11407, DOI: 10.1029/2004JB003054.

Brzezinski A., 2005, "Chandler wobble and free core nutation: observation, modeling and geophysical interpretation." Proc. *Earth rotation and satellite geodesy from astrometry to GNSS*, Warsaw, September 2003, *J. Planet. Geodesy, Artificial Satellites*, **40**(*1*), 21–33.

Buffett B.A., 1992, "Constraints on magnetic energy and mantle conductivity from the forced nutations of the Earth." *J. Geophys. Res.*, **97**, 19581–19597.

Buffett B.A., 1993, "Influence of a toroidal magnetic field on the nutations of the Earth." *J. Geophys. Res.*, **98**, 2105–2117.

Buffett B.A., Mathews P.M., Herring T.A., and Shapiro I.I., 1993, "Forced nutations of the Earth: contributions from the effects of ellipticity and rotation on the elastic deformations." *J. Geophys. Res.*, **98**(*B12*), 21659–21676, DOI: 10.1029/92JB01339.

Buffet B.A., 1996a, "Effects of a heterogeneous mantle on the velocity and magnetic fields at the top of the core." *Geophys. J. Int.*, **125**, 303–317.

Buffett B.A., 1996b, "A mechanism for decade fluctuations in the length of day." *Geophys. Res. Letters*, **23**(*25*), 3803–3806, DOI: 10.1029/96GL03571.

Buffett B.A., 1996c, "Numerical solution of boundary-layer problems." *Geophys. J. Int.*, **126**(*1*), 291–295.

Buffett B.A., 1997, "Geodynamic estimates of the viscosity of the Earth's inner core." *Nature*, **388**(*6642*), 571–573.

Buffett B.A., 2000a, "Dynamics of the Earth's core." In: '*Earth's Deep Interior: Mineral Physics and Tomography From the Atomic to the Global Scale*', Geophys. Monogr. Ser., **117**, Eds. S. Karato *et al.*, pp 37–62, AGU, Washington, D.C., DOI: 10.1029/GM117p0037.

Buffett B.A., 2000b, "Earth's core and the geodynamo." *Science*, **288**(*5473*), 2007–2012, DOI: 10.1126/science.288.5473.2007.

Buffett B.A., 2000c, "Sediments at the top of Earth's core." *Science*, **290**(*5495*), 1338–1342, DOI: 10.1126/science.290.5495.1338.

Buffett B.A. and Bloxham J., 2000, "Deformation of Earth's inner core by electromagnetic forces." *Geophys. Res. Letters*, **27**(*24*), 4001–4004, DOI: 10.1029/2000GL011790.

Buffett B.A., Mathews P.M., and Herring T.A., 2002, "Modeling of nutation and precession: effects of electromagnetic coupling." *J. Geophys. Res.*, **107**(*B4*), CI: 2070, DOI: 10.1029/2000JB000056.

Buffett B.A., 2002, "Estimates of heat flow in the deep mantle based on the power requirements for the geodynamo." *Geophys. Res. Letters*, **29**(*12*), 7-1–7-4, CiteID 1566, DOI: 10.1029/2001GL014649.

Buffett B.A., 2003, "A comparison of subgrid-scale models for large-eddy simulations of convection in the Earth's core." *Geophys. J. Int.*, **153**(*3*), 753–765, DOI: 10.1046/j.1365-246X.2003.01930.x.

Buffett B.A. and Mound J.E., 2005, "A Green's function for the excitation of torsional oscillations in the Earth's core." *J. Geophys. Res.*, **110**(*B8*), CiteID B08104, DOI: 10.1029/2004JB003495.

Buffett B.A. and Christensen U.R., 2007, "Magnetic and viscous coupling at the core–mantle boundary: inferences from observations of the Earth's nutations." *Geophys. J. Int.*, **171**(*1*), 145–152, DOI: 10.1111/j.1365-246X.2007.03543.x.

Buffett B.A., 2009a, "Onset and orientation of convection in the inner core." *Geophys. J. Int.*, **179**(*2*), 711–719, DOI: 10.1111/j.1365-246X.2009.04311.x.

Buffett B.A., 2009b, "Geodynamo: a matter of boundaries." *Nature Geoscience*, **2**(*11*), 741–742, DOI: 10.1038/ngeo673.

Buffett B.A. and Seagle C.T., 2010, "Stratification of the top of the core due to chemical interactions with the mantle." *J. Geophys. Res.*, **115**(*B4*), CiteID B04407, DOI: 10.1029/2009JB006751.

Buffett B.A., 2010a, "Chemical stratification at the top of Earth's core: Constraints from observations of nutations." *Earth Planet. Sci. Lett.*, **296**(*3–4*), 367–372, DOI: 10.1016/j.epsl.2010.05.020.

Buffett B.A., 2010b, "Tidal dissipation and the strength of the Earth's internal magnetic field." *Nature, Letter*, **468**, 952–955, DOI: 10.1038/nature09643.

Buffett B.A., 2012, "Geomagnetism under scrutiny." *Nature, Research*, **485**, 319–320, DOI: 10.1038/nature09643.

Bulirsch R. and Stoer J., 1966, "Numerical treatment of ordinary differential equation by extrapolation methods." *Numer. Math.*, **8**, 1–13.

Bullen K.E., 1940, "The problem of the Earth's density variation." *Bull. Seism. Soc. Amer.*, **30**, 235–250.

Burridge R., 1969, "Spherically symmetric differential equations, the rotation group, and tensor spherical functions." *Proc. Camb. Phil. Soc.*, **65**, 157–175.

Busse F.H., 1970, "The dynamical coupling between inner core and mantle of the Earth and the 24-year libration of the pole." In: *Earthquake Displacement Fields and the Rotation of the Earth*, pp 88–98, Eds. Mansinha L., Smylie D.E., and Beck A.E., D. Reidel, Dordrecht, the Netherlands.

Calkins M.A., Noir J., Eldredge J.D., and Aurnou J.M., 2010a, "Axisymmetric simulations of libration-driven fluid dynamics in a spherical shell geometry." *Physics of Fluids*, **22**(*8*), 086602–086602-12, DOI: 10.1063/1.3475817.

Calkins M.A., Noir J., Eldredge J.D., and Aurnou J.M., 2010b, "The effects of boundary topography on convection in Earth's core." *Geophys. J. Int.*, **189**(*2*), 799–814, DOI: 10.1111/j.1365-246X.2012.05415.x.

Capderou M., 2012, "*Satellites: De Kepler au GPS.*" Springer-Verlag, Paris, France, in French, 486 pp.

Capitaine N., 1982, "Effets de la non-rigidité de la Terre sur son mouvement de rotation: étude théorique et utilisation d'observations." Ph.D. thesis, Paris, France, in French, 205 pp.

Capitaine N., 1986, "The Earth rotation parameters: conceptual and conventional definitions." *Astr. Astrophys.*, **162**, 323–329.

Capitaine N., Souchay J., and Guinot B., 1986, "A non-rotating origin on the instantaneous equator: definition, properties and use." *Celest. Mech.*, **39**(*3*), 283–307, DOI: 10.1007/BF01234311.

Capitaine N. and Guinot B., 1988, "A non-rotating origin on the instantaneous equator." In: *The Earth's Rotation and Reference Frames for Geodesy and Geodynamics*, Eds. A.K. Babcock and G.A. Wilkins, IAU Publ., pp 33–38.

Capitaine N., 1990, "The celestial pole coordinates." *Celest. Mech. Dynam. Astr.*, **48**, 127–143.

Capitaine N. and Gontier A.-M., 1993, "Accurate procedure for deriving UT1 at a submilliarcsecond accuracy from Greenwich Sidereal Time or from the stellar angle." *Astr. Astrophys.*, **275**, 645–650.

Capitaine N., 1999, "Theory and observation of the 'Forced diurnal polar motion' from Fedorov's work to now." *Kinematika i Fizika Nebesnykh Tel, Prilozhenie*, (*1*), 25–31.

Capitaine N., 2000, "Definition of the celestial ephemeris pole and the celestial ephemeris origin." In: *Towards Models and Constants for Sub-microarcsecond Astrometry.*, IAU Colloquium 180, Eds. K. Johnson, D. McCarthy, B. Luzum, G. Kaplan, Washington, D.C., pp 153–163.

Capitaine N., Guinot B., and D.D. McCarthy, 2000, "Definition of the celestial ephemeris origin and of UT1 in the International Celestial Reference Frame." *Astr. Astrophys.*, **355**, 398–405.

Capitaine N., Gambis D., McCarthy D.D., *et al.*, 2002, "*Proceedings of the IERS Workshop on the Implementation of the New IAU Resolutions.*" IERS Technical Note No. 29, Ed. Verlag des Bundesamts für Kartographie und Geodäsie, Frankfurt am Main, 140 pp.

Capitaine N., 2002, "Comparison of 'Old' and 'New' concepts: the Celestial intermediate pole and Earth Orientation parameters." In: *Proc. IERS Workshop on the Implementation of the New IAU Resolutions*, Observatoire de Paris, Paris, France, 18–19 April 2002, Eds. N. Capitaine, D. Gambis, D.D. McCarthy, *et al.*, International Earth Rotation Service (IERS), Frankfurt am Main, Germany, Verlag des Bundesamtes für Kartographie und Geodäsie, ISBN 3-89888-866-5, pp 35–44.

Capitaine N., 2003, "Comparison of 'old' and 'new' concepts: The Celestial Intermediate Pole and Earth rotation parameters." IERS Technical Note No. 29, pp 35–44.

Capitaine N., Chapront J., Lambert S., and Wallace P., 2003a, "Expressions for the celestial intermediate pole and celestial ephemeris origin consistent with IAU2000A precession-nutation model." *Astr. Astrophys.*, **400**, 1145–1154, DOI: 10.1051/0004-6361:20030077.

Capitaine N., Chapront J., Lambert S., and Wallace P., 2003b, "Expressions for the coordinates of the CIP and the CEO using IAU2000 precession-nutation." IERS Technical Note No. 29, pp 89–91.

Capitaine N., Wallace P., and Chapront J., 2003c, "Expressions for IAU2000 precession quantities." *Astr. Astrophys.*, **412**, 567–586, DOI: 10.1051/0004-6361:20031539.

Capitaine N., Wallace P., and McCarthy D.D., 2003d, "Expressions to implement the IAU2000 definition of UT1." *Astr. Astrophys.*, **406**, 1135–1149, DOI: 10.1051/0004-6361:20030817.

Capitaine N., Wallace P., and Chapront J., 2004, "Comparison between high precision precession models for the ecliptic and the equator." *Astr. Astrophys.*, **421**, 365–379, DOI: 10.1051/0004-6361:20035942.

Capitaine N., Wallace P., and Chapront J., 2005a, "Improvement of the IAU 2000 precession model." *Astr. Astrophys.*, **432**(*1*), 355–367, DOI: 10.1051/0004-6361: 20041908.

Capitaine N., Hohenkerk C., Andrei A.H., *et al.*, 2005b, "Report of the IAU Division 1 Working Group on '*Nomenclature for Fundamental Astronomy (NFA)*'." In: *Proc. Journées Systèmes de Référence Spatio-temporels 2004*, Ed. N. Capitaine, Observatoire de Paris, pp 161–165.

Capitaine N., Andrei A.H., Calabretta M., *et al.*, 2005c, "Report of the IAU Division 1 Working Group on '*Nomenclature for Fundamental Astronomy (NFA)*'." In: *Reports on Astronomy* 2002–2005, IAU Transactions XXVIA, Ed. O. Engvold.

Capitaine N. and Wallace P., 2006, "High precision methods for locating the celestial intermediate pole and origin." *Astr. Astrophys.*, **450**, 855–872, DOI: 10.1051/0004-6361:20054550.

Capitaine N., Hohenkerk C., Andrei A.H., *et al.*, 2006a, "Latest proposals of the IAU WG on *Nomenclature for Fundamental Astronomy* (NFA)." In: *Proc. Journées Systèmes de Référence Spatio-temporels 2005*, Eds. A. Brzezinski, N. Capitaine, and B. Kolaczek, pp 143–146.

Capitaine N., Folgueira M., and Souchay J., 2006b, "Earth rotation based on the celestial coordinates of the celestial intermediate pole. I. The dynamical equations." *Astr. Astrophys.*, **445**(*1*), 347–360, DOI: 10.1051/0004-6361:20053778.

Capitaine N., Andrei A.H., Calabretta M., *et al.*, 2007, "Report of the IAU Division 1 Working Group on *Nomenclature for Fundamental Astronomy (NFA)*." In: Proc. 26th IAU General Assembly, IAU Transactions XXVIB, Ed. K.A. van der Hucht.

Capitaine N. and Wallace P.T., 2008, "Concise CIO based precession-nutation formulation." *Astr. Astrophys.*, **478**(*1*), 277–284, DOI: 10.1051/0004-6361:20078811.

Capitaine N., Mathews P.M., Dehant V., Wallace P.T., and Lambert S.B., 2009, "On the IAU 2000/2006 precession-nutation and comparison with other models and VLBI observations." *Celest. Mech. Dynam. Astr.*, **103**, 179–190, DOI: 10.1007/s10569-008-9179-9.

Capitaine N., 2012, "Micro-arcsecond celestial reference frames: definition and realization – Impact of the recent IAU Resolutions." *Research in Astronomy and Astrophysics*, **12**(*8*), 1162–1184, DOI: 10.1088/1674-4527/12/8/013.

Carrascal B., Estevez G.A., Lee P., and Lorenzo V., 1991, "Vector spherical harmonics and their application to classical electrodynamics." *Eur. J. Phys.*, **12**, 184–191.

Cartwright D.E. and Tayler R.J., 1971, "New computations in the tide-generating potential." *Geophys. J. Roy. Astron. Soc.*, **23**, 45–74.

Cébron D., Le Bars M., Noir J., and Aurnou J. M., 2012, "Libration driven elliptical instability." *Physics of Fluids*, **24**(*6*), 061703–061703-7, DOI: 10.1063/1.4729296.

Chao B.F. and Gross R.S., 1987, "Changes in the Earth's rotation and low-degree gravitational field induced by earthquakes." *Geophys. J. Roy. Astron. Soc.*, **91**, 569–596.

Chao B.F. and Au A.Y., 1991, "Atmospheric excitation of the Earth's annual wobble – 1980–1988." *J. Geophys. Res.*, **96**, 6577–6582.

Chao B.F., Ray R.D., Gipson J.M., Egbert G.D., and Ma C., 1996, "Diurnal/semidiurnal polar motion excited by oceanic tidal angular momentum." *J. Geophys. Res.*, **101**(*B9*), 20151–20164.

Chao B.F., Dehant V., Gross R.S., *et al.*, 2000, "*Space Geodesy Monitors Mass Transports in Global Geophysical Fluids*." EOS, AGU Publication, **81**(*22*), 247, 249, 250.

Chao B.F. and Chung W.Y., 2012, "Amplitude and phase variations of Earth's Chandler wobble under continual excitation." *Journal of Geodynamics*, **62**, 35–39, DOI: 10.1016/j.jog.2011.11.009.

Chapman S. and Lindzen R.S., 1970, "*Atmospheric Tides, Thermal and Gravitational*." D. Reidel Publishing Company, Dordrecht-Holland, 200 pp.

Chapront-Touzé M. and Chapront J., 1983, "The lunar ephemeris ELP-2000." *Astr. Astrophys.*, **124**(*1*), 50–62.

Chapront-Touzé M. and Chapront J., 1988, "ELP2000-85: a semi analytical lunar ephemeris adequate for historical times." *Astr. Astrophys.*, **190**, 342–352.

Chapront J., Chapront-Touzé M., and Francou G., 1999, "Determination of the lunar orbital and rotational parameters and of the ecliptic reference system orientation from LLR measurements and IERS data." *Astr. Astrophys.*, **343**, 624–633.

Chapront J., Chapront-Touzé M., and Francou G., 2002, "A new determination of lunar orbital parameters, precession constant and tidal acceleration from LLR measurements." *Astr. Astrophys.*, **387**, 700–709.

Chapront J. and Francou G., 2003, "The lunar theory ELP revisited. Introduction of new planetary perturbations." *Astr. Astrophys.*, **404**, 735–742, DOI: 10.1051/0004-6361:20030529.

Chapront J. and Francou G., 2006, "Lunar laser ranging: measurements, analysis, and contribution to the reference systems." In: *The International Celestial Reference System and Frame*, ICRS Center Report for 2001–2004, Eds. J. Souchay and M. Feissel-Vernier, IERS Technical Note No. 34, pp 97–116.

Charlot P., Campbell R.M., Alef W., *et al.*, 2002, "Improved positions of non-geodetic EVN telescopes." In: Proc. 6th European *VLBI Network* Symposium, Eds. E. Ros, R.W. Porcas, A.P. Lobanov, and J.A. Zensus, June 25–28 2002, Bonn, Germany.

Cheng M., Ries J.C., and Tapley B.D., 2011, "Variations of the Earth's figure axis from satellite laser ranging and GRACE." *Journal of Geophysical Research: Solid Earth*, **116**(*B1*), CiteID: B01409, DOI: 10.1029/2010JB000850.

Christensen U.R., Aubert J., Cardin P., *et al.*, 2001, "A numerical dynamo benchmark." *Phys. Earth Planet. Int.*, **128**(*1–4*), 25–34, DOI: 10.1016/S0031-9201(01)00275-8.

Christensen U.R. and Olson P., 2003, "Secular variation in numerical geodynamo models with lateral variations of boundary heat flow." *Phys. Earth Planet. Int.*, **138**(*1*), 39–54, DOI: 10.1016/S0031-9201(03)00064-5.

Christensen U.R. and Tilgner A., 2004, "Power requirement of the geodynamo from ohmic losses in numerical and laboratory dynamos." *Nature*, **429**(*6988*), 169–171, DOI: 10.1038/nature02508.

Christensen U.R., Aubert J., Cardin P., *et al.*, 2009, "Erratum to 'A numerical dynamo benchmark' [*Phys. Earth Planet. Int.* **128** (*1–4*) (2001) 25–34]." *Phys. Earth Planet. Int.*, **172**(*3–4*), 356–356, DOI: 10.1016/j.pepi.2008.09.014.

Christensen U.R., Aubert J., and Hulot G., 2010, "Conditions for Earth-like geodynamo models." *Earth Planet. Sci. Lett.*, **296**(*3–4*), 487–496, DOI: 10.1016/j.epsl.2010.06.009.

Christensen U.R., 2011, "Geodynamo models: tools for understanding properties of Earth's magnetic field." *Phys. Earth Planet. Int.*, **187**(*3*), 157–169, DOI: 10.1016/j.pepi.2011.03.012.

Christensen U.R., Wardinski I., and Lesur V., 2012, "Timescales of geomagnetic secular acceleration in satellite field models and geodynamo models." *Geophys. J. Int.*, **190**(*1*), 243–254, DOI: 10.1111/j.1365-246X.2012.05508.x.

Chulliat A. and Olsen N., 2010, "Observation of magnetic diffusion in the Earth's outer core from MagSat, Oersted, and CHAMP data." *Journal of Geophysical Research: Solid Earth*, **115**(*B5*), CiteID: B05105, DOI: 10.1029/2009JB006994.

Clairaut A.C., 1743, "*Théorie de la Figure de la Terre.*" Ed. Paris: David, XL, 305 pp.

Cottaar S. and Buffett B., 2012, "Convection in the Earth's inner core." *Phys. Earth Planet. Int.*, **198**, 67–78, DOI: 10.1016/j.pepi.2012.03.008.

Crossley D.J., Rochester M.G., and Peng Z.R., 1992, "Slichter modes and Love numbers." *Geophys. Res. Letters*, **19**(*16*), 1679–1682.

Dahlen F.A., 1968, "The normal modes of a rotating elliptical Earth." *Geophys. J. Roy. Astron. Soc.*, **16**, 329–367.

Dahlen F.A., 1972, "Elastic dislocation theory for a self-gravitating elastic configuration with an initial static stress field." *Geophys. J. Roy. Astron. Soc.*, **28**, 357–383.

Dahlen F.A., 1973, "Elastic dislocation theory for a self-gravitating elastic configuration with an initial static stress field-II: Energy release." *Geophys. J. Roy. Astron. Soc.*, **31**, 469–484.

Dahlen F.A., 1974, "On the static deformation of an Earth model with a fluid core." *Geophys. J. Roy. Astron. Soc.*, **36**, 461–485.

Dahlen F.A. and Smith M.L., 1975, "The influence of rotation on the free oscillations of the Earth." *Phil. Trans. Roy. Soc., Ser. A*, **279**, 583–624.

Dahlen F.A., 1976, "The passive influence of the oceans upon the rotation of the Earth." *Geophys. J. Roy. Astron. Soc.*, **46**, 363–406.

Dahlen F.A., 1982, "Variation of gravity with depth in the Earth." *Physical Review D (Particles and Fields)*, **25**(6), 1735–1736.

Dahlen F.A. and Tromp J., 1998, *"Theoretical Global Seismology."* Princeton University Press, Princeton, New Jersey, 1025 pp.

Davies R. and Whaler K., 1997, "The 1969 geomagnetic impulse and spin-up of the Earth's liquid core." *Phys. Earth Planet. Inter.*, **103**, 181–194.

Defraigne P., Dehant V., and Hinderer J., 1994, "Stacking gravity tide measurements and nutation observations in order to determine the complex eigenfrequency of the nearly diurnal free wobble." *J. Geophys. Res.*, **99**(B5), 9203–9213.

Defraigne P., Dehant V., and Hinderer J., 1995a, "Correction to 'Stacking gravity tide measurements and nutation observations in order to determine the complex eigenfrequency of the nearly diurnal free wobble'." *J. Geophys. Res.*, **100**(B2), 2041–2042.

Defraigne P., Dehant V., and Pâquet P., 1995b, "Link between the retrograde-prograde nutations in obliquity and longitude." *Celestial Mech. Dynam. Astr.*, **62**, 363–373.

Defraigne P., Dehant V., and Wahr J.M., 1996, "Internal loading of an homogeneous compressible Earth with phase boundaries." *Geophys. J. Int.*, **125**, 173–192.

Defraigne P. and Smits I., 1999, "Length of day variations due to zonal tides for an inelastic Earth in non-hydrostatic equilibrium." *Geophys. J. Int.*, **139**(2), 563–572.

Defraigne P., de Viron O., Dehant V., Van Hoolst T., and Hourdin F., 2000, "Mars rotation variations induced by atmospheric CO_2 and winds." *J. Geophys. Res. (Planets)*, **105**(E10), 24563–24570.

Defraigne P., Dehant V., and Van Hoolst T., 2001, "Steady state convection in Mars' mantle." *Planet. Space Sci.*, **49**, 501–509.

Defraigne P., Rivoldini A., Van Hoolst T., and Dehant V., 2003, "Mars nutation resonance due to free inner core nutation." *J. Geophys. Res. (Planets)*, **108**(E12), 5128, DOI: 10.1029/2003JE002145.

Degryse K. and Dehant V., 1993, "Analytical computation of modes using boundary layer theory." *Dynamics of Earth's Deep Interior and Earth Rotation*, Geophysical Monograph (IUGG/AGU Publication), **72**, IUGG Vol. 12, pp 69–80.

Degryse K. and Dehant V., 1995, "Analytical computation of modes for an Earth with viscous boundary layers, and influence of viscosity on non-rotating Slichter period." *Manuscripta Geodetica*, **20**, 498–514.

Degryse K. and Dehant V., 1996, "Are earthquakes responsible for the excitation of the FCN and/or the FICN?", *Phys. Earth Planet. Inter.*, **94**, 133–143.

Dehant V., 1986, "Intégration des equations différentielles aux déformations d'une terre ellipsoïdale, inélastique, en rotation uniforme, avec un noyau liquide." Ph.D. thesis of the Université Catholique de Louvain, in French.

Dehant V., 1987a, "Integration of the gravitational motion equations for an elliptical uniformly rotating Earth with an inelastic mantle." *Phys. Earth Planet. Inter.*, **49**, 242–258.

Dehant V., 1987b, "Tidal parameters for an inelastic Earth." *Phys. Earth Planet. Inter.*, **49**, 97–116.

Dehant V. and Ducarme B., 1987, "Comparison between the theoretical and observed tidal gravimetric factors." *Phys. Earth Planet. Inter.*, **49**, 192–212.

Dehant V., 1989, "Core undertones in an elliptical uniformly rotating Earth." In: Proc. Symp. U2, 19th General Assembly of the IUGG, Vancouver, Canada, August 1987, Geophysical Monograph 46, IUGG Vol. 1: *Structure and Dynamics of the Earth's Deep Interior*, Eds. D.E. Smylie and R. Hide, pp 29–34.

Dehant V. and Zschau J., 1989, "The effect of mantle inelasticity on tidal gravity: a comparison between the spherical and the elliptical Earth model." *Geophys. J.*, **97**, 549–555; and *Communication de l'Observatoire Royal de Belgique*, Série A no. 89, *Série Géophysique* no. 160.

Dehant V., 1990, "On the nutations of a more realistic Earth's model." *Geophys. J. Int.*, **100**, 477–483.

Dehant V. and Wahr J.M., 1991, "The response of a compressible, non-homogeneous Earth to internal loading: theory." *J. Geomag. Geoelectr.*, **43**, 157–178.

Dehant V., 1991, "Review of the Earth tidal models and contribution of Earth tides in geodynamics." *J. Geophys. Res.*, **96**(*B12*), 20235–20240.

Dehant V., 1993, "Free and forced oscillations in the core and the mantle." In: Proc. World Space Congress 92, COSPAR meeting, 28 August–9 Sept. 1992, Washington D.C., USA, Eds. J.M. Wahr and J. Dickey, *Adv. Space Res.*, **13**(*11*), 235–249; and *Communication de l'Observatoire Royal de Belgique*, Série A no. 111, *Série Géophysique* no. 184.

Dehant V., Hinderer J., Legros H., and Lefftz M., 1993a, "Analytical approach to the computation of the Earth, the outer core and the inner core rotational motions." *Phys. Earth Planet. Inter.*, **76**, 259–282.

Dehant V., Ducarme B., and Defraigne P., 1993, "New analysis of the superconducting gravimeter data of Brussels." *Dynamics of Earth's Deep Interior and Earth Rotation*, Geophysical Monograph (IUGG/AGU Publication), 72, IUGG Vol. 12, pp 35–44; and *Communication de l'Observatoire Royal de Belgique*, Série A no. 110, *Série Géophysique* no. 181.

Dehant V., Bizouard Ch., Hinderer J., Legros H. and Lefftz M., 1996, "On atmospheric pressure perturbations on precession and nutations." *Phys. Earth Planet. Inter.*, **96**, 25–39.

Dehant V. and Capitaine N., 1997, "On the luni-solar precession, the Tilt-Over-Mode, and the Oppolzer terms." *Celest. Mech. Dynam. Astr.*, **65**, 439–458.

Dehant V. and Defraigne P., 1997, "New transfer function formulations of a non-rigid Earth." *J. Geophys. Res.*, **102**, 27659–27688.

Dehant V., Wilson C.R., Salstein D.A., *et al.*, 1997a, "Study of Earth's rotation and geophysical fluids progresses." *EOS, Transactions, American Geophysical Union*, **78**(*34*), 357 and 360.

Dehant V., Feissel M., Defraigne P., Roosbeek F., and Souchay J., 1997b, "Could the energy near the FCN and the FICN be explained by luni-solar or atmospheric forcing?" *Geophys. J. Int.*, **130**, 535–546.

Dehant V., Arias F., Brzezinski A., *et al.*, 1999a, "Considerations concerning the non-rigid Earth nutation theory." *Celest. Mech. Dynam. Astr.*, **72**(*4*), 245–310.

Dehant V., Defraigne P., and Wahr J.M., 1999b, "Tides for a convective Earth." *J. Geophys. Res.*, **104**(*B1*), 1035–1058.

Dehant V., Defraigne P., and Van Hoolst T., 2000a, "Computation of Mars' transfer function for nutation tides and surface loading." *Phys. Earth Planet. Inter.*, **117**, 385–395.

Dehant V., Van Hoolst T., and Defraigne P., 2000b, "Comparison between the nutations of the planet Mars and the nutations of the Earth." *Survey Geophys.*, **21**(*1*), 89–110.

Dehant V. and de Viron O., 2002, "Earth rotation as an interdisciplinary topic shared by astronomers, geodesists and geophysicists." Proceedings of COSPAR meeting, Warsaw, Poland, *Adv. Space Res.*, **30**(*2*), 163–173.

Dehant V. and Mathews P.M., 2002, "Information about the core from nutation." In: AGU Monograph series, '*Earth's Core Dynamics, Structure and Rotation*', Geodynamics Series Volume 31, Eds. V. Dehant, K. Creager, S. Karato, S. Zatman, DOI: 10.1029/031GD18.

Dehant V., Feissel-Vernier M., de Viron O., *et al.*, 2003a, "Remaining error sources in the nutation at the sub-milliarcsecond level." *J. Geophys. Res. (Solid Earth)*, **108**(*B5*), 2275, DOI: 10.1029/2002JB001763.

Dehant V., Van Hoolst T., de Viron O., *et al.*, 2003b, "Can a solid inner core of Mars be detected from observations of polar motion and nutation of Mars?", *J. Geophys. Res. (Planets)*, **108**(*E12*), 5127, DOI: 10.1029/2003JE002140.

Dehant V., de Viron O., and Greff-Lefftz M., 2005a, "Atmospheric and oceanic excitation of the rotation of a three-layer Earth." *Astr. Astrophys.*, **438**, 1149–1161, DOI: 10.1051/0004-6361:20042210.

Dehant V., de Viron O., and Barriot J.-P., 2005b, "Geophysical excitation of the Earth orientation parameters and its contribution to GGOS." In: Proc. 2004 IUGG General Assembly, Sapporo, Japan, *J. Geodynamics*, **40**(*4–5*), Special Issue on The Global Geodetic Observing System, Ed. Hermann Drewes, Nov.–Dec. 2005, 394–399.

Dehant V. and Van Hoolst T., 2006, "Gravity, rotation, and interior of the terrestrial planets from planetary geodesy." In: Proc. IAG-IAPSO-IABO General Assembly on *Dynamic planet*, Cairns, Australia, Chapter 124, 887–894.

Dehant V. and Mathews M.P., 2007, "Earth rotation variations." In: *Treatise of Geophysics*, invited paper, Elsevier Publ., Vol. 3 *Geodesy*, Eds. T. Herring and J. Schubert, pp 295–349.

Dehant V., Lambert S.B., Rambaux N., Folgueira M., and Koot L., 2008, "Recent advances in modeling precession-nutation." Invited, in: *Proc. Journées Systèmes de Référence Spatio-temporels 2007*, Paris, France, September 2007, Ed. N. Capitaine, pp 82–87.

Dehant V., Folkner W., Renotte E., *et al.*, 2009, "Lander Radioscience for obtaining the rotation and orientation of Mars." *Planet. Space Sci.*, **57**, 1050–1067, DOI: 10.1016/j.pss.2008.08.009.

Dehant V., Le Maistre S., Rivoldini A., *et al.*, 2011, "Revealing Mars' deep interior: future geodesy missions using radio links between landers, orbiters, and the Earth." *Planet. Space Sci.*, Special Issue on *Comparative Planetology: Venus-Earth-Mars*, **59**(*10*), 1069–1081, DOI: 10.1016/j.pss.2010.03.014.

Dehant V., Banerdt B., Lognonné P., *et al.*, 2012a, "Future Mars geophysical observatories for understanding its internal structure, rotation, and evolution." *Planet. Space Sci.*, **68**(*1*), 123–145, DOI: 10.1016/j.pss.2011.10.016.

Dehant V., Folgueira M., and Puica M., 2012b, "Analytical computation of the effects of the core-mantle boundary topography on tidal length-of-day variations." In: Proc. *Journées Systèmes de Référence Spatio-temporels 2011*, Vienna, Austria, 113–116.

Dehant V., Lambert S., Koot L., Trinh A., and Folgueira M., 2013, "Recent advances in applications of geodetic VLBI to geophysics." In: Proc. IVS General Meeting 2012, *Launching the Next Generation IVS Network*," Madrid, Spain, Eds. D. Behrend and K.D. Baver, NASA/CP-2012-217504, pp 362–369, http://ivscc.gsfc.nasa.gov/publications/gm2012/Dehant.pdf.

Deleplace B. and Cardin P., 2006, "Visco-magnetic torque at the core mantle boundary." *Geophys. J. Int.*, **167**(*2*), 557–566.

Denis C., 1974, "Oscillations de configurations sphériques auto-gravitantes et applications à la Terre." Ph.D. thesis of the Université de Liège, in French, 359 pp.

Denis C., 1993, "Global deformations and evolution of the Earth." *Acta Geod. Mont. Hung.*, **28**(*1–2*), 15–131.

Denis C., Rogister Y., Amalvict M., *et al.*, 1997, "Hydrostatic flattening, core structure, and translational mode of the inner core." *Phys. Earth Planet. Inter.*, **99**, 195–206.

Denis C., Amalvict M., Rogister Y., and Tomecka-Suchoń S., 1998, "Methods for computing internal flattening, with applications to the Earth's structure and geodynamics." *Geophys. J. Int.*, **132**(*3*), 603–642, DOI: 10.1046/j.1365-246X.1998.00449.x.

de Sitter W., 1916a, "Space, time, and gravitation." *Observatory*, **39**, 412–419.

de Sitter W., 1916b, "A. Einstein's theory of gravitation and its astronomical consequences." *Month. Notice R. Astr. Soc.*, **76**, 699–728 (1916); **77**, 155–183 (1916); **78**, 3–28 (1917).

de Viron O. and Dehant V., 1999, "Earth's rotation and high frequency equatorial angular momentum budget of the atmosphere." *Survey in Geophysics*, **20**(*6*), 441–462.

de Viron O., Bizouard C., Salstein D. and Dehant V., 1999, "Atmospheric torque on the Earth and comparison with atmospheric angular momentum variations." *J. Geophys. Res.*, **104**(*B3*), 4861–4875.

de Viron O., Ponte R.M., and Dehant V., 2001a, "Indirect effect of the atmosphere through the ocean on the Earth's nutation by the torque approach." *J. Geophys. Res.*, **106**(*B5*), 8841–8851.

de Viron O., Marcus S.L., and Dickey J.O., 2001b, "Diurnal angular momentum budget of the atmosphere and its consequences for Earth's nutation." *J. Geophys. Res.*, **106**(*B11*), 26747–26759.

de Viron O., Marcus S.L., and Dickey J.O., 2001c, "Atmospheric torques during the winter of 1989: impact of ENSO and NAO positive phases." *Geophys. Res. Letters*, **28**(*10*), 1985–1988.

de Viron O., Dickey J.O., and Marcus S.L., 2002a, "Annual atmospheric torques: processes and regional contributions." *Geophys. Res. Letters*, **29**(*7*), DOI: 10.1029/2001GL013859.

de Viron O., Dehant V., and Goosse H., 2002b, "The 'hidden torque': the art, for a torque, to dominate everywhere and appear in no equation." In: IAG Symposia Proceedings series, **125**, *Vistas for Geodesy in the new Millennium*, pp 423–427.

de Viron O., Dehant V., Goosse H., Crucifix M., and the participating CMIP group, 2002c, "Effect of global warming on the length-of-day." *Geophys. Res. Letters*, **29**(*10*), 10.1029/2001GL013672.

de Viron O. and Dehant V., 2003, "Test on the validity of the atmospheric torques on Earth computed from model outputs." *J. Geophys. Res.*, **108**(*B2*), DOI: 10.1029/2001JB001196.

de Viron O., Boy J.-P., and Goosse H., 2004, "Geodetic effects of the ocean response to atmospheric forcing in an ocean general circulation model." *J. Geophys. Res.*, **109**(*B3*), CiteID B03411, DOI: 10.1029/2003JB002837.

de Viron O., Schwarzbaum G., Lott F., and Dehant V., 2005a, "Diurnal and sub-diurnal effects of the atmosphere on the Earth rotation and geocenter motion." *J. Geophys. Res.*, **110**(*B11*), B11404, DOI: 10.1029/2005JB003761.

de Viron O., Koot L., and Dehant V., 2005b, "Polar motion models: the torque approach." Workshop 'Forcing of polar motion in the Chandler frequency band: a contribution to understanding interannual climate variations', April 21–23, 2004, Luxembourg, *Cahiers du Centre Européen de Géodynamique et de Séismologie*, Vol. 24, Eds. H.P. Plag, B. Chao, R. Gross, and T. Van Dam, pp 9–14.

de Viron O., 2008, "Interaction Terre-atmosphère et rotation de la Terre." Habilatation (tenure) thesis, Institut de Physique du Globe de Paris, France, 107 pp.

de Vries D. and Wahr J.M., 1991, "The effects of the solid inner core and nonhydrostatic structure on the Earth's forced nutations and Earth tides." *J. Geophys. Res.*, **96**(*B5*), 8275–8293.

Dickey J.O. and de Viron O., 2009, "Leading modes of torsional oscillations within the Earth's core." *Geophys. Res. Letters*, **36**(*15*), CiteID: L15302, DOI: 10.1029/2009GL038386.

Dickey J.O., Marcus S.L., and de Viron O., 2010, "Closure in the Earth's angular momentum budget observed from subseasonal periods down to four days: no core effects needed." *Geophys. Res. Letters*, **37**(*3*), CiteID: L03307, DOI: 10.1029/2009GL041118.

Dickman S.R., 2003, "Evaluation of 'effective angular momentum function' formulations with respect to core-mantle coupling." *Journal of Geophysical Research (Solid Earth)*, **108**(*B3*), ETG 5-1, CiteID: 2150, DOI: 10.1029/2001JB001603.

Dickman S.R., 2005, "Rotationally consistent Love numbers." *Geophys. J. Int.*, **161**(*1*), 31–40, DOI: 10.1111/j.1365-246X.2005.02574.x.

di Virgilio A., Schreiber K.U., Gebauer A., *et al.*, 2010, "A laser gyroscope system to detect the gravito-magnetic effect on Earth." *International Journal of Modern Physics D*, **19**(*14*), 2331–2343, DOI: 10.1142/S0218271810018360.

Dobslaw H., Dill R., GröTzsch A., Brzezinski A., and Thomas M., 2010, "Seasonal polar motion excitation from numerical models of atmosphere, ocean, and continental hydrosphere." *Journal of Geophysical Research: Solid Earth*, **115**(*B10*), CiteID: B10406, DOI: 10.1029/2009JB007127.

Doodson A.T., 1921, "The harmonic development of the tide-generating potential." *Proc. Roy. Soc. A*, **100**, 305–329.

Dumberry M. and Buffett B.A., 1999, "On the validity of the geostrophic approximation for calculating the changes in the angular momentum of the core." *Phys. Earth Planet. Inter.*, **112**(*1–2*), 81–99, DOI: 10.1016/S0031-9201(98)00178-2.

Dumberry M. and Bloxham J., 2002, "Inner core tilt and polar motion." *Geophysical J. Int.*, **151**(*2*), 377–392, DOI: 10.1046/j.1365-246X.2002.01756.x.

Dumberry M. and Bloxham J., 2003, "Torque balance, Taylor's constraint and torsional oscillations in a numerical model of the geodynamo." *Phys. Earth Planet. Inter.*, **140**(*1–3*), 29–51, DOI: 10.1016/j.pepi.2003.07.012.

Dumberry M. and Bloxham J., 2004, "Variations in the Earth's gravity field caused by torsional oscillations in the core." *Geophysical J. Int.*, **159**(*2*), 417–434, DOI: 10.1111/j.1365-246X.2004.02402.x.

Dumberry M. and Bloxham J., 2006, "Azimuthal flows in the Earth's core and changes in length of day at millennial timescales." *Geophysical J. Int.*, **165**(*1*), 32–46, DOI: 10.1111/j.1365-246X.2006.02903.x.

Dumberry M., 2007, "Geodynamic constraints on the steady and time-dependent inner core axial rotation." *Geophysical J. Int.*, **170**(*2*), 886–895, DOI: 10.1111/j.1365-246X.2007.03484.x.

Dumberry M., 2008a, "Gravitational torque on the inner core and decadal polar motion." *Geophysical J. Int.*, **172**(*3*), 903–920, DOI: 10.1111/j.1365-246X.2007.03653.x.

Dumberry M., 2008b, "Decadal variations in gravity caused by a tilt of the inner core." *Geophysical J. Int.*, **172**(*3*), 921–933, DOI: 10.1111/j.1365-246X.2007.03624.x.

Dumberry M. and Mound J.E., 2008, "Constraints on core-mantle electromagnetic coupling from torsional oscillation normal modes." *J. Geophys. Res.*, **113**(*B3*), CiteID: B03102, DOI: 10.1029/2007JB005135.

Dumberry M., 2009, "Influence of elastic deformations on the inner core wobble." *Geophys. J. Int.*, **178**(*1*), 57–64, DOI: 10.1111/j.1365-246X.2009.04140.x.

Dumberry M., 2010a, "Gravitationally driven inner core differential rotation." *Earth Planet. Sci. Lett.*, **297**(*3–4*), 387–394, DOI: 10.1016/j.epsl.2010.06.040.

Dumberry M., 2010b, "Gravity variations induced by core flows." *Geophysical J. Int.*, **180**(*2*), 635–650, DOI: 10.1111/j.1365-246X.2009. 04437.x.

Dumberry M. and Mound J.E., 2010, "Inner core-mantle gravitational locking and the super-rotation of the inner core." *Geophysical J. Int.*, **181**(*2*), 806–817, DOI: 10.1111/j.1365-246X.2010.04563.x.

Dumberry M., 2011, "Earth's core: a new twist on inner-core spin." *Nature Geoscience*, **4**(*4*), 216–217, DOI: 10.1038/ngeo1091.

Dumberry M. and Koot L., 2012, "A global model of electromagnetic coupling for nutations." *Geophys. J. Int.*, **191**, 530–544, DOI: 10.1111/j.1365-246X.2012.05625.x.

Dumberry M., Rivoldini A., Van Hoolst T., and Yseboodt M., 2013, "The role of Mercury's core density structure on its longitudinal librations." *Icarus*, **225**, 62–74, DOI: 10.1016/j.icarus.2013.03.001.

Dziewonski A.M. and Gilbert F., 1972, "Observations of normal modes from 84 recordings of the Alaskan earthquake of 1964 March 28." *Geophys. J. Int.*, **27**(*4*), 393–446, DOI: 10.1111/j.1365-246X.1972.tb06100.x.

Dziewonski A.M. and Gilbert F., 1973, "Observations of normal modes from 84 recordings of the Alaskan earthquake of 1964 March 28-II. Further remarks based on new spheroidal overtone data." *Geophys. J. Int.*, **35**(*4*), 401–437, DOI: 10.1111/j.1365-246X.1973.tb00607.x.

Dziewonski A.M. and Gilbert F., 1976, "The effect of small, aspherical perturbations on travel times and a re-examination of the corrections for ellipticity." *Geophys. J. Int.*, **44**(*1*), 7–17, DOI: 10.1111/j.1365-246X.1976.tb00271.x.

Dziewonski A.D. and Anderson D.L., 1981, "Preliminary reference earth model." *Phys. Earth Planet. Inter.*, **25**, 297–356.

Eckert W.J., Walker M.J., and Eckert D., 1966, "Transformation of the lunar coordinates and orbital parameters." *Astron. J.*, **71**, 314–332.

Edmonds A.R., 1960, "*Angular Momentum in Quantum Mechanics*." Princeton University Press, Princeton, New Jersey.

Efroimsky M., 2000, "Inelastic dissipation in wobbling asteroids and comets." *Monthly Notices of the Royal Astronomical Society*, **311**(*2*), 269–278, DOI: 10.1046/j.1365-8711.2000.03036.x.

Efroimsky M., 2005, "Long-term evolution of orbits about a precessing oblate planet: 1. The case of uniform precession." *Celestial Mechanics and Dynamical Astronomy*, **91**(*1–2*), 75–108, DOI: 10.1007/s10569-004-2415-z.

Efroimsky M., 2006, "Long-term evolution of orbits about a precessing oblate planet. 2. The case of variable precession." *Celestial Mechanics and Dynamical Astronomy*, **96**(*3–4*), 259–288, DOI: 10.1007/s10569-006-9046-5.

Englich S., Snajdrova K., Weber R., and Schuh H., 2007, "High frequency variability in Earth rotation from VLBI and GNSS data." In: Proc. IAU Joint Discussion JD16 on *Nomenclature, precession, and new models in fundamental astronomy*, IAU XXVIth General, Prague, 2006.

Escapa A., Getino J., and Ferrándiz J.M., 2001, "Canonical approach to the free nutations of a three-layer Earth model." *J. Geophys. Res.*, **106**(*B6*), 11387–11397.

Escapa A., Getino J., and Ferrándiz J.M., 2002, "Indirect effect of the triaxiality in the Hamiltonian theory for the rigid Earth nutations." *Astr. Astrophys.*, **389**, 1047–1054.

Eubanks T.M., 1993, "Variations in the orientation of the Earth." In: *Space Geodesy to Geodynamics: Earth Dynamics*, AGU Geodynamics Series, **24**, 1–53.

Eubanks T.M., Matsakis D.N., Josties F.J., *et al.*, 1994, "Secular motions of extragalactic radio sources and the stability of the radio reference frame." Proc. IAU Symposium 166, Eds. E. Hog and P.K. Seidelmann, p. 283.

Euler L., 1744, "Theoria motuum planetarum et cometarum. Continens methodum facilem EX aliquot observationibus orbitas cum planetarum tum cometarum determinandi." Berolini, sumtibus Ambrosii Haude.

Euler L., 1747, "De attractione corporum sphaeroidico-ellipticorium." Commentarii Academiae Petropolitanae 1736, Vol. X.

Euler L., 1753, "Theoria motus lunae: exhibens omnes eius inaequalitates: in additamento HOC idem argumentum aliter tractatur simulque ostenditur quemadmodum motus lunae cum omnibus inaequalitatibus innumeris aliis modis repraesentari atque AD calculum reuocari possit." Petropolitanae: Impensis Academiae Imperialis Scientiarum; Berolini: Ex officina Michaelis.

Fedorov Ye.P., translated from Russian by Jeffreys B.S. and Jeffreys H., 1963, *"Nutation and Forced Motion of the Earth's Pole."* Pergamon Press, London, 150 pp.

Feissel M. and Mignard F., 1998, "The adoption of ICRF in January 1998: meaning and consequences." *Astr. Astrophys.*, **331**(*3*), L33–L36.

Feissel M., Gontier A.-M., and Eubanks T.M., 2000, "Spatial variability of compact extragalactic radiosources." *Astr. Astrophys.*, **359**, 1201–1204.

Feissel-Vernier M., 2003, "Selecting stable extragalactic compact radio sources from the permanent astrogeodetic VLBI program." *Astr. Astrophys.*, **403**, 105–110, DOI: 10.1051/0004-6361:20030348.

Feissel-Vernier M., Ma C., Gontier A.-M., and Barache C., 2005, "Sidereal orientation of the Earth and stability of the VLBI celestial reference frame." *Astr. Astrophys.*, **438**(*3*), 1141–1148.

Feissel-Vernier M., Ma C., Gontier A.-M., and Barache C., 2006, *"Analysis Strategy Issues for the Maintenance of the ICRF Axes."* Kluwer Academic Publishers, Dordrecht, The Netherlands.

Felli M. and Spencer R.E., 1989, *"Very Long Baseline Interferometry; Techniques and Applications."* Kluwer Academic Publishers, 456 pp.

Ferrándiz J.M., Escapa A., Navarro J.F., and Getino J., 2003, "Recent work on theoretical modeling of nutation." In: Proceedings of the IERS Workshop on *Combination Research and Global Geophysical Fluids*, Bavarian Academy of Sciences, Munich, Germany, 18–21 November 2002, Eds. B. Richter, W. Schwegmann, and W.R. Dick, International Earth Rotation and Reference Systems Service (IERS), IERS Technical Note No. 30, Frankfurt am Main, Germany: Verlag des Bundesamtes für Kartographie und Geodäsie, pp 163–167.

Ferrándiz J.M., Navarro J.F., Escapa A., and Getino J., 2004, "Precession of the nonrigid Earth: effect of the fluid outer core." *Astron. J.*, **128**(*3*), 1407–1411.

Fey A.L. and Charlot P., 1997, "VLBA observations of radio reference frame sources. II. Astrometric suitability based on observed structure." *The Astrophysical Journal Supplement Series*, **111**, 95–142.

Fey A.L. and Charlot P., 2000, "VLBA observations of radio reference frame sources. III. Astrometric suitability of an additional 225 sources." *The Astrophysical Journal Supplement Series*, **128**, 17–83.

Fey A.L., Boboltz D.A., Gaume R.A., and Johnston K.J., 2000, "Improving the ICRF using the radio reference frame image database." In: Proc. International VLBI Service for Geodesy and Astrometry 2000 General Meeting, Eds. N.R. Vandenberg and K.D. Baver, pp 285–287.

Fey A.L., Ma C., Arias E.F., *et al.*, 2004, "The second extension of the international celestial reference frame: ICRF-EXT.1." *Astron. J.*, **127**(*6*), 3587–3608, DOI: 10.1086/420998.

Fey A.L., Gordon D., and Jacobs C.S. (Eds), and IERS / IVS Working Group, 2009, "The second realization of the international celestial reference frame by very long baseline interferometry." IERS Technical Note No. 35, 204 pp.

Fienga A., Laskar J., Mache H., and Gastineau M., 2005, "IMCCE planetary ephemerides: present and future." In: Proc. Gaia Symposium *The Three-Dimensional Universe with Gaia* (ESA SP-576), Observatoire de Paris-Meudon, October 2004, Eds. C. Turon, K.S. O'Flaherty, M.A.C. Perryman, from http://www.rssd.esa.int/

index.php?project=Gaia&page=Gaia2004Proceedings, supersedes printed version, p 293.

Fienga A., Manche H., Laskar J., and Gastineau M., 2006, "INPOP06: a new planetary ephemeris." In: Proc. IAU Joint Discussion JD16 on 'Nomenclature, precession, and new models in fundamental astronomy', IAU XXVIth General, Prague, 2006.

Fienga A., Manche H., Laskar J., and Gastineau M., 2008, "INPOP06: a new numerical planetary ephemeris." *Astr. Astrophys.*, **477**(*1*), 315–327, DOI: 10.1051/0004-6361:20066607.

Fienga A., Laskar J., Morley T., *et al.*, 2009a, "INPOP08, a 4-D planetary ephemeris: from asteroid and time-scale computations to ESA Mars Express and Venus Express contributions." *Astr. Astrophys.*, **507**(*3*), 1675–1686, DOI: 10.1051/0004-6361/200911755.

Fienga A., Laskar J., Manche H., *et al.*, 2009b, "Gravity tests with the INPOP planetary ephemerides." American Astronomical Society, IAU Symposium 261 on *Relativity in fundamental astronomy: Dynamics, reference frames, and data analysis*, 27 April–1 May 2009, Virginia Beach, VA, USA, 6.02, *Bulletin of the American Astronomical Society*, **261**, 159–169.

Fienga A., Manche H., Kuchynka P., Laskar J., and Gastineau M., 2010, "Planetary and lunar ephemerides INPOP10a." Proc. Journées Systèmes de Référence Spatio-temporels 2010, Paris, September 2010, Ed. N. Capitaine.

Fienga A., Laskar J., Kuchynka P., *et al.*, 2011, "The INPOP10a planetary ephemeris and its applications in fundamental physics." *Celest. Mech. Dynam. Astr.*, **111**(*3*), 363–385, DOI: 10.1007/s10569-011-9377-8.

Fienga A., 2012, "How GAIA can improve planetary ephemerides?" *Planetary and Space Science*, **73**(*1*), 44–46, DOI: 10.1016/j.pss.2012.06.016.

Fienga A., Manche H., Laskar J., Gastineau M., and Verma A., 2014, "INPOP new release: INPOP13b." arXiv:1405.0484.

Folgueira M., Souchay J., and Kinoshita H., 1998a, "Effects on the nutation of the non-zonal harmonics of third degree." *Celest. Mech.*, **69**(*4*), 373–402, DOI: 10.1023/A:1008298122976.

Folgueira M., Souchay J., and Kinoshita H., 1998b, "Effects on the nutation of C4m and S4m harmonics." *Celest. Mech.*, **70**(*3*), 147–157, DOI: 10.1023/A:1008383423586.

Folgueira M., Bizouard C., and Souchay J., 2001, "Diurnal and subdiurnal luni-solar nutations: comparisons and effects." *Celest. Mech. Dynam. Astr.*, **81**(*3*), 191–217.

Folgueira M. and Souchay J., 2005, "Free polar motion of a triaxial and elastic body in Hamiltonian formalism: application to the Earth and Mars." *Astr. Astrophys.*, **432**(*3*), 1101–1113, DOI: 10.1051/0004-6361:20041312.

Folgueira M., Dehant V., Lambert S.B., and Rambaux N., 2007, "Impact of tidal Poisson terms to non-rigid Earth rotation." *Astr. Astrophys.*, **469**(*3*), 1197–1202, DOI: 10.1051/0004-6361:20066822.

Folkner W.M., Charlot P., Finger M.H., *et al.*, 1994, "Determination of the extragalactic-planetary frame tie from joint analysis of radio interferometric and lunar laser ranging measurements." *Astr. Astrophys.*, **287**(*1*), 279–289.

Folkner W.M., Williams J.G., and Boggs D.H., 2008, "The planetary and lunar ephemeris DE 421." IPN Progress Report 42–178, August 15, 2009, 31 pp, see http://ipnpr.jpl.nasa.gov/progress report/42-178/178C.pdf.

Folkner W.M., Williams J.G., Boggs D.H., Park R. S., and Kuchynka P., 2014, "The planetary and lunar ephemerides DE430 and DE431." The Interplanetary Network Progress Report, vol. 42–196, Jet Propulsion Laboratory, California Institute of Technology, pp 1–81.

Fomalont E., Johnston K., Fey A., *et al.*, 2011, "The position/structure stability of four ICRF2 sources." *Astron. J.*, **141**(*3*), ID. 91, 19 pp, DOI: 10.1088/0004-6256/141/3/91.

Forte A.M., Mitrovica J.X., and Woodward R.L., 1995, "Seismic-geodynamic determination of the origin of excess ellipticity of the core-mantle boundary." *Geophys. Res. Letters*, **22**(*9*), 1013–1016, DOI: 10.1029/95GL01065.

Fournier A., Hulot G., Jault D., *et al.*, 2010, "An introduction to data assimilation and predictability in geomagnetism." *Space Science Reviews*, **155**(*1–4*), 247–291, DOI: 10.1007/s11214-010-9669-4.

Frede V. and Dehant V., 1999, "Analytical versus semi-analytical determinations of the Oppolzer terms for a non-rigid Earth." *J. Geodesy*, **73**, 94–104.

Fricke W., 1982, "Determination of the equinox and equator of the FK5." *Astr. Astrophys.*, **107**(*1*), L13–L16.

Fricke W., Schwan H., Lederle T., *et al.*, 1988, *"Fifth Fundamental Catalogue (FK5), Part I: The Basic Fundamental Stars."* Veröffentlichungen Astronomisches Rechen-Institut Heidelberg, Verlag G. Braun, Karlsruhe, **32**, pp 1–106.

Frost H.J. and Ashby M.F., 1983, *"Deformation Mechanism Maps."* Pergamon Press, Oxford, 161 pages.

Fukushima T., 1991, "Geodesic nutation." *Astr. Astrophys.*, **244**, L11–L12.

Fukushima T., 2003a, "New precession formula." IERS Technical Note No. 29, 97–98.

Fukushima T., 2003b, "A new precession formula." *Astron. J.*, **126**, 494–534.

Furuya M., Hamano Y., and Naito I., 1996, "Quasi-periodic wind signal as a possible excitation of Chandler wobble." *J. Geophys. Res.*, **101**(*B11*), 25537–25546, DOI: 10.1029/96JB02650.

Gambis D., Johnson T., Gross R., and Vondrak J., 2003, "General combination of EOP series." In: Proceedings of the IERS Workshop on *Combination Research and Global Geophysical Fluids*, Bavarian Academy of Sciences, Munich, Germany, 18–21 November 2002, Eds. B. Richter, W. Schwegmann, and W.R. Dick, International Earth Rotation and Reference Systems Service (IERS), IERS Technical Note No. 30, Frankfurt am Main, Germany: Verlag des Bundesamtes für Kartographie und Geodäsie, pp 39–50.

Garnero E.J., Revenough J., Williams Q., Lay T., and Kellogg L.H., 1998, "Ultralow velocity zone at the core-mantle boundary." In: *The Core-Mantle Boundary Region*, Eds. M. Gurnis, M. E. Wysession, E. Knittle, and B. A. Buffett, Geodynamics Series, 28, American Geophysical Union, Washington, D.C., pp 319–334.

Gegout P., Hinderer J., Legros H., Greff M., and Dehant V., 1998, "Influence of atmospheric pressure on the free core nutation, precession and some forced nutational motions of the Earth." *Phys. Earth Planet. Inter.*, **106**(*3–4*), 337–351.

Getino J. and Ferrándiz J.M., 1991, "A Hamiltonian theory for an elastic Earth – Secular rotational acceleration." *Celest. Mech. Dynam. Astr.*, **52**(*4*), 381–396.

Getino J., 1993, "Perturbed nutations, Love numbers and elastic energy of deformation for Earth models 1066A and 1066B." *ZAMP Zeitschrift für angewandte Mathematik und Physik*, **44**(*6*), 998–1021, DOI: 10.1007/BF00942762.

Getino J. and Ferrándiz J.M., 1994, "A rigorous Hamiltonian approach to the rotation of elastic bodies." *Celest. Mech. Dynam. Astr.*, **58**(*3*), 277–295.

Getino J. and Ferrándiz J.M., 1995, "On the effect of the mantle elasticity on the Earth's rotation." *Celest. Mech. Dynam. Astr.*, **61**, 117–180.

Getino J. and Ferrándiz J.M., 1997, "A Hamiltonian approach to dissipative phenomena between the Earth's mantle and core, and effects on free nutations." *Geophys. J. Int.*, **130**(*2*), 326–334, DOI: 10.1111/j.1365-246X.1997.tb05650.x.

Getino J. and Ferrándiz J.M., 1999, "Accurate analytical nutation series." *Monthly Notices Roy. Astron. Soc.*, **306**(*4*), L45–L49.

Getino J., Martin P., and Farto J.M., 1999, "Improved nutation series for the non-rigid Earth with a precise adjustment of parameters with nonlinear dependence." *Celest. Mech. Dynam. Astr.*, **74**(*3*), 153–162, DOI: 10.1023/A:1008360303986.

Getino J. and Ferrándiz J.M., 2000, "Effects of dissipation and a liquid core on forced nutations in Hamiltonian theory." *Geophys. J. Int.*, **142**(*3*), 703–715.

Getino J., González A.B., and Escapa A., 2000, "The rotation of a non-rigid, non-symmetrical Earth II: Free nutations and dissipative effects." *Celest. Mech. Dynam. Astr.*, **76**(*1*), 1–21, DOI: 10.1023/A:1008373613208.

Getino J. and Ferrándiz J.M., 2001, "Forced nutations of a two-layer Earth model." *Monthly Notices Roy. Astron. Soc.*, **322**(*4*), 785–799.

Getino J., Ferrándiz J.M., and Escapa A., 2001, "Hamiltonian theory for the non-rigid Earth: semidiurnal terms." *Astr. Astrophys.*, **370**, 330–341.

Getino J., Escapa A., and Miguel D., 2010, "General theory of the rotation of the non-rigid Earth at the second order. I. The rigid model in Andoyer variables." *Astron. J.*, **139**(*5*), 1916–1934, DOI: 10.1088/0004-6256/139/5/1916.

Gibert D. and Le Mouël J.L., 2008, "Inversion of polar motion data: Chandler wobble, phase jumps, and geomagnetic jerks." *Journal of Geophysical Research: Solid Earth*, **113**(*B10*), CiteID: B10405, DOI: 10.1029/2008JB005700.

Gilbert F. and Dziewonski A.M., 1973, "The structure of the Earth retrieved from eigenspectra." *EOS, Trans. Am. Geophys. Un.*, **54**, 374.

Gilbert F. and Dziewonski A.M., 1975, "An application of normal mode theory to the retrieval of structural parameters and source mechanisms from seismic spectra." *Phil. Trans. Roy. Soc. A.*, **278**, 187–269.

Gipson J.M., 2006, "Correlation due to station dependent noise in VLBI." In: Proc. 2006 IVS Workshop, pp 286–290.

Glatzmaier G.A. and Roberts P.H., 1995, "A three-dimensional convective dynamo solution with rotating and finitely conducting inner core and mantle." *Phys. Earth Planet. Inter.*, **91**, 63–75.

Glatzmaier G.A. and Roberts, 1996, "An anelastic evolutionary geodynamo simulation driven by compositional and thermal convection." *Physica D*, **97**, 81–94.

Goldstein H., 1950, "*Classical Mechanics*." Addison-Wesley, Reading, Massachusetts.

Gontier A.-M., Le Bail K., and Feissel M., 2001, "Stability of the extragalactic reference frame." *Astr. Astrophys.*, **375**, 661–669.

Gontier A.-M., 2003, "An implementation of IAU2000 resolutions in VLBI analysis software." IERS Technical Note No. 29, 71–75.

Gontier A.-M. and Feissel-Vernier M., 2005, "Contribution of stable sources to ICRF improvements." In: *Highlights of Astronomy*, Vol. 13, Proc. XXVth General Assembly of the IAU–2003, Sydney, Australia, July 2003, Ed. O. Engvold, p 602.

González A.B. and Getino J., 1997, "The rotation of a non-rigid, non-symmetrical Earth I: Free nutations." *Celest. Mech. Dynam. Astr.*, **68**(*2*), 139–149.

Greenspan H., 1969, "*Theory of Rotating Fluids*." Cambridge University Press, New York.

Greff-Lefftz M. and Legros H., 1999a, "Correlation between some major geological events and resonances between the free core nutation and luni-solar tidal waves." *Geophys. J. Int.*, **139**, 131–151, DOI: 10.1046/j.1365-246x.1999.00935.x.

Greff-Lefftz M. and Legros H., 1999b, "Magnetic field and rotational eigenfrequencies." *Phys. Earth Planet. Inter.*, **112**(*1–2*), 21–41.

Greff-Lefftz M., Legros H., and Dehant V., 2000, "Influence of the inner core viscosity on the rotational eigenmodes of the Earth." *Phys. Earth Planet. Inter.*, **122**, 187–204.

Greff-Lefftz M., Dehant V., and Legros H., 2002, "Effect of inner core viscosity on gravity changes and spatial nutations induced by luni-solar tides." *Phys. Earth Planet. Inter.*, **129**(*1–2*), 31–41.

Greff-Lefftz M., Pais M. A., and Le Mouël J.-L., 2004, "Surface gravitational field and topography changes induced by the Earth's fluid core motions." *J. Geodesy*, **78**(*6*), 386–392, DOI: 10.1007/s00190-004-0418-x.

Greff-Lefftz M., Métivier L., and Legros H., 2005, "Analytical solutions of Love numbers for a hydrostatic ellipsoidal incompressible homogeneous Earth." *Celest. Mech. Dynam. Astr.*, **93**(*1–4*), 113–146, DOI: 10.1007/s10569-005-6424-3.

Greff-Lefftz M. and Legros H., 2007, "Fluid core dynamics and degree-one deformations: Slichter mode and geocenter motions." *Phys. Earth Planet. Inter.*, **161**(*3–4*), 150–160, DOI: 10.1016/j.pepi.2006.12.003.

Greff-Lefftz M., Métivier L., and Besse J., 2010, "Dynamic mantle density heterogeneities and global geodetic observables." *Geophys. J. Int.*, **180**(*3*), 1080–1094, DOI: 10.1111/j.1365-246X.2009.04490.x.

Greff-Lefftz M., 2011, "Length of day variations due to mantle dynamics at geological timescale." *Geophys. J. Int.*, **187**(*2*), 595–612, DOI: 10.1111/j.1365-246X.2011.05169.x.

Greff-Lefftz M. and Besse J., 2012, "Paleo movement of continents since 300 Ma, mantle dynamics and large wander of the rotational pole." *Earth Planet. Sci. Lett.*, **345**, 151–158, DOI: 10.1016/j.epsl.2012.06.017.

Greiner-mai H. and Barthelmes F., 2001, "Relative wobble of the Earth's inner core derived from polar motion and associated gravity variations." *Geophys. J. Int.*, **144**(*1*), 27–36, DOI: 10.1046/j.1365-246X.2001.00319.x.

Gross R.S., 1982, "A determination and analysis of polar motion." Ph.D. thesis, Colorado Univ., Boulder.

Gross R.S. and Chao B.F., 1985, "Excitation study of the LAGEOS-derived Chandler wobble." *Journal of Geophysical Research*, **90**, 9369–9380, DOI: 10.1029/JB090iB11p09369.

Gross R.S., 1992, "Correspondence between theory and observation of polar motion." *Geophys. J. Int.*, **109**, 162–170.

Gross R.S., 1993, "The effect of ocean tides on the Earth's rotation as predicted by the results of an ocean tide model." *Geophys. Res. Letters*, **20**(*4*), 293–296, DOI: 10.1029/93GL00297.

Gross R.S., 2000, "The excitation of the Chandler Wobble." *Geophys. Res. Letters*, **27**(*15*), 2329–2332.

Gross R.S., Fukumori I., and Menemenlis D., 2003, "Atmospheric and oceanic excitation of the Earth's wobbles during 1980–2000." *J. Geophys. Res.*, **108**(*B8*), CiteID: 2370, DOI: 10.1029/2002JB002143.

Gross R.S., Fukumori I., and Menemenlis D., 2004, "Atmospheric and oceanic excitation of length-of-day during 1980–2000." *J. Geophys. Res.*, **109**(*B1*), CiteID: B01406, DOI: 10.1029/2003JB002432.

Gross R.S., 2005a, "The observed period and Q of the Chandler wobble." Workshop *Forcing of polar motion in the Chandler frequency band: A contribution to understanding interannual climate variations*, April 21–23, 2004, Luxembourg, Cahiers du Centre Européen de Géodynamique et de Séismologie, Vol. 24, Eds. H.P. Plag, B. Chao, R. Gross, and T. Van Dam, pp 31–37.

Gross R.S., 2005b, "Oceanic excitation of polar motion: a review." Workshop *Forcing of polar motion in the Chandler frequency band: A contribution to understanding interannual climate variations*, April 21–23, 2004, Luxembourg, Cahiers du Centre Européen de Géodynamique et de Séismologie, Vol. 24, Eds. H.P. Plag, B. Chao, R. Gross, and T. Van Dam, pp 89–102.

Gross R.S., Fukumori I., and Menemenlis D., 2005, "Atmospheric and oceanic excitation of decadal-scale Earth orientation variations." *Journal of Geophysical Research: Solid Earth*, **110**(*B9*), CiteID: B09405, DOI: 10.1029/2004JB003565.

Groten E., Lenhardt H., and Molodenskii S.M., 1991, "On the period and damping of polar motion." In: Proc. Conference on *Mathematical Geophysics*, 18th, Jerusalem, Israel, June 1990, *Journal of Geophysical Research*, **96**, 20241–20255, DOI: 10.1029/91JB01134.

Gubanov V.S., 2009, "Dynamics of the Earth's core from VLBI observations." *Astronomy Letters*, **35**(*4*), 270–277, DOI: 10.1134/S1063773709040070.

Gubanov V.S., 2010, "New estimates of retrograde free core nutation parameters." *Astronomy Letters*, **36**(*6*), 444–451, DOI: 10.1134/S1063773710060083.

Gubbins D. and Roberts P., 1987, "Magnetohydrodynamics of Earth's core." In: *Geomagnetism 2*, Ed. J.A. Jacobs, Academic Press, London, pp 1–183.

Gubbins D., Masters G., and Nimmo F., 2008, "A thermochemical boundary layer at the base of Earth's outer core and independent estimate of core heat flux." *Geophys. J. Int.*, **174**(*3*), 1007–1018, DOI: 10.1111/j.1365-246X.2008.03879.x.

Guinot B., 1979, "Basic problems in the kinematics of the rotation of the Earth." In: *Time and the Earth's Rotation*, Eds. D.D. McCarthy and P.D. Pilkington, IAU Publ., Reidel Publishing Company, pp 7–18.

Guinot, B., 2001, "Time systems and time frames, the epochs." Proceedings Journées Systèmes de Référence Spatio-temporels 2000, Ed. N. Capitaine, Observatoire de Paris, pp 209–213.

Guinot B., 2003a, "Celestial ephemeris origin (CEO), terrestrial ephemeris origin (TEO)." IERS Technical Note No. 29, pp 99–100.

Guinot B., 2003b, "Comparison of 'old' and 'new' concepts: celestial ephemeris origin (CEO), terrestrial ephemeris origin (TEO), Earth rotation angle (ERA)." IERS Technical Note No. 29, 45–46.

Guo J.Y. and Ning J.S., 2002, "Influence of inner core rotation and obliquity on the inner core wobble and the free inner core nutation." *Geophys. Res. Letters*, **29**(*8*), 45-1–45-4, CiteID: 1203, DOI: 10.1029/2001GL014058.

Guo J.Y., Mathews P.M., Zhang Z.X., and Ning J.S., 2004, "Impact of the inner core rotation on outer core flow: the role of outer core viscosity." *Geophys. J. Int.*, **159**(*1*), 372–389, DOI: 10.1111/j.1365-246X.2004.02416.x.

Guo J.Y., Greiner-Mai H., and Ballani L., 2005a, "A spectral search for the inner core wobble in Earth's polar motion." *J. Geophys. Res.*, **110**(*B10*), CiteID: B10402, DOI: 10.1029/2004JB003377.

Guo J.Y., Greiner-Mai H., Ballani L., Jochmann H., and Shum C.K., 2005b, "On the double-peak spectrum of the Chandler wobble." *J. Geodesy*, **78**(*11–12*), 654–659, DOI: 10.1007/s00190-004-0431-0.

Guo J.Y., Li Y.B., Dai C.L., and Shum C.K., 2013, "A technique to improve the accuracy of Earth orientation prediction algorithms based on least squares extrapolation." *Journal of Geodynamics*, **70**, 36–48, DOI: 10.1016/j.jog.2013.06.002.

Gurfil P., Lainey V., and Efroimsky M., 2007, "Long-term evolution of orbits about a precessing oblate planet: 3. A semianalytical and a purely numerical approach." *Celest. Mech. Dynam. Astr.*, **99**(*4*), 261–292, DOI: 10.1007/s10569-007-9099-0.

Gwinn C.R., Herring T.A., and Shapiro I.I., 1986, "Geodesy by radio interferometry: studies of the forced nutations of the Earth. II – Interpretation." *J. Geophys. Res.*, **91**, 4755–4766, DOI: 10.1029/JB091iB05p04755.

Haas R., 2004, "Analysis strategies and software for geodetic VLBI." In: Proc. 7th European VLBI Network Symp., Eds. R. Bachiller, F. Desmurs, J.F. de Vicente, Toledo, Spain, pp 297–301.

Hager B.H., Clayton R.W., Richards M.A., Comer R.P., and Dziewonski A.M., 1985, "Lower mantle heterogeneity, dynamic topography and the geoid." *Nature*, **313**, 541–545, DOI: 10.1038/313541a0.

Han S.-C., Ray R.D., and Luthcke S.B., 2010, "One centimeter-level observations of diurnal ocean tides from global monthly mean time-variable gravity fields." *J. Geodesy*, **84**(*12*), 715–729, DOI: 10.1007/s00190-010-0405-3.

Hartmann T. and Soffel M., 1994, "The nutation of a rigid Earth model: direct influences of the planets." *Astron. J.*, **108**, 1115–1120.

Hartmann T. and Wenzel H.-G., 1995a, "Catalogue HW95 of the tide generating potential." *Bulletin d'Informations Marées Terrestres*, **123**, 9278–9301.

Hartmann T. and Wenzel H.-G., 1995b, "The HW95 tidal potential catalogue." *Geophys. Res. Letters*, **22**(*24*), 9278–9301.

Hartmann T., Soffel M., and Ron C., 1999, "The geophysical approach towards the nutation of a rigid Earth." *Astronomy and Astrophysics Supplement*, **134**, 271–286.

Hass R. and Schuh H., 1996, "Determination of frequency dependent Love and Shida numbers from VLBI data." *Geophys. Res. Letters*, **23**(*12*), 1509–1512, DOI: 10.1029/96GL00903.

Heinkelmann R., Boehm J., Schuh H., *et al.*, 2007, "Combination of long time-series of troposphere zenith delays observed by VLBI." *J. Geodesy*, **81**(*6–8*), 483–501, DOI: 10.1007/s00190-007-0147-z.

Heinkelmann R., Boehm J., Bolotin S., *et al.*, 2011, "VLBI-derived troposphere parameters during CONT08." *J. Geodesy*, **85**(*7*), 377–393, DOI: 10.1007/s00190-011-0459-x.

Herring T.A., 1986, "Precision of vertical position estimates from very long baseline interferometry." *J. Geophys. Res.*, **91**, 9177–9182, DOI: 10.1029/JB091iB09p09177.

Herring T.A., Gwinn C.R., and Shapiro I.I., 1986a, "Geodesy by radio interferometry: studies of the forced nutations of the Earth. I – Data analysis." *J. Geophys. Res.*, **91**, 4745–4755, DOI: 10.1029/JB091iB05p04745.

Herring T.A., Shapiro I.I., Clark T.A., Ma C., and Ryan J.W., 1986b, "Geodesy by radio interferometry – Evidence for contemporary plate motion." *J. Geophys. Res.*, **91**, 8341–8347, DOI: 10.1029/JB091iB08p08341.

Herring T.A., Buffett B.A., Mathews P.M., and Shapiro I.I., 1991, "Forced nutations of an Earth: influence of inner core dynamics 3. Very long interferometry data analysis." *J. Geophys. Res.*, **96**(*B5*), 8259–8273.

Herring T.A., 1991, "The rotation of the Earth." *Rev. Geophys. Supplement*, **29**, part 1, 172–175.

Herring T.A. and Dong D., 1994, "Measurement of diurnal and semidiurnal rotational variations and tidal parameters of Earth." *J. Geophys. Res.*, **99**(*B9*), 18051–18071, DOI: 10.1029/94JB00341.

Herring T.A., Mathews P.M., and Buffett B., 2002, "Modeling of nutation-precession of a non-rigid Earth with ocean and atmosphere." *J. Geophys. Res.*, **107**(*B4*), ETG4-1–12, CI: 2069, DOI: 10.1029/2001JB000165.

Hide R. and Dickey J.O., 1992, "Analyses and forecasts of fluctuations in the angular momentum of the atmosphere and changes in the Earth's rotation." *Meteorological Magazine*, **121**, 22–26.

Hill E.L., 1954, "The theory of vector spherical harmonics." *Am. J. Phys.*, **22**, 211–214.

Hinderer J., Legros H., and Amalvict M., 1987a, "Tidal motions within the Earth's fluid core: resonance process and possible variations." *Phys. Earth Planet. Inter.*, **49**(*3–4*), 213–221.

Hinderer J., Legros H., Gire C., and Le Mouël J.L., 1987b, "Geomagnetic secular variation, core motions and implications for the Earth's wobbles." *Phys. Earth Planet. Inter.*, **49**(*1–2*), 121–132, DOI: 10.1016/0031-9201(87)90136-1.

Hinderer J., Boy J.-P., Gegout P., *et al.*, 2000, "Are the free core nutation parameters variable in time?" *Phys. Earth Planet. Inter.*, **117**, 37–49.

Holme R., 1998a, "Electromagnetic core-mantle coupling; I: Explaining decadal changes in the length of day." *Geophys. J. Int.*, **132**(*1*), 167–180, DOI: 10.1046/j.1365-246x.1998.00424.x.

Holme R., 1998b, "Electromagnetic core-mantle coupling; II: Probing deep mantle conductance." In: *The Core-Mantle Boundary Region*, Geodynamics Series, volume 28, Eds. M. Gurnis, M.E. Wysession, E. Knittle, and B.A. Buffett, AGU Geophysical Monograph, Washington D.C., pp 139–152.

Holme R., 2000, "Electromagnetic core-mantle coupling; III: Laterally varying mantle conductance." *Phys. Earth Planet. Inter.*, **117**, 329–344.

Holme R. and de Viron O., 2013, "Characterization and implications of intradecadal variations in length of day." *Nature*, **499**, 202–204, DOI: 10.1038/nature12282.

Höpfner J., 2003a, "Parameter variability of the observed periodic oscillations of polar motion with smaller amplitudes." *J. Geodesy*, **77**(*7*), 388–401, DOI: 10.1007/s00190-003-0342-5.

Höpfner J., 2003b, "Polar motions with a half-Chandler period and less in their temporal variability." *Journal of Geodynamics*, **36**(*3*), 407–422, DOI: 10.1016/S0264-3707(03)00059-0.

Höpfner J., 2003c, "Chandler and annual wobbles based on space-geodetic measurements." *Journal of Geodynamics*, **36**(*3*), 369–381, DOI: 10.1016/S0264-3707(03)00056-5.

Hopkins W., 1839, "On the phenomena of precession and nutation, assuming the fluidity of the interior of the Earth." *Phil. Trans. Roy. Soc. London*, **129**, 381–423.

Hori G.I., 1966, "Theory of general perturbations with unspecified canonical variables." *Publ. Astron. Soc. Japan*, **18**(*4*), 287–296.

Hough S.S., 1895, "The oscillations of a rotating ellipsoidal shell containing fluid." *Philos. Trans. Roy. Soc., Ser. A*, **186**, 469–506.

Huang C.L., 1999, "The scalar boundary conditions for the motion of the elastic Earth to second order in ellipticity." *Earth, Moon, and Planets*, **84**(*3*), 125–141, DOI: 10.1023/A:1018999708294.

Huang C.L., Liao X.H., Zhu Y.Z., and Jin W.J., 2001a, "The derivation of a scalar boundary condition for the motion of the elastic and slightly elliptical Earth." *Acta Astronomical Sinica*, **42**(*1*), 81–87.

Huang C.L., Jin W.J., and Liao X.H., 2001b, "A new nutation model of a non-rigid Earth with ocean and atmosphere." *Geophys. J. Int.*, **146**(*1*), 126–133, DOI: 10.1046/j.1365-246X.2001.00429.x.

Huang C., Dehant V., and Liao X., 2004, "The explicit equations of infinitesimal elastic-gravitational motion in the rotating, slightly elliptical fluid outer core of the Earth." *Geophys. J. Int.*, **157**(*2*), 831–837, DOI: 10.1111/j.1365-46X.2004.02238.x.

Huang C.L., Dehant V., Liao X.H., Van Hoolst T., and Rochester M.G., 2011, "On the coupling between magnetic field and nutation in a numerical integration approach." *J. Geophys. Res.*, **116**, B03403, DOI: 10.1029/2010JB007713.

IAU 1976, *"Proceedings of the Sixteenth General Assembly."* Ed. E. Muller, *Trans. Int. Astron. Union*, Vol. 16B, Kluwer Academic Publishers, 547 pp.

IAU 1992, *"Proceedings of the Twenty-First General Assembly."* Ed. J. Bergeron, *Trans. Int. Astron. Union*, Vol. 21B, Kluwer Academic Publishers.

IAU 1998, *"Proceedings of the Twenty-Third General Assembly."* Ed. J. Anderson, *Highlights of Astronomy* Vol. 11, *Trans. Int. Astron. Union*, Vol. 23B, Kluwer Academic Publishers.

IAU 2000a, *"Proceedings of the IAU Symposium 180 'Towards Models and Constants for Sub-Microarcsecond Astrometry'."* Washington, D.C., USA, March 27–30, 2000, Eds. K.J. Johnston, D.D. McCarthy, B.J. Luzum, and G.H. Kaplan.

IAU 2000b, *"Proceedings of the Twenty-Fourth General Assembly."* Ed. H. Rickman, *Highlights of Astronomy* Vol. 12, *Trans. Int. Astron. Union*, Vol. 24B, Kluwer Academic Publishers.

IAU 2000c, Approved Resolutions, *"Proceedings of the Twenty-Fourth General Assembly."* Ed. H. Rickman, *Highlights of Astronomy* Vol. 12, *Trans. Int. Astron. Union*, Vol. 24B, Kluwer Academic Publishers, pp 34–58.

IAU 2003, *"Proceedings of the Twenty-Fifth General Assembly."* Ed. O. Engvold, *Highlights of Astronomy* Vol. 13, *Trans. Int. Astron. Union*, Vol. 25B, Kluwer Academic Publishers.

IAU 2006, *"Proceedings of the Twenty-Sixth General Assembly."* Ed. K.A. van der Hucht, *Highlights of Astronomy* Vol. 14, *Trans. Int. Astron. Union*, Vol. 26B, Kluwer Academic Publishers.

IERS annual report 1988–1999 and their annexes.

IERS annual report 2000–2009, available online on IERS website.

IERS Conventions 1996, IERS Technical Note No. 21, 98 pp, Ed. D.D. McCarthy, see also McCarthy (1996).

IERS Conventions 2000, published in 2003, IERS Technical Note No. 32, Eds. D.D. McCarthy and G. Petit, Frankfurt am Main, Verlag des Bundesamts für Kartographie und Geodäsie, 127 pp, see also McCarthy and Petit (2003).

IERS Conventions 2010, published in 2011, IERS Technical Note No. 36, Eds. G. Petit and B. Luzum, Frankfurt am Main, Verlag des Bundesamts für Kartographie und Geodäsie, IERS Conventions Center, 179 pp, see also Petit and Luzum (2011).

IERS Standards 1989, IERS Technical Note No. 3, Ed. D.D. McCarthy.

IERS Standards 1992, IERS Technical Note No. 13, Ed. D.D. McCarthy.

IUGG 1987 (Vancouver), IUGG Resolutions, available online on the IUGG website.

IUGG 1991 (Vienna), IUGG Resolutions, available online on the IUGG website, also, 1992, IUGG Resolutions, *Bull Géodésique*, **66**, 128–129.

IUGG 1995 (Boulder), IUGG Resolutions, available online on the IUGG website.

IUGG 1999 (Birmingham), IUGG Resolutions, available online on the IUGG website.

IUGG 2003 (Sapporo), IUGG Resolutions, available online on the IUGG website.

IUGG 2007 (Perugia), IUGG Resolutions, available online on the IUGG website.

Jackson A. and Livermore P., 2009, "On Ohmic heating in the Earth's core I: nutation constraints." *Geophys. J. Int.*, **177**(2), 367–382, DOI: 10.1111/j.1365-246X.2008.04008.x.

Jackson A., Livermore P.W., and Ierley G., 2011, "On Ohmic heating in the Earth's core II: Poloidal magnetic fields obeying Taylor's constraint." *Geophys. J. Int.*, **187**(3), 322–327, DOI: 10.1016/j.pepi.2011.06.003.

Jacobs C.S., Sovers O.J., Williams J.G., and Standish E.M., 1993, "The extragalactic and solar system celestial frames: accuracy, stability, and interconnection." *Adv. Space Res.*, **13**(*11*), 161–174, DOI: 10.1016/0273-1177(93)90219-2.

Jacobs C.S., Sovers O.J., Naudet C.J., Branson R.P., and Coker R.F., 1998, "The JPL extragalactic radio reference frame: astrometric results of 1978–1996 deep space network VLBI." In: *The Telecommunications and Data Acquisition Progress Report* 42–133, Jet Propulsion Laboratory, California Institute of Technology, 48 pp.

Jault D. and Le Mouël J.L., 1993, "Circulation in the liquid core and coupling with the mantle." *Adv. Space Res.*, **13**(*11*), 221–233, DOI: 10.1016/0273-1177(93)90225-Z.

Jeanloz R., 1990, "The nature of the Earth's core." *Ann. Rev. Earth Planet. Sci.*, **18**, 357–386, DOI: 10.1146/annurev.ea.18.050190.002041.

Jeanloz R.,1993, "Chemical reactions at the Earth's core-mantle boundary: summary of evidence and geomagnetic implications." In: *Relating Geophysical Structures and Processes: the Jeffreys Volume*, Eds. K. Aki and R. Dmowska, AGU/IUGG, 121–127, DOI: 10.1029/GM076p0121.

Jeffreys H. and Vicente R.O., 1957a, "The theory of nutation and the variation of the latitude." *Mon. Not. R. Astron. Soc.*, **117**, 142–161.

Jeffreys H. and Vicente R.O., 1957b, "The theory of nutation and the variation of the latitude: the Roche model cor." *Mon. Not. R. Astron. Soc.*, **117**, 162–173.

Jeffreys H., 1958, "Nutation: comparison of theory and observations." *Mon. Not. R. Astron. Soc.*, **119**, 75–80.

Jeffreys H., 1959a, "Nutation." *Geophys. J. Int., The Observatory*, **79**, 222–222.

Jeffreys H., 1959b, "Nutation: comparison of theory and observations." *Mon. Not. R. Astron. Soc.*, **119**(*2*), 75–80.

Jeffreys H., 1978, "Some difficulties in the theory of nutation." *Geophys. J. Int.*, **54**(*3*), 711–712, DOI: 10.1111/j.1365-246X.1978.tb05505.x.

Jeffreys H., 1979, "On imperfection of elasticity in the Earth's interior." *Geophys. J. Int.*, **55**(*2*), 273–281, DOI: 10.1111/j.1365-246X.1978.tb04271.x.

Jeffreys H., 1983, "Earth models." *Geophys. J. Int.*, **74**(*1*), 5, DOI: 10.1111/j.1365-246X.1983.tb01867.x.

Johnson T., Kammeyer P., and Ray J., 2001, "The effects of geophysical fluids on motions of the Global Positioning System satellites." *Geophys. Res. Letters*, **28**, 17, DOI: 10.1029/2001GL013180, issn: 0094-8276.

Kakuta C., Okamoto I., and Sasao T., 1975, "Is the nutation of the solid inner core responsible for the 24-year libration of the pole?" *Publ. Astron. Soc. Jpn.*, **27**, 357–365.

Kalarus M., Schuh H., Kosek W., *et al.*, 2010, "Achievements of the Earth orientation parameters prediction comparison campaign." *J. Geodesy*, **84**(*10*), 587–596, DOI: 10.1007/s00190-010-0387-1.

Kammeyer P., 2000, "A UT1-like quantity from analysis of GPS orbit planes." *Celest. Mech. Dynam. Astr.*, **77**, 241–272.

Kanamori H. and Anderson D.L., 1977, "Importance of physical dispersion in surface wave and free oscillation problems: review." *Rev. Geophys. Space Phys.*, **15**(*1*), 105–112.

Kaplan G.H., 1998, "High-precision algorithms for astrometry – a comparison of two approaches." *Astron. J.*, **115**, 361–372.

Kaplan G.H., 2003, "Another look at non-rotating origins." In: Proc. IAU 25th General Assembly, Sidney, Australia, JD 16 on '*The International Celestial Reference System, Maintenance and Future Realizations*', Ed. O. Engvold, *Highlights of Astronomy* Vol. 13, *Trans. Int. Astron. Union*, Vol. 25B, Kluwer Academic Publishers, 612–613.

Kaplan G.H., 2005a, "The IAU resolutions on astronomical reference systems." *Time Scales and Earth Rotation Models*, USNO Circular No. 179, U.S. Naval Observatory, 118 pp.

Kaplan G.H., 2005b, "Celestial pole offsets: conversion from (dX,dY) to (dψ,dϵ)." U.S. Naval Observatory technical note, 4 pp.

Karatekin O., Duron J., Rosenblatt P., *et al.*, 2005, "Mars time-varying gravity and its determination; simulated geodesy experiments." *J. Geophys. Res., Planets*, **110**, Cite ID: E06001, DOI: 10.1029/2004JE002378.

Karatekin O., Dehant V., and Van Hoolst T., 2006a, "Martian global-scale CO_2 exchange from time-variable gravity measurements." *J. Geophys. Res.*, **111**(*E6*), CiteID: E06003, DOI: 10.1029/2005JE002591.

Karatekin O., Van Hoolst T., Tastet J., de Viron O., and Dehant V., 2006b, "The effects of seasonal mass redistribution and interior structure on length-of-day variations of Mars." *Adv. Space Res.*, **38**(*4*), 739–744, DOI: JASR-D-04-01301R1.

Karatekin O., de Viron O., Lambert S., *et al.*, 2011, "Atmospheric angular momentum variations of Earth, Mars and Venus at seasonal time scales." *Planet. Space Sci.*, Special Issue on *Comparative Planetology: Venus-Earth-Mars*, **59**(*10*), 923–933, DOI: 10.1016/j.pss.2010.09.010.

Karato S.-I., 2008, *"Deformation of Earth Materials. An Introduction to the Rheology of Solid Earth."* Cambridge University Press, USA, New York, 463 pp.

Kennett B.L.N., Engdahl E.R., and Buland R., 1995, "Constraints on seismic velocities in the Earth from traveltimes." *Geophys. J. Int.*, **122**, 108–124, DOI: 10.1111/j.1365-246X.1995.tb03540.x.

King E.M., and Buffett B.A., 2013, "Flow speeds and length scales in geodynamo models: the role of viscosity." *Earth Planet. Sci. Lett.*, **371**, 156–162, DOI: 10.1016/j.epsl.2013.04.001.

King R.W., 1987, "The rotation of the Earth." *Rev. Geophys.*, **25**, 871–874, DOI: 10.1029/RG025i005p00871.

Kinoshita H., 1975, *"Formulas for Precession."* Smithsonian Astrophys. Obs. Special Report, 364.

Kinoshita H., 1977, "Theory of the rotation of the rigid Earth." *Celest. Mech.*, **15**, 277–326.

Kinoshita H., Nakajima K., Kubo Y., *et al.*, 1979, "Note on nutation in ephemerides." *Publ. Int. Lat. Obs. Mizusawa*, **12**, pp 71–108.

Kinoshita H. and Souchay J., 1990, "The theory of the nutation for the rigid Earth model at the second order." *Celest. Mech.*, **48**, 187–265, DOI: 10.1007/BF02524332.

Kinoshita H. and Souchay J., 1993, "Erratum: 'The theory of the nutation for the rigid Earth model at the second order' [*Celest. Mech. Dynam. Astr.*, **48**(*3*), 187–265 (1990)]." *Celest. Mech. Dynam. Astr.*, **57**(*3*), 515.

Klioner S.A. and Voinov A.V., 1993, "Relativistic theory of astronomical reference systems in closed form." *Physical Review D (Particles, Fields, Gravitation, and Cosmology)*, **48**(*4*), 1451–1461, DOI: 10.1103/PhysRevD.48.1451.

Klioner S.A., 1998, "Astronomical reference frames in the PPN formalism." Journées Systèmes de Référence Spatio-temporels 1997, September 1997, Prague, Czech Rep., Eds. J. Vondrák and N. Capitaine, 32–37.

Klioner S.A. and Soffel M.H., 2000, "Relativistic celestial mechanics with PPN parameters." *Physical Review D (Particles, Fields, Gravitation, and Cosmology)*, **62**(*2*), ID. 024019, DOI: 10.1103/PhysRevD.62.024019.

Klioner S.A., 2003, "A practical relativistic model for microarcsecond astrometry in space." *Astron. J.*, **125**, 1580–1597.

Koelemeijer P.J., Deuss A., and Trampert J., 2012, "Normal mode sensitivity to Earth's D'' layer and topography on the core-mantle boundary: what we can and cannot see." *Geophys. J. Int.*, **190**(*1*), 553–568, DOI: 10.1111/j.1365-246X.2012.05499.x.

Koning A. and Dumberry M., 2013, "Internal forcing of Mercury's long period free libra-tions." *Icarus*, **223**(*1*), 40–47, DOI: 10.1016/j.icarus.2012.11.022.

Konopliv A.S., Yoder C.F., Standish E.M., Yuan D.-N., and Sjogren W.L., 2006, "A global solution for the Mars static and seasonal gravity, Mars orientation, Pho-bos and Deimos masses, and Mars ephemeris." *Icarus*, **182**(*1*), 23–50, DOI: 10.1016/j.icarus.2005.12.025.

Konopliv A.S., Asmar S.W., Folkner W.M., *et al.*, 2011, "Mars high resolution gravity fields from MRO, Mars seasonal gravity, and other dynamical parameters." *Icarus*, **211**(*1*), 401–428, DOI: 10.1016/j.icarus.2010.10.004.

Koot L., 2006, "Etude de la structure interne de la Terre à partir des observations de la précession et des nutations." Ph.D. thesis, in French, 230 pp.

Koot L., de Viron O., and Dehant V., 2006, "Atmospheric angular momentum time-series: characterization of their internal noise and creation of a combined series." *J. Geodesy*, **79**, 663–674, ISSN: 0949–7714, DOI: 10.1007/s00190-005-0019-3.

Koot L., Rivoldini A., de Viron O., and Dehant V., 2008, "Estimation of Earth interior parameters from a Bayesian inversion of VLBI nutation time series." *J. Geophys. Res.*, **113**(*B8*), CiteID B08414, DOI: 10.1029/2007JB005409.

Koot L., Dumberry M., Rivoldini A., de Viron O., and Dehant V., 2010, "Constraints on the coupling at the core-mantle and inner core boundaries inferred from nutation observa-tions." *Geophys. J. Int.*, **182**, 1279–1294, DOI: 10.1111/j.1365-246X.2010.04711.x.

Koot L. and Dumberry M., 2011, "Viscosity of the Earth's inner core: con-straints from nutation observations." *Earth Planet. Sci. Lett.*, **308**, 343–349, DOI: 10.1016/j.epsl.2011.06.004.

Koot L. and de Viron O., 2011, "Atmospheric contributions to nutations and implications for the estimation of deep Earth's properties from nutation observations." *Geophys. J. Int.*, **185**, 1255–1265, DOI: 10.1111/j.1365-246X.2011.05026.x.

Koot L. and Dumberry M. 2013, "The role of the magnetic field morphology on the electromagnetic coupling for nutations." *Geophys. J. Int.*, **195**(*1*), 200–210, DOI: 10.1093/gji/ggt239.

Kopal Z., 1968, "The precession and nutation of deformable bodies." *Astrophys. Space Sci.*, **1**(*1*), 74–91, DOI: 10.1007/BF00653847.

Kopal Z., 1969a, "The precession and nutation of deformable bodies, II." *Astrophys. Space Sci.*, **4**(*3*), 330–364, DOI: 10.1007/BF00661824.

Kopal Z., 1969b, "The precession and nutation of deformable bodies, III." *Astrophys. Space Sci.*, **4**(*4*), 427–458, DOI: 10.1007/BF00651349 .

Kosek W., 2004, "Possible excitation of the Chandler wobble by variable geophysical annual cycle." In: Proceedings of the seminar *Earth rotation and satellite geodesy from astrometry to GNSS*, Warsaw, September 18–19, 2003, Part II: "Earth rotation, geodynamics, and dynamics of the satellite motion." *Journal of Planetary Geodesy*, **39**(*2*), 135–145.

Kouba J., 2005, "Comparison of polar motion with oceanic and atmospheric angular momentum time series for 2-day to Chandler periods." *J. Geodesy*, **79**(*1–3*), 33–42, DOI: 10.1007/s00190-005-0440-7.

Kouba J., 2008, "Implementation and testing of the gridded Vienna mapping function 1 (VMF1)." *J. Geodesy*, **82**(*4–5*), 193–205, DOI: 10.1007/s00190-007-0170-0.

Kovalevsky J., 1967, "*Introduction to Celestial Mechanics*." Springer, Series: Astrophysics and Space Science Library, 134 pp.

Kovalevsky J. and Mueller I.I., 1989, "Introduction to reference frames in astronomy and geophysics." In: *Reference Frames in Astronomy and Geophysics*, Eds. J. Kovalevsky, I.I. Mueller and B. Kolaczek, Astrophys. Space Sc. Lib., Kluwer Acad. Publ., pp 1–12.

Kovalevsky J., Lindegren L., Perryman M.A.C., *et al.*, 1997, "The HIPPARCOS catalogue as a realisation of the extragalactic reference system." *Astr. Astrophys.*, **323**, 620–633.

Kovalevsky J., 2003, "Comparison of 'old' and 'new' concepts: reference systems." IERS Technical Note No. 29, pp 31–34.

Kovalevsky J. and Seidelmann P.K., 2004, *"Fundamentals of Astrometry."* Cambridge University Press.

Krásná H., Boehm J., and Schuh H., 2013, "Free core nutation observed by VLBI." *Astron. Astrophys.*, **555**(*A29*), 5, DOI: 10.1051/0004-6361/201321585.

Krügel M., Thaller D., Tesmer V., *et al.*, 2007, "Tropospheric parameters: combination studies based on homogeneous VLBI and GPS data." *J. Geodesy*, **81**(*6–8*), 515–527, DOI: 10.1007/s00190-007-0127-8.

Kuang W.J. and Bloxham J., 1997, "An Earth-like numerical dynamo model." *Nature*, **389**, 371–374.

Kuang W.J. and Bloxham J., 1999, "Numerical modelling of magnetohydrodynamic convection in a rapidly rotating spherical shell: weak and strong field dynamo action." *J. Comput. Phys.*, **153**, 51–81.

Kuang W.J., Tangborn A., Wei Z., and Sabaka T., 2009, "Constraining a numerical geodynamo model with 100 years of surface observations." *Geophys. J. Int.*, **179**(*3*), 1458–1468, DOI: 10.1111/j.1365-246X.2009.04376.x.

Kubo Y., 1979, "A core-mantle interaction in the rotation of the Earth." *Celest. Mech.*, **19**(*3*), 215–241, DOI: 10.1007/BF01230216.

Kubo Y. and Fukushima T., 1988, "Numerical integration of precession and nutation of the rigid Earth." In: Proc. of the 128th IAU Symp. on *Earth's Rotation and Reference Frames for Geodesy and Geodynamics*, Washington, USA, 1986, Eds. A.K. Babcock and G.A. Wilkins, Reidel Publ. Comp., pp 331–340.

Kubo Y., 1991, "Solution to the rotation of the elastic Earth by method of rigid dynamics." *Celest. Mech. Dynam. Astr.*, **50**(*2*), 165–187.

Kubo Y., 2009, "Rotation of the elastic Earth: the role of the angular-velocity-dependence of the elasticity-caused perturbation." *Celest. Mech. Dynam. Astr.*, **105**(*4*), 261–274, DOI: 10.1007/s10569-009-9225-2.

Kubo Y., 2011, "Kinematical modeling of the Earth rotation, focusing on the Oppolzer terms in a rigid Earth and the Oppolzer-like terms in an elastic Earth." *Celest. Mech. Dynam. Astr.*, **110**(*2*), 143–168, DOI: 10.1007/s10569-011-9347-1.

Kuchynka P. and Folkner W.M., 2013, "A new approach to determining asteroid masses from planetary range measurements." *Icarus*, **222**(*1*), 243–253, DOI: 10.1016/j.icarus.2012.11.003.

Kuchynka P., Folkner W.M., Konopliv A.S., *et al.*, 2014, "New constraints on Mars rotation determined from radiometric tracking of the Opportunity Mars Exploration Rover." *Icarus*, **229**, 340–347, DOI: 10.1016/j.icarus.2013.11.015.

Kudryashova M., Snajdrova K., Weber R., Heinkelmann R., and Schuh H., 2008, "Combination of nutation time series derived from VLBI and GNSS." IVS General Meeting Proceedings, pp 240–245.

Kudryashova M., Lambert S., Dehant V., Bruyninx C., and Defraigne P., 2011, "Determination of nutation offsets by combining VLBI/GPS-produced normal equations." In: Proc. Journées Systèmes de Reférence, Observatoire de Paris, 20–22 September 2010, pp 202–203.

Lamb H., 1932, *"Hydrodynamics."* 6th ed. Dover, New York, 738 pp.

Lambeck K., 1980, *"The Earth's Variable Rotation: Geophysical Causes and Consequences."* Cambridge University Press, Cambridge Monographs on Mechanics and Applied Mathematics, 449 pp.

Lambert S.B. and Bizouard C., 2002, "Positioning the terrestrial ephemeris origin in the international terrestrial reference frame." *Astr. Astrophys.*, **394**, 317–321, DOI: 10.1051/0004-6361:20021139.

Lambert S.B., 2003, "Analyse et modélisation de haute précision pour l'orientation de la Terre." Ph.D. thesis, Observatoire de Paris, France, in French.

Lambert S.B. and Capitaine N., 2004, "Effects of zonal deformations and the Earth's rotation rate variations on precession-nutation." *Astr. Astrophys.*, **428**, 255–260, DOI: 10.1051/0004-6361:20035952.

Lambert S.B., 2006, "Atmospheric excitation of the Earth's free core nutation." *Astr. Astrophys.*, **457**(*2*), 717–720, DOI: 10.1051/0004-6361:20065813.

Lambert S.B., Bizouard C., and Dehant V., 2006, "Rapid variations in polar motion during the 2005–2006 winter season." *Geophys. Res. Letters*, **33**(*13*), CiteID: L03303, DOI: 10.1029/2006GL026422.

Lambert S.B. and Mathews P.M., 2006, "Second-order torque on the tidal redistribution and the Earth's rotation." *Astr. Astrophys.*, **453**(*1*), 363–369.

Lambert S.B. and Dehant V., 2007, "The Earth's core parameters as seen by the VLBI." *Astr. Astrophys.*, **469**, 777–781, DOI: 10.1051/0004-6361:20077392.

Lambert S.B. and Mathews P.M., 2008, "Second-order torque on the tidal redistribution and the Earth's rotation." *Astr. Astrophys.*, **481**, 883–884, DOI: 10.1051/0004-6361:20054516e.

Lambert S.B., Dehant V., and Gontier A.-M., 2008, "Celestial frame instability in VLBI analysis and impact on geophysics." *Astr. Astrophys.*, **481**(*3*), 535–541, DOI: 10.1051/0004-6361:20078489.

Langel R.A. and Estes R.H., 1982, "A geomagnetic field spectrum." *Geophys. Res. Letters*, **9**, 250–253.

Lanzano P., 1984, "Earth tides and polar motion." *J. Geodyn.*, **1**, 121–142, DOI: 10.1016/0264-3707(84)90024-3.

Lanzano P., 1984, "*Deformations of an Elastic Earth.*" International Geophysics Series, Volume 31, Academic Press, 221 pp.

Laplace P.S., 1775, "Recherches sur plusieurs points du système du monde." Mémoires de l'Académie Royale des Sciences de Paris, **88**, pp 75–182; reprinted in *Oeuvres Completes de Laplace*, Gauthier-Villars, Paris, 1893.

Laplace P.S., 1776, "Recherches sur plusieurs points du système du monde." Mémoires de l'Académie Royale des Sciences de Paris, **89**, 177–264; reprinted in *Oeuvres Completes de Laplace*, Gauthier-Villars, Paris, 1893.

Laskar J., 1986, "Secular terms of classical planetary theories using the results of general theory." *Astr. Astrophys.*, **157**(*1*), 59–70.

Laskar J., Jontel F., and Boudin F., 1993, "Orbital, precessional, and insolation quantities for the Earth from −20 Myr to +10 Myr." *Astr. Astrophys.*, **270**, 522–533.

Laskar J., Levrard B., and Mustard J.F., 1993, "Orbital, precessional, and insolation quantities for the Earth from –20 MYR to +10 MYR." *Astr. Astrophys.*, **270**(*1–2*), 522–533.

Laskar J., Levrard B., and Mustard J.F., 2002, "Orbital forcing of the Martian polar layered deposits." *Nature*, **419**(*6905*), 375–377.

Laskar J., Correia A.C.M., Gastineau M., *et al.*, 2004, "Long term evolution and chaotic diffusion of the insolation quantities of Mars." *Icarus*, **170**(*2*), 343–364, DOI: 10.1016/j.icarus.2004.04.005.

Lay T., Williams Q., and Garnero E.J., 1998, "The core-mantle boundary layer and deep Earth dynamics." *Nature*, **392**, 461–468, DOI: 10.1038/33083.

Lefftz M., Legros H., and Hinderer J., 1991, "Non-linear equations for the rotation of a viscoelastic planet taking into account the influence of a liquid core." *Celest. Mech. Dynam. Astr.*, **52**(*1*), 13–43, DOI: 10.1007/BF00048585.

Lefftz M. and Legros H., 1992a, "Some remarks about the rotations of a viscous planet and its homogeneous liquid core – Linear theory." *Geophys. J. Int.*, **108**, 705–724, DOI: 10.1111/j.1365-246X.1992.tb03463.x.

Lefftz M. and Legros H., 1992b, "Influence of viscoelastic coupling on the axial rotation of the Earth and its fluid core." *Geophys. J. Int.*, **108**, 725–739, DOI: 10.1111/j.1365-246X.1992.tb03464.x.

Lefftz M. and Legros H., 1993, "On the fluid and viscoelastic deformations of the planets." *Celest. Mech. Dynam. Astr.*, **57**(*1–2*), 247–278, DOI: 10.1007/BF00692477.

Legros H., Hinderer J., Lefftz M., and Dehant V., 1993, "The influence of the solid inner core on gravity changes and spatial nutations induced by luni-solar tides and surface loading." *Phys. Earth Planet. Inter.*, **76**, 283–315.

Legros H., Greff-Lefftz M., and Tokieda T., 2006, "Physics inside the earth: deformation and rotation." In: *Dynamics of Extended Celestial Bodies and Rings*, Lecture Notes in Physics Volume 682, Ed. Jean Souchay, Springer, Berlin, Germany, 23 pp., DOI: 10.1007/3-540-32455-02.

Le Maistre S., Rosenblatt P., Rivoldini A., *et al.*, 2012, "Lander radio science experiment with a direct link between Mars and the Earth." *Planet. Space Sci.*, **68**(*1*), 105–122, DOI: 10.1016/j.pss.2011.12.020.

Le Mouël J.L., Narteau C., Greff-Lefftz M., and Holschneider M., 2006, "Dissipation at the core-mantle boundary on a small-scale topography." *J. Geophys. Res.*, **111**(*B4*), CiteID: B04413, DOI: 10.1029/2005JB003846.

Lhuillier F., Aubert J., and Hulot G., 2011, "Earth's dynamo limit of predictability controlled by magnetic dissipation." *Geophys. J. Int.*, **186**(*2*), 492–508, DOI: 10.1111/j.1365-246X.2011.05081.x.

Lieske J.H., 1976, this reference is often used in the literature to refer to the IAU resolution on Precession, adopted in 1976; the associated paper is Lieske *et al.*, 1977.

Lieske J.H., Lederle T., Fricke W., and Morando B., 1977, "Expressions for the precession quantities based upon the IAU(1976) system of astronomical constants." *Astr. Astrophys.*, **58**, 1–16.

Lieske J.H., 1979, "Precession matrix based on IAU /1976/ system of astronomical constants." *Astr. Astrophys.*, **73**(*3*), 282–284.

Liu H.P., Anderson D.L., and Kanamori H., 1976, "Velocity dispersion due to anelasticity; implications for seismology and mantle composition." *Geophys. J. Roy. Astron. Soc.*, **47**, 41–58.

Liu J.-C., Capitaine N., Lambert S.B., Malkin Z., and Zhu Z., 2012, "Systematic effect of the Galactic aberration on the ICRS realization and the Earth orientation parameters." *Astr. Astrophys.*, **548**(*A50*), 9, DOI: 10.1051/0004-6361/201219421.

Longuet-Higgins M.S., 1968, "The eigenfunctions of Laplace's tidal equations over a sphere." *Phil. Trans. R. Soc. London A*, **262**, 511–607.

Lott F., de Viron O., Viterbo P., and Vial F., 2008, "Axial AAM budget at diurnal and sub-diurnal time scale." *J. Atmospheric Sci.*, **65**(*1*), 156–171.

Love A.E.H., 1909a, "The yielding of the Earth to disturbing forces." *Mon. Not. R. Astron. Soc.*, **69**, 476–479.

Love A.E.H., 1909b, "The yielding of the Earth to disturbing forces." *Proc. R. Soc. London A*, **82**, 73–88.

Love A.E.H., 1911, "*Some Problems of Geodynamics.*" Cambridge University Press, Cambridge.

Lubkov M., 2005, "The evaluation of the effects of visco-elastic mantle on luni-solar nutations." *Kinematika i Fizika Nebesnykh Tel, Suppl*, **5**, 355–358.

Luh P.C., 1973, "Free oscillations of the laterally inhomogeneous Earth. Quasi-degenerate multiplet coupling." *Geophys. J. R. Astron. Soc.*, **32**, 187–202.

Luzum B., Capitaine N., Fienga A., *et al.*, 2011, "The IAU 2009 system of astronomical constants: the report of the IAU working group on numerical standards for Fundamental Astronomy." *Celest. Mech. Dynam. Astr.*, **110**(4), 293–304, DOI: 10.1007/s10569-011-9352-4.

Lyard F., Lefevre F., Letellier T., and Francis O., 2006, "Modelling the global ocean tides: modern insights from FES2004." *Ocean Dynam.*, **56**(5–6), 394–415.

Ma C. and Shaffer D.B., 1991, "Stability of the extragalactic reference frame realized by VLBI." In: *Reference Systems*, Hughes J.A., Smith C.A., and Kaplan G.H. (Eds.), Proceedings of IAU Colloquium 127, U. S. Naval Observatory, Washington, pp 135–144.

Ma C., Arias E.F., Eubanks T.M., *et al.*, 1997, "The international celestial reference frame realized by VLBI." In: *Definition and Realization of the International Celestial Reference System by VLBI Astrometry of Extragalactic Objects*, IERS Technical Note No. 23, Part II, Observatoire de Paris, Eds. C. Ma and M. Feissel.

Ma C., Arias E.F., Eubanks T.M., *et al.*, 1998, "The international celestial reference frame as realized by very long baseline interferometry." *Astron. J.*, **116**, 516–546, DOI: 10.1086/300408.

Ma C., 2000, "The international celestial reference frame (ICRF) and the relationship between frames." In: Proc. International VLBI Service for Geodesy and Astrometry 2000 General Meeting, Session 1: 'Highlights and challenges of VLBI', pp 52–56.

Ma C., 2005, "Potential refinement of the ICRF." In: *Highlights of Astronomy*, Vol. 13, Proc. XXVth General Assembly of the IAU, Ed. O. Engvold, San Francisco, CA, Astronomical Society of the Pacific, 631 pp.

Ma C., Arias F., Bianco G., *et al.*, 2009, "The second realization of the international celestial reference frame by very long baseline interferometry; Presented on behalf of the IERS/IVS Working Group." IERS Technical Note No. 35, 105 pp.

Malkin Z., 2001, "On computation of combined IVS EOP series." In: Proc. 15th Workshop Meeting on *European VLBI for Geodesy and Astrometry*, Institut d'Estudis Espacials de Catalunya, Consejo Superior de Investigaciones Científicas, Barcelona, Spain, September 2001, Eds. D. Behrend and A. Rius, pp 55–63.

Malkin Z., 2007, "Empiric models of the Earth's free core nutation." *Solar System Research*, **41**(6), 492–497, DOI: 10.1134/S0038094607060044.

Malkin Z., 2008, "On the accuracy assessment of celestial reference frame realizations." *J. Geodesy*, **82**(6), 325–329, DOI: 10.1007/s00190-007-0181-x.

Malkin Z., 2009, "On comparison of the Earth orientation parameters obtained from different VLBI networks and observing programs." *J. Geodesy*, **83**(6), 547–556, DOI: 10.1007/s00190-008-0265-2.

Malkin Z., 2010, "Analysis of the accuracy of prediction of the celestial pole motion." *Astronomy Reports*, **54**(11), 1053–1061, DOI: 10.1134/S1063772910110119.

Malkin Z. and Miller N., 2010, "Chandler wobble: two more large phase jumps revealed." *Earth, Planets and Space*, **62**(12), 943–947, DOI: 10.5047/eps.2010.11.002.

Malkin Z., 2011a, "Study of astronomical and geodetic series using the Allan variance." *Kinematics and Physics of Celestial Bodies*, **27**(1), 42–49, DOI: 10.3103/S0884591311010053.

Malkin Z., 2011b, "Erratum to: On comparison of the Earth orientation parameters obtained from different VLBI networks and observing programs." *J. Geodesy*, **85**(8), 551, DOI: 10.1007/s00190-011-0483-x.

Malkin Z., 2011c, "The influence of galactic aberration on precession parameters determined from VLBI observations." *Astronomy Reports*, **55**(*9*), 810–815, DOI: 10.1134/S1063772911090058.

Malkin Z., 2011d, "The impact of celestial pole offset modelling on VLBI UT1 intensive results." *J. Geodesy*, **85**(*9*), 617–622, DOI: 10.1007/s00190-011-0468-9.

Malkin Z., 2013a, "Impact of seasonal station motions on VLBI UT1 intensive results." *J. Geodesy*, **87**(*6*), 505–514, DOI: 10.1007/s00190-013-0624-5.

Malkin Z., 2013b, "Free core nutation and geomagnetic jerks." *Journal of Geodynamics*, **72**, 53–58, DOI: 10.1016/j.jog.2013.06.001.

Manga M. and Jeanloz R., 1996, "Implications of a metal-bearing chemical boundary layer in D'' for mantle dynamics." *Geophys. Res. Letters*, **23**(*22*), 3091–3094, DOI: 10.1029/96GL03021.

Marcus S.L., Ghil M., and Dickey J.O., 1996, "The extratropical 40-day oscillation in the UCLA general circulation model. Part II: Spatial structure." *J. Atmospheric Sc.*, **53**(*14*), 1993–2014.

Marcus S.L., Chao Yi, Dickey J.O., and Gegout P., 1998, "Detection and modeling of nontidal oceanic effects on Earth's rotation rate." *Science*, **281**(*5383*), 1656–1659.

Marcus S.L., de Viron O., and Dickey J.O., 2004, "Atmospheric contributions to Earth nutation: geodetic constraints and limitations of the torque approach." *J. Atmosph. Sc., Notes and Correspondence*, **61**, 352–356.

Marcus S.L., de Viron O., and Dickey J.O., 2010, "Interannual atmospheric torque and El Nino-Southern Oscillation: Where is the polar motion signal?" *Journal of Geophysical Research: Solid Earth*, **115**(*B12*), CiteID: B12409, DOI: 10.1029/2010JB007524.

Marcus S.L., Dickey J.O., and de Viron O., 2011, "Abrupt atmospheric torque changes and their role in the 1976–1977 climate regime shift." *Journal of Geophysical Research: Atmospheres*, **116**(*D3*), CiteID: D03107, DOI: 10.1029/2010JD015032.

Marcus S.L., Dickey J.O., Fukumori I., and de Viron O., 2012, "Detection of the Earth rotation response to a rapid fluctuation of Southern Ocean circulation in November 2009." *Geophys. Res. Letters*, **39**(*4*), CiteID: L04605, DOI: 10.1029/2011GL050671.

Markowitz W., 1960, "Latitude and longitude, and the secular motion of the pole." *Methods and Techniques in Geophysics*. Ed. S.K. Runcorn, Vol. I, John Wiley and Sons-Interscience, New York.

Marty J.C., Balmino G., Duron J., *et al.*, 2009, "Martian gravity field model and its time variations." *Planetary Space Sci.*, **57**(*3*), 350–363, DOI: 10.1016/j.pss.2009.01.004.

Mathews P.M. and Venkatesan K., 1976, second edition, 2010, "*A Textbook of Quantum Mechanics*." Tata McGraw-Hill, New Delhi, 387 pp.

Mathews P.M., Buffett B.A., Herring T.A. and Shapiro I.I., 1991a, "Forced nutations of the Earth: influence of inner core dynamics. I. Theory." *J. Geophys. Res.*, **96**(*B5*), 8219–8242.

Mathews P.M., Buffett B.A., Herring T.A., and Shapiro I.I., 1991b, "Forced nutations of the Earth: influence of inner core dynamics. II. Numerical results and comparisons." *J. Geophys. Res.*, **96**(*B5*), 8243–8258.

Mathews P.M. and Shapiro I.I., 1992, "Nutations of the Earth." *Annu. Rev. Earth Planet. Sci.*, **20**, 469–500.

Mathews P.M., Buffett B.A., and Shapiro I.I., 1995a, "Love numbers for a rotating spheroidal Earth: new definitions and numerical values." *Geophys. Res. Letters*, **22**(*5*), 579–582, DOI: 10.1029/95GL00161.

Mathews P.M., Buffett B.A., and Shapiro I.I., 1995b, "Love numbers for diurnal tides: relation to wobble admittances and resonance expansions." *J. Geophys. Res.*, **100**(*B6*), 9935–9948, DOI: 10.1029/95JB00670.

Mathews P.M., Dehant V., and Gipson J.M., 1997, "Tidal station displacements." *J. Geophys. Res.*, **102**, 20469–20477.

Mathews P.M., 2001, "Time based on Earth rotation." In: Proc. Journées Systèmes de Référence Spatio-temporels 2001, Ed. N. Capitaine, Paris Observatory, Paris, pp 180–184.

Mathews P.M., Herring T.A., and Buffett B.A., 2002, "Modeling of nutation and precession: new nutation series for nonrigid Earth and insights into the Earth's interior." *J. Geophys. Res.*, **107**(*B4*), CI: 2068, DOI: 10.1029/2001JB000390.

Mathews P.M. and Bretagnon P., 2003, "Polar motions equivalent to high frequency nutations for a nonrigid Earth with anelastic mantle." *Astr. Astrophys.*, **400**(*3*), 1113–1128.

Mathews P.M. and Guo J.Y., 2005, "Viscoelectromagnetic coupling in precession-nutation theory." *J. Geophys. Res.*, **110**(*B02402*), DOI: 10.1029/2003JB002915.

Mathews P.M. and Lambert S.B., 2009, "Effect of mantle and ocean tides on the Earth's rotation rate." *Astr. Astrophys.*, **493**, 325–330, DOI: 10.1051/0004-6361: 200810343.

Matsui H. and Buffett B.A., 2012, "Large-eddy simulations of convection-driven dynamos using a dynamic scale-similarity model." *Geophysical and Astrophysical Fluid Dynamics*, **106**(*3*), 250–276, DOI: 10.1080/03091929.2011.590806.

Matsui H. and Buffett B.A., 2013, "Characterization of subgrid-scale terms in a numerical geodynamo simulation." *Phys. Earth Planet. Int.*, **223**, 77–85, DOI: 10.1016/j.pepi.2013.08.004.

McCarthy D.D. and Luzum B.J., 1991, "Observations of luni-solar and free core nutation." *Astron. J.*, **102**, 1889–1895, DOI: 10.1086/116011.

McCarthy D.D., 1996, "IERS Conventions 1996." IERS Technical Note No. 21, 98 pp.

McCarthy D.D. and Petit G., 2003, "IERS Conventions 2000." IERS Technical Note No. 32, Frankfurt am Main, Verlag des Bundesamts für Kartographie und Geodäsie, 127 pp.

McCarthy D.D. and Capitaine N., 2003a, "Practical consequences of Resolution B1.6 'IAU2000 precession-nutation model,' Resolution B1.7 'Definition of the Celestial Intermediate Pole,' and Resolution B1.8 'Definition and use of Celestial Ephemeris Origin and Terrestrial Ephemeris Origin'." IERS Technical Note No. 29, pp 9–18.

McCarthy D.D. and Capitaine N., 2003b, "Compatibility with past observations." IERS Technical Note No. 29, pp 85–88.

McCarthy D.D. and Luzum B.J., 2003, "An abridged model of the precession-nutation of the celestial pole." *Celest. Mech. Dynam. Astr.*, **85**, 37–49.

McCarthy D.D. and Petit G., 2004, "Conventions 2003." IERS Technical Note No. 32, Publ. Frankfurt am Main: Verlag des Bundesamts für Kartographie und Geodäsie, 127 pp.

MacMillan D.S. and Ma C., 2007, "Radio source instability in VLBI analysis." *J. Geodesy*, **81**(*6–8*), 443–453, DOI: 10.1007/s00190-007-0136-2.

MacCullagh J., 1843, "On surfaces of the second order." *Proceedings of the Royal Irish Academy*, 429–494.

MacCullagh J., 1853, "On the attraction of ellipsoids, with a new demonstration of Clairaut's theorem, being an account of the late Professor MacCullagh's lectures on those subjects." *R. Irish Acad.*, **22**(*139*), 139–135, published by G.J. Allman.

MacCullagh J., 1880, "*The Collected Work of James MacCullagh.*" Eds. J.H. Jellet, S. Haughton, Hodges, Figgis & Co, Dublin.

Manga M. and Jeanloz R., 1996, "Implications of a metal-bearing chemical boundary layer in D″ for mantle dynamics." *Geophys. Res. Letters*, **23**, 3091–3094.

Melchior P. and Georis B., 1968, "Earth tides, precession-nutation and the secular retardation of Earth's rotation." *Phys. Earth Planet. Inter.*, **1**, 267–287.

Melchior P., Crossley D., Dehant V., and Ducarme B., 1989, "Have inertial waves been identified from the Earth's core?" In: Proc. Symp. U2, 19th General Assembly of the IUGG, Vancouver, Canada, August 1987, Geophysical Monograph 46, IUGG Vol. 1: *Structure and Dynamics of the Earth's Deep Interior*, Eds. D.E. Smylie and R. Hide, pp 1–12.

Mendes Cerveira P.J., Boehm J., Schuh H., *et al.*, 2009, "Earth rotation observed by very long baseline interferometry and ring laser." *Pure and Applied Geophysics*, **166**(*8–9*), 1499–1517, DOI: 10.1007/s00024-004-0487-z.

Merrian J.B., 1984, "Tidal terms in universal time: effects of zonal winds and mantle Q." *J. Geophys. Res.*, **89**(*B12*), 10109–10114, DOI: 10.1029/JB089iB12p10109.

Métivier L., Greff-Lefftz M., and Diament M., 2005, "A new approach to computing accurate gravity time variations for a realistic Earth model with lateral heterogeneities." *Geophys. J. Int.*, **162**(*2*), 570–574, DOI: 10.1111/j.1365-246X.2005.02692.x.

Métivier L., Greff-Lefftz M., and Diament M., 2006, "Mantle lateral variations and elastogravitational deformations – I. Numerical modelling." *Geophys. J. Int.*, **167**(*3*), 1060–1076, DOI: 10.1111/j.1365-246X.2006.03159.x.

Métivier L., Greff-Lefftz M., and Diament M., 2007, "Mantle lateral variations and elastogravitational deformations - II. Possible effects of a superplume on body tides." *Geophys. J. Int.*, **168**(*3*), 897–903, DOI: 10.1111/j.1365-246X.2006.03309.x.

Métivier L., Karatekin O.E., and Dehant V., 2008, "The effect of the internal structure of Mars on its seasonal loading deformations." *Icarus*, **194**(*2*), 476–486, DOI: 10.1016/j.icarus.2007.12.001.

Métivier L., Greff-Lefftz M., and Altamimi Z., 2011, "Erratum to 'On secular geocenter motion: The impact of climate changes' [*Earth Planet. Sci. Lett.* **296** (2010) 360–366]." *Earth Planet. Sci. Lett.*, **306**(*1–2*), 136–136, DOI: 10.1016/j.epsl.2011.03.026.

Minster J.B. and Anderson D.L., 1980, "Dislocation and nonelastic processes in the mantle." *J. Geophys. Res.*, **85**(*B11*), 6347–6352.

Minster J.B. and Anderson D.L., 1981, "A model of dislocation-controlled rheology for the mantle." *Phil. Trans. Roy. Soc. London, Series A, Math. Phys. Sci.*, **299**(*1449*), 319–356.

Misner C.W., Thorne K.S., and Wheeler J.A., 1973, "*Gravitation*." Freeman publ., San Francisco, Sec. 40.7.

Mitrovica J.X., Davis J.L., Mathews P.M., and Shapiro I.I., 1994, "Determination of tidal h Love number parameters in the diurnal band using an extensive VLBI data set." *Geophys. Res. Letters*, **21**(*8*), 705–708.

Mocquet A., Rosenblatt P., Dehant V., and Verhoeven O., 2011, "The deep interior of Venus, Mars, and the Earth: a brief review and the need for planetary surface-based measurements." *Planet. Space Sci.*, Special Issue on *Comparative Planetology: Venus-Earth-Mars*, **59**(*10*), 1048–1061, DOI: 10.1016/j.pss.2010.02.002.

Moffat H.K., 1978, "*Magnetic Field Generation in Electrically Conducting Fluids*." Cambridge University Press, New York.

Moisson X., 2000, "Intégration du mouvement des planètes dans le cadre de la relativité générale." Ph.D. thesis, Observatoire de Paris, 165 pp.

Moisson X. and Bretagnon P., 2001, "Analytical solution. VSOP2000." *Celest. Mech. Dyn. Astron.*, **80**, 205–213.

Molodensky M.S., 1961, "The theory of nutation and diurnal Earth tides." *Commun. Observ. Roy. Belgique*, **188**, 25–56.

Molodensky M.S. and Groten E., 1996, "On 'pathological' oscillations of rotating fluids in the theory of nutation." *J. Geodesy*, **70**(*10*), 603–621, DOI: 10.1007/s001900050051.

Molodensky S.M., 2004, "Tides and nutation of the Earth: I. Models of an Earth with an inelastic mantle and homogeneous, inviscid, liquid core." *Solar System Research*, **38**(*6*) 476–490, DOI: 10.1007/s11208-005-0019-0.

Molodensky S.M., 2005, "Tides and nutation of the Earth: II. Realistic models of the liquid core." *Solar System Research*, **39**(*1*), 54–72, DOI: 10.1007/s11208-005-0005-6.

Molodensky S.M., 2006, "On the effect of oceanic tides on the inertial coupling between the liquid core and the mantle." *Izvestiya, Physics of the Solid Earth*, **42**(*5*), 357–361, DOI: 10.1134/S1069351306050016.

Molodensky S.M. and Molodenskaya M.S., 2009, "On mechanical Q parameters of the lower mantle inferred from data on the Earth's free oscillations and nutation." *Izvestiya, Physics of the Solid Earth*, **45**(*9*), 744–752, DOI: 10.1134/S1069351309090031.

Molodensky S.M., 2010, "Correctives to the scheme of the Earth's structure inferred from new data on nutation, tides, and free oscillations." *Izvestiya, Physics of the Solid Earth*, **46**(*7*), 555–579, DOI: 10.1134/S1069351310070013.

Molodensky S.M., 2011a, "Models of density and mechanical Q-factor distributions according to new data on the nutations and free oscillations of the Earth: 1. Ambiguity in the inverse problem." *Izvestiya, Physics of the Solid Earth*, **47**(*4*), 263–275, DOI: 10.1134/S1069351311040069.

Molodensky S.M., 2011b, "Models of density and mechanical Q-factor distributions according to new data on the nutations and free oscillations of the Earth: 2. Comparison with astrometric data." *Izvestiya, Physics of the Solid Earth*, **47**(*7*), 559–574, DOI: 10.1134/S1069351311070056.

Molodensky S.M. and Molodenskaya M.S., 2012, "The admissible interval for the Q factor in the solid inner core: constraints from new data on the nutation and free oscillations of the Earth." *Izvestiya, Physics of the Solid Earth*, **48**(*7–8*), 562–571, DOI: 10.1134/S106935131207004X.

Molodensky S.M. and Molodenskaya M.S., 2013a, "On the admissible range for the mass and inertia moments of the liquid core: I. Solving the inverse problem of nutations and free oscillations of the Earth by decomposition of mechanical parameters in an orthogonalized basis." *Izvestiya, Physics of the Solid Earth*, **49**(*4*), 449–458, DOI: 10.1134/S1069351313030099.

Molodensky S.M. and Molodenskaya M.S., 2013b, "On the admissible range for the mass and inertia moments of the liquid core: II. Results of numerical calculations." *Izvestiya, Physics of the Solid Earth*, **49**(*4*), 459–463, DOI: 10.1134/S1069351313030105.

Moor A., Frey S., Lambert S.B., Titov O.A., and Bakos J., 2011, "On the connection of the apparent proper motion and the VLBI structure of compact radio sources." *Astron. J.*, **141**(*6*), ID. 178, 7 pp, DOI: 10.1088/0004-6256/141/6/178.

Morelli A. and Dziewonski A.M., 1987, "Topography of the core-mantle boundary and lateral heterogeneity of the inner core." *Nature*, **325**, 678–683.

Moritz H. and Mueller I.I., 1987, *"Earth Rotation: Theory and Observation."* The Ungar Publishing Company, New York, 617 pp.

Moucha R., Forte A.M., Mitrovica J.X., and Daradich A., 2007, "Lateral variations in mantle rheology: implications for convection related surface observables and inferred viscosity models." *Geophys. J. Int.*, **169**(*1*), 113–135, DOI: 10.1111/j.1365-246X.2006.03225.x.

Mound J.E. and Buffett B.A., 2003, "Interannual oscillations in length of day: implications for the structure of the mantle and core." *J. Geophys. Res.*, **108**(*B7*), ETG 2–1, CiteID 2334, DOI: 10.1029/2002JB002054.

Mound J.E., 2005, "Electromagnetic torques in the core and resonant excitation of decadal polar motion." *Geophys. J. Int.*, **160**(*2*), 721–728, DOI: 10.1111/j.1365-246X.2004.02495.x.

Mound J.E. and Buffett B.A., 2005, "Mechanisms of core-mantle angular momentum exchange and the observed spectral properties of torsional oscillations." *J. Geophys. Res.*, **110**(*B8*), CiteID B08103, DOI: 10.1029/2004JB003555.

Mound J.E. and Buffett B.A., 2006, "Detection of a gravitational oscillation in length-of-day." *Earth Planet. Sci. Lett.*, **243**(*3–4*), 383–389, DOI: 10.1016/j.epsl.2006.01.043.

Mound J.E. and Buffett B.A., 2007, "Viscosity of the Earth's fluid core and torsional oscillations." *J. Geophys. Res.*, **112**(*B5*), CiteID B05402, DOI: 10.1029/2006JB004426.

Mueller I.I., 1981, "Reference coordinate systems for Earth dynamics: a preview." In: *Reference Coordinate Systems for Earth Dynamics*, Eds. E.M. Gaposchkin and B. Kolaczek, Reidel Publ. Company, pp 1–22.

Munk W.H. and McDonald G.J.F., 1960, "*The Rotation of the Earth – A Geophysical Discussion.*" Cambridge University Press, (reprinted with corrections 1975).

Murphy F.E., Neuberg J.W., and Jacob A.W.B., 1997, "Alternatives to core-mantle boundary topography." *Phys. Earth Planet. Int.*, **103**, 349–364, DOI: 10.1016/S0031-9201(97)00052-6.

Murray C.A., 1978, "On the precession and nutation of the Earth's axis of figure." *Mon. Not. R. Astron. Soc.*, **183**, 677–685.

Nafisi V., Madzak M., Boehm J., Ardalan A.A., and Schuh H., 2012, "Ray-traced tropospheric delays in VLBI analysis." *Radio Science*, **47**(*2*), CiteID: RS2020, DOI: 10.1029/2011RS004918.

Nakada M., 2009, "Earth's rotational variations by electromagnetic coupling due to core surface flow on a timescale of 1 yr for geomagnetic jerk." *Geophys. J. Int.*, **179**(*1*), 521–535, DOI: 10.1111/j.1365-246X.2009.04256.x.

Nakada M., 2011, "Earth's rotational variations due to rapid surface flows at both boundaries of the outer core." *Geophys. J. Int.*, **184**(*1*), 235–246, DOI: 10.1111/j.1365-246X.2010.04862.x.

Nakada M., Iriguchi C., and Karato S.I., 2012, "The viscosity structure of the D″ layer of the Earth's mantle inferred from the analysis of Chandler wobble and tidal deformation." *Phys. Earth Planet. Int.*, **208**, 11–24, DOI: 10.1016/j.pepi.2012.07.002.

Nakada M. and Karato S.I., 2012, "Low viscosity of the bottom of the Earth's mantle inferred from the analysis of Chandler wobble and tidal deformation." *Phys. Earth Planet. Int.*, **192**, 68–80, DOI: 10.1016/j.pepi.2011.10.001.

Nastula J. and Ponte R.M., 1999, "Further evidence for oceanic excitation of polar motion." *Geophys. J. Int.*, **139**, 123–130.

Navarro J.F. and Ferrándiz J.M., 2002, "A new symbolic processor for the Earth rotation theory." *Celest. Mech. Dynam. Astr.*, **82**(*3*), 243–263, DOI: 10.1023/A:1015059002683.

Newcomb S., 1895, "Tables of motion of the Earth on its axis and around the Sun." In: "*Tables of the Four Inner Planets*," Vol. 6, Part 1 of *Astronomical Papers* prepared for the use of the American *Ephemeris and Nautical Almanac*, pp 1–169.

Newhall X.X., Standish E.M., and Williams J.G., 1983, "DE102 – A numerically integrated ephemeris of the Moon and planets spanning forty-four centuries." *Astr. Astrophys.*, **125**, 150–167.

NFA Chart of transformation process, 2006a, see NFA WG website http://syrte.obspm.fr/iauWGnfa/.

NFA Explanation behind the terms, 2006b, see NFA WG website http://syrte.obspm.fr/iauWGnfa/.

NFA Final recommendations, 2006c, see NFA WG website http://syrte.obspm.fr/iauWGnfa/.

NFA IAU 2006 Glossary, 2006d, see NFA WG website http://syrte.obspm.fr/iauWGnfa/.

NFA List of terms by category, 2006e, see NFA WG website http://syrte.obspm.fr/iauWGnfa/.

NFA List of acronyms and symbols, 2006f, see NFA WG website http://syrte.obspm.fr/iauWGnfa/.

NFA References, 2006g, see NFA WG website http://syrte.obspm.fr/iauWGnfa/.

Niell A. and Tang J., 2002, "Gradient mapping functions for VLBI and GPS." In: Proc. IVS 2002 General Meeting, Tsukuba, Japan, February 4–7 June 2002, Eds. N.R. Vandenberg and K.D. Baver, pp 215–218.

Nilsson T., Boehm J., and Schuh H., 2011, "Universal time from VLBI single-baseline observations during CONT08." *J. Geodesy*, **85**(7), 415–423, DOI: 10.1007/s00190-010-0436-9.

Noir J., Cardin P., Jault D., and Masson J.-P., 2003, "Experimental evidence of non-linear resonance effects between retrograde precession and the tilt-over mode within a spheroid." *Geophys. J. Int.*, **154**(2), 407–416, DOI: 10.1046/j.1365-246X.2003.01934.x.

Noir J., Hemmerlin F., Wicht J., Baca S.M., and Aurnou J.M., 2009, "An experimental and numerical study of librationally driven flow in planetary cores and subsurface oceans." *Phys. Earth Planet. Int.*, **173**(1–2), 141–152, DOI: 10.1016/j.pepi.2008.11.012.

Noir J., Calkins M.A., Lasbleis M., Cantwell J., and Aurnou J.M., 2010, "Experimental study of libration-driven zonal flows in a straight cylinder." *Phys. Earth Planet. Inter.*, **182**(1–2), 98–106, DOI: 10.1016/j.pepi.2010.06.012.

Noir J., Cébron D., Le Bars M., Sauret A., and Aurnou J.M., 2012, "Experimental study of libration-driven zonal flows in non-axisymmetric containers." *Phys. Earth Planet. Inter.*, **204**, 1–10, DOI: 10.1016/j.pepi.2012.05.005.

Nothnagel A., 2009, "Conventions on thermal expansion modeling of radio telescopes for geodetic and astrometric VLBI." *J. Geodesy*, **83**, 787–792, DOI: 10.1007/s00190-008-0284-z.

Nothnagel A. and Schnell D., 2008, "The impact of errors in polar motion and nutation on UT1 determinations from VLBI intensive observations." *J. Geodesy*, **82**(12), 863–869, DOI: 10.1007/s00190-008-0212-2.

Ohta K., Onoda S., Hirose K., *et al.*, 2008, "The electrical conductivity of post-perovskite in Earth's D'' layer." *Science*, **320**, 89–91.

Ohta K., Hirose K., Masahiro I., *et al.*, 2010, "Electrical conductivities of pyrolitic mantle and MORB materials up to the lowermost mantle conditions." *Earth Planet. Sci. Lett.*, **289**, 497–502, DOI: 10.1016/j.epsl.2009.11.042.

Ohta K., Takashi Y., Taketoshi N., *et al.*, 2012, "Lattice thermal conductivity of $MgSiO_3$ perovskite and post-perovskite at the core-mantle boundary." *Earth Planet. Sci. Lett.*, **349**, 109–115, DOI: 10.1016/j.epsl.2012.06.043.

Olson P. and Christensen U.R., 2002, "The time-averaged magnetic field in numerical dynamos with non-uniform boundary heat flow." *Geophys. J. Int.*, **151**(3), 809–823, DOI: 10.1046/j.1365-246X.2002.01818.x.

Ooe M., 1973a, "On the nearly diurnal free nutation." *Publ. Int. Lat. Obs. Mizusawa*, **IX**(*1*), 133–160.

Ooe M., 1973b, "Note on the retrograde sway." *Publ. Int. Lat. Obs. Mizusawa*, **IX**(*1*), 161–166.

Ozawa H., Hirose K., Mitome M., *et al*., 2009, "Experimental study of reaction between perovskite and molten iron to 146 GPa and implications for chemically distinct buoyant layer at the top of the core." *Phys. Chem. Miner.*, **36**, 355–363, DOI: 10.1007/s00269-008-0283-x.

Palmer A. and Smylie D.E., 2005, "VLBI observations of free core nutations and viscosity at the top of the core." *Phys. Earth Planet. Int.*, **148**(*2–4*), 285–301, DOI: 10.1016/j.pepi.2004.09.003.

Panafidina N., Malkin Z., and Weber R., 2006, "A new combined European permanent network station coordinates solution." *J. Geodesy*, **80**(*7*), 373–380, DOI: 10.1007/s00190-006-0076-2.

Pany A., Boehm J., MacMillan D., *et al*., 2011, "Monte Carlo simulations of the impact of troposphere, clock and measurement errors on the repeatability of VLBI positions." *J. Geodesy*, **85**(*1*), 39–50, DOI: 10.1007/s00190-010-0415-1.

Petit G. and Klioner S.A., 2008, "Does relativistic time dilation contribute to the divergence of universal time and ephemeris time?" *Astron. J.*, **136**(*5*), 1909–1912, DOI: 10.1088/0004-6256/136/5/1909.

Petit G. and Luzum B. (Eds.), 2011, "IERS Conventions 2010." IERS Technical Note No. 36, Frankfurt am Main, Verlag des Bundesamts für Kartographie und Geodäsie, IERS Conventions Center, 179 pp.

Petrov L., 2007, "The empirical Earth rotation model from VLBI observations." *Astr. Astrophys.*, **467**(*1*), 359–369, DOI: 10.1051/0004-6361:20065091.

Petrov L., Gordon D., Gipson J., *et al*., 2009, "Precise geodesy with the Very Long Baseline Array." *J. Geodesy*, **83**(*9*), 859–876, DOI: 10.1007/s00190-009-0304-7.

Petrov S., Brzezinski A., and Gubanov V., 1996, "A stochastic model for polar motion with application to smoothing, prediction, and combining." *Artificial Satellites, Planetary Geodesy 26*, **31**(*1*), 51–70.

Phinney R.A. and Burridge R., 1973, "Representation of the elastic-gravitational excitation of a spherical Earth model by generalized spherical harmonics." *Geophys. J. Roy. Astron. Soc.*, **34**, 451–487.

Pil'Nik G.P., 1976, "On the study of forced nutation." *Soviet Astr.*, **20**, 502–507.

Pil'Nik G.P., 1979, "Tides and nutation in time service observations." *Soviet Astr.*, **23**(*1*), 100–104.

Pitjeva E.V., 2001a, "Modern numerical ephemerides of planets and the importance of ranging observations for their creation." *Celest. Mech. Dynam. Astr.*, **80**(*3–4*), 249–271.

Pitjeva E.V., 2001b, "Progress in the determination of some astronomical constants from radiometric observations of planets and spacecraft." *Astr. Astrophys.*, **371**, 760–765.

Pitjeva E.V., 2005, "High-precision ephemerides of planets – EPM and determination of some astronomical constants." *Solar System Research*, **39**(*3*), 176–186, DOI: 10.1007/s11208-005-0033-2.

Pitjeva E.V. and Standish E.M., 2009, "Proposals for the masses of the three largest asteroids, the Moon–Earth mass ratio and the Astronomical Unit." *Celest. Mech. Dynam. Astr.*, **103**(*4*), 365–372, DOI: 10.1007/s10569-009-9203-8.

Pitjeva E.V., 2013, "Updated IAA RAS planetary ephemerides-EPM2011 and their use in scientific research." *Solar System Research*, **47**(*5*), 386–402, DOI: 10.1134/S0038094613040059.

Poincaré H., 1910, "Sur la précession des corps déformables." *Bull. Astr.*, **27**, 321–356.

Poirier J.-P. and Peyronneau J., 1992, "Experimental determination of the electrical conductivity of the material of the Earth's lower mantle." In: *High-Pressure Research: Applications to Earth and Planetary Sciences*, AGU Geophysical Monograph Series 67, Eds. Y. Syono and M.H. Manghnani, pp 77–87.

Poirier J.-P., 1993, "Core-infiltrated mantle and the nature of the D″ layer." *J. Geomag. Geoelectr.*, **45**, 1221–1227.

Ponsar S., Dehant V., Holme R., *et al.*, 2002, "The core and fluctuations in the Earth rotation." In: AGU Monograph series, *Earth's Core Dynamics, Structure and Rotation*, Geodynamics Series Volume 31, Eds. V. Dehant, K. Creager, S. Karato, S. Zatman, DOI: 10.1029/031GD17.

Ponte R.M. and Rosen R.D., 1994, "Angular momentum and torques in a simulation of the atmosphere's response to the 1982–83 El Niño." *J. Climate*, **7**(*4*), 538–550.

Ponte R.M., 1997, "Oceanic excitation of daily to seasonal signals in Earth rotation: results from a constant-density numerical model." *Geophys. J. Int.*, **95**, 11 369–11 376.

Ponte R.M., Stammer D., and Marshall J., 1998, "Oceanic signals in observed motions of Earth's pole of rotation." *Nature*, **391**, 476–479.

Ponte R.M. and Rosen R.D., 1999, "Torques responsible for evolution of atmospheric angular momentum during the 1982–83 El Niño." *J. Atm. Sciences*, **56**(*19*), 3457–3462.

Ponte R.M. and Stammer D., 1999, "Role of ocean currents and bottom pressure variability on seasonal polar motion." *J. Geophys. Res. (Oceans)*, **104**(*C10*), 23393–23409.

Ponte R.M. and Stammer D., 2000, "Global and regional axial ocean angular momentum signals and length-of-day variations (1985–1996)." *J. Geophys. Res. (Oceans)*, **105**(*C7*), 17161–17171.

Ponte R.M., Rajamony J., and Gregory J.M., 2002, "Ocean angular momentum signals in a climate model and implications for Earth rotation." *Climate Dynamics*, **19**(*2*), 181–190, DOI: 10.1007/s00382-001-0216-6.

Pozzo M., Davies C., Gubbins D., and Alfè D., 2012, "Thermal and electrical conductivity of iron at Earth's core conditions." *Nature*, **485**, 355–358, DOI: 10.1038/nature11031.

Rambaux N., Van Hoolst T., Dehant V., and Bois E., 2005, "Earth librations due to coremantle coupling." In: Proc. Journées Systèmes de Référence Spatio-temporels 2004, Paris, France, September 2004, pp 150–151.

Ray R., 1999, "A global ocean tide model from Topex/Poseidon altimetry: GOT99.2." NASA Tech Memo 209478, 58 pp.

Ray J., Kouba J., and Altamimi Z., 2005, "Is there utility in rigorous combinations of VLBI and GPS Earth orientation parameters?", *J. Geodesy*, **79**(*9*), 505–511, DOI: 10.1007/s00190-005-0007-7.

Ray R.D. and Zaron E.D., 2011, "Non-stationary internal tides observed with satellite altimetry." *Geophys. Res. Lett.*, **38**(*17*), CiteID: L17609, DOI: 10.1029/2011GL048617.

Ray R.D. and Egbert G.D., 2012, "Fortnightly Earth rotation, ocean tides and mantle anelasticity." *Geophys. J. Int.*, **189**(*1*), 400–413, DOI: 10.1111/j.1365-246X.2012.05351.x

Ray R.D. and Egbert G.D., 2013, "Reply to comments by S.R. Dickman on 'Fortnightly Earth rotation, ocean tides and mantle anelasticity'." *Geophys. J. Int.*, **192**(*3*), 1055–1058, DOI: 10.1093/gji/ggs078.

Richter B., Engels J., and Grafarend E., 2010, "Transformation of amplitudes and frequencies of precession and nutation of the Earth's rotation vector to amplitudes and frequencies of diurnal polar motion." *J. Geodesy*, **84**(*1*), 1–18, DOI: 10.1007/s00190-009-0339-9.

Rioja M., Dodson R., Asaki Y., Hartnett J., and Tingay S., 2012, "The impact of frequency standards on coherence in VLBI at the highest frequencies." *Astron. J.*, **144**(*4*), ID. 121, 11 pp, DOI: 10.1088/0004-6256/144/4/121.

Rivoldini A., Van Hoolst T., Verhoeven O., Mocquet A., and Dehant V., 2011, "Geodesy constraints on the interior structure of Mars." *Icarus*, **213**, 451–472, DOI: 10.1016/j.icarus.2011.03.024.

Roberts P.H., 1972, "Electromagnetic core-mantle coupling." *J. Geomag. Geoelectr.*, **24**, 231–259.

Robertson D.S., Carter W.E., and Wahr J.M., 1986, "Possible detection of the Earth's free-core nutation." *Geophys. Res. Letters*, **13**, 949–952, DOI: 10.1029/GL013i009p00949.

Rochester M.G., 1960, "Geomagnetic westward drift and irregularities in the Earth's rotation." *Phil. Trans. Roy. Soc. London, Ser. A*, **252**, 531–555.

Rochester M.G., 1969, "Core-mantle interactions: geophysical and astronomical consequences." *Phys. Earth Planet. Int.*, **2**(*5*), 381–381, DOI: 10.1016/0031-9201(69)90033-8.

Rochester M.G., 1970, "Polar wobble and drift: a brief history, earthquake displacement fields and the rotation of the Earth." Eds. Mansinha L., *et al.*, Springer-Verlag, New York, pp 3–13.

Rochester M.G., 1973, "The Earth's rotation." *EOS*, **54**, 769–780.

Rochester M.G., 1975, "Chandler wobble and viscosity in the Earth's core." *Nature*, **255**(*5510*), 655, DOI: 10.1038/255655a0.

Rochester M.G., 1981, "Comments on the paper by Y. Kubo: 'A core-mantle interaction in the rotation of the Earth'." *Celest. Mech.*, **24**(*3*), 231–234, DOI: 10.1007/BF01229554.

Rochester M.G., 1984, "Causes of fluctuations in the rotation of the Earth." In: 'Discussion on rotation in the solar system', London, England, Royal Society (London), *Philosophical Transactions, Series A*, **313**(*1524*), 95–105, DOI: 10.1098/rsta.1984.0086.

Rochester M.G. and Crossley D.J., 2009, "Earth's long-period wobbles: a Lagrangian description of the Liouville equations." *Geophys. J. Int.*, **176**, 40–62, DOI: 10.1111/j.1365-246X.2008.03991.x.

Roden R., 1963, "Electromagnetic core-mantle coupling." *Geophys. J. Roy. Astron. Soc.*, **7**, 361–374.

Rogister Y., 2001, "On the diurnal and nearly diurnal free modes of the Earth." *Geophys. J. Int.*, **144**(*2*), 459–470, DOI: 10.1046/j.1365-246x.2001.00359.x.

Rogister Y., 2003, "Splitting of seismic-free oscillations and of the Slichter triplet using the normal mode theory of a rotating, ellipsoidal Earth." *Phys. Earth Planet. Inter.*, **140**(*1–3*), 169–182, DOI: 0.1016/S0031-9201(03)00171-7.

Rogister Y. and Rochester M.G., 2004, "Normal-mode theory of a rotating Earth model using a Lagrangian perturbation of a spherical model of reference." *Geophys. J. Int.*, **159**(*3*), 874–908, DOI: 10.1111/j.1365-246X.2004.02447.x.

Rogister Y. and Valette B., 2009, "Influence of liquid core dynamics on rotational modes." *Geophys. J. Int.*, **176**(*2*), 368–388, DOI: 10.1111/j.1365-246X.2004.02447.x.

Rogister Y., 2010, "Multiple inner core wobbles in a simple Earth model with inviscid core." *Phys. Earth Planet. Int.*, **178**(*1–2*), 8–15, DOI: 10.1016/j.pepi.2009.08.012.

Ronnang B., 1989, "*Very Long Baseline Interferometry*." Eds. M. Felli and R.E. Spencer, Kluwer, Chapter 16.

Roosbeek F., 1996, "RATGP95: a harmonic development of the tide-generating potential." *Geophys. J. Int.*, **126**, 197–204.

Roosbeek F. and Dehant V., 1998, "RDAN97: an analytical development of rigid Earth nutation series using the torque approach." *Celest. Mech. Dynam. Astr.*, **70**, 215–253.

Roosbeek F., 1999, "Diurnal and subdiurnal terms in RDAN97 series." *Celest. Mech. Dynam. Astr.*, **74**(*4*), 243–252.

Roosbeek F., Defraigne P., Feissel M., and Dehant V., 1999, "The free core nutation period is between 431 and 434 sidereal days." *Geophys. Res. Letters*, **26**(*1*), 131–134.

Roosbeek F., 2000, "Analytical developments of rigid Mars nutation and tide generating potential series." *Celest. Mech. Dynam. Astr.*, **75**, 287–300.

Rosat S. and Lambert S.B., 2009, "Free core nutation resonance parameters from VLBI and superconducting gravimeter data." *Astr. Astrophys.*, **503**(*1*), 287–291, DOI: 10.1051/0004-6361/200811489.

Rosat S., Florsch N., Hinderer J., and Lubes M., 2009, "Estimation of the free core nutation parameters from SG data: sensitivity study and comparative analysis using linearized least-squares and Bayesian methods." *Journal of Geodynamics*, **48**(*3*), 331–339, DOI: 10.1016/j.jog.2009.09.027.

Rosat S. and Rogister Y., 2012, "Excitation of the Slichter mode by collision with a meteoroid or pressure variations at the surface and core boundaries." *Phys. Earth Planet. Int.*, **190**, 25–33, DOI: 10.1016/j.pepi.2011.10.007.

Rosat S., Boy J.-P., and Rogister Y., 2014, "Surface atmospheric pressure excitation of the translational mode of the inner core." *Phys. Earth Planet. Int.*, **227**, 55–60, DOI: 10.1016/j.pepi.2013.12.005.

Rosenblatt P., Marty J.C., Perosanz F., *et al.*, 2004, "Numerical simulations of a Mars geodesy network experiment: effect of orbiter angular momentum desaturation on Mars' rotation estimation." *Planet. Space Sci.*, **52**(*11*), 965–975.

Rosenblatt P. and Dehant V., 2010, "Mars geodesy and rotation." *Research in Astronomy and Astrophysics (RAA)*, **10**(*8*), 713–736, DOI: 10.1088/1674-4527/10/8/002, impact factor 0.856.

Rothacher M., Beutler G., Herring T.A., and Weber R., 1999a, "Estimation of nutation using the Global Positioning System." *J. Geophys. Res.*, **104**, 4835–4859.

Rothacher M., Beutler G., Weber R., and Hefty J., 1999b, "High-frequency variations in Earth rotation from Global Positioning System data." *J. Geophys. Res.*, **106**(*B7*), 13711–13738.

Rummel R., Rothacher M., and Beutler G., 2005, "Integrated global geodetic observing system (IGGOS)-science rationale." *Journal of Geodynamics*, **40**(*4–5*), 357–362, DOI: 10.1016/j.jog.2005.06.003.

Sabadini R. and Vermeersen B., 2004, *Global Dynamics of the Earth; Applications of Normal Mode Relaxation Theory to Solid-Earth Geophysics*. G. Kluwer Academic Publishers, Dordrecht, The Netherlands, 330 pp.

Sailor R.V. and Dziewonski A.M., 1978, "Measurements and interpretation of normal mode attenuation." *Geophys. J. Roy. Astron. Soc.* **53**, 559–581.

Salstein D.A., Kann D.M., Miller A.J. and Rosen R.D., 1993, "The sub-bureau for atmospheric angular momentum of the International Earth Rotation Service: a meteorological data center with geodetic applications." *Bull. Amer. Meteorol. Soc.*, **74**, 67–80.

Salstein D.A., de Viron O., Yseboodt Y., and Dehant V., 2001, "High frequency geophysical fluid modeling necessary to understand Earth rotation variability." *EOS, AGU Publication*, **82**(*21*), 237–238.

Sasao T., Okamoto I., and Sakai S., 1977, "Dissipative core-mantle coupling and nutational motion of the Earth." *Publ. Astron. Soc. Japan*, **39**, 83–105.

Sasao T., Okubo S., and Saito M., 1980, "A simple theory on the dynamical effects of a stratified fluid core upon nutational motion of the Earth." In: *Nutation and the Earth's Rotation*, Proc. IAU Symposium, Kiev, Ukrainian SSR, May 23–28, 1977, Dordrecht, D. Reidel Publishing Co., 165–183.

Sasao T. and Wahr J.M., 1981, "An excitation mechanism for the free core nutation." *Geophys. J. Roy. Astron. Soc.*, **64**, 729–746.

Savcenko R. and Bosch W., 2008, "EOT08a – Empirical ocean tide model from multi-mission satellite altimetry." DGFI Report no 81.

Schastok J., Soffel M., Ruder H., and Schneider M., 1986, "Newton's law of gravity modified? Celestial mechanical consequences." *Phys. Letters A*, **118**(*1*), 8–10, DOI: 10.1016/0375-9601(86)90523-2.

Schastok J., Soffel M., and Ruder H., 1990, "Numerical derivation of forced nutation terms for a rigid Earth." *Celest. Mech. Dynam. Astr.*, **47**(*2*), 219–223, DOI: 10.1007/BF00051206.

Schastok J., Soffel M., and Ruder H., 1994, "A contribution to the study of fortnightly and monthly zonal tides in UT1." *Astr. Astrophys.*, **283**(*2*), 650–654.

Schastok J., 1997, "A new nutation series for a more realistic model Earth." *Geophys. J. Int.*, **130**(*1*), 137–150, DOI: 10.1111/j.1365-246X.1997.tb00993.x.

Schlüter W. and Behrend D., 2007, "The International VLBI Service for Geodesy and Astrometry (IVS): current capabilities and future prospects." *J. Geodesy*, **81**(*6–8*), 379–387, DOI: 10.1007/s00190-006-0131-z.

Schreiber K.U., Klügel T., Wells J.-P.R., Hurst R.B., and Gebauer A., 2011, "How to detect the Chandler and the annual wobble of the Earth with a large ring laser gyroscope." *Physical Review Letters*, **107**(*17*), CiteID: 173904, DOI: 10.1103/PhysRevLett.107.173904.

Schuh H., Nagel S., and Seitz T., 2001, "Linear drift and periodic variations observed in long time series of polar motion." *J. Geodesy*, **74**(*10*), 701–710, DOI: 10.1007/s001900000133.

Schuh H. and Behrend D., 2012, "VLBI: a fascinating technique for geodesy and astrometry." *Journal of Geodynamics*, **61**, 68–80, DOI 10.1016/j.jog.2012.07.007.

Seidelmann P.K., 1982, "1980 IAU theory of nutation: the final report of the IAU working group on nutation." *Celest. Mech.*, **27**, 79–106.

Seidelmann P.K., Abalakin V.K., Bursa M., *et al.*, 2002, "Report of the IAU/IAG Working Group on 'Cartographic coordinates and rotational elements of the planets and satellites: 2000'." *Celest. Mech. Dynam. Astr.*, **82**(*1*), 83–111.

Seidelmann P.K. and Kovalevsky J., 2002, "Application of the new concepts and definitions (ICRS, CIP, and CEO) in fundamental astronomy." *Astr. Astrophys.*, **392**, 341–351.

Seidelmann P.K., 2003, "Comparison of 'old' and 'new' concepts: astrometry." IERS Technical Note No. 29, 51–52.

Seitz F., Stuck J., and Thomas M., 2004, "Consistent atmospheric and oceanic excitation of the Earth's free polar motion." *Geophys. J. Int.*, **157**(*1*), 25–35, DOI: 10.1111/j.1365-246X.2004.02208.x.

Seitz F., Stuck J., and Thomas M., 2005, "White noise Chandler wobble excitation." Workshop "Forcing of polar motion in the Chandler frequency band: a contribution to understanding interannual climate variations," April 21–23, 2004, Luxembourg, Cahiers du Centre Européen de Géodynamique et de Séismologie, Vol. 24, Eds. H.P. Plag, B. Chao, R. Gross, and T. Van Dam, 15–21.

Seitz F. and Schmidt M., 2005, "Atmospheric and oceanic contributions to Chandler wobble excitation determined by wavelet filtering." *Journal of Geophysical Research: Solid Earth*, **110**(*B11*), CiteID: B11406, DOI: 10.1029/2005JB003826.

Seitz M., Angermann D., Blossfeld M., Drewes H., and Gerstl M., 2012, "The 2008 DGFI realization of the ITRS: DTRF2008." *J. Geodesy*, **86**(*12*), 1097–1123, DOI: 10.1007/s00190-012-0567-2.

Selim H.H., 2007, "Hamiltonian of a second order two-layer Earth model." *J. Korean Astron. Soc.*, **40**(2), 49–60.

Shaffer D., 1995, "*Very Long Baseline Interferometry and the VLBA.*" Eds. J.A. Zensus, P.J. Diamond, P.J. Napier, ASP Conference Series, Vol. 82, Chapter 18, 345–361.

Shankland T.J., Peyronneau J., and Poirier J.P., 1993, "Electrical conductivity of the Earth's lower mantle." *Nature*, **366**, 453–455.

Shen P.Y. and Mansinha L., 1976, "Oscillation, nutation and wobble of an elliptical rotating Earth with liquid outer core." *Geophys. J. Roy. Astron. Soc.*, **46**, 467–496.

Shida T. and Matsuyama M., 1912, "Change of the plumb line referred to the axis of the Earth as found from the result of the international latitude observations." *J. Mem. College Sci. Eng., Kyoto Imp. Univ.*, **4**, 277–284.

Shirai T., Fukushima T., and Malkin Z., 2005, "Detection of phase disturbances of free core nutation of the Earth and their concurrence with geomagnetic jerks." *Earth, Planets Space*, **57**, 151–155.

Shum C.K., Woodworth P.L., Andersen O.B., *et al.*, 1997, "Accuracy assessment of recent ocean tide models." *J. Geophys. Res.*, **103**(*C11*), 25 173–25 194.

Sidorenkov N.S., 2000, "Chandler wobble of the poles as part of the nutation of the atmosphere–ocean–earth system." *Astronomy Reports*, **44**(6), 414–419, DOI: 10.1134/1.163865.

Simmons N.A., Forte A.M., and Grand S.P., 2006, "Constraining mantle flow with seismic and geodynamic data: a joint approach." *Earth Planet. Sci. Lett.*, **246**(*1–2*), 109–124, DOI: 10.1016/j.epsl.2006.04.003.

Simmons N.A., Forte A.M., and Grand S.P., 2009, "Joint seismic, geodynamic and mineral physical constraints on three-dimensional mantle heterogeneity: implications for the relative importance of thermal versus compositional heterogeneity." *Geophys. J. Int.*, **177**(*3*), 1284–1304, DOI: 10.1111/j.1365-246X.2009.04133.x.

Simon J.L., Bretagnon P., Chapront J., *et al.*, 1994, "Numerical expressions for precession formulae and mean elements for the Moon and the planets." *Astr. Astrophys.*, **282**, 663–683.

Simon J.-L., Francou G., Fienga A., and Manche H., 2013, "New analytical planetary theories VSOP2013 and TOP2013." *Astr. Astrophys.*, **557**(*A49*), DOI: 10.1051/0004-6361/201321843.

Sipkin S.A. and Jordan T.H., 1980, "Frequency dependence of Q_{ScS}." *Bull. Seism. Soc. Am.*, **70**, 1071–1102.

Smith M., 1974, "The scalar equations of infinitesimal elastic-gravitational motion for a rotating, slightly elliptical Earth." *Geophys. J. Roy. Astron. Soc.*, **37**, 491–526.

Smith M.L. and Dahlen F.A., 1981, "The period and Q of the Chandler wobble." *Geophys. J. Roy. Astron. Soc.*, **64**, 223–282.

Smith D.E., Zuber M.T., Torrence M.H., *et al.*, 2009, "Time variations of Mars' gravitational field and seasonal changes in the masses of the polar ice caps." *J. Geophys. Res.*, **114**(*E5*), CiteID: E05002, DOI: 10.1029/2008JE003267.

Smylie D.E., Szeto A.M.K., and Sato K., 1990a, "Elastic boundary conditions in long-period core oscillations." *Geophys. J. Int.*, **100**(2), 183–192, DOI: 10.1111/j.1365-246X.1990.tb02478.x.

Smylie D.E., Szeto A.M.K., and Sato K., 1990b, "Addendum and corrigendum to 'Elastic boundary conditions in long-period core oscillations'." *Geophys. J. Int.*, **105**(2), 553–553, DOI: 10.1111/j.1365-246X.1991.tb06734.x.

Smylie D.E., Jiang X., Brennan B.J., and Sato K., 1992, "Numerical calculation of modes of oscillation of the Earth's core." *Geophys. J. Int.*, **108**(2), 465–490, DOI: 10.1111/j.1365-246X.1992.tb04629.x.

Smylie D.E., 1992, "The inner core translational triplet and the density near Earth's center." *Science*, **255**, 1678–1682, DOI: 10.1126/science.255.5052.1678.

Smylie D.E. and McMillan D.G., 1998, "Viscous and rotational splitting of the translational oscillations of Earth's solid inner core." *Phys. Earth Planet. Int.*, **106**, 1–18, DOI: 10.1016/S0031-9201(97)00114-3.

Smylie D.E., 1999, "Viscosity near Earth's solid inner core." *Science*, **284**(*5413*), 461, DOI: 10.1126/science.284.5413.461.

Smylie D.E. and McMillan D.G., 2000, "The inner core as a dynamic viscometer." *Phys. Earth Planet. Int.*, **117**(*1–4*), 71–79, DOI: 10.1016/S0031-9201(99)00088-6.

Smylie D.E., Brazhkin V.V., and Palmer A., 2009, "Methodological notes: Direct observations of the viscosity of Earth's outer core and extrapolation of measurements of the viscosity of liquid iron." *Physics Uspekhi*, **52**(*1*), 79–92, DOI: 10.3367/UFNe.0179.200901d.0091.

Smylie D.E., 2013, "*Earth Dynamics; Deformations and Oscillations of the Rotating Earth.*" Cambridge University Press.

Snajdrova K., Boehm J., Willis P., Haas R., and Schuh H., 2006, "Multi-technique comparison of tropospheric zenith delays derived during the CONT02 campaign." *J. Geodesy*, **79**(*10–11*), 613–623, DOI: 10.1007/s00190-005-0010-z.

Soffel M., Schastok J., Ruder H., and Schneider M., 1985, "Relativistic astrometry." *Astrophys. Space Sci.*, **110**(*1*), 95–101, DOI: 10.1007/BF00660610.

Soffel M., Wirrer R., Schastok J., Ruder H., and Schneider M., 1988, "Relativistic effects in the motion of artificial satellites. I. The oblateness of the central body." *Celest. Mech.*, **42**(*1–4*), 81–89, DOI: 0.1007/BF01232949.

Soffel M. and Klioner S., 1998, "The present status of Einstein relativistic celestial mechanics." Journées Systèmes de Référence Spatio-temporels, September 1997, Prague, Czech Rep., Eds. J. Vondrák and N. Capitaine, pp 27–31.

Soffel M., Klioner S., Petit G., *et al.*, 2003, "Explanatory supplement for the IAU 2000 resolutions for astrometry, celestial mechanics, and metrology in the relativistic framework." *Astron. J.*, **126**, 2687–2706.

Sohl F. and Spohn T., 1997, "The interior structure of Mars: implications from SNC meteorites." *J. Geophys. Res.*, **102**, 1613–1635.

Sokolova J. and Malkin Z., 2007, "On comparison and combination of catalogues of radio source positions." *Astr. Astrophys.*, **474**(*2*), 665–670, DOI: 10.1051/0004-6361:20077450.

Soldati G., Boschi L., and Forte A.M., 2012, "Tomography of core-mantle boundary and lowermost mantle coupled by geodynamics." *Geophys. J. Int.*, **189**(*2*), 730–746, DOI: 10.1111/j.1365-246X.2012.05413.x.

Soldati G., Koelemeijer P., Boschi L., and Deuss A., 2013, "Constraints on core-mantle boundary topography from normal mode splitting." *Geochemistry, Geophysics, Geosystems*, **14**(*5*), 1333–1342, DOI: 10.1002/ggge.20115.

Somenzi L., Fienga A., Laskar J., and Kuchynka P., 2010, "Determination of asteroid masses from their close encounters with Mars." *Planetary and Space Science*, **58**(*5*), 858–863, DOI: 10.1016/j.pss.2010.01.010.

Souchay J. and Kinoshita H., 1991, "Comparison of the new nutation series with numerical integration." *Celest. Mech.*, **52**, 45–55, DOI: 10.1007/BF00048586.

Souchay J. and Kinoshita H., 1993, "Erratum: 'The theory of the nutation for the rigid Earth model at the second order' [*Celest. Mech. Dynam. Astr.*, **48**(*3*), 187–265 (1990)]." *Celest. Mech.*, **57**(*3*), 515.

Souchay J., 1993, "Comparison between theories of nutation for a rigid-Earth model." *Astr. Astrophys.*, **276**, 266–277.

Souchay J., Feissel M., Bizouard Ch., Capitaine N., and Bougeard M., 1995, "Precession and nutation for a non-rigid Earth: comparison between theory and VLBI observations." *Astr. Astrophys.*, **299**, 277–287.

Souchay J. and Kinoshita H., 1995, "Recent results of the nutation theory for a rigid Earth." *Highlights of Astronomy*, **10**, 237.

Souchay J., Feissel M., and Ma C., 1996, "Precise modeling of nutation and VLBI observations." *Astr. Astrophys. Supplement*, **116**, 473–481.

Souchay J. and Kinoshita H., 1996, "Corrections and new developments in rigid Earth nutation theory: I. Lunisolar influence including indirect planetary effects." *Astr. Astrophys.*, **312**, 1017–1030.

Souchay J. and Kinoshita H., 1997, "Corrections and new developments in rigid Earth nutation theory: II. Influence of second-order geopotential and direct planetary effect." *Astr. Astrophys.*, **318**, 639–652.

Souchay J., 1998, "Comparisons of the REN-2000 tables with numerical integration and other recent analytic tables." *Astron. J.*, **116**(*1*), 503–515, DOI: 10.1086/300401.

Souchay J. and Folgueira M., 1998, "The effect of zonal tides on the dynamical ellipticity of the Earth and its influence on the nutation." *Earth Moon Planets*, **81**, 201–216, DOI: 10.1023/A:1006331511290.

Souchay J., Loysel B., Kinoshita H., and Folgueira M., 1999, "Corrections and new developments in rigid Earth nutation theory: III. Final tables 'REN-2000' including crossed-nutation and spin-orbit coupling effects." *Astr. Astrophys. Supplement*, **135**, 111–131, DOI: 10.1051/aas:1999446.

Souchay J., Folgueira M., and Bouquillon S., 2003a, "Effects of the triaxiality on the rotation of celestial bodies: application to the Earth, Mars and Eros." *Earth Moon Planets*, **93**(*2*), 107–144, DOI: 10.1023/B:MOON.0000034505.79534.01.

Souchay J., Ma C., and Feissel-Vernier M., 2003b, "Celestial reference frame issues." IERS Technical Note No. 30, 35–38.

Souchay J., Lambert S.B., and Le Poncin-Lafitte C., 2007, "A comparative study of rigid Earth, non-rigid Earth nutation theories, and observational data." *Astr. Astrophys.*, **472**(*2*), 681–689, DOI: 10.1051/0004-6361:20077065.

Souchay J. and Capitaine N., 2013, "Precession and nutation of the Earth." *Tides in Astronomy and Astrophysics*, Lecture Notes in Physics, Volume 861, Springer-Verlag Berlin Heidelberg, DOI: 10.1007/978-3-642-32961-64, 115 p.

Sovers O.J., Fanselow J.L., and Jacobs C.S., 1998, "Astrometry and geodesy with radio interferometry: experiments, models, results." *Reviews of Modern Physics*, **70**(*4*), 1393–1454.

Sovers O.J., Charlot P., Fey A.L., and Gordon D., 2002, "Structure corrections in modeling VLBI delays for RDV data." In: Proc. IVS 2002 General Meeting, Tsukuba, Japan, February 4–7 June 2002, Eds. N.R. Vandenberg and K.D. Baver, pp 343–347.

Stacey F.D., Paterson M.S., and Nicholas A., 1981, "*Anelasticity in the Earth.*" American Geophysical Union, Washington, D.C.

Stacey F.D., and Anderson O.L., 2001 "Electrical and thermal conductivities of Fe-Ni-Si alloy under core conditions." *Phys. Earth Planet. Int.*, **124**(*3–4*), 153–162, DOI: 10.1016/S0031-9201(01)00186-8.

Standish E.M., 1981, "Two differing definitions of the dynamical equinox and the mean obliquity." *Astr. Astrophys.*, **101**, L17–18.

Standish E.M., 1982a, "The JPL planetary ephemerides." *Celest. Mech.*, **26**, 181–186, DOI: 10.1007/BF01230883.

Standish E.M., 1982b, "Orientation of the JPL ephemerides, DE200/LE200, to the dynamical equinox of J2000." *Astr. Astrophys.*, **114**, 297–302.

Standish E.M., 1985, "On the orientation of ephemeris reference frames." *Celest. Mech.*, **37**, 239–242, DOI: 10.1007/BF02285048.

Standish E.M., 1990a, "The observational basis for JPL's DE200, the planetary ephemerides of the Astronomical Almanac." *Astr. Astrophys.*, **233**(*1*), 252–271.

Standish E.M., 1990b, "An approximation to the outer planet ephemeris errors in JPL's DE 200." *Astr. Astrophys.*, **233**(*1*), 272–274.

Standish E.M., Newhall X.X., Williams J.G., and Folkner W.F., 1995, "JPL planetary and lunar ephemerides, DE403/LE403." JPL Memorandum IOM (Inter Office Memorandum), 314.10-127.

Standish E.M., 1998a, "JPL planetary and lunar ephemerides, DE405/LE405." JPL Memorandum IOM (Inter Office Memorandum), 312.F-98-048.

Standish E.M., 1998b, "JPL long ephemeris; JPL planetary and lunar ephemerides DE406/LE406." Available on CD-ROM at http://www.willbell.com/software/jpl.htm.

Standish E.M., 1998c, "Linking the dynamical reference frame to the ICRF." In: *Highlights of Astronomy* Vol. 11A, as presented at Joint Discussion 14 of the XXIIIrd General Assembly of the IAU, 1997. Edited by Johannes Andersen, Kluwer Academic Publishers, 310–311.

Standish E.M., 1998d, "Time scales in the JPL and CfA ephemerides." *Astr. Astrophys.*, **336**, 381–384.

Standish E.M. and Fienga A., 2002, "Accuracy limit of modern ephemerides imposed by the uncertainties in asteroid masses." *Astr. Astrophys.*, **384**, 322–328, DOI: 10.1051/0004-6361:20011821.

Standish E.M., 2004, "An approximation to the errors in the planetary ephemerides of the Astronomical Almanac." *Astr. Astrophys.*, **417**, 1165–1171, DOI: 10.1051/0004-6361:20035663.

Standish E.M., 2005, "Relating the dynamical reference frame and the ephemerides to the ICRF." In: *Highlights of Astronomy*, Vol. 13, Proc. XXVth General Assembly of the IAU, Ed. O. Engvold, San Francisco, CA, Astronomical Society of the Pacific, p. 609.

Steigenberger P., Tesmer V., Krügel M., *et al.*, 2007, "Comparisons of homogeneously reprocessed GPS and VLBI long time-series of troposphere zenith delays and gradients." *J. Geodesy*, **81**(*6–8*), 503–514, DOI: 10.1007/s00190-006-0124-y.

Stelzer Z. and Jackson A., 2013, "Extracting scaling laws from numerical dynamo models." *Geophys. J. Int.*, **193**(*3*), 1265–1276, DOI: 10.1093/gji/ggt083.

Stewart D.N., Busse F.H., Whaler K.A., and Gubbins D., 1995, "Geomagnetism, Earth rotation and the electrical conductivity of the lower mantle." *Phys. Earth Planet. Int.*, **92**, 199–214.

Stix M. and Roberts P.H., 1984, "Time-dependent electromagnetic core-mantle coupling." *Phys. Earth Planet. Int.*, **36**, 49–60.

Sylvander M., Ponce B., and Souriau A., 1997, "Seismic velocities at the core-mantle boundary inferred from P waves diffracted around the core." *Phys. Earth Planet. Int.*, **101**, 189–202, DOI: 10.1016/S0031-9201(97)00006-X.

Sze E.K.M. and van der Hilst R.D., 2003, "Core mantle boundary topography from short period PcP, PKP, and PKKP data." *Phys. Earth Planet. Int.*, **135**(*1*), 27–46, DOI: 10.1016/S0031-9201(02)00204-2.

Takeuchi H., 1950, "On the Earth tide of the compressible Earth of variable density and elasticity." *Trans. Amer. Geophys. Union*, **31**, 651–689.

Tamura Y., 1987, "A harmonic development of the tide-generating potential." *Bull. Info. Marées Terrestres*, **99**, 6813–6855.

Teke K., Boehm J., Nilsson T., *et al.*, 2011, "Multi-technique comparison of troposphere zenith delays and gradients during CONT08." *J. Geodesy*, **85**(7), 395–413, DOI: 10.1007/s00190-010-0434-y.

Tesmer V., Boehm J., Heinkelmann R., and Schuh H., 2007, "Effect of different tropospheric mapping functions on the TRF, CRF and position time-series estimated from VLBI." *J. Geodesy*, **81**(6–8), 409–421, DOI: 10.1007/s00190-006-0126-9.

Thaller D., Krügel M., Rothacher M., *et al.*, 2007, "Combined Earth orientation parameters based on homogeneous and continuous VLBI and GPS data." *J. Geodesy*, **81**(6–8), 529–541, DOI: 10.1007/s00190-006-0115-z.

"*The Astronomical Almanac 2006*," Nautical Almanac Office, US Naval Observatory, HM Nautical Almanac Office, Rutherford Appleton Laboratory, published by US Government Printing Office and in the UK by TSO, 2005.

Thompson A.R., Moran J.M., and Swenson G.W. Jr., 1986, "*Interferometry and Synthesis in Radio Astronomy*." First Edition, John Wiley and Sons, New York, 1986. Reprinted by Krieger Publishing Co, Malabar FL, 1991, third printing 1994, 534 pp, also Second Edition, John Wiley and Sons, New York, 2001, and Wiley-VCH, Berlin, 2004, 692 pp.

Tisserand F., 1891, "*Traité de Mécanique Céleste*." Tome II 'Théorie de la figure des corps célestes et de leur mouvement de rotation', 552 pp.

Titov O. and Malkin Z., 2009, "Effect of asymmetry of the radio source distribution on the apparent proper motion kinematic analysis." *Astr. Astrophys.*, **506**(3), 1477–1485, DOI: 10.1051/0004-6361/200912369.

Toomre A., 1966, "On the coupling of the Earth's core and mantle during the 26 000-year precession." In: "*The Earth–Moon System*," 33–45, Eds. Marsden B.G. and Cameron A.G.W., Plenum Press, New York.

Toomre A., 1974, "On the 'Nearly diurnal wobble' of the Earth." *Geophys. J. Roy. Astron. Soc.*, **38**, 335–348.

Trinh A., Rivoldini A., Van Hoolst T., and Dehant V., 2011, "The librations of a triaxial, synchronously rotating planetary satellite." EPSC-DPS Joint Meeting 2011, La Cité Internationale des Congrès Nantes Métropole, Nantes, France, 3–7 October 2011, extended abstract, EPSC Proceedings, **6**, EPSC-DPS2011-1514, 2 pp.

Trinh A., 2013, "Précession, nutations et librations d'une planète triaxiale." Ph.D. thesis, Université catholique de Louvain.

Tromp J. and Dahlen F.A., 1990, "Summation of the Born series for the normal modes of the Earth." *Geophys. J. Int.*, **100**, 527–533.

Van den Acker E., Van Hoolst T., de Viron O., *et al.*, 2002, "Influence of the winds and of the CO_2 mass exchange between the atmosphere and the polar ice caps on Mars' rotation." *J. Geophys. Res.*, **107**(E7), 10.1029/2000JE001539.

Van Hoolst T., Dehant V., and Defraigne P., 2000a, "Sensitivity of the free core nutation and the Chandler wobble to changes in the interior structure of Mars." *Phys. Earth Planet. Int.*, **117**, 397–405.

Van Hoolst T., Dehant V., and Defraigne P., 2000b, "Chandler wobble and free core nutation for Mars." *Planet. Space Sci.*, **48**, 12–14, 1145–1151.

Van Hoolst T., Dehant V., de Viron O., Defraigne P., and Barriot J.-P., 2002, "Degree-one displacements on Mars." *Geophys. Res. Letters*, **29**(11), 1511, 10.1029/2002GL014711.

Van Hoolst T. and Dehant V., 2002, "Influence of triaxiality and second-order terms in flattenings on the rotation of terrestrial planets: I. Formalism and rotational normal modes." *Phys. Earth Planet. Int.*, **134**, 17–33.

Van Hoolst T., Dehant V., Roosbeek F., and Lognonné P., 2003, "Tidally induced surface displacements, external potential variations, and gravity variations on Mars." *Icarus*, **161**, 281–296, DOI: 10.1016/S0019-1035(2)00045-2.

Van Hoolst T., 2007, "The rotation of the terrestrial planets." Treatise on Geophysics, Volume 10 *"Planets and Moons,"* 123–164, DOI: 10.1007/s11214-007-9202-6.

Van Hoolst T., Baland R.-M., and Trinh A., 2013, "On the librations and tides of large icy satellites." *Icarus*, **226**, 299–315.

Varadi F., Musotto S., Moore W., and Schubert G., 2005, "Normal modes of synchronous rotation." *Icarus*, **176**(*1*), 235–249, DOI: 10.1016/j.icarus.2005.01.002.

Verhoeven O., Rivoldini A., Vacher P., *et al.*, 2005, "Interior structure of terrestrial planets. I. Modelling Mars' mantle and its electromagnetic, geodetic and seismic properties." *J. Geophys. Res. Planets*, **110**(*E4*), E04009, DOI: 10.1029/2004JE002271.

Verhoeven O., Mocquet A., Vacher P., *et al.*, 2009, "Constraints on thermal state and composition of the Earth's lower mantle from electromagnetic impedances and seismic data." *J. Geophys. Res. (Planets)*, **114**(*B3*), CiteID B03302, DOI: 10.1029/2008JB005678.

Verma A., Fienga A., Laskar J., Manche H., and Gastineau M., 2014, "Use of MESSENGER radioscience data to improve planetary ephemeris and to test general relativity." *Astr. Astrophys.* **561**(*A115*), DOI: 10.1051/0004-6361/201322124.

Vicente R.O., 1962, "The values of the nutations and of the period of the variation of latitude." *Vistas in Astronomy*, **5**(*1*), 1–10, DOI: 10.1016/0083-6656(62)90001-6.

Vondrák J., Ron C., and Weber R., 2003, "Combined VLVI/GPS series of precession-nutation and comparison with IAU2000 model." *Astr. Astrophys.*, **397**, 771–776.

Vondrák J., Weber R., and Ron C., 2005, "Free core nutation: direct observations and resonance effects." *Astr. Astrophys.*, **444**(*1*), 297–303.

Vondrák J. and Ron C., 2005, "Solution of Earth orientation parameters in the frame of new Earth orientation catalogue." *Kinematika i Fizika Nebesnykh Tel, Suppl*, **5**, 305–310.

Vondrák J., Capitaine N., and Wallace P., 2011, "New precession expressions, valid for long time intervals (Corrigendum)." *Astr. Astrophys.*, **534**(*ID A22*), 19 p., DOI: 10.1051/0004-6361/201117274.

Vondrák J., Capitaine N., and Wallace P., 2012, "New precession expressions, valid for long time intervals (Corrigendum)." *Astr. Astrophys.*, **541**(*ID C1*), 1 p., DOI: 10.1051/0004-6361/201117274e.

Voorhies C.V., Sabaka T.J., and Purucker M., 2002, "On magnetic spectra of Earth and Mars." *J. Geophys. Res. (Planets)*, **107**(*E6*), 5034, DOI: 10.1029/2001JE001534.

Wahr J.M., 1979, "The tidal motions of a rotating, elliptical, elastic, and oceanless Earth." Ph.D. thesis, University of Colorado.

Wahr J.M., 1981a, "The forced nutations of an elliptical, rotating, elastic and oceanless Earth." *Geophys. J. Roy. Astron. Soc.*, **64**, 705–727.

Wahr J.M., 1981b, "Body tides on an elliptical, rotating, elastic and oceanless Earth." *Geophys. J. Int.*, **64**(*3*), 677–703.

Wahr J.M., 1981c, "A normal mode expansion for the forced response of a rotating Earth." *Geophys. J. Roy. Astron. Soc.*, **64**, 651–675.

Wahr J.M., 1981d, "A review of: 'The Earth's variable rotation: Geophysical causes and consequences'." *Geophys. Astrophys. Fluid Dynam.*, **18**(*3*), 321–322.

Wahr J.M. and Sasao T., 1981, "A diurnal resonance in the ocean tide and in the Earth's load response due to the resonant free core nutation." *Geophys. J. Roy. Astron. Soc.*, **64**, 747–765.

Wahr J.M., Smith M.L., and Sasao T., 1981, "Effect of the fluid core on changes in the length of day due to long period tides." *Geophys. J.*, **64**, 635–650.

Wahr J.M., 1982, "The effects of the atmosphere and oceans on the Earth's wobble and the seasonal variations in the length of day. I. Theory." *Geophys. J. Roy. Astron. Soc.*, **70**, 349–372.

Wahr J.M. and Bergen Z., 1986, "The effects of mantle anelasticity on nutations, Earth tides, and tidal variations in the rotation rate." *Geophys. J. Roy. Astron. Soc.*, **87**, 633–668.

Wahr J.M., 1988, "The Earth's rotation." *Annual Rev. Earth Planet. Sci.*, **16**, 231–249, DOI: 10.1146/annurev.ea.16.050188.001311.

Wahr J.M., 2005, "Polar motion models: angular momentum approach." Workshop "Forcing of polar motion in the Chandler frequency band: A contribution to understanding interannual climate variations," April 21–23, 2004, Luxembourg, Cahiers du Centre Européen de Géodynamique et de Séismologie, Vol. 24, Eds. H.P. Plag, B. Chao, R. Gross, and T. Van Dam, pp 1–8.

Walker A.D. and Backus G.E., 1997, "A six-parameter statistical model of the Earth's magnetic field." *Geophys. J. Int.*, **130**, 673–700.

Wallace P., 2003, "Software for implementing the IAU2000 resolutions." IERS Technical Note No. 29, pp 65–70.

Wallace P.T. and Capitaine N., 2006, "Precession-nutation procedures consistent with IAU 2006 resolutions." *Astr. Astrophys.*, **459**(*3*), 981–985, DOI: 10.1051/0004-6361:20065897.

Wicht J. and Christensen U.R., 2010, "Torsional oscillations in dynamo simulations." *Geophys. J. Int.*, **181**(*3*), 1367–1380, DOI: 10.1111/j.1365-246X.2010.04581.x.

Williams J.G., 1994, "Contributions to the Earth's obliquity rate, precession and nutation." *Astron. J.*, **108**, 711–724.

Williams J.G., 1995, "Planetary–induced nutation of the Earth: direct terms." *Astron. J.*, **110**, 1420–1426.

Wilson C.R., 1979, "Estimation of the parameters of the Earth's polar motion." IAU Publ., Eds. D.D McCarthy and J.D. Pilkington, *"Time and the Earth's Rotation,"* pp 307–312.

Wilson C.R. and Vicente R.O., 1981, "Estimates of Chandler's component of polar motion as derived from various data sets." *Astronomische Nachrichten*, **302**(*5*), 227–232.

Wilson C.R. and Vicente R.O., 1990, "Maximum likelihood estimates of polar motion parameters." Geophys. Monograph, IUGG Volume 9, *"Variations in Earth Rotation,"* IUGG and AGU Publ., Eds. D.D McCarthy and W.E. Carter, 151–155.

Woodhouse J.H. and Dziewonski A.M., 1989, "Seismic modelling of the Earth's large-scale three-dimensional structure." *Phil. Trans. Roy. Soc. of London. Series A, Mathematical and Physical Sciences*, **328**(*1599*), 291–308, DOI: 10.1098/rsta.1989.0037.

Woolard E.W., 1953, "Theory of the rotation of the Earth around its center of mass." *Astron. Papers for Amer. Ephemeris and Nautical Almanac XV*, Part I, US Government Printing Office, Washington D.C., pp 1–165.

Woolard E.W., 1953, "Redevelopment of the theory of nutation." *Astron. J.*, **58**, 1–3.

Wresnik J., Haas R., Boehm J., and Schuh H., 2007, "Modeling thermal deformation of VLBI antennas with a new temperature model." *J. Geodesy*, **81**(*6–8*), 423–431, DOI: 10.1007/s00190-006-0120-2.

Wu X. and Wahr J.M., 1997, "Effects of non-hydrostatic core-mantle boundary topography and core dynamics on Earth rotation." *Geophys. J. Int.*, **128**, 18–42.

Wünsch J., 2002, "Oceanic and soil moisture contributions to seasonal polar motion." *Journal of Geodynamics*, **33**(*3*), 269–280, DOI: 10.1016/S0264-3707(01)00070-9.

Xu C., Wu X., Soffel M., and Klioner S.A., 2003, "Relativistic theory of elastic deformable astronomical bodies: perturbation equations in rotating spherical coordinates and junction conditions." *Physical Review D*, **68**(*6*), ID. 064009, DOI: 10.1103/Phys-RevD.68.064009.

Xu S. and Szeto A.M.K., 1998, "The coupled rotation of the inner core." *Geophys. J. Int.*, **133**(*2*), 279–297, DOI: 10.1046/j.1365-246X.1998.00495.x.

Xu S., Crossley D., and Szeto A.M.K., 2000, "Variations in length of day and inner core differential rotation from gravitational coupling." *Phys. Earth Planet. Int.*, **117**(*1–4*), 95–110.

Yoder C.F. and Standish E.M., 1997, "Martian precession and rotation from Viking lander range data." *J. Geophys. Res.*, **102**(*E2*), *1242*, 4065–4080.

Yoder C.F., Konopliv A.S., Yuan D.N., Standish E.M., and Folkner W.M., 2003, "Fluid core size of Mars from detection of the solar tide." *Science*, **300**(*5617*), 299–303.

Yseboodt M., de Viron O., Chin T.M., and Dehant V., 2002, "Atmospheric excitation of the Earth nutation: comparison of different atmospheric models." *J. Geophys. Res.*, **107**(*B2*), 2036, 10.1029/2000JB000042.

Yseboodt M., Barriot J.-P., and Dehant V., 2003, "Analytical modeling of the Doppler tracking between a lander and a Mars orbiter in term of rotational dynamics." *J. Geophys. Res.*, **108**(*E7*), 5076, DOI: 10.1029/2003JE002045.

Yseboodt M., Rivoldini A., Van Hoolst T., and Dumberry M., 2013, "Influence of an inner core on the long-period forced librations of Mercury." *Icarus*, **226**(*1*), 41–51, DOI: 10.1016/j.icarus.2013.05.011.

Yuen D.A. and Peltier W.R., 1982, "Normal modes of the viscoelastic Earth." *Geophys. J. Roy. Astron. Soc.*, **69**, 495–526.

Zensus J.A. and Napier P.J. (Eds), 1995, "Very long baseline interferometry and the VLBA." ASP Conference Series, **82**, 453 pp.

Zerhouni W. and Capitaine N., 2009, "Celestial pole offsets from lunar laser ranging and comparison with VLBI." *Astr. Astrophys.*, **507**(*3*), 1687–1695, DOI: 10.1051/0004-6361/200912644.

Zhang Keke, 1992, "On inertial waves in the Earth's fluid core." *Geophys. Res. Letters*, **19**(*8*), 737–740, DOI: 10.1029/92GL00357.

Zhang Keke and Fearn D.R., 1994, "Hydromagnetic waves in rapidly rotating spherical shells generated by magnetic toroidal decay modes." *Geophys. Astrophys. Fluid Dynamics*, **77**(*1*), 133–157, DOI: 10.1080/03091929408203679.

Zhang Keke, Earnshaw P., Liao X., and Busse F.H., 2001, "On inertial waves in a rotating fluid sphere." *Journal of Fluid Mechanics*, **437**(*1*), 103–119.

Zhang Keke, Liao X., and Li L., 2003, "Differential rotation driven by precession." *Celest. Mech. Dynam. Astr.*, **87**(*1*), 39–51.

Zhang Keke, Kong D., and Liao X., 2010a, "On fluid flows in precessing narrow annular channels: asymptotic analysis and numerical simulation." *Journal of Fluid Mechanics*, **656**, 116–146, DOI: 10.1017/S0022112010001059.

Zhang Keke, Chan K.H., and Liao X., 2010b, "On fluid flows in precessing spheres in the mantle frame of reference." *Physics of Fluids*, **22**(*11*), 116604–116604-8, DOI: 10.1063/1.3515344.

Zharkov V.N., Molodensky S.M., Brzezinski A., Groten E., and Varga P., 1996, *"The Earth and its Rotation: Low Frequency Geodynamics."* Wichmann Publication, 497 pp.

Zharkov V.N. and Molodensky S.M., 1996, "On the Chandler wobble of Mars." *Planet. Space Sci.*, **11**, 1457–1462.

Zhu S.Y. and Groten E., 1989, "Various aspects of numerical determination of nutation constants. I - Improvement of rigid-Earth nutation." *Astron. J.*, **98**, 1104–1111, DOI: 10.1086/115201.

Zotov L.V., 2010, "Dynamical modeling and excitation reconstruction as fundamental of Earth rotation prediction." *Artificial Satellites*, **45**(2), 95–106, DOI: 10.2478/v10018-010-0010-y.

Zotov L.V. and Bizouard C., 2012, "On modulations of the Chandler wobble excitation." *J. Geodynamics*, **62**, 30–34, DOI: 10.1016/j.jog.2012.03.010.

Index

Printed in the United States
By Bookmasters